Advances in Intelligent and Soft Computing 128

Editor-in-Chief: J. Kacprzyk

Advances in Intelligent and Soft Computing

Editor-in-Chief

Prof. Janusz Kacprzyk
Systems Research Institute
Polish Academy of Sciences
ul. Newelska 6
01-447 Warsaw
Poland
E-mail: kacprzyk@ibspan.waw.pl

Further volumes of this series can be found on our homepage: springer.com

Vol. 111. L. Jiang (Ed.)
Proceedings of the 2011 International Conference on Informatics, Cybernetics, and Computer Engineering (ICCE 2011) November 19-20, 2011, Melbourne, Australia, 2011
ISBN 978-3-642-25187-0

Vol. 112. L. Jiang (Ed.)
Proceedings of the 2011 International Conference on Informatics, Cybernetics, and Computer Engineering (ICCE 2011) November 19-20, 2011, Melbourne, Australia, 2011
ISBN 978-3-642-25193-1

Vol. 113. J. Altmann, U. Baumöl, and B.J. Krämer (Eds.)
Advances in Collective Intelligence 2011, 2011
ISBN 978-3-642-25320-1

Vol. 114. Y. Wu (Ed.)
Software Engineering and Knowledge Engineering: Theory and Practice, 2011
ISBN 978-3-642-03717-7

Vol. 115. Y. Wu (Ed.)
Software Engineering and Knowledge Engineering: Theory and Practice, 2011
ISBN 978-3-642-03717-7

Vol. 116. Yanwen Wu (Ed.)
Advanced Technology in Teaching - Proceedings of the 2009 3rd International Conference on Teaching and Computational Science (WTCS 2009), 2012
ISBN 978-3-642-11275-1

Vol. 117. Yanwen Wu (Ed.)
Advanced Technology in Teaching - Proceedings of the 2009 3rd International Conference on Teaching and Computational Science (WTCS 2009), 2012
ISBN 978-3-642-25436-9

Vol. 118. A. Kapczynski, E. Tkacz, and M. Rostanski (Eds.)
Internet - Technical Developments and Applications 2, 2011
ISBN 978-3-642-25354-6

Vol. 119. Tianbiao Zhang (Ed.)
Future Computer, Communication, Control and Automation, 2011
ISBN 978-3-642-25537-3

Vol. 120. Nicolas Loménie, Daniel Racoceanu, and Alexandre Gouaillard (Eds.)
Advances in Bio-Imaging: From Physics to Signal Understanding Issues, 2011
ISBN 978-3-642-25546-5

Vol. 121. Tomasz Traczyk and Mariusz Kaleta (Eds.)
Modeling Multi-commodity Trade: Information Exchange Methods, 2011
ISBN 978-3-642-25648-6

Vol. 122. Yinglin Wang and Tianrui Li (Eds.)
Foundations of Intelligent Systems, 2011
ISBN 978-3-642-25663-9

Vol. 123. Yinglin Wang and Tianrui Li (Eds.)
Knowledge Engineering and Management, 2011
ISBN 978-3-642-25660-8

Vol. 124. Yinglin Wang and Tianrui Li (Eds.)
Practical Applications of Intelligent Systems, 2011
ISBN 978-3-642-25657-8

Vol. 125. Tianbiao Zhang (Ed.)
Mechanical Engineering and Technology, 2011
ISBN 978-3-642-27328-5

Vol. 126. Khine Soe Thaung (Ed.)
Advanced Information Technology in Education, 2011
ISBN 978-3-642-25907-4

David Jin and Sally Lin (Eds.)

Advances in Multimedia, Software Engineering and Computing Vol.1

Proceedings of the 2011 MSEC International Conference on Multimedia, Software Engineering and Computing, November 26–27, Wuhan, China

Editors

Prof. David Jin
International Science & Education
Researcher Association
Wuhan Section
Special No.1, Jiangxia Road of Wuhan
Wuhan
China, People's Republic
E-mail: dayang1818@163.com

Dr. Sally Lin
ISER Association
Guangshan Road, Baoli Garden 1
430072 Wuhan
China, People's Republic
E-mail: 1375170731@qq.com

ISBN 978-3-642-25988-3 e-ISBN 978-3-642-25989-0

DOI 10.1007/978-3-642-25989-0

Advances in Intelligent and Soft Computing ISSN 1867-5662

Library of Congress Control Number: 2011943102

Typeset by Scientific Publishing Services Pvt. Ltd., Chennai, India

Printed on acid-free paper

5 4 3 2 1 0

springer.com

Preface

International Science & Education Researcher Association (ISER) puts her focus on studying and exchanging academic achievements of international teaching and scientific research, and she also promotes education reform in the world. In addition, she serves herself on academic discussion and communication too, which is beneficial for education and scientific research. Thus it will stimulate the research interests of all researchers to stir up academic resonance.

MSEC2011 is an integrated conference concentrating its focus upon Multimedia, Software Engineering, Computing and Education. In the proceeding, you can learn much more knowledge about Multimedia, Software Engineering, Computing and Education of researchers all around the world. The main role of the proceeding is to be used as an exchange pillar for researchers who are working in the mentioned field. In order to meet high standard of Springer, AISC series, the organization committee has made their efforts to do the following things. Firstly, poor quality paper has been refused after reviewing course by anonymous referee experts. Secondly, periodically review meetings have been held around the reviewers about five times for exchanging reviewing suggestions. Finally, the conference organization had several preliminary sessions before the conference. Through efforts of different people and departments, the conference will be successful and fruitful.

MSEC2011 is co-sponsored by International Science & Education Researcher Association, Beijing Gireida Education Co.Ltd and Wuhan University of Science and Technology,China. The goal of the conference is to provide researchers from Multimedia, Software Engineering, Computing and Education based on modern information technology with a free exchanging forum to share the new ideas, new innovation and solutions with each other. In addition, the conference organizer will invite some famous keynote speaker to deliver their speech in the conference. All participants will have chance to discuss with the speakers face to face, which is very helpful for participants.

During the organization course, we have got help from different people, different departments, different institutions. Here, we would like to show our first sincere thanks to publishers of Springer, AISC series for their kind and enthusiastic help and best support for our conference. Secondly, the authors should be thanked too for their enthusiastic writing attitudes toward their papers. Thirdly, all members of program chairs, reviewers and program committees should also be appreciated for their hard work.

In a word, it is the different team efforts that they make our conference be successful on November 26–27, Wuhan, China. We hope that all of participants can give us good suggestions to improve our working efficiency and service in the future. And we also hope to get your supporting all the way. Next year, In 2012, we look forward to seeing all of you at MSEC2012.

September, 2011 ISER Association

Organizing Committee

Honor Chairs

Prof. Chen Bin	Beijing Normal University, China
Prof. Hu Chen	Peking University, China
Chunhua Tan	Beijing Normal University, China
Helen Zhang	University of Munich, China

Program Committee Chairs

Xiong Huang	International Science & Education Researcher Association, China
LiDing	International Science & Education Researcher Association, China
Zhihua Xu	International Science & Education Researcher Association, China

Organizing Chair

ZongMing Tu	Beijing Gireida Education Co.Ltd, China
Jijun Wang	Beijing Spon Technology Research Institution, China
Quanxiang	Beijing Prophet Science and Education Research Center, China

Publication Chair

Song Lin	International Science & Education Researcher Association, China
Xionghuang	International Science & Education Researcher Association, China

International Committees

Sally Wang	Beijing normal university, China
LiLi	Dongguan University of Technology, China
BingXiao	Anhui University, China
Z.L. Wang	Wuhan University, China
Moon Seho	Hoseo University, Korea
Kongel Arearak	Suranaree University of Technology, Thailand
Zhihua Xu	International Science & Education Researcher Association, China

Co-sponsored by

International Science & Education Researcher Association, China
VIP Information Conference Center, China

Reviewers of MSEC2011

Chunlin Xie	Wuhan University of Science and Technology, China
LinQi	Hubei University of Technology, China
Xiong Huang	International Science & Education Researcher Association, China
Gangshen	International Science & Education Researcher Association, China
Xiangrong Jiang	Wuhan University of Technology, China
LiHu	Linguistic and Linguidtic Education Association, China
Moon Hyan	Sungkyunkwan University, Korea
Guangwen	South China University of Technology, China
Jack H. Li	George Mason University, USA
Marry Y. Feng	University of Technology Sydney, Australia
Feng Quan	Zhongnan University of Finance and Economics, China
PengDing	Hubei University, China
XiaoLie Nan	International Science & Education Researcher Association, China
ZhiYu	International Science & Education Researcher Association, China
XueJin	International Science & Education Researcher Association, China
Zhihua Xu	International Science & Education Researcher Association, China
WuYang	International Science & Education Researcher Association, China

QinXiao	International Science & Education Researcher Association, China
Weifeng Guo	International Science & Education Researcher Association, China
Li Hu	Wuhan University of Science and Technology, China
ZhongYan	Wuhan University of Science and Technology, China
Haiquan Huang	Hubei University of Technology, China
Xiao Bing	Wuhan University, China
Brown Wu	Sun Yat-Sen University, China

Contents

Model Retrieval Based on Three-View 1
FuHua Shang, YaDong Zhou, HongTao Xie

A Bayesian Network Approach in the Relevance Feedback of Personalized Image Semantic Model 7
Lei Huang, Jian-guo Nan, Lei Guo, Qin-ying Lin

An Empirical Study on Elastic Effect of Fiscal Expenditure to Household Consumption in China 13
Jianbao Chen, Huanjun Zhu, Tingting Cheng

Extraction and Recognition of Bridges over Water in High Resolution SAR Image 19
XiongMei Zhang, JianShe Song, ZhaoXiang Yi, JunHui Xu

Unsupervised SAR Imagery Segmentation Based on SVDD 25
XiongMei Zhang, JianShe Song, ZhaoXiang Yi, RuiHua Wang

An Approach of Building Areas Segmentation of SAR Images Based on the Level Set Method 33
Ruihua Wang, Jianshe Song, Xiongmei Zhang, Yibing Wu

SAR Image Classification in Urban Areas Using Unit-Linking Pulse Coupled Neural Network 39
Ruihua Wang, Jianshe Song, Xiongmei Zhang, Yibing Wu

The Research of Intrusion Detection System in Cloud Computing Environment 45
Huaibin Wang, Haiyun Zhou

Zero-Voltage-Switching Voltage Doubled SEPIC Converter 51
Hyun-Lark Do

The Study on Application of Computer Assisted-Language Learning in College English Reading Teaching 55
Haiquan Huang

Single-Switch Step Up/Down Converter with Positive and Negative Outputs 61
Hyun-Lark Do

Energy Recovery Address Driver with Reduced Circulating Current 65
Hyun-Lark Do

A Scheme of Feasibility with Static Analysis in Software Testing Environment 69
ManSha Lu

Active Clamped Resonant Flyback with a Synchronous Rectifier 75
Hyun-Lark Do

A Robust Digital Watermarking Algorithm Based on Integer Wavelet Matrix Norm Quantization and Template Matching 79
Xin Liu, XiaoQi Lv, Qiang Luo

Semi-supervised Metric Learning for 3D Model Automatic Annotation 85
Feng Tian, Xu-kun Shen, Xian-mei Liu, Kai Zhou

ZVS Synchronous Buck Converter with Improved Light Load Efficiency and Ripple-Free Current 93
Hyun-Lark Do

A Prediction Model Based on Linear Regression and Artificial Neural Network Analysis of the Hairiness of Polyester Cotton Winding Yarn 97
Zhao Bo

Predicting the Fiber Diameter of Melt Blowing through BP Neural Network and Mathematical Models 105
Zhao Bo

Humanizing Anonymous More Sensitive Attributes Privacy Protection Algorithm 111
GuoXing Peng, Qi Tong

Research of Music Class Based on Multimedia Assisted Teaching in College 117
Hongbo Zhang

Analysis and Design of E Surfing 3G Mobile Phone's Payment Schemes of China Telecom Based on BP Neural Network 123
Yan Shen, Shuangshuang Sun, Bo Zhang

Soft-switching High Step-Up DC-DC Converter with Single Magnetic Component 129
Hyun-Lark Do

Research on Association Rule Algorithm Based on Distributed and Weighted FP-Growth 133
Huaibin Wang, Yuanchao Liu, Chundong Wang

The Research of PWM DC-DC Converter Based on TMS320F28335 139
Jinhua Liu, Xuezhi Hu

The New Method of DDOS Defense 145
Shuang Liang

Research of Blind Signal Separation Algorithm Based on ICA Method 153
Xinling Wen, Yi Ru

Research of Intelligent Physical Examination System Based on IOT 159
Wei Han

Research of Intelligent Campus System Based on IOT 165
Wei Han

Training Skills of Autonomous Learning Abilities with Modern Information Technology 171
Xianzhi Tian

Medical Image Processing System for Minimally Invasive Spine Surgery . . . 177
YanQiu Chen, PeiLli Sun

Multi-scale Image Transition Region Extraction and Segmentation Based on Directional Data Fitting Information 183
Jianda Wang, Yanchun Wang

Analysis and Design of Digital Education Resources Public Service Platform in Huanggang 191
Yun Cheng, Yanli Wang, SanHong Tong, ZhongMei Zheng, Feng Wang

Study on E-commerce Course Practice and Evaluation System under Bilingual Teaching 199
Yiqun Li, Quan Jin, Xuezheng Zhang

Non-termination Analysis of Polynomial Programs by Solving Semi-Algebraic Systems 205
Xiaoyan Zhao

Construction of Matlab Circuit Analysis Virtual Laboratory 213
Shoucheng Ding

Identification and Evaluation Methods of Expert Knowledge Based on Social Network Analysis 219
Guoai Gu, Wanjun Deng

Effect of Ammonium and Nitrate Ratios on Growth and Yield of Flowering Chinese Cabbage 227
Shiwei Song, Lingyan Yi, Houcheng Liu, Guangwen Sun, Riyuan Chen

Research of Rapid Design Method for Munitions Based on Generalized Modularity 233
Chao Wang, Chunlan Jiang, Zaicheng Wang, Ming Li

Structured Evaluation Report of an Online Writing Resource Site 239
Xing Zou

Inversion Calculation and Site Application for High-Resolution Dual Laterolog (HRDL) Tool 245
Zhenhua Liu, Jianhua Zhang

GPU Accelerated Target Tracking Method 251
Jian Cao, Xiao-fang Xie, Jie Liang, De-dong Li

Web Database Access Technology Based on ASP.NET 259
Jin Wang

A Novel Approach for Software Quality Evaluation Based on Information Axiom 265
Houxing You

Research on Model of Ontology-Based Semantic Information Retrieval 271
Yu Cheng, Ying Xiong

Using Dynamic Fuzzy Neural Networks Approach to Predict Ice Formation 277
Qisheng Yan, Muhua Ding

The Design of Simple Experiment System for Automatic Control System . . . 285
YiMing Li

Deformation Calculation for Uplift Piles Based on Generalized Load Transfer Model 291
Wenjuan Yao, Shangping Chen, Shengqing Zhu

Teaching Linear Algebra: Multimedia, Strategies, Methods and Computing Technology Development 299
Jing Zhang

Research on Keyword Information Retrieve Based on Semantic 305
Xin Li, Wanxin Dong

The Implementation of a Specific Algorithm by Traversing the Graph Stored with Adjacency Matrix . . . 311
Min Wang, YaoLong Li

Research of Key Technologies in Development in Thesis Management System . . . 317
Huadong Wang

Evaluation Model of Business Continuity Management Environment on E-learning . . . 323
Gang Chen

Knowledge Discovering and Innovation Based on General Education Curriculum Reform around Classical Reading in Colleges . . . 329
ShuQin Wu

Based on Embedded Image Processing Technology of Glass Inspection System . . . 337
Lian Pan, Xiaoming Liu, Cheng Chen

Wavelet Weighted Multi-Modulus Blind Equalization Algorithm Based on Fractional Lower Order Statistics . . . 343
Fang Xu, Yecai Guo, Jun Guo

Orthogonal Wavelet Transform Blind Equalization Algorithm Based on Chaos Optimization . . . 349
Yecai Guo, Wencai Xu, Fang Xu, Jun Guo

Research on Soil Heavy Metal Pollution Assessment System Based on WebGIS . . . 355
He Shen, TingYan Xing, ShengXuan Zhou, YongHong Mi, Xian Feng

The Impact of Visual Transcoder on Self-similarity of Internet Flow . . . 361
JingWen Zhu, Jun Steed Huang, BoTao Zhu

The Research and Design of NSL-Oriented Automation Testing Framework . . . 367
Chongwen Wang

Based on Self-manufactured Experimental Equipments, the Curriculums of Microcomputer Experiment Teaching Mode Reform and Innovation . . . 375
Jiangchun Xu, Jiande Wu, Nan Lin, Ni Liu, Xiaojun Xue, Yan Chen

"Flexibility" of Software Development Method . . . 383
Yu Gao, Yong-hua Yang

Direction of Arrival (DOA) Estimation Algorithm Based on the Radial Basis Function Neural Networks . . . 389
Hong He, Tao Li, Tong Yang, Lin He

Interference Analysis of WCDMA Base from TD-SCDMA Terminals 395
Hong He, Xin Yin, Tong Yang, Lin He

Research of Dispatching Method in Elevator Group Control System Based on Fuzzy Neural Network 401
Jun Wang, Airong Yu, Lei Cao, Fei Yang

Modal of Dynamic Data Collection Based on SOA 409
Airong Yu, Jun Wang, Lei Cao, Yihui He

A Distributed Website Anti-tamper System Based on Filter Driver and Proxy 415
Jun Zhou, Qian He, Linlin Yao

The Application of TDL in the Windows Intelligent Control 423
Wang Na, Yanxia Pang

The System Design and Edge Detection on Island and Reef Based on Gabor Wavelet 429
XiJun Zhu, QiuJu Bai, GuiFei Liu, Ke Xu

The Designs of the Function and Structure of the New Hydraulic Cylinder Test-Bed 435
DongHai Su, ZhengHui Qu

A Positioning Algorithm for Node of WSN Based on Received Signal Strength 441
Xiang Yang, Wei Pan, Yuanyi Zhang

Linear Generalized Synchronization between Two Complex Networks 447
Qin Yao, Guoliang Cai, Xinghua Fan, Juan Ding

The Application of Digital Media Technology in Art Design 453
Qiang Liu, Lixin Diao, Guangcan Tu, Linlin Lu

A Metadata Architecture Based on Combined Index 459
Li Cai, JianYing Su

Two-Stage Inventory Model with Price Discount under Stochastic Demand 465
Yanhong Qin, Xinghong Qin

Study a Text Classification Method Based on Neural Network Model 471
Jian Chen, Hailan Pan, Qinyun Ao

Security Design and Implementation of Yangtze Gold Cruises Website 477
Huashan Tan, You Yang, Ping Yu

The Effect of Multimedia CAI Courseware in the Modern Art Teaching 485
Guangcan Tu, Qiang Liu, Linlin Lu

A Virtual Laboratory Framework Based on Mobile Agents 491
Chao Yang, Gang Liu

Research and Improvement of Resource Scheduling Mechanism in Alchemi Desktop Grid 497
CaiFeng Cao, DeDong Jiang

Customers Evaluation Effects of Brand Extension towards Brand Image of Chinese Internet Companies 503
Ming Zhou

Research on the Fault Diagnosis of Excess Shaft Ran of Electric Submersible Pump 509
Fengyang Tao, Guangfu Liu, Wenjing Xi

An Empirical Research of Effects of Realness on Microblogging Intention Model 515
Zhijie Zhang, Haitao Sun, Huiying Du

A Study on the Relation between Moldingroom's Temperature and the Cock Issue in FDM Techniques 521
Hao Wu, Zheng Yang

The Gracefulness of a Kind of Unconnected Graphs 525
Yan-Hua Yu, Wen-Xiang Wang, Li-xia Song

Six-Order Symplectic Integration in Quasi-classical Trajectory Computation 533
Xian-Fang Yue

Identity for Sums of Five Squares 539
Guowen Xi

The Optimal Traits of Semiorthogonal Trivariate Matrix-Valued Small-Wave Wraps and Trivariate Framelets 545
Delin Hua

The Nice Features of Two-Direction Poly-Scale Trivariate Small-Wave Packages with Finite Support 553
Bingqing Lv, Jing Huang

The Excellent Traits of Multiple Dual-Frames and Applications 561
Kezhong Han

Finite-Time Control of Linear Discrete Singular Systems with Disturbances 569
Yaning Lin, Fengjie An

Construction of Gray Wave for Dynamic Fabric Image Based on Visual Attention 575
Zhe Liu, Xiuchen Wang

Theoretical Framework of Responsible Leadership in China 581
XiaoLin Zhang, YangHua Lu

Research on Dual-Ring Network Based TDM Ring and CSMA Ring 587
Haiyan Chen, Li Hua, Donghua Lu

Improved Authentication Model Based on Kerberos Protocol 593
Xine You, Lingfeng Zhang

Predicting Arsenic Concentration in Rice Plants from Hyperspectral Data Using Random Forests 601
Jie Lv, Xiangnan Liu

Design and Implementation of Online Experiment System Based on Multistorey Architectures 607
Wei Li, WenLong Hu, YiWen Zhang

CT-PCA Algorithm Based on Contourlet Transform and KPCA for Image Enhancement 615
Qisong Chen, Maonian Wu, Xiaowei Chen, Yanlong Yang

A Robust STBC-OFDM Signal Detection with High Move Speed Environments 623
Yih-Haw Jan

Application in Evaluating Driving Fatigue Influence Factors by Grey Interval Theory 629
Jixuan Yuan, Zhumei Song, Shiqiong Zhou

An Improved Semantic Differential Method and Its Application on Sound Quality 635
Jixuan Yuan, Zhumei Song, Shiqiong Zhou

Design and Application of Linux-Based Embedded Systems 641
Chunling Sun

Detection of Human Movement Behavior Rules Using Three-Axis Acceleration Sensor 647
Hui Xu, Lina Zhang, Wenting Zhai

Research on Forecasting Models for Material Consumption 653
Qingtian Han, Wenqiang Li, Wenjing Cao

Study on Optimal Model and Algorithm of Equipment Spares 659
Qingtian Han, Wenjing Cao, Wenqiang Li

Research on Role and Context-Based Usage Control Model 665
HaiYing Wu

A No Interference Method for Image Encryption and Decryption by an Optical System of a Fractional Fourier Transformation and a Fourier Transformation 671
Huaisheng Wang

A Minimum Spanning Tree Problem in Uncertain Networks 677
FangGuo He, GuiMing Shao

Author Index 685

Model Retrieval Based on Three-View

FuHua Shang, YaDong Zhou, and HongTao Xie

School of Computer and Information Technology, Northeast Petroleum University, Daqing Heilongjiang 163318, China
Shangfh@163.com, zhouyadong@sogou.com

Abstract. In this paper we propose a new 3D shape retrieval method based on three views. In our approach, top view front view left / right view projection approach is represented by a set of depth images captured uniformly from three view for 3D model matching. Then 2D shape descriptors. dynamic programming distance (DPD) is used to compare the depth line descriptors. The DPD leads to an accurate matching of sequences even in the presence of local shifting on the shape. Experimental results show that it can quickly and efficiently retrieval 3D model.

Keywords: three view, retrieval, depth image, matching, DPD.

1 Introduction

Presently, as the huge demand of computer cartoon and three-dimensional game manufacture, immense amounts of 3D shapes are created and stored.Consequently a 3D shape database system is required in order to make us retrieve the shapes we needed conveniently and accurately[1]. And this leads to the experimental research on 3D retrieval search engine bases on shape, such as the Ephesus search engine at the National Research Council of Canada, the 3D model search engine at Princeton University, the 3D model retrieval system at the National Taiwan University, the Ogden IV system at the National Institute of Multimedia Education, Japan, the 3D search system at the Informatics and Telematics Institute, Greece, the 3D model similarity search engine at the University of Konstanz, and the 3D retrieval engine at Utrecht University[2].

The core problem of model library is the retrieval problem, and using search method based on view to conduct 3D shape retrieval is better than other methods[3]. This article just adopted this method to carry out a research on the model retrieval and proposed a model retrieval method based on three-view.

2 Model Retrieval Based on Three-View

Presently retrieval can be divided into two categories:text-based retrieval and content-based retrieval, and the latter contains retrieval based on two-dimensional projection and retrieval based on shape. This article adopted retrieval based on shape. Shape-based description can be divided into four categories: view-based, image-based, statistical

D. Jin and S. Lin (Eds.): Advances in MSEC Vol. 1, AISC 128, pp. 1–5.
springerlink.com

data-based. The current study shows that search method based on view to conduct 3D shape retrieval is better than other methods and it can also be applied in query search interface based on binary image or 2D sketch[4]. This article presented a new search method based on three-view which can save search time, improve the retrieval efficiency.

Shape-based 3D object recognition is the core issue of computer vision. And there are many ways to classify the existing methods which can be divided into three categories:(1)feature-based approach, generally described by utilizing statistical properties, global and local features, histogram or combination of the above[5], (2)graph-based approach, mainly described by utilizing topological characteristics of objects[6], (3)geometry-based approach, in addition to multi-view 3D search method, there are approach based on deformation and approach based on the volume error[7].

The current content-based 3D model retrieval technology generally includes three steps: preprocessing, feature extraction and similarity matching[8].

And in these three steps, preprocessing is the basis of the other two subsequent steps and should be carried out firstly.

Ⅰ. Preprocessing

The role of preprocessing is to conduct normalization on different coordinate systems in order to make 3D models under different coordinate systems have the same similarity and improve the efficiency, accuracy of feature extraction and similarity matching[9]. As the location, size, orientation of various 3D models are often inconsistent in 3D space, we adopted PCA to carry out preprocessing on models which avoided the inconsistency influence on feature extraction of models.

Ⅱ. Feature Extraction and Similarity Matching

Range images were generated according to 3D model boundary cuboids.

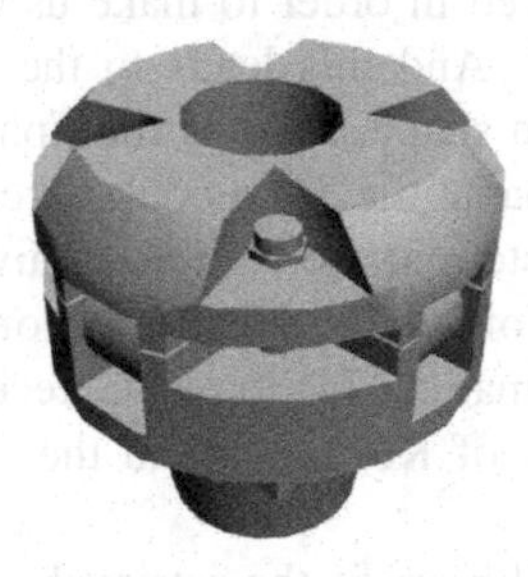

Image 1. Roller kelly bushing of drilling model

Image 2. Range images of roller kelly bushing projection on outsourcing cuboids

In order to compare the two range images I^1 and I^2, 2×N observation sequences $S_{i,j}^1$ and $S_{i,j}^2$ were generated to the two range images, and then we calculated the dynamic programming distance (DPD) between sequences of I^1 and corresponding sequences of I^2. Finally, we made the calculated sum total of DPD of all sequences the dissimilarity value of the two range images.

Take the front view as an example.

```
ooooooooooooooooooooooooo/------\ooooooooooooooooooooooooo
oooooooooooooooooooooooo////----\\\\oooooooooooooooooooooooo
oooooooooooooooooooooo////////---\\\\\\\\oooooooooooooooooooo
ooooooooooooooooooo///////////---\\\\\\\\\\\oooooooooooooooooo
oooooooooooooooo/////////////-----------\\\\\\\\\\\\\ooooooooooooooo
ooooooooooooooo////////------------------------\\\\\\\\oooooooooooooo
ooooooooooooo/////////-/----------------------\-\\\\\\\\\oooooooooooo
ooooooooooooo//////////--/--------------------\--\\\\\\\\\\ooooooooooooo
ooooooooooooo/////////----\-/--------------\-/----\\\\\\\\\ooooooooooooo
oooooooooooo///////////-----\-/-----------\-/-----\\\\\\\\\\\oooooooooooo
oooooooooooo//////////////-----\---/----\---/----\\\\\\\\\\\\\\ooooooooooo
ooooooooooooo//////////////(--//--/---------\--\\--)\\\\\\\\\\\\\\oooooooooooo
oooooooooo/////////////--/--\--/-\--/--\c/--\--/-\--/--\--\\\\\\\\\\\\\oooooooooo
oooooooooo////////////----\-/--\-/--/cccc\--\-/--\-/----\\\\\\\\\\\\oooooooooo
ooooooooo/////////////-------------/ccccccc\-------------\\\\\\\\\\\\\ooooooooo
oooooooo/////////////------------/cccccccccccc\------------\\\\\\\\\\\\\oooooooo
ooooooo//////////////-----------/ccccccccccccccc\-----------\\\\\\\\\\\\\\ooooooo
oooooo//////////------------/cccccccccccccccccc\------------\\\\\\\\\\oooooo
ooooo//////////------------/ccccccccccccccccccccc\------------\\\\\\\\\\ooooo
```

Image 3. Level sequence generated according to depth line

'o', 'c', '/', '\', '-'denotes respectively the external background, the internal background, depth increasing, depth reducing and depth unchanging.

If we simply compared the Hamming Distance of the two sequences, it is likely that we would obtain a large value than the true value. So we can consider the application of DPD to obtain the minimum distance of each pair of sequences.

The formula to calculate DPD between sequences of I^1 and corresponding sequences of I^2 is as follows:

$$d\left(I^1, I^2\right) = \sum_{i=r,c} \sum_{j=0}^{N-1} DPD\left(S_{i,j}^1, S_{i,j}^2\right)$$

'i' denotes if the depth line is a row or a column, and 'j' denotes orders of lines in the depth image.

Here the similarity between two 3D models, O_1 and O_2, is calculated by summing up the similarity of all corresponding images, the formula is as follows:

$$\Delta\left(O^1, O^2\right) = \sum_{i=1}^{3} d\left(I_i^1, I_i^2\right)$$

'i' denotes the value of the rendered image. Here we set the maximum value for i 3 which means that we match through using range images projected by only three views(front view, top view, left/right view). Though views are less, we can get good search results for the lightweight model library and save search time greatly.

3 Analysis of Experimental Result

Matching capabilities of 3D model based on shape is showed by precision/recall curve[10].

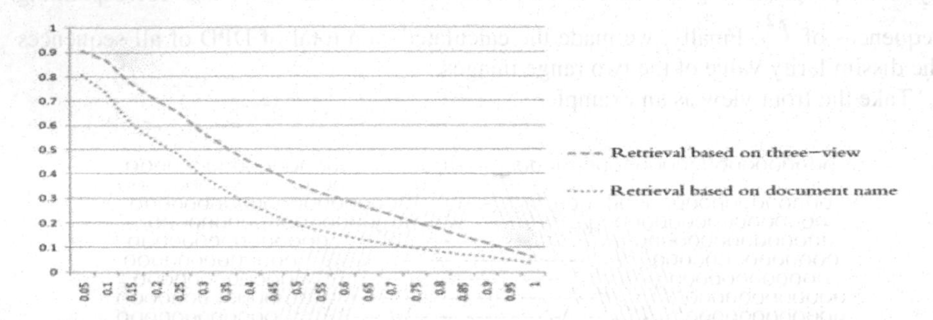

Image 4. Analysis of Experimental Result

After a test comparison, the search results based on document name were much lower than the search results based on three-view. As the model library was self-developed and finished, some of the model names were non-standarded, and the model number was large.While this complexity reflected the advantages of using three-view to search and it was in line with the mechanism of finding models on web.

4 Conclusion

3D drilling model library system based on web can realize the reusability of models and improve the model resource sharing between developers. By increasing model reusability and resource sharing, many invalid development, which was caused by poor communication between developers and incomplete grasp to the model information, can be avoided, thus, improve development efficiency, accelerate the project progress[11].

This paper briefly described the design and implementation of the drilling model library, emphasized the design of retrieval module, especially the new model retrieval method based on three-view. With the abundant development, sharing of 3D models, network 3D model resources will grow rapidly. How to find the required network 3D models is the problem to be solved in the future. 3D model retrieval approach has two main directions: text-based model retrieval and content-based model retrieval. And the research focus of text-based model retrieval is the semantic annotation, which is also the direction to explore on the basis of the model library.

References

1. Okada, Y.: 3D MODEL DATABASE system by hand sketch query, vol. 1, pp. 889–892 (2002)
2. Tangelder, O.W.H., Veltkamp, R.C.: A Survey of Content Based 3D Shape Retrieval Methods. J. of Institute of Information and Computing Sciences (2008)

3. Wagan, A.I., Bres, S., Emptoz, H., Godil, A.: 3D shape retrieval based on Multi-scale Integral Orientations. INSA de Lyon, France National Institute of Standards and Technology (2010)
4. Wagan, A.I., Bres, S., Emptoz, H., Godil, A.: 3D shape retrieval based on Multi-scale Integral Orientations. INSA de Lyon, France National Institute of Standards and Technology (2010)
5. Mahmoudi, S., Benjelloun, M., Ansary, T.F.: 3D Objects Retrieval using Curvature Scale Space and Zernike Moments. Journal of Pattern Recognition Research 6(1) (2011)
6. Tierny, J., Vandeborre, J.-P., Daoudi, M.: Partial 3D shape retrieval by reeb pattern unfolding. Computer Graphics Forum 28 (March 2009) (to appear)
7. Tangelder, J.W.H., Veltkamp, R.C.: A survey of content based 3d shape retrieval methods. Multimedia Tools and Applications 32(3), 441–471 (2008)
8. Zhen, C.L., Jiang, H., Liu, H.T.: Wireless sensor network node level, an effective strategy for energy research. System Simulation 19(10), 2351–2356 (2007)
9. Tang, Y.: Improved 3D model retrieval PCA pre-processing algorithm. System Simulation 20(11) (2008)
10. Shilane, P., Min, P., Kazhdan, M., Funkhouser, T.: The Princeton Shape Benchmark. Proceedings of Shape Modeling Applications 2, 3, 167–178 (2004)
11. Wu, Y.L., Qiu, X.G., Liu, B.H.: Web-based simulation model library system's overall framework. Ordnance Industry Automation Network Information Technology 25(1) (2006)

A Bayesian Network Approach in the Relevance Feedback of Personalized Image Semantic Model

Lei Huang[1,2], Jian-guo Nan[2], Lei Guo[1], and Qin-ying Lin[1,2]

[1] Department of automatic, Northwest Polytechnical University, Xi'an, P.R. China
leiguo@gmail.com
[2] College of Engineering, Air Force Engineering University, Xi'an, P.R. China
leillon@gmail.com, njgzhj@163.com,
qinying@gmail.com

Abstract. Based on a natural and friendly human-computer interaction, relevance feedback is used to determine a user's requirement s and narrow the gap between low-level image features and high-level semantic concepts in order to optimize query result s and perform a personalized search. In this paper, we proposed a novel personalized approach for image semantic retrieval based on PISM (Personalized image semantic model), which use the user queries related to the image of feedback mechanism, dynamic image adjustment semantic similarity of the distribution, and fuzzy clustering analysis, PISM training model to make it more accurate expression of semantic image to meet the different needs of the user's query. And the limitations of image-based semantic memory of learning algorithm, the initial experimental system developed by a number of user feedback to participate in relevant training, which analyzes the performance of the algorithm, the experiments show that the algorithm is a viable theory, with a value of the application.

Keywords: Personalized, image retrieval, image semantic, relevant feedback, Bayesian Networks.

1 Introduction

Human-computer interaction plays an important role in image searches. Based on a natural and friendly human-computer interaction, relevance feedback is used to determine a user's requirement s and narrow the gap between low-level image features and high-level semantic concepts in order to optimize query result s and perform a personalized search. With a number of users participating in relevant feedback training, CBIR system will aggregate abundance of semantic [1, 2]. In a way, CBIR system provides a better result for users' follow-up retrieval. Though CBIR system overcomes the shortage of text-based systems to a certain degree, there still exist some limitations. Because of personalized characteristics of users' relevant feedback, the result from CBIR system always reflects users' common understanding of image semantic. The result is common not special to one user, which leads to great difference between retrieval result and users' expect .In other words, CBIR system cannot provide one user with the result according to his or her personalized needs [3].

D. Jin and S. Lin (Eds.): Advances in MSEC Vol. 1, AISC 128, pp. 7–12.
springerlink.com

In this paper, we proposed a novel image retrieval method based on personalized semantic model. By mining user query logs of different information, this method builds Personalized Image Semantic Model (PISM) for users. Combination of the same images for different people to understand the semantic differences arising analyzed. Use of semantic clustering and relevance feedback learning, build PISM of personalized semantic factor smaller, making semantics with the speed and accuracy of training and gradually growing to be consistent with human subjective inquiry necessary to judge semantics, semantic clustering of shared and binding of a balance in PISM. User information for different query strategy, meet the user needs to return retrieval results.

2 Related Work

Personalized search is an important item in the field of Human-computer interaction research. The correlativity used in Generic image search engine that the all user unanimously approves is thought be the single user's, and The correlativity that adoption unifies method. But the correlativity used in personalized search that adoption the unique computing method for each user. Personalized search can not only raise retrieval precision, but also provide the certain user with the better method that inspects intention. Therefore, personalized search is the purpose that carries out friendly retrieval.

The key of Personalized search that is user Personalized analysis,and establish Personalized profile.the accessing interest can be reasoning and classify by user web behavior mining,including user register information, the interested vocabulary, server logs and cookies logs...etc. the user accessing information can be obtained by the dynamic increment clustering arithmetic and parallel arithmetic.

Personalized search is to take user as the retrieval of centre. according to the user's participation degree can be classify as following:

1) user participation: The user initiative provides his own interest instruction to the system,such as the sort of interest,web page,and keyword.the high quality of user information can be obtained by such mode,and ambiguity less. But there are high request to the user,for example, The consumer wants apprehensibility so feedback of purpose, also the patience participate in system to the show type earth.

2) non-user participation: The user doesn't need show the type of interest direct. Personalized system Statistics a behavior method for user to use Internet,and Discover the consumer's interest from it,such as web usage mining, context search, and query session.

Bayesian networks are directed acyclic graphs, which allow for an efficient representation of the joint probability of a set of random variables [4] [5]. Each vertex in a Bayesian network represents a random variable $X_i \in \{X_1, X_2,\ldots, X_n\}$,$1 \leq i \leq n$, and edges represent dependencies between the variables. It is common to denote $P(X_i=x)$ as the probability of X_i assuming one of its possible values x_i.

The joint probability of a set of random variables can be defined as a function $P(X_1,\ldots,X_n)$.The conditional probability of X_i, given a second variable X_j, is the probability of X_i conditioned on the fact that X_j assumes a value x_j. It is denoted as $P(X_i=x_i/X_j=x_j)$ .The structure of the network implies several conditional independence statements. The random variables X_i and X_j are conditionally independent given a third random variable X_z when $P(X_i/X_j, X_z)= P(X_i/X_z)$.

Bayesian networks reduce the number of parameters needed to characterize a joint probability, which enable the encoding of ratiocinative knowledge and causal relationships in the model, and facilitate efficient computation of posterior probabilities given evidence.

3 Image Retrieval with Semantic Feedback

Inquiring request analyzing data's origin includes inquiring input data, which is provided by users and include extracted features and semantic key words about inquiring image. This can be divided into three processes, validating a user's identity, stylebook training and semantic network analysis. The purpose of validating a user's identity is to ensure whether the user will have the authority to use protected PISM in power. Sample training is based on inquiring and Sample image visual characteristic matching and semantic clustering. Sample training analyses output data by connecting semantic network. Combining the two processes, semantic network analysis shows similarity weight distribute standard. And it can output retrieval conclusion through features matching index and relevant feedback from users. At the same time, it can record feedback conclusion, study personalized semantic, return feedback information to feature data-base, divide weight of semantic similarity in a new way, and replace data to the semantic network node where the user is.

The feedback arithmetic, which is based on personalized image semantic model PISM and personalized semantic, is as follows:

1. Show semantic gathering about inquiring request in measure. Collect image visual features, and then choose what is affiliated with *N-M* sequence from semantic features with *N* observing number by using Bayesian classifier. And by using Bayesian rule [6, 7], we distribute them to effective sort sequence mapping in order to build semantic feature gathering Q_{S_L}. Use inquiring key words to connect semantic network for the purpose of building semantic feature gathering Q_{Sk}. Specially, if $Q_S \neq \phi$, evaluate $w_O = 0, w_U = 0$, perform the fourth step. In other words, execute image similarity matching from low visual feature, or perform the second step.

2. Build inquiring image personalized semantic gathering Q_L on the basis of checking a user's identity.

$$Q_L = \{X_L \mid X_L = f(U_0, Q_S)\} \tag{1}$$

L is personalized semantic feature vector quantities; *f* is the function of X_L, a user's identity and inquiring request semantic adjusting collection. It is obtained by judging the similarity of semantic feature vector X and personalized semantic vector X_L. Perform the third step to finish creating Q_L.

3. Map and build all the image semantic feature collection H in image warehouse. And judge $X_H(X_H \in H)$ from the average clustering arithmetic K-means so that X_H can share in some clustering of $H \cap Q_S$. In other words, there is vector $X_H \in (H \cap Q_S)$, which can be used to create personalized semantic collection Q_{ngL} with the help of Q_S.

4. According to elements in Q_L, that is, image Q personalized semantic vector, we calculate the similarity of elements in Q_L and semantic feature vector of an image I, that is, personalized semantic similarity S_U. Calculate the similarity between users' inquiring image Q and image I from $S(Q, I) = w_L S_L + w_O S_O + w_U S_U$.

w_L、 w_O、 w_U each stands for the weight of low feature similarity S_L between image Q and image I, common semantic similarity S_O, personalized semantic similarity S_U, and satisfy the following $w_L + w_O + w_U = 1$.

Different feedback from the same user and different feedback from different users synthesize the above process, and finally make both inquiring image Q and image I totally have high quality $w_O S_O$ and low quality $w_U S_U$. According to the user, $w_{Ui} S_{Ui} > w_U S_U$ under usual condition. This makes the inquiring of image tally with users' needs. Adjust the value of according to methods chosen by users, in order to display different inquiring results.

5. Output results, rebuild the value range of collection Q_L and create new personalized semantic Q_L'.

4 Experimental Results and Discussions

In order to measure the accuracy of the proposed methods, we have implemented a demo system on a Pentium PC platform running Windows XP. We use a collection of 1240 images which belong to 22 groups, which contain a variety of images with various contents and textures. The experimental results and evaluations are shown and discussed as following.

The comparison is made through single user query in terms of vary w_O and w_U. The average accuracy is then plotted against the number of user feedbacks. The result is shown in Figure 1. As we can see from the results, as users of the distribution of w_O and w_U different values, resulting in images of different semantic similarity on the distribution of output and thus significantly different. The result proves the PISM have the stronger capability of feedback learning.The result proves with the numbers of feedback increasing, the performances of Query by weight method is more than the others query methods. Since it is a better description of the user's information need the retrieval result is improved. For each query,3 feedback iterations are run and the results are reported in Table 1, in comparison with the same user query by using Query by weight method with the vary Similarity weight. As we can see from the results, with the numbers of w_U increasing, although the response time is increasing, the Precision and recall will be improved.

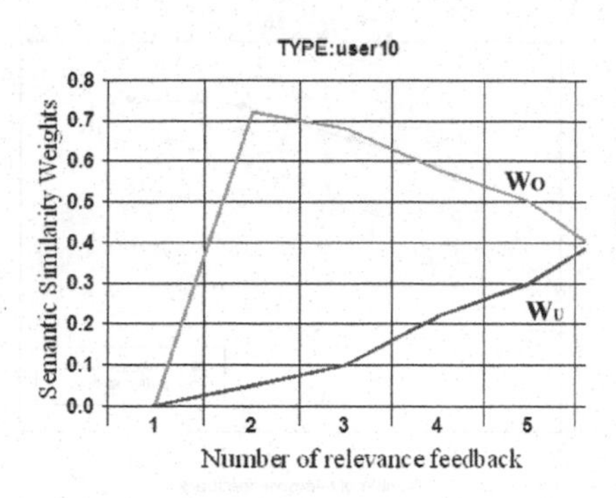

Fig. 1. w_O、 w_U Sampling Distributing Chart

To verifying the effectiveness of our system through the performance measure, we have compared it against other state of the image retrieval systems. We have chosen to compare our method with the retrieval technique MARS used in the [1]. The comparison is made through 5 sets of random queries with 4 feedback iterations for each set of query and the number of correctly retrieved images is counted after each user feedback. The average accuracy is then plotted against the number of user feedbacks.

Table 1. Stat. Result of Query

Similarity weight			Semantic	Recall	Precision	Response Time
W_L	W_O	W_U				
0.6	0.3	0.1	flower	0.969	0.97	1.3s
0.5	0.3	0.2	Grass	0.951	0.955	1.2s
0.2	0.3	0.3	Butterfly	0.831	0.917	2.8s
0.4	0.2	0.4	Human	0.94	0.961	2.3s
0.3	0.2	0.5	Kinetic	0.86	0.884	8.6s
0.5	0.3	0.6	Construction	0.831	0.827	1.6s
0.2	0.3	0.7	sunset	0.588	0.63	2.4s
0.4	0.2	0.8	car	0.618	0.587	2.6s

In Figure 2, we show performances of the Proposed and MARS methods. As we see in the figure, the proposed method has the better performance in these two methods. We performed four random queries on our system. As we can see from the results, our system achieves on average 50% retrieval accuracy after just 4 user feedback iterations and over 95% after 8 iterations for any given query. In addition, we can clearly see that more relevant images are being retrieved as the number of user feedbacks increase. Unlike MARS methods where more user feedback may even lead to lower retrieval accuracy, our method proves to be more stable. It is easily seen from the result that by combining PISM with low-level feature feedback, the retrieval accuracy is improved substantially.

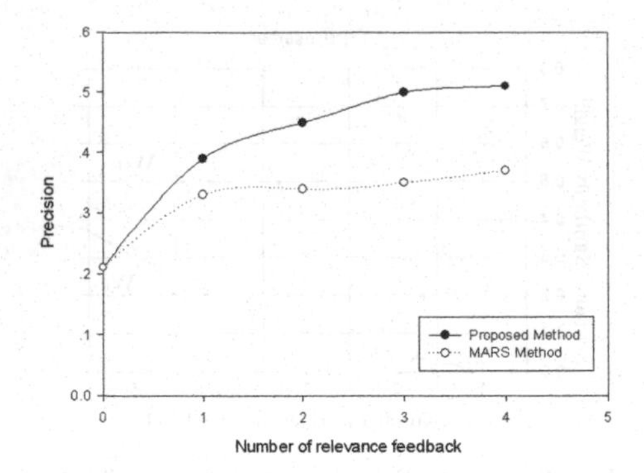

Fig. 2. Comparison of Precision

5 Conclusion

Through the realization of the feedback learning algorithm based on personalized image semantic, the purpose of user relevance feedback was recorded and was dynamic allocated in PISM model, therefore, the system whole retrieval more efficient. And with the increase in the number of feedback, the system capacity about clustering analysis for different semantic would be enhancement. In particular, according to the different user ID, query for the same parameters will retrieve the results of a high degree of support to users. Comparison with other typical methods, the PISM model can be used to improve the retrieval accuracy. Due to the semantic richness and PISM repeatedly called constraints, after two rounds of relevance feedback in the response time is slower. So this will be our next step will focus on solving problems.

References

1. Rui, Y., Huang, T.S., Mehrotra, S.: Relevance feedback: a powerful tool in interactive content-based image retrieval. IEEE Trans. on CSVT 8(5), 644–655 (1998)
2. Rui, Y., Huang, T.S.: Optimizing learning in image retrieval. In: IEEE Conf. on CVPR, South Carolina, USA (2000)
3. Cox, I.J., Miller, M.L., Minka, T.P., et al.: The Bayesian image retrieval system PicHunter: theory, implementation and psychophysical experiments. IEEE Trans. On Image Processing 9(1), 20–37 (2000)
4. Souafi-Bensafi, S., Parizeau, M., Lebourgeois, F., Emptoz, H.: Logical labeling using Bayesian networks. In: Proceedings of IEEE ICDAR, pp. 832–836 (2001)
5. Zhang, R., Zhang, Z.: Addressing cbir efficiency, effectiveness, and retrieval subjectivity simultaneously. In: ACM Multimedia 2003 Multimedia Information Retrieval Workshop, Berkeley, CA (November 2003)
6. Souafi-Bensafi, S., Parizeau, M., Lebourgeois, F., Emptoz, H.: Bayesian networks classifiers applied to documents. In: Proc. IEEE ICPR, vol. 1, pp. 483–486 (2002)
7. Su, Z., Zhang, H.J.: Relevance feedback using Bayesian classifier in content-based image retrieval. In: SPIE (2001)

An Empirical Study on Elastic Effect of Fiscal Expenditure to Household Consumption in China

Jianbao Chen, Huanjun Zhu, and Tingting Cheng

Department of Statistics, School of Economics
Xiamen University, Xiamen 361005, P.R. of China
jbjy2006@126.com, 42111872@qq.com, tingting.cheng@monash.edu

Abstract. By using a series of modern econometric methods, we studied the relationship between fiscal expenditure and household consumption in China. Our results are summarized as follows: (a) there exists a long term equilibrium relationship between them; (b) fiscal expenditure has positive effect to household consumption but there exists a decreasing marginal effect; (c) error spatial auto-regression model is a good choice to study the relation between them, and the household consumption level of local province is not only affected by that of adjacent provinces but also by the fiscal expenditure of local and adjacent provinces.

Keywords: Household Consumption, Fiscal Expenditure, Elastic Effect, Empirical Study.

1 Introduction

In recent three decades, China has attained a great achievement in economic development with a remarkable average growth speed about 10%. However, the low household consumption rate has been a key factor retraining the Chinese economic development. It is essential to study the relationship between the fiscal expenditure and household consumption, the main aim is to propose some appropriate financial policies and enhance household consumption level effectively.

As regard to the relationship between fiscal expenditure and household consumption, many scholars have done a lot of research work and obtained some valuable results. The papers were written by Feldstein (1982)[1], Karras (1994)[2], Hjelm(2002)[3], Li(2005)[4], Zhou and Lu(2005)[5], Zhang and Wu(2007)[6] and etc.. Nevertheless, most of the researches above are confined to discuss whether the relationship is substituted or complementary in terms of empirical study with time series data. This paper tries to study the cointegration relation, spatial dependence relation respectively via time series and sectional data.

2 Cointegration Analysis, Error Correction Model and VAR Analysis

2.1 Co-integration Analysis

The time series data are considered from 1978-2007 in China and collected from Chinese Economy database. Let *LNCO* and *LNFO* be the logarithm forms of

D. Jin and S. Lin (Eds.): Advances in MSEC Vol. 1, AISC 128, pp. 13–18.
springerlink.com

household consumption and fiscal expenditure per capita respectively. By using Engle-Granger test, we easily find that both *LNCO* and *LNFO* are first order unit root series under test level 5%. Therefore, the following fitted linear regression can be obtained:

$$LNCO_t = 1.61 + 0.88 LNFO_t + \varepsilon_t, \qquad t = 1,2,\cdots,20 \tag{1}$$

Its residuals are a white-noise series at test level 5%. This implies that there exists a long-term equilibrium relationship between *LNCO* and *LNFO*, and the long term elastic coefficient of fiscal expenditure with respect to household consumption is 0.88.

2.2 Error Correction Model Analysis

Next, we study the short term elastic coefficient between two variables. It is not hard to establish an error correction model (ECM) as follows:

$$\Delta LNCO_t = 0.45 \Delta LNFO_t - 0.02 ECM_{t-1} \qquad t = 2,3,\cdots,20 \tag{2}$$

where $\Delta LNCO_t = LNCO_t - LNCO_{t-1}$, $\Delta LNFO_t = LNFO_t - LNFO_{t-1}$. Equation (2) indicates the short term elastic coefficient of *LNCO* with respect to *LNFO* is 0.45 which is much smaller than the long term effect of financial policy. The negative coefficient of EMC is coordinate with the reverse adjustment mechanism.

2.3 Vector Auto-Regression (VAR) Analysis

The VAR analysis can tell us the path and intensity of the effect caused by financial policy to the household consumption level. The VAR analysis follows the next three steps.

Step 1. Fix the optimal lagged order of the VAR model. By using AIC and SC criterion, we find that the optimal lagged order of the VAR model is 1 to 2, i.e., VAR(2) model. The estimation results are as follows:

$$\underset{}{LNCO_t} = \underset{(5.67)}{1.15 LNCO_{t-1}} \underset{(-0.90)}{- 0.18 LNCO_{t-2}} \underset{(1.74)}{+ 0.16 LNFO_{t-1}} \underset{(-1.57)}{- 0.15 LNFO_{t-2}} \underset{(1.23)}{- 0.11} \tag{3}$$

$$LNFO_t = \underset{(-0.46)}{-0.17 LNCO_{t-1}} \underset{(0.80)}{+ 0.29 LNCO_{t-2}} \underset{(8.89)}{+ 1.51 LNFO_{t-1}} \underset{(-3.31)}{- 0.58 LNFO_{t-2}} \underset{(-1.96)}{- 0.33} \tag{4}$$

Here the data in brackets represent t-statistics. The coefficient 1.15 reveals how many percentages of $LNCO_t$ will increase when $LNCO_{t-1}$ increase 1%. The t-statistics in equation (4) show that $LNFO_t$ is mainly affected by $LNFO_{t-1}$ and $LNFO_{t-2}$ with inhibitory action. That is, when $LNFO_{t-2}$ increase 1%, current $LNFO_t$ will decrease 0.58 percentages. However, 1% increase of $LNFO_{t-1}$ will lead to 1.51% positive change of $LNFO_t$.

Step 2. Analyze of the impulse response function curve. The impulse response function curve of model VAR(2) demonstrates the interactional path between the fiscal expenditure and household consumption per capita, which is shown in Fig. 1. The left part of Fig.1 shows the reaction of both *LNCO* and *LNFO* to a standard innovation change of *LNCO* ; the right part of Fig.1 shows the reaction of both *LNCO* and *LNFO* to a standard innovation change of *LNFO* .

On the basis of analyzing the curves, we find that the increase of fiscal expenditure will generally lead to the improvement of household consumption and the increasing household consumption can also enhance the fiscal expenditure in a long period as it boosts the economic development. Such an interaction between them is helpful to the aim of expanding domestic demand.

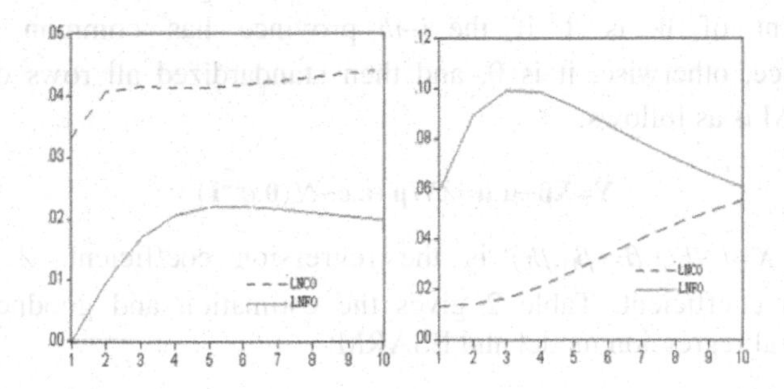

Fig. 1. The reaction of *LNCO* and *LNFO* to innovation change

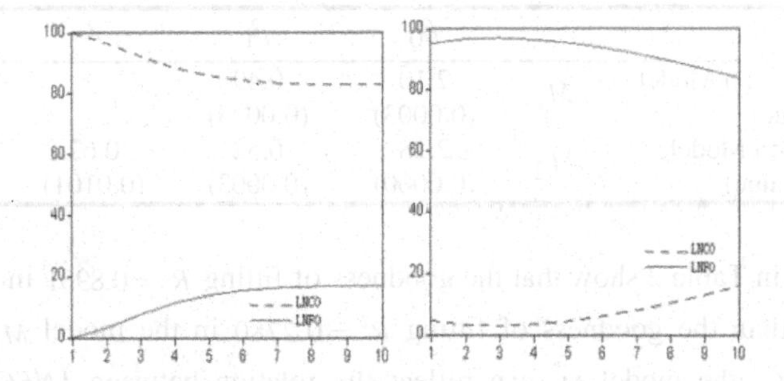

Fig. 2. Variance decomposition of predictive error of *LNCO* and *LNFO*

Step 3. Decompose the expected variance of predictive error of *LNCO* and *LNFO* . Fig.2 indicates that the influences of *LNCO* and *LNFO* chiefly appear in medium to long term while influences of *LNCO* and *LNFO* mainly occur in a long term period.

3 Spatial Dependence Relation

Next, we will use the sectional spatial econometric models (see Anselin[7], 1988; Lesage[8], 1997) to study the spatial dependence relation between *LNCO* and *LNFO* , the data of two variables are sample data from 31 provinces of China in 2007.

The spatial auto-correlation test of regression residuals demonstrates the existence of spatial correlation and suggests an error spatial auto-regression model (ESARM) to be used. The test results are listed in Table 1.

Table 1. The Spatial autocorrelation test of regression residuals

	Moran I Statistics	LR	Wald Value	LM
Indicator values	-2.23	5.01	9.66	4.03
p-value	0.0330	0.0251	0.0019	0.0446

We use W to denote the spatial structure matrix among all provinces of China, the (i,j)–*th* element of W is 1 if the *i–th* province has common border with the *j–th* province, otherwise, it is 0, and then standardized all rows of W as 1. A general ESARM is as follows:

$$\mathbf{Y}=\mathbf{X}\boldsymbol{\beta}+\boldsymbol{\mu},\boldsymbol{\mu}=\lambda\mathbf{W}\boldsymbol{\mu}+\boldsymbol{\varepsilon},\boldsymbol{\varepsilon}\sim N(\mathbf{0},\sigma^2\mathbf{I}) \tag{5}$$

Here $Y=LNCO,X=LNFO,\beta=(\beta_0,\beta_1)'$ is the regression coefficient, λ is the error autocorrelation coefficient. Table 2 gives the estimation and goodness of fitting results of general regression model and ESARM.

Table 2. Parameter estimation and goodness of fitting

		β_0	β_1	λ	R^2
General Regression Model	M_1	2.10	0.49		0.2780
(p-value)		(0.0003)	(0.0023)		
Error Regression Model	M_2	2.28	0.52	0.62	0.8931
(p-value)		(0.0000)	(0.0003)	(0.0101)	

The results in Table 2 show that the goodness of fitting $R^2=0.8931$ in model M_2 is much higher than the goodness of fitting $R^2=0.2780$ in the model M_1. Compared with model M_1, the model M_2 can reflect the relation between *LNFO* and *LNCO* much better. The elastic coefficient in the model M_2 is 0.52. By using generalized difference method and estimation results in Table 2, we can rewritten the model (5) as:

$$Y=0.62\mathbf{W}Y+X\beta-0.62\mathbf{W}X\beta+\varepsilon \tag{6}$$

Equation (6) reveals that the provincial *LNCO* is not only affected by the household consumption level of adjacent provinces but also by the *LNFO* of local and adjacent provinces. Except *LNFO* of its adjacent province is negative, others are positive. This implies that the change of resource allocation caused by fiscal expenditure may affect economic development and consumption level.

4 Analysis of Current Elastic Coefficient

The current elastic coefficients of *LNFO* with respect to *LNCO* in the models M_1 and M_2 are 0.49 and 0.52 respectively, both values are much smaller than the long term elastic coefficient 0.88 in the cointergation analysis, and bigger than short term elastic coefficient 0.45 in the ECM. Therefore, we need to explore whether there exists a decreasing marginal elastic coefficient between them by adding the quadratic term of *LNFO* into the above mentioned models, the corresponding estimation and goodness of fitting results are listed in Table 3.

Table 3. Parameter estimation and goodness of fitting

		β_0	β_1	β_2	λ	R^2
General regression model	M_3	–	1.64	-0.068		0.2733
(p-value)			(0.0000)	(0.0004)		
ESRM	M_4	–	1.88	-0.47	0.62	0.8705
(p-value)			(0.0000)	(0.0011)	(0.0101)	

Remark: β_2 represents the coefficient of the quadratic term of *LNFO*.

From Table 3, we find that the goodness-of-fitting of models M_3 and M_4 have not much changes compared with the models M_1 and M_2, this implies their fitting effects are almost same. However, the two coefficients of quadratic term of *LNFO* in models M_3 and M_4 are negative values and significant under test level 5%. The negative values indicate that there exists an inverted-U relationship between *LNCO* and *LNFO*, i.e., the marginal elastic coefficient of household with respect to fiscal expenditure is decreasing. It is not hard to find that all values of *LNFO* for provinces in China are located at the right side of the inverted-U curve. This is a contradict result considered from its outside surface. In fact, the fiscal expenditures in all provinces of China are used to support government consumption, the ratio of government consumption to fiscal expenditure is too high, and it is not helpful to promote household consumption level. In order to promote household consumption level, we should decrease the ratio of government consumption to fiscal expenditure and increase people's income in China.

5 Summary

In this paper, by using a series of econometric methods, which include cointegration analysis, ECM, VAR model, Impulse response function, variance decomposition and sectional ESARMs, we find the following results: (a) there exists a long term equilibrium relationship between fiscal expenditure and household consumption per capita; (b) fiscal expenditure has positive effect to household consumption but there exists a decreasing marginal effect; (c) ESARM is a good choice to study the relation between fiscal expenditure and household consumption, and the local household consumption level is not only affected by the household consumption level of adjacent provinces but also by the financial expenditure of local and adjacent provinces. Because there exists an inverted-U shape between fiscal expenditure and household consumption, we suggest our government to allocate more fiscal expenditure to promote people's income; the aim is to expand household consumption.

Acknowledgements. This work was partly supported by the MOE Project of Key Research Institute of Humanities and Social Sciences at Universities (07JJD790145), and the Key Research Projects of Social Sciences of China (08&ZD034 and 09AZD045).

References

1. Feldstein, M.: Journal of Monetary Economics 9(1), 1–20 (1982)
2. Karras, G.: Journal of Money: Credit and Banking 26(1), 9–22 (1994)
3. Hjelm, G.: Journal of Economics 24(7), 17–39 (2002)
4. Li, G.Z.: World Economics 28(5), 38–45 (2005) (in Chinese)
5. Zhou, Y.F., Lu, Y.H.: Statistic Research 12(10), 67–71 (2005) (in Chinese)
6. Zhang, Z.J., Wu, D.Y.: The Journal of Quantitative & Technical Economics 14(5), 53–61 (2007) (in Chinese)
7. Anselin, L.: Spatial Econometrics: Methods and Models. Academic Publishers, Kluwer (1988)
8. Lesage, J.P.: International Regional Science Review 20(1), 113–129 (1997)

Extraction and Recognition of Bridges over Water in High Resolution SAR Image

XiongMei Zhang, JianShe Song, ZhaoXiang Yi, and JunHui Xu

Xi'an Research Institute of High Tech
Xi'an 710025, China
zxw.ok@163.com

Abstract. Based on the characteristics of bridges over water in high resolution SAR image, a novel method for bridge extraction and recognition by combining the multi-scale decomposition and region analysis is proposed. Firstly, the nonsubsampled pyramid (NSP) transform is employed to denoise the SAR image. And then, by using the information provided by the multi-scale subbands and analyzing the region characteristics in the segmentation results, the contour of water region is extracted. Finally, the bridge is detected and recognized according to the knowledge of bridges over water. Experimental results obtained on real SAR images confirm the effectiveness of the proposed method.

Keywords: SAR image, bridge, multi-scale decomposition, nonsubsampled pyramid (NSP), image segmentation.

1 Introduction

With the development of Synthetic Aperture Radar (SAR) technique, the acquisition of high quality and high spatial resolution SAR images becomes available. Accordingly, the recognition of bridges over water in high resolution SAR images has received an increasing amount of attention from the image processing community and many methods have been proposed over the last few years [1-3]. However, due to the complex nature of SAR image as well as the complicated terrain appearances, the above methods may fail in many cases.

As an important component of nonsubsampled contour transform (NSCT) [4], the nonsubsampled pyramid (NSP) transform is a fully shift-invariant and multi-scale expansion. When used to image, it can filter the noises as well as the high frequency information to obtain low-pass subbands of the same size to the original one. All the above characteristics enable it to deal efficiently with images having smooth contours, thereby providing robust performance when used in image processing. In this paper, based on the aforementioned NSP and the region analysis, a novel method for extraction and recognition of bridges over water is proposed (see Fig.1). The ideal is to automatically obtain the number of bridges in the coarse subbands of NSP which is fed back to the procedure of the fine subbands to extract the bridges. Experiments carried out on real SAR images show that the new method is able to extract bridges precisely and effectively.

D. Jin and S. Lin (Eds.): Advances in MSEC Vol. 1, AISC 128, pp. 19–24.
springerlink.com

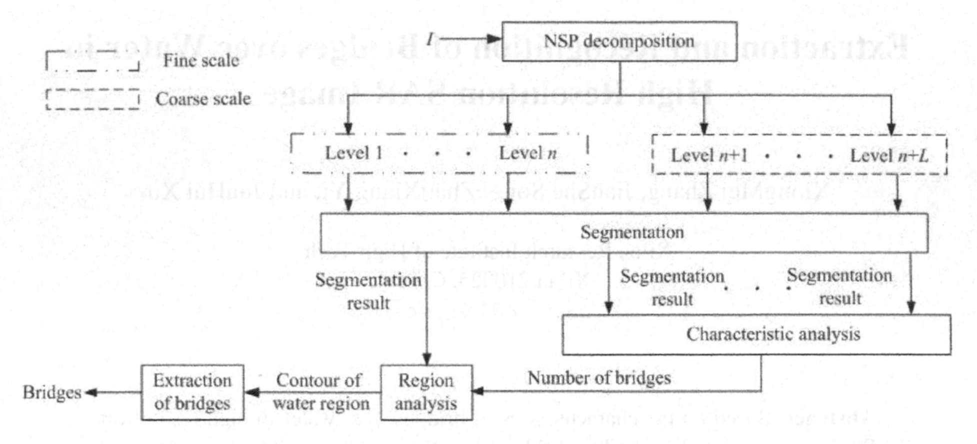

Fig. 1. Block scheme of the proposed method

2 Image Denoise Based on NSP

As a foundational component of NSCT, NSP provides multi-scale decomposition and can be iterated repeatedly on the low-pass subband output of previous stage. The building block of NSP is a two-channel nonsubsampled filter bank (NSFB). Since NSFB has no downsampling or upsampling, it is shift-invariant. The perfect reconstruction condition is given as

$$H_0(z)G_0(z)+H_1(z)G_1(z)=1 \tag{1}$$

Particularly, the $G_0(z)$ and $G_1(z)$ are low-pass and high-pass. Thus, they can filter certain parts of the noise spectrum in the processed pyramid coefficients. The filters for subsequent stages are obtained by upsampling the filters of the first stage.

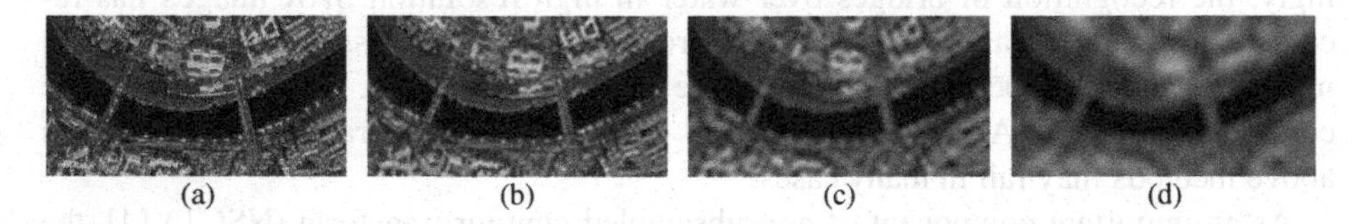

Fig. 2. SAR image and its NSP decomposition: (a) the original SAR image; subband of (b) stage1; (c) stage 2; (d) stage 3

The NSP is shift-invariant such that each pixel of the transform subbands corresponds to that of the original image in the same location. Therefore, we employ the NSP to denoise the original SAR images. A SAR image including bridges over water as well as its 3-stage decomposition of NSP are shown in Fig.2. It can be observed that with the increase of stage, though more noises are removed, the contour and edge are blurred, which is unfavorable in bridges extraction. An effective solution is to obtain the overall information in coarse stages and detailed information such as edges and parameters in fine stages.

3 Water Region Extraction

Due to the intuitive properties and simplicity of implementation, Otsu [5] is adopted here to segment the low-pass subbands of NSP. To improve the segmentation performance, we enhance the original SAR image before the NSP analysis. Fig.3 displays the segmentation results of NSP subbands of enhanced image of Fig.2 (a). As the figure shows, in the coarse stages, though the contours of bridges and water are distorted and blurred, it is easy to locate bridges, while in the fine stages, though there are lots of irrelevant information, the contour and shape of bridges and water are well retained, which is propitious to extract the bridges accurately and precisely.

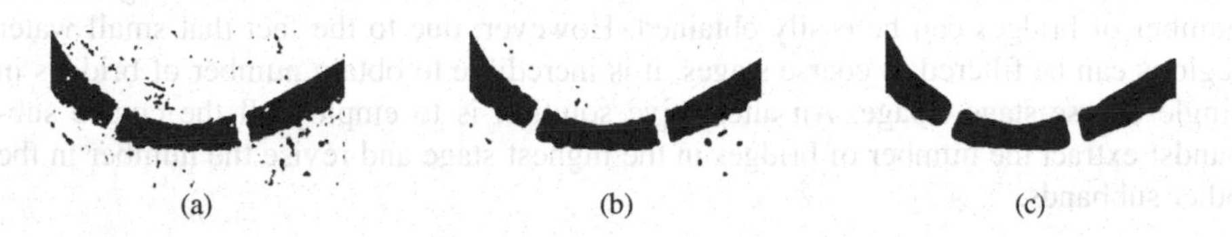

Fig. 3. Segmentation of subbands of enhanced SAR image: segmentation result of (a) stage1; (b) stage 2; (c) stage 3

3.1 Characteristic Analysis in Coarse Stages

The challenge here is to select the features and rules that allow one to represent and extract the objects of interest. Thus, a large set of features have been tested over a variety of data. Here we selected the ones that have a simple implementation and good performance.

1) Eccentricity: The eccentricity of a boundary is defined as the ratio of the major to the minor axis. Compared to the minor axis of water region, the minor axis of false alarm is usually shorter. However, the minor axis of water region between two adjacent bridges is also short. Therefore, we use the minor and major axis together to filer false alarm. Let max_*ax* be the maximum minor axis, ma_*ax*(*i*) and mi_*ax*(*i*) the major and minor axis of region *i*, respectively. The water region can be extracted by

$$\begin{cases} \mathrm{mi\_}ax(i)/\mathrm{max\_}ax \geq a \\ b \leq \mathrm{ma\_}ax(i)/\mathrm{max\_}ax \leq c \end{cases} \tag{2}$$

where *a*, *b* and *c* are constant.

2) Compactness: The compactness of a region is defined by $C=P^2/A$, where A is the number of pixels in the region and P is the length of its boundary. Compared to the region of false alarm, the water region is generally bigger and smoother, therefore leading to a higher value in C. So, we can extract water region by

$$\begin{cases} C_i/C_{\max} \leq T_c & i \in water \\ C_i/C_{\max} > T_c & else \end{cases} \tag{3}$$

where C_i is the compactness of region i, $C_{\max}$ is the maximum compactness of all regions and $0< T_c <1$ is the threshold.

3) Proximity interaction: The proximity interaction $D_{i,j}$ of regions R_i and R_j is defined as the minimum distance between their extremity points [6]. In the high resolution SAR images, though the bridges can not be simplied to two parallel lines, they really take up small regions and the breadth of bridge is just about few pixels. So, there is no bridge between two regions with high value in D and the extraction rule can be defined as

$$D_{i,j} > T_D \times \max\_ax, \quad j \in [1,...,n] \tag{4}$$

where $0<T_D<1$ is the threshold and n is the number of regions in the binary image.

After the above procedure, there will be only water regions in the binary image. According to the fact there exists one bridge between two adjacent water regions, the number of bridges can be easily obtained. However, due to the fact that small water regions can be filtered in coarse stages, it is incredible to obtain number of bridges in single coarse stage image. An alternative solution is to employ all the coarse subbands: extract the number of bridges in the highest stage and revise the number in the other subbands.

3.2 Region Analysis in Fine Stages

As the Fig.4 shows, compared to water regions, the land object which is wrongly assigned to water regions is small. Accordingly, compared to the land target, the objects such as islands and ships are much smaller. Based on this observation, the water region can be extracted by:

$$\begin{cases} A_i / A_r \geq T_s & i \in water \\ A_i / A_r < T_s & else \end{cases}, \quad i \in [1,...,n] \tag{5}$$

where A_i is the area of region i , A_r is the area of reference region.

The land region can also be extracted by (5). As to the water region, the reference region is defined as the smallest water region. As to the land region, due to the fact that the bridges are connected with the land while the islands as well as ships are usually isolated, small region in water, the reference region is defined as the biggest region in land.

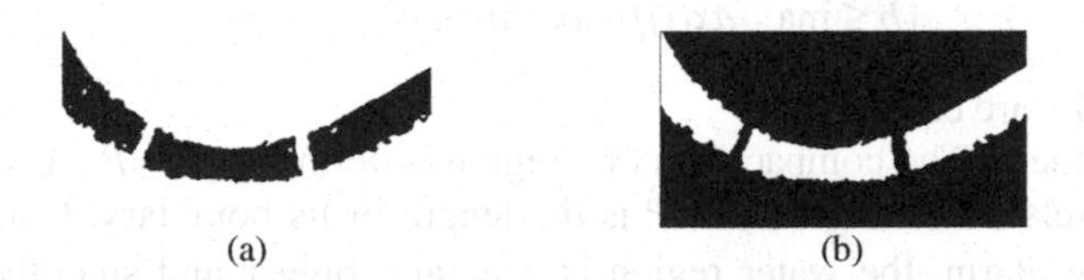

Fig. 4. Filtering of false alarms: (a) land object; (b) water regions

Fig.4 displays the result of Fig.3 (a) using (5). It can be observed that this technique can effectively filter the false alarms while keep contour and edge of bridges intact.

4 Bridge Extraction

According to [3], in the neighborhood of bridge, the points in the contour of water regions are almost in two parallel lines with shortest distance. However, it is time consuming to search and validate all the points in the boundary of water regions one by one. An effective solution is to extract the extreme point, convex point of two adjacent water regions *A* and *B* firstly, denoted as S_A, S_B. Then extract the bridge edges from these candidate points.

1) Extract the minor axis of the largest water region, denoted as *W* and set *W* be the upper bound of the length of bridge;

2) Compute the shortest and secondly shortest distance between *A* and *B*, denoted as min_*dist*, smin_*dist* and let *Pa*_min, *Pb*_min, *SPa*_min, *SPb*_min be the endpoints;

3) Extract the subsets of S_A, S_B, i.e., S_{A1} and S_{B2}:

$$S_{A1}= \{p \mid dist(p, q) \leqslant \mathrm{min\_}dist + \lambda,\ p \in S_A,\ q \in S_B\} \tag{6}$$

$$S_{B1}= \{q \mid dist(q, p) \leqslant \mathrm{min\_}dist + \lambda,\ q \in S_B,\ p \in S_A\} \tag{7}$$

where $1 < \lambda < 3$ is the admitted error.

4) Extract the two points with the highest curvature in S_{A1}, S_{B2}, respectively, denoted as P_{A1}, P_{A2}, and P_{B1}, P_{B2};

5) Draw up the edge lines (i.e., *L*1, *L*2) of bridge based on the points *Px*_min, *SPx*_min and P_{x1}, P_{x2} ($x \in (A, B)$);

6) If *L*1and *L*2 satisfy the following constraints, accept them.

a) *L*1 and *L*2 are almost parallel;

b) *L*1 and *L*2 are almost same in length and shorter than *W*;

c) The furthest distance between *L*1 and *L*2 is less than *W*/2.

5 Experimental Results and Analysis

Several experiments have been organized and carried out on different SAR images. Here, we just show three of them. All the results are shown in Fig.5. Fig.5(a), (b), (c) and (d) are the original SAR images including ridges over water, the obtained bridges, the superposition of bridges over the water contours and the superposition of bridges over the original SAR images, respectively. In the second image, the ship is just under the bridge, making the bridge extraction and recognition more difficult. Due to the employment of information in multi-scale subbands, our method extracted the bridge without disturbance. In the third image, the bridges in the northeast are close to each other, forming a relatively small water region. However, due to the denoising procedure using NSP, the large false alarms are removed while the small water region is retained intact, making the extraction of the adjacent bridges practicable. As to the extracted bridge lines in Fig.5 (b), though they are not perfect in parallelism and usually wider than the real ones, due to the influence of attachment (e.g., street lamps and billboards), the obtained results is acceptable.

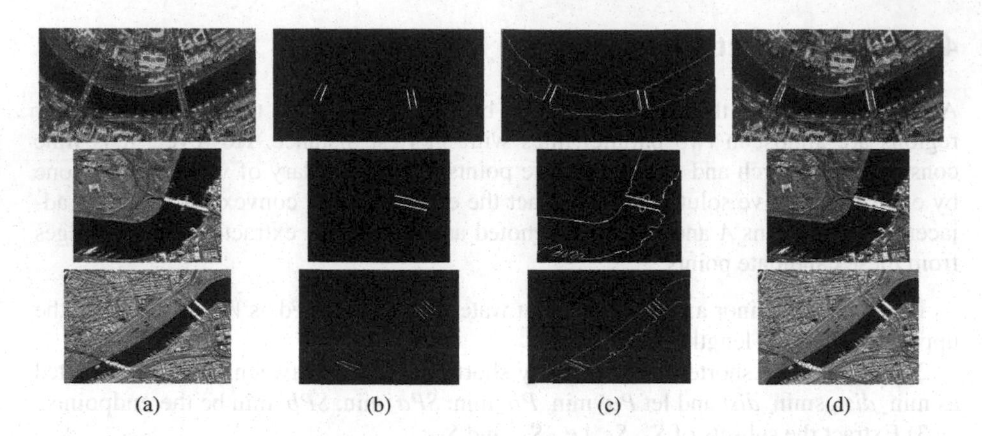

Fig. 5. Experiments: (a)original SAR images; (b)the extracted bridges; (c)bridges superposed over edge maps of water regions; (d)bridges superposed over the original SAR images

6 Conclusion

In this paper, a novel method for extraction and recognition of bridge over water in high resolution SAR image is presented. The ideal is to employ the multi-scale information in the subbands of NSP as well as the specific characteristics of bridges over water in SAR image to extract and recognize bridges. Experimental results show that the new method can be applied to SAR image without complicated pre-processing operations and can extract the bridges exactly. Further work in this field will focus on generalizing the proposed method to images captured by other sensors.

Acknowledgments. This paper is supported by the NSFC of China (61072141).

References

1. Wang, Z., Li, Z., Su, Y.: SAR image bridge detection using regularized heat diffusion equation. Modern Radar 31, 46–49 (2009)
2. Cheng, H., Yu, Q., Liu, J., Tian, J.: SVM-based detection of bridge object over water. Journal of Astronautics 26, 600–605 (2005)
3. Min, L., Tang, Y., Shi, Z.: Segmentation and recognition of bridge over water based on Mumford-Shah model. Infrared and Laser Engineering 35, 499–504 (2006)
4. Da Cunha, A.L., Zhou, J., Do, M.N.: The nonsubsampled contourlet transform: theory, design, and applications. IEEE Trans. on Image Processing 15, 3089–3101 (2006)
5. Zhang, J.: The study of methods of SAR image segmentation. Computer Knowledge and Technology 7, 648–650 (2011)
6. Bernad, G.P., Denise, L., Réfrégier, P.: Hierarchical Feature-Based Classification Approach for Fast and User-Interactive SAR Image Interpretation. IEEE Geoscience and Remote Sensing Letters 6, 117–121 (2009)

Unsupervised SAR Imagery Segmentation Based on SVDD

XiongMei Zhang, JianShe Song, ZhaoXiang Yi, and RuiHua Wang

Xi'an Research Institute of High Tech
Xi'an 710025, China
zxw.ok@163.com

Abstract. Based on the extraction of texture features, the Bayesian decision rule is employed to identify the decision threshold that separates the target from the background in the magnitude image. Then, the training samples for the SVDD classifier are automatically selected and used to train the classifier. Finally, the trained SVDD classifier is used to classify the rest pixels of the thresholding process. Experimental results obtained on real and simulated SAR imageries demonstrate the effectiveness of the proposed method.

Keywords: SAR imagery, support vector domain description(SVDD), unsupervised segmentation, texture feature, Bayesian decision.

1 Introduction

The segmentation of synthetic aperture radar (SAR) imagery is still one of the most challenging tasks in image processing and has received an increasing amount of attention from the image processing community. Many segmentation methods for SAR images have been proposed over the last few years. Among them, the thresholding method and method of integrating with some specific theories are widely used and studied[1].

Due to the intuitive properties and simplicity of implementation, image thresholding enjoys a central position in image segmentation[2]. However, because of the complex nature of SAR images, none of the obtained results is desirable. To fully exploit its complexity, the Support Vector Machine (SVM)[3] was introduced, which can fully utilize the information of intensity and texture and achieve better performance than threshodling method. But in these methods, SVM is usually used in supervised way and time consuming in training phase due to the nature of SVM. Recently, another kernel technique, the support vector domain description (SVDD)[4], was introduced in image processing. SVDD aims at mapping the data into a high dimensional feature space where a hypersphere encloses most of the patterns belonging to the target class and rejecting the rest. In [5] and [6], SVDD was demonstrated to be effective in solving classification and change detection problem.

Based on the above analysis, in this paper, a novel unsupervised segmentation method for SAR image is proposed. Both the SVDD and threshodling method are included, aiming at separating the target from the background. To properly constrain the

D. Jin and S. Lin (Eds.): Advances in MSEC Vol. 1, AISC 128, pp. 25–31.
springerlink.com

learning process, an unsupervised way for identifying examples is adopted, which is based on the selective Bayesian thresholding[7] of the magnitude image. Due to the specific nature of segmentation problem, this procedure leads to the identification of both positive and negative examples. The negative examples are also included in the training of SVDD to improve the description capability. Segmentation result is achieved by applying the SVDD classifier to the rest pixels of the thresholding processing. Experimental results confirm the effectiveness of the proposed method.

2 Proposed Methodology

Let I be SAR image of size $M \times N$. The proposed method consists of two steps (see Fig.1):1) initialization of method using a Baysian thresholding of the magnitude image; 2) generation of segmentation result using SVDD classifier including negative examples. These two steps are described in detail in the following sections.

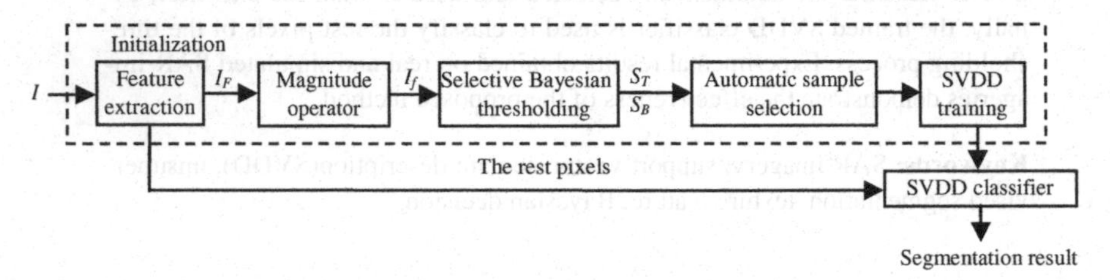

Fig. 1. Block scheme of the proposed method

2.1 Initialization

Initialization aims to segment the image roughly and identify the sets S_T and S_B of target and background pixels to be used as seeds for training SVDD classifier. In our method, the sets S_T and S_B would be directly merged into the final result, and the reliability of sets S_T and S_B would affect the reliability of classifier trained on them, so we should improve the accuracy of this procedure to make sure pixels in S_T and S_B are associated with high probability to belong to target and background areas.

Thresholding is the simplest way to segment images and the selection of proper threshold is of fundamental importance. In the literature, several threshold-selection methods have been proposed. Among them, we adopt the threshold-selection method based on the Bayesian decision theory, which is shown to be effective in many segmentation scenarios. However, due the complexity nature of SAR images, the threshold obtained from the original images can not separate the target from the background correctly. An effective solution is to estimate threshold from the magnitude image. Here, we first extract the Gabor texture of each pixel in a sliding window as well as the mean value to obtain a multidimensional feature vector. Then the magnitude of each feature vector is computed. Finally, the selective Bayesian thresholding is applied to the magnitude image I_f.

The threshold value T obtained from the above approach can separate the target from the background in I_f. However, due to the loss of information of magnitude operator, the result is affected by high uncertainty and sometimes incorrect when the magnitude of target and background is interlaced. Nevertheless, the value T represents a relative reasonable threshold to identify the sets S_T and S_B. According to this observation and the following F. Bovolo [6], the offset constants δ_1, δ_2 are introduced and the final threshold result is defined as (see Fig.2(a)):

$$S_B = \left\{ \boldsymbol{x}_n \in \Re^d \left| i_n^f \le T - \delta_1 \right. \right\}_{n=1}^{M \times N} \tag{1}$$

$$S_T = \left\{ \boldsymbol{x}_n \in \Re^d \left| i_n^f \ge T + \delta_2 \right. \right\}_{n=1}^{M \times N} \tag{2}$$

where $\boldsymbol{x}_n$ is a d-dimensional feature vector of the nth pixel in I_F and i_n^f is the magnitude of the nth pixel in I_f.

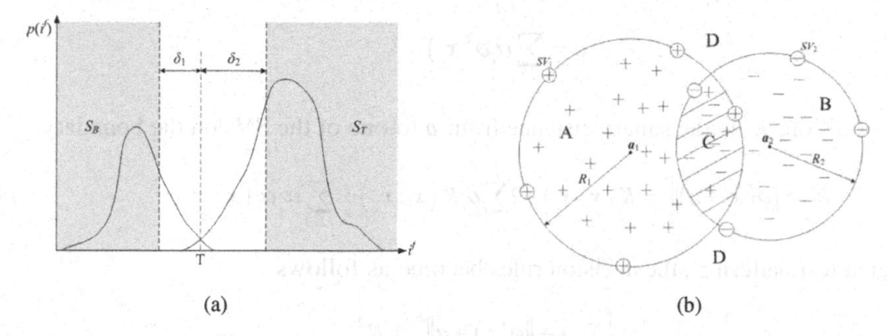

Fig. 2. (a)thresholding segmentation of introducing offsets; (b)revised SVDD including positive and negative examples

Depending on the size of input image, the cardinality of S_T and S_B can be high. To decrease the computational cost, we randomly select part of S_T and S_B as the training examples for the SVDD classifier.

2.2 Image Segmentation Base on SVDD Including Negative Examples

In this part, we employ the SVDD technique to classify the rest patterns into target class and background class. Given a set $X = \{\boldsymbol{x}_i\}_{i=1}^k$, SVDD aims at defining a minimum volume hypersphere in the kernel space to include all (or most) of the target patterns in X by minimizing a cost function. To reduce the effect of outliers, we introduce slack variable ξ_i and the minimization problem of the minimum enclosing ball (MEB) with center $\boldsymbol{a}$ and radius R ($R > 0$) can be defined as:

$$\min_{R,\boldsymbol{a},\xi_i}\left(R^2+C\sum_{i=1}^{k}\xi_i\right) \tag{3}$$

$$\text{s.t. } \left\|\phi(\boldsymbol{x}_i)-\boldsymbol{a}\right\|^2 \le R^2+\xi_i,\ \xi_i \ge 0,\ \forall i=1,\cdots,k \tag{4}$$

where C is regularization parameter that gives the tradeoff between the volume of the description and the errors. The above problem is usually solved by:

$$\max_{\alpha_i}\left\{\sum_{i=1}^{k}\alpha_i K(\boldsymbol{x}_i,\boldsymbol{x}_i)-\sum_{i,j=1}^{k}\alpha_i\alpha_j K(\boldsymbol{x}_i,\boldsymbol{x}_j)\right\} \tag{5}$$

$$\text{s.t. } \sum_{i=1}^{k}\alpha_i=1,\ 0\le\alpha_i\le C,\ \forall i=1,\cdots,k \tag{6}$$

In the minimization of the above problem, a large fraction of the weights become 0 and only a few patterns $\boldsymbol{x}_i$ with non-zero α_i which are called as support vectors (SVs). Then the center $\boldsymbol{a}$ is expressed as a linear combination of objects with non-zero α_i :

$$\boldsymbol{a}=\sum_{i=1}^{k}\alpha_i\phi(\boldsymbol{x}_i) \tag{7}$$

By definition, R^2 is the square distance from $\boldsymbol{a}$ to one of the SVs on the boundary:

$$R^2=\left\|\phi(\boldsymbol{x}_i-a)\right\|^2=K(\boldsymbol{x}_i,\boldsymbol{x}_i)-2\sum_{j=1}^{k}\alpha_i K(\boldsymbol{x}_i,\boldsymbol{x}_j)+\sum_{j,l=1}^{k}\alpha_j\alpha_l(\boldsymbol{x}_j,\boldsymbol{x}_l) \tag{8}$$

Given a test pattern z , the decision rules become as follows:

$$z\in\begin{cases} S_T \leftrightarrow \left\|\phi(z)-\boldsymbol{a}\right\|^2 \le R^2 \\ S_B \leftrightarrow \left\|\phi(z)-\boldsymbol{a}\right\|^2 > R^2 \end{cases} \tag{9}$$

The above SVDD classifier involves only target examples in the definition of the cost function. As a result, the obtained decision boundary is usually secund to the target and can not classify patterns precisely. To improve the data description of the SVDD, we involve both positive and negative examples and reformulate the problem as follows (see Fig.2(b)).

Employ the technique described in previous section on the negative and positive examples, respectively, to obtain the center and radius of MEB as well as the support vectors, i.e., ($\boldsymbol{a}_1$, R_1 , SV_1) and ($\boldsymbol{a}_2$, R_2 , SV_2). Given a pattern z , evaluate the distances, i.e., d_1 and d_2 , from z to the centers $\boldsymbol{a}_1$ and $\boldsymbol{a}_2$, respectively, then the decision rules can be reformulated as: 1)If $d_1 < R_1$ and $d_2 > R_2$, z belongs to A and will be assigned to the target class; 2)If $d_1 > R_1$ and $d_2 < R_2$, z belongs to B and will be assigned to the background class; 3)If $d_1 < R_1$ and $d_2 < R_2$ or $d_1 > R_1$ and $d_2 > R_2$, z belongs to $C \cup D$, then the minimum distance classifier based on SV_1 and SV_2 will be adopted to decide its final assignment.

The d_1 and d_2 in the above rules are obtained by:

$$d_i = \left\| \phi(z) - \boldsymbol{a}_i \right\|, \quad i = 1,2 \tag{10}$$

3 Experimental Results and Analysis

To test the validity of the proposed method, several experiments have been organized on real and simulated SAR images and compared to the classical Otsu method and SVM method. Here, we just show three of them. In all of the experiments, we use the same kind of features as well as the same kernel function (Gaussian kernel function) for the SVM and our methods and the width σ of the employed Gaussian kernel function and the parameter C are set with respect to different images. The sizes of sliding widow are5×5 and7×7for real and simulated SAR images, respectively. The offset constants δ_1, δ_2 are the 0.2 times of the threshold value to the minimal and maximum magnitude.

Experiment 1(see Fig.3) was carried out on real SAR images. Fig.3(a) are the original real SAR images, Fig.3(b), (c) and (d) are the results of the Otsu method, the SVM method and our method, respectively. It can be observed that in Fig.3(b), the pixels of target and background intermix with each other and cannot offer accurate and meaningful information about the target. Though Fig.3(c) and Fig.3(d) cannot present the exact boundaries, they obviously achieved better performance both in retaining coherence and uniformity. Compared to the SVM method, our method achieves better performance in keeping details.

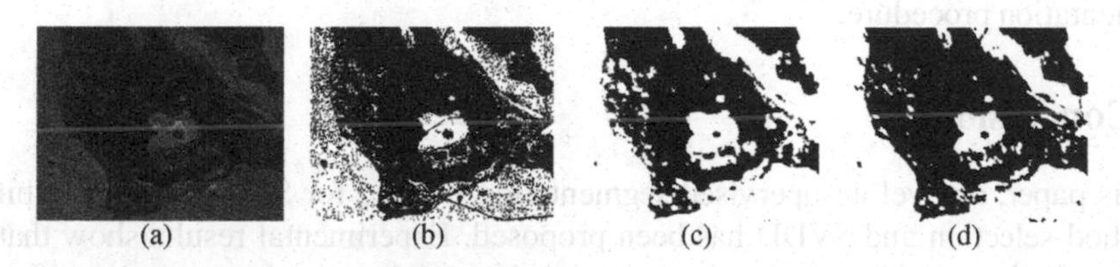

Fig. 3. Experiment 1:(a)the real SAR images;(b)segmentation of Otsu method;(c)segmentation of SVM method;(d)segmentation of our method

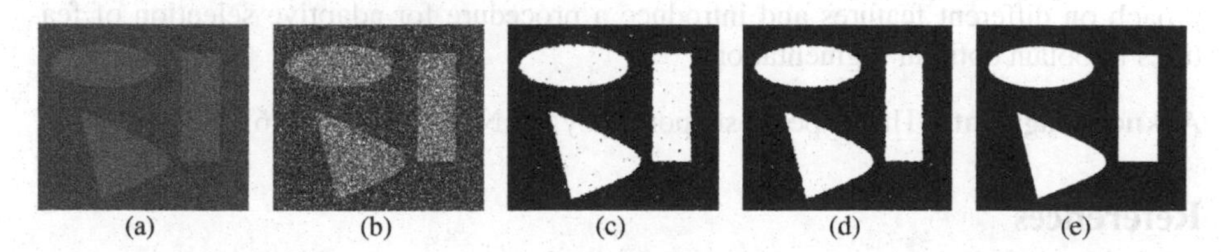

Fig. 4. Experiment2: (a) the simulated SAR image; (b)segmentation of Otsu method; (c) segmentation of SVM method; (d) segmentation of our method; (e) the reference result

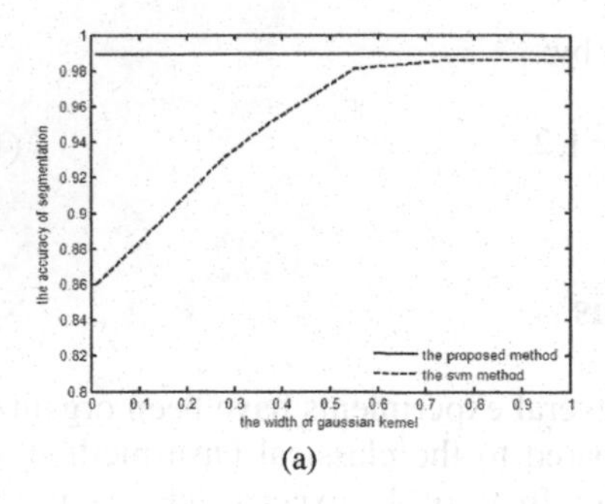

(a)

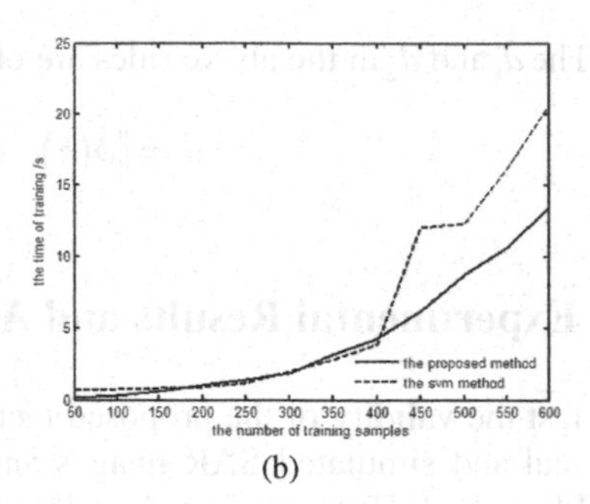

(b)

Fig. 5. Experiment 3: (a) accuracy varying with the width σ of the Gaussian kernel function ; (b) training time varying with the number of train samples

Experiment 2(see Fig.4) was carried out on simulated SAR image. Fig.4(a) is the original image, Fig.4(b), (c) and (d) are the results of the Otsu method, the SVM method and our method, respectively. The reference segmentation of Fig.4(a) is presented in Fig.4(e). It is obvious that the SVM method and our method are more effective than Otsu in dealing with severely noised image. From Fig.4(c), (d) and (e), we can clearly observe that compared to the SVM method, our method is able to obtain result more analogical to the reference one.

Experiment 3(see Fig.5) was carried out on Fig.4(a) to provide quantitative compare of the SVM and our method. Fig.5(a) clearly shows that compared to the SVM method, our method is more robust against varying parameters and achieve higher accuracy. Fig.5(b) explicitly indicates that our method can definitely accelerate the training phase when dealing a large set of training samples (the number of training samples increases with the size of input image), and therefore to accelerate the whole segmentation procedure.

4 Conclusion

In this paper, a novel unsupervised segmentation method for SAR images combining threshod-selection and SVDD has been proposed. Experimental results show that the new method can achieve better segmentation in retaining consistency and uniformity while compared to the Otsu method, and achieve stronger robustness against varying parameters while compared to the SVM method. Future work aims to test our approach on different features and introduce a procedure for adaptive selection of features to obtain optimal segmentation.

Acknowledgments. This paper is supported by the NSFC of China (61072141).

References

1. Li, Y., Ji, K., Su, Y.: Surveys on SAR image segmentation algorithms. Journal of Astronautics 29(2), 407–412 (2008)
2. Gonzalez, R.C., Woods, R.E.: Digital Image Processing, 2nd edn. Publishing House of Electronics Industry (2006)

3. Mantero, P., Moser, G., Serpico, S.B.: Partially Supervised Classification of Remote Sensing Images through SVM-Based Probability Density Estimation. IEEE Trans. Geosci. Remote Sensing 43(3), 559–570 (1999)
4. Tax, D., Duin, R.: Support vector domain description. Pattern Recognition Letters 20, 1191–1199 (1999)
5. Liu, W., Liu, S., Xue, Z.: Separation hyperplane based on support vector domain description. Systems Engineering and Electronicx 30, 748–751 (2008)
6. Bovolo, F., Camps-Valls, G., Bruzzone, L.: A support vector domain method for change detection in multitemporal images. Pattern Recognition Letters 31, 1148–1154 (2010)
7. Zhu, Z., Miao, J.: The Application of Bayes Decision. Journal of Taiyuan University 8, 74–76 (2007)

An Approach of Building Areas Segmentation of SAR Images Based on the Level Set Method

Ruihua Wang, Jianshe Song, Xiongmei Zhang, and Yibing Wu

Xi'an Research Institute of High-tech
710025, Xi'an, China
wrh_1208@163.com

Abstract. Because of the imaging characteristic of Synthetic Aperture Radar (SAR) building areas, an approach based on the level set approach is used to segment building areas. We use the edge feature based on exponential wavelet and texture feature based on variogram to build the energy criterion model, and then the evolutive curve is implemented by minimizing the energy criterion model via variational level set approach to detect the building areas. Compared to the traditional energy criterion model based on the gray information, experimental results show that the proposed model is effective and building areas of large SAR image is detected accurately.

Keywords: SAR, Image segmentation, Building Areas, Level set, Variogram.

1 Introduction

Because of the unique advantages of Synthetic Aperture Radar (SAR), SAR has become an indispensable method to obtain multi-band and multi-polar high resolution images. Therefore, the segmentation of Man-Made targets, such as buildings, bridges in large SAR images are aroused widespread concern. This paper mainly studies the building areas of large SAR image segmentation.

In recent years, image segmentation based on the level set method has vigorous development. Chan and Vese put forward Chan-Vese model[1] based on the Mumford-Shah model[2]. Paragios proposed Geodesic Active Regions model [3] based on the gray probability distribution. But because of the speckle noises, the above models which only rely on the gray information cannot achieve good results. SAR images have much texture information, so the proposed model in this paper is based on the texture feature. And the results show that it can achieve perfect segmentation.

2 The Proposed Energy Criterion Model

2.1 Texture Feature Based on Variogram

In large SAR images, building areas change dramatically in gray level, and have arrangement rules and certain structural, which is an effect way to distinguish building areas. The variogram function[4,5] can describe the similarity between image pixel

D. Jin and S. Lin (Eds.): Advances in MSEC Vol. 1, AISC 128, pp. 33–38.
springerlink.com

and its neighbor pixels, so it can reflect the structural and statistical characteristics of images. In this paper, texture feature $R(h)$ of image $f(x)$ is computed by

$$R(h)=\frac{1}{2N(h)}\sum_{i=1}^{N(h)}\left(f(x_i)-f(x_i+h)\right)^2 \tag{1}$$

$N(h)$ is the number of pixels which have an interval of h, so we need to fix on the interval directions. In this paper, we select four directions: 0°, 45°, 90°, 135°. Therefore, for a window image $w(x,y)$, in which (x_0,y_0) is the centre, the texture features of four directions are:

$$R^{o}(h)=\frac{1}{2N(h)}\sum_{x=x_0-d}^{x_0+d}\sum_{y=y_0-d}^{y_0+d-h}\left(w(x,y)-w(x,y+h)\right)^2 \tag{2}$$

$$R^{45}(h)=\frac{1}{2N(h)}\sum_{x=x_0-d}^{x_0+d-h}\sum_{y=y_0-d}^{y_0+d-h}\left(w(x,y+h)-w(x+h,y)\right)^2 \tag{3}$$

$$R^{90}(h)=\frac{1}{2N(h)}\sum_{x=x_0-d}^{x_0+d-h}\sum_{y=y_0-d}^{y_0+d}\left(w(x,y)-w(x+h,y)\right)^2 \tag{4}$$

$$R^{135}(h)=\frac{1}{2N(h)}\sum_{x=x_0-d}^{x_0+d-h}\sum_{y=y_0-d}^{y_0+d-h}\left(w(x,y)-w(x+h,y+h)\right)^2, \tag{5}$$

and the size of the window image is $W\times W$, $W=2\bullet d+1$, so the texture feature of the widow image is

$$R(h)=\left(R^{o}(h)+R^{45}(h)+R^{90}(h)+R^{135}(h)\right)/4. \tag{6}$$

We can see that the texture feature $R(h)$ is influenced by the interval h and the size of window W. If h is too small, the noises will be detected as the targets. And if h is too big, the backgrounds of the building areas will be more.

2.2 Edge Function Based on Exponential Wavelet

Because of the speckle noise, the traditional edge detection operator cannot get nice results. And exponential wavelet deriving from Gaussian function can not only enhance the edge of image, but also smooth the even areas, so we can detect edge accurately based on the exponential wavelet.

The Gaussian function of two dimensions in s scale is defined as follow:

$$G(x,y,s)=\exp\left(-\left(x^2+y^2\right)/\left(2s^2\right)\right). \tag{7}$$

The two deriving wavelet functions with two directions x and y are:

$$W_0(x,y,s)=\frac{\partial G(x,y,s)}{\partial x}=-\frac{x}{s}\exp\left(-\left(x^2+y^2\right)/\left(2s^2\right)\right) \tag{8}$$

$$W_{90}(x,y,s)=\frac{\partial G(x,y,s)}{\partial y}=-\frac{y}{s}\exp\left(-\left(x^2+y^2\right)/\left(2s^2\right)\right). \quad (9)$$

$$WT_0(x,y,s)=W_0(x,y,s)*f(x,y) \quad (10)$$

$$WT_{90}(x,y,s)=W_{90}(x,y,s)*f(x,y). \quad (11)$$

The grads modulus of image $f(x,y)$ is defined as:

$$M(x,y,s)=\sqrt{WT_0(x,y,s)+WT_{90}(x,y,s)}. \quad (12)$$

And the edge function g can be defined as:

$$g(M)=1/\left(1+M^2\right). \quad (13)$$

2.3 The Proposed Energy Criterion Model

The proposed energy model is

$$E=\alpha E_{edge}+\beta E_{area}. \quad (14)$$

E_{edge} is edge energy model, and E_{area} is area energy model.

For every pixel of the image, the texture feature of its window image is its texture value, so we can obtain the texture image $R(x,y)$ corresponding to the image $f(x,y)$. And the area energy model based on texture feature is defined as

$$E_{area}=vS(C)+\lambda_1\int_{inside(C)}\left|R(x,y)-Rc_1\right|^2 dxdy+\lambda_2\int_{outside(C)}\left|R(x,y)-Rc_2\right|^2 dxdy. \quad (15)$$

$s(c)$ is the neighborhood area of the evolutive curve C, v, λ_1, λ_2 are weight coefficients. Rc_1, Rc_2 are the mean texture values inside and outside the curve C in texture image $R(x,y)$.

For the evolutive curve $C(p)$, the edge energy model can be defined as:

$$E_{edge}=\int_C g(M)\,dp. \quad (16)$$

g is the edge function, $g\in[0,1]$, and if g is close to 0, C is close to edge of the image. So, the proposed energy model is:

$$E=\alpha vS(C)+\alpha\lambda_1\int_{inside(C)}\left|R(x,y)-Rc_1\right|^2 dxdy+\alpha\lambda_2\int_{outside(C)}\left|R(x,y)-Rc_2\right|^2 dxdy+\beta\int_C g(M)\,dp \quad (17)$$

According to the level set method, for image field Ω, the curve C can be as the zero level set of a Lipschitz function $\phi:\Omega\to R$, and the level set function $\phi(x,y)$ is defined as[1]

$$\phi(x,y)\begin{cases}>0, & (x,y) \quad is \quad inside \quad C, \quad (x,y)\in\Omega\\ =0 & (x,y) \in C, \quad (x,y)\in\Omega\\ <0 & (x,y) \quad is \quad outside \quad C, \quad (x,y)\in\Omega\end{cases} \quad . \tag{18}$$

The Heaviside function and Dirac function

$$H(z)=\begin{cases}1 & z\ge 0\\ 0 & z<0\end{cases}, \quad \delta(z)=\frac{d}{dz}H(z) \tag{19}$$

are used to change Integral field to image field. So,

$$\begin{aligned}E(\phi)&=\alpha v\int_\Omega H(\phi)\,dxdy+\alpha\lambda_1\int_\Omega\left|R(x,y)-Rc_1\right|^2 H(\phi)\,dxdy\\&+\alpha\lambda_2\int_\Omega\left|R(x,y)-Rc_2\right|^2(1-H(\phi))\,dxdy+\beta\int_\Omega g\delta(\phi)\left|\nabla\phi\right|dxdy\end{aligned} \tag{20}$$

The texture image $R(x,y)$ can be defined as

$$R(x,y)=Rc_1H(\phi(x,y))+Rc_2(1-H(\phi(x,y))) \tag{21}$$

Make ϕ unchanged, and minimize the energy function to get

$$Rc_1=\frac{\int_\Omega R(x,y)H(\phi)\,dxdy}{\int_\Omega H(\phi)\,dxdy}, \quad Rc_2=\frac{\int_\Omega R(x,y)(1-H(\phi))dxdy}{\int_\Omega(1-H(\phi))dxdy} \tag{22}$$

In practice, we use regularizations of Heaviside function and Dirac function as follows :

$$H_{2\varepsilon}(z)=\frac{1}{2}\left(1+\frac{2}{\pi}\arctan\left(\frac{\pi z}{\varepsilon}\right)\right), \quad \delta_{2\varepsilon}(z)=\frac{d}{dz}H_{2\varepsilon}(z)=\frac{1}{\pi}\bullet\frac{\varepsilon}{\varepsilon^2+z^2} \tag{23}$$

Keeping the Rc_1, Rc_2 fixed, and minimizing the model with respect to ϕ, we deduce the associated Euler-Lagrange equation for ϕ :

$$\frac{\partial\phi}{\partial t}=-\frac{\partial E}{\partial\phi} \tag{24}$$

$$\frac{\partial\phi}{\partial t}=\delta_{2\varepsilon}(\phi)\left(\beta div\left(g\frac{\nabla\phi}{|\nabla\phi|}\right)-\alpha v-\alpha\lambda_1\left(R-Rc_1\right)^2+\alpha\lambda_2\left(R-Rc_2\right)^2\right) \tag{25}$$

Parameterizing the descent direction by an artificial time $t \geq 0$, we use a finite differences implicit scheme to discretize the equation in $\phi(x, y, t)$.

Supposed Δh as discrete step, so

$$\phi_{i,j}^{n} = \phi(i\Delta h, j\Delta h, n\Delta t)\ ,\ \phi_{i,j}^{n+1} = \phi_{i,j}^{n} + V\Delta t \tag{26}$$

$$V = -\alpha\nu - \alpha\lambda_1 (R - Rc_1)^2 + \alpha\lambda_2 (R - Rc_2)^2 + \beta div\left(g\frac{\nabla\phi}{|\nabla\phi|} \right) \tag{27}$$

Fig.1 shows the flow of the proposed algorithm based on the level set method.

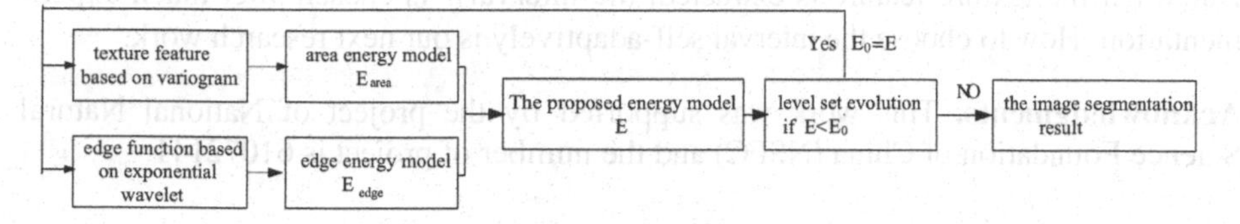

Fig. 1. The proposed algorithm

3 Experiment Results and Discussion

We use two SAR images to test the proposed method. Fig.2 shows the segmentation result of the interest building areas. Fig.2 (a) 1 is the SAR image which has more buildings, and Fig.2 (a) 2 is the SAR image which has a few of buildings. Fig.2 (b) 1~2 are the image segmentation results based on the traditional CV model, and we can see almost all background is also segmented. So, the model based on the gray information is not suitable for SAR images. Fig.2 (c) 1~2 shows the image segmentation results based on the proposed model, and we can see the building areas are segmented

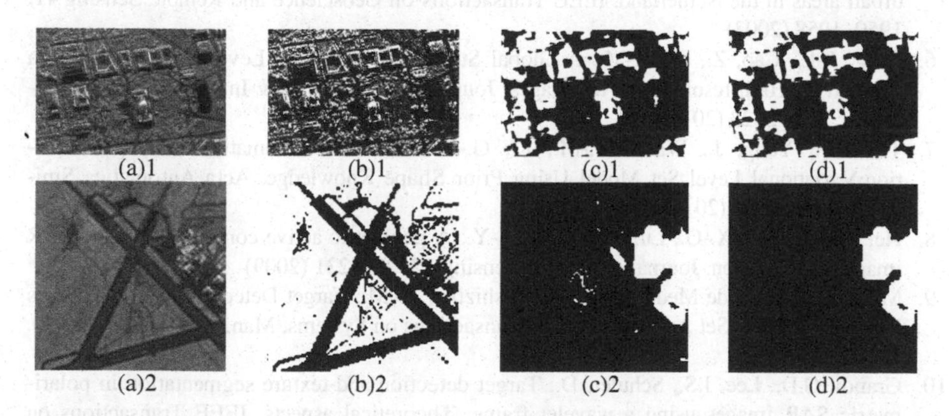

Fig. 2. The segmentation results of the interest building areas. (a)1~2 the orginal SAR images; (b)1~2 the segmentation results based on CV model; (c)1~2 the segmentation results based on the proposed model; (d)1~2 the segmentation results after removing flase areas.

accurately and effectively. But some false areas are also segmented, and we can remove these false areas according to their area sizes. Fig.2 (d) 1~2 are the results after removing the false areas. Here, when computing, $h = 3$, $W = 11$, $\nu = \lambda_1 = \lambda_2 = 1$, $\alpha = \beta = 1$.

4 Conclusion

In this paper, an approach of building areas segmentation of SAR images based on the level set method is presented. Compared to the energy model based on the gray information, the proposed model based on the texture feature can segment more exactly. But when the texture feature is extracted, the interval h is chosen after much experimentation. How to choose the interval self-adaptively is our next research work.

Acknowledgments. This work was supported by the project of National Natural Science Foundation of China (NSFC) and the number of project is 61072141.

References

1. Chan, T.F., Vese, L.A.: Active Contours Without Edges. IEEE Transctions on Image Processing 10, 266–277 (2001)
2. Vese, L.A., Chan, T.F.: A Multiphase Level Set Framework for Image Segmentation Using the Mumford and Shah Model. International Journal of Computer Vision 50, 271–293 (2002)
3. Paragios, N.: Geodesic Active Regions and Level Set Methods for Supervised Texture Segmentation. International Journal of Computer Vision 46, 223–247 (2002)
4. Miranda, F.P., Fonseca, L.E.N., Carr, J.R.: Semivariogram textural classification of JERS-1(Fuyo-1)SAR data obtained over a flooded area of the Amazon rainforest. International Journal of Remote Sensing 19, 549–556 (1998)
5. Decker, R.J.: Texture analysis and classification of ERS SAR images for map updating of urban areas in the Netherland. IEEE Transactions on Geoscience and Remote Sensing 41, 1950–1958 (2003)
6. Feng, J.-L., Cao, Z., Pi, Y.-M.: A Global Stationary Minimum Level Set Segmentation Method for High-resolution SAR Images. Journal of Electronics & Information Technology 32, 2618–2623 (2010)
7. Tian, H., Yang, J., Wang, Y.-M., Li, G.-H.: Towards Automatic Building Extraction:Variational Level Set Model Using Prior Shape Knowledge. Acta Automatica Sinica 36, 1502–1511 (2010)
8. He, Z.-G., Zhou, X.-G., Lu, J., Kuang, G.-Y.: A geometric active contour model for SAR image segmentation. Journal of Remote Sensing 13, 224–231 (2009)
9. Marques, R.C.P., de Medeiros, F.N.S., Ushizima, D.M.: Target Detection in SAR Images Based on a Level Set Approach. IEEE Transactions on Systems, Man, and Cybernetics 39, 214–222 (2009)
10. Grandi, G.D., Lee, J.S., Schuler, D.: Target detection and texture segmentation in polarimetric SAR images using a wavelet frame: Theoretical aspects. IEEE Transactions on Geoscience and Remote Sensing 45, 3437–3453 (2007)
11. Liu, D., Zhang, G.: Special target detection of the SAR image via exponential wavelet fractal. Journal of Xidian University 37, 366–373 (2010)

SAR Image Classification in Urban Areas Using Unit-Linking Pulse Coupled Neural Network

Ruihua Wang, Jianshe Song, Xiongmei Zhang, and Yibing Wu

Xi'an Research Institute of High-tech
710025, Xi'an, China
wrh_1208@163.com

Abstract. A method for synthetic aperture radar (SAR) image classification in urban areas based on modified Unit-linking pulse coupled neural networks (Unit-linking PCNN) and texture feature is presented. Unit-linking PCNN is modified to be two levels in order to make it classify more classes. The primary level corresponds to determining the initial threshold value of the secondary level, and in the secondary level, the similar neurons are captured using Unit-linking PCNN. Because of the imaging characteristic of SAR building areas, the texture feature of the neuron's $n \times n$ window image is used as the input pulse signal. Experimental results show that the proposed method is effective.

Keywords: Synthetic aperture radar (SAR), Image classification, Unit-linking PCNN, Urban areas.

1 Introduction

Pulse Coupled Neural Network (PCNN) is a result of the research that focused on the development of artificial neuron models that are capable of emulating the behavior of cortical neurons observed in the visual cortices of cats [1]. Because of the one-to-one correspondence between image pixels and network neurons, and the ability of capturing the linking neurons, PCNN is widely used in image segmentation [1~3]. But the parameters of PCNN models which influence the segmentation result are complex. Conversely, in Unit-linking PCNN model proposed in reference [4], the parameter number decreases from 6 to 3.

Because SAR images have speckle noise and complex grey distribution, the traditional Unit-linking PCNN can't obtain good results. So, in this paper, the modified Unit-linking PCNN and texture feature are used to classify SAR images in urban areas to more classes, and the results show that it can achieve perfect classification.

2 Unit-Linking PCNN

2.1 Unit-Linking PCNN

Fig.1 is a simple unit-linking pulse coupled neuron (unit-linking PCNN) model, and all parts are described as follows:

D. Jin and S. Lin (Eds.): Advances in MSEC Vol. 1, AISC 128, pp. 39–44.
springerlink.com

(1)Feeding receptive field: for the neuron (i, j), in the n-th iteration, the receptive field has two parts:

a) the input impulse signal: $F_{ij}[n] = S_{ij}$ (1)

b) the linking input: In PCNN, L_{ij} of neuron (i, j) is complex, because it relates not only to the number of neuron fired in its neighborhood, but also to the weight coefficients. In fact, the impulse wave should transmit as long as there are suitable channels, just like water. So, in Unit-linking PCNN, as long as there are neurons fired in its neighborhood, L_{ij} is equal to 1.

$$L_{ij}[n] = \begin{cases} 1, & \sum_{k \in N(i,j)} Y_k[n-1] > 0 \\ 0, & else \end{cases} \tag{2}$$

$N(i, j)$ is the neighborhood of neuron (i, j), which includes 8 neurons(see Fig.2). So, in Unit-linking PCNN, the impulse transmission wave is easy to analyze and control.

(2)Internal activity: the neuron (i, j)'s internal activity

$$U_{ij}[n] = F_{ij}[n](1 + \beta L_{ij}[n]) . \tag{3}$$

β, a positive constant, is referred to as the linking coefficient of the neuron.

(3) Pulse generator: the neuron (i, j)'s pulse output

$$Y_{ij}[n] = \begin{cases} 1, & if \quad U_{ij}[n] \geq \theta_{ij}[n] \\ 0, & if \quad U_{ij}[n] < \theta_{ij}[n] \end{cases}, \tag{4}$$

and $\theta_{ij}[n]$ is the threshold which is defined as follows:

$$\theta_{ij}[n] = e^{-\alpha_\theta} \theta_{ij}[n-1] + V_\theta Y_{ij}[n] \tag{5}$$

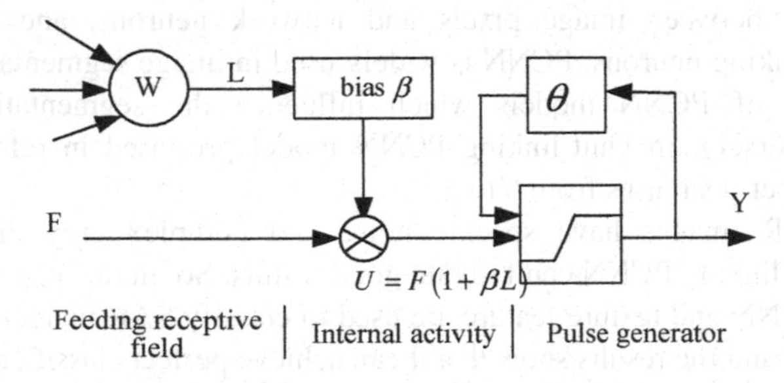

Fig. 1. Simple pulse coupled neuron model

When PCNN is used in image processing, the number of network neurons is equal to the number of image pixels. And the initial values of U , L and θ are assumed to be zeros. In the first iteration, U is equal to F , θ of all neurons decrease from the initial value. If θ_{ij} of neuron (i, j) decrease to be equal to or less than corresponding U_{ij}, the neuron

(i, j) is fired, and its output Y_{ij} is equal to 1.Simultaneously, θ_{ij} of the fired neuron increase fast. When θ_{ij} decreases to be equal to or less than U_{ij} again, the neuron fires again. In the above process, the fired neurons communicate with their neighbors, and the output of each neuron affects the outputs of its neighbors. The result is an auto wave that expands from an active pixel to the whole region, which can realize image classification.

2.2 Modified Unit-Linking PCNN

The main idea of modified Unit-linking PCNN is dividing Unit-linking PCNN iteration into two levels: the primary level corresponds to determining the initial threshold value of the secondary level, and in the secondary level, the similar neurons are captured using Unit-linking PCNN. The whole steps are described as follows: the input impulse signal, the linking input, and the neuron's internal activity are the same as traditional Unit-linking PCNN's. But the two levels have their own thresholds θ_p and θ_s . The initial values of U , L and Y are assumed to be zeros. When starting the primary level, let $\theta_p = F_{\max}$, and $F_{\max}$ is the maximum of the input impulse signals. The neurons whose input impulse signals are equal to $F_{\max}$ will pulse naturally. Then the secondary level starts, θ_s decreases from θ_p in order to capture neurons. Until no neuron can be captured, the secondary level ends. Then the thresholds of neurons fired are made large to avoid to be fired again and all outputs Y of the network are initialized to be zeros. Get in the primary level again, and θ_p decreases until some neurons can fire naturally. Then go to the secondary level to cluster. The two levels are iterated until all neurons in the network are fired.

Fig.3 is the detailed steps of modified Unit-linking PCNN.

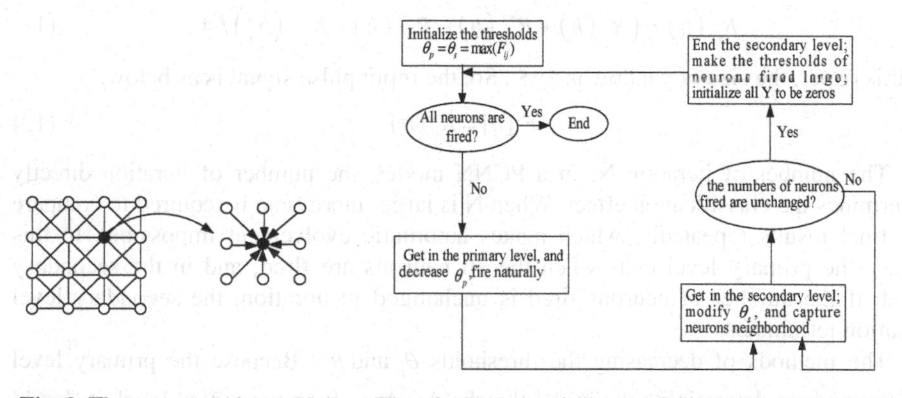

Fig. 2. The structure of each Unit-linking PCN

Fig. 3. The detailed steps of modified Unit-linking PCNN

3 SAR Image Classification Using Modified Unit-Linking PCNN

(1) The input impulse signal: because of the speckle noises, the grey distribution of SAR image is uneven. And SAR images in urban areas have much unique texture

information. So, we use the texture feature of the neuron's $n \times n$ window image based on the variogram function as the input impulse signal.

The variogram function[5,6] can describe the similarity between image pixel and its neighbor pixels, so it can reflect the structural and statistical characteristics of images. In this paper, texture feature $R(h)$ of image $f(x)$ is computed by

$$R(h) = \frac{1}{2N(h)} \sum_{i=1}^{N(h)} \left(f(x_i) - f(x_i + h) \right)^2 \tag{6}$$

$N(h)$ is the number of pixels which have an interval of h, so we need to fix on the interval directions. In this paper, we select four directions: 0^0, 45^0, 90^0, 135^0. Therefore, for the neuron (i, j), the texture features of its window image $w(x, y)$ are:

$$R^o(h) = \frac{1}{2N(h)} \sum_{x=i-d}^{i+d} \sum_{y=j-d}^{j+d-h} \left(w(x, y) - w(x, y+h) \right)^2 \tag{7}$$

$$R^{45}(h) = \frac{1}{2N(h)} \sum_{x=i-d}^{i+d-h} \sum_{y=j-d}^{j+d-h} \left(w(x, y+h) - w(x+h, y) \right)^2 \tag{8}$$

$$R^{90}(h) = \frac{1}{2N(h)} \sum_{x=i-d}^{i+d-h} \sum_{y=j-d}^{j+d} \left(w(x, y) - w(x+h, y) \right)^2 \tag{9}$$

$$R^{135}(h) = \frac{1}{2N(h)} \sum_{x=i-d}^{i+d-h} \sum_{y=j-d}^{j+d-h} \left(w(x, y) - w(x+h, y+h) \right)^2, \tag{10}$$

and the size of the window image is $W \times W$, $W = 2 \bullet d + 1$, so the texture feature of the widow image is

$$R_{ij}(h) = \left(R^o(h) + R^{45}(h) + R^{90}(h) + R^{135}(h) \right) / 4. \tag{11}$$

In this paper, the size of window is 5×5. So, the input pulse signal is as below:

$$F_{ij}[n] = R_{ij}(h) \tag{12}$$

(2) The number of iteration N: in a PCNN model, the number of iteration directly determines the classification effect. When N is large, more time is required to compare the final results repeatedly, which makes automatic evolvement impossible. In this paper, the primary level ends when the all neurons are fired, and in the secondary level, if the number of neurons fired is unchanged in iteration, the secondary level iteration terminates.

(3) The methods of decreasing the thresholds θ_p and θ_s : Because the primary level corresponds to determining the initial threshold value of the secondary level, it should be defined as:

$$\theta_p[n_1] = \max(F_{pq}, (p, q) \notin fire) \tag{13}$$

The threshold θ_s of the secondary level which should be decreased slowly to capture similar neurons is defined as below:

$$\theta_s[n_2] = Fm_s[n_2-1]\ ,\ Fm_s[n_2-1] = \frac{1}{m}\sum_{p,q\in fire} F_{pq}[n_2-1] \quad (14)$$

$Fm_s[n-1]$ is the mean of F of neurons fired at n-1-th iterative time in the secondary level.

(4) The linking coefficient β : β can control the influence of the linking input on the internal activity. The bigger β is, the more neurons captured to be fired are. According to the proposed method, in the primary level, $\beta = 0$, and in the secondary level, because different classes are clustered, different linking coefficient β should be used.

4 Experiment Results and Discussion

Two SAR images are used to test the validity of the proposed method. The experimental results of SAR image classification using the proposed method and the traditional Unit-linking PCNN are shown to be compared. Fig.4 (b) 1~2 are the results of image classification using the traditional Unit-linking PCNN. We can see that the classification images are binary, so the urban areas are mixed in the background and can't be recognized. Fig.4 (c) 1~2 are the results of image classification using the proposed method. In Fig.4(c) 1, we set the linking coefficients of the secondary level $\beta_1 = 0.1$ and $\beta_2 = 0.15$, and the urban areas, background and the water are classified effectively and clearly. In Fig.4(c) 2, the linking coefficients of the secondary level are $\beta_1 = 0.5$ and $\beta_2 = 0.8$. From the classification result, we can see the urban areas, background, and the road are classified nicely.

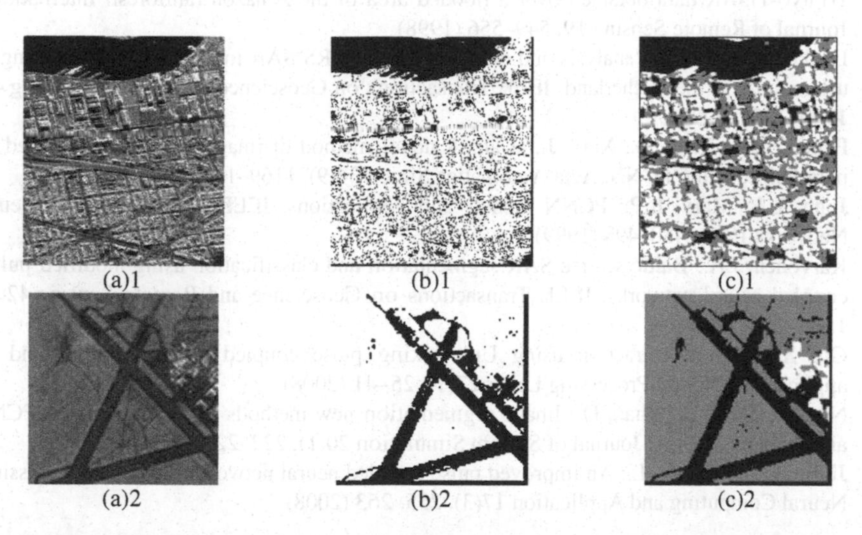

Fig. 4. The results of image classification.(a)1~(a)2: Original SAR images; (b)1~(b)2: The results of image classification using the troditional Unit-linking PCNN; (c)1~(c)2: SAR classification images using the proposed method (class=3)

5 Conclusion

In this paper, the modified Unit-linking PCNN is proposed to classify the SAR images in urban area. As results above show, the SAR image classification using modified Unit-linking PCNN is effective. However, the linking coefficients β in the secondary level are chosen after many experiments. How to choose β self-adaptively according to image is vital to Unit-linking PCNN for achieving the best effect. And it's the key point of research in the future.

Acknowledgments. This work was supported by the project of National Natural Science Foundation of China (NSFC) and the number of project is 61072141.

References

1. Kuntimad, G., Ranganath, H.S.: Perfect Image Segmentation Using Pulse Coupled Neural Networks. IEEE Transactions on Neural Networks 10(3), 591–598 (1999)
2. Zhao, S.-J., Zhang, T.-W., Zhang, Z.H.: A study of a new image segmentation algorithm based on PCNN. Acta Electronica Sinica 33(7), 1342–1344 (2005)
3. Waldemark, K., Lindblad, T., et al.: Patterns from the sky Satellite image analysis using pulse coupled neural networks for pre-processing, segmentation and edge detection. Pattern Recognition Letters 21(3), 227–237 (2000)
4. Gu, X.-D., Zhang, L.-M., Yu, D.-H.: Automatic image segmentation using Unit-linking PCNN without choosing parameters. Journal of Circuits and Systems 12(6), 54–59 (2007)
5. Miranda, F.P., Fonseca, L.E.N., Carr, J.R.: Semivariogram textural classification of JERS-1(Fuyo-1)SAR data obtained over a flooded area of the Amazon rainforest. International Journal of Remote Sensing 19, 549–556 (1998)
6. Decker, R.J.: Texture analysis and classification of ERS SAR images for map updating of urban areas in the Netherland. IEEE Transactions on Geoscience and Remote Sensing 41, 1950–1958 (2003)
7. Peng, Z.-M., Jiang, B., Xiao, J., et al.: A novel method of image segmentation based on parallelized firing PCNN. Acta Automatica Sinica 34(9), 1169–1173 (2008)
8. John, L.J., Mary, L.P.: PCNN models and applications. IEEE Transactions on Neural Networks 10(3), 480–498 (1999)
9. Karvonen, J.A.: Baltic sea ice SAR segmentation and classification using modified pulse-coupled neural networks. IEEE Transactions on Geoscience and Remote Sensing 42(7), 1566–1574 (2004)
10. Gu, X.: Feature extraction using Unit-linking pulse coupled neural network and its applications. Neural Processing Letter 27(1), 25–41 (2008)
11. Nie, R., Zhou, D., Zhao, D.: Image segmentation new methods using unit-linking PCNN and image's entropy. Journal of System Simulation 20(1), 222–227 (2008)
12. Ji, L., Yi, Z., Shang, L.: An improved pulse coupled neural networks for image processing. Neural Computing and Application 17(3), 255–263 (2008)

The Research of Intrusion Detection System in Cloud Computing Environment

Huaibin Wang and Haiyun Zhou

Key Laboratory of Computer Vision and System, Ministry of Education
Tianjin University of Technology, 300191 Tianjin, China
haiyun313@163.com

Abstract. A variety of convenient services have been provided by cloud computing to end users. So providing safe and reliable service assurance is quiet important. Many Network-based intrusion detection systems (NIDS) are used to obtain the packets from the cloud. It has lower detection rate, higher false-positive rate and is unable to resist the single point attack of failure. In this paper, multiple intrusion detection systems (IDSs) are deployed in each layer of cloud infrastructure for protecting each Virtual Machine (VM) against threats. We also propose the cloud alliance concept by the communication agents exchanging the mutual alerts to resist the single point attack of failure. The simulation results indicate that the proposed system have a higher detection rate, lower false-positive rate and can resist the single point attack of failure.

Keywords: Cloud Computing, Detection Rate, VM-based IDS, Cloud Alliance.

1 Introduction

According to different deployment mechanisms, IDS can be divided into software-based IDS, hardware-based IDS and VM-based IDS [1]. In the proposed virtual cloud infrastructure, due to the highly heterogeneous architecture, the VM-based IDS new structure is the core of the paper.

Owing to the combination of the means of cloud service and the different deployment of cloud computing, new security challenges emerge such as how to resolve the deployment of the virtual infrastructure in cloud platform when virtual technology provides the flexible deployment of resource for cloud computing platform. So it's necessary to deploy IDS sensor to monitor the separated VM at each layer which is controlled by the VM management unit. In order to integrate and analyze the alerts generated by multiple distributed sensors deployed in cloud, a plug-in-concept is proposed in core management unit .Furthermore, in the paper, the single point attack of failure must be considered, that is to say, to realize cloud alliance concept by the communication agents. All above are proposed in proof-of-concept to realize the architecture.

D. Jin and S. Lin (Eds.): Advances in MSEC Vol. 1, AISC 128, pp. 45–49.
springerlink.com

2 The Related Work

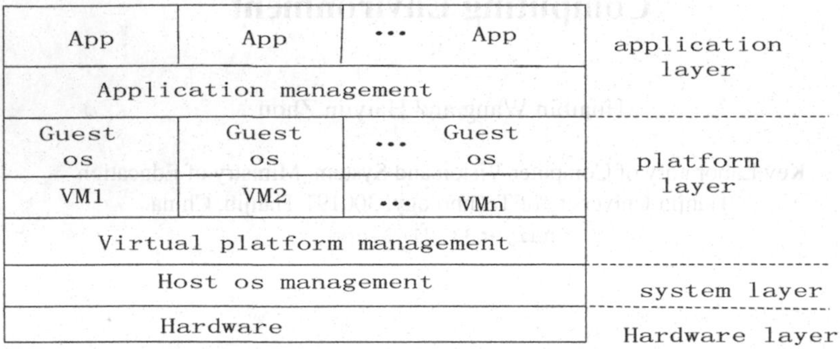

App | App | ... App — application layer
Application management
Guest os | Guest os | ... Guest os — platform layer
VM1 | VM2 | ... VMn
Virtual platform management
Host os management — system layer
Hardware — Hardware layer

Fig. 1. Architecture of cloud computing

2.1 The Related Cloud Model

As shown in Fig. 1, the cloud architecture can be divided into specific four layers.

The hardware layer which doesn't join the cloud directly we don't introduce specifically here.

Cloud computing software as a service (SaaS) is provided by the cloud provider and could be accessed by the interfaces of a variety of clients. The underlying infrastructure of the cloud including networks ,servers, operating systems, storage or even a single application functionality need not to be managed by the user. SaaS's case is such as Google's app engine, Alibaba, Babai-off and so on.

Cloud computing platform as a service (PaaS), the application development environment created by the tools (such as Java, python, .Net) provided by provider are automatically developed to the cloud computing infrastructure. The underlying cloud infrastructure including networks, servers, operating systems and storage needn't to be managed and managed by the user. The consumer could control and deploy the application and environment. For example, this type of service could be provided by Windows Azure.

Cloud infrastructure as a service (IaaS), computing power, storage capacity, network rental provided by provider are available to users. Any software including operating system and application configuration could be deployed by users. The underlying infrastructure are not be controlled or managed by users .Amazon is a typical IaaS service provider.

So as [2] described, each layer is extremely likely suffered from attacks.

2.2 Denial of Service (DoS) and Distributed Denial of Service(DDoS) Attack

In order to make the computer and network denied serve the normal service by consuming bandwidth and host system resources, mining program defects and providing false DNS information. For example, network communication is blocked and access to service is denied, server crashes or service have been damaged. The denial service capability of DDoS [3] is increased by depending on client and server technology with multiple computers together as an attack platform to launch DoS attack to one or more targets and generate more attack traffic than DoS.

3 Proposed Architecture and Simulation

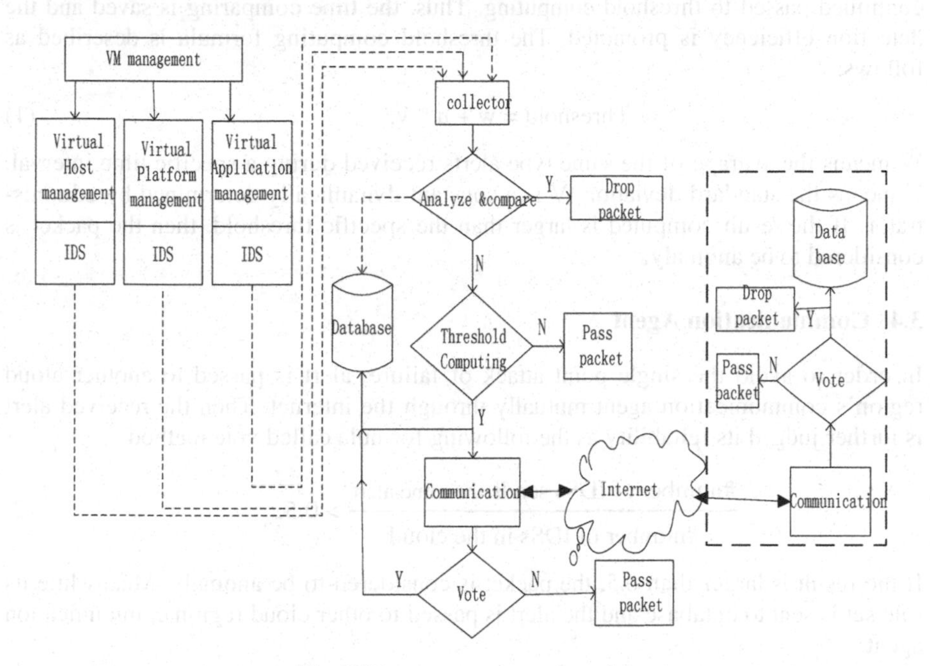

Fig. 2. The core management unit

3.1 VM Management

As shown in Fig 2, Network-based or host-based IDS sensors of different layers in virtual environment are managed by VM management unit which is a part of core management unit. The state of VM such as start, shutdown, stop, continue, reset or update and weather the VM is running, how its platform are involved in virtual environment information. The attacks related to the virtual component could be recognized by the provider with the VM management unit.

3.2 Collector

Alerts generated from each layer multiple sensors are collected by the component. Then the alert with the format of IDMEF [4] has been proposed as a standard to enable interoperability among different IDS approaches. A message with the type of IDMEF message is the part of IDMEF library which is based IDMEF XML scheme by RFC [5]. However due to the highly heterogeneous architecture, especially VM-based new structure IDMEF is inevitable.

3.3 Analyze and Compare and Threshold Computing

The alerts unified by the front component are passed to database to compare with the signatures. If the type of packet is correspondence with the one listed in the database, then the alert is considered to be anomaly and to be dropped. Otherwise the alert is continued passed to threshold computing. Thus, the time comparing is saved and the detection efficiency is promoted. The threshold computing formula is described as follows:

$$\text{Threshold} = w + u * v. \quad (1)$$

W means the average of the same type alerts received during a specific time interval. U means the standard deviation. V is a constant dynamically determined by administrator. If the result computed is larger than the specific threshold, then the packet is considered to be anomaly.

3.4 Communication Agent

In order to avoid the single point attack of failure, alert is passed to another cloud region's communication agent mutually through the internet. Then the received alert is further judged its reliability as the following formula called vote method.

$$\frac{\#\text{number of IDSs sends the same alert}}{\#\text{number of IDSs in the cloud}} > 0.5. \quad (2)$$

If the result is larger than 0.5, the packet is considered to be anomaly. Meanwhile its rule set is sent to database and the alert is passed to other cloud region communication agent.

3.5 Simulation

To test the feasibility of the above mentioned architecture, we simulate the experiment. It consists of two servers A and B in different cloud regions. A executes two VMs including F-Secure sensor and snort sensor. B executes three VMs including two Samhain [6] sensors and one snort sensor. Attacker firstly launches attacks like TCP/IP packets, SYN flooding [7] to the servers in the format NMAP to scan the ports. We also compare with the NIDS system deployed in the cloud. The experiment result is as follows.

Table 1. The detection condition of SYN flooding

Simulate systems	Detection rate	False-positive rate	Negative rate
NIDS system	82%	0.8%	1.12%
Proposed system	94.27%	0.55%	0.52%

The architecture's simulation results prove that the proposed system has higher detection rate, lower false-positive rate and negative rate.

4 Conclusion

Considering the complexity of the cloud security architecture, a extensible VM-based multiple IDS is deployed in each layer to monitor specific virtual component and the core management unit are constructed by multiple plugs with the IDMEF standard to realize the ideas of virtualization and cloud alliance which is mainly used for avoiding the single point attack of failure through the communication agent.

Acknowledgments. This work was supported by the Foundation of Tianjin for Science and Technology Innovation (No.10FDZDGX004 00&11ZCKFGX00900), Education Science and Technology Foundation of Tianjin (No.SB20080053 & SB20080055).

References

1. Laureano, M., Maziero, C., Jamhour, E.: Protecting Host-based Intrusion Detectors through Virtual Machines. Computer Networks 51, 1275–1283 (2007)
2. Kandukuri, B.R., Paturi, V., Rakshit, A.: Cloud Security Issues. In: 38th IEEE International Conference on Services Computing, Bangalore, India, pp. 517–520 (2009)
3. Wei, L., Xiang, L., Derek, P., Bin, L.: Collaborative Distributed Intrusion Detection System. In: Second International Conference on Future Generation Communication and Networking, pp. 172–177 (2008)
4. Debar, H., Curry, D., Feinstein, B.: The Intrusion Detection Message Exchange Format, Internet Draft Technical Report, IETF Exchange Format Working Group (July 2004)
5. Laureano, M., Maziero, C., Jamhour, E.: Intrusion detection in virtual machine environments. In: EUROMICRO Conference, Brazil, pp. 520–525 (2004)
6. F-Secure Linux Security, http://www.f-secure.com
7. Moore, D., Shannon, C., Brown, D.J., et al.: Inferring Internet Denial-of-Service Activity. In: ACM Transactions on Computer Systems (TOCS), Berkeley, pp. 115–139 (2006)

Zero-Voltage-Switching Voltage Doubled SEPIC Converter

Hyun-Lark Do

Department of Electronic & Information Engineering,
Seoul National University of Science and Technology, Seoul, South Korea
hldo@seoultech.ac.kr

Abstract. A zero-voltage-switching (ZVS) voltage doubled single-ended primary inductor converter (SEPIC) is presented in this paper. An active clamp circuit is added to the conventional isolated SEPIC converter to clamp the voltage across the switches and provide ZVS operation. Voltage doubler is adopted as an output stage to provide high voltage gain and confine the voltage stress of output diodes to the output voltage. Moreover, the reverse-recovery problem of the output diodes is alleviated due to the leakage inductance of the transformer. This converter besides an electrical isolation provides higher efficiency due to soft-switching commutations of the power semiconductor devices. The operation principle and steady-state analysis of the proposed converter are provided. A prototype of the proposed converter is developed, and its experimental results are presented for validation.

Keywords: Zero-voltage-switching, DC-DC converter, SEPIC.

1 Introduction

SEPIC converter has been adopted for many applications such as power factor correction [1], photovoltaic system [2], and LED lighting [3]. However, it has several drawbacks. Its switching loss is large because of its hard-switching operation. To reduce the volume and weight of a converter, switching frequency needs to be raised. High frequency operation allows reduction of the volume and weight of magnetic components and capacitors. However, switching losses and electromagnetic interference noises are significant in high frequency operation. Another drawback is its high voltage stress. In conventional isolated SEPIC converter, the voltage stresses across switches and output diodes are serious and additional snubbers are required to suppress them.

In this paper, a ZVS voltage doubled SEPIC converter is proposed. In order to clamp the voltage across the switches and provide ZVS function, an active clamp circuit consisting of a clamp switch and a capacitor is added to the conventional SEPIC converter. In addition, a voltage doubler is adopted as an output stage to provide high voltage gain and confine the voltage stress of output diodes to the output voltage. Moreover, the reverse-recovery problem of the output diodes is alleviated due to the leakage inductance of the transformer.

D. Jin and S. Lin (Eds.): Advances in MSEC Vol. 1, AISC 128, pp. 51–54.
springerlink.com

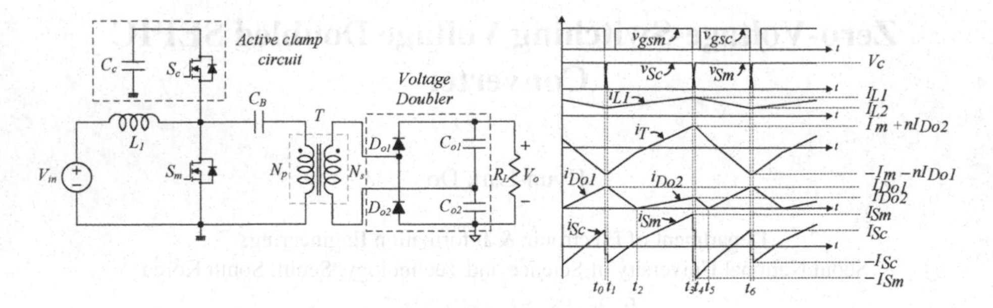

Fig. 1. Circuit diagram and key waveforms of the proposed converter

2 Analysis of the Proposed Converter

Fig. 1 shows the circuit diagram and key waveforms of the proposed converter. The switches S_m and S_c are operated asymmetrically and the duty ratio D is based on S_m. The transformer T has a turn ratio of 1:n ($n= N_s/N_p$). To simplify the steady-state analysis, it is assumed that the voltages across the capacitors C_c, C_B, C_{o1}, and C_{o2} are constant.

2.1 Operation

The operation of the proposed converter in one switching period T_s can be divided into six modes. Before t_0, the switch S_c and the diode D_{o1} are conducting.

Mode 1 [t_0, t_1]: At t_0, S_c is turned off. Then, the energy stored in the magnetic components starts to charge/ discharge the parasitic capacitances of S_m and S_c. Therefore, the voltage v_{Sm} across S_m starts to fall and the voltage v_{Sc} across S_c starts to rise. Since the parasitic capacitances of the switches are very small, the transition interval is very short and it can be neglected. Therefore, all currents can be assumed as constant during this mode.

Mode 2 [t_1, t_2]: At t_1, the voltage v_{Sm} becomes zero and the body diode of S_m is turned on. Then, the gate signal is applied to S_m. Since the current has already flown through the body diode and v_{Sm} becomes zero before S_m is turned on, zero-voltage turn-on of S_m is achieved. Since the voltage across L_1 is V_{in}, the current i_{L1} is linearly increasing from its minimum value I_{L2}. Since the primary voltage v_p is $-V_{in}$ and it is reflected at the secondary side, the magnetizing current i_m increases linearly from its minimum value $-I_m$ and the diode current i_{Do1} decreases linearly from its maximum value I_{Do1}.

Mode 3 [t_2, t_3]: At t_2, the secondary current changes its direction. The diode current i_{Do1} decreases to zero and D_{o1} is turned off. Then, D_{o2} is turned on and its current increases linearly. Since the changing rate of i_{Do1} is controlled by the leakage inductance of T, its reverse-recovery problem is alleviated significantly. Since v_p is maintained as $-V_{in}$, i_m decreases with the same slope.

Mode 4 [t_3, t_4]: At t_3, S_m is turned off. Then, the energy stored in the magnetic components starts to charge/ discharge the parasitic capacitances of the switches. Therefore, the voltages v_{Sm} and v_{Sc} start to rise and fall with a similar manner in Mode 1. With the same reason, this transition interval is very short and it can be neglected.

Mode 5 [t_4, t_5]: At t_4, the voltage v_{Sc} becomes zero and the body diode of S_c is turned on. Then, the gate signal is applied to S_c. Since the current has already flown through the body diode and v_{Sc} becomes zero before S_c is turned on, zero-voltage turn-on of S_c is achieved. Since the voltage across L_1 is $-(V_{in}/(1-D)-V_{in})$, the current i_{L1} is linearly decreasing from its maximum value I_{L1}. Since v_p is $DV_{in}/(1-D)$, the current i_m decreases linearly. The current i_{Do2} decreases linearly from its maximum value I_{Do2}.

Mode 6 [t_5, t_6]: At t_5, the secondary current changes its direction. The diode current i_{Do2} decreases to zero and D_{o2} is turned off. The reverse-recovery problem of D_{o2} is also alleviated due to the leakage inductance of T. Then, the diode D_{o1} is turned on and its current linearly increases. Since vp is $DV_{in}/(1-D)$, i_m increases with the same slope and it approaches to its maximum value I_m at the end of this mode.

2.2 Voltage Gain

Since the average value of the output diode current is equal to the average output current Io and the average voltage across the leakage inductance L_k of T should be zero at steady-state operation, the voltage gain of the proposed converter is given by

$$\frac{V_o}{V_{in}} = \frac{nD(1-2d)}{(D-d(2D-1))(1-D+d(2D-1))}, \tag{1}$$

$$d = \frac{1}{2}\left(1-\sqrt{1-\frac{8L_k I_o}{nV_{in}DT_s}}\right). \tag{2}$$

3 Experimental Results

The prototype is implemented with specifications of V_{in}=24V, V_o=160V, P_o=100W. The circuit parameters are L_1=320uH, C_c=C_B=13.2uF, C_{o1}=C_{o2}=220uF, n=3, leakage inductance=70uH, magnetizing inductance=180uH, f_s=100kHz. Fig. 2 shows the experimental waveforms of the prototype of the proposed converter. It can be seen that the experimental waveforms agree with the theoretical analysis. The input current is continuous. It is clear that the reverse-recovery problem of the output diodes is alleviated dramatically by the leakage inductance of T. The ZVS of Sm and Sc is achieved. The proposed converter exhibits an efficiency of 93.2% at full load condition. Due to its soft-switching characteristic and alleviated reverse-recovery problem, the efficiency was improved by around 2% compared with the conventional isolated SEPIC converter.

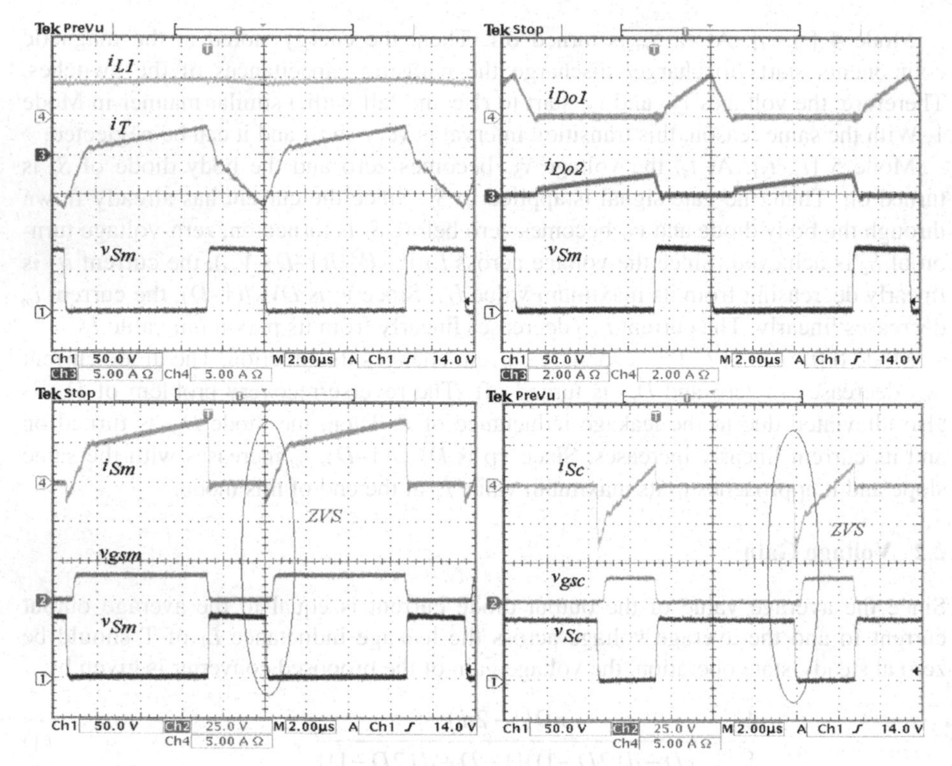

Fig. 2. Experimental waveforms

4 Conclusion

A ZVS voltage doubled SEPIC converter has been proposed. It provides a continuous input current and minimizes the voltage stresses of the switches. Due to soft commutation of semiconductor devices, higher efficiency can be obtained compared to the conventional isolated SEPIC converter.

References

1. Melo, P.F., Gules, R., et al.: A modified SEPIC converter for high-power-factor rectifier and universal input voltage applications. IEEE Trans. Power Elec. 25, 310–321 (2010)
2. Chiang, S.J., et al.: ZIB Structure Prediction Pipeline: Modeling and control of PW charger system with SEPIC converter. IEEE Trans. Ind. Elec. 56, 4344–4353 (2009)
3. Ye, Z., et al.: Design considerations of a high power factor SEPIC converter for high brightness white LED lighting applications. In: Proc. IEEE PESC, pp. 2657–2663 (2008)

The Study on Application of Computer Assisted-Language Learning in College English Reading Teaching

Haiquan Huang

Foreign Language School, Hubei University of Technology, Wuhan, Hubei, China
Huanghaiquan824@163.com

Abstract. Nowadays, the research and practice of CALL have been paid great attention in the foreign language learning and teaching. But the research on the application of CALL into reading teaching is less developed. In this paper, the author attempts to use empirical study to illustrate how to improve the teaching of reading and classroom efficiency by multimedia and Internet.

Keywords: multimedia, Internet, reading of reading, assisted teaching.

1 Introduction

Computer-Assisted Language Learning (CALL) is a new mode of modern educational technology. It has been an inevitable outcome of the combination of multi-media technique and educational theories since it entered the field of education. It has been the main current of the development of education modernization to use the technique into the field of education. As a teaching method, CALL is widely used in classroom teaching of college English, it helps to create a pleasant communication environment, stimulate students' motivation and enthusiasm to learn English, construct their cognitive schemata and cultivate their cross-culture awareness and ability. Besides, it plays a positive role in the respects of developing English teaching resources, opening study outlets, enlarging the classroom teaching capacity, improving learning methods, developing students' thinking abilities, and raising teaching efficiency.

Besides, reading comprehension is a very important part of college English teaching, so it is necessary to explore the teaching method to improve students' reading comprehension. The author believes that applying multimedia and Internet to improve students' reading ability will be of great practical significance to both the students and teachers. This paper will try to do such a research in order to prove that CALL has its advantages in reading teaching.

2 Methodology

2.1 Hypothesis

As for the present study, the author wants to find out whether CALL can help students improve their English. So, the author put forward the following hypothesis:

D. Jin and S. Lin (Eds.): Advances in MSEC Vol. 1, AISC 128, pp. 55–59.
springerlink.com

CALL is a more effective method of teaching college English in China. That is to say, CALL can help to arouse the learner's interest; CALL can help to improve the learner's reading ability.

The results of the questionnaire and the marks of the two examinations will be collected and analyzed to find out whether or not.

2.2 Subjects

The participants are the students from two intact classes of the first year in Hubei University of Technology. In 2010, the author has been assigned to teach Civil Engineering Class3 and Computer Engineering Class5, with each class having exactly 65 students. The average score of the two classes in the English test in the College Entrance Examination are nearly the same, besides; the students who have scored above 105 in these two classes are nearly equal in number. The real situation of subjects can be seen in the following table .Accordingly, Civil Engineering Class3 was randomly chosen as the experimental class (EC) and Computer Engineering Class5 as the control class (CC). What's more, during this research these two classes were taught by the author in person.

Table 1.

Class	Average age	Mean score	Highest Score	Lowest Score	Above105
3	19	94.5	128	80	38
5	19	95.3	127	79	37

2.3 Instruments

During the experiment, the author designed the pre-test and the post-test.

2.3.1 The Pre-test

At the beginning of the experiment, to compare the reading ability of the students in the EC and in the CC, all the students were required to take part in the pre-test. The test paper was specially designed for these two classes about reading, which is intended to test their reading ability. After the test was finished, students' test papers were gathered and corrected, and then the results of the test were collected, analyzed and discussed.

2.3.2 The Post-test

The post-test was used to analyze achievement differences between the two classes after CALL was applied to the EC's classroom teaching.

2.4 Procedures

The experiment was carried out within sixteen weeks, lasting from September 2010 to January 2011.The Experiment was done in about sixteen weeks in two classes. One is

Experimental Class (EC) and another is Control Class (CC).The English textbook being used is New Horizon College English (the second edition), which focuses on reading because there are one or two reading passages in each unit. How to teach reading well has been as issue facing challenges. Now, I'd like to say how to use multimedia while teaching reading.

2.4.1 Experimental Class

Using Multimedia Can Demonstrate Wide and Relevant Background Knowledge

Background knowledge is a critical component of text comprehension (Anderson, etal.,1985;McNeil,1992; Stanovich,1994).Research has shown that when background knowledge is provided before reading(for instance, through video clips),story comprehension improves(Kinzer&Leu,1997,Sharpetal,1995).According to Schema theory, schema that a reader owns in his head has a great effect on his reading comprehension. Multimedia has obvious advantages over other media. Multimedia technology seems more suitable to prepare students by "helping them build background knowledge on the topic prior to reading, through appropriate pre-reading activities (Carrell 1988b:245).

Multimedia and Network Can Bring about a New Teaching Mode

Multimedia and network technology not only provide friendly and audio-visual mutual learning environment in the form of hypertext and hyper-chain, but also provide lovely pictures, pleasant sounds or colorful tableaus to stimulate students' senses. In network environment, interaction between teachers and students, students and students, students and computers, is becoming more plentiful.

The Internet Makes Extensive Reading Facilitating

Intensive reading is inadequate large amounts of self-selected, easy and interesting reading called extensive reading should be the underpinning of all foreign language reading instruction. Extensive reading is generally associated with reading large amounts with the aim of getting an overall understanding of the material.

All the participants can study in the autonomous language learning center of Hubei University of Technology. They are exposed to large amount of English reading materials in the Internet.

2.4.2 Control Class

In the control class, the author adopts a traditional way of the teaching in reading Grammar-Translation Method.

3 Analysis and Discussion of the Results

3.1 Data Collection

The result of pre-test is shown as follows:

Table 2.

Class	Number of students	Mean score	Highest score	Lowest score	Above 105
EC	65	94.5	129	82	39
CC	65	95.2	128	80	38

The post-test was done in the EC and CC at the end of the experiment in order to get comparable result.

Table 3.

Class	Number of students	Mean score	Highest score	Lowest score	Above 105
EC	65	99.4	132	90	45
CC	65	96.2	129	86	40

3.2 Analysis

From the above two tables, we can see clearly that the mean score of the EC has been significantly increased, while the mean score of the CC has almost stayed the same. Anyway, it might be safe to say that the reading ability of the EC has been greatly improved because of the application of CALL. The students in EC are more involved and motivated in today's class. They feel more relaxed and are willing to have English class. After analyzing the data, it can be clearly seen that 86% of the students like to challenge CALL. They are willing to be on the Internet or use other sources to search for some information for their project. And they find their reading ability has been improved.

3.3 Discussion

From the research results, we can see that the EC under CALL outperforms the CC in reading ability. Therefore, it can be concluded that CALL is more effective in teaching English. After implementing CALL, we find out that CALL can also help students learn to communicate, learn to think about a question, learn to make a decision and deal with an emergency. Furthermore, with the help of CALL, students can also experience the sense of achievement and realize their own value. In addition, CALL can greatly improve the integrated ability of a student, which is in accordance with the general target of the empowerment education in China.

4 Conclusion

This paper insists that CALL is good for improving teaching of reading in college English. Utilization of multimedia and the Internet in the teaching of reading is becoming wider and more efficient. Students can be more capable of obtaining information or knowledge not only from books but also from the Internet so as to fit in with society. Such a teaching reform is the only way to carry out quality-oriented education which concentrates on students' development in an all-round way and stresses that students are the autonomous learners. Although a lot of research into CALL has been done, thing will change with the rapid development of computer technology. Some research will be out of date soon and ways to improve the teaching of reading are under discussion. Therefore, the study of how to utilize computers to assist language teaching or learning is well worth studying.

References

1. Ahmad, K., Greville, C., Rogers, M., Sussex, R.: Computers, Language Learning and Language Teaching. Cambridge University Press (1985)
2. Berge, Z., Collins, M.: Computer-mediated communication and the online classroom in distance learning. Hampton Press, Cresskill (1995)
3. Chapelle, C.A.: Multimedia CALL: Lessons to be Learned from Research on Instructed SLA. Language Learning & Technology (1998)
4. Chapelle, C.A.: Computer Applications in Second Language Acquisition: Foundations for Teaching, Testing and Research. Cambridge University Press (2001)
5. Canale, M., Wain, M.S.: Theoretical Bases of Communicative Approaches to Second Language Teaching and Testing. Applied Linguistics 1, 1–47 (1980)
6. Hubbard, P.: A Methodological Framework for CALL Courseware Development. In: Pennington and Stevens (1992)
7. Higgins, J.: Language, Learners and Computers. Longman, London (1988)
8. Warschauer, M., Healey, D.: Computers and Language Learning: An Overview Language Teaching (1998)
9. Warschauer, M.: E-Mail for English teaching. Teachers of English to Speakers of Other Languages, Washington, D. C (1995)
10. Taylor, R.P.: The computer in the School: Tutor, Tool, Tutee. Teacher's College, Columbia University, Teacher's College Press, New York (1980)

Single-Switch Step Up/Down Converter with Positive and Negative Outputs

Hyun-Lark Do

Department of Electronic & Information Engineering,
Seoul National University of Science and Technology, Seoul, South Korea
hldo@seoultech.ac.kr

Abstract. A single-switch step up/down converter with positive and negative outputs is proposed in this paper. In the porposed converter, a single-ended primary inductor converter (SEPIC) and a Cuk converter are merged. The SEPIC converter stage provides a positive step up/down output and the Cuk converter stage provides a negative step up/down output. An input stage consisting of a switch and an inductor is shared. The proposed converter has a continuous input current, step up/down capability, dual outputs, and good cross regulation. The operation principle and steady-state analysis of the proposed converter are provided. A prototype of the proposed converter is developed, and its experimental results are presented for validation.

Keywords: Step up/down capability, SEPIC converter, Cuk converter, input current ripple.

1 Introduction

Recently, DC-DC power conversion techniques have been researched due to their increasing demands in applications such as fuel cells and photovoltaic systems [1]. Generally, input sources in these applications have low voltages and are easy to vary. Also, the input current ripple is an important factor in theses applications. Large input current ripple may shorten the lifetime of the sources and input filter capacitors used in the applications. Especially in the fuel cell systems, reducing the input current ripple is very important because the large current ripple shortens fuel cell lifetime as well as decreasing performances [2]. SEPIC and Cuk converters are suitable for these applications due to their input stages consisting of an inductor and a switch like a boost converter and step up/down capability.

In this paper, a single-switch step up/down converter with positive and negative outputs is proposed. A SEPIC converter and a Cuk converter are merged in the proposed converter. Since a SEPIC converter and a Cuk converter have step up/down capability, each output of the proposed converter has step up/down capability. The shared input stage provides a continuous input current. Moreover, the proposed converter shows good cross regulation. Also, two outputs can be used as a single output. Then, the voltage gain of the proposed converter doubles.

D. Jin and S. Lin (Eds.): Advances in MSEC Vol. 1, AISC 128, pp. 61–64.
springerlink.com

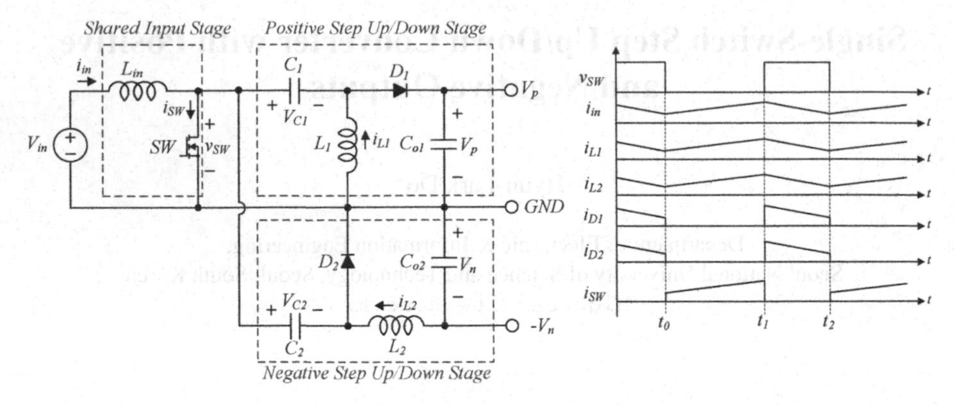

Fig. 1. Circuit diagram and key waveforms of the proposed converter

2 Analysis of the Proposed Converter

Fig. 1 shows the circuit diagram and key waveforms of the proposed converter. To simplify the steady-state analysis, it is assumed that the voltages across the capacitors C_1, C_2, C_{o1}, and C_{o2} are constant and all the semiconductor devices are ideal.

2.1 Operation

The operation of the proposed converter in one switching period T_s can be divided into two modes. Before t_0, the diodes D_1 and D_2 are conducting.

Mode 1 [t_0, t_1]: At t_0, the switch *SW* is turned on. Then, D_1 and D_2 are turned off. In this mode, the input voltage V_{in} is applied to the inductor L_{in}. Therefore, the input current i_{in} increases from its minimum value with the slope of V_{in}/L_{in}. The capacitor voltage V_{C1} is applied to L_1. Then, i_{L1} increases linearly with the slope of V_{C1}/L_1. Similarly, V_{C2}-V_n is applied to L_2. Then, i_{L2} increases linearly with the slope of $(V_{C2}$-$V_n)/L_2$. At the end of this mode, all the inductor currents arrive at their maximum values.

Mode 2 [t_1, t_2]: At t_1, the switch *SW* is turned off. Then, D_1 and D_2 are turned on. In this mode, the input voltage $-(V_{C2}-V_{in})$ is applied to the inductor L_{in}. Therefore, the input current i_{in} decreases from its maximum value with the slope of $-(V_{C2}-V_{in})/L_{in}$. The capacitor voltage $-V_p$ is applied to L_1. Then, i_{L1} decreases linearly with the slope of $-V_p/L_1$. Similarly, $-V_n$ is applied to L_2. Then, i_{L2} decreases linearly with the slope of $-V_n/L_2$. At the end of this mode, all the inductor currents arrive at their maximum values.

2.2 Design Parameters

By applying volt-second balance law to the voltage waveforms across the inductors, the capacitor voltages can be easily obtained by V_{C1}=V_{in}, V_{C2}=V_{in}/(1-D),

V_p=V_n=$DV_{in}/(1\text{-}D)$. When two outputs are used as a single output, the voltage gain of the proposed converter is given by

$$M = \frac{V_p + V_n}{V_{in}} = \frac{2D}{1-D} \quad . \tag{1}$$

Input current ripple ΔI_{in} is given by

$$\Delta I_{in} = \frac{V_{in} DT_s}{L_{in}} \quad . \tag{2}$$

Minimum value of L_{in} for a continuous input current is given by

$$L_{in,\min} = \frac{\eta V_{in}^2 DT_s}{2P_o} \quad , \tag{3}$$

where P_o is total output power and η is the efficiency of the converter. The ripple component ΔI_{L1} of i_{L1} is given by

$$\Delta I_{L1} = \frac{V_{in} DT_s}{L_1} \quad . \tag{4}$$

Minimum value of i_{L1} is given by

$$I_{L1,\min} = I_{o,positive} - \frac{V_{in} DT_s}{2L_1} \quad . \tag{5}$$

3 Experimental Results

A prototype is implemented with specifications of V_{in}=24V, V_p(=V_n)=50V, P_o=100W. The circuit parameters are L_{in}=200uH, C_1=C_2=13.2uF, C_{o1}=C_{o2}=220uF, L_1=L_2=120uH, f_s=100kHz. The experimental waveforms of the prototype are shown in Fig. 2. The measured duty cycle is 0.682. The voltage stress of *SW* is around 75V. It can be seen that the experimental waveforms agree with the theoretical waveforms and the analysis. The input current is continuous. By adjusting the inductance of L_{in}, the input current ripple can be controlled. The cross regulation of two outputs is also shown in Fig. 2. When the negative output is loaded with 4W, the variation of V_n is around 1.2%. The proposed converter exhibits an efficiency of 92.3% at full load condition.

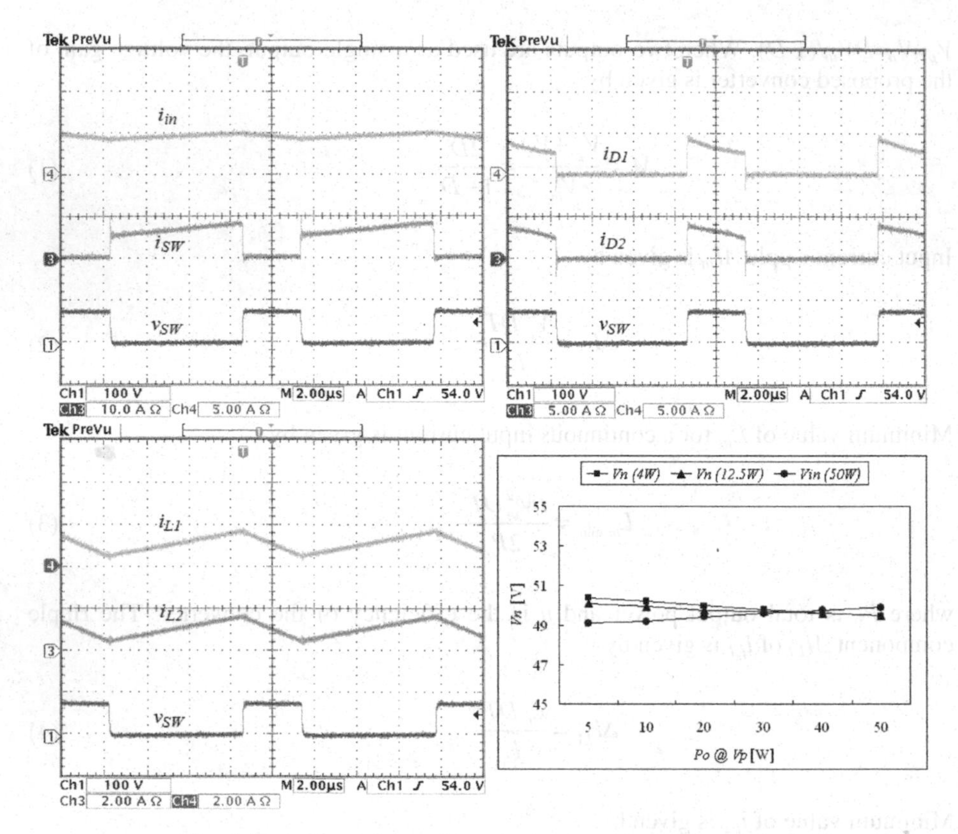

Fig. 2. Experimental waveforms and measured cross regulation

4 Conclusion

A single-switch step up/down converter with positive and negative outputs has been proposed. It features a continuous input current, step up/down capability, dual outputs, good cross-regulation, and a single power switch. Due to these features, the proposed converter can be a good candidate for non-isolated step up/down DC-DC converter.

References

1. Do, H.L.: A zero-voltage-switching DC-DC converter with high voltage gain. IEEE Trans. Power Elec. 26, 1578–1586 (2011)
2. Kong, X., Khambadkone, A.M.: Analysis and Implementation of a High Efficiency, Interleaved Current-Fed Full Bridge Converter for Fuel Cell System. IEEE Trans. Power Elec. 22, 543–550 (2007)

Energy Recovery Address Driver with Reduced Circulating Current

Hyun-Lark Do

Department of Electronic & Information Engineering,
Seoul National University of Science and Technology, Seoul, South Korea
hldo@seoultech.ac.kr

Abstract. An energy recovery address driver for plasma display panel (PDP) is proposed. It utilizes a resonant circuit consisting of panel capacitance and external inductors to reduce the voltage across the data drive IC when its output stages change their status. The resonant circuit reduces the power consumption and relieves the thermal problems of the drive ICs. Moreover, circulating current is significantly reduced in the proposed address driver. It also has load adaptive characteristic (LAC). Then, low efficiency problem at light load does not exist. Compared to the conventional LC resonant driving method, the component count is significantly reduced.

Keywords: Energy recovery, address driver, resonance, circulating current.

1 Introduction

In PDPs, an energy recovery circuit (ERC) has been used to recover the reactive power from PDP on the sustain electrodes in [1] and [2] and address electrodes in [3]. Most PDPs utilize an LC resonant ERC to obtain high efficiency. The conventional LC resonant address driving method shown in Fig. 1 uses the resonance between the panel capacitance and an external inductor. It provides high efficiency at heavy data loads which have a lot of data switchings. However, it shows relatively low efficiency at light loads such as full white image which has no data switching. It is because it has no load adaptive characteristic. Low efficiency at light load can be overcome simply by turning off the ground clamping switch S_4 during the address period or replacing the switch S_4 with the clamping diode [2]. However, many components and complex control circuits are still required. In addition, it has a large circulating current which occurs due to the reverse recovery of D_1. The large circulating current causes additional power loss [1].

An energy recovery address driver with reduced circulating current for address drive ICs is proposed. Components count is significantly reduced compared to that of the conventional LC resonant driving method. Moreover, it has intrinsic LAC without ground clamping. In addition, the circulating current is significantly reduced. Compared to the conventional method, the component count is reduced and the cost is cut down. Reduced power consumption can also lower the cost of power supplies by cutting down the maximum power consumption.

D. Jin and S. Lin (Eds.): Advances in MSEC Vol. 1, AISC 128, pp. 65–68.
springerlink.com

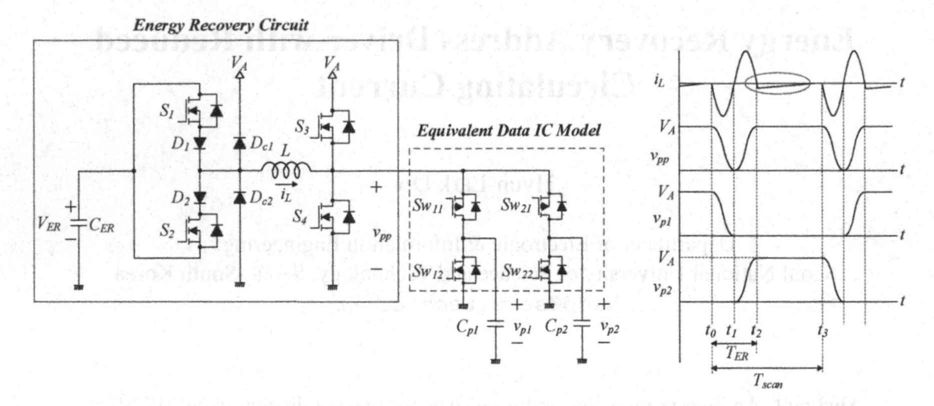

Fig. 1. Circuit diagram and key waveforms of conventional resonant address driver

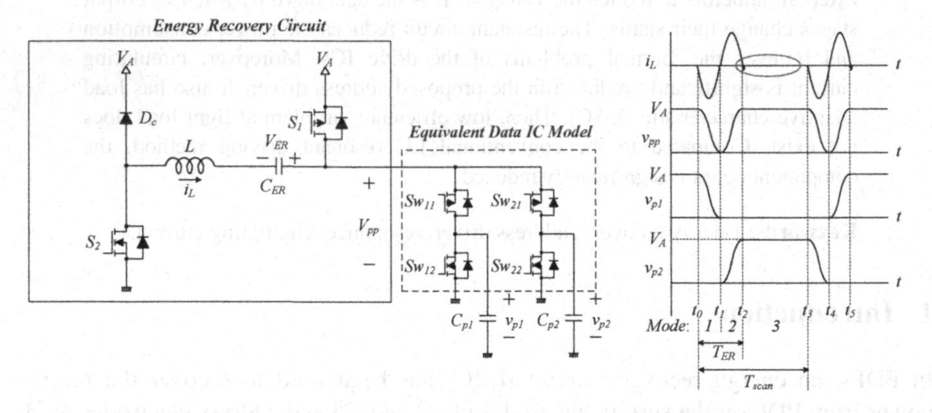

Fig. 2. Circuit diagram and key waveforms of proposed address driver

2 Analysis of the Proposed Address Driver

Prior to Mode 1, the panel voltage v_{p1} is maintained as the address voltage V_A with S_{W11} conducting and the panel voltage v_{p2} is clamped as zero with S_{W22} conducting.

Mode 1 [t_0, t_1]: At t_0, the switch S_2 is turned on and the resonance occurs between the external inductor L and the panel capacitance C_{p1}. The panel starts to be discharged with a resonant manner through the body diode of S_{W11}, the inductor L, the energy recover capacitor C_{ER}, and the switch S_2. The energy stored in the panel is transferred to the energy recovery capacitor C_{ER}.

Mode 2 [t_1, t_2]: At t_1, the switch S_{W12} is turned on and the panel voltage v_{p1} is clamped as zero. The switch S_{W21} is turned on and the panel voltage v_{p2} starts to rise with a resonant manner. The resonance occurs between the inductor L and the panel capacitance C_{p2}. The panel starts to be charged through the body diode of S_2, the inductor L, the capacitor C_{ER}, and S_{W21}. The energy stored in C_{ER} is transferred to the panel.

Mode 3 [t_2, t_3]: At t_2, the switch S_1 is turned on and the panel voltage v_{p2} is clamped as the address voltage V_A. The panel voltage v_{p1} is maintained as zero with S_{W12} conducting. Then, the current i_L changes its direction and linearly decreased with the same slope. When the reverse recovery of the body diode of S_2 ends, the energy stored in L is transferred to C_{ER}. Since $-V_{ER}$ is applied to L, the inductor current i_L decreases to zero rapidly.

In the proposed address driver, the voltage V_{ER} across the energy recovery capacitor naturally varies according to N_{data} which represents the number of data on single vertical line. Since the voltage V_{ER} increases as N_{data} increases, the energy recovery operation is weakened and the circulating current is reduced. When the full white image is displayed, the voltage V_{ER} arrives at V_A and the energy recovery operation stops naturally. The voltage V_{ER} is calculated as

$$V_{ER} = V_A - \frac{V_A}{\sqrt{2\left(\frac{N_{data} - N_{data\_chg}}{N_{data\_chg}} \cdot \frac{\omega^2}{\frac{R^2}{4L^2} + \omega^2} \cdot \left(1 - e^{-\frac{\pi R}{\omega L}}\right) + 2\right)}}, \tag{1}$$

where N_{data_chg} is the number of address data switchings, R is the inevitable parasitic resistances, and ω is given by

$$\omega = \sqrt{\frac{1}{LC_p} - \left(\frac{R}{2L}\right)^2}. \tag{2}$$

3 Experimental Results

The prototype address driver is tested on a single-scan 42-inch XGA ac PDP (Ncell=768×1024×3). The panel capacitances C_a and C_m are about 40pF and 30pF, respectively. The energy recovery capacitor C_{ER} is 0.22uF. There are 16 data drivers of 192 outputs. The proposed ERC block is designed to control two data driver ICs. Therefore, eight ERC blocks are required. The address voltage V_A is set to 60V. T_{frame} is 16.67ms and 10 subfields are used. The scan period T_{scan} is 1.2μs. Fig. 3 shows the experimental waveforms and power consumption comparison. Since N_{data} is equal to N_{data_chg} at one-dot on/off pattern which is the heaviest load, V_{ER} is around $V_A/2$. Since the data are not changed at full white pattern, the voltage V_{ER} goes up to V_A and the energy recovery operation stops. Power consumption of the proposed address driver is reduced by 71% at one-dot on/off pattern compared with the address driver without an ERC. The proposed method shows slightly higher efficiency than the conventional LC resonant driving method with LAC. It is simply because the proposed method has less components and low circulating current.

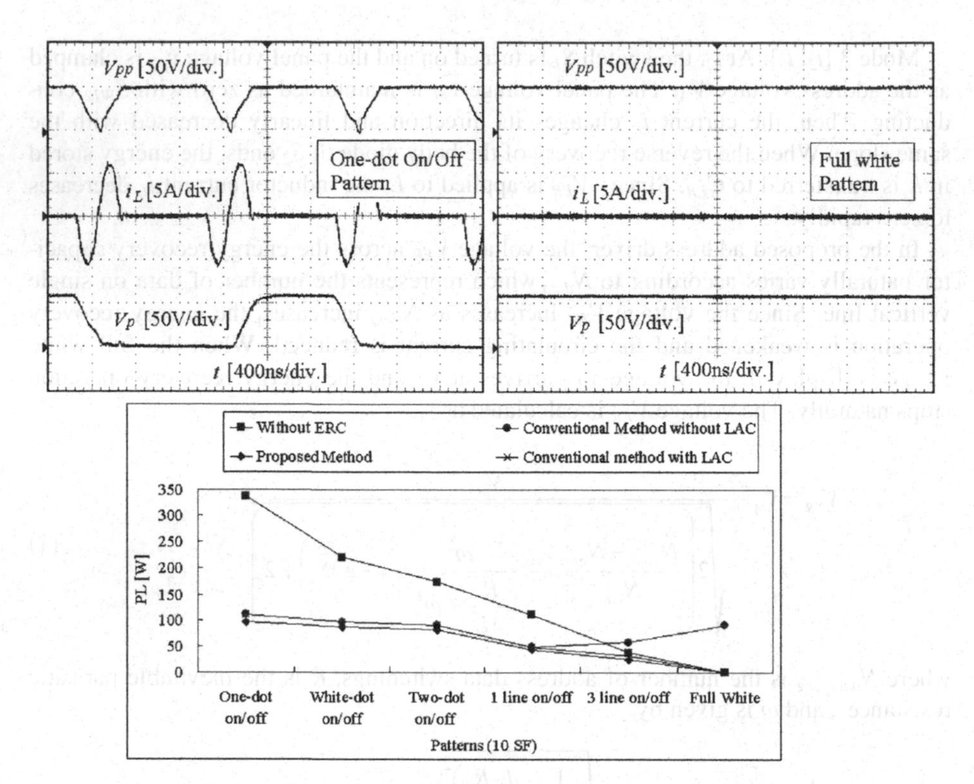

Fig. 3. Experimental waveforms and measured power consumption

4 Conclusion

A new energy recovery address driver with reduced circulating current for PDP has been proposed. It has LAC and low circulating current. It showed a higher efficiency compared to the conventional method. Moreover, it has less components.

References

1. Do, H.L.: Energy recovery sustain driver with low circulating current. IEEE Trans. Ind. Elec. 57, 801–804 (2010)
2. Hsu, H.B., et al.: Regenerative power electronics driver for plasma display panel in sustain-mode operation. IEEE Trans. Ind. Elec. 47, 1118–1124 (2000)
3. Do, H.L.: An efficient driving method for a high-voltage CMOS driver IC. International Journal of Electronics 97, 883–896 (2010)

A Scheme of Feasibility with Static Analysis in Software Testing Environment

ManSha Lu

College of Computer and Communication Engineering,
Changsha University of Science and Technology, ChangSha, HuNan, China
lumansha8832@163.com

Abstract. The static analysis in the software testing is to abstract the necessary data from the source code, which is used in the analysis of control flow and syntax analysis. It is used to automatically generate the graph, list and it is the first step to insert testing code in the program. This paper mainly analyzes the details of pertaining modules of source code, project file, code insertion and automatic generation of graph and list. And it gives a feasible method.

Keywords: Static analysis, program structured graph, insertion.

The software testing consists of static analysis, dynamic analysis and finally generates the testing report. The static analysis is the analysis of syntax of software, which generates static data file and inserted source code according to testing requirement. It also supports the successive procedure of dynamic analysis.

1 Introduction

The tool of software testing requires internal logic of source code analysis. It can identify the syntax of source code and understand program logic. An application usually consists of many files. The project file manages these files. The function of static analysis module is to analyze the project file. It obtains the original files according to project file, and then analyzes the syntax of source files. In this procedure the code insertion is completed. Finally the new files are compiled and linked and the executive is generated. The static analysis procedure is to abstract the key data, such as function name, class name of program in OOP and line number etc. It also analyzes the control flow of source code, block source code and then stores the data in a certain file format. These files are used to automatically generate the graph. The code insertion is the procedure which the specific code is inserted in specific location of source code. The executives, generated from the inserted codes can generate the dynamic data when it executes. These data can be used in the procedure of overcast analysis automation and dynamic tracing. In the procedure of testing only static analysis module directly interacts with source code. The automation module only interacts with data files and the executive but language. The static analysis module can support many different languages.

D. Jin and S. Lin (Eds.): Advances in MSEC Vol. 1, AISC 128, pp. 69–73.
springerlink.com

2 Implementation of the Analysis Module of Source Code

An application many contains many source files with different types. The syntax and structure of these source files is not same. Sometimes lexical analyzer and syntax analyzer is called the front end of compiler. The program analysis module can be generated from source code analysis with the method of compiling procedure. But it is different from compiler, e.g. the interim code generation. The interim code from program analysis is a text. It is the input of successive modules, but compiler only analyzes syntax of program and finds syntax errors. Also complier does not care about calling procedure of function. It only requires the syntax of program to comply with language rules. The source analysis module learns the calling procedure of function and generates the graph of relationship among the functions. In source analysis module the lexical analysis and syntax analysis are used as front end and storage of calling relationship trailing end. The structure of source analysis module is as figure 1.

The front end of source analysis module mainly finds the all functions in the source code with lexical analyzer and syntax analyzer. It stores the information, i.e. function information list. It also analyzes the logic of functions, separates the blocks in these functions, labels these blocks and stores the information, i.e. block information list in the data files. The function information list is the input of function calling and control flow and the block information list the input of program insertion

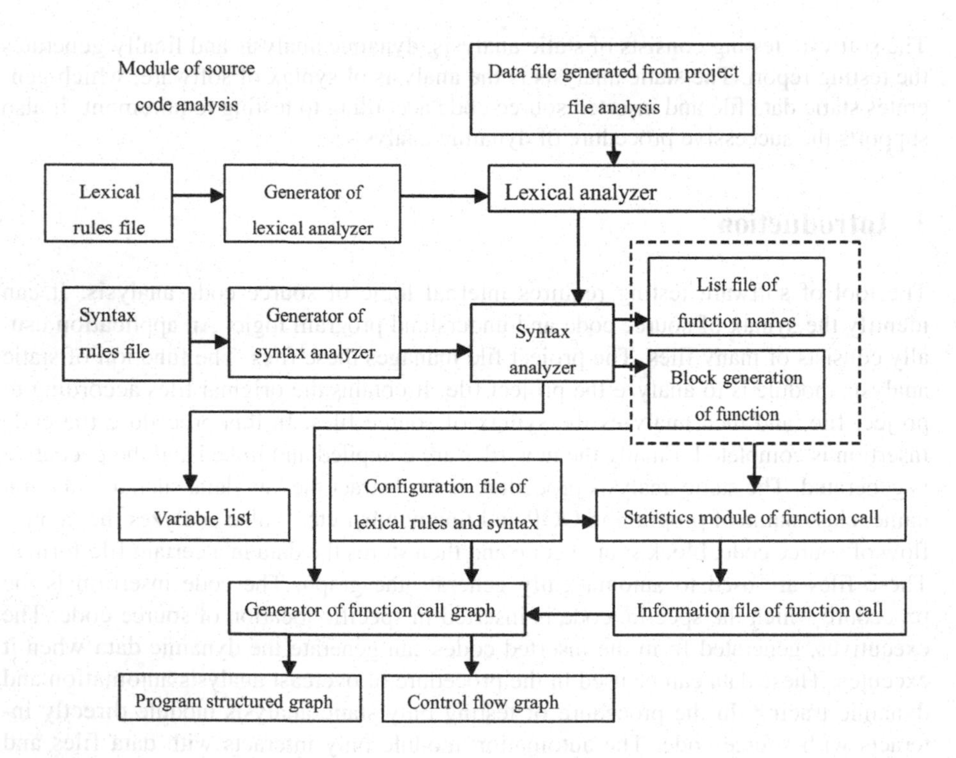

Fig. 1. Structured graph of source analysis module

3 Implementation of the Analysis Module of Project File

An application may contain hundreds or thousands of source files. It is very complicated. The project file can provide the automatic management. Its functions are as follows:

(1) to provide the calling relationship among the source files
(2) to provide management of date of file modification
(3) to provide the compatibility of program in different language

Since an application many contain source code in different language the analysis of project is a very complicated procedure. So it is required to understand:type of source file, lexical rules of project file, syntax rules of project file.

As a management file, the project file is written in printable characters. The target of project analysis module is to find all source files, file attributes, file size and use condition of file (i.e. which conditions are used) in the project and generate a data file. The data file consists of many items, such as file name, file type, the located directory (including relative path and absolute path), file size, use condition and included header files

The project analysis module is required to identify the file type since the source files may be written in different languages. The project analysis module is as figure2

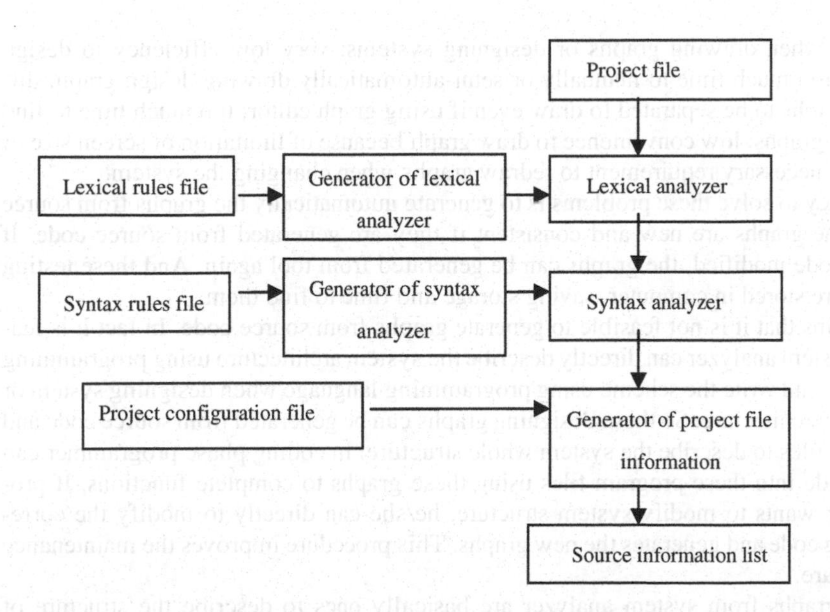

Fig. 2. Graph of project analysis module

4 Implementation of Program Insertion Module

The program insertion module is to insert required information in a specified location in source code. It is impossible to automatically insert the required information since the information which programmer requires is vast. The insertion module is to insert template codes in the specific location of program according to user's requirement, and control the switch between controlling debug version and distributed version with pre-compiled control code. The user needs to direct the service of insertion module, such as a certain variable value, return value and parameter value of function when called. The program insertion module inserts the code in corresponding location and notifies user the required information with a certain mechanism. The program insertion contains the following three phases:

1. phase of pre-processing: in this phase it is to insert the specific code in source file. The handling procedure of this phase can progress with static testing. It inserts specific code in specific location when processing syntax analysis
2. phase of compilation and execution: in this phase compilation and execution is completed
3. phase of handling: the data from execution is used with source code to generated graph, report, including overcastting report

5 Implementation of Automatic Generation Module of Graphics

It exists when drawing graphs or designing systems: very low efficiency to design system; too much time to manually or semi-automatically drawing design graph; different graphs to be separated to draw even if using graph editor; too much time to find different graphs; low convenience to draw graph because of limitation of screen size of monitor; necessary requirement to redraw graphs when changing the system.

The key to solve these problems is to generate automatically the graphs from source code. The graphs are new and consistent if they are generated from source code. If source code modified, the graphs can be generated from tool again. And these testing graphs are stored in computer, saving storage and time to find them.

It seems that it is not feasible to generate graphs from source code. In fact it is feasible. System analyzer can directly describe the system architecture using programming language and write the scheme using programming language when designing system or software architecture. So these designing graphs can be generated from source code and program files to describe the system whole structure. In coding phase programmer can insert code into these program files using these graphs to complete functions. If programmer wants to modify system structure, he/she can directly to modify the corresponding code and generates the new graphs. This procedure improves the maintenance of software.

The graphs from system analyzer are basically ones to describe the structure of program. It is considered here to automatically generates these graphs. Two graphs are important: function call graph and project file graph

(1) function call graph
The function call graph describes the calling relationship among calling functions and called functions. There are non-member functions and member functions of source code in the function call graph. The frame of non-member function uses the real line; the frame of member function contains class name and function name; the line between frames is as calling relationship. The strip graph in the frame and digits represent ratio of testing overcastting, complexity and module size.

(2) project file graph
There are many files in an application. The programmer cannot easily locate the file with complexity of calling relationship among functions even if the function call graph can directly describe the relationship and details among internal modules(procedure-oriented) or components (object-oriented). The programmer cannot directly describe the relationship among modules or components in some condition, e.g. strongly-coupled or weakly-coupled or weakly-coupled among the calling relationship or message passing when coding. The programmer can locate the file which the function locates with navigation with these project file graph. And he/she can easily master the relationship among these files, e.g. file A including a call to func B, and func B included in file B The programmer can find the definition file of the function. The project file graph is generated from calling information of functions.

It is required to analyze syntax of source code to generate the above two graphs. The data can be abstracted in the procedure of syntax analysis and then the program structured graph is generated.

[Project name]:research on automation of object-oriented software testing
[Project number]:10C0366

References

1. Pressman, R.S.: Software Engineering: A Practitioner's Approach, 5th edn. McGraw-Hill Companies, New York (2007)
2. Minkiewicz, A.F.: In Measuring Objectoriented Software with Predictive Object Points. Price Systems, L.L.C, California (2006)
3. Kuchana, P.: Software Architecture Design Patterns in Java. Auerbach (2006)
4. McGregor, J.D., Sykes, D.A.: A Practical Guide to Testing Object-Oriented Software, 1st edn. Addison-Wesley Professional (2006)
5. Fewster, M., Graham, D.: Software Test Automation: Effective Use of Test Execution Tools. Addison-Wesley Professional (2005)

Active Clamped Resonant Flyback with a Synchronous Rectifier

Hyun-Lark Do

Department of Electronic & Information Engineering,
Seoul National University of Science and Technology, Seoul, South Korea
hldo@seoultech.ac.kr

Abstract. A zero-voltage-switching (ZVS) resonant flyback converter with a synchronous rectifier is presented in this paper. An active clamp circuit is added to the conventional flyback converter to clamp the voltage across the switches and provide ZVS operation. By utilizing a resonance between the leakage inductance of a transformer and a clamp capacitor, the energy stored in the primary side is transferred to the secondary side with a resonant manner. Therefore, the output rectifier is turned off softly. Moreover, a synchronous rectifier is adopted at the secondary side to improve the efficiency. The operation principle and steady-state analysis of the proposed converter are provided. A prototype of the proposed converter is developed, and its experimental results are presented for validation.

Keywords: Zero-voltage-switching, flyback converter, active clamp, synchronous rectifier.

1 Introduction

Due to its simplicity when compared with other topologies, a flyback converter has been adopted in many applications such as single-stage power converters, bidirectional power conversion for electric vehicles, photovoltaic systems, and lighting systems [1]. However, a switch in the flyback converter operates at hard switching and it causes several problems such as high switching loss, low efficiency, and high voltage spikes across the switch and the output diode. To overcome these problems, various soft-switching techniques have been presented [2] and [3]. Most of them require more than one magnetic component. It raises the volume and size.

In order to overcome these problems, an active clamped resonant flyback converter with synchronous rectifier is proposed. Its circuit diagram and key waveforms are shown in Fig. 1. There is only one magnetic component in the proposed converter. An active clamp circuit consisting of a clamp capacitor and an auxiliary switch is added to the conventional flyback converter. By utilizing the resonance between the leakage inductance of transformer and the clamp capacitor, both main and auxiliary switches can operate with ZVS. Moreover, a synchronous rectifier is adopted at the secondary side to improve the efficiency. Due to these features, the proposed converter shows high efficiency. The theoretical analysis is verified by an experimental prototype.

D. Jin and S. Lin (Eds.): Advances in MSEC Vol. 1, AISC 128, pp. 75–78.
springerlink.com

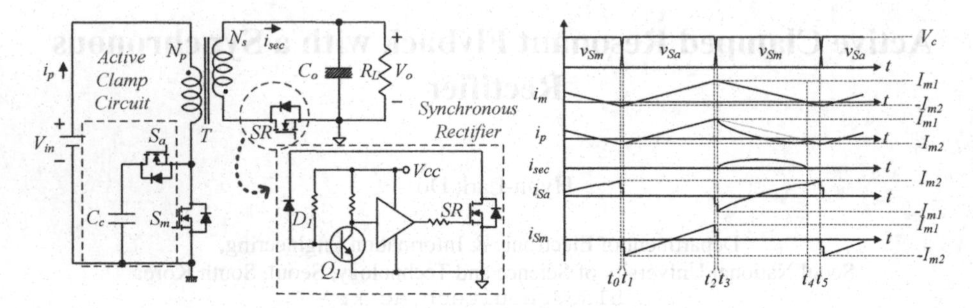

Fig. 1. Circuit diagram and key waveforms of the proposed converter

2 Analysis of the Proposed Converter

By utilizing the active clamp circuit, the energy trapped in the leakage inductance L_{lk} of T is recycled and the voltage across the main switch S_m is clamped as the clamp capacitor voltage V_c. Due to the resonance between the leakage inductance and the clamp capacitor, ZVS is achieved in both main and auxiliary switches. Without any additional magnetic component, soft-switching operation is achieved. The switches S_m and S_a are operated asymmetrically and the duty ratio D is based on S_m. The transformer T has a turn ratio of 1:n ($n= N_s/N_p$). To simplify the steady-state analysis, it is assumed that the voltages across the capacitors C_c and C_o are constant as V_c and V_o, respectively.

2.1 Operation of the Active Clamped Resonant Flyback Converter

The operation of the proposed converter in one switching period T_s can be divided into five modes. Before t_0, S_a is conducting and SR is turned off. The magnetizing current i_m flowing through the magnetizing inductance L_m of T decreases linearly and approaches to its minimum values $-I_{m2}$.

Mode 1 [t_0, t_1]: At t_0, S_a is turned off. Then, the energy stored in the magnetizing inductance and the leakage inductance starts to charge/ discharge parasitic capacitances of the switches. Therefore, v_{Sa} across S_a starts to rise from zero and v_{Sm} across S_m starts to fall from V_C. Since these capacitances are very small, this time interval is very short.

Mode 2 [t_1, t_2]: At t_1, v_{Sm} becomes zero and the body diode of S_m is turned on. Then, the gate signal is applied to S_m. Since the current has already flown through the body diode and v_{Sm} is zero before the gate signal is applied, zero-voltage turn-on of S_m is achieved. In this mode, SR is turned off. Then, the input current i_{in} is equal to the current i_m. Since V_{in} is applied to the primary inductance, i_m increases linearly.

Mode 3 [t_2, t_3]: S_m is turned off at t_2. Similar to mode 1, the parasitic capacitances are charged/ discharged at the same time. This time interval is also very short.

Mode 4 [t_3, t_4]: At t_3, the voltage v_{Sa} across S_a becomes zero and the body diode of S_a is turned on. Similar to mode 2, zero-voltage turn-on of S_a is achieved. With the turn-on of S_a, SR starts to conduct and the resonance between the leakage inductance of T and C_c occurs. Since the reflected output voltage is applied to L_m, the current i_m decreases linearly in this mode.

Mode 5 [t_4, t_5]: At t_4, the current i_{sec} decreases to zero with a resonant manner and the SR is turned of softly. Since S_a is still on, i_m decreases linearly.

Since the average inductor voltage must be zero under a steady-state, the average voltage of v_{Sm} is equal to V_{in} and the average capacitor voltage V_c is $V_{in}/(1\text{-}D)$. By applying the volt-second balance law to the voltage across L_m, the voltage gain is obtained as follows:

$$\frac{V_o}{V_{in}} = \frac{L_m}{L_m + L_{lk}} \cdot \frac{D}{1-D} \cdot \frac{1}{n} \approx \frac{D}{1-D} \cdot \frac{1}{n} \quad . \tag{1}$$

The maximum voltage stress of a synchronous switch is about V_o+V_{in}/n.

2.2 Synchronous Rectifier

At t_3, the body diode of SR is turned on at first. Then, the voltage of drain terminal of SR is lower than the source terminal. Since the base charge of Q_1 is removed by turn-on of d_1, Q_1 is turned off. Then, "High" signal is applied to a gate driver and the gate signal is applied to SR. Now, SR is fully turned on and all the current is flowing through the channel of SR. At t_4, i_{sec} decreases to zero. If i_{sec} changes its direction, the voltage of drain terminal of SR is higher than the source terminal. Then, d_1 is turned off and Q_1 is turned on. Then, "Low" signal is applied to the gate driver and SR is turned off rapidly.

3 Experimental Results

A prototype is implemented and tested with specifications of V_{in}=48V, V_o=150V, f_s=73kHz, $P_{o,max}$=45W. According to (1), n is selected as 0.5 D_{max} of 0.6. L_m is chosen as 83uH and L_k is 800nH. C_c is chosen as 0.47uF. Fig. 2 shows the experimental waveforms and measured efficiency. The switch voltages are well clamped as V_c. Both S_m and S_a operate with ZVS and SR is turned off softly. It agrees with the theoretical analysis. Fig. 2 shows the measured efficiency of the proposed converter and it is compared with the conventional flyback converter and the resonant flyback converter without synchronous rectification. The power consumed in the control circuit is not included in the efficiency curve. The maximum efficiency of the proposed converter is 94.8%. Due to its soft-switching characteristic of both S_a and S_m and soft commutation of SR, the proposed converter shows higher efficiency.

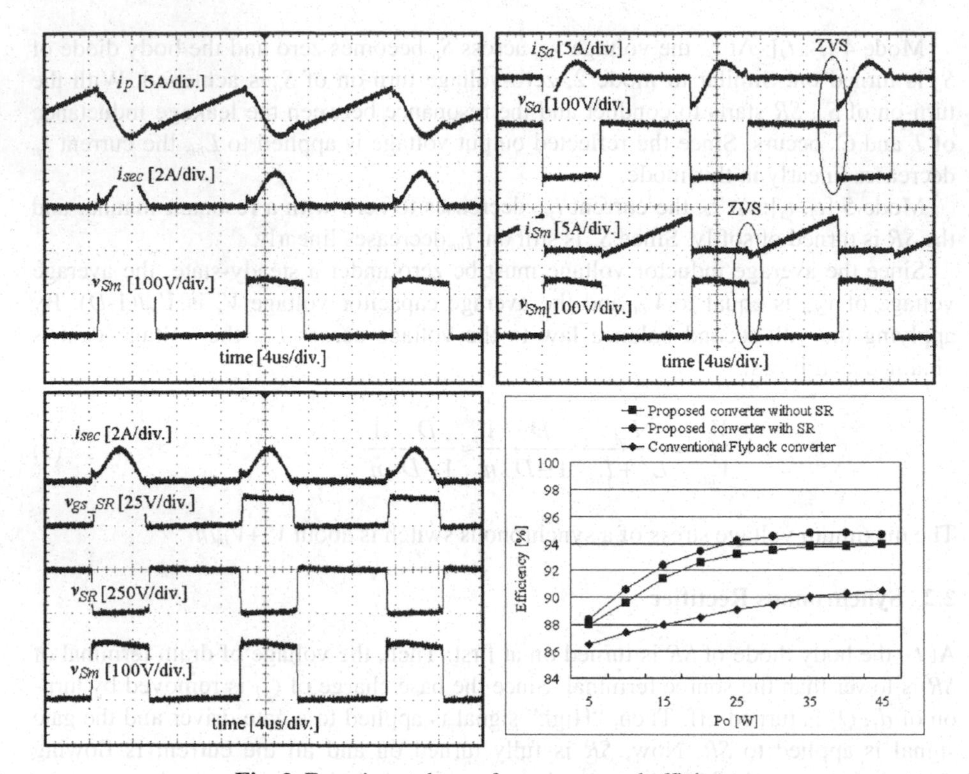

Fig. 2. Experimental waveforms measured efficiency

4 Conclusion

An active clamped resonant flyback converter with synchronous rectifier has been proposed. By utilizing the active clamp circuit and the resonance between the leakage inductance and the clamp capacitor, both main and auxiliary switches can operate with ZVS and soft commutation of a synchronous rectifier. Due to these features, the proposed converter shows high efficiency.

References

1. Chiu, H.J., et al.: A Single-stage soft-switching flyback converter for power-factor-correction applications. IEEE Trans. Ind. Elec. 57, 2187–2190 (2010)
2. Wang, C.M., et al.: A novel ZCS-PWM flyback converter with simple ZCS-PWM commutation cell. IEEE Trans. Ind. Elec. 55, 749–757 (2008)
3. Watson, R., et al.: Utilization of an active-clamp circuit to achieve soft switching in flyback converters. IEEE Trans. Power Elec. 11, 162–169 (1996)

A Robust Digital Watermarking Algorithm Based on Integer Wavelet Matrix Norm Quantization and Template Matching

Xin Liu, XiaoQi Lv, and Qiang Luo

School of Information Engineering, Inner Mongolia University of Science and Technology, Baotou014010, China

Abstract. As a protection and authentication technology, digital watermarking technology has been used in digital image field. A robust digital watermarking algorithm is proposed based on integer wavelet transform to estimate integrity and authenticity of digital images. Using matrix norm quantization, it embeds watermarks into medium-frequency and high-frequency detail sub-bands of digital images' integer wavelet domain. The scheme realizes blind watermark extraction without any extra information. By shaping the watermarked image before extraction, the precision of the extracted watermark is enhanced. Attacking experiments show that the algorithm not only has both robustness and sensitivity, but also exactly locates distorted area, so it is an effective robust digital watermarking algorithm for digital image field.

Keywords: Robust Digital Watermarking for Digital Image Field, Integer Wavelet Transform, Matrix Norm Quantization, Template Matching, SVD.

1 Introduction

Along with fast development of digital technology, digital images' integrity and authenticity are transmitted conveniently but queried more widely. Digital watermarking technology has been used to digital image field, while as an important branch of authentication watermarking technology, robust watermarking technology is attached importance to protect copyright and integrity of multimedia products[1-5]. Wong proposed that original images were blocked, the LSB of each block was set zero and encrypted by XOR with watermarks after scrambled by HASH-function[1]. Barni M proposed a better robust watermarking algorithm which showed watermarks were quantitatively embedded into DWT domain[2]. According to JPEG lossy compression, Dong Gang proposed a DCT-domain watermarking algorithm[3]. Chen Fan proposed a robust watermarking algorithm based on image singular value in which watermarks were selected from singular value corresponding with image content[4]. When verified, the tempered intensity of distorted parts was reflected from singular value difference. So the robust watermarking technology can be used to protect integrity and authenticity of digital images.

Robust watermarks require not only certain robustness but also sensitivity, so this paper proposes a robust watermarking algorithm based on matrix norm quantization and

D. Jin and S. Lin (Eds.): Advances in MSEC Vol. 1, AISC 128, pp. 79–84.
springerlink.com

template matching, in which watermarks are separately embedded into medium-frequency and high-frequency in digital images' integer wavelet domain. The scheme realizes blind watermark extraction without any extra information. By shaping the watermarked image before extraction, the precision of the extracted watermark is enhanced. The algorithm can not only test tempered intensity, but also locate tempered area of digital images.

2 Singular Value Decomposition and Matrix Norm

Matrix singular value decomposition(SVD) is an orthogonal transform and can diagonalize matrix. That is, if matrix $A \in R^{m \times n}$ is nonnegative, thus

$$A = USV^T \tag{1}$$

$$U^T AV = S = diag(\sigma_1, \sigma_2, \sigma_3, \cdots\cdots, \sigma_p) \tag{2}$$

where p is minimum of n and m, σ_i is singular value of matrix A, and $\sigma_1 \geq \sigma_2 \geq \sigma_3 \geq \cdots\cdots \geq \sigma_p$, and U 、 V is separately left and right singular value vector.

Spectral norm and F-norm of nonnegative matrix A's express as follows:

$$\|A\|_2 = \sqrt{\lambda_{\max}(A^T A)} = \sigma_1 \tag{3}$$

$$\|A\|_F = \sqrt{\sum_{i=1}^{m}\sum_{j=1}^{n} a_{ij}^{\ 2}} = \sum_{i=1}^{p} \lambda_i = \sigma_1^2 + \sigma_2^2 + \cdots\cdots + \sigma_p^2 \tag{4}$$

As it is known, matrix's spectral norm is maximal singular value, F-norm is square sum of all singular values, and there is certain corresponding relationship between matrix norm and singular values.

Recently, SVD is applied to digital watermarking technology[10], which depends on following theories:

(1) SVD can't finite image size;

(2) Stability of image singular value is good, that is, when images are disturbed slightly, singular value won't change violently.

Watermarks are embedded through changing singular values, which can improve watermarking robustness, but SVD calculating will increase algorithm's complexity and computing cost.

Viewed in formula (3) and (4), changing matrix singular will relevantly change matrix spectral norm and F-norm, so changing matrix norm can implement watermark embedding instead of changing matrix singular values.

3 A Robust Watermarking Algorithm Based on Matrix Norm Quantization and Template Matching

In order to avoid rounding error of floating point, host digital images are transformed by integer wavelet; whereas the wavelet low-frequency coefficients have more stronger robustness and high-frequency coefficients have more fragile sensitivity and medium-frequency coefficients lie in them, so watermarks are embedded into medium-frequency and high-frequency sub-bands in host digital image's integer wavelet domain, which will guarantee both robustness and sensitivity; in order to locate tempered area, it blocks each sub-band and calculate spectral norm and embeds watermarks through norm quantification; in order to improve watermark information's security, Logistic chaotic sequences are used to scramble watermarks. Embedding process as Figure 1 displays.

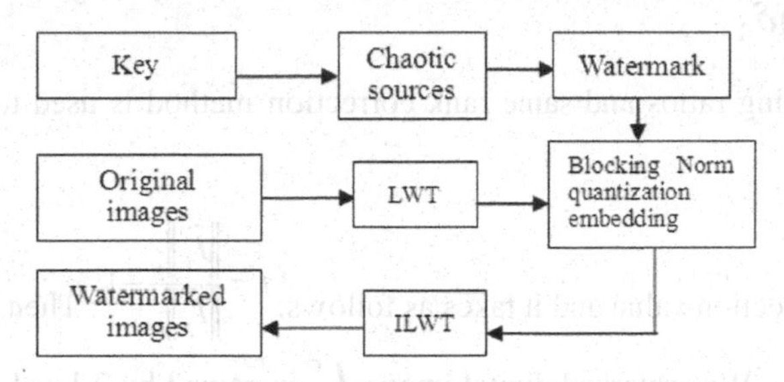

Fig. 1. Embedding watermarks process

The first step: Watermarks' encryption by chaotic scrambling

Logistic chaotic system: $x_{k+1} = \mu x_k(1-x_k)$

Where bifurcation parameter $\mu \in (3.569945, 4]$ and $x_k \in (0,1)$.

Initial value x_0 and bifurcation parameter μ are used as key space $K(x_0, \mu)$ to generate chaotic sequences X which scramble watermarks W. Concrete contents is as follows:

(1)Chaotic sequences are rearranged according to ascending order, obtaining ascending order chaotic sequence X' and index sequence l;

(2)Index sequence l is converted by column into two-dimensional matrix L ;

(3)Original watermark W is rearranged by scrambled index matrix L to generate scrambled watermarks W';

The second step: Integer wavelet transform

Original digital image I is transformed by 2-level integer wavelet transformation to generate seven sub bands, where 9-7 orthogonal base is used as wavelet base. Medium-frequency fragile watermarks are embedded into HL2, LH2 and HH2 detail sub-bands; high-frequency fragile watermarks are embedded into HL1, LH1 and HH1 detail sub-bands.

The third step: Embedding watermarks

(1)Each layer detail sub-band is blocked to several $q \times q$ cubic blocks which are denoted as $I_g \in R^{q \times q}$. Calculate spectral norm $\|I_g\|_2$, quantization parameter $\lambda = \left\lfloor \frac{\|I_g\|_2}{\delta} \right\rfloor$ and control ratio $\gamma = \|I_g\|_2 - \lambda\delta$, where δ is quantization step based on the feature of visual masking of human vision system;

(2)Modify matrix spectral norm:If $(\lambda + W_1') \bmod 2 = 1$ and $\gamma < \frac{\delta}{2}$, $\|\widetilde{I_g}\|_2 = (\lambda - \frac{1}{2})\delta$;If $(\lambda + W_1') \bmod 2 = 1$ and $\gamma \geq \frac{\delta}{2}$, $\|\widetilde{I_g}\|_2 = (\lambda + \frac{3}{2})\delta$;Else, $\|\widetilde{I_g}\|_2 = (\lambda + \frac{1}{2})\delta$;

(3)Equal jigging ratios and same rank correction method is used to modify I_g as $\widetilde{I_g} = I_g + tI_g$

Where t is correction value and it takes as follows: $t = \frac{\|\widetilde{I_g}\|_2}{\|I_g\|_2} - 1$; Then $\widetilde{I_g} = \frac{\|\widetilde{I_g}\|_2}{\|I_g\|_2} I_g$;

The forth step: Watermarked digital image I' is gained by 2-level integer wavelet inverse transformation.

The scheme realizes blind watermark extraction without any extra information. By shaping the watermarked image before extraction, the precision of the extracted watermark is enhanced, as follows:

The first step: Integer wavelet transformation

Image for testing I'' is transformed by 2-level integer 9-7 orthogonal wavelet base to generate six detail sub-bands;

The second step: Extracting watermarks

Every detail sub-band is blocked into $q \times q$ cubic blocks, every block is regarded as I_g''. Calculate its spectral norm $\|I_g''\|_2$ and quantization ratio $\lambda_i'' = \left\lfloor \frac{\|I_g''\|_2}{\delta} \right\rfloor$;If $\lambda_i'' \bmod 2 = 1$, extracted watermark $w_i'' = 1$;If $\lambda_i'' \bmod 2 = 0$, extracted watermark $w_i'' = 0$ .Then gain scrambled watermarks W'';

The third step: Decrypting watermarks and testing

Scrambled watermarks W'' can be decrypted through index sequences which Logistic chaotic sequences are rearranged by ascending order to generate, gaining watermarks W^*. Watermarks W^* not only can be judged subjectively, but also is tested through correlating detection function (SIM) and temper assessment function (T_{AF}), which can certificate multimedia datum completely.

4 Experiment Simulation and Analysis

Aiming at some digital images, the algorithm is simulated by means of MATLAB. Some simulation datum about a CT 256-level gray image display as follows. Figure 2(a) is original CT host image, figure 2(b) is binary watermarks about patients' information, in sequence of medium-frequency robust watermarks and high-frequency robust watermarks, figure 2(c) is scrambled watermarks, figure 2(d) is watermarked digital image.

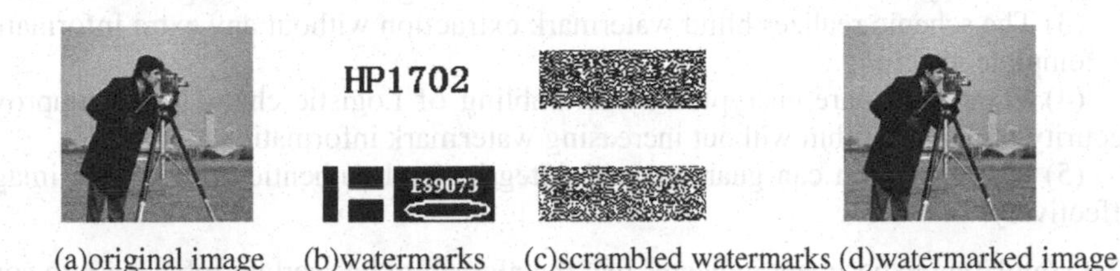

(a)original image (b)watermarks (c)scrambled watermarks (d)watermarked image

Fig. 2. Original digital image, watermarks and watermarked image

Attacking watermarked digital images can test effectiveness of the algorithm. Attacking methods include altering (Fig.3(a) and Fig.3(b)), shearing(Fig.3(c)), JPEG compressing(Fig.3(d)), adding noise(Fig.3(e) and Fig.3(f)), etc..

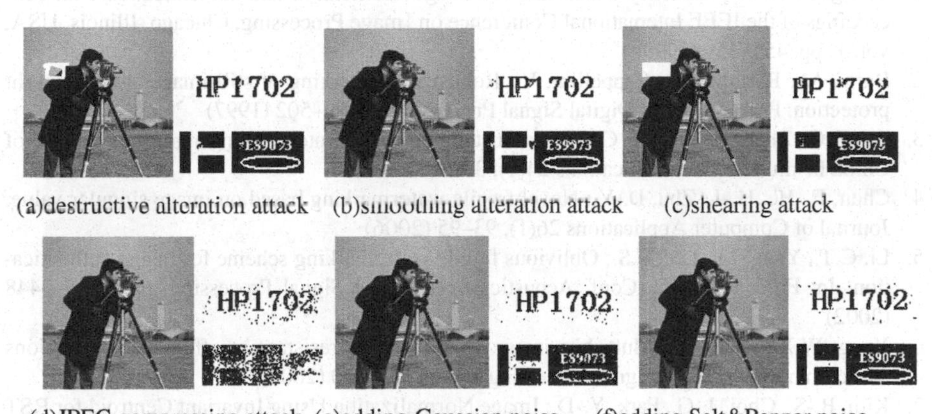

(a)destructive alteration attack (b)smoothing alteration attack (c)shearing attack

(d)JPEG compressing attack (e)adding Gaussian noise (f)adding Salt&Pepper noise

Fig. 3. Extracted watermarks after attacking

As it is known from above attack detection, the algorithm can reflect tempered area and frequency characteristics of watermarked digital images, and it has certain robustness. But with the increase of attack intensity, watermarks distortion is greater and greater. As regards the same attack, high-frequency sub-band is more sensitive than medium-frequency.

5 Conclusions

This paper puts forward a robust digital watermarking algorithm which is used to certificate integrity of digital images. Its characteristics as follows:

(1) Watermarks are embedded by stratification into detail sub-bands of digital images' integer wavelet domain, which can avoid rounding error of floating point and extract watermarks according to requirement;

(2) It proposes a embedding method of matrix norm quantization which not only is simple and easy to calculate, but also doesn't need original digital images when testing;

(3) The scheme realizes blind watermark extraction without any extra information by template matching.;

(4) Watermarks are encrypted by scrambling of Logistic chaos, which improves security of the algorithm without increasing watermark information amount;

(5) The algorithm can guarantee the integrity and authenticity of digital images effectively.

The simulating experiment indicates that the algorithm has certain robustness to some common attack and locates tempered area of digital images exactly. It is an effective robust digital watermarking algorithm for digital image field.

References

1. Wong, P.W.: A public key watermark for image verification and authenticationJ. In: Proceedings of the IEEE International Conference on Image Processing, Chicago, Illinois, USA, vol. 1, pp. 455–459 (1998)
2. Barni, M., Bartolini, F., Cappellini, V.: Robust watermarking of still images for copyright protection. Proceedings of Digital Signal Processing 2, 499–502 (1997)
3. Dong, G., Zhang, L., Zhang, C.-T.: A robust digital imagewatermarking algorithm. Journal of China Institute of Communications 24(1), 33–38 (2003)
4. Chen, F., He, H.-J., Zhu, D.-Y.: Novel fragile watermarking based on image singular value. Journal of Computer Applications 26(1), 93–95 (2006)
5. Li, C.T., Yang, M., Lee, C.S.: Oblivious fragile watermarking scheme for image authentication. In: Proc. IEEE Int. Conf. Acoustics Speech and Signal Processing, pp. 3445–3448 (2002)
6. Yang, W.X., Zhao, Y.: Multi-bits image watermarking resistant to affine transformations based on normalization. Signal Processing 20(3), 245–250 (2004)
7. Kim, B.-S., Choi, J.-G., Park, Y.-D.: Image Normalization Using Invariant Centroid for RST Invariant Digital Image Watermarking. In: Petitcolas, F.A.P., Kim, H.-J. (eds.) IWDW 2002. LNCS, vol. 2613, pp. 202–211. Springer, Heidelberg (2003)
8. Lee, S.-W., Choi, J.-G., Kang, H.-S., Hong, J.-W., Kim, H.-J.: Normalization Domain Watermarking Method Based on Pattern Extraction. In: Kalker, T., Cox, I., Ro, Y.M. (eds.) IWDW 2003. LNCS, vol. 2939, pp. 390–395. Springer, Heidelberg (2004)
9. Zhou, B., Chen, J.: A geometric distortion resilient image watermarking algorithm based on SVD. Image and Graphics 9(41), 506–512 (2004)

Semi-supervised Metric Learning for 3D Model Automatic Annotation

Feng Tian[1,2], Xu-kun Shen[1], Xian-mei Liu[2], and Kai Zhou[2]

[1] State Key Lab. of Virtual Reality Technology and Systems, BeiHang University Beijing, China

[2] School of Computer and Information Technology, Northeast Petroleum University Daqing, China

Tianfeng80@gmail.com

Abstract. Automatically assigning relevant text labels to 3D model is an important problem. For this task we propose a semi-supervised measure learning method. Labels of 3D models are predicted using a graph-based semi-supervised method to exploit labeled and unlabeled 3D models. In this manner, we can get semantic confidence of labels. An improved relevant component analysis method is also proposed to learn a distance measure based on label's semantic confidence. A novel approach based on the semantic confidence and the distance is applied on multi-semantic automatic annotation task. We investigate the performance of our method and compare to existing work. The experimental results demonstrate that the method is more accurate when a small amount of labels were given.

Keywords: automatic annotation, 3D model retrieval, semi-supervised learning.

1 Introduction

In recent years, 3D scanning equipment, modeling tools and Internet technology have led to a large number of 3D models. 3D model retrieval has become a research hotspot. Several 3D model search engines have been developed [1-3]. These search engines are all include two search types. One is using traditional text-based retrieval which keywords are extracted from captions, titles, etc. The other type is using content-based retrieval method which search sample is 2d image or 3d object. The text-based retrieval provides users with a simple and natural interface, so it is friendlier for the user, but the text labels is required. In order to improve the retrieval effectiveness and capture the user's semantic knowledge, the semantic automatic annotation technique has been introduced to the 3D model retrieval broadly in recent years [4-6]. Most current automatic annotation methods need a large number of models hand tagged with text labels, so the training sample size and quality are in high demand [7]. At the same time manually annotation brought tedious workload, which made the label results imperfect, inaccurate and subjective. Figure 1 shows some hand tagged models and their labels. In this paper, we present a method called 3D model multiple semantic automatic annotation based on semi-supervised metric learning (MS3ML) to label 3D models, which has achieved a better annotation result when a small amount of labels were given.

D. Jin and S. Lin (Eds.): Advances in MSEC Vol. 1, AISC 128, pp. 85–91.
springerlink.com

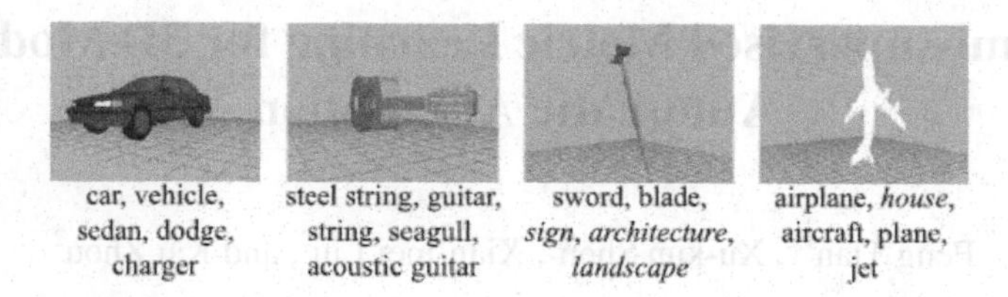

Fig. 1. Four hand tagged models

2 The Overview of MS3ML

The process of MS3ML is shown in Figure 2. The corpus is comprised of a small amount of hand tagged models and a large number of unlabeled models. Firstly, the feature of 3D models was extracted, and the process of dimension deduction is needed. Secondly, we make full use of the unlabeled models to expand the training dataset (known as label propagation) and the label confidence was computed. Thirdly, a new distance metric considered label confidence as well as the correlation between features is learned. Lastly, for each model needing labels, we label it by multiple semantic annotation strategy.

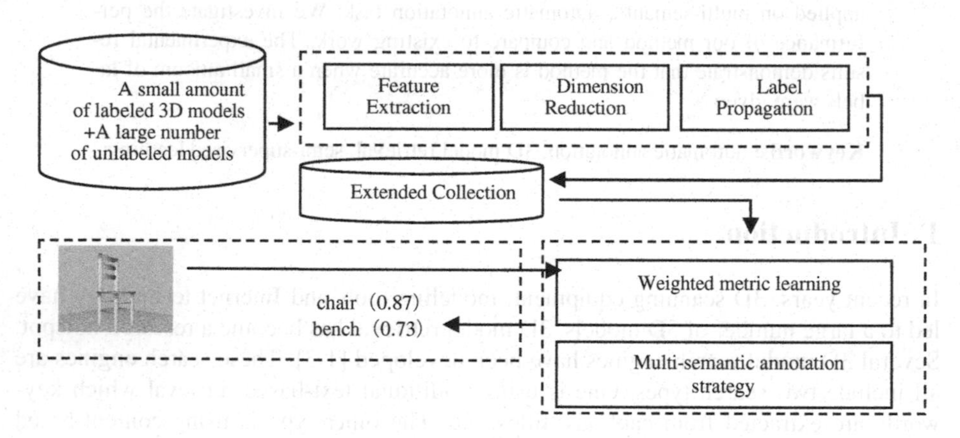

Fig. 2. The process of MS3ML

3 The Process of Automatic Annotation

We take full use of labelled and unlabeled models to expand the amount of labelled models. The graph-based semi-supervised learning has become the mainstream of semi-supervised learning because of its efficiency [8]. To do this, we use a corpus of known hand tagged models $L = \{(x_1, y_1) \cdots (x_{|L|}, y_{|L|})\}$ where x_i denotes the model and y_i denotes *i-th* model's label collection, $y_i \subset T$, $T = \{\lambda_1 \cdots \lambda_{|T|}\}$ denotes the collection consisting of all labels. $U = \{x_{|L|+1}, \cdots, x_n\}$ denotes the unlabeled model. The model x_i is represented by the point x_i in feature space. Define a graph $G=\{V,E\}$, each vertices corresponds to model from $L \cup U$, its weighted edge reflects similarity between adjacent models. So $n \times n$ similarity matrix W can denote the graph. Each element in W can be formally defined by RBF kernel function as follows:

$$w_{ij} = \exp\left(-\frac{\| x_i - x_j \|^2}{\alpha^2}\right) \tag{1}$$

where w_{ij} denotes the similarity between model x_i and x_j .A $n \times n$ matrix P is defined to represent the edge propagation probability of label information to the neighbour node:

$$P_{ij} = P(j \to i) = \frac{w_{ij}}{\sum_{k=1}^{n} w_{ik}} \tag{2}$$

where P_{ij} denotes probability that x_i learns labels from x_j .The labels of model x_i is expressed by $1 \times |T|$ row vector f_i, if $x_i \in L$, the *j-th* element is defined as follows:

$$f_{ij} = \begin{cases} 1, & j \in y_i \\ 0 & j \notin y_i \end{cases} \tag{3}$$

That is, the *j-th* elements of f_i is 1 if the *j-th* label in T is one of the model's label, and the rest are zero. If $x_i \in U$, $f_{ij} \in [0,1]$. Define $|L| \times |T|$ matrix f_L to denote the label semantic matrix, the $|U| \times |T|$ matrix f_U to denote unlabeled semantic matrix. Define f_X to denote matrix of all the data as follows:

$$f_X = \begin{pmatrix} f_L \\ f_U \end{pmatrix} \tag{4}$$

The data's label is propagated from the neighbors, that is

$$f_X^{(i)} = P \times f_X^{(i-1)} \tag{5}$$

We summarize the standard process of label propagation algorithm as follows:

1) $i=0$, Initialize $f_U^{(i)}=0$;
2) Calculate P;
3) $i=i+1$, get $f_U^{(i)}$ by $f_X^{(i)} = P \times f_X^{(i-1)}$;
4) Repeat step 3 until convergence;
5)Define f_{U_i} as i-th row vector of f_U, each elements of f_{U_i} has been assigned a real-value which is used to measure the confidence of *i-th* model label in U .

The final state of label propagation is that all the vectors of unlabeled data are no longer changed, which means semantic labels have achieved smooth distribution in unlabeled data. So we expand the manually labeled data set L to $L \cup U$, meanwhile for each label we assigned a confidence value which we interpret as the probability that the label is relevant to the model. Now we can learn a new distance metric from $L \cup U$. RCA (Relevant Component Analysis) is an effective distance metric learning method [9]. But we found that when the amount of labelled information is insufficient, the results got from traditional RCA will bias. So we propose a method called weighted RCA. The extended labelled dataset and labelling confidence are a guarantee of the algorithm's validity. We firstly normalize the label confidence of each label in T, and a $|T| \times |T|$ diagonal matrix of confidence is generated:

$$W = Diagonal[w_1, w_2, w_3, ..., w_{|T|}] \tag{6}$$

where w_i is mean confidence of all the models described by *i-th* label in T. So we can use weighted covariance matrix instead of centralized covariance matrix of RCA:

$$C = \frac{1}{n} \sum_{c=1}^{|T|} \sum_{i=1}^{n_i} (x_{c,i} - x_{c-mean}) W (x_{c,i} - x_{c-mean})^T \tag{7}$$

where $x_{c,i}$ denotes *i-th* model in feature space described by *c-th* label. x_{c-mean} is mean point in feature space described by *c-th* label. We calculate C^{-1} as a mahalanobis distance metric, and then we get the weighted-RCA distance:

$$d_{weighted-RCA}(x_1, x_2) = (x_1 - x_2)^T C^{-1} (x_1 - x_2) \tag{8}$$

Given an unlabeled 3D model X_{new}, we wish assign labels from the set of all possible labels $T = \{\lambda_1 ... \lambda_{|T|}\}$ to X_{new}. Specifically, for each label we wish to assign a confidence value which we interpret as the probability that λ_i is a relevant label for X_{new}. So we start with a shape similarity metric and find the neighbors of X_{new} within some distance threshold. Note that the distance threshold is allowed to be a function of the model, which allows for adaptively defining the threshold based on the density of models in a given portion of the descriptor space. We take

$$\mathrm{P}(X_{new} \approx X_{neighbour-i}) = (1 - d_{weighted-RCA}(X_{new}, X_{neighbour-i}))^2 \tag{9}$$

to be an estimate of the probability that X_{new} and $X_{neighbour-i}$ represent the same type of model and therefore should have similar text labels. Then given unlabelled model X_{new}, a possible text label λ_i, and a neighbor $X_{neighbour-i}$ from the extended labeled data set $L \cup U$, the probability that X_{new} should have the label is

$$C(\lambda^i, X_{new}) = P(X_{new} \approx X_{neighbour-i}) \wedge C(\lambda^i, X_{neighbour-i}) \tag{10}$$

Where $C(\lambda^i, X_{new})$ denotes the confidence of label λ_i. Intuitively this means that the confidence that λ_i is appropriate for X_{new} is the confidence that it is appropriate for $X_{neighbour-i}$ and that X_{new} and $X_{neighbour-i}$ are similar enough to share labels. $C(\lambda^i, X_{new})$ can be thought of as measuring how much we trust the original annotation on $X_{neighbour-i}$. Considered over full set of k neighbors this generalizes to

$$C(\lambda^i, X_{new}) = \bigcup_{j=1}^{k} P(X_{new} \approx X_{neighbour-j}) \wedge C(\lambda^i, X_{neighbour-j}) \tag{11}$$

By analogy to the *TF-IDF* method from text processing we reweight these probabilities such that:

$$C_{tf-idf}(\lambda^i, X_{new}) = \frac{C(\lambda^i, X_{new})}{\sum_j C(\lambda^j, X_{new})} \cdot \log \frac{|L \cup U|}{\sum_{X_k \in L \cup U} C(\lambda^i, X_k)} \tag{12}$$

For each unlabeled model, we get a vector of probabilities for each semantic label. We choose the *TOP-N* labels to describe the model.

4 Experiments

To evaluate the proposed method, our experiments were performed on a database containing 1125 3D models, which were collected from the Princeton Shape Benchmark (PSB) [1], and 725 models were semantically hand tagged with text labels. Figure 3 shows some automatic annotation samples. In the experiment, we mainly use the depth buffer method to extract the 3D models' feature (438-dimensional feature vector) [10]. We performed PCA over the descriptors, and kept the top 20 dimensions.

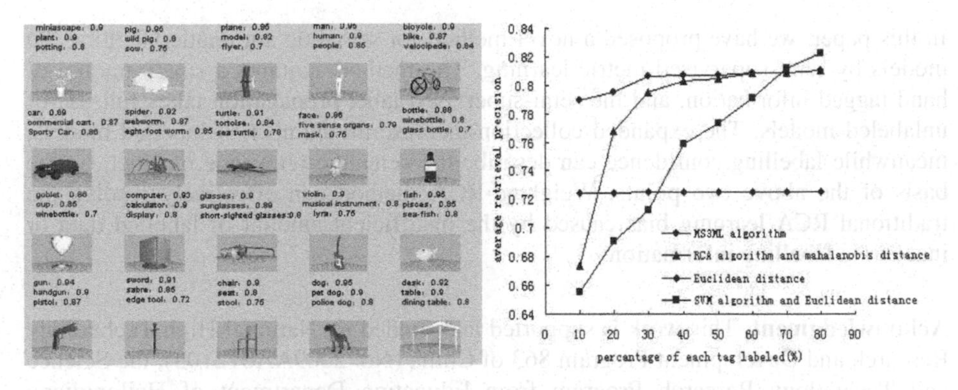

Fig. 3. Automatic labeling samples **Fig. 4.** Comparison of retrieval precision

In this paper, we use " Average Precision" VS "Percentage of each tag labeled" to evaluate both automatic annotation and retrieval process, figure 4 lists the average retrieval precision of five times. These types of labeling methods includes: Euclidean distance metric method, typical supervised classification learning method (SVM algorithm and the Euclidean distance), RCA distance metric method and MS3ML.Results show that MS3ML has higher labelling precision when there is a small amount of label information, and the kernel function of the supervised labelling method SVM adopted RBF kernel [11]; the distance metric function used Euclidean distance. Since

SVM requires a large number of training data, if the training set's label is insufficient, the result is not accurate. Table 1 shows the average retrieval efficiency of various methods in the case of insufficient labelled data (label 1, 2, 3 and 4 models for each label), and only the first 16 retrieval results is taken into account. Results show that MS3ML has a better retrieval result.

Table 1. Comparison of the retrieval effectiveness with a small amount of labels

Methods		Labeled models per label	Precision (%)	Recall (%)
Supervised method	SVM and Euclidean distance	1	17.31	5.51
		2	34.47	10.64
		3	49.52	16.82
		4	64.67	22.93
Semi-supervised method	RCA	1	38.92	13.12
		2	44.02	14.65
		3	46.41	15.92
		4	66.51	23.42
	MS3ML	1	74.11	26.51
		2	75.53	26.97
		3	77.55	27.78
		4	78.07	28.08

5 Conclusion

In this paper, we have proposed a novel method for semantic automatic labelling 3D models by semi-supervised metric learning. The method acquires a small amount of hand tagged information, and the semi-supervised label propagation takes full use of unlabeled models. The expanded collection increase the number of labelled models; meanwhile labelling confidence can describe the semantic relevance of label, on the basis of the above two points, Weighted-RCA method can effectively resolve the traditional RCA learning bias caused by the insufficient amount of labelled data or inaccurate labelling information.

Acknowledgment. This work is supported and funded by National High Technology Research and Development Program 863 of China (No. 2009AA012103), the Science and Technology Research Program from Education Department of Heilongjiang Province of China (No. 12511011).

References

1. Philip, S., Patrick, M., Michael, K.: The Princeton Shape Benchmark. Shape Modeling International, 388–399 (2004)
2. Chen, D.-Y., Tian, X.-P., Shen, Y.-T.: On visual similarity based 3D model retrieval. Eurographics, 22–26 (2003)
3. Paquet, E., Murching, A., Naveen, T., Tabatabai, A., Rioux, M.: Description of shape information for 2-D and 3-D objects. Signal Process. Image Commun., 103–122 (2000)

4. Min, P., Kazhdan, M., Funkhouser, T.: A Comparison of Text and Shape Matching for Retrieval of Online 3D Models. In: Heery, R., Lyon, L. (eds.) ECDL 2004. LNCS, vol. 3232, pp. 209–220. Springer, Heidelberg (2004)
5. Meng, Z., Atta, B.: Semantic-associative visual content labeling and retrieval: A multimodal approach. Signal Processing: Image Communication, 569–582 (2007)
6. Funkhouser, T., Min, P., Kazhdan, M., Chen, J., Halderman, A., Dobkin, D., Jacobs, D.: A Search Engine for 3D Models. ACM Transactions on Graphics 22(1), 83–105 (2003)
7. Datta, R., Ge, W., Li, J.: Toward bridging the annotation-retrieval gap in image search by a generative modeling approach. In: Proceedings of the 14th Annual ACM International Conference on Multimedia, Santa Barbara, CA, USA, pp. 23–27 (2006)
8. Zhou, D., Bousquet, O., Lal, T., Weston, J., Scholkopf, B.: Ranking on data manifolds. In: Proceedings of the 18th Annual Conference on Neural Information Processing System, pp. 169–176 (2003)
9. Shental, N., Hertz, T., Weinshall, D., Pavel, M.: Adjustment Learning and Relevant Component Analysis. In: Heyden, A., Sparr, G., Nielsen, M., Johansen, P. (eds.) ECCV 2002, Part IV. LNCS, vol. 2353, pp. 776–790. Springer, Heidelberg (2002)
10. Heczko, M., Keim, D., Saupe, D.: Methods for similarity search on 3D databases. Datenbank-Spektrum, 54–63 (2002) (in German)
11. Hou, S., Lou, K., Ramanik: SVM based semantic clustering and retrieval of a 3D model database. Journal of Computer Aided Design and Application, 155–164 (2005)

ZVS Synchronous Buck Converter with Improved Light Load Efficiency and Ripple-Free Current

Hyun-Lark Do

Department of Electronic & Information Engineering,
Seoul National University of Science and Technology, Seoul, South Korea
hldo@seoultech.ac.kr

Abstract. A zero-voltage-switching (ZVS) synchronous buck converter with improved light load efficiency and ripple-free current is proposed in this paper. An auxiliary circuit is added to a conventional buck converter. The auxiliary circuit provides a ripple-free current and ZVS feature. Due to the ZVS feature, the switching loss is significantly reduced. However, the conduction loss due to the auxiliary circuit increases. Especially at light loads, the conduction loss is much larger than the switching loss. Therefore, by disabling the synchronous switch at light loads, the circulating current in the auxiliary circuit is reduced and the efficiency at light loads is improved. The operation principle and steady-state analysis of the proposed converter are provided. A prototype of the proposed converter is developed, and its experimental results are presented for validation.

Keywords: Zero-voltage-switching, buck converter, ripple-free, synchronous buck converter.

1 Introduction

The buck converter is the simplest solution to realize step-down DC-DC power conversion. In order to increase the system efficiency, a synchronous buck converter is sometimes adopted. However, due to its hard switching operation, the switching loss is large [1]. In order to remedy these problems, ZVS control scheme for a pulse-width modulation buck converter under discontinuous conduction mode/ continuous conduction mode boundary was suggested in [2]. ZVS control scheme can reduce the switching loss. However, it significantly increases the ripple component of the inductor current. Therefore, the output voltage ripple is increased. Interleaving technique can be adopted to reduce input/output current ripple and simplify the filter stage [3]. However, the multi-channel interleaved structure has many components. As the channel increases, more components are required.

A ZVS synchronous buck converter with ripple-free inductor current is proposed. It utilizes an auxiliary circuit consisting of an additional winding of the filter inductor, an auxiliary inductor and a capacitor to provide ZVS of the power switches and ripple-free inductor current. The ZVS feature solves the reverse recovery problem of the

D. Jin and S. Lin (Eds.): Advances in MSEC Vol. 1, AISC 128, pp. 93–96.
springerlink.com

anti-parallel body diode of the synchronous switch. The ripple component of the filter inductor current is cancelled out effectively due to the auxiliary circuit. The theoretical analysis is provided in the following section. The theoretical analysis is verified by a 110W experimental prototype with 100V-to-48V conversion.

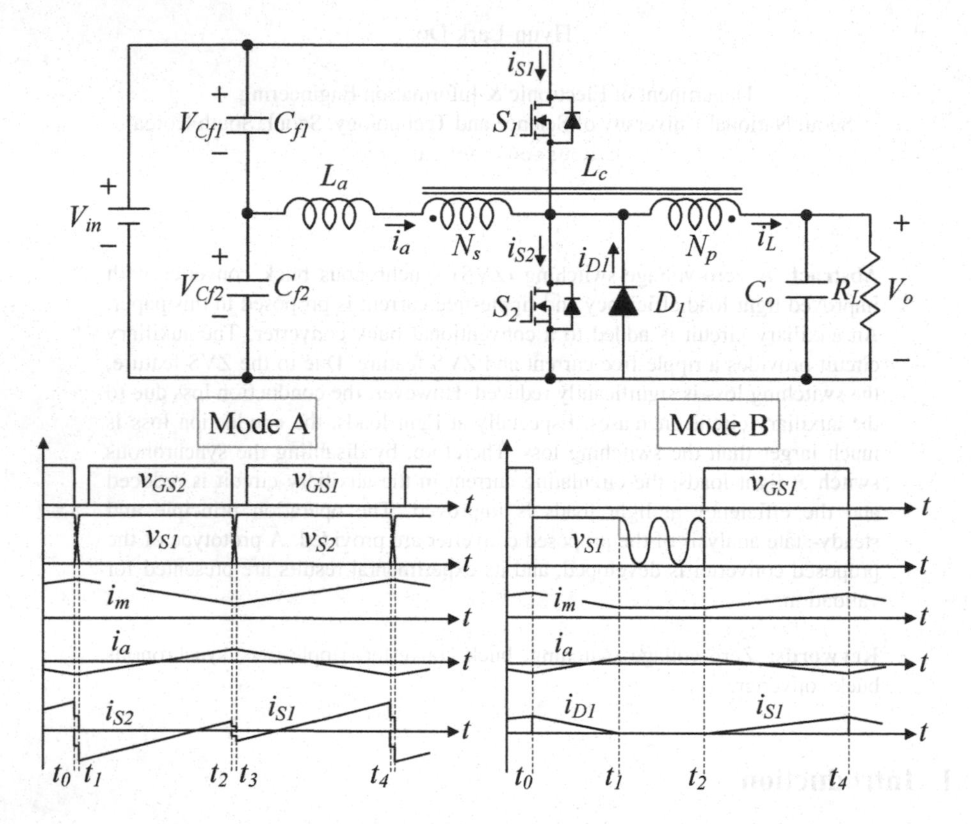

Fig. 1. Circuit diagram and key waveforms of the proposed converter

2 Analysis of the Proposed Converter

Fig. 1 shows the circuit diagram and key waveforms of the proposed converter. The proposed converter operates with two operation scheme. Under medium and heavy loads, its operation is equal to the conventional synchronous buck converter—mode A. Under light load, the synchronous switch is disabled to reduce the circulating current caused by the additional circuit—mode B.

The switches S_1 and S_2 are operated asymmetrically and the duty ratio D is based on S_1. The coupled inductor L_c has a turn ratio of 1:n ($n = N_s/N_p$). D_1 is mainly used in stead of the body diode of S_2 in the disabled synchronous operation mode. In both modes, the magnetizing current i_m of L_c increases linearly when S_1 is on. At that time, the current i_a decreases linearly. By adjusting the slope of i_a properly, the ripple component of i_m can be perfectly removed. The ripple-free condition is given by

$$L_a = n(1-n)L_m \ . \quad (1)$$

In mode A, when S_1 is turned off, S_2 is turned on. The current i_m decreases linearly and i_a increases linearly. With the same condition of (1), i_L can be ripple-free. In mode B, when S_1 is turned off, D_1 is turned on. Similar to mode A, i_m decreases and i_a increases. Also, in this time interval, i_L can be ripple-free with the condition of (1). When D_1 is turned off and S_1 is not turned on yet in mode B, the voltages across L_m and L_a are zero and the currents i_a and i_m are constant. The current i_L is definitely ripple-free. When a loosely coupled inductor L_c is used, the auxiliary inductor L_a can be replaced with the leakage inductance of L_c can be used as L_a. As you can see in Fig. 1, if the peak value of i_a is larger than i_L which is ripple-free and equal to the output current, ZVS operation of both switches S_1 and S_2 is easily achieved. Since the average inductor voltage must be zero under a steady-state, in mode A, the average voltage across C_{f2} is equal to V_o and the average voltage across C_{f1} is V_{in}-V_o. By applying the volt-second balance law to the voltage across L_m, the voltage gain in mode A is obtained as follows:

$$\frac{V_o}{V_{in}} = D \ . \quad (2)$$

The voltage gain in mode B is given by

$$\frac{V_o}{V_{in}} = \frac{D}{D-d_1} \ , \quad (3)$$

where d_1 is $(t_2 - t_1)/T_s$ in Fig. 1.

3 Experimental Results

The prototype of the proposed buck converter is implemented with specifications and parameters of n=0.3, V_{in}=100V, V_o=48V, L_m=155uH, L_s=32uH, C_a=6.6uF f_s=100kHz, $P_{o,max}$=110W. Fig. 2 shows the experimental waveforms and measured efficiency. The conventional synchronous buck converter is also implemented with the same circuit parameters except for the auxiliary circuit. It can be seen that the experimental waveforms agree with the theoretical analysis. The ripple component of the inductor current i_L is effectively removed regardless of load condition. The proposed converter provides almost ripple-free inductor current without raising the inductance significantly. The efficiency of the proposed converter is compared with the conventional synchronous buck converter and the proposed converter which always operates in mode A. The power consumed in the control circuit is ignored in the efficiency curve. Due to its soft-switching characteristic, the efficiency of the proposed ZVS synchronous buck converter is significantly improved at heavy load compared with the conventional one. Since the conventional synchronous buck converters can operate with ZVS below a specific load level, they can provide high efficiency below such a load level. At light load, the conduction loss is reduced and the efficiency is improved by disabling the synchronous switch.

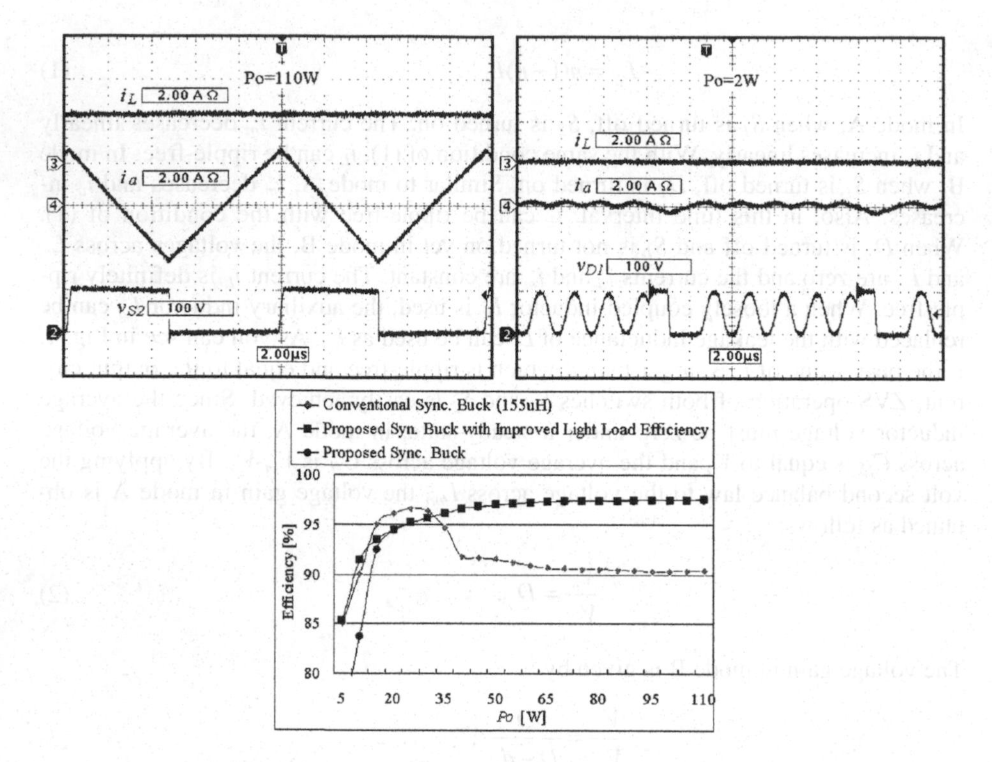

Fig. 2. Experimental waveforms measured efficiency

4 Conclusion

A ZVS synchronous buck converter has been proposed. By utilizing the auxiliary circuit, both switches can operate with ZVS and ripple-free inductor current is achieved. Soft-switching operation improves the efficiency at medium and heavy loads. By disabling the synchronous switch, the light load efficiency is improved.

References

1. Abedinpour, S., et al.: A multistage interleaved synchronous buck converter with integrated output filter in 0.18 um SiGe process. IEEE Trans. Power Elec. 22, 2164–2175 (2007)
2. Chiang, C.Y., et al.: Zero-voltage-switching control for a PWM buck converter Under DCM/ CCM boundary. IEEE Trans. Power Elec. 24, 2120–2126 (2009)
3. Li, W., et al.: A family of interleaved boost and buck converters with winding-cross-coupled inductors. IEEE Trans. Power Elec. 23, 3164–3173 (2008)

A Prediction Model Based on Linear Regression and Artificial Neural Network Analysis of the Hairiness of Polyester Cotton Winding Yarn

Zhao Bo*

College of Textiles, Zhongyuan University of Technology, Henan, Zhengzhou, 450007, People's Republic of China
zhaobohenan@sina.com zhaobohenan@163.com

Abstract. The polyester/cotton blended yarn hairiness of winding is related to the winding processing parameters (winding tension, winding speed, balloon position controller, ring yarn hairiness, and ring yarn twist). However, it is difficult to establish physical models on the relationship between the processing parameters and the winding yarn hairiness. Due to the ANN model has excellent abilities of nonlinear mapping and self-adaptation. Therefore, it can be well used to predict yarn properties quantitatively. In this research, two modeling methods are used to predict the hairiness of polyester/cotton winding yarn. The results show that ANN model is more effective than linear regression model, which is an excellent approach for predictors.

Keywords: winding yarn, hairiness, polyester/cotton, linear regression model, artificial neural network, prediction.

1 Introduction

Winding yarn hairiness [1-6] is primarily determined by fiber properties, spinning processing parameters and winding processing parameters etc. In recent years, in order to increase the hairiness of winding polyester/cotton yarn, many researchers have studied the hairiness of winding yarn, and there are many modeling methods for predicting yarn hairiness. However, the reported methods were quite elementary, they can not associate with real-world process dynamics because it is difficult to determine yarn structural parameters, and it is rather difficult to establish a physics model to describe the process-property relation of the winding yarn hairiness.

Due to the ANN model has excellent abilities of nonlinear mapping and self-adaptation. Therefore, it can be well used to predict yarn properties quantitatively. Presently, it is widely used in many engineering fields to predict material properties. Within the textile industry alone, numerous applications have been reported. They indicated the ANN model can provide a very good and reliable reference for yarn performances [7-10]. However, the applications of ANN for predicting the yarn hairiness of ring spinning are very scanty. In this work, we attempt to predict winding yarn hairiness

* Corresponding author.

D. Jin and S. Lin (Eds.): Advances in MSEC Vol. 1, AISC 128, pp. 97–103.
springerlink.com

with ANN and linear regression, and to compare the prediction results with each other. The results show that ANN model is more effective than linear regression model.

2 Experimental

2.1 Experiment Variety

The polyester/cotton 65/35 yarn with linear density of 13tex was manufactured with the use of the Espero cone winding machine.

2.2 Experiment Machine Type

Espero cone winding machine is used in this work, speed of the winding is 900-1600r/min, and winding tension of the winder is 80-210 mbar.

2.3 Experiment Conditions

Test samples were conditioned in a standard atmosphere of 25℃ temperature and 65% relative humidity for minimum of 24 hours.

2.4 Experiment Tester

The 2mm hairiness tests of polyester/cotton 65/35 13tex blended yarns were carried out in YG172A yarn hairiness tester with 30m/min.

2.5 Experimental Program

Table 1. Experimental program and results

No.	Winding tension (mbar) (A)	Winding speed (r/min) (B)	Balloon position controller (mm) (C)	Ring yarn hairiness (2mm) (Num./10m) (D)	Ring yarn twist (twist/10cm) (E)
1	173	900	25	61. 5	90
2	96	1300	35	83.0	95
3	114	1460	30	72. 5	90
4	158	1000	25	72. 5	92.5
5	196	1500	30	80.0	87.0
6	106	1400	40	80. 5	83.5
7	96	1200	35	91. 0	90
8	86	1250	30	79. 50	95
9	107	1300	29	89. 00	91.5
10	92	1400	2 6	82. 0	92.5
11	104	1100	32	92. 0	90
12	108	1200	3 4	92. 5	93.5
13	100	1270	28	81. 5	90
14	162	1350	38	61. 5	90
15	180	1250	3.2	72. 5	88.0
16	168	1350	29	80.0	100
17	180	1320	30	72. 5	92.0
18	178	950	35	60.0	95.0

3 Artificial Neural Network Model

Artificial neural network models [11] have been applied to different textile problems. They are designed to simulate the way in which the human brain processes information, and gather their knowledge by detecting relationships and patterns in data and hence they are able to learn from experience, differently from common programming. There are many types of neural networks, but all of them have the same basic principle. Each neuron in the network is able to receive input signal. Each neuron is connected to another neuron and each connection is represented by a real number called weight coefficient that reflects the degree of importance of the given connection to the neural network. Among many network schemes, feed-forward neural networks with back-propagation learning algorithms based on gradient descent have been widely used, because they offer unlimited approximation power for non-linear mappings. Therefore, in this work, an ANN was used with the back-propagation learning scheme to predict the hairiness of winding polyester/cotton yarn.

3.1 The Advantage of Artificial Neural Network Model

The main advantage of ANN is the fact that they are equipped with non-linear algorithms of regression, and possess the ability to model multidimensional systems with maximum flexibility under the consideration to learn them. The main advantage of ANN is that they are able to use some information apparently hidden in data. The process of capturing the unknown information is done during the training step of the ANN, when one may say that the ANN, when one may say that the ANN is learning how to output a satisfactory response for an input dataset. In mathematic language, the learning process is the adjustment of the set weight coefficient in such a way that some conditions are fulfilled. The main advantage of ANN over classic algorithmic methods lies in the fact that full knowledge of the issue of the model is necessary in the classic methods, permitting the formulation of fixed rules of inferring, while the ANN possesses the ability to programme itself on the basis of examples fed to the set-up.

3.2 The Feed-Forward Neural Network Model

A feed-forward neural network model is proposed for predicting the hairiness of winding polyester/cotton yarn. It consists of a number of simple neuron-like processing elements (neurons). It is organized in layers that are classified as an input layer, a hidden layer, and an output layer. Every unit within a layer is connected with all of the units in the previous layer. These connections are not all equals, each connections has a different strength or weight. The weights of these connections encode the knowledge of the network. Data enters at the input layer and passes through the network, layer by layer, until it arrives at the output layer. During normal operation, there is no feedback between layers, and hence it is called a feed-forward neural network.

3.3 The Back Propagation Neural Network Model Algorithm

The most popular ANN is the back-propagation feed-forward type, whose architecture is based on an input layer containing all the entrance variables that feed the network and an output layer that contains the response of the ANN to the desired problem. All the layers between the input and the output are called hidden layers. There is no limit to the number of hidden layers, but one hidden layer with an arbitrary large number of processing elements is enough to solve the majority of problems, although some rare functions require two hidden layers to be modeled.

The output value of the *i*th neuron x_i is determined by Equation 1 and 2, which holds:

$$x_i = f(y_i) \tag{1}$$

$$y_i = b_i + \sum_{j=1}^{n} w_j x_j \tag{2}$$

Where y_i is the potential of the *i*th neuron, b_i is the bias coefficient and can be understood as a weight coefficient of the connection formally added to the neuron, *n* is the number of input connections on the *i*th neuron, w_j is the weight coefficient of the connection between the input *j* and the *i*th neuron, and x_i is the value of the input *j*. The function $f(y_i)$ is the so-called transfer function.

3.4 The Training of Neuron Network Model

One of the most crucial aims of the back propagation neuron network is to minimize the error function by using the gradient steepest descent method. The error function is

$$E = \frac{1}{2}\sum_k (Y_{dk} - Y_k)^2 \tag{3}$$

Where Y_{dk} is the desired output, and Y_k the calculated output value of the output layer, respectively. The weights updated themselves by using error function as

$$\Delta w_{ij} = -\eta \frac{\partial E}{\partial w_{ij}} \tag{4}$$

Here η is the learning and determines the performance of the learning capability of network,

3.5 Error Measurement

In the learning network, the degree of convergence can be repressed in a mean square error (MSE), as follows

$$MSE = \frac{1}{N}\sum_{i=1}^{N}(Y_i - X_i)^2 \tag{5}$$

Where N is the number of objects, Y_i the neural network predicted values, and X_i the actual output values. Where n is the number of units processed by the output layer. The values of MSE lie within the range of [0, 1]. If the MSE converges to less than 0.005, a good convergence effect is obtained, and the network learning is satisfactory. The variation of learning cycle with MSE is represented in Figure 1. Here, the maximum epoch is determined as 20 and the training least mean square error is obtained as 0.001336, which is a satisfactory.

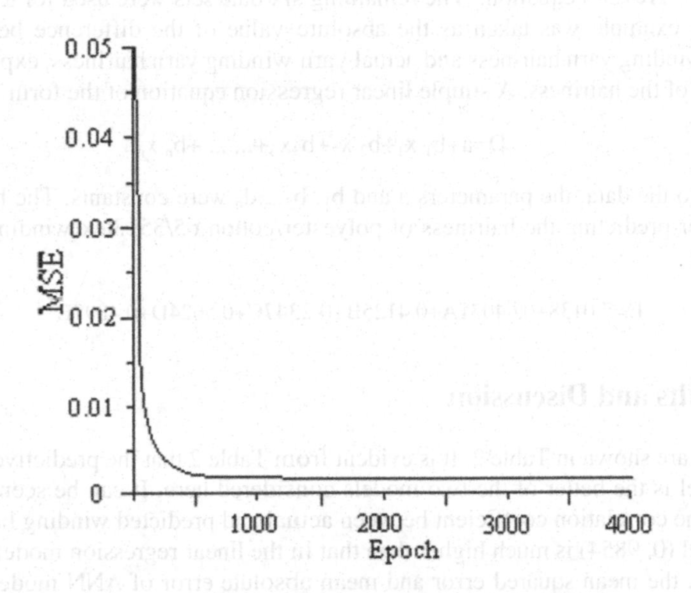

Fig. 1. Variation of error mean square with epoch

3.6 The Practicing of Artificial Neural Network

The neural network model practice is guided by back-propagation training algorithm. We use 12 groups of datum as direct practice and six groups as test and verification. The transfer function between input layer and hidden layer is as following:

$$f(x) = 2/(1+e^{-2x}) - 1 \tag{6}$$

which functions to transform the output value of neurons in the hidden layer to be ranged between -1 and 1 and the transfer function between the hidden layer and output layer is

$$f(x) = x \tag{7}$$

The learning rate is set to 0.01. The iterations are broken when the epoch reached 4000. All connection weights and biases are initialized randomly between -1 and 1. All the samples indexes are scale to be in the set of [-1, 1] before training. The program is written in Matlab 7.0.

4 Linear Regression Method

Linear regression analysis has been used to establish a quantitative relationship of hairiness of polyester/cotton blended winding yarn with respect to winding tension (A), winding speed(B) , balloon position controller (C), ring yarn hairiness(D) , and ring yarn twist(E), which is very frequently used to predict the winding yarn properties. Out of the eighteen data set, twelve were randomly chosen and used for constructing the best fitting regression equation. The remaining six data sets were used for testing. The error of an example was taken as the absolute value of the difference between the predicted winding yarn hairiness and actual yarn winding yarn hairiness, expresses as a percentage of the hairiness. A simple linear regression equation of the form

$$D=a+b_1 x_1+b_2 x_2+b_3x_3+\ldots\ldots +b_n x_n \quad (8)$$

was fitted to the data, the parameters a and b_1, $b_2 \ldots d_n$ were constants. The best fitting equation for predicting the hairiness of polyester/cotton 65/3513tex winding yarns is found to be

$$D=7.0138+0.64031A+0.4125B+0.2347C+0.5624D+0.4897E \quad (9)$$

5 Results and Discussion

The results are shown in Table 2. It is evident from Table 2 that the predictive power of ANN model is the better of the two models considered here. It can be seen from this table that the correlation coefficient between actual and predicted winding hairiness of ANN model (0. 9854) is much higher than that in the linear regression model (0.8780). In addition, the mean squared error and mean absolute error of ANN model (0. 1351 and 4. 1970%, respectively) is smaller than that in the linear regression model(0. 2551 and 7. 5818%, respectively)..It is also worthwhile to note that the maximum error and minimum error of ANN model (10.1689%and 0.0409%, respectively) is smaller than that in the linear regression model (16.9219%and 0. 1202%, respectively.). In all the cases, the error in the ANN model is smaller than that in the linear regression model. As it can be seen, the ANN model can yields more accurate and stable predictions than the linear regression model.

Table 2. Comparison of prediction performance of two models

Statistical parameter	Linear regression model	ANN model
Correlation coefficient , R	0.8780	0. 9854
Mean squared error	0. 2551	0. 1351
Mean absolute error %	7. 5818	4. 1970
Cases with more than 10% error	4	1
Maximum error %	16.9219	10.1689
Minimum error %	0. 1202	0.0409

6 Conclusions

In this study, we predicted the hairiness of polyester/cotton blended winding yarn with both ANN and linear regression models. On the basis of the results obtained, with the aid of both ANN and linear regression analysis, we can predict the winding yarn hairiness easily and accurately. The results show that the performance of ANN seems to be better than that of the linear regression model, which is a useful alternative approach for the development of spinning prediction models.

References

1. Barella, A.: Yran hairiness. J. Text. Inst. 13, 39–49 (1983)
2. Barella, A.: The hairiness of yarns. J. Text. Inst. 24, 41–50 (1993)
3. Lang, J.: Frictional behavior of synthetic yarns during processing. Textile Res. J. 73, 1071–1078 (2003)
4. Wang, X.: The effect of testing speed on the hairiness of ring-spun and sirospun yarns. J. Text. Inst. 88, 99–106 (1997)
5. Wang, X.: Effect of speed and twist level on the hairiness of worsted yarns. Textile Res. J. 69, 889–894 (1999)
6. Wang, X.: A study on the formation of yarn hairiness. J. Text. Inst. 90, 555–561 (1997)
7. Guba, A.: Predicting yarn tenacity: A comparison of mechanistic, statistical and neural network models. J. Text. Inst. 92, 139–145 (2001)
8. Zeng, Y.C.: Predicting the tensile properties of air-jet spun yarns. Textile Res. J. 74, 689–694 (2004)
9. Ertugrul, S.: Predicting bursting strength of cotton plain knitted fabrics using intelligent technique. Textile Res. J. 70, 845–851 (2000)
10. Yao, G.F.: Predicting the warp breakage rate in weaving by neural network techniques. Textile Res. J. 75, 274–278 (2005)
11. Hagan, M.T.: Neural Network Design. China Machine Press, Beijing (2007)

Predicting the Fiber Diameter of Melt Blowing through BP Neural Network and Mathematical Models

Zhao Bo*

College of Textiles, Zhongyuan University of Technology, Henan, Zhengzhou, 450007, People's Republic of China
zhaobohenan@sina.com, zhaobohenan@163.com

Abstract. Artificial neural network (ANN) model and mathematical method of the air drawing model are established and utilized for predicting the fiber diameter of melt blowing nonwovens from the processing parameters. A mathematical method provides a useful insight into the process parameters and fibber diameter. By analyzing the results of the mathematical model, the effects of process parameters on fiber diameter of melt blowing nonwovens can be predicted. A artificial neural network model provides quantitative predictions of fiber diameter. The results reveal that the ANN model produces a very accurate prediction.

Keywords: artificial neural network model, mathematical method, melt blowing, nonwoven, fiber diameter.

1 Introduction

Melt blowing is used commercially as a one-step process for converting polymer resin directly into a nonwoven mat of ultrafine fibers. In the melt blowing process, high velocity hot air streams impact on a stream of molten polymer as the polymer issues from a fine capillary. Many researchers have been devoting much attention to the study of the air drawing models of polymers [1-2]. In previous years [2], the air drawing model of polymers based on the numerical computation results of the air drawing model of melt blowing are established. The predicted fiber diameter gives good agreement with the experimental results.

In the recent years, the ANN models have been widely used to predict textiles problems [1-3].However, there is death of published papers that encompasses the scope of predicting the fiber diameter of melt blowing nonwovens with ANN models [4-7]. In this article, we attempt to predict the fiber diameter of melt blowing nonwovens using mathematical method and ANN models. The results show the ANN model produces more accurate and stable predictions than mathematical method.

* Corresponding author.

D. Jin and S. Lin (Eds.): Advances in MSEC Vol. 1, AISC 128, pp. 105–110.
springerlink.com

2 Experimental

2.1 Materials

The polymer used in the experimental runs was 36 MFR (melt flow rate) YMC 9747 polypropylene pellet. The polymer has a *Mn* of 36000.

2.2 Sample Preparation and Test Conditions

Images of the collected webs were captured by a 3D video microscope and the diameters of fiber in the webs were measured using Image-Pro software. All the tests were performed under a standard atmosphere of 22℃and 65% RH.

Table 1. Experimental program

No.	Polymer flow rate (g/s)	Polymer temperature (℃)	Initial polymer velocity (m/s)
1	0.0024	270	162
2	0.0016	250	109
3	0.0028	290	98
4	0.0028	290	162
5	0.0024	250	162
6	0.0016	270	109
7	0.0016	250	109
8	0.0016	290	98
9	0.0016	270	98
10	0.0024	270	109
11	0.0024	250	109
12	0.0028	290	162
13	0.0028	290	162
14	0.0016	270	98
15	0.0028	270	162

2.3 Program and Orthogonal Table

In this work, the orthogonal table of $L_9(3^4)$ is used. To investigate the main effects of each processing parameter, single factor experiments are also performed. The experimental programme is shown in the table 1.

3 Mathematical Model of the Air Drawing Model

The air drawing model of polymers consists of a continuity equation, a momentum equation, an energy equation, and constitutive equation [1-3].

3.1 Continuity Equation

$$W = \frac{\pi}{4} D^2 V \rho_f \tag{1}$$

Where W is the polymer mass flow rate, D the fiber diameter, V the fiber velocity and ρ_f the polymer density.

3.2 Momentum Equation

$$\frac{dF_{rheo}}{dz} = \frac{\pi}{2} j \rho_a C_f (V_a - V)^2 D + W \frac{dv}{dz} - \rho_f \frac{\pi}{4} D^2 g \tag{2}$$

Where F_{rheo} =rheological force, z =axial direction, ρ_a =air density, ρ_f =polymer density, V =fiber velocity, V_a =air velocity, g =gravitational acceleration, C_f =air drawing coefficient.

The rheological force is as following.

$$F_{rheo} = \frac{\pi}{4} D^2 (\tau_{xx} - \tau_{yy}) \tag{3}$$

Where τ_{xx} is the axial tensile stress of polymer, and τ_{yy} the transversal tensile stress of polymer.

3.3 Energy Equation

$$\frac{dT}{dz} = \frac{\pi D h (T - T_a)}{\rho_f W C_{pf}} \tag{4}$$

Where T =polymer temperature, T_a =air temperature, h =heat transfer coefficient, C_{pf} =specific heat capacity of the polymer at constant pressure.

3.4 Constitutive Equation

As is commonly known, the simplest constitutive equation is used in our work.

$$\tau_{xx} = 2\eta \frac{dV}{dz}$$

$$\tau_{yy} = -\eta \frac{dV}{dz} \tag{5}$$

Where η is shear viscosity.

4 Artificial Neural Network Model

Artificial neural network [8] is a parallel running nonlinear system which simulates the human brain construction and encouraging behavior. It has some handling units like neuron and gets the solution through expressing the problem as the weighted value between the neurons. Because the encouraging function usually is a nonlinear function, when many neurons connect a network and operate in a motive manner, it will constitute a nonlinear power system. Artificial neural network is model for computational systems, either in hardware or in software, which imitate the behavior of biological nervous in human brain by using a lot of structural artificial neurons. Artificial neural network can learn from the past and predict the future, extract rules for models that are unavailable or too cumbersome. The nonlinear relationships based on neural networks can be enhanced by learning and training. Artificial neural network is an algorithm to describe a special relationship between input variables and target values by the weights. It is utilized to study the nonlinear relationships between input variables and target values.

4.1 BP Neural Network Model

There are many kinds of neural networks, in which Back-propagation (BP) network model is one of the most researched and applied model. Back-propagation network is a feed-forward connecting model, which contains input layer, hidden layer and output layer, as shown in Fig. 1. Neurons in the same layer do not link each other, but neurons in neighboring layer are connected by the weight. Back-propagation network algorithm is the common calculations whose special features are distribute memory and parallel handling. As a forecast model, BP neural network is a kind of widely used artificial neural network in industry and it employs the BP algorithm. BP neural network consists of input layers, output layers and one or many hidden layers. Each layer has several neural units, which are related to those in the layers by nodes. BP neural network is a front-connection model and composed of input layer, output layer and some hidden layers. The neurons in the same layer have no connection with each other, but they are connected in the neighbor layers by individual weights.

According to experimental requirement, the BP neural network has three neural units for input, and has 1 neural unit for output. But in hidden layer, the number of neural units which can affect practical result is alterable in great range. The number can be defined according to the deviation produced in practice on every neural unit of BP. The practice result shows that the optimal number of units in hidden layer is equal to the average of unit number in output layer and input layer.

4.2 Back-Propagation Algorithm

Back-propagation algorithm is a neural network training methods mostly in use. It is simply a gradient descent method to minimize the total squared error of the output computed by the net. The algorithm has three steps. Firstly, initiate weight and bias. Secondly, amend weight and bias, input training data, calculate error between output and expected output; evaluate the correction of weight and bias, correct weight and bias. Thirdly, go on dong. Fourth, until error is less than expected value.

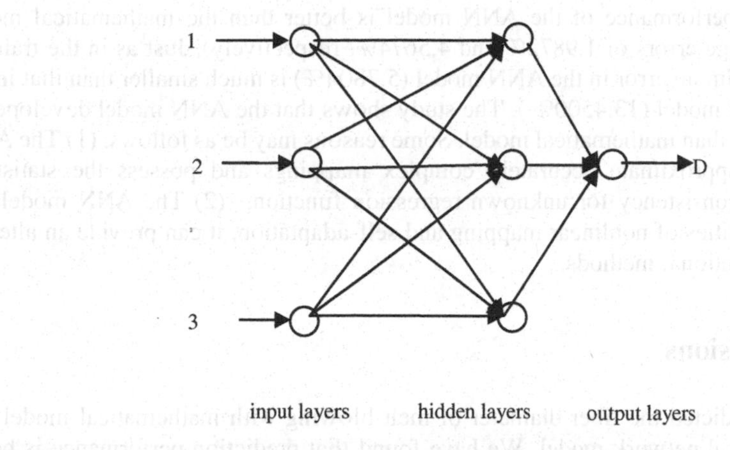

Fig. 1. Structure and architecture of an ANN model

4.3 Architecture and Designing of Neural Network Model

In this work, thirteen pieces of nonwoven samples are chosen as specimens. The results are input-to-output layer as model parameter, a network with three layers is sufficient for most practical applications. The number of input neurons normally corresponds to the number of input variables of the process to be modeled. In selecting the output neurons, note that it is generally inadvisable to train a network for several tasks. No corresponding rule is known for selecting the number of neurons of the second hidden layer, so purely empirical approach is necessary here. Therefore, we have selected a network with three layers: an input layer with 3 neurons, hidden layers with 3 neurons, and an output layer with 1 neuron. All combinations of 10 and 3 data points are used to train and test the ANN.

4.4 Practicing of Neural Network Model

The neural network model practice is guided by BP realization processing system. The algorithm of model is momentum regulation and reading ratio is adjusted to itself. When the BP network is being practiced, its parameters could be defined according to the structure. There are 3 nodes for input here, and the number of hidden layers is 3, the minimum velocity is 0.1, the momentum factor is 0.8, the sigmoid is 0.9, the acceptable deviation is 0.0001, the maximum times of practice is 6000, meanwhile, the datum in input nodes are standardized.

5 Results and Analysis

The average error of the mathematical model (2.3481%) is much larger than that in the ANN model (0.0013%).It is worthwhile to note that the maximum error is much smaller in the ANN model(0.0045%)than that in the mathematical model(5.6231%). In all the cases, the error in the ANN model is smaller than that in the mathematical model. The performance of the ANN model is better than the mathematical model (giving average errors of 1.9872% and 4.5674%, respectively). Just as in the training sets, the maximum error in the ANN model (5.7804%) is much smaller than that in the mathematical model (13.4500%). The study shows that the ANN model developed is more reliable than mathematical model. Some reasons may be as follows: (1) The ANN model can approximate accurately complex mappings and possess the statistical property of consistency for unknown regression functions. (2) The ANN model has excellent abilities of nonlinear mapping and self-adaptation, it can provide an alternative to conventional methods.

6 Conclusions

We have predicted the fiber diameter of melt blowing with mathematical model and artificial neural network model. We have found that prediction performance is better for the ANN model than mathematical model. The fiber diameter of melt blowing nonwoven is mostly influenced by processing parameters. The ANN model provides quantitative predictions of fiber diameter. The predicted and measured results agree well, indicating that the ANN model is an excellent method.

References

1. Chen, T.: Modeling polymer air drawing in the melt blowing nonwoven process. Textile Res. J. 73, 651–658 (2003)
2. Uyttendaele, M.A.J.: Melt blowing: General equation development and experimental verification. AIChE J. 36, 175–181 (1990)
3. Chen, T.: Predicting the fiber diameter of melt blown nonwovens: comparison of physical, statistical and artificial neural network models. Modeling Simul. Mater. Sci. Eng. 13, 575–584 (2005)
4. Guba, A.: Predicting yarn tenacity: A comparison of mechanistic, statistical and neural network models. J. Text. Inst. 92, 139–145 (2001)
5. Zeng, Y.C.: Predicting the tensile properties of air-jet spun yarns. Textile Res. J. 74, 689–694 (2004)
6. Ertugrul, S.: Predicting bursting strength of cotton plain knitted fabrics using intelligent technique. Textile Res. J. 70, 84–851 (2000)
7. Yao, G.F.: Predicting the warp breakage rate in weaving by neural network techniques. Textile Res. J. 75, 27–278 (2005)
8. Hagan, M.T.: Neural Network Design. China Machine Press, Beijing (2007)

Humanizing Anonymous More Sensitive Attributes Privacy Protection Algorithm

GuoXing Peng * and Qi Tong

School of Computer and Communication, Hunan University of Technology,
Zhuzhou, Hunan, China
pengguoxing168@163.com

Abstract. Paper first analyzes the major technical database of existing privacy and privacy requirements in more sensitive attribute data released under the scenarios studied in the publishing process more sensitive data privacy issues, pointing out that a single individual more than the corresponding records and the corresponding personal data published anonymously in the privacy protection model number of key issues need to be addressed, including information disclosure risk measure, information loss measure and control the dynamic semantic tree, and induction, summed up the release of existing multi-attribute data privacy-sensitive technology characteristics.

Keywords: personal anonymity, status maintained, semantic classification trees, multi-sensitive properties.

1 Introduction

Personal information privacy is about your personal identifiable information to control and use the removal of unlawful interference. With the rapid development of information technology, people enjoy the convenience and the resulting fast, but the information technology of the "double-edged sword" makes the collection of personal information is becoming increasingly easy to improper use of such information or be personal property will result in an open and spiritual loss. Therefore, the protection of personal privacy can not just stop at the so-called protection of the right alone, but should be moving in the direction of the protection of personal information. Privacy from the traditional "individual to live in peace without interference," the negative rights of the evolution of positive significance for the modern "information privacy." Performance of personal privacy on private affairs and control over the private information.

Demand from the perspective of privacy protection, privacy protection can be divided into for the current user's privacy and data-oriented privacy protection. Privacy protection for users from the user point of view, the main information on the protection of personal privacy, which is involved in protecting the privacy of the database user-related sensitive data and some of the sensitive behavior (such as query or delete

---

* GuoXing Peng (1968.4-): nationality: Hunan province, PRC; degree: M.S.; professional title: senior engineer; research field: data mining and computer science and its applications.

D. Jin and S. Lin (Eds.): Advances in MSEC Vol. 1, AISC 128, pp. 111–116.
springerlink.com

some data.) Privacy protection for users with different users, the specific application environment, a variety of laws and regulations have a close relationship. Comprehensive privacy protection for users needs include four aspects: the user's anonymity, pseudonym of the user, observation of user behavior is not.

Data-oriented database, to consider how to protect the privacy of data mainly expressed in the privacy of information about some of the privacy protection of sensitive data used to eliminate data access caused by the leakage of privacy issues, generally through the adding of these data some of the labels or by some special treatment to achieve the above objectives. Privacy protection for data privacy protection is the focus of the study, this discussion is for data privacy protection, to consider how to protect stored in a database of personal privacy information, that can directly or indirectly reflect the private information of the data to establish some privacy information on the relevant mechanisms.

2 Related Technologies

During the general algorithm on minimum loss, because of the record alone during the merge, the resulting loss of information beyond the scope of minimum information loss, the loneliness it was recorded as an independent element group.
General algorithm:

1. In the table to meet the rest of the records to find properties in this category, generalized disorder of records, from the disorder of the cost of the property value within the group, consider ordering property records, find records that meet the minimum distance join group
2. If there is no record of the disorder to meet the group of generalized attribute generalization, to continue the generalization properties of the disorder in this group, repeat one, until the meet the l-diversity
3. Alone tuple tuples ordered directly through the minimum distance to find the corresponding broad group, but if the combination of individual and group alone element caused by the loss of information when a great degree, you do not have to merge the records, saying the record is Independent alone tuple, its operations remain in the multi-sensitive property sheet to reconsider the merger.

3 Almost One-Dimensional Method

Alignment of generalized method of generalized identity property, have a relational table, which identifies the value of the property the same quasi-tuples form a QI group. In order to reduce the level of generalization anonymous table can be used to allow the right amount of tuples inhibition (suppression) of the method, but also reduced the tuple release table suppression of information validity.

Almost re-encoding method into global and local recoding categories. Requirements of the global re-encoded on the same quasi-identifier property values are generalized to almost the same level of the hierarchy tree. [3] proposed a global re-encoding algorithm, it can also be applied to maintain the identity of the generalization method.

The algorithm is simple, it is time to select a quasi-generalized identity property, until the meet the needs of anonymity. Global re-encoding is usually over-generalization.

Partial re-encoding does not require the same values of the property to the same level of generalization. Wong et al proposed a top-down partial re-encoding algorithm [5]. This method will first of all tuples are generalized to a wide Overview of the group, and then in maintaining the principle of anonymity, based on the specialization of tuples (specialization). This cycle continues until the special needs of anonymity until the break. This paper proposes a bottom-up partial re-encoding algorithm.

In order to merge into a single multi-dimensional table table produce the least sensitive property of the duplicate records, this article limits the sense of property of the traditional single Weimin anonymous conditions. For each single Weimin sense that the property sheet, in addition to a single record corresponding to the record number of sensitive attributes and the last shall be independent of tuple alone, the rest of the record must meet the traditional anonymous methods. Here we first summarize the new approach through the sense of a single Weimin property sheet to meet the constraints of the anonymous condition.

Table 1. Overview of the Disease attribute table after an anonymous (to meet the limit of 3 - Anonymous and 3 - diversification)

Goup ID	NO	Age	Postcode	Disease
1	1	55	10085	Bronchitis
1	2	55	10085	Coryza
1	3	55	10085	Flu
2	4	43	10075	Pneumonia
2	5	48	10075	Coryza
3	6	[45~47]	1007*	Gastritis
3	8	[45~47]	1007*	Gastric ulcer
3	9	[45~47]	1007*	Dyspepsia
4	7	[48~49]	100**	Dyspepsia
4	10	[48~49]	100**	Flu
4	11	[48~49]	100**	Bronchitis
4	12	[48~49]	100**	Flu
4	13	[48~49]	100**	Bronchitis
5	14	5	20084	Bronchitis

Described in Table 1, the first corresponding to a single record for multiple sensitive attributes impoverished group, and then a single case from the first record of a non-start, first of all minimal generalized disorder properties, then the last chapter summarizes grouping algorithm 1, the remaining records 12 and 13 can not meet the 3 - tuple diversity of anonymity for the lonely, according to algorithm 1.1, the first record minimum of 12 properties disorder generalized as 1007 *, corresponding to the third group, then consider ordering attribute values, but the record of 12 48 Ordered attribute value, not the third group, ordered range and are therefore incorporated into the record 12, the fourth group. Similarly, Algorithm 1 was recorded there should also be incorporated into the fourth group of 13. Hop into any of the records of 14 information packet loss caused by too much, so lonely as an independent element group 5.

Table 2. Overview of Medicine attributes of the table after an anonymous (to meet the limit of 3 - Anonymous and 3 - diversification)

Goup ID	NO	Age	Postcode	Medicine
1	1	55	10085	An
1	2	55	10085	Bn
1	3	55	10085	Gn
2	4	43	10075	Cn
2	5	48	10075	En
3	6	[45~48]	1007*	Dn
3	8	[45~48]	1007*	Fn
3	9	[45~48]	1007*	Dn
3	12	[45~48]	1007*	Fn
4	7	[48~49]	100**	En
4	10	[48~49]	100**	Fn
4	11	[48~49]	100**	Gn
4	13	[48~49]	100**	An
5	14	5	20084	An

3 describes the general process that, according to algorithm 1.1, the first three records will be assigned to a 6,8,9 record group, this time recording the minimum order is [45 47], but at this point does not meet the three principles of diversity, Therefore, the minimum distance algorithm to find records based on 12 into the group to meet after the anonymous request, so this range of four after an orderly record of generalization into one group. Similarly, alone recorded 13 into group 4. Hop into any of the records of 14 information packet loss caused by too much, so lonely as an independent element group 5.

4 The General Algorithm for Anonymous Individuals

Personal and more sensitive to general algorithm description attribute Anonymous

1. Individual requirements of privacy protection through the sensitive property value, found in the semantic tree leaf node corresponding

2. From the leaf node (sensitive property value) be traced back to the root
3. If (the second from the root of the left sub-tree exists) / / false to 4 conditions
3.1. Remove the root of the tree sub-set of A
3.2. if (protection of sensitive attribute values in the set A,) / / false to 4 conditions
3.2.1. While (not present in the second sub-tree traversal of the leaf clusters)
3.2.1.1. if (protection of sensitive attribute values in the first i-leaf clusters)
3.2.1.1.1. Remove the leaves of sensitive attribute values of another cluster
3.2.1.1.2. Sensitivity obtained by the new property value, in the corresponding one-dimensional classification tree, whichever is the root node
3.2.1.1.3. The release of the person found in the table correspond to the two different sensitive attribute values of multiple records, each sensitive attribute generalization is the root of the corresponding
3.2.1.2. else continue to traverse to the next leaf cluster
4. While (third from left sub-tree beginning to end, there is not a sub-tree traversal of the sub-tree)
4.1. while (i-sub-tree exists)
4.1.1. Remove the root of the tree sub-set of A
4.1.2. if (protection of sensitive attribute values in the set A,)
4.1.2.1. While (the semantic sub-tree traversal of the leaves are not present in the cluster)
4.1.2.1.2. if (protection of sensitive attribute values in the first i-leaf clusters)
4.1.2.1.2.1. Remove the leaves of sensitive attribute values of another cluster
4.1.2.1.2.2. Sensitivity obtained by the new property value, in the corresponding one-dimensional classification tree, whichever is the root node
4.1.2.1.2.3. Found in the published table in the person's records are in different dimensions in different sensitive attribute values for the protection of privacy values and be sensitive to the new property value, respectively generalization
4.1.2.1.3. else continue to traverse to the next leaf cluster
4.1.3. else continue to traverse a sub-tree, until finally a semantic subtree
4.2 in the release table to find the person is protecting sensitive attributes node multiple records, these records of sensitive attribute values corresponding to the root of the generalization is
5. Return

Mike more sensitive properties through a personal anonymous algorithm summary, due to their personal privacy and property values Bronchitis Bronchitis medicine An attribute has an associated dimension, so to change the value of the corresponding private property the same dimension also changes the sensitive records of the corresponding attributes of other dimensions value, in this case is the generalization for the respiratory tablet. In addition, private property in the same dimension, the cluster identified by the leaf node property values and personal privacy of the property value associated with flu, flu the same reason the properties of this algorithm is summarized respiratory infection.

Generalization based on anonymous mike after the individual data sheets for the SST `, note 1 GroupID and Name the first and third records.

Data Sheet for the SST `

5 Conclusions

Mike more sensitive properties through a personal anonymous algorithm summary, due to their personal privacy and property values Bronchitis Bronchitis medicine An attribute has an associated dimension, so to change the value of the corresponding private property the same dimension also changes the sensitive records of the corresponding attributes of other dimensions value, in this case is the generalization for the respiratory tablet. In addition, private property in the same dimension, the cluster identified by the leaf node property values and personal privacy of the property value associated with flu, flu the same reason the properties of this algorithm is summarized respiratory infection.

Acknowlegement. I would extend my great thanks to the assistance and help of the project team by college teaching reform research of the Hunan province in 2011.

References

1. Li, T., Tang, C., Wu, J., Zhou, M.: Two clustering based on k-anonymity privacy protection. Jilin University (Information Science) 02 (2009)
2. Han, J., Cen, T., Yu, H.: Data table k-anonymous study of the micro-clustering algorithm. Electronics 10 (2008)
3. Li, J., Lv, Y., Zhao, W., Liu, G., Li, H.J.: Barrels of K-based anonymity multidimensional table incremental update algorithm. Yanshan University 05 (2009)
4. Qi, R., Wang, K., Guo, X., Li, J., Tang, J., Liu, G.: Priority strategy based on the maximum Ye Zizi tree protection method of multi-sensitive properties. Yanshan University 05 (2009)
5. Teng, J., Zhong, C.: Anonymous method based on data published privacy leaks Control Technologies. Guangxi Academy of Sciences 04 (2009)

Research of Music Class Based on Multimedia Assisted Teaching in College

Hongbo Zhang

Department of Music Education, Chifeng College, Chifeng, 024000, China
Happyzg3@163.com

Abstract. Information technologies are played the important role and are effect our concepts and teaching methods in music education from MIDI, computer music, multimedia music to computers and network-based teaching. During the music education in college, the configuration and application of multimedia music learning system in higher music education, as the modernization of teaching methods, are to achieve the best solutions. This paper introduced the advantages of music appreciation class based on multimedia assisted teaching in college and introduced some hardware and software configuration of the teaching system and teaching in music, writing, research and management application. In the end, several issues were suggested on music appreciation class based on multimedia-assisted teaching.

Keywords: Multimedia music instruction, Computer music, Teaching methods.

1 Introduction

College music appreciation class is a larger capacity, wider knowledge of the course [1, 2]. In addition to enable students to appreciate music, they also need to explain the music of the background of the times, the author's life, work style, the band works of the musical composition and analysis, display cases and related music scores pictures. However, it is difficult to complete these within the limited hours of teaching content using the traditional teaching methods. Multimedia technology in the field of music teaching in a wide range of applications, greatly expanding the teaching capacity of music to enrich the teaching methods and teaching resources to enhance student interest in music. Especially for the teaching of music appreciation has brought a new situation, greatly easing the small hours of the teaching content, and more conflicts, effectively promote the improvement of teaching quality. CAI has many advantages over traditional teaching, so more and more teachers and students alike, and gradually become the mainstream of college music appreciation teaching aids [3] multimedia music instruction can Help play music files quickly easily, help improve teaching efficiency , help students to start a rich auditory association ,help to improve the efficiency of teachers ,help communicate and interaction between teachers and students and so on.This paper suggested the multi-media music education system and introduced some hardware and software configuration of the teaching system and teaching in music, writing, research and management application. In the end, several

D. Jin and S. Lin (Eds.): Advances in MSEC Vol. 1, AISC 128, pp. 117–121.
springerlink.com

issues were suggested on music appreciation class based on multimedia-assisted teaching [4].

2 Multi-media Music Education System

Multi-media music education system is a multimedia computer system with a combination of computer music. The multimedia computer system to the multimedia information processing and music equipment, music software combined with intuitive, concrete, show the image of the multimedia music instruction content. Application of multimedia teaching system for teaching music, management and other activities, can significantly improve the teaching quality and efficiency. The components of multi-media music education system are showed in Figure 1.

2.1 The Hardware of Multimedia Music Teaching System

Multi-media music education system, equipment can be divided into the following types, it showed in Figure2.

Computer: Multimedia Computer (including sound card, TV card, etc.).
Projection equipment: projectors, video showcase.
Video equipment: VCR[5].
Audio equipment: recording deck, amplifier, speakers.
Music equipment: electronic synthesizer.

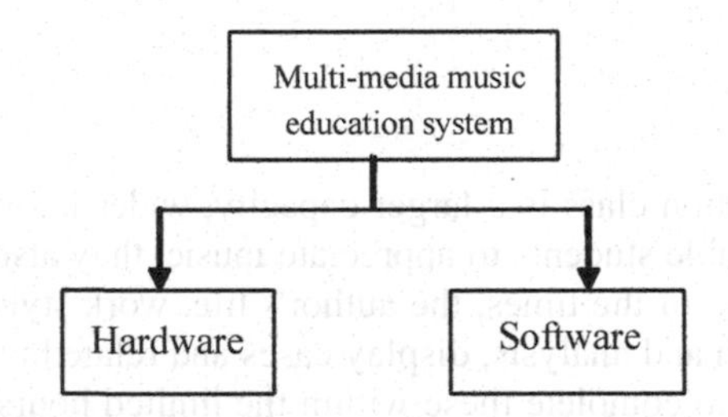

Fig. 1. The components of multi-media music education system

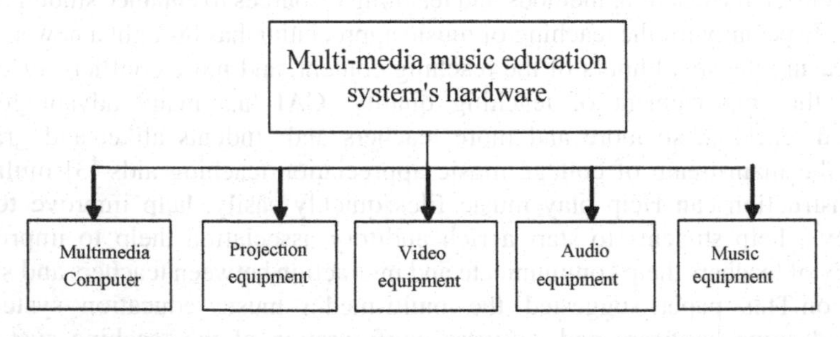

Fig. 2. The components of multi-media music education system's hardware

Multimedia computer is to handle text, graphics, images, sound, animation and film and other media information, that information can be converted into digital signals and to edit and modify.

, The projector is to put the content on the computer screen clear and large screen to be displayed, so that more people can also see the computer screen to display a variety of text and image information. Video Showcase is to sent directly the image information to the projector, according to the need to highlight and amplify local image.

TV receiving car is to to watch television (CCTV, satellite TV or outdoor antenna) program, can be directly on the big screen or TV display. The image quality, sound quality are better than ordinary TV[6].

Sound card is to processaudio signal via computer .it can take advantage of computer music creation and performance, with a MIDI interface and the corresponding software adapter produce electronic music.

Recording card is to play and record music tapes, all of the system to the transcription of audio tape.

PA is to accept the system and all of the audio signal for power amplifier and then transmitted to the speakers, sound.

Speaker is to transmit power amplifier receiving the signal, the signal is reduced to the sound broadcast.

Electronic synthesizer is to be Built-in computer interface electronic synthesizers and computers can be connected directly, which combines music and audio for the whole keyboard, the keyboard can play music directly to the computer input of information, its source can provide high-quality, multi-tone and the multi-voice polyphony sound.

2.2 The Software of Multimedia Music Teaching System

Generally use Windows 95 or 98 operating system and Microsoft Office business software, use of multimedia tools related software, music software and audio and video editing software. it showed in Figure3.

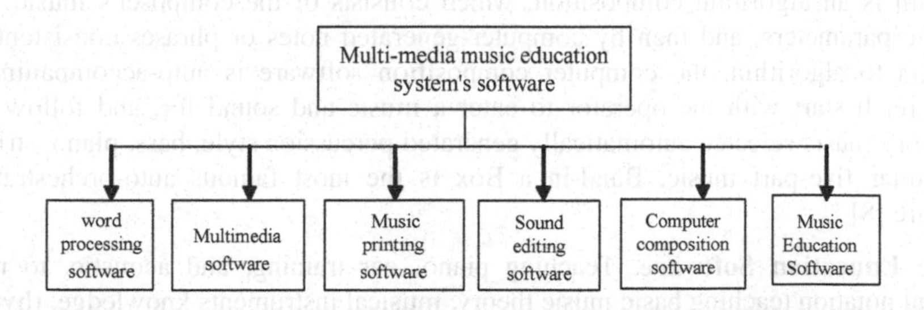

Fig. 3. The components of multi-media music education system's hardware

Word Processing Software. Use in Microsoft Office Word for word processing, the textbooks in the text, images, music, photos and other content via the keyboard or scanner into the computer, and then need to edit the text or image zoom and special effects processing, you can the images, sounds, MIDI and other files placed in the text of which can also be inserted with a miniature icon in the text which need to click on the icon with the mouse when you can see images or hear sounds.

Multimedia Software. Using Microsoft Office Power-point slide to display the content in book on the big screen and preparation of their own software using multimedia music courseware with "Director" program production and broadcasting systems, and multimedia editing text, images, sounds, animation production.

Music Software. Sequencer software is the most common music software, which can record a variety of data input into the computer, and then modify the data, shift, delete, stitching, generation, and editing, and finally replay of these data, the sequencer software similar to multi-track digital recorder. The most famous sequence software is Cakewalk, Cakewalk sequencer software can be used for music production, digital recording (hard disk recording) and audio editing. Other sequencer software as well as Microsoft's "Music Producer", "Computer Music Master", "Computer Band" (musical notation) [7] and "shell sound professional music system" (musical notation) and so on.

Music Printing Software. You can use the synthesizer keyboard or mouse input music, then music editing, layout, add music to input Chinese characters and symbols, print out the standard of professional music. Encore is the most famous music printing software. MIDI Scan software: you can use a scanner to read music and then scanned into image files into MIDI files, sheet music for music playback or printing.

Sound Editing Software. Sound editing software is divided into sound database, software sound synthesis and sample waveform editing software. Sound synthesis software is the electronic voice synthesizer or audio synthesis parameters of items made of software on the computer to process data or graphical way to adjust the instrument parameters in the synthesis of new sounds. Sample waveform editing software sampler or a waveform RAM using the software of electronic musical instruments, it is mainly through the MIDI cable input waveform sampling data collected, and then edit, synthesize new sounds.

Computer Composition Software. Computer composition software is used to help the composer editing and recording their work, and computer composition software program is an algorithm composition, which consists of the composer's music that specific parameters, and then by computer-generated notes or phrases consistent. In addition to algorithm, the computer composition software is auto-accompaniment software. It start with the operator to enter a music and sound for, and follow the operator's music selected automatically generated percussion style, bass, piano, strings and guitar five-part music. Band-in a Box is the most famous auto-orchestration software [8].

Music Education Software. Teaching piano, ear training, and acoustic, to read musical notation teaching basic music theory, musical instruments knowledge, rhythm training and educational software. Some special famous musicians, such as Mozart, Beethoven, Schubert and other famous musicians can be introduced via multimedia software.

3 Conclusions

Music CAI has many advantages, but inevitably there are some shortcomings and deficiencies. After all, a multimedia-aided teaching is still in development stage, is not satisfactory in many ways, until the maturity needed to use a process. CAI is the trend of educational development. We should be clear to their strengths, and on this basis, according to university music appreciation course in the use of multimedia in teaching the problems, weaknesses of the measures designed to explore a music appreciation class and efficient development of modern teaching methods update the road.

References

[1] Zhao, Y.S., Lu, L.C.: Music multimedia courseware, p. 17. China Central Conservatory of Music Press, Beijing (2004)
[2] Liang, M., Guo, G.: Li Jie of multimedia CAI courseware and case-based tutorial, vol. 11, pp. 1–4. Electronic Industry Press, Beijing (2006)
[3] Huang, X.: Analysis of the music teaching values education. Jilin Education 2 (2010)
[4] Hu, Y.: On the lower grades under the new curriculum music teaching. China Ti Weiyi Education 01 (2010)
[5] Huang, Y.: Relying on the media to build a variety of music classes. School Audio-Visual 2 (2010)
[6] Zhang, S.: Happy music class. New Courses (Teachers) 1 (2010)
[7] Zhao, C.Y.: Bit about music appreciation teaching. Exam (Academic Version) 02 (2010)
[8] Le, M.: New Curriculum Innovation Methods of teaching high school musical. Secondary School Teaching Reference 6 (2010)

Analysis and Design of E Surfing 3G Mobile Phone's Payment Schemes of China Telecom Based on BP Neural Network

Yan Shen*, Shuangshuang Sun, and Bo Zhang

Harbin Engineering University, College of Science, Harbin 150001, China
shenyan@hrbeu.edu.cn

Abstract. In order to let the mobile phone operators and the consumers both maximize profits by way of launching plans of payment schemes and choosing the one which is suitable for consumers' own consumption characteristics correspondingly, this paper analyzes and compares the current mobile phone's payment schemes by means of adopting BP neural network to establish the relationship between multivariable and single-objective, and launches new design plan of payment schemes. It is demonstrated that the consumers suffer less total consumption in the new payment scheme than that of previous ones on the basis of guaranteeing operators' former profits.

Keywords: 3G mobile phone's payment schemes, BP Neural Network, model predictive.

1 Introduction

With the quick development of mobile communication technology going by, consumers suffer tough choices among kinds of payment schemes carried out by mobile phone operators. In consideration of common users of mobile phones, when changing the payment schemes, how to choose the suitable operator to meet own consuming characteristics as well as the plans of payment schemes, how to target the consuming groups, how to price the highly pertinent payment schemes to improve market competitiveness and expand market share are crucial problems that operators should focus on during the update from 2G to 3G.

At present, the analysis and design of relevant business are mostly carried out by adopting statistics, in combination with some tests of MATLAB. Artificial neural networks method is applied in the research and prediction of flow of mobile phone short message service[1], mobile phone virus detection[2], applied research on mobile phone traffic[3], prediction of multimedia messaging service[4], credit analysis of segmentation of mobile communication users and so forth[5]. This paper employs artificial neural networks method to analyze e surfing 3G mobile phone's payment schemes of China Telecom, and launches a new design plan of payment schemes.

* Corresponding author.

D. Jin and S. Lin (Eds.): Advances in MSEC Vol. 1, AISC 128, pp. 123–128.
springerlink.com

2 BP Neural Network

The principle of BP arithmetic: It has two processes that forward propagation of signals and backward propagation of error signals [6]. The algorithm flow of BP is as follows:

1) Initialize the network
2) Input sample vector, and calculate the output of each layer

$$y_j = f\left(V_j^T X\right) \quad j = 1,2,\cdots,m \tag{1}$$

$$o_k = f\left(W_j^T Y\right) \quad k = 1,2,\cdots,l \tag{2}$$

3) Calculate error of network output

$$E^p = \sqrt{\sum_{k=1}^{l}\left(d_k^p - o_k^p\right)^2} \tag{3}$$

4) Calculate error signal of each layer

$$\delta_k^o = \left(d_k - o_k\right)\left(1 - o_k\right)o_k \quad k = 1,2,\cdots,l \tag{4}$$

$$\delta_j^y = \left(\sum_{k=1}^{l}\delta_k^o w_{ij}\right)\left(1 - y_j\right)y_j \quad j = 1,2,\cdots,m \tag{5}$$

5) Adjust weight of each layer

$$w_{jk} \Leftarrow w_{jk} + \eta\delta_k^o y_j \quad j = 1,2,\cdots,m \;\; k = 1,2,\cdots,l \tag{6}$$

$$v_{ij} \Leftarrow v_{ij} + \eta\delta_j^y x_i \quad i = 1,2,\cdots,n \;\; j = 1,2,\cdots,m \tag{7}$$

6) Check whether the system has been completed once training;
7) Check whether the total precision of network is in the allowable range of error[6].

3 The Empirical Analysis Is Based on the E Surfing 3G Mobile Phone's Payment Schemes of China Telecom

This paper choses the e surfing 3G campus mobile phone's payment schemes of China telecom as examples. Aim to analyze and design the payment schemes of the selected samples. Sample data are the detailed list of consumers from May 2010 to August 2010 which is selected from the China telecom database of a certain area. These sample data are consuming charges with the unit of fen.

3.1 Screening of Sample Data

Firstly, selecting 100 thousand users' the data randomly from the database as the experiment database of this paper.

Secondly, according to the concrete contents of mobile phone's payment schemes of China telecom, select the sample data items of neural network training.

Thirdly, according to the consuming characteristics of users, set the minimum value of selected the sample data items and screen data which do not meet the conditions manually. Then 1439 original experimental items can be obtained.

Finally, cluster the selected 1439 experimental items based on the distance of link. Have an operation of intersection on the data of five clustering for 5 data items. Then 39 sample data can be obtained at last.

3.2 Optimal Structure Design of Network

Determine the node number of hidden layer by the principle of rms error between actual output and expected output as small as possible, shown in Table 1:

Table 1. The relationship between node number of hidden layer and rms error

Node number of hidden layer	2	3	4	5	6	7	8
Rms error	0.483	0.656	0.421	0.173	0.074	0.064	0.026

According to Table 1, there are 8 nodes of hidden layer.

3.3 Analysis and Design of E Surfing 3G Campus Payment Schemes of China Telecom

Network Training. Set up a BP neutral network structure with three inputs, one output and one hidden layer. Select anterior 30 group data from 39 group sample data as training samples, and the rest 9 group data as testing samples. Function service charge, data service charge, and month total consumption as the input sample items. The output sample item of point-to-point short message charge is network I, and the output sample item of local telephone charge is network II.

The magnitude of rms error of network I and network II are respectively 10^{-9} and 10^{-7} when they reach stable. The magnitude of rms error between actual output and expected output are 10^{-14} and 10^{-12} when they reach stable by testing with 9 group samples respectively. Thus the two network accord with the network requirements.

Analysis and design of e surfing 3G campus payment schemes of China telecom. There are three consumption plans of e surfing 3G campus payment schemes of China telecom, shown in Table 2:

Table 2. E surfing 3G campus payment schemes of China telecom

Scheme	Basic consumption[fen]	Charge of short message[fen]	Charge of talking [fen]	Charge of surfing flow[fen]
Talking scheme	1900	3000	1000	615
Music scheme	3900	5000	3500	1230
WAP scheme	5900	8000	4750	2460

This paper adopts average of each consuming item of current sample population as input vector, and corresponding output can be obtained respectively.

For network I, the output point-to-point short message charge is 1.0018×10^3, which is equivalent to1002 by the rule of "round". For network II, the output local phone charge is 6.3936×10^3, which is equivalent to 6934 by the rule of "round". The above two output data is consumption charge with the unit of fen.

Contrast the obtained data with the data of exiting schemes under the standard of expected consumption. The comparison results are shown in Table 3:

Table 3. Comparison Results

Scheme	The balance of short message charge[fen]	The balance of talking charge[fen]	total consumption [fen]
Talking scheme	1998	-5394	7294
Music scheme	3998	-2894	6794
WAP scheme	6998	-1644	7544

The Table 3 shows that the cost of music scheme of e surfing 3G campus plan of China telecom is the lowest one in the sample population.

The basis and principle for designing a new scheme for the sample population are as follows:

From the perspective of operators, on the basis of keeping the original monthly mean of consumers consumption, set the month total consumption average of the sample data as the basic consumption of new plan. Considering the charging standard of exiting plan, adjust 4396 to 4500.

From the perspective of consumers, consulting to the charging standard of exiting plan, then setting 4250 as local telephone charge and 2000 as point-to-point short message charge. Setting 1383 as surfing flow charge. The consuming scheme of the samples is showed in Table 4:

Table 4. New designing payment schemes

Scheme	Basic consumption[fen]	Charge of short message[fen]	Charge of talking [fen]	Charge of surfing flow[fen]
New scheme	4500	2000	4250	1383

According to the consumption characteristics of the consumers, the consuming list of users when they chose new payment scheme is shown in Table 5:

Table 5. Charge of new payment schemes

Scheme	The balance of short message charge[fen]	The balance of talking charge[fen]	total consumption [fen]
New scheme	998	-2144	6644

From the comparison between Table 5 and Table 3, it shows that the consumers suffer less total consumption under the new payment scheme than that of the previous one on the basis of guaranteeing operators' original profits.

4 Conclusion and Prospect

In this study, we firstly divide the sample data into two classes--one class as training sample and the other one as the texting sample. Secondly, we train the two BP networks respectively. At last, we take the average of sample data as the input and obtained point-to-point short message charge and local telephone charge. It shows that the consumers suffer less total consumption under the new payment scheme than that of the previous one on the basis of guaranteeing operators' original profits.

We only adopt BP neutral network to set up a network model as to analyze and design payment schemes in this paper. Combining some intelligent algorithm with BP algorithm to realize complementary integration, whether the result will be better need further research.

References

1. Wan, N.-H.: Research and prediction on the flow of Mobile Phone Short Message Service based on BP neural network. Zhejiang Normal University, Jinhua (2008)
2. Mei, H.W., Zhang, M.Q.: On Mobile Phone Viruses Detection Based On BP Neural Network. Computer Applications and Software 27, 283–300 (2010)
3. Tao, N.Y., Jiang, J.Z., Ze, C.X.: The application of neural network in Mobile Communication Traffic Prediction. Shandong Communication Technology 28, 9–12 (2008)
4. Yu, Z.F., Guo, M.S.: Prediction Model for Multimedia Messaging Service based on improved BP neural network. Mobile Communication 33, 88–91 (2009)
5. Shen, C.: Research on Market Segmentation of rural Mobile Communication. Nanjing University of Posts and Telecommunications, Nanjing (2007)
6. Han, L.Q.: Theory, Design and Application of Artificial Neural Network, 2nd edn. Chemical Industry Press, Beijing (2007)

Soft-switching High Step-Up DC-DC Converter with Single Magnetic Component

Hyun-Lark Do

Department of Electronic & Information Engineering,
Seoul National University of Science and Technology, Seoul, South Korea
hldo@seoultech.ac.kr

Abstract. A soft-switching high step-up DC-DC converter with single magnetic core is presented in this paper. By utilizing the leakage inductance of a transformer, power switches operate with soft-switching and reverse-recovery problem of output diodes is dramatically alleviated. The two-stage voltage doubler which is adopted at the secondary side can raise the voltage gain with a relatively low turn ratio of a transformer. Also, voltage stresses of output diodes are confined to half of the output voltage. A prototype of the proposed converter is developed, and its experimental results are presented for validation.

Keywords: Soft-switching, DC-DC converter, reverse-recovery.

1 Introduction

Recently high step-up DC-DC converters have been widely employed in many applications such as electric vehicles, fuel cells, and photovoltaic systems [1]. In those applications, DC-DC converters act as an interface system between the low voltage sources and the load which requires higher voltage. Thus, a step-up converter is required to boost low input voltage to a sufficiently high and constant level. A conventional boost converter can be used in those applications. It is because it has many advantages such as its simple structure and low cost. However, it requires an extreme duty cycle to obtain high voltage gain and its voltage gain is limited due to its parasitic components [2]. In order to increase voltage gain, high step-up DC-DC converter using coupled inductors have been suggested in [3]. It can provide high voltage gain but it has several disadvantages such as electrical non-isolation, large voltage ringing across the semiconductor devices, and low efficiency.

In order to remedy these problems, a soft-switching high step-up DC-DC converter with single magnetic core is proposed. Its circuit diagram and key waveforms are shown in Fig. 1. The proposed converter features high voltage gain, fixed switching frequency, soft-switching operations of all power switches and output diodes, and clamped voltage across power switches and output diodes without any clamping circuits. The reverse-recovery loss of the output diodes is significantly reduced due to the leakage inductance. Thus, zero-current-switching (ZCS) operation of the output diodes is achieved.

D. Jin and S. Lin (Eds.): Advances in MSEC Vol. 1, AISC 128, pp. 129–132.
springerlink.com

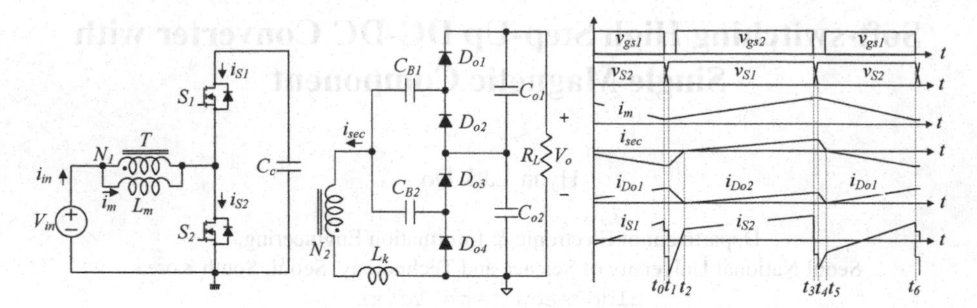

Fig. 1. Circuit diagram and key waveforms of the proposed converter

2 Analysis of the Proposed Converter

The operation of the proposed converter during a switching period T_s $(=t_6 - t_0)$ is divided into six modes. The switch S_1 and the switch S_2 are operated asymmetrically and the duty cycle D $(=(t_3 - t_0)/T_s)$ is based on S_2. The voltages across S_1 and S_2 are confined to the voltage V_c across C_c. The two-stage voltage doubler adopted at the secondary side raises the voltage gain with a relatively low turn ratio n $(= N_2/N_1)$ of T. L_m is the magnetizing inductance and L_k is the leakage inductance. Before t_0, the switch S_1, D_{o1}, and D_{o3} are conducting. At t_0, i_m and i_{sec} arrive at their minimum values, respectively.

Mode 1 [t_0, t_1]: At t_0, S_1 is turned off. Then, the energy stored in the magnetic components starts to charge/discharge the parasitic capacitances of S_1 and S_2. Therefore, the voltage v_{S1} starts to rise from zero and the voltage v_{S2} starts to fall from V_c. Since the parasitic capacitances are small, this time interval can be ignored.

Mode 2 [t_1, t_2]: At t_1, v_{S2} arrives at zero. Then, the body diode of S_2 is turned on. After that, the gate signal is applied to S_2. Since v_{S2} is zero before S_2 is turned on, zero-voltage turn-on of S_2 is achieved. The currents i_m and i_{sec} increase linearly.

Mode 3 [t_2, t_3]: At t_2, the currents i_{Do1} and i_{Do3} arrive at zero and D_{o1} and D_{o3} are turned off. Then, i_{sec} changes its direction and D_{o2} and D_{o4} are turned on. Since the changing rates of i_{Do1} and i_{Do3} are controlled by L_k, their reverse-recovery is significantly alleviated. In this mode, i_{sec} increases linearly. At the end of this mode, i_m and i_s arrive at their maximum values, respectively.

Mode 4 [t_3, t_4]: At t_3, S_2 is turned off. Similar to mode 1, the parasitic capacitances of S_1 and S_2 are charged/discharged. Also, this time interval can be ignored.

Mode 5 [t_4, t_5]: At t_4, v_{S1} becomes zero and its body diode is turned on. Then, the gate signal is applied to S_1. Since v_{S1} is zero before S_1 is turned on, zero-voltage turn-on of S_1 is achieved. With the turn-on of S_1, i_m and i_{sec} decrease linearly from their maximum values.

Mode 6 [t_5, t_6]: Similar to mode 3, i_{Do2} and i_{Do4} arrive at zero and D_{o2} and D_{o4} are turned off. Then, the output diode D_{o1} and D_{o3} are turned on and their currents increase linearly. Since the diode currents are controlled by L_k, its reverse-recovery problem is significantly alleviated. At the end of this mode, i_m and i_{sec} arrive at their minimum values, respectively.

By applying the volt-second balance law to the voltage across L_m, V_c is derived by $V_{in}/(1\text{-}D)$. The voltages across the power switches are confined to V_c. Since V_c depends on D, the voltage stresses of S_1 and S_2 can be varied according to D and V_{in}. The voltage gain M of the proposed converter is given by

$$M=\frac{V_o}{V_{in}}=\frac{2n(1-2k)D}{(D+(1-2D)k)(1-D-(1-2D)k)}, \tag{1}$$

$$k=\frac{1}{2}\left(1-\sqrt{1-\frac{8L_kI_o}{nDV_{in}T_s}}\right). \tag{2}$$

Due to the two stage voltage doubler structure, the voltages across all the output diodes are confined to $V_o/2$.

The ZVS condition for S_1 is given by $I_{m1}+2nI_{Do2}>0$, where I_{m1} is the maximum value of i_m and I_{Do2} is the maximum value of i_{Do2}. Since n, I_{m1}, and I_{Do2} always have positive values, it can be easily seen that the inequality is satisfied. Similarly, the ZVS condition for S_2 is given by $-I_{m2}+2nI_{Do1}>0$, where I_{m2} is the minimum value of i_m and I_{Do1} is the maximum value of i_{Do1}. Since I_{m2} approximates to $nI_o/(1-D) - V_cDT_s/2L_m$ and I_{Do1} approximates to $I_o/(1-D)$, the ZVS condition for S_2 is satisfied.

3 Experimental Results

The performance of the proposed converter was verified on a 110W prototype. The prototype was designed to operate from a 48V input voltage and provide 820V output voltage. Its operating frequency was 75kHz. The turn ratio n was selected as 3. The proposed converter provides the voltage gain of 17 with the turn ratio of 3. The inductance L_m was selected as 132uH. The leakage inductance L_k was 140uH. The capacitor C_c was chosen as 220uF. The blocking capacitors C_{B1} and C_{B2} were selected as 6.6uF. The output capacitors C_{o1} and C_{o2} were selected as 22uF. Fig. 2 shows the measured key waveforms of the proposed converter. It agrees with the experimental result. The measured maximum voltage stresses of S_1 and S_2 are around 160V, which agrees with the theoretical analysis. The ZVS operations of S_1 and S_2 are also shown in Fig. 2. Since the voltages across the switches go to zero before the gate pulses are applied to the switches, the ZVS turn-on of the switches is achieved. The ZCS of the output diodes are also shown in Fig. 2. After the diode currents fall to zero, the voltages across the diodes rise toward $V_o/2$. The voltages v_{Do1} and v_{Do2} are confined to $V_o/2$ without a clamping circuit. The proposed converter exhibits the maximum efficiency of 94.2% at full load. Due to its soft-switching characteristic and alleviated reverse-recovery problem, it shows a higher efficiency than other isolated high step-up DC-DC converters.

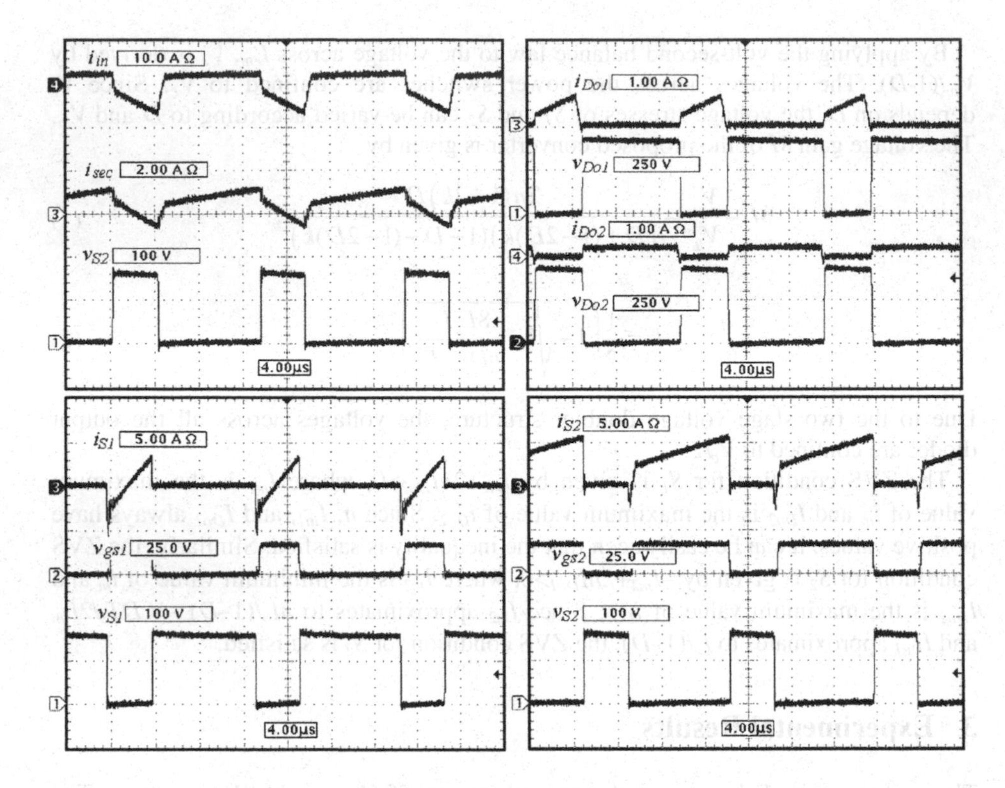

Fig. 2. Experimental waveforms

4 Conclusion

A soft-switching high step-up DC-DC converter with single magnetic component has been proposed. Both power switches operate with soft-switching and reverse-recovery problem of output diodes is dramatically alleviated by the leakage inductance. With the transformer turn ratio of 3, the voltage gain of 17 has been achieved. Due to its soft commutation of power semiconductor devices, it shows a high efficiency.

References

1. Wai, R.J., et al.: High-performance stand-alone photovoltaic generation system. IEEE Trans. Ind. Elec. 55, 240–250 (2008)
2. Yang, L.S., et al.: Transformerless DC-DC converters with high step-up voltage gain. IEEE Trans. Ind. Elec. 56, 3144–3152 (2009)
3. Grant, D.A., et al.: Synthesis of tapped-inductor switched-mode converters. IEEE Trans. Power Elec. 22, 1964–1969 (1997)

Research on Association Rule Algorithm Based on Distributed and Weighted FP-Growth

Huaibin Wang[1,2,*], Yuanchao Liu[1,2], and Chundong Wang[1,2]

[1] Key Laboratory of Computer Vision and System (Tianjin University of Technology)
[2] Tianjin Key Laboratory of Intelligence Computing and Novel Software Technology Tianjin University of Technology, 300191, Tianjin, China
Michael3769@163.com

Abstract. In this paper a distributed weight mining algorithm is proposed based on FP-growth. As an important part of network fault management, the association rule takes effect on eliminating redundant alarms and preventing alarm storm. In traditional association rules the importance of each item is seen as equivalent during mining which is not realistic. By considering the different weights of the items, the AHP approach is introduced in the paper. Without any candidate generation process FP-growth performs well in mining alarm records. The distributed architecture of master and slave site can effectively reduced the complexity of the algorithm. The experimental results and comparison with other algorithms prove the validity of this proposed algorithm and good performance of decreasing the run-time.

Keywords: Association Rule, Distributed, Weighted, FP-growth.

1 Introduction

With the increasing expansion of the scales of communication networks, Modern telecom system becomes more and more complicated and provides more new functions. It inevitably generates a large number of alarms. The most destructive factor to the network is alarm storm. Alarm correlation analysis can help network administrators to delete redundant alarms, locate faults and prevent alarm storm.

In the correlation analysis, the importance of each item in database is often seen as equivalent during in the traditional association rules. Actually the algorithm has two assumptions: (1) the item in database has the same nature and role, (2) the distribution of item in database is the uniform distribution. But in the real mobile network, the important of each item is often different [1]. For example, a critical alarm has more destructive effect to the system than normal alarm of a hundred times. Therefore the importance of each alarm should be considered. The communication network is a distributed surrounding and the alarm information is distributed stored in database. So the research of distributed weighted association rules is consistent with the characteristics of communication network.

* Corresponding author.

D. Jin and S. Lin (Eds.): Advances in MSEC Vol. 1, AISC 128, pp. 133–138.
springerlink.com

2 Related Works

2.1 The AHP Approach

Alarm information with different properties or different levels means that the records should not be treated equally while mining association rules. To solve this problem, assigning weights for alarm information is considered. Weights of alarm information can directly reflect the relative importance of the alarm. It has significant impact for data mining.

As a method of decision-making, the AHP method has been widely used in various disciplines [2].

While using the AHP method, it can be roughly divided into the following steps:

Step 1, establishing a hierarchical structure of the system according to the relationship between various factors.

Step 2, generating the judgment matrix while comparing the importance of elements on the same level.

Step 3, calculating the relative value of each element relatively to the upper standard.

Step 4, calculating the synthetic weights of each elements towards to the system target.

Step 5, the consistency check. Sorting criterion in the calculation of a single weight vector must also be consistency test.

The hierarchical structure of alarm weights is shown in Fig.1.

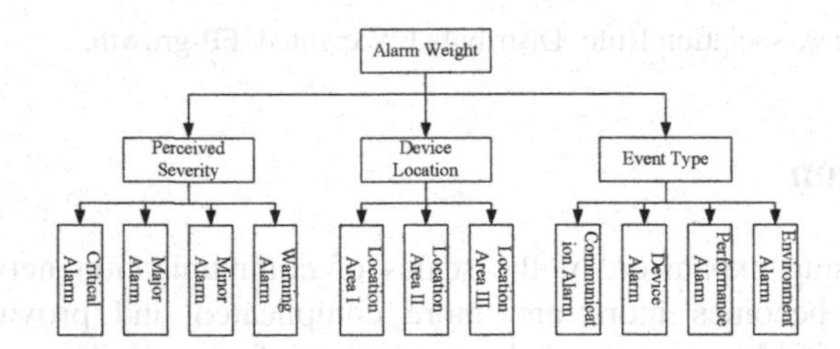

Fig. 1. Hierarchical Structure of Alarm Weights

2.2 The FP-Growth Algorithm

The FP-growth algorithm was put forward by Han. It generates frequent item sets without any candidate generation process based on a prefix tree representation of the given database of transactions. The algorithm can be divided into two phases: the construction of FP-tree and mining frequent patterns from FP-tree [3].

(1) Construction of FP-tree

The construction of FP-tree requires two scans on transaction database. First scan accumulates the support of each item and then selects items that satisfy minimum support. The items are sorted in frequency descending order to form F-list. The second scan constructs FP-tree.

While non-frequent items are stripped off, the transactions are recorded according to F-list. Recorded transactions are inserted into FP-tree. The order of items is important because in FP-tree itemset with same prefix shares same nodes. If the node corresponds to the items in transaction exists the count of the node is increased, otherwise a new node is generated and the count is set to 1.

(2) FP-growth
FP-tree and the minimum support is the input of FP-growth algorithm. To find all frequent patterns whose support are higher than minimum support, FP-growth traverses nodes in the FP-tree starting from the least item in F-list. The node-link originating form each item in the frequent-item header table connects the same item in FP-tree. While visiting each node, FP-growth also collects the prefix-path of the node. FP-growth also stores the count on the node as the count of the prefix path [4].

The FP-growth create small FP-tree from the conditional pattern based conditional FP-tree. The process is recursively iterated until no conditional pattern base can be generated and all frequent patterns are discovered. The same iterative process is repeated for other frequent items in the F-list.

3 Alarm Association Rules

Here is briefly recall the problem of frequent item set mining. Let I = $\{i_1, ..., i_m\}$ be the set of items in a database DB consisting of transactions T = (tid,X) where tid is transaction identifier and $X \subseteq I$. Suppose: W= $\{w_1,..., w_m\}$ is set of weights of each items. The weights can be calculated by the AHP method introduced in 2.1 [5].

According to the description of classical association rules algorithm, the weight support is defined as follows:

$$\text{wsupport}(x) = \text{weight}(x) \times \text{support}(x), \tag{1}$$

$$\text{support}(x) = \frac{\text{support - count}(x)}{|DB|} \times 100\%, \tag{2}$$

It is the classical support, |DB| is the number of transactions, while support-count(x) is the number of transactions including x. The weight confidence of rule x=>y is

$$\text{wconfidence}(x \Rightarrow y) = \frac{\text{wsupport}(x \cup y)}{\text{wsupport}(x)} \times 100\% \tag{3}$$

To analyze alarm more effectively, a good model for data mining of alarm is established, as shown in Fig 2.

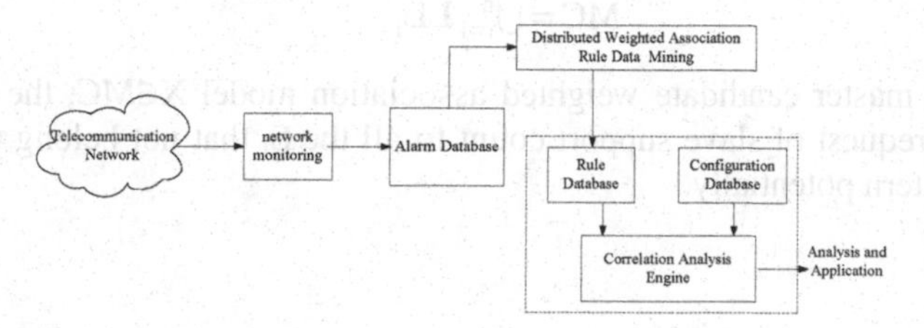

Fig. 2. Model for Data Mining of Alarm

Here are some definitions of symbols used in the algorithm shown in Table 1.

Table 1. Meaning of Symbol

Symbol	Meaning	Symbol	Meaning
DB_i	the slave database of site S_i	$Tree_i$	weighted association pattern tree in S_i
SP_i	slave weighted potential 1-pattern in S_i	SL_i	slave collection of association pattern in S_i
MCP	master collection of candidate potential 1-pattern	ML	master collection of candidate association pattern
MP	master collection of 1-pattern	MR	master collection of association rule
MC	master candidate associa-tion pattern		

The proposed algorithm in the slave site of the operation as follows [6]:

(1).By scanning the slave database DB_i, it gets the slave weighted 1-pattern potentially SP_i and its corresponding support counter. If SP_i=0, the slave mining would be finished. But it still responds to the request of queries of slave support counter of master site.

(2) To any X∈SP_i, X and its slave support counts X should be sent to the master site.

(3) S_i receives the request of queries of slave support counter send by master site. The queried object is collections of master candidate and weighted 1-pattern belongs to MCP but not SP_i included. If P =0, Then jump to step (5).

(4) While re-scanning the slave database DB_i, calculate the support counter of 1-pattern in slave. Then the slave support count will be sent to the master site.

(5) If it receives MP broadcasted by the master site, S_i receives messages sent by the master site. Then the slave pattern tree is generated according to MP and the potential 1-pattern is updated $SP_i = SP_i \cap MP$. Otherwise, if the site receives messages of ending mining sent by the master, the slave mining should be finished.

The proposed algorithm in the master site is described as follows:

(1) Receiving the slave weighted potential 1-pattern and its corresponding support counter submitted by the slave site S_i. Then it generates a master candidate weighted 1-pattern potentially.

$$MC = \cup_{i=1}^{n} LL_i \tag{4}$$

(2) For any master candidate weighted association model X∈MC, the master site sends a query request of slave support count to all the S_i that not belong to the slave weighted 1-pattern potentially.

(3) According to the support count returned by the slave site, thc master support count of X is calculated. If the weighted support of X meets the requirements of wminsupport, X will be added to the master weighted association pattern ML.

$$ML=ML+X \tag{5}$$

(4) With the master weighted association pattern, a master weighted association rules MR is generated.

(5) The master weighted association rule is over, and the master site informs the slave site that the mining is end.

This algorithm uses a hierarchical structure. Each site does not need data exchange which can effectively reduce the amount of data traffic. Furthermore the FP-Tree is used in generating slave frequent itemsets which do not generate candidate itemsets. Slave site only need to manage data on the site, it effectively reduced the redundancy calculating.

4 Experimental Analysis

To evaluate the performance of the proposed algorithm, several experiments have been performed. The program was written in Java 6.0 and run with Windows XP operating system on a Pentium Dual Core 1.8GHz CPU with 2GB memory. One computer is the master site while two computers are the slave ones. The 100000 alarm records come from the local mobile communications. The algorithm compared with the classical FP-growth algorithm. The run-time and the number of frequent pattern are selected to evaluate the two algorithms. The wminsupport is set as 0.25% [7].

Fig.3 shows the comparison of the two algorithms. While wminsupport added, the run-time of two algorithms is decreased rapidly. But the proposed algorithm is decreased more effectively and the distinction is more evident when the value of wminsupport is small.

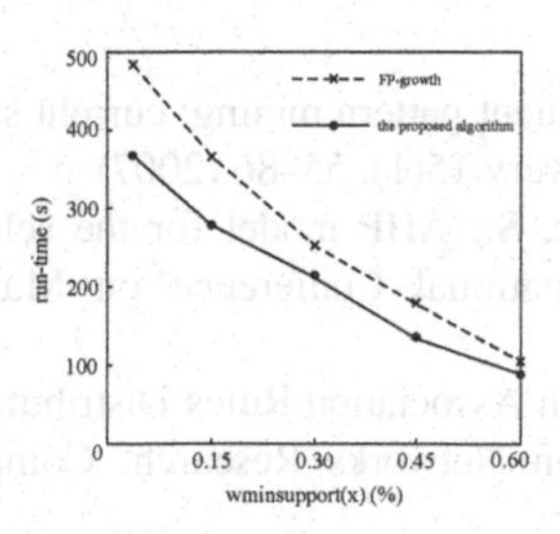

Fig. 3. Comparison of two algorithms with different wminsupport

Fig.4 shows the comparison of the run-time with different number records. It is clearly that the run-time is linearly varied with the increasing of records. When the number of records is higher, the proposed algorithm is better than FP-growth.

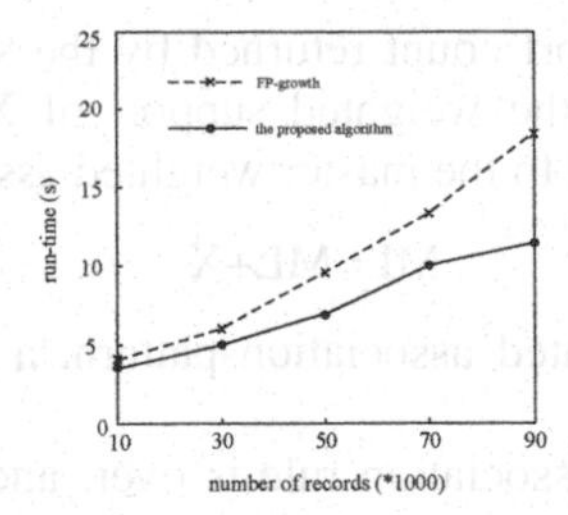

Fig. 4. Run-time of two algorithms with different records

5 Conclusion and Future Work

Applying the mining association rules to network fault management system can make good use of alarm information and prevent alarm storm. In this paper we present a distributed weight mining algorithm based on FP-growth. The paper combined three approaches which enhanced the efficiency of data mining. The AHP approach is introduced to mark the importance of different item. The FP-growth algorithm without any candidate generation process combined distributed architecture of master and slave site effectively reduced the complexity of the algorithm. Experimental results and comparison with other algorithms prove the validity of this proposed algorithm.

Since the alarm database is constantly updated, it is unrealistic for mining when database updates each time. Therefore, the future work should focus on improving the algorithm for incremental frequent-pattern mining.

Acknowledgement. This work was supported by the Foundation of Tianjin for Science and Technology Innovation (No.10FDZDGX004 00&11ZCKFGX00900), Education Science and Technology Foundation of Tianjin(No.SB20080053 & SB20080055).

References

1. Han, J., Cheng, H., et al.: Frequent pattern mining: current status and future directions. Data Mining and Knowledge Discovery 15(1), 55–86 (2007)
2. Sari, B., Sen, T., Engin Kilic, S.: AHP model for the selection of partner companies in virtual enterprises. In: International Conference on Manufacturing Research, vol. 38, pp. 367–376 (2008)
3. Jian, W., Ming, L.X.: Based on Association Rules Distributed Mining Algorithm for Alarm Correlation in Communication Networks Research. Computer Science 36(11), 204–207 (2009)
4. Silva, A., Antunes, C.: Pattern Mining on Stars with FP-Growth. In: Torra, V., Narukawa, Y., Daumas, M. (eds.) MDAI 2010. LNCS, vol. 6408, pp. 175–186. Springer, Heidelberg (2010)
5. Yun, U.: An efficient mining of weighted frequent patterns with length decreasing support constraints. Knowledge-Based Systems 21(8), 741–752 (2008)
6. Zhang, K.: Application of Based on Association Rules Data Mining in Telecommunication Alarm Management. University of Electronic Science and Technology, Chengdu (2006)
7. Li, H., Wang, Y., Zhang, D., et al.: PFP: Parallel FP-Growth for Query Recommendation. In: Proceedings of the 2008 ACM Conference on Recommender Systems, pp. 107–114 (2008)

The Research of PWM DC-DC Converter Based on TMS320F28335

Jinhua Liu[1] and Xuezhi Hu[2]

[1] College of Mechatronics and Control Engineering
Hubei Normal University, Huangshi Hubei, China
hbhsljh999@sina.com
[2] School of Electric and Electronic Information Engineering
Huangshi Institute of Technology, Huangshi Hubei, China
tlmj6688@163.com

Abstract. Transformer original edges circulation of traditional phase-shifted full bridge ZVS PWM converter caused larger loss of conduction and reduced the system efficiency. In this paper a phase-shifted full bridge hybrid ZVZCS PWM converter is proposed to overcome these weakness. The realization of leading-leg is based on the non-mutation theory of capacitor voltages. And the realization of zero current (ZCS) switching of lagging-leg benefits from the saturated inductance which makes the original edge current reset to zero during transformer original edge voltage cross zero and clamp in zero current. This paper presents the operational principle and implementation process of the proposed coverter. Finally, the experiment results verify the feasibility of the proposed method.

Keywords: Converter, Full-bridge DC-DC converter, Zero voltage and zero current switches.

1 Introduction

The phase-shifted full bridge ZVS DC/DC converter is a kind of phase-shifted full bridge converter with excellent performance , which has got extensive research and application in the high power DC/DC converter. It makes the power switch tube realize turn-on and turn-off of the soft switch through the phase shifting control mode , reduces the switch loss and the current and voltage stress of switch tube and has high efficiency; But it exists the weakness that ZVS of the lagging-leg doesn't easily satisfy ZVS conditions as a result of the limit of load range. In order to solve this problem, this paper proposes an ZVZCS PWM converter, which removes parallel capacitors of the lagging-leg in main circuit of the traditional phase-shifted full bridge ZVS DC - DC converter and adds saturated inductance and blocking capacitance . Its feature is that the leading-leg use the non-mutations theory of capacitor voltages to realize zero voltage (ZVS) switching, and zero current (ZCS) switching of the lagging-leg uses saturated inductance during transformer original edge voltage across zero to make the original edge current reset to zero and briefly clamp in zero current. This paper analyzes the operational principle of converter and TMS320F28335 is used

D. Jin and S. Lin (Eds.): Advances in MSEC Vol. 1, AISC 128, pp. 139–144.
springerlink.com

as the main control chip to realize the closed-loop control of the system and completes the transformation of digital realization. the experiment done presents the main circuit parameters and experimental result,and also proves the correctness of the related theory of the converter.

2 Converter Basic Working Principle

Figure 1 shows full bridge converter schematic diagram of phase-shifted full bridge ZVZCS PWM DC/DC ,the circuit adopts phase shifting control mode, which VT_1 and VT_4 constitute the leading-leg , VT_2 and VT_3 constitute the lagging-leg. C_1 and C_4 is VT_1 parasitic capacitance and VT_4 external capacitor respectively and $C1 = C_4$, TR vice party and DR1, DR2 constitute full wave rectifying circuit, L_f and C_f constitute output filter. The leading-leg uses the non-mutations theory of capacitor voltages to realize zero voltage (ZVS) switching; Transformer original edges in main circuit cascade a DC-blocked capacitor C_b , which can avoid the phenomenon of transformer saturation caused by DC biasing for the reason of asymmetry of device characteristics on the one hand , and make the lagging-legs realize zero current (ZCS) switching with saturated inductance L_s on the other hand ; VT_1 and VT_3 advance VT_4 and VT_2 respectively a phase, namely phase-shifting angle , and we can adjust the output voltage by adjusting the size of phase-shifting angle . Here, assume that all of switch tubes and diodes devices are ideal and the inductance of saturated inductance is infinite before saturation and zore after saturation . The circuit topology in half cycle are divided into six working patterns, the circuit waveform is showed in figure 2 .

Mode 1 $[t_0 \sim t_1]$: VT_1 and VT_4 conduct, energyof this stage transmits to load through the transformer, at this stage, the voltage of inductor saturation and blocking capacitance C_b increases linearly from negative maximum ; When VT_1 turns off , the operating mode enters into mode 2 $[t_1 \sim t_2]$ at the moment of t_1 , the original edge current transfers to the C_1, C_3 branch , charges C_1 and discharges C_3 at the same time ,and U_{AB} declines linearly. The voltage of switch VT_1 rises, VT_1 is a soft shutoff ; When the voltage of C_1 rises up to U_{in} and the voltage of C_3 drops to zero, D_3 conducts naturally ,voltage clamp of VT_3 is in zero and VT_3 realizes zero voltage turning-on. The converter enters into mode 3$[t_2 \sim t_3]$, from the moment of t2 U_{AB} is clamped in zero, at this moment because saturated inductance Ls is still in the saturated state, the voltage of blocking capacitance C_b will completely add to both ends of the transformer leakage inductance, and the transformer leakage inductance is very small, so the original edge current IP will decrease linearly ;When IP drops to zero, the operating mode enters mode 4 $[t_3 \sim t_4]$,at the moment of t_3 because saturated inductance L_s has quit saturation, it stops the reverse growth of the current . At this stage, blocking capacitor voltages will be constant and add to both ends of saturated inductance completely, there is no current in VT_4 . When VT_4 is turned off , the operating mode enters into the mode 5 $[t_4 \sim t_5]$, at this stage, I_P is still zero, VT_4 is turned off so that the ZCS can be achieved.. Enter into mode 6 $[t_5, t_6]$,at circulation end t_5, because of the existence of saturated inductance L_s and the non-mutation of original edge current , VT_2 can also achieve ZCS.

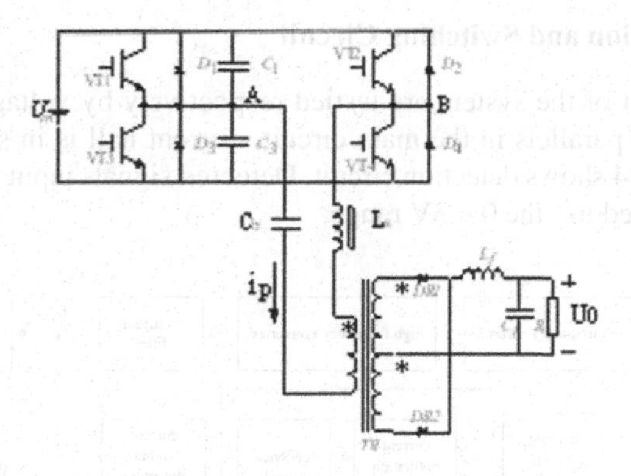

Fig. 1. ZVZCS full bridge converter topology graph

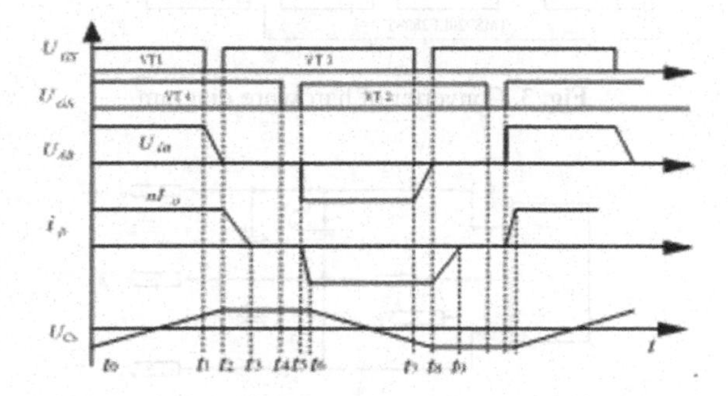

Fig. 2. ZVZCS full bridge converter working waveform

3 System Control Strategy and Digital Realization

3.1 The System Hardware Structure

This converter uses TMS320LF28335 as the controller chip to realize digital control of the converter. The hardware composition of converter is showed in figure 3. As the figure shows ,the load voltage and current are detected by detection circuit and translate into the corresponding voltage and current signal and go to filter, then the feedback signals enter into digital signal processor (DSP) and do double closed loops control operation, and the inner and outer loops of this system use PI adjusters, the digital PI adjuster based on the given value and the feedback signal does deviation adjustment, so the outcome determines the size of the PWM waveform phase-shifting angle between the leading-leg and the lagging-leg and the controlled makes track for the quantitative , and the driving signal sent out by DSP is converted by the level translator ,then the signal is sent to driver chip EXB240 ,eventually becomes IGBT driver signal.

3.2 Signal Detection and Switching Circuit

Voltage and current of the system are testied respectively by voltage hall and current hall , voltage hall parallels in the main circuit, current hall is in series and tests in main circuit, figure 4 shows detection circuit. Detected signals input to a DSP ADC and should be converted to the 0 ~ 3V range.

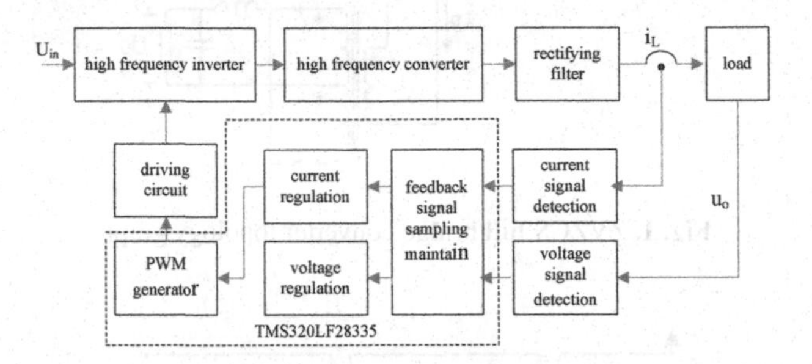

Fig. 3. Converter of hardware diagram

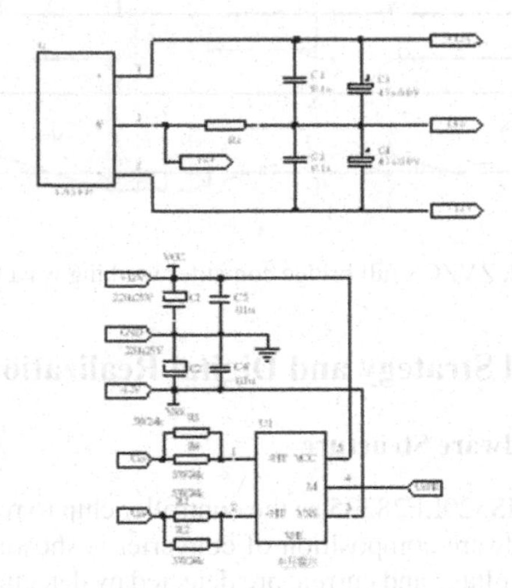

Fig. 4. Detection circuit. (a) Current detection circuit (b) Voltage detection circuit.

3.3 System Software Design

System software can be divided into two parts, namely main program and interrupt service routine. The main program mainly completes system initialization and sets various function modules for TMS320F28335 and the workings of system function modules ;The switch is detected , and then the mode enters the main program and cycles and waits for interrupt. The interrupt service routines include cycle interrupt routine and underflow interruption program, etc. The cycle interrupt routine completes the reading of voltage and current sample value and digital filtering, starts A/D conversion and A/D calibration, and adjusts operation program to implement control algorithm ,eventually produces PWM signal. If the system malfunctions, the external hardware generates signals to blockade pulse amplification and shaping circuit and the signals are sent to DSP to produce interrupt and blockade pulse output.

4 The Analysis of Experimental Results

According to the theoretical analysis, we have designed the experiment circuit. Main parameters of the experiment circuit : DC input voltage, U_{in} = 180V; DC output voltage,U_0 = 60V; f = 40kHz ; I_0 = 30A ; Transformer former vice edge turn ratios, N_1 = 25,N_2 = 10 ; Shunt capacitance, C_1 = C_3 = 10nF; Saturated inductance, L_s = 3mH; blocking capacitance, C_b = 3 u F; Output filter inductance, L_f = 1.8 mH, Output filter capacitance, C_f = 22000uF. The main circuit uses PM200DSA120 as the main switch device, the control device adopts TMS320F28335 chip and the system adopts voltage and current dual-loop control . Figure 5 shows the leading-legs IGBT soft switch waveform in the case of 20A load . Channel 1 represents the driver signal waveform of the leading-legs IGBT ,and channel 2 represents the tube voltage drop waveform of the leading-leg IGBT . From the graph , we can see that the tube voltage drop of the leading-legs switch tube has been zero and realized ZVS before the driver waveform rising edge arrives. Figure 6shows the lagging-leg ZCS experimental waveform with the output current being 25A . Channel 1 shows the driver waveform of the lagging-leg switch tube and channel 2 the transformer original edge current waveform. As can be seen from the graph, when the switch tube is turned off, the original edge current has basically been zero and realized ZCS; When the switch tube is turned on, the original edge current goes up again after a certain delay and realizes zero current turning-on.

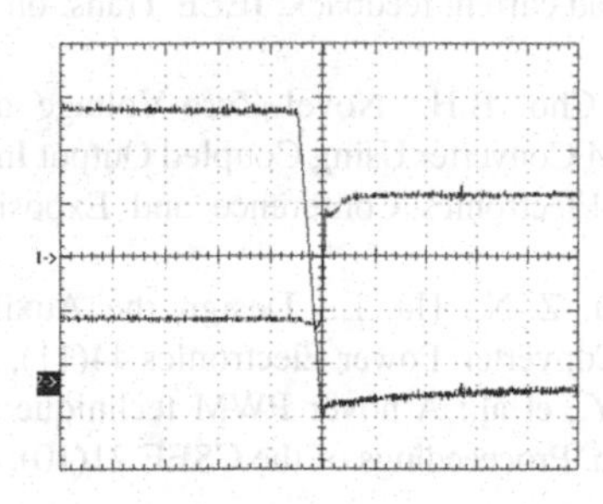

Fig. 5. Leading-legs ZVS IGBT soft switch waveform. (Channe 1: the horizontal axis: the 2 us/div longitudinal axis: 10 V/divChannel 2: the horizontal axis: the 2 us/div longitudinal axis: 50V/div)

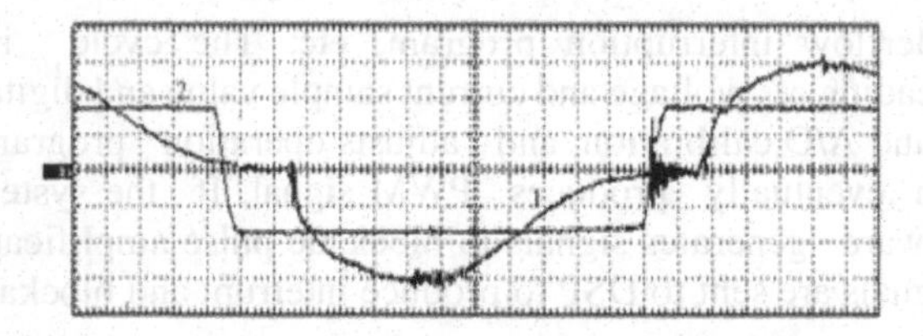

Fig. 6. Lag- legs ZCS experimental waveform.(Channe 1: the horizontal axis: 2 u s/div longitudinal axis: 2A/div;Channel 2: the horizontal axis: 2 u s/div longitudinal axis: 6V/div)

5 Conclusion

This paper puts forward a new phase-shifted full bridge ZVZCSSDC/DC converter and analyzes its basic principle, and we has built the experiment system and the experimental results shows the control circuit of soft switch DC/DC converter using TMS320F283352 as the main control chip increases the stability of output voltage , reduces the degree of the output voltage waveform distortion , realizes effectively ZVS of the leading-leg switches and ZCS of the lagging-leg switches and obtains the high quality control waveform; In addition, from this experimental implement ,we can see that digital control is more intelligent than analog control ,and the design of regulator is more convenient..

References

1. Sun, T., Wang, G., Tang, P., et al.: Novel full-bridge phase-shift ZVZCS PWM DC-DC converter based on DSP control. In: Proceedings of the CSEE, vol. 18(9), pp. 46–50 (2005)
2. Chenjian Power Electronics-power Electronics Transformation and Control Technology. China Higher Education Press, Beijing (2002)
3. Zhao, Q., Xu, M., Lee, F.C., et al.: Single-switch parallel power factor correction AC-DC converters with inherent load current feedback. IEEE Trans. on Power Elec. 19(4), 928–936 (2004)
4. Choi, H.-S., Kim, J.W., Cho, B.H.: Novel Zero-Voltage and Zero-Current-Switching (ZVZCS) Full-Bridge PWM Converter Using Coupled Output Inductor. In: Sixteenth Annual IEEE on Applied Power Electronics Conference and Exposition, APEC 2001, vol. (2), pp. 967–973 (2001)
5. Qian, L., Tao, G.-S., Zhu, Z.-N., Hu, L.: Design the Auxiliary Circuit for a ZVZCS Phase-shifted Full-bridge Converter. Power Electronics 44(11), 30–32 (2010)
6. Peng, L., Lin, X., Kang, Y., et al.: A nover PWM technique in high-frequency converter controlled by digital system. Proceeedings of the CSEE 21(10), 47–51 (2001)

The New Method of DDOS Defense

Shuang Liang

Polytechnic School of Shenyang Ligong University, Liaoning, Fushun
ls_happiness@163.com

Abstract. Currently, IP tracking based on packet marking and attacking package recognition technology is one of the main means for effective against distributed denial of service attack. This paper has proposed a defense new method based on determined package marking, which increased Tacking Server, altered EPS coding in subnet, tracked and identified attack package by Border Router. Experimental results have shown that the method has such advantages that tracking the large number of attack sources, no false alarm, identifying the attack packets, tracking single package and effective protecting the network topology etc..

Keywords: DDOS, Attacking, Defense, Packet Marketing, Certainty, IP Tracking, Attack Packet Identification.

1 DDOS Attack Principle and Prevention

1.1 DDOS Attack Principle

DDOS, which is Distributed Denial of Service, DDOS attack is the further development of DOS attack [1]. DoS attack is to make the target computer as a lot of system resources are occupied, resulting in system cannot run properly, so that external services failed. DoS attack has a variety of ways, but the most basic way is through a reasonable request to make the computer a large number of service resources are occupied, however , legitimate users cannot get effective services [2].

It can be seen from Figure 1, DDoS attacks mainly be divided into three layers: the attacker, the host computer and meat machine. The three layers play different roles in the attacks.

(1) The Attacker: The attacker is the attack source during the attacking process, which is responsible for handling all the attacking process, is responsible for issuing various control command for attacking to the host computer. It could be any host on the network, even a notebook PC.

(2) The Host Computer: The host computer s are computers that were invaded and was controlled illegally by the attcker. There computers were installed a specific application software in some ways, the computers can accept commands sent by the attack, and these instructions can be sent to the meat machine, which can realize the attacking of target.

(3) Meat Machine: Meat machine is also a series of compute which was invaded and controlled by the attacker. The attacker installed the same specific application on it. It

D. Jin and S. Lin (Eds.): Advances in MSEC Vol. 1, AISC 128, pp. 145–151.
springerlink.com

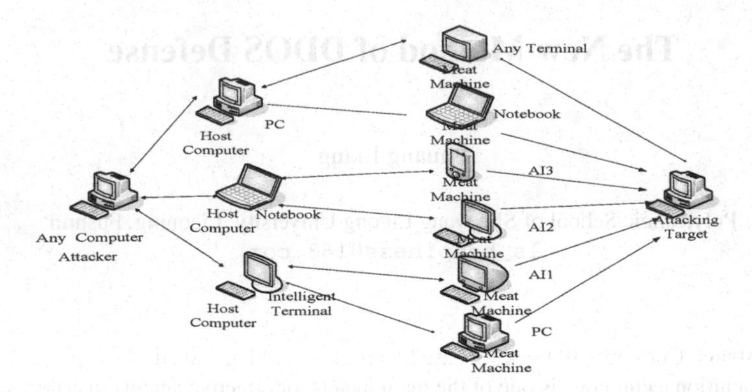

Fig. 1. Attack Principle

can accept and run the command sent by the host computer, which has realized the attacking for the target computer. As far as I am concerned, "If there are more meat machines which are controlled by the attacker, his attacking team will be bigger, the impact will be even stronger to the target."

In the course of the DDOS attack, the attacker is not directly attack the target computer, but use other computers to complete the task, so it will not be tracked and monitored during the attack process, his identity is not easy to find.

1.2 DDOS Common Attack Methods

Denial of service attacks mainly have the following ways:

① Attacker controls the meat machines to generate lots of useless data through the host computer continuely, resulting in network congestion, so that the target machine cannot properly communicate with the outside world since the data transmission channel is occupied.

② The defects of using service by the targetcomputer,the attacker repeated high-speed issued the fixed services request to the target computer. Leading to the target computer is busy dealing with some useless service requests, however, some of the normal service request does not work in time.

③ Attacker uses the defects of target host about providing services in data processing way, keep sending invalid data to the target host service program causes the target host run error, which a lot of system resources have been occupied, even bring about a crash situation.

Overall, DDOS attack is a method to damage the target host network services, in fact, DDOS fundamental aim is to target host or network lost ability of receiving and processing external service request in time , so that normal service requests cannot get response.

1.3 The Common Defense Measures of DDOS

Effective means to prevent DDOS attacks mainly includes three aspects, namely using network intrusion detection tool to detect the operation of network, using the router defense and using the firewall defense [2]. I believe that preventive measures in the above 3, the intrusion detection tools is lagging behind, the firewall defense need additional settings tools, so the best defense is the router. The usually router defense achieves the purpose of DDoS defense through the router's access control and service quality settings. The IP tracking technology based on packet marking and attack packet identification technology which are in current research focus[3]. They use the features that routers connect the entire Internet and have proposed the means to prevent DDOS attacks.

IP tracking technology based on packet marking , the router marks the IP data packets when data packets from one to another network. Through such IP tag, we can easily know the data packets from which sub network ,so can to quickly identify the host sending the packet. The DDOS attack to defense, the main method is to find out the origin of the package, thus shielding them from the address to send packets ,for achieve the purpose of protection of other hosts. IP tracking technology based on packet marking can track well the birthplace of the packet[4].

Attack packet identification technology is mainly used to add the packet mark to identify the different sources of data packets[5].

Most existing schemes use the router's IP address as the tag content, this marker is simple and effective, but the biggest problem is the attacker is very easy to get the structure of entire network through the IP address, thus bringing the entire network in front of the attacker.

It aims at the existing problems of IP tracking technology based on packet marking. This paper has proposed a new router defenses - Tracking Packet Marking (TPM).Through the ingress router marks the packets and signature packets can easily to find and protect against DDOS attacks.

2 The New Method of Router Defense

2.1 Network Divided into Several Subnets, an Ingress Router to Determine the Packet Marking

According to the router playing the different role in data forward, routers can be divided into two categories: edge routers and intermediate routers. According to the needs described in this article, first introduced the following definition:

Definition 1: Border Router (BR): They are the router in the entire network or a subnet boundary.

Definition 2: Intermediate Router(IR): They are the routers In addition to border router.

Definition 3: Entry Point (EP): Packets from one subnet to another subnet go through the first router.

Definition 4: Entry Point Signature (EPS): The packets add the uniquely entry point identity string in entry point (instead of the usual IP address).

Definition 5: Tracking Server (TS): They are responsible for tracking the request authentication and tracking services within a subnet, as well as the collaboration between subnets.

Description: Tracking packet marking only execute the determined packet marking on EP, which is EPS, it don't process any intermediate routers. Once EPS is confirmed, the EP identification information string is not changing during the entire data transfer process. That is the signature is not only the basis of tracking data, but also the basis of data packets identification.

The basic principles of tracking packet marking method shown in Figure 2:

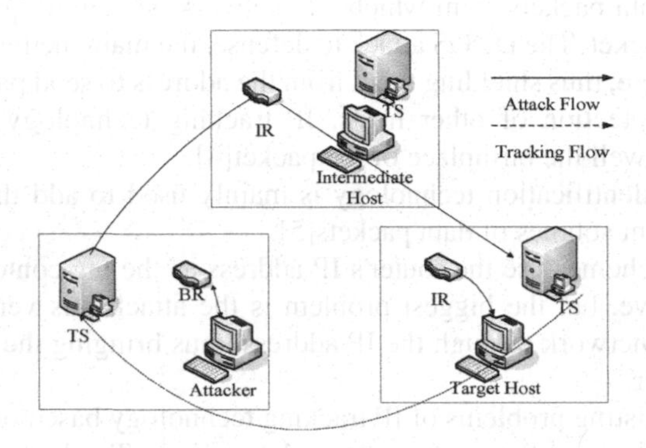

Fig. 2. The Basic Principles of TPM

Since each host will record border router in EP when it request to send data packet, so once they have been found to attack the target host, they can send characteristics attack packet with EPS to the TS, as long as TS validate host legitimate, TS can perform ingress router anti-identification method, the border routers quickly are found, to find further the attack host.

2.2 EPS Encoding Change to Adapt to the New Router Defenses

In order to uniquely identify each router, the easiest way is to divide the network into subnets, recording each border router. According to authorities CAIDI measurement statistics, the vast majority of border router subnet number are less than 1000, only a small part of large subnet, its border routers generally do not exceed 2000. According to this number statistics, we choose 11 to express 2048 router address (with RouterID). In order to adapt to the larger network, it can set aside a part of the router address, so we can choose 13 to describe each subnet router address. A non-zero 13-bit router ID can express 8192 router address, such scope is often enough for a common network.

RouterID can uniquely identify the router in subnet, but how to distinguish different subnets, you must define an identifier for each subnet, which is the subnet ID, which can be reflected by EPS. We define a 16-bit ID to describe the entry point signature identification (EPSID) for each subnet EP.

EPSID can ensure that any one host cannot receive the tag in package to determine IP address, thereby have protected the confidentiality of network topology. EPS encoding shown in Figure 3:

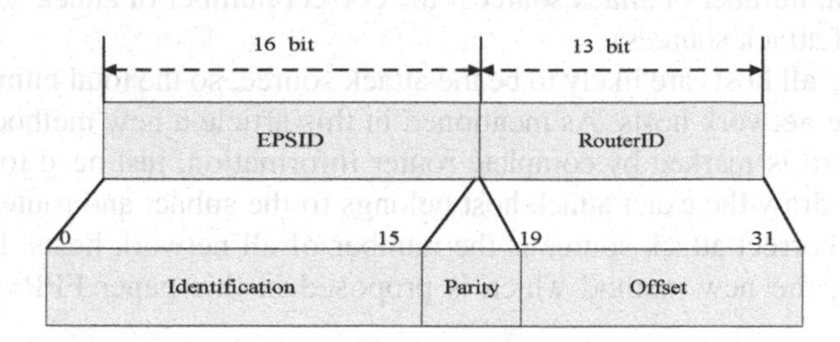

Fig. 3. EPS code

16 bits express EPSID,13 bits express the offset of RouterID.Were signed with the 16bits entry point that with the 13bits ID and the offset that RouterID, the Parity of 3bits express identification check in them, it may verify whether the information that TS stored is correct.

There are the correspondence between RouterID and IP address in TS, which can simply realize tracking query.

2.3 Ingress Router Anti-identification Methods

When data packets are from one network to another network through the entry point, the entry point respectively write the packet's RouterID and EPSID into RouterID and EPSID fields. Packets across the network transmission process, the intermediate routers don't make any mark about the packets. Entry point marking algorithm is as follows:

For each incoming Packet P
P.ESPID=Q.ESPID
P.RouterID=Q.RouterID

Entry point address reduction is very simple. Because each tag data packet has recorded the complete router information, as long as an attack packet can be get, it can be clearly seen from the packet where the subnet EPSID and his RouterID, which can easily be drawn the attacker's IP address.

2.4 New Methods of Defense Processes

The target host can judge the attacks from the same subnet or from other subnets according mark packets' EPSID field. If the attack comes from the internal network, TS can directly find the attack host according to the mapping table which saved router and its IP address, accordingly to take some defensive measures about the host. If the attack comes from the other network, TS need to inform TS which is from attack source the entry points RouterID, at the same time it is responsible for the defense of corresponding entry point.

3 Experimental Testing

(1) FalseReport Rate

FRR=(The total number of attack sources- the correct number of attack sources)/(The total number of attack sources)

In a network, all hosts are likely to be the attack source, so the total number of attack source is all the network hosts. As mentioned in this article a new method of defense, the packet record is marked by complete router information, just need to get a single packet, we can draw the exact attack host belongs to the subnet and router address, so the number of correct attack source is the number of all network hosts. Based on the above analysis, the new method which is proposed in this paper FPR=0, that is no false reports.

(2) The maximum number of tracking attack sources (N)

In this paper, 16bits identifies EPS, which is used to distinguish the various subnets with 16bits, 13bits identifies the router ID. That is this method which proposed in this article can theoretically track the number of $2^{16} * 2^{13}$ attack sources, and this number in real life we can be close to any large-scale DDos attack.

The proposed method in Reference 3 is called DPM, which is a very classical deterministic packet marking method. The proposed method in Reference 1 is called DPM-RD, which is an improved method based on classic deterministic packet marking method.

Table 1. Comparison between TPM and other deterministic packet marking method

scheme	FRR	N	Packet Identification	Single Package Tracking	Confidentiality of Network Topology
DPM	≤0.01	2048	no	no	no
DPM-RD	≤0.08	2048	no	no	no
TPM	0	2^{29}	yes	yes	yes

As can be seen from the table, regardless of from the false report rate of the packet or the maximum number of tracking data source of view, the proposed method are significantly better than other methods. In addition, this method also has a data packet identification, single packet track and effective protection the confidentiality of network topology and so on. It is shown that the new method which is proposed in this paper has some practical significance.

4 Conclusions

Aimed at DDOS attack, this paper has proposed a new defense method based on determined packet marking, the method is that the network is divided into several subnets, increase TS in each subnet, change the EPS encoding, and track and identify data packet through the border router. Experiments have shown that the method has such advantage as tracking a large number of attack sources, no false report rate, realizing attack packet identification, single packet tracking and effective protection confidentiality of a network topology, etc.. The method has some promotional value.

References

1. Yang, B., Hu, H., Guo, S.: Cost-oriented task allocation and hardware redundancy policies in heterogeneous distributed computing systems considering software reliability. Computers & Industrial Engineering 56(4), 1687–1696 (2009)
2. Herzog, P.: Open-source security testing methodology manual [EB/OL] (December 10, 2009), http://isecom.securentled.com/osstmm.es.2.1.pdf
3. Jin, G.: The Research of DDOS defense technology based on data packet mark. Zhejiang University, Hangzhou (2008)
4. Savage, S., Wetherall, D., Karlin, A., et al.: Network Support for IPTraceback. ACM SIGCOMM Computer Communication Review 30(4), 295–306 (2000)
5. Belenky, A., Ansari, N.: IP Traceback with Deterministic PacketMarking. IEEE Communication Letters 7(4), 162–164 (2003)

Research of Blind Signal Separation Algorithm Based on ICA Method

Xinling Wen and Yi Ru

Zhengzhou Institute of Aeronautical Industry Management,
450015 Zhengzhou, China
wenxinling@zzia.edu.cn

Abstract. This paper studies a kind of blind source separation algorithm-independent component analysis (ICA) and the application in blind source signal separation. First of all, we introduce the blind source separation in theory, and then introduce the specific algorithm analysis, as well as analyze the mathematical model and principle of the ICA in detail. This paper discusses the different independence standards and the existing ICA several main algorithm, focus on the Fast ICA algorithm. Fast ICA algorithm is based on the maximum principle of non-gaussian and adopt Newton iterative algorithm, which is a fast algorithm in ICA and has wide application prospect.

Keywords: BSS, ICA, projection tracking, negative entropy, Fast ICA.

1 Introduction

In real life, we observed signals are often unknown multi-input and multi-output linear system signal. If having no other priori information, only according to the independent character of source signal, and using statistical observations to separate the mixed signal through the source signal apart, which called blind source separation (BSS). The term 'blind source' has two meanings: (1) the source signal cannot be observation; (2) how to mix the source signal is unknown. Obviously, it is hard to set up the math model when from the transmission properties between source and sensor, or prior knowledge can't get, the blind source separation is a very natural choice. the core of Blind source separation (or solution mixed) is matrix learning algorithm, its basic idea is statistically independent feature extraction acting as input expression, and the representation of information is not lost.

Blind source separation is a powerful signal processing method. Blind source separation of the research in the early 1980's, due to the development and application of blind source separation in wireless communication, image processing, earthquake, sonar, language and the biomedical sciences in the past.

2 Standards Introduction of ICA

Independent component analysis (ICA) essentially is an optimization problem, which is an maximize approximation problem that how to separate by algorithm between the

D. Jin and S. Lin (Eds.): Advances in MSEC Vol. 1, AISC 128, pp. 153–158.
springerlink.com

independent component and each source signal. Independent component analysis mainly includes two aspects: optimization algorithm and optimization criterion (objective function). The right target function has the right choice. Output components statistical independence, in a point of view, the mutual information for zero of the output component of is one of the most basic standard.

2.1 Non-gassian Maximization

According to the center limit theorem, many numbers of the independent random variables and the probability density distribution asymptoticly obey the gaussian distribution. Therefore, the linear superposition distribution of the independent component usually more close to the gaussian distribution than the original distribution of weight, which can act the criterion of independence as the non-gaussian characters of components. Then each component is more independent, the gaussian character is stronger. When the source signal is assumed to be independent and do not have the time structure, which usually adopt the higher order statistics as the measure of the gaussian character. But in this case, the source signal there can be at most only one gauss signals. Now that the gaussian character acts as the measure of independence, there must be a standard to measure the gaussian character. The key of independent component analysis model to estimate is gaussian character. The size of the gaussian character usually can use negative entropy (ngeeniorpy) and kurtosis (kurtosis) to measure. When the data has non-gaussian character maximum, which can obtain the each independent component source, one of the most basic theory is center limit theorem. There are two measurement method about the non-gaussian character:

The Non-gaussian Maximization Based on Negative Entropy

By the method of the information theory, it is known that the entropy value is related to the information amount of observation data, in all having the random variable variance, the gaussian character is stronger, the gaussian distribution information entropy is smaller. Often this means using entropy can measure the gaussian character. Negative entropy is a kind of the information theory amount based on the differential entropy, the normalized of negative entropy is shown as follows formula (1):

$$J(\mathrm{x}) = H(\mathrm{x}_{gauss}) - H(\mathrm{x}) \tag{1}$$

Among it,

$$H(\mathrm{x}) = \int f(\mathrm{x}) \log f(x) dx \tag{2}$$

x_{gauss} is gaussian random variable, which has same covariance with x. It remains the same to x's any linear transformation, this is an important characteristic of negative entropy. Negative entropy usually always the negative, only in x is gaussian distribution, negative entropy just is zero. Usually, in order to simplify the calculation in real application, negative entropy taken an approximate value.

$$J(\mathrm{x}) \propto [E\{G_i(\mathrm{x})\} - E\{v\}]^2 \tag{3}$$

Among them, the v is a gaussian random vector with zero mean and unit variance, the x's mean is zero, and the x is the unit variance. We take $G(\bullet)$ as a quadratic function such as $G_1(\mathrm{u}) = \frac{1}{a_1}\log\cos a_1 u$, $1 \le a_1 \le 2$, or $G_1(\mathrm{u}) = \exp(-u^2/2)$, etc. In the gaussian character measure, it is negative entropy, which is a good compromise between negative entropy and classical kurtosis. The approximate characteristics is quickly in concept calculations, simple and has good robustness.

The Gaussian Maximization Based on kurtosiS (kurtosiS)
The kurtosis to random data x is defined as:

$$\mathrm{kurt}(\mathrm{x}) = E(\mathrm{x}^4) - 3(E\{\mathrm{x}^2\})^2 = C_4[\mathrm{x}^4] \tag{4}$$

It has the linear properties. The gaussian character of observation signal is stronger than the source signal, this is because it is the linear combination by many independent source signal. In other words, the observation signal non-gaussian character is weaker than the source signal , yet the gaussian character is stronger, it is more independent. For the vast majority of the gaussian random variable concerned, the kurtosis have positive or negative, but to gaussian random variable is concerned, their kurtosis is usually equal to zero.In order to get the independent component of non-gaussian, we usually extract each independent component through the maximization or minimize kurtosis serial, and finally find each of kurtosis local maximum. This metric method has very simple features in calculated and the theory, so, it is widely used in independent component analysis and related fields.

2.2 Mutual Information Minimal

K-L degree is the best measure of statistical independence, which making the output signal x (t) as possible independence. Random vector $\mathrm{x} = (\mathrm{x}_1, \mathrm{x}_2, ..., x_n)^T$ of each element of the mutual information between each elements can be expressed as:

$$I(x_1, x_2, ..., x_n) = \sum_{i=1}^{n} H(x_i) - H(x) \tag{5}$$

The mutual information of random variable is a natural measure to the relevance chaaracter. It is always the negative, and mutual information has a very important properties is $I(x_1, x_2, ..., x_n) \ge 0$, when and only when each component of the random vector x is independent, it put a zero. Using the definition of negative entropy, and the component hat not relevant, the expression can be further expressed as:

$$I(x_1, x_2, ..., x_n) = C - \sum_{\mathrm{i}} J(x_i) \tag{6}$$

Among them, C is the constant nothing to matrix, show that a basic relations between the negative entropy and the mutual information. From which we can see, minimization mutual information rough is equivalent to maximizing negative entropy. Because negative entropy is measure non-gaussian character, so, this shows that the mutual information minimization for independent component analysis estimates is equivalent to the maximum sum of gaussian character. In the framework of the mutual information, the authors give the maximization of the principle of the gaussian heuristic explained. By the expression we can further get formual (7).

$$I(x_1, x_2,..., x_n) = -\sum_i E\{\varphi(x_i)\} - \log|\det(W)| - H(y) \tag{7}$$

When the estimate amount is irrelevant, independent component analysis model is estimated through the mutual information to minimize, which is equivalent to the sum of the non-gaussian maximize estimated. If we can find the minimize mutual information of reversible matrix form, then we can find out the direction of negative entropy maximization. In other words, it is equivalent to find one dimension sub-space, and the subspace have the projection in largest negative entropy.

3 The ICA Optimization Algorithm

Independent component analysis methods include two aspects: determining the objective function, choice of optimization algorithm. In establishing the objective function, the problem is how to get the solution of the optimal objective function. In the development process of the independent component analysis, there have been some good algorithm, mainly including natural gradient method, gradient method and the designated iterative method and relative gradient method.

3.1 Gradient Method

People usually optimize the objective function through adopting general gradient method before putting forward relative gradient and natural gradient. Gradient method is actually the process of solving extreme value to maximum likelihood estimation objective function based on the gradient algorithm, this is a kind of common and very effective method. Make the objective function for *H* (*x*, *W*), the general gradient method to the objective function optimization, when it take the maximum gives separable matrix. First of all to get the gradient for target function *H* (*x*,*w*) about *W* , so, adaptive learning algorithm structure is show as follows:

$$W(k+1) = W(k) + \mu \frac{\partial H(x, W(k))}{\partial W(k)} \tag{8}$$

3.2 Newton Method

Newton method is based on the second order derivatives, and in front of the introduction of target function gradient method is based on the first derivative. Newton method usually need to estimate the Hessian matrix and its inverse matrix, to order number is lower matrix, the algorithm has certain feasibility. We develop the objective function with Taylor series, which is as follows:

$$J(W(\mathrm{t})) = J(W(t-1) + \Delta W) \approx J(W(t-1)) + g_{t-1}^T \Delta W + \frac{1}{2} \Delta W H_{t-1} \Delta W$$

$$g_{t-1} = \frac{\partial J(W)}{\partial (W)}\Big|_{W-W(t-1)}, \quad H_{t-1} = \frac{\partial^2 J(W)}{\partial^2 (W)}\Big|_{W-W(t-1)} \tag{9}$$

Among them, the gradient (first derivative) of objective function is g_{t-1}, and the second order derivatives of objective function is H_{t-1}, it is the Hessian matrix. We derivative to ΔW, and for a derivative is 0, then, we can get $g_{t-1} + H_{t-1}\Delta W = 0$ and then $\Delta W = -H_{t-1}^{-1} g_{t-1}$.

4 Fast ICA Method

Fast ICA algorithm acts as one of the most popular algorithm of independent component analysis algorithm, it can start from observation signal and estimate source signals with little known information, and get the approximation of original signal independent each other. Fast ICA algorithm is a kind of fixed point iteration method, which in order to find out the maximum of Gaussian character, and use formula $J(y) = [E\{G_i(y)\} - E\{G(v)\}]^2$ to measure its independence, and can also approximately derive it by Newton iterative method.

First, we notice that the maximum value of approximation negative entropy of $w^T x$ is gotten through $E\{G(w^T x)\}$ optimized. According to the conditions of Kuhn-Tucker, the most optimized point of $E\{G(w^T x)\}$ is obtained in meeting formula (10) under the restriction of $E\{G(w^T x)^2\} = \|w\|^2 = 1$.

$$E\{xg(w^T x)\} - \beta w = 0 \tag{10}$$

In formula (10), β is a constant value and can easy to get through $\beta = E\{w_0{}^T x g(w_0{}^T x)\}$. And w_0 is the initial boundary value of w. If we assume we had solve the equations by the Newton method, and express the left of function formula (10) by F, then we can get the Jacobian matrix $JF(w)$ as bellow formula (11).

$$JF(w) = E\{xx^T g'(w^T x)\} - \beta I \tag{11}$$

In order to simplify the calculation of transposed matrix, we take the approximation value of first item and the reasonable estimate is shown as formula (12).

$$E\{xx^T g'(w^T x)\} \approx E\{xx^T\} E\{g(w^T x)\} = E\{g'(w^T x)\} I \tag{12}$$

Jacobian matrix is the Lord diagonal matrix and singular, and easy to deferring relatively, the approximate iteration formula is shown as formula (13).

$$w_{k+1} = w_k - \frac{E\{xg(w_k{}^T x)\} - \beta w_k}{E\{g'(w_k{}^T x)\} - \beta} \tag{13}$$

Among formula (13), $\beta = E\{xg(w_k{}^T x)\}$. In order to enhance the stability of the iterative algorithm, after iteration, we use formula $w_{k+1} = w_{k+1} / \| w_{k+1} \|$ to normalization *w*, and in formula (13), we multiplied $\beta - E\{g^{'}(w_k{}^T x)\}$ on both sides, then, the algorithm will be further simplified, which will get protection iteration algorithm formula is shown as formula (14).

$$w_{k+1} = E\{xg(w_K{}^T x)\} - E\{xg^{'}(w_K{}^T x)\}w_K \tag{14}$$

5 Conclusions

Fast ICA blind separation technology can be used in language recognition, communications, and image processing, etc, and has wide application prospects. According to its application field points mainly in the following aspects: the voice, image recognition and strengthen, array antenna processing and data communication, biomedical image processing, the feature extraction and recognition, and denosing the noise.

Acknowledgments. This paper is supported by the Aeronautical Science Foundation in China. (No.2010ZD55006).

References

1. Parra, L., Spence, C.: Convolutive blind source separation of nonstationary sources. IEEE Trans. on Speech Audio Processing (10), 320–327 (2000)
2. Jutten, C., Herault, J.: Blind separation ofsources, partl: An adaptive algorithm basedon neuromimetic architecture. Signal Processing 24, 1–10 (1991)
3. Bi, Y.: Research and Application of BBS Algorithm Based on Fast ICA. Xian university of science and technology (2007)
4. Cao, H., Zhang, B., Ma, L.: The Separation of Mixed Images Based on Fast ICA Algorithm. Computer Study 1, 45 (2004)
5. Ma, J.: Blind Signal Processing. Defense industry press (2006)
6. Zhang, X., Bao, Z.: Blind Signal Separation. Chinese Journal of Electronics 29(12), 1766–1771 (2001)
7. Liu, J., Lu, Z., He, Z., Mei, L.: Blind source separation method Based on information theory standards. Applied Science Journal 17(6), 156–162 (1999)
8. Yang, F., Hongbo: Principle and application of independent component analysis. Tstinghua University Press (2006)
9. Wang, L.: Research of BBS Method Based on ICA. Lanzhou University of Technology (2010)

Research of Intelligent Physical Examination System Based on IOT

Wei Han

School of mathematics, Chifeng College, Chifeng 024000, China
Happyzg3@163.com

Abstract. Considering the shortcomings of the current Physical examination, the novel methods of Physical examination based on Internet of Things (IOT) is proposed in the paper. The design scheme of the system is suggested. Meanwhile, the topology structure and components of system are given. In the end, the development of this technology was prospected, which can create a new platform for economic growth and greatly improve our lives and serve users better. In the end, the development of this technology was prospected, which can provide mankind with the better services of personal healthy and make people live more and more health, more effective and more comfortable.

Keywords: Internet of Things (IOT), Intelligent physical examination, Body chip.

1 Introduction

The Internet of Things is a network of Internet-enabled objects together with web services that interact with these objects [1, 2]. Underlying the Internet of Things are technologies such as RFID (radio frequency identification), sensors, and smart phones. The Internet fridge is probably the most oft-quoted example of what the Internet of Things will enable. The Internet of Things comprises a digital overlay of information over the physical world. Objects and locations become part of the Internet of Things in two ways. Information may become associated with a specific location using GPS coordinates or a street address. Alternatively, embedding sensors and transmitters into objects enables them to be addressed by Internet protocols, and to sense and react to their environments, as well as communicate with users or with other objects. Imagine a refrigerator that monitors the food inside it and notifies you when you're low on milk. It also perhaps monitors all of the best food websites, gathering recipes for your dinners and adding the ingredients automatically to your shopping list. This fridge knows what kinds of foods you like to eat based on the ratings you have given to your dinners. Indeed the fridge helps you take care of your health, because it knows which foods are good for.

The application area of IOT is quite wide, such as supply chain in manufacturing, equipment monitor and management in industry, intelligent control on home electronic appliances [3, 4]. There are a large number of applications that can be included as Internet of Things services, and these can be classified according to different criteria.

D. Jin and S. Lin (Eds.): Advances in MSEC Vol. 1, AISC 128, pp. 159–163.
springerlink.com

According to technical features, Internet of Things services can be divided into 4 types: identity-related services[5],information aggregation services, collaborative -aware services, and ubiquitous services[6].It is generally agreed that an inevitable trend for the Internet of Things will be its development from information aggregation to collaborative awareness and ubiquitous convergence, and that not all services of the Internet of Things will develop to the stage of ubiquitous convergence. Many applications and services only require information aggregation, and are not intended for ubiquitous convergence as the information is closed, confidential, and applicable only to a small group [6].

However, with the continuous development of science and technology, people's work pressures are becoming stronger and stronger. Meanwhile, their own physical conditions are more and more important to their own physical condition. People eager to clearly understand their own physical conditions timely without spend time to go to the hospital for examinations [7].

Considering the shortcomings of the current Physical examination, the novel methods of Physical examination based on Internet of Things (IOT) is proposed to solve current Physical examination in the paper. The design scheme of the system is suggested. Meanwhile, the topology structure and components of system are given. The development of this technology was prospected, which can create a new platform for economic growth and greatly improve our lives and serve users better. In the end, the development of this technology was prospected, which can provide mankind with the better services of personal healthy and make people live more and more health, more effective and more comfortable.

2 The System Overview of Intelligent Physical Examination System Based on IOT

To achieve this design requires the use of implanted chips to send as a data collection terminal, the collected data via TCP / IP protocol sent by the GPRS module to the GPRS public networks. A server send data receives from the body of the chip using TCP / IP and upload data to the Internet. the data processed is sent back to the user timely after finished analysis of the data processed via GPRS.

Electronic health records systems through reliable portal focused medical integration and sharing, so that various therapeutic activity may not the hospital administrative boundaries form an integrative perspective. With the electronic health record system, the hospital can be accurately and smoothly moving patients to other outpatient or other hospitals, patients can readily understand their illness, the doctor can pass a reference in patients with complete history for the accurate diagnosis and treatment.

People can install different sensors, on human health parameters are monitored and transmitted in real time to the related medical care center, if there is abnormal, health center through mobile phone, to remind you to go to the hospital to check the body. The system overview of intelligent physical examination system based on IOT is showed in Figure 1. The workflow of intelligent physical examination system is showed in Figure 2.

3 The Structure of Intelligent Physical Examination System

3.1 Terminal Data Sending and Acquisition via the Chip

The chip is a very small chip, can be easily implanted in the body underneath the skin, a record of personal information. With a particular machine can display the contents. In fact it is a use of radio frequency identification technology developed can be implanted in the body of the chip is fitted inside the chip, antenna and an information transmitting device, corresponding to different receiving device outside the body. The data terminal structure is showed in Figure 3.

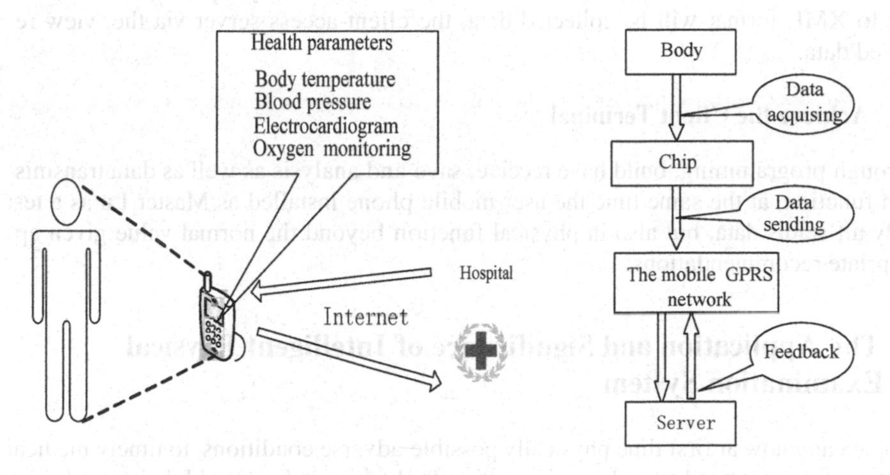

Fig. 1. The construction of the system **Fig. 2.** The workflow of the system

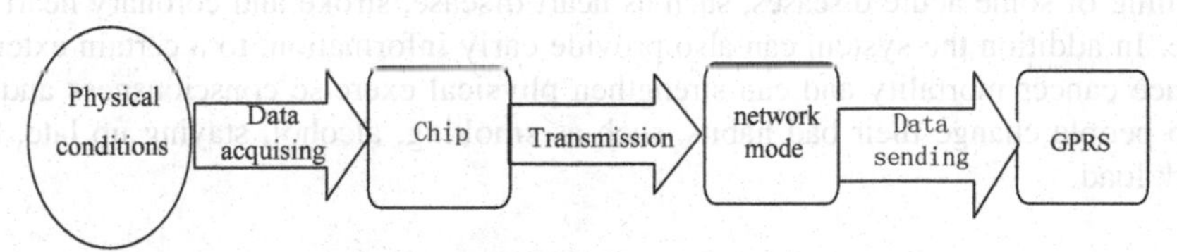

Fig. 3. The data terminal structure

3.2 Data Acquising

Implantation of human chip collecting human data, such as blood pressure, blood glucose, cholesterol, body basic physical data transferred via GPRS to the GPRS network, the server receives the data after the database of normal human data analysis, is returned to the user (mobile phone), and grant proposals.

3.3 Data Transmission and Sender(Embedded Operating System)

The real-time data compression tasks are completed by the DSP in the data transfer bus converter. ARM S3C4510B between PC and Ethernet communication and its software implementation of the requirement of real-time, reliability and complexity makes the choice of a TCP / IP protocol packet embedded real-time operating system is necessary, and C Linux is a complete TCP / IP protocol in C Linux operating system, adding real-time RT-Linux module of embedded operating system in order to meet the real-time requirement.

3.4 The Mobile GPRS Network

GPRS network mode is used in hospital public network mode. Hospital center station configured with a fixed IP address, and the remote terminal to implement dynamic IP address allocation. Remote terminal starting, active connection server, data collection, terminal module to obtain an IP address automatically, actively reported to the server, and to XML format will be collected data, the client access server via the, view received data.

3.5 Achieve the Client Terminal

Through programming, build have receive, save and analysis as well as data transmission function, at the same time the user mobile phone installed as Master Lu as a test body all status data, but also in physical function beyond the normal value given appropriate recommendations.

4 The Application and Significance of Intelligent Physical Examination System

People can know at first time physically possible adverse conditions, to timely medical treatment, do not need to make an inspection to the hospital can hold their own health through the physical examination of the Internet of things. Also, the system could early warning of some acute diseases, such as heart disease, stroke and coronary heart disease. In addition the system can also provide early information, to a certain extent to reduce cancer mortality and can strengthen physical exercise consciousness and can help people change their bad habits, such as smoking, alcohol, staying up late, high work load.

5 Conclusions

This paper introduces a novel methods of Physical examination based on Internet of Things (IOT) to solve the shortcomings of the current Physical examination. The design scheme and the topology structure and components of system of the system are suggested. The development of this technology was prospected, which can create a new platform for economic growth and greatly improve our lives and serve users better. In the end, The schemes can collected timely data information through body chip via a GPRS network wireless transmission to the server, and upload them to Internet. The system embodies the human body through the chip enable networking based on the Internet realization feasibility. the development of this technology was prospected, which can provide mankind with the better services of personal healthy and make people live more and more health, more effective and more comfortable.

References

[1] Liu, Q., Cui, L., Chen, H.M.: The Internet of Things and application of key technologies. Computer Science (06) (2010)
[2] Shen, S.-B., Mao, Y., et al.: The conceptual model and architecture of IOT. The Journal of Nanjing Posts and Telecommunications University (04) (2010)
[3] Junhua: Based Intelligent the Internet of Things Digital Campus Research and Design. Wuzhou University (03) (2010)
[4] Sun, X., Guo, C.: The Internet of Things networking technology and application. Science and Technology Information (26) (2010)
[5] Ho, F.: The Internet of Things architecture analysis and research. Guangdong University (04) (2010)
[6] Jie, F., Zhen, X.: The Internet of Things to explore network architectures. Information and Computer (08) (2010)
[7] Zhang, F., Zhang, X., Wu, G.: The airport baggage handling system and its application based on the Internet. Computer Applications (10) (2010)

Research of Intelligent Campus System Based on IOT

Wei Han

School of mathematics, Chifeng College, Chifeng 024000, China
Happyzg3@163.com

Abstract. Information technologies are played the important role and are leading to profound changes in education and management, such as the education ideas, models, methods, and concepts. This paper introduced the advantages and functions of intelligent campus system. Considering the shortcomings of education and management, the novel methods of intelligent campus system based on Internet of Things (IOT) is proposed in the paper. The topology structure and components of system are given. Meanwhile, the design scheme based on intelligent management and control is suggested. In the end, the applications of this technology were prospected, which can create a new platform for improving the applications of various materials and resources of education and greatly improve our study and serve users better and make education progress more effective and more comfortable.

Keywords: Internet of Things (IOT), intelligent campus system, education and management.

1 Introduction

With the popularity of computers and networks, the digital technologies and digital products are bringing people a new, more colorful way of life called "digital life"[1]. The campus is also facing the same trends. Digital technology will lead to profound changes in education and management, such as the education ideas, models, methods, and concepts.

Digital Campus is a network (the campus network and Internet) based on the use of advanced information technology means and tools to achieve from the environment (including equipment, classrooms, etc.), resources (such as books, courseware, materials, etc.[2]), to the activities (including teaching , science, management, service, office, etc.) all digital, on the basis of the traditional campus to build a digital space, so as to enhance the efficiency of the traditional campus, expanding the traditional campus function, and ultimately the overall educational process information. To put it simply, "digital campus" is set against the backdrop of campus teaching, learning, management, entertainment, a new digital work, study and living environment[3].

This paper introduced the advantages and functions of intelligent Campus system. Considering the shortcomings of education and management, the novel methods of intelligent Campus system based on Internet of Things (IOT) is proposed in the paper. The topology structure and components of system are given. Meanwhile, the design scheme based on intelligent management and control is suggested.. At present, China is

D. Jin and S. Lin (Eds.): Advances in MSEC Vol. 1, AISC 128, pp. 165–169.
springerlink.com

pushing forward the pace of information technology in education, the Ministry of Education in October 2000[8,9], made the decision to build the campus network project. And we have various information to exploer in order to create a new platform for improving the applications of various materials and resources of education and greatly improve our study and serve users better and make education progress more effective and more comfortable.

2 The System Overview of Intelligent Campus System Based on Internet of Things (IOT)

Digital Campus building is a long and arduous task, must be unified planning and step by step. First, make a long-term and overall system planning is the implementation of any necessary steps and systems engineering practice. Second, the digital campus building is a long-term task that requires step by step, gradually implemented at different levels, and gradually improved. Meanwhile, the digital campus is an extremely difficult and complex task that requires the school staff and students great efforts. The system overview of intelligent physical examination system based on IOT is showed in Figure 1. The workflow of intelligent physical examination system is showed in Figure 2.

3 The Structure of Intelligent Education and Management System

The structure of intelligent education and management system include the following aspects: Hardware: scanner, camera, light sensor actuator temperature sensors (fans, windows and doors), etc. it is showed in Figure 2.The work modules of intelligent management and control system is showed in Figure 3.

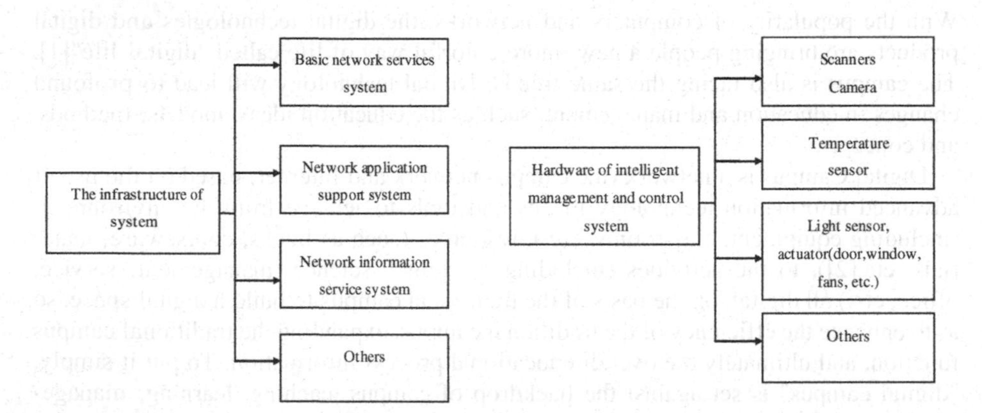

Fig. 1. The construction of system **Fig. 2.** Work modules of system

Scan Module. First need to install the scanner is a reasonable position in the Education, then based in accordance with the infrared scanner to scan the frequency of the entire classroom, then get the data to the core processor, CPU instructions to make judgments and then sent to the actuator, actuator to make the final action.

Light Control Module. First optical sensor placed in the classroom reasonable reasonable position, the light sensor will collect classroom light intensity data to the CPU, CPU instructions and then make judgments issued actuator, the actuator to make the final action.

Thermostat Module. First temperature sensor placed in the classroom reasonable reasonable position, temperature sensor will collect high and low temperature to teach data to CPU, CPU instructions to make judgments and then distributed to the actuator, the actuator to make the final action.

Alarm Module. This module mainly depends on CPU instruction, according to the temperature and light control, and scanner data to judge, and then do further processing. The alarm module structure is showed in Figure 4.

Practice and principle:

Smart Classroom. The management of the classroom so three modes, namely, school patterns and self-study mode, as well as conference mode.

First introduce Mode. There will be one of the CPU on all the classroom curriculum, for example it took one of the classroom! Teachers and classmates scanners will scan the distribution of seats, the results to CPU, CPU will be based on the distribution of teachers and students to the next step. According to the next step is the so-called light sensor and temperature sensor data dimming thermostat. The instructions sent to the actuator, which is the light curtain doors and windows so air-conditioning fans.

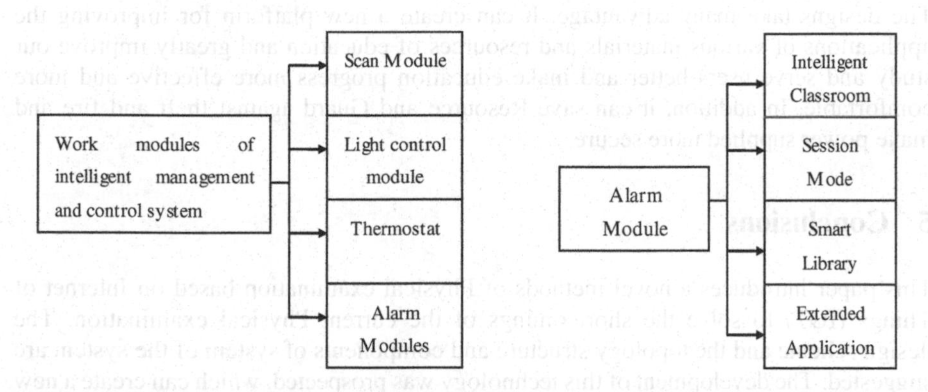

Fig. 3. The control system module structure **Fig. 4.** The alarm module structure

Reintroduce the self-study mode. That CPU is not room here to determine when the class curriculum, scanner to scan to determine whether someone is, no one is close doors and windows, and all equipment. Some people are into self-study mode. Into the self-study mode, CPU will be identified by the scanner as the set was, and only the locked position dimming thermostat.

Finally introduce session mode, which is referred to the class or student organization activities, classes will be so. Before entering this mode, the organizational meeting must make an application There are many ways to apply, can be written, you can log in the bedroom, the campus computer network applications, can also be applied in the school room, more convenient is that you can school-specific format of messages sent by the server, the server will automatically provide you the information you arrange the classroom.

Smart Library. In fact similar to library management and classroom management. In many schools will have a seat on the phenomenon, especially in the end would be particularly acute. With this management can reduce or even prevent such phenomenon, that is, you want to have to apply for self-study, you can book a few days in advance you can temporarily go to the library application (that is, to the campus entrance swipe card). If the book or at the time an application you can not be overhauled to ten minutes before the time to unsubscribe, the time to do the bit when the scanner will be scanned by frequency to determine the time, did not you come over, the latest can not late more than ten minutes, picking up and leaving count unsubscribe. If you come late or not, CPU will reduce your credibility, credibility that you are down to a certain extent you will have no right to self-study, after waiting some time the value of credibility credible back up again before booking.

Eextended application. This management system can also be used for other purposes. Such as supermarkets, shopping malls housing, etc.

4 Evaluation

The designs take many adwantage. It can create a new platform for improving the applications of various materials and resources of education and greatly improve our study and serve users better and make education progress more effective and more comfortable. In addition, it can save Resource and Guard against theft and fire and make power supplied more secure.

5 Conclusions

This paper introduces a novel methods of Physical examination based on Internet of Things (IOT) to solve the shortcomings of the current Physical examination. The design scheme and the topology structure and components of system of the system are suggested. The development of this technology was prospected, which can create a new platform for economic growth and greatly improve our lives and serve users better. In the end, The schemes can collected timely data information through body chipl via a GPRS network wireless transmission to the server, and upload them to Internet. The system embodies the human body through the chip enable networking based on the Internet realization feasibility. the development of this technology was prospected, which can provide mankind with the better services of personal healthy and make people live more and more health, more effective and more comfortable.

References

[1] Liu, Q., Cui, L., Chen, H.M.: The Internet of Things and application of key technologies. Computer Science (06) (2010)
[2] Zhou, Q., Liang, Y.: In order to optimize the out-patient-centered processes. Journal of Hospital Administration 8(20), 491–493 (2004)
[3] Hu, Y.: IT process optimization on out-patient medical treatment and practice. Science and Technology Information 27(23), 473 (2009)
[4] Memorial. Development of The Internet of Things. Nanjing University of Posts and Telecommunications (Social Science Edition) (02) (2010)
[5] Junhua: Based Intelligent the Internet of Things Digital Campus Research and Design. Wuzhou University (03) (2010)
[6] Gao, J., Liu, F., Ning, H., et al.: RFID coding, Name and information service for internet of things. In: Proc Wireless, Mobile and Sensor Network, CCWMSN 2007, IEEE Press, Shanghai (2007)
[7] Ho, F.: The Internet of Things architecture analysis and research. Guangdong Radio & TV University (04) (2010)
[8] Jie, F., Zhen, X.: The Internet of Things to explore network architectures. Information and Computer (Theory) (08) (2010)
[9] Peng, X.: The Internet of Things technology development and application prospect of. Shantou Technology (01) (2010)
[10] Zhang, F., Zhang, X., Wu, G.: Based on integration of The Internet of Things airport baggage handling system and its application. Computer Applications (10) (2010)

Training Skills of Autonomous Learning Abilities with Modern Information Technology

Xianzhi Tian

Wuchang University of Technology
Julia030712@163.com

Abstract. With the development of modern information technology, more and more studies are put on the application of it . In college study,especially college English study, autonomous learning abilities are very important for the improvement of learning achievements or learning efficiencies. In order to make learners to have good autonomous learning abilities, the author has studied some training skills of autonomous learning abilities with modern information technology. In order to show the views very clearly, the author has compared the different training skills and given some examples to analyse the study points. From the analysis, the training skills can be seen clearly. And from the study, the author wants to improve autonomous learning abilities of learners in their autonomous learning time.

Keywords: training skills, autonomous learning abilities, modern information technology.

1 Introduction

In the modern learning course, learners can not only learn by the guide of teachers,but also learn by autonomous learning by themselves. Generally speaking, these two ways are mixed with each other. In the long run, autonomous learning abilities play an important role in learning efficiency of a person. In the past, autonomous learning abilities are hidden in the long run learning course. The autonomous learning abilities can be seen and realized through the learning achievements. Learners can learn by themselves through some learning materials such as learning books, learning exercises ect. In modern society, autonomous learning abilities can be realized by some network sources in a large degree.

In fact, in recent years, the application of modern information technology is very wide. As for autonomous learning abilities, modern information technology can save time and improve learning efficiencies. On the one hand, learners can get enough learning resources from the web, on the other hand, learners can get some efficient learning methods from the web, and the learning methods can guide learners to learn efficiently.

In modern time, teachers and students rely greatly on web and information of web. They not only includes learning skills, learning contents, learning methods and learning experiences of others. People can not be separated from information technology everyday. So they must put their status very well to get effective

D. Jin and S. Lin (Eds.): Advances in MSEC Vol. 1, AISC 128, pp. 171–175.
springerlink.com

information to serve their learning courses. In this paper, the author tries her best to build up an effective learning mode for autonomous learning, which will be helpful for learners in their learning courses.

2 Types of Autonomous Learning in Information Time

Generally speaking, autonomous learning means learning by learners themselves. All types of autonomous learning are closely connected with learners' learning type. Owning to the development of information technology, autonomous learning begins to be connected with information technology. At the same time, autonomous learning turns to appear in different types.

According to the investigation, the author finds out that there are several types of autonomous learning as shown in the table.

Table 1. Main types of autonomous learning

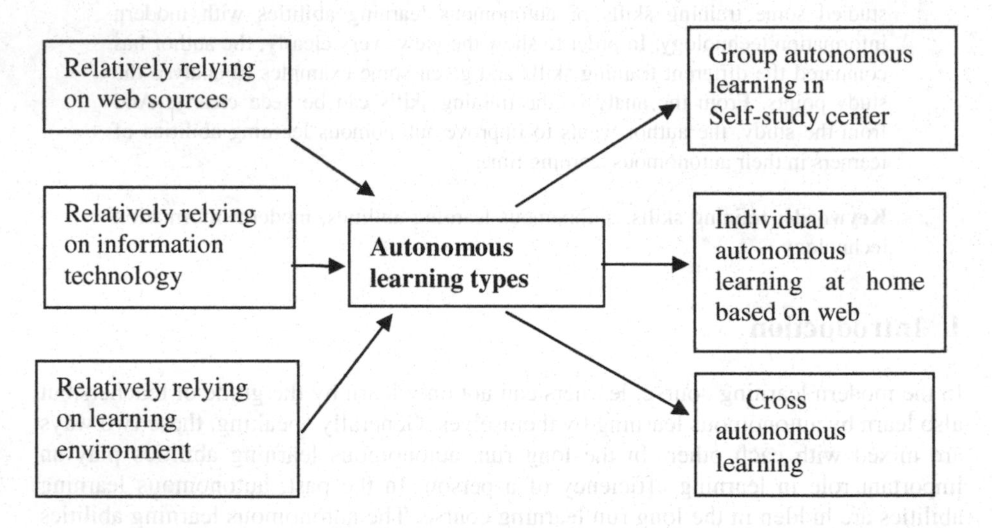

From the above table, we can see that autonomous learning types includes three main types. The first one is the group autonomous learning in self-study center. That is to say, this type of autonomous learning can be finished among group learning with group cooperation and group task. Nowadays, this type of learning appears very frequently in many universities. Many universities has built self-study center based on information technology. In self-study center, most learning groups come to the center with a natural class. And they have monitor to adjust group discipline,group learning tasks, even phases of examination and comparison. Therefore, group autonomous learning in self-study center occupies most of effective autonomous learning types. In addition, the second main one is individual autonomous learning type at home, which also occupies some parts of effective autonomous learning. But this type relies highly on personal self-consciousness and scientific learning plans, or it will have no effect on learning achievements. Recently, with the development of web sources, this type is also closely related with information technology. Finally, the last one, cross autonomous learning combines the first one and the second one. Learners can learn by themselves, but sometimes, they can also cooperate with their partners for some tasks. All of the three main autonomous learning types closely rely on web sources for learning materials, learning skills ect. They also rely on information technology for effective learning such as PPT, multimedia, video ect.

3 Training Skills of Autonomous Learning Abilities with Modern Information Technology

As for autonomous learning abilities,they have been closely connected with many elements. But the main element is on individual effort. In recent years, many researchers find out a new question. That is, the autonomous learning abilities, in fact, can be improved or developed by the scientific training skills of teachers or trainers. So in this perspective, the author wants to show her new research on training skills of autonomous learning abilities with modern information technology.

3.1 Knowing Learning Motivation Is the Base of Training Skills

Learning motivation is the internal factor for stimulating learners' learning. It is just like the actuator of a machine. The machine can work only with the normal working of the actuator. When the trainer knows the learning motivation of the learners, then the autonomous learning abilities can be easily stimulated, which just like "suit the remedy to the case".

According to the investigation, the author finds that different learners have different learning motivation. As for the learners in universities. The learning motivations are mainly listed as follows.

Table 2. Main learning motivation and their percentages

Main Learning motivations and their percentages					
Only for certain examinations	For improving certain skills to get better jobs in future	Only for spending spare time	Only for finishing certain tasks	For defeating the competitors	others
45%	32%	4%	12%	4%	3%

From the above table, it is easy to see that most learning motivation is on certain examination for Chinese authors. And then the second one is on improving certain skills to get better jobs. In fact, these two kinds of learning motivation both serve for future jobs. So autonomous learning abilities must be trained for future jobs. Only do some trainers know these key points, can they know how to train autonomous learning abilities of learners.

3.2 Basic Principles of Training Skills

As for the training skills, the author thinks that there must be some basic principles of training skills. Generally speaking, there mainly includes practical principle, planing principle, target principle and effective principle.

Firstly, practical principle refers to the fact that all learners must know practical knowledges in their autonomous learning course. That is to say, their learning materials must be useful for their future application such as their future job, their future communication ect. If the learners don't know how to judge whether their learning knowledges are practical or not , they can get information or reference suggestions from their trainers, teachers or partners. According to the different suggestions, they can have judgements on their learning contents and then adjust their learning plans. Secondly, planning principle is also very useful for every learner. Good plan will be effective for autonomous learning results. So learners must learn how to adjust their learning plans according to their learning arrangements and learning targets. In this course, trainers can give autonomous learners some suggestions according to the period achievement of learner. So trainers must learn how to evaluate scientificalness of learning plans and then they know how to give suggestions to learners to adjust their learning course. Thirdly, target principle is also very important for autonomous learners. Autonomous learners must have their learning target, or they will miss their learning motivation. Learning target can guide autonomous learners increase or decrease their learning contents. Finally, all autonomous learners and trainers must grasp effective principle, which will decide the effectiveness of learning results of learners.

In fact, as a trainer, he/she must be above of learners in psychological layer and teaching or guiding layer. The trainer must grasp all of the above principles very well and then he can guide the learner.

3.3 Training Skills with Modern Information Technology

As for training skills, trainers must know several "combination" as follows:

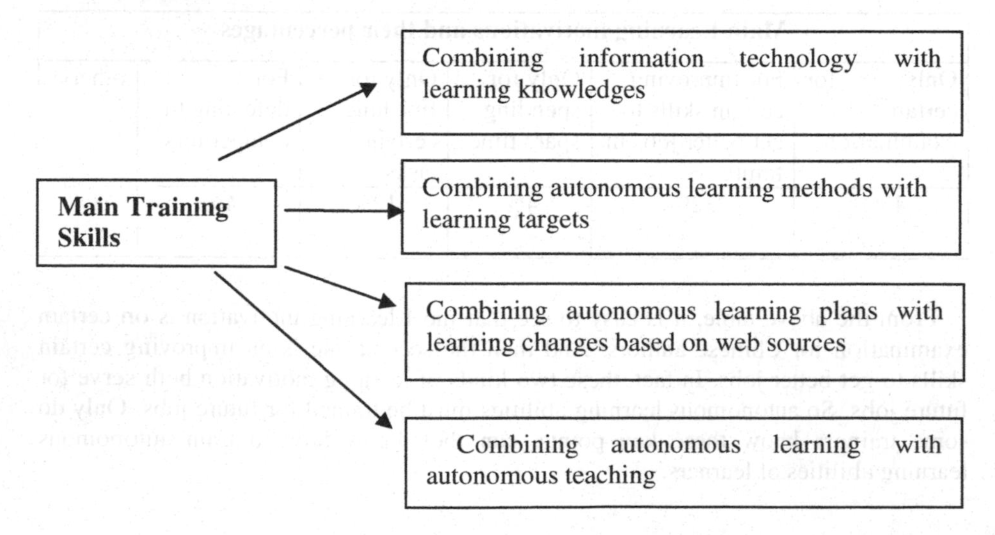

From the above table, we can see that there are several main training skills with modern information technology. Firstly, teachers can combine information technology with learning knowledges in class such as using PPT,video ect. To promote learning autonomous abilities. Furthermore, teachers can combine learning methods with learning targets. Learning targets can urge learners to make new plans and adjust their learning contents during learning courses. Therefore, learning plans and learning changes can be adjusted personally and swiftly according to some problems appearing in their learning course. In fact, the most important one is on combing autonomous learning with autonomous teaching. Only through creative autonomous teaching, can learners learn much more from their teachers. It is the necessary condition for the improvement of students. In recent teaching reform, teachers actually apply many different autonomous teaching methods in their teaching course such as exploration-internalization teaching method. The exploration-internalization methods will be researched in the next paper of the author, which is a useful teaching method to improve autonomous learning abilities of students.

In the course of training abilities of students, learners must know how to explore their potential abilities in their learning courses. They must learn how to learn actively and independently. At the same time, recent age is an age full of information technology. Learners must learn how to get enough knowledges from web or by different kinds of information technology. As for learning skills and methods, the author will study in her further research.

4 Conclusion

All in all,with development of information technology, different measures of training skills have been used to improve autonomous learning abilities of students. And teachers have also adopted many different teaching modes to help students develop their learning skills. But the achievements of students are from the efforts of the combination between teachers and learners. In the further study, the teachers will learn how to apply useful learning and teaching modes in autonomous learning courses.

References

1. Holec, H.: On Autonomy: Some Elementary Concepts. In: Riley, P. (ed.) Discourse and Learning, Longman, New York (1985)
2. Littlewood, W.: Defining and developing autonomy in East Asian contexts. Applied Linguistics (1999)
3. Smith, R.C.: Starting with ourselves: Teacher-learner autonomy in language learning. In: Sinclair, B., McGrath, I., Lamg, T. (eds.) Learner Autonomy, Teacher Autonomy: Future Directions, Longman, London (2000)

Medical Image Processing System for Minimally Invasive Spine Surgery

YanQiu Chen[1] and PeiLli Sun[2]

[1] Department of Computer Science, Dalian Neusoft Institute of Information, Dalian, P.R. China
chenyanqiu@neusoft.edu.cn
[2] Dalian Ocean University, Dalian, P.R. China
sunpeili@dlou.edu.cn

Abstract. A conceptual medical robotic system applicable for establishing surgical platform in the process of minimally invasive spine surgery (MISS) is proposed in this paper. According to the requirement of MISS operation, the surgical navigation system is established. The navigation system proposed a set of image intensification algorithm, which can enhance the images visual effect and surgery precision. At last, the experimental results made for the prototype illustrate the system well. This research will lay a good foundation for the development of a medical robot to assist in MISS operation.

Keywords: medical robotics, minimally invasive spine surgery, surgical navigation, image enchancement.

1 Introduction

Minimally invasive surgery (MIS) is a type of surgery performed through several small incisions (usually less than one inch in diameter), or puncture sites. It features some outstanding advantages compared to traditional open surgery with significantly reduce tissue traumas and decrease recovery time. Therefore, this procedure has become a common method during the last decades.

Minimally invasive procedure is an important developing trend of the spinal surgery. In the process of spinal puncture doctors determine the sick location of spine with multi-slice images acquired from the computed tomography (CT) scanner pre-operation, and plan the route of puncture needle guided by information of CT images with their experience. However, it is difficult to target accurately during deep percutaneous interventions because of soft tissue deformations. When doctor discovers there may be errors, the position and orientation of needle should be adjusted through re-scanning patient with CT scanner. This process may be repeated several times. Figure 1 shows the schematic of MISS operation [1].

The medical robot technology is the key to solving this problem. Robots can perform accurate positioning under manual or automatic operation. Moreover, it can be used as external holder of surgical instruments after robot joints are locked, which provides a stationary and reliable platform for doctors to operate minimally invasive surgery.

D. Jin and S. Lin (Eds.): Advances in MSEC Vol. 1, AISC 128, pp. 177–182.
springerlink.com

2 System Overview

2.1 System Structure

In view of the above problems, we have developed a minimally invasive spine surgery robotic system which mainly consists of three parts, as shown in Figure 1.

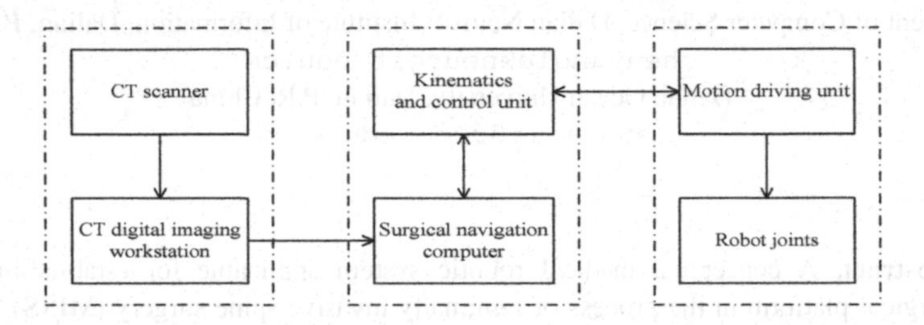

Fig. 1. Structure diagram of the medical robotic system

2.2 Main Working Process

Here, the main working process of this system is generally introduced. Firstly, doctors make treatment plan on the basis of focus information from CT imaging instrument, then input it to surgical navigation computer through graphical user interface (GUI). Secondly, the computer image processing and navigation system builds virtual surgical environment, in which motion parameters will be calculated. Thirdly, doctors evaluate motion path and improve surgical procedure, and after surgical plan is finally determined, the relevant movement information is entered to the robot motion system. Finally, according to the received orientation and movement data, robotic system carries on the initial localization of the surgical tools and controls motion of treatment tools.

In above process, the surgical navigation system is very critical. In order to help doctors to locate focus position precisely, this navigation system proposed a set of image intensification algorithm, which can enhance the images visual effect and surgery precision through improving edges、outline、contrast and so on.

3 Surgical Navigation System Design

3.1 Overrall Design

The surgical navigation system assists doctors in locating exact points within the body during surgery. Using of image information to establish a virtual environment, which provides evaluating treatment platform for doctors. According to computer graphics theory, virtual 3D image will be constructed on the basis of patient's CT images. Focuses can be identified and marked through feature extraction and pattern recognition. This is a very difficult work because the human spinal physiological anatomy has a very complex anatomical anatomic structure. Therefore, the system requires marking the color of important biological organizations significantly and should be designed to provide multi-angle, multi-scale visualization of the results.

In order to model three-dimensional human organs from CT datasets, we can reconstruct anatomical surface model by a certain 3D reconstruction algorithm, such as contour based reconstruction and volume based reconstruction [4]-[6]. We have programmed using Microsoft Visual C++ on the Windows XP platform. The proposed GUI of the surgical navigation system, as shown in Figure 2, provides the 2D and 3D spinal images, medical image processing unit, surgical marker's space coordinates, SpineNav robot self-test, system control function, etc.

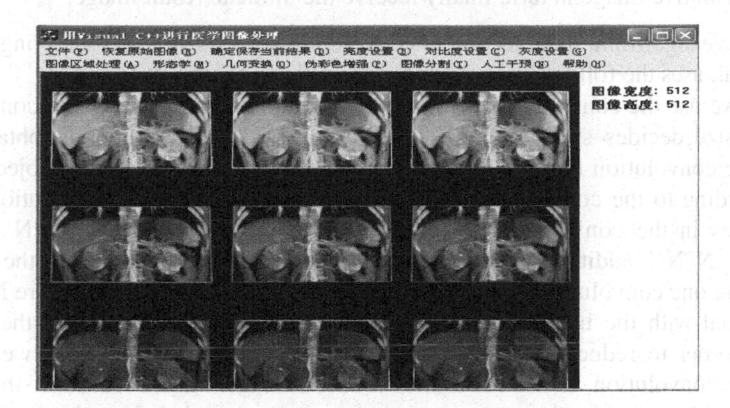

Fig. 2. Graphical user interface of the navigation program

3.2 Image Enhancement Algorithm Design

At present there are many algorithms which can realize the image intensification like gray level transformation algorithm 、 histogram equalization algorithm and so on. But these methods are only useful to global equalization, and are unable to realize details and organizations strengthen. Therefore, the method of enhancing image details and tissue edges becomes the urgent demand in medical image processing field. This paper presents a new image processing algorithm, which can enhance objects of different spatial frequency band, highlight organize edges and detail information, assist doctors to gain more useful information from the CT image.

This new algorithm first filters image through the low-pass filter for smoothing the image, obtains the low-pass smoothing image, and then subtracts the low-pass smoothing image with primitive image, extracts high frequency component which will be superimposed again to the primitive image, that may strengthen size close objects with the convolution kernel of the low-pass filter, and enhance image organize edges and details information.

The concrete step is as follows:

- Firstly, using low-pass filter with appropriate size convolution kernel filters the primitive image, obtains a low-pass image of including low frequency component, and then subtracts the low-pass image with primitive image, obtains high frequency component of the primitive image.

- Secondly, carries on the dot sampling in horizontal and vertical direction, obtains a reduced image; using low-pass filter with appropriate size convolution kernel filters the reduced image, obtains a low-pass image of including the reduced image's low frequency component, and then subtracts the low-pass image with reduced image, obtains high frequency component which will be enlarged to the primitive image size.
- Thirdly, repeat step2 with low-pass image of including reduced image's low frequency component.
- Lastly, high frequency component of each level will be overlapped to the primitive image in turn, finally receive the ultimate result image.

In the above algorithm, the selection of convolution kernel's size and sampling interval is essential, uses the following method to process these questions.

First, we use the convolution method to realize the low-pass filter. The convolution kernel's size decides strengthened spatial frequency band. In order to obtain ideal effect, the convolution kernel's size should be adjusted to the interested object's size. But according to the convolution algorithm principle, a size of N convolution kernel participates in the convolution, for each pixel point, it needs to do N*N multiply operation, N*N-1 additive operation and 1 division operation, therefore the time of completing one convolution operation is proportional with the time of square N, and is proportional with the image size, obviously the speed is unable to meet the clinical need. In order to reduce the computation and imagery processing time, we use the fixed size convolution kernel convolutes reduced image. Also because of the bigger convolution kernel size, the more computation time is needed. In order to guarantee one convolution operation be processed in 2 seconds, the convolution kernel's size should be smaller or equal to 5 pixel point, but the smaller convolution kernel's size, the more noises will be introduced to image, therefore we use fixed size of 5 pixel point.

Next, the goal of dot sampling is to withdraw objects of different spatial frequency band, it is possible to separate one pixel, two pixels and even several pixels. Sampling of separating one pixel reduces 1/2 images, Sampling of separating two pixels reduces 1/3 images, Sampling of separating three pixels reduces 1/4 images, which will be performed until the reduced image size is smaller than low-pass filter convolution kernel's size.

Finally, there is an intensification factor when each high frequency component overlaps with the primitive image. We discover that intensification factor selection is related with outline clarity and noise size: If the outline is not clear and the noise is big, the intensification factor is 0, if the outline is clear and the noise is small, the intensification factor is 1.

With the above processing method, result image simultaneously strengthened object of different space frequency band effectively including organize edge 、 detail information etc. Moreover the method of reducing image causes convolution kernel invariable, which reduced the computing time and meets the clinical demand.

We treated a chest's CT with above algorithm, the effect as shown in Figure 3, in which the first image is primitive image, the second one is 1/2 reduced image, the last one is 1/4 reduced image. Figure 4 is the final effect image.

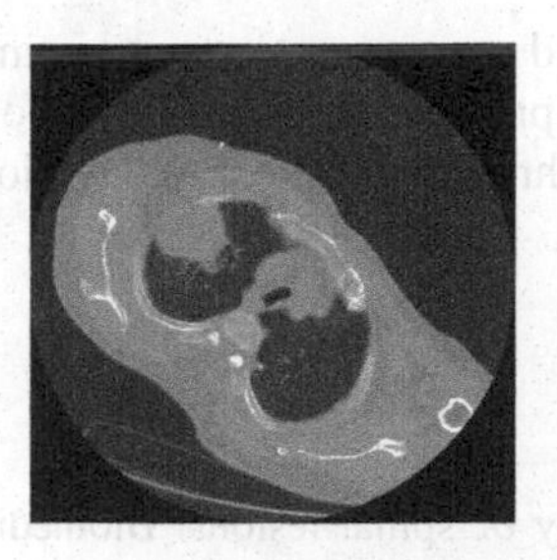
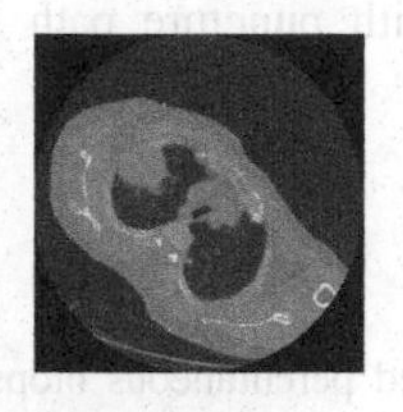
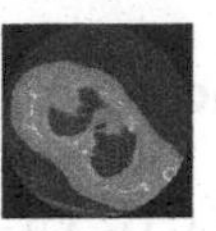

Fig. 3. Reduced image after dot sampling

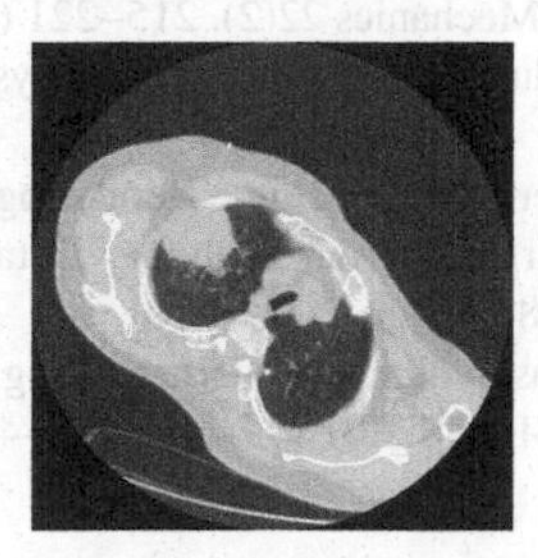

Fig. 4. Result image after enhancement processing

We may see from Figure 4 that detail and big organization edge are well strengthened after using our method, which proves that our algorithm is quite effective.

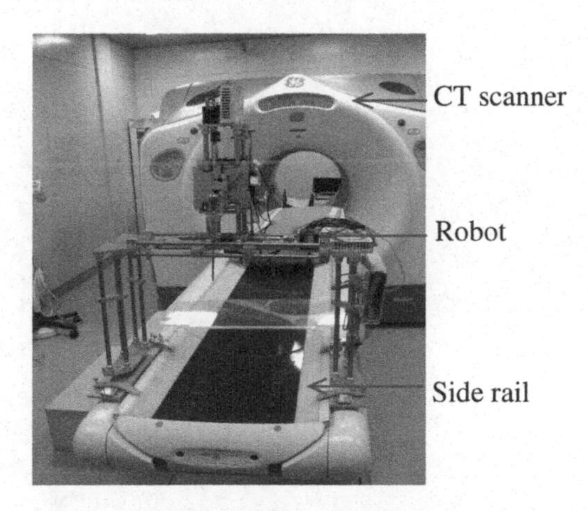

Fig. 5. Photograph of the testing environment

4 Conclusion

In this paper, a novel concept of employing a medical robot applicable to establish surgical platform in the process of MISS operation has been proposed. We have developed a testing environment and carried on scene test many times. The actual testing environment shows as Figure 5. At present, the surgical robot can assist the positioning operation with puncture path through surgical navigation and surgical physician intervention.

References

1. Peh, W.C.G.: CT-guided percutaneous biopsy of spinal lesions. Biomedical Imaging and Intervention Journal 3(3) (2006)
2. Denavit, J., Hartenberg, R.S.: A kinematic notation for lower-pair mechanisms based matrices. Journal of Applied Mechanics 22(2), 215–221 (1955)
3. Niku, S.B., Sun, F.C.: Introduction to Robotics - Analysis, Systems, Applications, pp. 60–68. Phei Press (2004)
4. Johnson, C., Hansen, C.: Overview of Volume Rendering, pp.1-26. Academic Press (2005)
5. Levoy, M.: Display of Surfaces from Volume Data. IEEE Computer Graphics and Applications 8(3), 29–37 (1988)
6. Lacroute, P., Levoy, M.: Fast volume rendering using a shear-warp factorization of the viewing transformation. In: SIGGRAPH 1994, pp. 451–458 (1994)

Multi-scale Image Transition Region Extraction and Segmentation Based on Directional Data Fitting Information

Jianda Wang and Yanchun Wang

Information Science and Technology College
Dalian Maritime University
Dalian, China
wangjd8971@hotmail.com, yanchwang@yahoo.com.cn

Abstract. This paper presents multi-scale image transition region extraction and segmentation algorithm based on directional data fitting information. Using direction-data fitting information measure, image transition region will be gained by modulating the wavelet scale. An optimal segmentation threshold will be obtained from the transition region histogram. This method has gone beyond the limitation that traditional methods of image transition region extraction are sensitive to noises and depend on clip limits L_{low} and L_{high} Analyses and experiments show that the proposed method is characterized by speed, high robustness and good anti-noise performance.

Keywords: Information Measure, Wavelet Transformation, Transition Region.

1 Introduction

Image segmentation refers to the extraction of targets from an image, whose results greatly affect the succeeding image processing. Weszka et al proposed Average Gradient method [1]. Gerbrands described the features of transition regions [2]. Zhang et al first applied transition region extraction technique the Effective Average Gradient (EAG) and Clip Transformation to image segmentation [3]. Groenewald et al compared EAG method with Average Gradient (AG) method [4], and proved that EAG is actually the smooth style of AG, advancing windowed average gradient method. In the previous transition region-determined methods, there is a default, namely, the transition region is a region where the change of grey level is the most exquisite in the whole image, while the other region of the image, even if there is some change in grey scales, its result can be ignored, which in a way overlooked the effect of noises. Typical gradient-based transition region extraction method employed effective mean gradient and the clipping of grey scale [5].

In order to improve the deficiency of literature [5, 6] introduced weighted gradient operator, which in a way restrains random noises and makes $EAG(L) \sim L$ smoother. Literature [7] did one-dimensional fitting to the original image, reducing the effect of the noises in the image on $EAG(L) \sim L$, making it easier to extract L_{low} and L_{high} . The defect of this algorithm lies in the fact that the extraction of

D. Jin and S. Lin (Eds.): Advances in MSEC Vol. 1, AISC 128, pp. 183–190.
springerlink.com

tradition region is dependent on L_{low} and L_{high}, but the sensitivity of gradient operator to noises may result in the excursion of L_{low} and L_{high}, even in the extremeness that $L_{low} = 0, L_{high} = 255$. In addition, there may exist the situation that $L_{low} > L_{high}$, in which case transition region may not be extracted. To reduce the effect of noises on transition region, and to make the transition region get rid of the dependence on L_{low} and L_{high}, Literature [8, 9] proposed a transition region extraction method based on local complexity and local entropy, which overcame the sensitivity of gradient based method to pepper & salt noises, but they are still not very efficient for resisting gauss noises.

This paper presents a directional data fitting information based image transition region extraction and segmentation method. Based on directional data fitting information measure, transition region can be obtained by adjusting wavelet scale, which effectively eliminates the influence of pepper & salt noises and Gauss noises on the transition region extraction, consequently the transition region will distributes well around the target.

2 Image Transition Region Extraction and Segmentation Method Based on Directional Data Fitting Information

Owing to the deficiency of the typical transition region extraction methods, this paper proposes an image transition region extraction and segmentation method based on directional data fitting information, which consists of two parts: directional data fitting information measure, wavelet transformation scale self-adaptive modulation.

2.1 Transition Region Directional Data Fitting Information Measure

Transition region means there exist grayscale changes at certain direction, Fig1(a) and (b) are the two-dimensional performance of transition region and smooth region when noise is added. Fig1(c) and (d) are unidimensional sampled data when noise is added. When the unidimensional data of the sampled image is analyzed with monolinear regression, the fitting straight line slope will change with the transforming of sampled data length if these data generate in the transition region, and a sampled length also maximized linearity. Maximum fitting degree makes the straight line fitting slope maximum. If these data generated in the noised smooth region, they can be fitted with a straight line. When the length of the sampled data changes, the slope of the fitted straight line will not change much, and the slope is minor. Since the slope of the fitted straight line is only related to the distribution of the integrated sampled data, it is less related to the single sampled data, so it can reflect the statistic characteristics of the transition region and the smooth region neighborhood gray value in the noised condition.

Set the coordinate of the current pixel point (m,n), its neighborhood $R = \{(i,j) \mid |i-m| \le L, |j-n| \le L\}$, L is half of the variable neighborhood length. When the given neighborhood length is L, in the neighborhood R, in the direction of

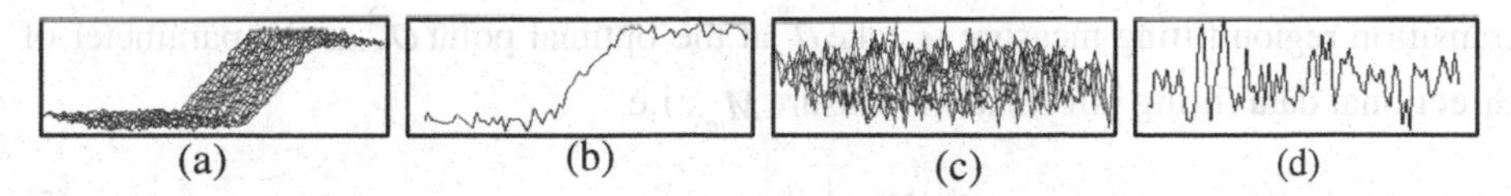

Fig. 1. Unidimensional and two dimensional sampled data demonstration of transition region and smooth region:(a) two dimensional demonstration of transition region; (b) sampled data in the direction of transition region; (c) two dimensional demonstration of smooth region; (d) sampled data of smooth region

transition region θ, taking the current pixel point as the center, collect $2L+1$ pixel points, its gray data sequence of the pixel points is $x_{-L}, x_{-L+1}, \cdots, x_{L-1}, x_L$, to fit straight line with monolinear regression to the this data sequence.

$$x = ai + b . \tag{1}$$

Where, $i \in \{-L, -L+1, \cdots, L-1, L\}$, $x \in \{x_{-L}, x_{-L+1}, \cdots, x_{L-1}, x_L\}$.

Suppose the image has Gauss noises, the experimental observation values $x_{-L}, x_{-L+1}, \cdots, x_{L-1}, x_L$ meet $\hat{x} = ai + b + \varepsilon$, of which a, b are constants, ε is subject to normal distribution $N(0, \sigma^2)$, from monolinear regression analysis, we know that the estimated value of a, σ^2 is

$$a = \sum_{i=-L}^{L} i(x_i - \bar{x}) / \sum_{i=-L}^{L} i^2 . \tag{2}$$

$$\sigma^2 = \frac{1}{2L+1}\left[\sum_{i=-L}^{L}(x_i - \bar{x})^2 - a^2 \sum_{i=-L}^{L} i^2\right] . \tag{3}$$

Where. $\bar{x} = \sum_{i=-L}^{L} x_i \Big/ 2L+1$.

The slope of the fitted straight line a demonstrates statistically the gray mutation within the neighborhood. The more intense the transition region, the larger the slope a is, and vice versa. When the transition region has the same intensity, it is less wide, and the slope a is larger; the wider the transition region, the smaller the slope a . So slope a can show both the intensity and width of the transition region. σ^2 reflects the linearity of the sampled data sequence. The bigger σ^2 is, the lower the linearity is, and vice versa. σ^2 is in smallest level when linearity is optimal. The optimal σ_*^2 can be obtained when the length of sampled data L is changed. viz.

$$\sigma_*^2 = \min_{3 \le l \le 2L+1} \sigma_l^2 . \tag{4}$$

Where, σ_l^2 is σ^2 of fitted straight line when the sampled data length is l. As straight line fitting is mainly relevant to data distribution, the value of slope a has a better noise restriction. Slope a of the fitted straight line can denote the possible measure that there exists transition region within the neighborhood, namely, the parameter of transition region fitting measure M_a take a^* at the optimal point σ_*^2 as the parameter of directional data fitting information measure M_a, i.e.

$$M_a = |a^*| \cdot \tag{5}$$

Analyzing the relationship of the region where the current pixel point is located and directional data fitting information measure M_a:

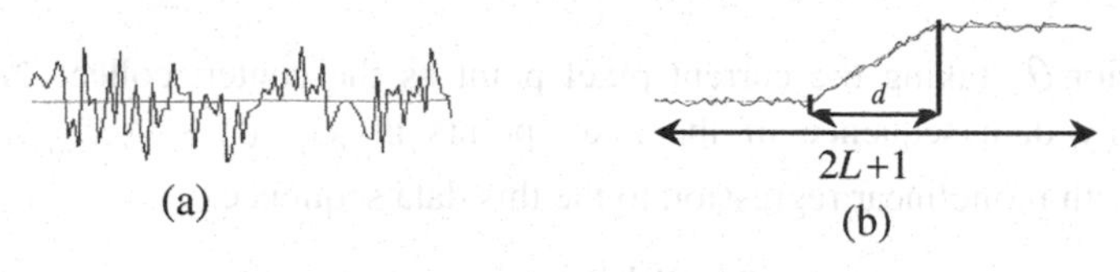

Fig. 2. Sampled data distribution and unidimensional fitting of transition region and smooth region: (a) smooth region; (b) transition region

(1) Smooth Region
As shown in Figure 2(a), in the smooth region, the gray distribution does not mutate statistically, therefore, the fitted straight line slope a is less valued, the value of M_a is less accordingly.
(2) Transition Region
As shown in Figure 2(b), in the transition region, the gray distribution has mutation statistically, consequently, when the sampled data length changes to d , the linear fitting is higher, σ^2 has a less value, the slope of straight line a is larger, i.e. the value of M_a is bigger; when the sampled data length is $2L+1$, the linearity of the fitted straight line is relatively poor, the value of σ^2 is bigger, the linear fitting degree is lower. Comprehensive analysis shows that directional data fitting information measure M_a may reflect the possibility that if transition region exists in the area.

2.2 Self-adaptive Modulation of Filter Scale

Self-adaptive modulation filter scale refers to the method that large scale filtering is conducted in smooth area, while small scale filtering is done in transition area so that noises are restricted and the weak intensity transition region is also better preserved. This paper adopts wavelet method to detect transition region.

The binary wavelet transform of two-dimensional signal (image) $f(x, y) \in L^2(R^2)$ is defined as follows:

$$\begin{aligned} W_{2^j}^1 f(x,y) &= f(x,y) * \psi_{2^j}^1(x,y). \\ W_{2^j}^2 f(x,y) &= f(x,y) * \psi_{2^j}^2(x,y) \end{aligned} \tag{6}$$

Write as vector:

$$\begin{bmatrix} W_{2^j}^1 f(x,y) \\ W_{2^j}^2 f(x,y) \end{bmatrix} = 2^j \begin{bmatrix} \frac{\partial}{\partial x}(f * \theta_{2^j}(x,y)) \\ \frac{\partial}{\partial y}(f * \theta_{2^j}(x,y)) \end{bmatrix} = 2^j \nabla f * \theta_{2^j}(x,y) \tag{7}$$

Two-dimensional binary wavelet transform $f(x,y)$ is actually total differential of $f(x,y)$ after being smoothed by $\theta_{2^j}(x,y)$, whose smooth function of wavelet transform adopts detachable cubic B-spline functions $\theta(x) = B^3(x)$, $\theta(y) = B^3(y)$, where $s = 2^j, j \in Z^+$, what needs self-adaptive modulation is filter scale s.

Both the filter scale s and the data fitting parameter d are relevant to the width of transition region. Suppose the transition region width is d, and transition region intensity is $\eta = \eta_2 - \eta_1$, as shown in Fig.3. From the triangle relationship we gain,

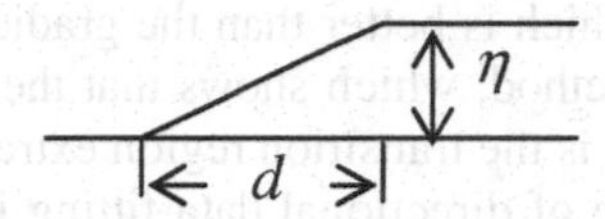

Fig. 3. The relationship between transition region and intensity

$$a = \frac{\eta}{d}. \tag{8}$$

The above formula shows that under the same intensity of the transition region, the bigger the value of a, the narrower the transition region d, and vice versa. Thus the straight line fitting measure M_a determined by parameter a carries width information of the transition region, the wider the transition region is, the smaller the value of M_a is, and vice versa. Hence, the value of straight line fitting measure M_a may be used to modulate filter scale. Fig.4 is the transition region extracted by multi-scale method where directional data fitting information is employed to the image. The transition region is relatively precisely distributed around the target.

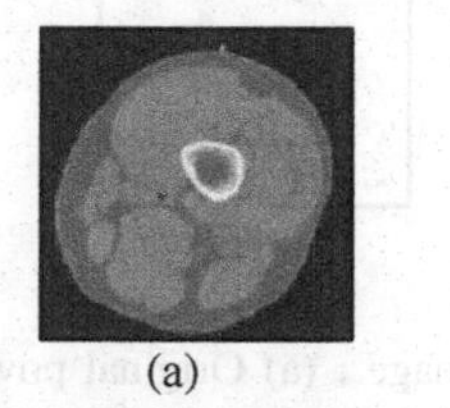

(a)

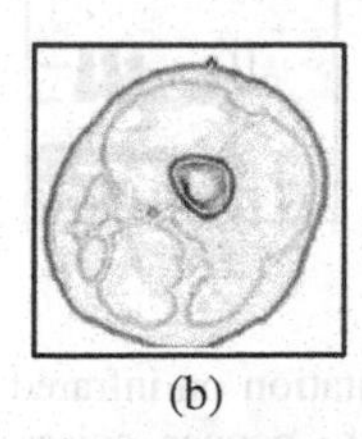

(b)

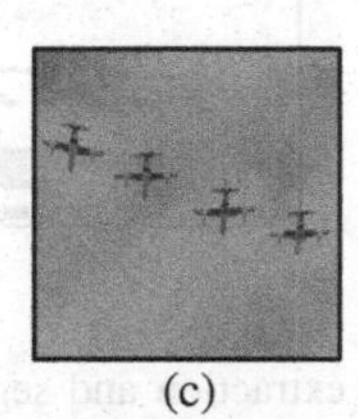

(c)

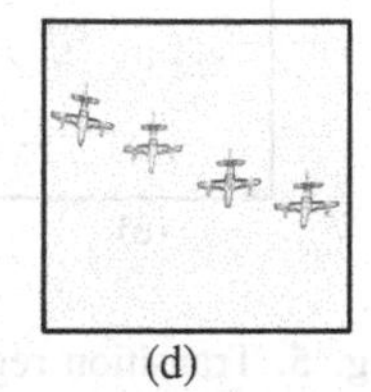

(d)

Fig. 4. Transition region extraction of original image: (a) original image; (b) transition region extracted by multi-scale; (c) original region; (d) transition region extracted by multi-scale

3 Results of Experiments and Analyses

To demonstrate the efficiency of our algorithm, we did experiments on transition region extraction and segmentation by adding pepper & salt noises of different intensity and gauss noises of different variances to different images, and compared respectively the segmentation effect with the traditional weighted effective average gradient algorithm (W-EAG) and local complexity-based transition region extraction method (C-TREM).

Fig 5(a) is the original infrared image of a power station, Fig 5(b) is the image after adding salt & pepper noises of intensity $D = 0.035$. From 5 (c) we can see that the range of the transition region extracted by W-EAG is smaller (framed in the Fig), it fails to distribute around the object, and a majority of the transition region around the cooling tower is not extracted, the inaccuracy of the transition region results in the dissatisfaction of the segmentation in Fig5 (f), as a result, the two chimneys are incompletely segmented. Fig5 (d) is the transition region extracted by local complexity algorithm, the effect of which is better than the gradient method. Fig 5(g) is segmentation effect by the latter method, which shows that the chimney on the left is not completely segmented. Fig 5(e) is the transition region extracted by our method, the region where the least multi-scale of directional data fitting information (already counter-colored) is around house and the cooling tower, where the comparability of the pixels in the image is the lowest, and the variation of the grayscale level is the most distinct. The image shows that the transition region has been well extracted and accurately distributed around the object. Thanks to the accuracy of the transition region extraction, the segmentation result of Fig 5(h) evidently excels that of Fig 5(f) and Fig 5(g).

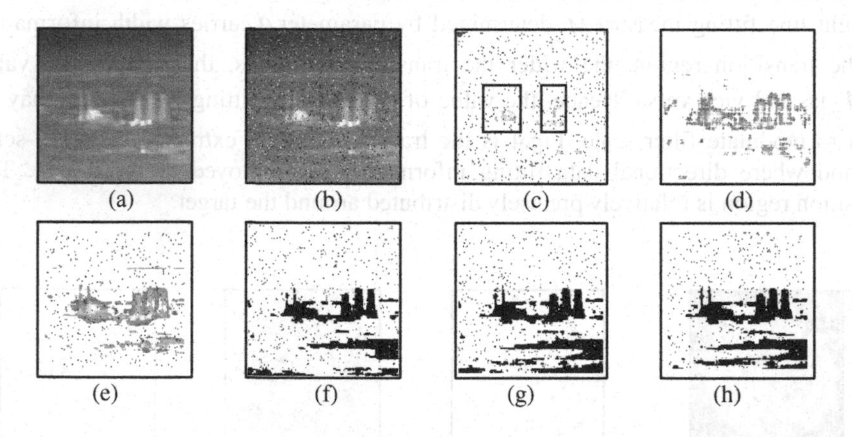

Fig. 5. Transition region extraction and segmentation on infrared image : (a) Original power station infrared image; (b)After adding salt & pepper noises of intensity $D = 0.035$;(c) Transition region extracted by W-EAG; (d) Transition region extracted by local complexity algorithm; (e) Transition region extracted by our method; (f) result by W-EAG; (g) result by local complexity; (h) result by our method

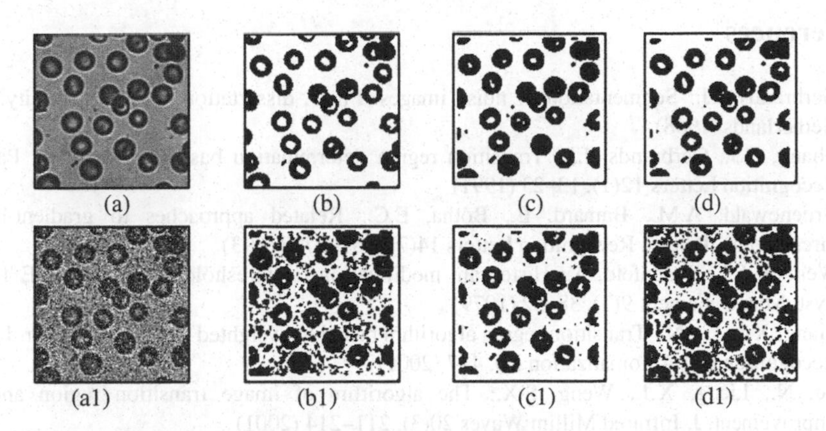

Fig. 6. Micro-image segmentation comparison experiments by three threshold methods: (a) original microimages; (b) segmentation result by weighted gradient; (c) segmentation result by local complexity; (d) segmentation result by our method (a1) noises with V=0.03, D=0.1 (b1) segmentation result by weighted gradient (c1) segmentation result by local complexity ; (d1) segmentation result by our method

To test the anti-mixed noises (gauss noises and pepper & salt noises) capability of our method, the experiment is shown in microimages. Fig 6(a) is the original image without any noises. The segmentation effects of Fig 6(b) and (d) are almost the same, while Fig6(c) shows that the two cells combines into one owing to the error of segmentation (the framed region).

When adding gauss noises with V=0.03 and pepper & salt noises with intensity D=0.2, the cells of Fig 6(b1) and (c1) becomes agglutinate (the framed region), and the background is greatly noised, so the segmentation effect is very poor. Fig 6(b1) and (c1) demonstrate that the segmentation effect of our method excels that of the other two methods.

4 Conclusions

Transition region extraction is the core of threshold segmentation, yet traditional transition region extraction methods are sensitive to noises and depend on clip limits L_{low} and L_{high}. This paper presents multi-scale image transition region extraction and segmentation algorithm based on directional data fitting information. Experiments demonstrate that our method, which shows better segmentation ability, outperforms the methods in existence. The target can be segmented even though gauss noises with variance 0.03 and pepper & salt noises with intensity 0.2 are added. This algorithm is robust and resistant to noises.

References

1. Gerbrands, J.J.: Segmentation of noise images. Ph D. dissertation, Delft University, The Netherlands (1988)
2. Zhang, Y.J., Gerbrands, J.J.: Transition region determination based thresholding. Pattern Recognition Letters 12(1), 13–23 (1991)
3. Groenewald, A.M., Bamard, E., Botha, E.C.: Related approaches to gradient-based thresholding. Pattern Recognition Letters 14(7), 567–572 (1993)
4. Weszka, J.S., Rosenfeld, A.: Histogram modification for threshold selection. IEEE Trans. Syst. Man Cybement 9(1), 38–52 (1979)
5. Liang, X.J., Le, N.: Transition region algorithm based on weighted gradient operator. Image Recognition and Automatization (1), 4–7 (2001)
6. Le, N., Liang, X.J., Weng, S.X.: The algorithm of image transition region and its improvement. J. Infrared Millim Waves 20(3), 211–214 (2001)
7. Chen, C.P., Qin, W., Fang, Z.F.: Infrared Image Transition Region Extraction and Segmentation Based on Local Definition Cluster Complexity. In: International Conference on Computer Application and System Modeling, pp. V3-50–V3-54(2010)
8. Yan, C.X., Sang, N., Zhang, T.X., Zeng, K.: Image Transition region extraction and segmentation based on local complexity. Infrared Millim Waves 24(4), 312–316 (2005)
9. Yan, C.X., Sang, N., Zhang, T.: Local entropy-based transition region extraction and thresholding. Pattern Recognition Letters 24, 2935–2941 (2003)

Analysis and Design of Digital Education Resources Public Service Platform in Huanggang

Yun Cheng[1,2], Yanli Wang[1], SanHong Tong[1,*], ZhongMei Zheng[1], and Feng Wang[1]

[1] Department of Educational Science and Technology
HuangGang Normal University, HuangGang, Hubei, China, 438000
[2] Department of Information Technology
Central China Normal University, WuHan, Hubei, China, 430079
yunccnu@126.com, 741645187@qq.com

Abstract. The regional digital educational resources sharing and public service platform is designed to collect and integrate all digital educational resources in certain region and provide an open, flexible and online information sharing and public service platform for all users to acquire quality, efficient, and personalized learning resources and services. This will be an effective way to improve rural education service system. This paper tried to discuss the regional digital educational resources sharing and service in Huanggang from the perspective of "regional service", and designed an educational resources public service platform mainly from three aspects: its system architecture, the framework module and the function of core module. This will be an effective and meaningful practice about regional digital educational resources sharing and service.

Keywords: Rural education, Digital educational resources, regional resources sharing, the public service platform.

1 Introduction

Educational resources construction always has been a very important factor in the process of educational informatization. Digital educational resources break multiple constraints of traditional educational resources, with features of multimedia, network, open and easy to update and management. However, due to the large difference in the distribution of educational resources, educational fund, infrastructure and equipments between towns and big cities, towns and rural areas can not enjoy good quality education resources and services. How to integrate township featured resources and strengthen the sharing of resources between different regions is very worthy of concern. National long-term education reform and development plan (2010-2020) has referred that "enhance the development and application of high-quality educational resources, enhance the construction of online teaching resource library, establish open and flexible educational resources public service platforms and promote the popularity and share of high-quality educational resources to speed up the process of

* Corresponding author.

D. Jin and S. Lin (Eds.): Advances in MSEC Vol. 1, AISC 128, pp. 191–198.
springerlink.com

educational informatization".[1] Therefore, based on the township's infrastructure and educational condition, integrating local featured resources and establishing a regional educational resource public service platform has very important significance for rural educational informatization.

2 Educational Resource and Its Construction and Share

According to Chinese E-Learning Technology Standardization Committee, educational resources refer to the educational information transmitted over the Internet through digital signal model. The construction of teaching resources has four levels: teaching material, online course construction, the evaluation of resource development and the development of educational resources management system. [2] Currently, the vast amounts of digital resources always exist with many problems, such as the lack of quality resources, resource "island", featureless, inconformity for rural teaching, especially how to effectively share resources between towns and cities and promote resources application in rural teaching still has no completely effective solution.

Construction of educational resources needs the joint efforts of our nation, province, municipality, county and schools or other educational departments. To avoid unnecessary duplication, every region or school should focus on the development of featured courses and school-based resource libraries. And different regions or schools can share their resources to form a greater resource library for wider range of users. So development and share of regional educational resources has particular significance for the application and popularization of resources in townships and rural areas.

3 Regional Educational Resources Public Service Platform and Its Role

The educational resource sharing service platform usually refers to the online software system which is built to support resources retrieving, downloading and sharing to enable more users to faster access to quality educational resources through Internet. The information sharing platform of educational resources combines with network technology, database and browser technology for the release of educational resources, and is the most direct information platform of resource sharing system, with the features of highly targeted, interactive and services. Users can quickly and easily search for the needed information and get them from the platform, upload new resources, track resource updating and receive the catalog of new information resources. [3]

So regional educational resource information sharing and public service platform will help to meet users' demands for high-quality, subject featured or regional featured educational resources, strengthen the sharing of educational resources between different regions, and enhance the accessibility and availability of digital educational resources for more users. It is an effective way to make resources development more high-quality, efficient and sustainable to support local education development. Based on the ideas of regional resources sharing and service, we are trying to analyze and design a regional educational resources public service platform according to the actual educational condition of Huanggang city, Hubei Province, aimed to promote the construction, sharing, unified management and service of digital resources in Huanggang.

4 The Framework of the Resources Sharing and Service in Huanggang

Huanggang city has one district, two county-level cities, seven counties, and a county-level farm. Most regions are rural areas. So the infrastructure is not so good, and lack of educational resources, low utilization rate and unevenly distributed exist. But Huanggang, known as a famous basic education area, also has plenty of educational resources in basic education, vocational education and college education. So collecting all kinds of digital educational resources to share and serve for all users in this region has great importance. So according to the need of regional resources sharing and service, we provide the overall framework of resource sharing and service in Huanggang, as shown in Figure 1.

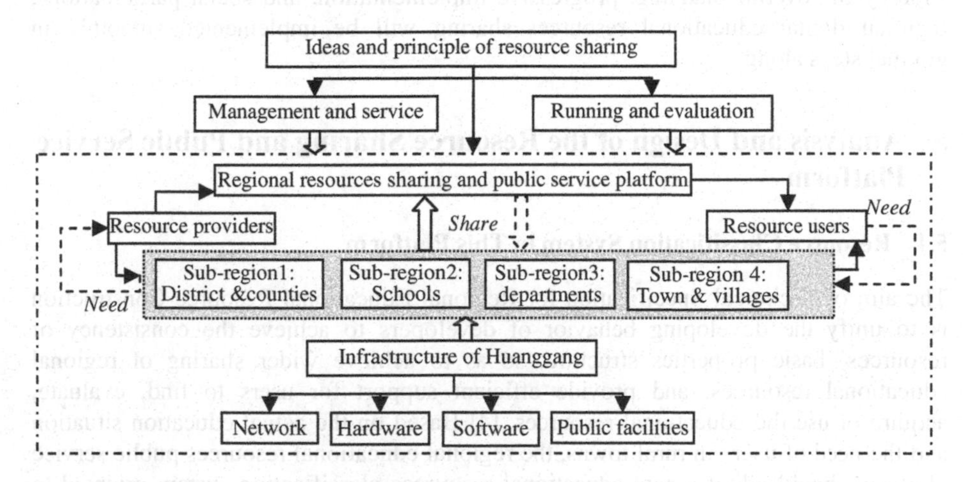

Fig. 1. The overall framework of resource sharing and service in Huanggang

1) The framework is based on the digital infrastructure of Huanggang area, including all kinds of network like Internet, satellite, digital television and telecommunication network, the hardware facilities and software services.

2) According to the geographical location, the whole region can be divided into more sub-regions, such as the audio-visual education office, district, counties, different departments and schools, towns or rural villages. The sub-regions are the main body of developing and using the resources. They collect useful educational resources from local area, or develop featured resources for local use, then share and exchange their resources through regional resources sharing and public service

platform. A sub-region also can develop an independent resources service platform on the local area network, which also can link to the center platform.

3) Regional digital educational resources sharing and public service platform provides a unified portal interface for regional resource publishing, sharing, service and management.

4) Resource providers can be all level schools, training institutions, education companies, county/town government or other institutions and individual users, while the resources users can be teachers, rural schools, training personnel, institutions and community. They interact with each other through the public service platform and sub-regions.

5) And resource management and services includes all management associated with the resources sharing, like costs, personnel, project, production and implementation. Huanggang government and the education department make unified plan for whole process, and provide policy support and supervision. The Education Information Center is responsible for the platform's daily running, and provides technical support, service and evaluation for resource management. Huanggang normal university can also provide some technical support and decision-making for whole resources sharing system.

6) With the guidance of scientific ideas and principles of resources sharing and the strategy of "overall planning, progressive implementation, and social participation", regional digital educational resources sharing will be implemented smoothly in gradual steps along.

5 Analysis and Design of the Resource Sharing and Public Service Platform

5.1 Resource Classification System in This Platform

The aim of Technical Specification for Regional Educational Resource Construction is to unify the developing behavior of developers to achieve the consistency of resources' basic properties structure; so as to achieve wider sharing of regional educational resources, and provide efficient support for users to find, evaluate, acquire or use the educational resources. [4] Based on the actual education situation and the need of users in rural towns, the regional educational resources public service platform should adopt a new educational resources classification system oriented to townships. According to the research of professor Liu Qingtang, Central China Normal University, educational resources classification model oriented to towns and villages include six aspects: its application field classification system, its subjects and curriculum classification system, its application objects and users classification, the version of textbooks, resources format and resource types. [5]Each classification also has its sub-classification and the described methods and systems. For example, the application field classification system can be divided into basic education, training services, vocational education, higher education and other educational areas. So we take this six-dimensional model classification system to classify the digital educational resources in the platform.

5.2 The System Architecture of the Educational Resources Public Service Platform

The platform in Huanggang provides "one-stop" resources and learning support service for users. So internally it can realize a variety of resources processing and management, such as resources classification, resources display/download/upload, resources retrieval, configuration, statistics, monitoring and evaluation. And externally users can realize functional interaction with the platform through web portal to achieve a variety of service, including user registration, resources service, learning support, information and communication service. The platform adopts three-tier architecture to achieve maximally "high cohesion and low coupling"[6], as shown in Fig.2.

The top Performance layer is the web portal of the platform to provide users with an interactive operation interface to achieve a variety of service. The business logic layer in the middle is responsible for all digital resources processing and management associated with the resources classification, display, upload and download, retrieval, configure, evaluation and management. The underlying database is to provide all data service for the upper business logic layer.

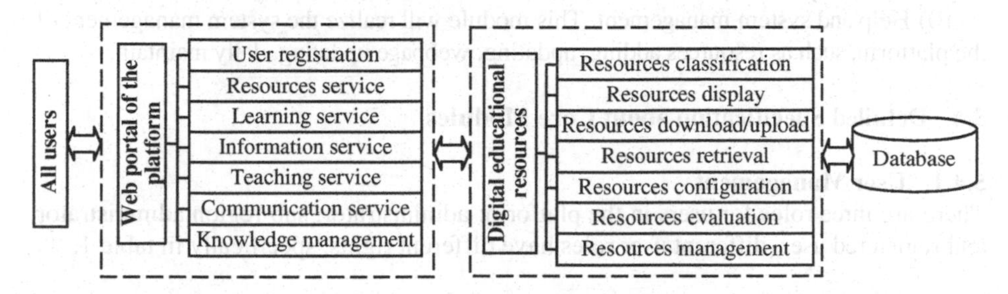

Fig. 2. The system architecture of the platform

5.3 The Framework Module Design of the Platform

The platform is designed to encourage all people and communities in Huanggang to participate in the activities of educational resources construction, sharing and application. So the framework module of the platform is shown as follows:

1) User management. This module mainly can realize user registration/accreditation, personal information revise, login / logout, credits, role management, right distribution and other personal information management.

2) Classified Resources display. All resources will be classified by the application field, the subjects and curriculum, the application objects, textbooks version, resources format and resource types. Then classified resources will be listed in the main web portal so that users can browse and search needed resources easily.

3) Recommendation of the newest/hottest resources. This module will display the top 10 new resources by their adding time, or display the top 10 hot resources by their downloading times, so that users will know the newest/hottest resources in this platform for their choice.

4) User evaluation and communication. Users can give their comments or opinions about each resource to show its quality, application effect or suggestion. This will provide reference for other users' choosing and using them in teaching.

5) Experience of teaching and resources application. This module will publish some articles about informationized teaching, experience of resources application in teaching and learning, or ideas about resources construction and sharing, as a space for users to exchange knowledge and experience in the process of resources construction and application.

6) Blog group of teachers. This module will organize some teachers' personal Blog as a group to form a learning community, which will promote local digital educational resources sharing and application in teaching practice.

7) Special subjects resources of Huanggang High School. This module will share the special subject resources from Huanggang High School; so that the high quality featured resources can be shared or used by other school.

8) News/Announcement management. This module will publish some news or announcements about the resources construction in Huanggang.

9) Other resource links. This module will provide some quick links to other educational resources service platforms or libraries, such as national educational resources library for basic education and so on.

10) Help and system management. This module will realize the system management of the platform, such as resources adding, updating, webpage updating, daily maintain.

5.4 Detailed Specification about Core Modules

5.4.1 User Management

There are three roles for users in the platform: administrator, sub-region administrator and registered user, different user roles have different rights, specifically in table 1.

Table 1. User roles and operation rights in the platform

User roles	Operation rights
Administrator	User management (user registration and approval, user information management, user rights assignment, group management), Announcement management (publish and update public news), sub-region management (sub-region approve, information/status view, user statistics), the resources management (resources classification, uploading, retrieving, modify and publishing information, etc.), users' comment management, system management, help.
Sub-region administrator	User management, news and articles management and resources management of local region.
Registered user	Personal information revise, browse, search, download and upload resources, publishing articles and teaching experiences, manage their own articles and resources, select group, give comments on resources.

It is noticed that all the users in this platform should be registered in real information, including real names, regions, units and other information, so that the administrator can know the number and state of users in different sub-region. The system will give reward credits to excellent users according to their login times, uploading/downloading resources times, comments and application situation.

5.4.2 Resources Uploading and Classification Display

All the resources uploaded to the platform should be met the standardization requirement, and users need to fill with the description information, including authors, application field, objects which is suitable for, subject and curriculum, textbooks, resource format, resource type, resource description and other information. This will be helpful for users to quickly and accurately manage, retrieve, browse and use their needed resources. According to six-dimension classification system for township resources, all registered resources will be organized in multilevel directory tree, which is ordered by application areas - for objects – the subject and curriculum - resource format. Each level also can be categorized in different sub-trees according to school-age section or subjects and corresponding resources will be showed in a list. Here the application areas are divided into basic education (high school section, middle school section, primary school section, pre-school), vocational education, higher education, training (teacher training, rural party member training, worker training, community service), other education (all kinds of public resources and introduced resources). Take basic education as an example, the directory tree is ordered by school segment – grade–subject–resource format. The tree can be fully expanded or partly expanded, and accordingly different resources list will be showed. Select the needed resource, users will see detail information, can download it or give comments about it.

5.4.3 Evaluation and Feedback of Educational Resources

The evaluation mechanism of regional digital educational resources sharing refers to using teaching evaluation criteria to conduct quality certification, progress check, efficiency assessment for digital educational resources, so as to ensure the quality of the shared educational resources and increase the benefits of resources sharing. [7]In this platform, we design the "User evaluation and communication" module, as an effective way to reflect the resources quality. Administrators regularly collect and analyze the evaluation information to find out some problems timely, and correct them to improve the quality and level of regional educational resource construction and sharing.

In the same time, as the feedback of evaluation result, we can set up some reward measures for sub-regions, departments or individual which have good performance in the resources construction process to stimulate all users' enthusiasm to develop and apply the digital resources. For example, the government can give formal reward to certain sub-region or individual users for their outstanding performance each year according to the comprehensive performance like users' number, login times, uploading and downloading resources, application situation and evaluation.

6 Conclusion

The joint construction and sharing of regional digital educational resources is a systematic project. Every region should pay attention to it, make scientific planning and strategy, establish a standard, advanced and practical educational resources public services platform to enhance communication and collaboration between different regions. This paper discussed the overall framework of digital educational resources sharing and service in Huanggang, and designed a resource public service platform to collect, publish and share various types of educational resources and provide relevant service to all users. In the future, this platform should promote the development of online courses, multimedia courses, and other online resources for users' online learning, and promote resource sharing in wider range and for more users.

Acknowledgments. This research is supported by the Humanities and Social Sciences Research Project of HuangGang Normal University (NO. 2011CB098); Key Research Institute of Humanities and Social Sciences in Universities of Hubei Province-Research Centre for Education and Culture of East Hubei Province (2011).

References

[1] National long-term education reform and development plan (2010-2020), http://www.gov.cn/jrzg/2010-07/29/content_1667143.htm
[2] Chinese E-Learning Technology Standardization Committee. Technical Specification for Educational Resource Construction. CELTS-41.1 (2002)
[3] Cheng, J.-J., Huang, J.-J., Pan, Y.: Theoretical and practical study on the sharing approach of regional digital educational resources. Modern Education Technology 21(4), 125–129 (2011)
[4] Liu, Q., Liu, M., Xie, Y., Li, H., Hu, M.: Research and Application of Educational Resource Classification System Oriented to Townships. E-education Research, 57-63 (December 2010)
[5] Li, H.: Design and Implementation of Rural Educational Resource Regional Online Service Platform. Thesis of M.S. Degree, Central China Normal University, Wuhan (2010)

Study on E-commerce Course Practice and Evaluation System under Bilingual Teaching

Yiqun Li, Quan Jin, and Xuezheng Zhang

International Business School, Shanghai Institute of Foreign Trade, SIFT,
1900 Wenxiang Road, 201620 Shanghai, China
{liyq,jinquan,zxz}@shift.edu.cn

Abstract. Based on the undergraduate program in e-commerce, this paper mainly discusses how to establish the reasonable bilingual teaching course program and strengthen students' cross-cultural communication ability of e-commerce training and practices. Combined with the common goal of bilingual teaching in higher education, it is proposed from the perspectives of e-commerce professional knowledge system and talent cultivation target that the solid specialized theory and rich practical innovation ability should be emphasized to construct e-commerce bilingual teaching courses program based on the social demand. Furthermore, the e-commerce course program of bilingual teaching is put forward including three objectives as knowledge, quality and ability, covering four areas of course, such as disciplines basics, professional cores, professional electives and practical education, and some teaching highlights. From understanding of the connotation and the bilingual teaching mode and combining with the features of Shanghai Institute of Foreign Trade e-commerce English teaching practice, it is proposed in this paper a reasonable evaluation system of bilingual teaching model based on four perspectives, the characteristics of teachers and students, course design and, curriculum organization and learning outcome based on comprehensive, motivational, subjective and instructive principle.

Keywords: e-commerce, bilingual teaching, course practice, evaluation system.

1 Introduction

Bilingual teaching in China is formulated by economic globalization and the development of modern educational technologies and it is a teaching system which its environment of teaching language, curriculum and teaching resources, training objectives and so on are continuously innovated. At present, the researches of bilingual teaching are always focus on the form and content of teaching, based on the undergraduate program in e-commerce, this paper mainly discusses how to establish the reasonable bilingual teaching course program and strengthen students' cross-cultural communication ability of e-commerce training and practices.

This paper began with the higher education course program and analyzed the three elements of it , there are “objectives curriculum and teaching process” , and it combined with the common goal of bilingual teaching in higher education, it is

D. Jin and S. Lin (Eds.): Advances in MSEC Vol. 1, AISC 128, pp. 199–204.
springerlink.com

proposed from the perspectives of e-commerce professional knowledge system and talent cultivation target that the solid specialized theory and rich practical innovation ability should be emphasized to construct e-commerce bilingual teaching courses program based on the social demand. Furthermore, the e-commerce course program of bilingual teaching is put forward including three objectives as knowledge, quality and ability, covering four areas of course, such as disciplines basics, professional cores, professional electives and practical education, and some teaching highlights, such as case study, group practice, network platform and special lectures.

2 E-commerce Bilingual Teaching Programs

Since the Ministry of Education allowed thirteen high schools to establish e-commerce professional in 2000 till to now, nearly 400 universities has established e-commerce professional in whole country. Through the relevant statistical from the Electronic Commerce Steering Committee member of university, the arrangement of core theory courses of these three types shows the cross-feature of e-commerce subject, and it covers economics, management, information management and computer applications and other disciplines.

According to the approval from China's Electronic Commerce Steering Committee of University, "Higher Education undergraduate e-commerce knowledge of professional education" published in 2008 (trial), e-commerce professional knowledge system can be divided into some distribution like tree of four levels as knowledge areas, knowledge modules, knowledge units and knowledge points according to classification of professional disciplines and content of knowledge, and knowledge points will be described by curriculum system. [1] (Table 1.)

Core knowledge unit is a public portion of the content which all the students of this subject direction must master. Optional knowledge unit is a type of professional knowledge that it is mastered by some one subject direction, but is only optional content of knowledge for other subject directions. Core knowledge unit itself can not be the whole of integral professional knowledge system, and it must be combined with related optional knowledge unit to build up integral knowledge system. Each university can select related optional knowledge unit according to the features of its subjects and the direction of professional training.

Table 1. E-commerce professional knowledge system

Knowledge area	Number of Modules	Number of Units	
		Required	Optional
E-commerce technology	4	7	3
E-commerce economy	5	15	18
E-commerce management	9	31	13
Integrated e-commerce	6	21	13
Total	**24**	**74**	**47**

The e-commerce major in Shanghai Institute of Foreign Trade was established in 2002, approved by State Education Commission. During the development of professional training model and training system, we mainly aimed at talents, multi-target adaptive, and build talent system mainly from three parts of knowledge, ability and quality. In the process of curriculum construction, attention to improve the teaching materials, strengthen bilingual teaching, rich the content of practical course, support teachers' training for innovative curriculum and curriculum construction, fully mobilize the advantages of network teaching of e-commerce courses, comprehensively promote that teaching materials electronic, teaching methods multimedia, teaching methods open, teaching platform networking, continuously strengthen teacher-student interaction and school-enterprise cooperation. (Figure 1)

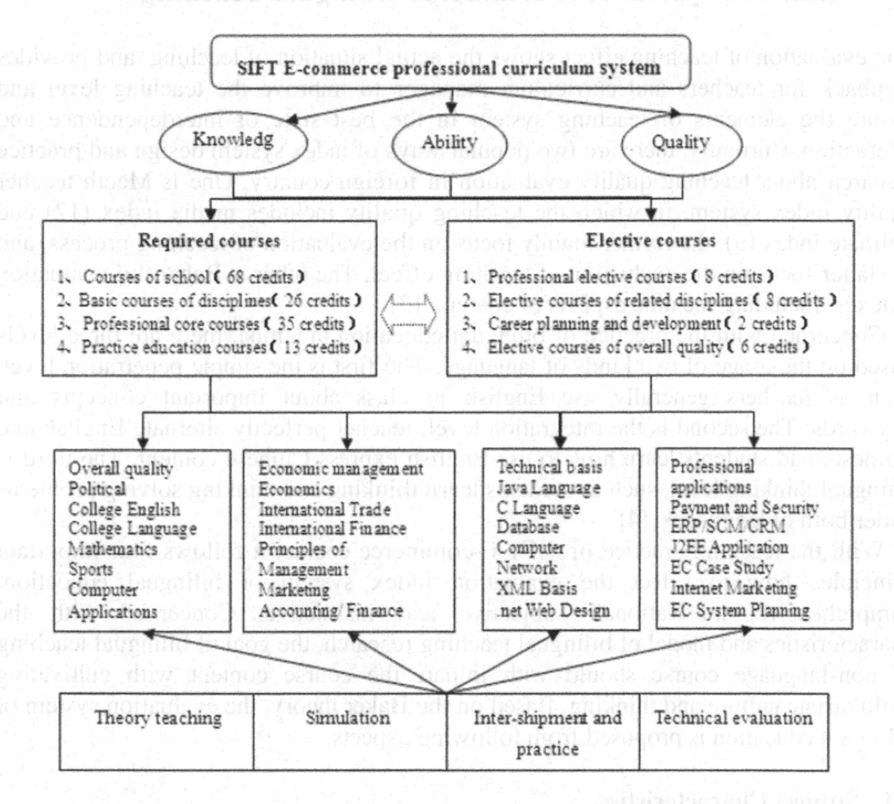

Fig. 1. The existing curriculum system of e-commerce courses in SIFT

Combined with the current curriculum construction of e-commerce courses and the teacher conditions and the content of professional training, we could clear the main courses of bilingual teaching system are: basic courses of disciplines, professional core courses, professional elective courses and practice education courses, and extended based on different curriculum. Through the comparative better education of bilingual curriculum system of e-commerce courses, it trains the students to get management theory and basement of analysis, and makes them to understand the reasons, technical background, scope of business application and characteristics, basement of industry's support etc of the emergence and rapid development of international e-commerce at home and abroad; With a comparative strong practice ability of professional English application, it is familiar with the operation of domestic and international e-commerce and master the basic management decision-making skills, explores e-commerce business strategy in different areas; and thus improves their ability of using information technology to carry out the capacity of the network, and obtaining the required professional knowledge and skills of the implementation of e-commerce under the background of cross-cultural.

3 Evaluation System of E-commerce Bilingual Teaching

The evaluation of teaching effect shows the actual situation of teaching, and provides feedback for teachers and curriculum manager to improve the teaching level and ensure the elements of teaching system in the best state of interdependence and interaction. Currently, there are two popular ways of index system design and practice research about teaching quality evaluation in foreign country. One is Meeth teacher quality index system, in which the teaching quality includes media index (12) and ultimate index (9), the former mainly focus on the evaluation of teaching process, and the latter focus on the evaluation of teaching effect. The other is Babanskii evaluation indexes, including the nine aspects of content. [3]

Concerned with the practice of bilingual education in China, there are three levels based on the usage of two kinds of language. The first is the simple penetration level, such as teachers generally use English in class about important concepts and keywords. The second is the integration level, teacher perfectly alternate English and Chinese, and students learn how to use English express Chinese content. The third is bilingual thinking level, such as students learn thinking and thinking solving problems under both of languages. [4]

With the teaching practice of SIFT E-commerce course, it follows four important principles how to select the evaluation index system of bilingual education: comprehensive, motivational, subjective and instructive. Concerned with the characteristics and model of bilingual teaching research, the goal of bilingual teaching of non-language course should with impart the course content with cultivating multivariate culture and thinking. Based on the Baker theory, the evaluation system of bilingual education is proposed from following aspects.

3.1 Subject Characteristics

Bilingual education is a complicated process, and the effect of bilingual teaching firstly depends on the quality of input in the teaching process, which mainly depends on the quality of teachers and students' ability and motivation. Therefore, it should begin from the quality of input elements to control and improve the quality of bilingual teaching. For teachers, both of professional ability and language skills should be assessed. It should take initiatives to encourage teachers to open bilingual course and ensure the quality of the bilingual education. For students, the language ability will be assessed to accept the bilingual courses under some conditions. In addition, the investigation of students' study enthusiasm should be necessary to determine the scope of bilingual education. The class of bilingual education without a comprehensive promotion can be further considered as different level and stratified types.

3.2 Course Design

Both of general courses and bilingual courses all need to be designed to plan teaching stages and the course planned activities will affect the quality of teaching quality. Therefore, the teaching quality of bilingual education will focus on evaluation of the curriculum design. In addition, some specific factors of bilingual teaching also should be combined with the school's objectives and nature of the course itself. Therefore the design of teaching evaluation can be used by teaching management department to assess the course level, namely whether the bilingual curriculum design meets the school requirement and the course characteristics. Also it can combine with the teaching object to evaluate students' feelings, whether students think that bilingual course design can help them understand the course content and form the critical thinking of particular field.

3.3 Curriculum Organization

The teaching process of bilingual course, especially the curriculum organization in class, has directly influence on teaching quality of bilingual education. As mentioned before, the teaching process of bilingual course not only includes knowledge and language learning, but also affects on students' thinking and values, cultural diversity and a multi-angle thinking through the process of classroom interaction between teachers and students. Therefore the evaluation of bilingual courses firstly should pay more attention to the organization ability of classroom interaction, both teachers and students, and students and students. Secondly, it should focus on whether teaching material is effective and reasonable use in class, and how teachers guide students to use the teaching materials reasonably, and to learn from out of class. It also should be one of the evaluation indexes of curriculum organization to cultivate and improve students' learning ability to achieve the goal of bilingual courses.

3.4 Learning Outcome

In particular model of bilingual education, the evaluation of teaching effect on bilingual courses not only is same as the general course, but also includes specific characteristic. Fox example, the students' ability of using foreign language should be taken into the evaluation index of teaching effect. The higher goal of bilingual education is the forming of multicultural understanding and critical thinking. Thus the evaluation of bilingual teaching effect also need includes students' values and thinking mode change. And the study of bilingual teaching system should extend beyond the course, and also need further evaluation and learn whether the learning in bilingual education is changed their long-term behavior, such as whether to use more foreign language than other student, whether to use multiple thinking and multicultural to study, work and life of the other aspects and so on.

4 Conclusion

Through several years constructing in special courses and practical teaching, the E-commerce major in Shanghai Institute of Foreign Trade has formed the preliminary bilingual course system, set the training objective for the bilingual teaching clearly. Teaching circumstance and effect has gradually become better. Bilingual's faculty has got improved steadily. In this paper, only a preliminary design on the evaluation system of bilingual teaching effect is proposed from four perspectives combined with the teaching practice of SIFT E-commerce course. In fact, it needs long-term research work of continuing investigation and extensive experiments to design a comprehensive evaluation system. At present, the promotion of bilingual education in colleges and universities is higher requirement for teachers and students, which needs both of them constantly exploring and progressing in teaching and learning practice. In this process, an active evaluation system provides an effective incentive and feedback mechanism to expand bilingual education.

Acknowledgements. This work is supported by the 2009 English–teaching Model Course Project of Shanghai Municipal Education Commission and the 085 Knowledge Innovation project (R08509003) of Shanghai Institute of Foreign Trade. The authors are grateful for the anonymous reviewers who made constructive comments.

References

1. E-commerce Program Education Instruction Committee, Ministry of Education. E-commerce Undergraduate Knowledge System (trial). Higher Education Publish press (2008)
2. Mike, M.F., et al.: Outline of the Bilingual Education, Guang Ming Daily (1989)
3. Ma, Y.N.: Research on the Bilingual Education in the High Education School. Peking University Journal (Philosophy and Social Scientist) (2007)
4. Wang, X.X.: Research on Bilingual Education in China, Da Liang University of Technique and Science (2006)
5. Baker, C.: Key Issues in Bilingualism and Bilingual Education, Clevedon, England. Multilingual Matters 35, 222 (1988)
6. Li, Y., Zhang, X.: Study on The Construction of Stereoscopic Teaching Model——The Practice Training of E-commerce Discipline in SIFT. In: International Colloquium on Computing, Communication, Control, and Management (CCCM), pp.114–118 (2009)

Non-termination Analysis of Polynomial Programs by Solving Semi-Algebraic Systems

Xiaoyan Zhao

Staff Room of Mathematics and Computer
North Sichuan Medical College, Nanchong, China
Zhaoxiaoyan79@163.com

Abstract. This paper reduces the non-terminating determination for polynomial programs to semi-algebraic system solving, and demonstrates how to apply the symbolic computation tool DISCOVERER to some interesting examples. This method proceeds in three phases, by first constructing a semi-algebraic system according to the given polynomial program, and then calling DISCOVERER to solve frontal semi-algebraic system, finally analyzing the result from DISCOVERER with the initial state of polynomial program to determine whether the program terminates or not.

Keywords: semi-algebraic system, polynomial program, non-termination.

1 Introduction

It is indisputable that the main value of verification tools lies in the discovery of feasible bugs, not in the correctness proof of program [1]. Presenting a bug to a programmer is often a more convincing demonstration of the utility of a tool than a proof of correctness.

Since techniques to prove termination are always incomplete, failure to prove termination does not immediately indicate the existence of a non-terminating execution. Therefore, a failed termination proof produces a possible counterexample. Currently, these counterexamples must be inspected manually to determine if they are indeed bugs. Furthermore, classical objects of study in temporal verification are auxiliary assertions geared towards proving temporal properties. Auxiliary assertions that demonstrate existence of non-terminating executions have not received adequate attention [2]. Therefore, dual to the search for termination proofs, we should also develop tools that demonstrate feasible non-terminating executions.

We present a method that analysis non-terminating program execution by solving semi-algebraic transition system (SATS), which is an extension of algebraic transition systems in [3]. The SATS is used to represent polynomial programs [4]. Our algorithm proceeds in three phases. For a loop in a given polynomial program, we can first translate it to a SATS in regard with its program variables. In order to construct the corresponding SATS, we need to get the current state and its next state of loop conditional expression according to its loop body, and prove that the loop is termination if and only if each of the SATS has real solutions for any input, non-termination otherwise. After the translation we apply the function of root

D. Jin and S. Lin (Eds.): Advances in MSEC Vol. 1, AISC 128, pp. 205–211.
springerlink.com

classification for SAS [5] of DISCOVERER to each of the SAS to generate conditions on solution. The second phase solves SATS for possible non-termination. Finally, the result from DISCOVERER was analyzed with the initial state of polynomial program to determine whether the program terminates or not. In general, the method is incomplete, as not all non-terminating program executions are solved precisely. However, we can concentrate on finding the most common non-termination bugs quickly.

This method has two advantages. First, the execution of the loop is guaranteed to be feasible. Second, such SATS is easy to be generated. We formulate and solve the existence of a real solution as a constraint satisfaction problem. The constraint satisfaction problem turns out to be equivalent to constraint systems for SATS solving [3, 6].

The rest of this paper is structured as follows: Section 2 presents a brief review of the theories and tools of semi-algebraic systems, and their implementations in the computer algebra tool DISCOVERER; The notion of algebraic transition systems of [3] was extended to semi-algebraic transition system to represent polynomial programs in Section 3; In Section 4, we use examples to illustrate the non-termination analysis of polynomial loops by SAS solving; and Section 5 draws a summary and discusses future work.

2 Theories and Tools on Solving Semi-Algebraic Systems

In this section, we introduce the theories of semi-algebraic system and the tool DISCOVERER on solving SATS [4].

2.1 Semi-Algebraic Systems

Let K be a field, $X=\{x_1,x_2,\cdots,x_n\}$ a set of indeterminates, and $K[x_1,x_2,\cdots,x_n]$ the ring of polynomials in the n indeterminates with coefficients in K, ranged over $p(x_1,x_2,\cdots,x_n)$ with possible subscription and superscription. Let the variables be ordered as $x_1 \prec x_2 \prec \cdots \prec x_n$. Then, the leading variable of a polynomial p is the variable with the biggest index which indeed occurs in p. If the leading variable of a polynomial p is x_k, p can be collected w.r.t its leading variable as $p=c_m x_k^m+\cdots+c_0$ where m is the degree of p w.r.t. x_k and c_i s are polynomials in $K[x_1,x_2,\cdots,x_{k-1}]$. We call $c_m x_k^m$ the leading term of p w.r.t. x_k and c_m the leading coefficient.

Anatomic polynomial formula over $K[x_1,x_2,\cdots,x_n]$ is of the form $p(x_1,x_2,\cdots,x_n) \rhd 0$, where $\rhd \in \{=,>,\geq,\neq\}$, while a polynomial formula over $K[x_1,x_2,\cdots,x_n]$ is constructed from atomic polynomial formulae by applying the logical connectives. Conjunctive polynomial formulae are those that are built from atomic polynomial formulae with the logical operator $\wedge$. We will denote by $PF(\{x_1,x_2,\cdots,x_n\})$ the set of polynomial formulae and by $CPF(\{x_1,x_2,\cdots,x_n\})$ the set of conjunctive polynomial formulae, respectively.

In what follows, we will use Q to stand for rationales and R for real, and fix K to be Q. In fact, all results discussed below can be applied to R.

In the following, the n indeterminates are divided into two groups: $u=(u_1,u_2,...,u_d)$ and $x=(x_1,x_2,\cdots,x_s)$, which are called parameters and variables, respectively, and we sometimes use "," to denote the conjunction of atomic formulae for simplicity.

Definition 1. A semi-algebraic system is a conjunctive polynomial formula of the following form:

$$\begin{cases} p_1(u,x)=0,...,p_r(u,x)=0, \\ g_1(u,x)\geq 0,...,g_k(u,x)\geq 0, \\ g_{k+1}(u,x)>0,...,g_t(u,x)>0, \\ h_1(u,x)\neq 0,...,h_m(u,x)\neq 0. \end{cases} \quad (1)$$

Where $r > 1$, $t \geq k \geq 0$, $m \geq 0$ and all p_i's, g_i's and h_i's are in $Q[u,x]\setminus Q$. An SAS of the form (1) is called parametric if $d \neq 0$, otherwise constant.

An SAS of the form (1) is usually denoted by a quadruple [P, G1, G2, H], where P $=[p_1,p_2,...,p_r]$, G1$=[g_1,g_2,...,g_k]$, G2$=[g_{k+1},...,g_t]$ and H$=[h_1,...,h_m]$.

2.2 DISCOVERER

In this section, we will give a short description of the main functions of DISCOVERER, which includes an implementation of the algorithms presented in the previous subsection with Maple. The reader can refer to [7, 8] for details.

The prerequisite to run the package is Maple 7.0 or a later version of it.

For a parametric SAS T of the form (1) and an argument N, where N is one of the following three forms:

– a non-negative integer b;
– a range b..c, where b, c are non-negative integers and b < c;
– a range b..w, where b is a non-negative integer and w is a name without value, standing for $+\infty$.

DISCOVERER can determine the conditions on u such that the number of the distinct real solutions of T equals to N if N is an integer, otherwise falls in the scope N. This is by calling tofind([P], [G1], [G2], [H], $[x_1,...,x_s]$, $[u_1,...,u_d]$,N), and results in the necessary and sufficient condition as well as the border polynomial BP of T in u such that the number of the distinct real solutions of T exactly equals to N or belongs to N provided $BP \neq 0$. If T has infinite real solutions for generic value of parameters, BP may have some variables.

Then, for the "boundaries" produced by "tofind", i.e. $BP = 0$, we can call Tofind([P,BP], [G1], [G2], [H], $[x_1,...,x_s]$,$[u_1,...,u_d]$,N) to obtain some further conditions on the parameters.

3 Polynomial Programs

A polynomial program takes polynomials of $R[x_1,...,x_n]$ as its only expressions, where $x_1,...,x_n$ stands for the variables of the program. Polynomial programs include expressive class of loops that deserves a careful analysis.

For technical reason, similar to [3], we use algebraic transition systems (ATSs) to represent polynomial programs. An ATS is a special case of standard transition system, in which the initial condition and all transitions are specified in terms of polynomial equations. We extend the notion of algebraic transition systems in [3] by associating with each transition a conjunctive polynomial formula as guard and allowing the initial condition possibly to contain polynomial inequalities. We call such an extension semi-algebraic transition system (SATS). It is easy to see that ATS is a special case of SATS.

Definition 2. A semi-algebraic transition system is a quintuple $<V,L,T,l_0,\Theta>$, where V is a set of program variables, L is a set of locations, and T is a set of transitions. Each transition τ∈T is a quadruple $<l_1,l_2,\rho_\tau,\theta_\tau>$, where l_1 and l_2 are the pre- and post- locations of the transition, $\rho_\tau \in$CPF(V, V') is the transition relation, and $\theta_\tau \in$CPF(V) is the guard of the transition. Only if θ_τ holds, the transition can take place. Here, we use V' (variables with prime) to denote the next-state variables. The location l_0 is the initial location; and Θ∈CPF (V) is the initial condition.

A state is an evaluation of the variables in V and all states are denoted by Val (V). Without confusion we will use V to denote both the variable set and an arbitrary state, and use F (V) to mean the (truth) value of function (formula) F under the state V. The semantics of SATSs can be explained through state transitions as usual.

A transition is called separable if its relation is a conjunctive formula of equations which define variables in V' equal to polynomial expressions over variables in V. It is easy to see that the composition of two separable transitions is equivalent to a single separable one. An SATS is called separable if each transition of the system is separable. In a separable system, the composition of transitions along a path of the system is also equivalent to a single separable transition. We will only concentrate on separable SATSs as any polynomial program can easily be represented by a separable SATS [9]. Any SATS in the rest of the paper is always assumed separable.

For convenience, by $l_1 \xrightarrow{\rho_\tau,\theta_\tau} l_2$ we denote the transition $\tau=(l_1,l_2,\rho_\tau,\theta_\tau)$, or simply by $l_1 \xrightarrow{\tau} l_2$. A sequence of transitions $l_{11} \xrightarrow{\tau_1} l_{12} \cdots l_{n1} \xrightarrow{\tau_n} l_{n2}$ is called composable if $l_{i2}=l_{(i+1)1}$ for $i=1,\cdots,n-1$, and written as $l_{11} \xrightarrow{\tau_1} l_{12}(l_{21}) \xrightarrow{\tau_2} \cdots \xrightarrow{\tau_n} l_{n2}$. A composable sequence is called transition circle at l_{11}, if $l_{11}=l_{n2}$. For any composable sequence $l_0 \xrightarrow{\tau_1} l_1 \xrightarrow{\tau_2} \cdots \xrightarrow{\tau_n} l_n$, it is easy to show that there is a transition of the form $l_0 \xrightarrow{\tau_1;\tau_2;...;\tau_n} l_n$ such that the composable sequence is equivalent to the transition, where $\tau_1;\tau_2;...;\tau_n, \rho_{\tau_1;\tau_2;...;\tau_n}$ and $\theta_{\tau_1;\tau_2;...;\tau_n}$ are the compositions of $\tau_1,\tau_2,...,\tau_n, \rho_{\tau_1},\rho_{\tau_2},...,\rho_{\tau_n}$ and $\theta_{\tau_1},\theta_{\tau_2},...,\theta_{\tau_n}$, respectively. The composition of transition relations is defined in the standard way, for example, $x'=x^4+3; x'=x^2+2$ is $x'=(x^4+3)^2+2$; while the composition of transition guards have to be given as a conjunction of the guards, each of which takes into account the past state transitions. In the above example, if we assume the first transition with the guard $x+7=x^5$, and the second with the guard $x^4=x+3$, then the composition of the two guards is $x+7=x^5 \wedge (x^4+3)^4=(x^4+3)+3$. That is,

Theorem 1. For any composable sequence $l_0 \xrightarrow{\tau_1} l_1 \xrightarrow{\tau_2} \cdots \xrightarrow{\tau_n} l_n$, it is equivalent to the transition $l_0 \xrightarrow{\tau_1;\tau_2;\ldots;\tau_n} l_n$.

Example 1. Consider the SATS:

$$P \triangleq \begin{Bmatrix} V=\{x\}, \\ L=\{l_0,l_1\}, \\ T=\begin{Bmatrix} \tau_1=<l_0,l_1,x'=x^2+7,x=5>, \\ \tau_2=<l_1,l_0,x'=x^3+12,x=12> \end{Bmatrix}, \\ l_0,\theta=x=5 \end{Bmatrix}$$

According to the definition, P is separable and $l_0 \xrightarrow{\tau_1} l_1 \xrightarrow{\tau_2} l_0$ is a composable transition circle, which is equivalent to $<l_0,l_0,x'=(x^2+7)^3+12, x=5 \wedge x^2+7=12>$.

4 Non-termination Analyses of Polynomial Loop Programs

For a given program, if its conditional expression is always been met, this program is non-terminated. If there is a state V' such that its conditional expression is not met, the program must terminate. So we can verifying the polynomial programs whether it has a state V, which meet the program condition, but its next state V' will not meet the program condition. If there isn't such state V, we can immediately conclude that this program is not terminates. Combined with the theories of SAS, we can construct a SATS according to the polynomial programs, and determine it terminates or not according to the following algorithm.

Step 1: construct a SATS according to the polynomial programs.

Example 2. Consider the program P0 shown as following:

P0: While(*x-y*>0) { *x*=-*x*+*y*;}

We can immediately get the corresponding SATS:

$$P=\{V=\{x,y\},L=\{l_0\},T=\{\tau_1\},\Theta=\{\}\}$$

$$where$$

$$\tau_1 :< l_0, l_0, x'+x-y=0, y'-y=0,$$

$$x-y>0, -x'+y'>0>$$

Step 2: Calling DISCOVERER to solve this SATS.

Example 3. For program P0, we should call DISCOVERER as follow:

tofind([$x'+x-y$, $y'-y$],[],[$x-y$, $-x'+y'$], [],[x' , y'],[x, y],1..n);

Step 3: Analyzing the results obtained from DISCOVERER.

Firstly, if the result shows that there is no solution of the SATS, so the polynomial is non-termination.

Example 4. Considering following program.

P1 : While(*x*>0) { *x*=*2x*; }

The corresponding SATS is:

$$P = \{V = \{x\}, L = \{l_0\}, T = \{\tau_1\}, \Theta = \{\}, where$$
$$\tau_1 :< l_0, l_0, x'-2x = 0, x > 0, -x' > 0 >\}$$

Calling: tofind([$x'-2x$],[],[x, x'],[], [x'],[x],1..n);

Its result shows that there is no real solution, so program P1 is not terminates.

Secondly, if the result shows that there has solution of the SATS for any input, so the polynomial program will terminate under any conditions.

Example 5. Let's considering following program.

P2 : While(*x*>0) { *x*=*x*-*1*; }

The corresponding SATS is :

$$P = \{V = \{x\}, L = \{l_0\}, T = \{\tau_1\}, \Theta = \{\}, where$$
$$\tau_1 :< l_0, l_0, x'-x+1 = 0, x > 0, -x' > 0 >\}$$

Calling: tofind([$x'-x+1$],[],[x , $-x'$],[],[x'], [x],1..n);

Its result shows that there is real solution for any input, so the program P2 will terminate under any condition.

Finally, the result shows that there has solution of the SATS under some conditions. In this case, we should analyses the result in company with the initial conditions.

Example 6. Let's considering following program.

P3 : While(*x*-*y*>0) { *x*=*x*+*1*; *y*=*x*+*y*; }

The corresponding SATS is :

$$P = \{V = \{x, y\}, L = \{l_0\}, T = \{\tau_1\}, \Theta = \{\}, where$$
$$\tau_1 :< l_0, l_0, x'-x-1 = 0, y'-x'-y = 0, x-y > 0, -x'+y' > 0 >\}$$

Calling: tofind([$x'-x-1$, $-x'+y'-y$],[], [$x-y$, $-x'+y'$],[],[x' , y'],[x , y],1..n);

Its result shows that there is real solution if and only if *y*>0. We can get that variable *y*'s state, after looping n times, can be expressed by $y(n) = y(0) + n \times x(0) + \frac{n(n-1)}{2}$, where $x(0)$ and $y(0)$ is the initial value of variable x and y respectively. On the basis of analysis to this program, we can easily get that variable y will be greater than 0 as the program executed continually, so this program terminates.

5 Conclusions

This paper uses the techniques on solving semi-algebraic systems to analyze non-termination of polynomial programs, and also shows how to use computer algebra tools DISCOVERER to solve semi-algebraic systems for some interesting programs.

References

1. Godefroid, P., Klarlund, N., Sen, K.: DART: Directed Automated Random Testing. In: Sarkar, V., Hall, M.W. (eds.) Proc. of the ACM SIGPLAN 2005 Conference on Programming Language Design and Implementation (PLDI 2005), pp. 213–223. ACM (2005)
2. Gupta, A.K., et al.: Proving Non-Termination. In: POPL 2008, San Francisco, California, USA, January 7-12 (2008)
3. Sankaranarayanan, S., Sipma, H.B., Manna, Z.: Non-linear loop invariant generation using Gröbner bases. In: ACM POPL 2004, pp. 318–329 (2004)
4. Chen, Y., Xia, B., Yang, L., Zhan, N., Zhou, C.: Discovering Non-linear Ranking Functions by Solving Semi-algebraic Systems. In: Jones, C.B., Liu, Z., Woodcock, J. (eds.) ICTAC 2007. LNCS, vol. 4711, pp. 34–49. Springer, Heidelberg (2007)
5. Yang, L., Hou, X., Xia, B.: A complete algorithm for automated discovering of a class of inequality-type theorems. Sci. in China (Ser. F) 44, 33–49 (2001)
6. Colón, M.A., Sankaranarayanan, S., Sipma, H.B.: Linear Invariant Generation Using Non-linear Constraint Solving. In: Hunt Jr., W.A., Somenzi, F. (eds.) CAV 2003. LNCS, vol. 2725, pp. 420–432. Springer, Heidelberg (2003)
7. Yang, L., Hou, X., Xia, B.: A complete algorithm for automated discovering of a class of inequality-type theorems. Sci. in China (Ser. F) 44, 33–49 (2001)
8. Yang, L., Xia, B.: Real solution classifications of a class of parametric semialgebraic systems. In: Proc. of Int'l. Conf. on Algorithmic Algebra and Logic, pp. 281–289 (2005)
9. Manna, Z., Pnueli, A.: Temporal Verification of Reactive Systems: Safety. Springer, Heidelberg (1995)

Construction of Matlab Circuit Analysis Virtual Laboratory

Shoucheng Ding[1,2]

[1] College of Electrical and Information Engineering, Lanzhou University of Technology, Lanzhou, China

[2] Key Laboratory of Gansu Advanced Control for Industrial Processes, Lanzhou, China

dingsc@lut.cn

Abstract. The Matlab virtual laboratory was composed of the management platform and experimental platform. Management platform used Microsoft Access database, and it was conducive to dynamic data organization, with little redundancy, data independence high and easy scalability and so on. System by the CGI interface to link the pilot designed project would to complete the virtual simulation. Practice shows that the virtual laboratory makes up the lack of traditional resources, and makes circuit analysis practical teaching more lively and flexible.

Keywords: circuit analysis, Matlab, virtual laboratory, practical teaching.

1 Introduction

The web-based virtual laboratory was the use of computer technology, network communication technology, multimedia technology for information processing and related technology to simulate the real experiment, determined in accordance with certain principles of experimental procedures and experimental rules to establish a test to be consistent with real the virtual environment of the experimental teaching in time and space are extended. In recent years, many domestic and foreign research institutions and universities are on the virtual experiment technology for a large number of useful attempts. Clarkson university, the United States to use Java Applet basic circuit design curriculum laboratory electronic teaching assistant; U.S. Illinois University Nmrscope system, through internet research can be used anywhere in Illinois at the University of instruments such as NMR instrument; Peking University computer www-based system developed by the virtual laboratory 3WNVLAB, the system has achieved the CACHE preliminary design and assembly line design experiments [1].

More common now virtual experimental platform technologies FLASH, JAVA, VRML and MATLAB, etc., each technology had its own advantages, but Matlab in circuit analysis, electric machinery and electrical control field simulation applications, making the theory of these courses experimental study of the problem opens up new trends. However, a Matlab web-based virtual laboratory system should include what kind of function, the function structure of how to match between modules and the field was still no uniform standard. However, Matlab provided a networking tool bag

D. Jin and S. Lin (Eds.): Advances in MSEC Vol. 1, AISC 128, pp. 213–218.
springerlink.com

of Matlab web server, development and construction of simulation based on Matlab platform for virtual experiments generally optimistic outlook. Web-based distance education can make full use of text, graphics, images, sounds and other digital media technology to improve student learning enthusiasm, promote student self-thinking, innovative and comprehensive development of freedom, and even the modern education system to promote change, however, The resource constraints of modern experimental education and other characteristics, making some of the requirements of teaching practical hands-on experimental courses is restricted, as the realization of experimental teaching of modern polytechnic schools a bottleneck or difficulty [2].

With the virtual instruments and virtual reality technology, through the network to build a virtual experiment teaching system is already technically possible, Web-based virtual experiment to become an important aspect of distance learning, and different ways more and more rich. However, a Web-based virtual laboratory system should include what kind of function, the functional structure of modules should be how to match, the field is still no uniform standard. In the paper, the characteristics of professional science and engineering and electrical electronics based on the problems facing the experimental courses, and built a network virtual laboratory, and break the traditional experimental teaching means, and built a new experimental teaching model. It injected new vitality for the experimental teaching of distance education to carry out not only the students in the experimental time and space are to maximize the extension, but also to some extent, inadequate funding settlement experiment, experiment insecurity issues.

2 System Mode

Construction of a remote virtual laboratory was a complicated systematic project, in order to ensure the quality and operation of the system after the operational efficiency, system operation must have an appropriate network structure. Common network teaching system currently mainly had two application modes, one traditional client / server (client /server was referred to as C / S mode) model structure; the other is the browser / server (browser / server was referred to as B / S mode) model structure, and this model was mainly suitable for wide area network. The server running the data load heavy security control capacity in the information relative to C / S mode was weak, poor stability of the system, which predictably, in the continuous development of broadband technology, B / S mode will be greatly improved these deficiencies [3]. Practical application should be according to their needs, determine the mode by which, this design used a B / S mode that was, web-based system model. In the B / S mode, a virtual experiment system consists of front and back office server, the browser two parts. Front of the main implementation technology 4: Java, QTVR, VRML, active control technology. The most common back-end servers had three kinds of dynamic web languages: JSP, ASP and PHP.

To Internet as a virtual processing environment of network computing model allowed a variety of scientific computing, simulation, information processing had been greatly improved, network resources were fully utilized. Then, the network can provide people with Matlab computing services. The answer was yes, it can install Matlab software component that came with the help of Matlab web server and web design and browsing technology, and achieved the so-called browser / server (B/ S) computing model.Clearly, the virtual laboratory using Matlab simulation, using Matlab web server components and web browser built B / S network model to implement virtual reality, and constructs a shared, multi-user process, experimental platform, to solve the current science and engineering school of electrical based in the plight of the experiment [3]. Using Matlab web server will extend the application of Matlab to the network, remote visual modeling and simulation. It was for the development of virtual electric and electronic experiment system had created better conditions. Matlab -based virtual experiment platform features powerful as Matlab itself, in data processing and analysis showed strong advantage.Virtual laboratory used active control technology and the Matlab web server. Through ASP web page programming, it implemented a virtual experiment system.

3 MATLAB Web Server

Matlab web server included Matlab server, matweb and matweb.m file. The web application by the Matlab web server and the Matlab web service agent composed of two parts. Matlab server.exe was running Matlab application server environment, responsible for managing web application and the communication between Matlab. Matlab web service agent was an executable program matweb.exe. It was the Matlab web server's TCP / IP client, but also web-CGI extensions. For Matlab's request it will redirect to Matlabserver.exe processing. It was a multi-threaded TCP / IP server, and you can configure in the matweb.conf to any legitimate TCP / IP port. Matlab server to handle web page by calling matweb.m implied by the specified field mlmfile M files, Web pages, MATLAB, M linkages among documents.

Matlab web server was the core of matweb.exe, responsible for interpreting HTML pages sent by the client's request, and converted to run Matlab applications required parameters, and then started a Matlab process, and the specified Matlab applications and their parameters to the calculation process. After calculation, Matlab program also responsible for the results to HTML pages of the way through matweb output to the client browser. Matlab web server was developed under Matlab toolbox. It can be used by the client browser running on the server side of the Matlab simulation program. The simulation results pre-set by the output HTML file outputted through the web browser to the client. Obviously, based on Matlab web server developed a complete test consists of three parts: client; web server; Matlab simulation server, which users enter data in the browser, Matlab submitted to the server to calculate the results show in the browser [4].

3.1 Server Configurations

Microsoft's Windows NT (Windows 2000 in IIS5. 0 version) was for the system development platform. It installed Matlab 6.5 and Matlabserver. Create a new Web site, the root directory is wsdemos (defined as needed), this directory can be under the \ toolbox \ webserver, and also copy it elsewhere. Web server specific steps to set the following four:

Create the virtual or physical cgi-bin directory under the wsdemos, and copy matweb.exe and matweb.conf to this directory and set permissions for this directory can execute scripts and applications.

In the IIS web service extension set to add a web service extension allowed, and require the file name is matweb.exe, take a random extension name. CGI is set to execute permission. Matlabserver way through the CGI work, in order to allow IIS to use CGI programs, the system must open the CGI permissions.

Under the wsdemos create the virtual or physical icons directory, and the image file of the wsdemos directory is added to this directory, and set this directory has write permissions.

In the website, increase the default index.html to the content of documents.

3.2 Matlab.conf and Matlabserver.conf Configuration

Configuration Matlab.conf and Matlabserver.conf file was as follows:

In <MATLAB> \ Webserver directory, installer file to create a matlabserver.conf. Matlabserver.conf consists of two parts: the port number p, which can run concurrently maximum number of threads m. After installation, the file only a single line:-m 1, it means that the default port number 8888, while the maximum concurrent number of threads to 1, according to need to change their number. If the port number is changed, the matweb.conf configuration file to set the port number must be changed accordingly, so that the two ports the same.

Each additional Matlab Web applications, it need to add a configuration in matweb.conf. In matweb.conf add the following:

[File name] / * M file name * /

mlserver = 127.0.0.1 / * server IP address * /

mldir = c: / matlab / toolbox / webserver / wsdemos / * Matlab program and save the picture * /

Note that the path is set to modify matweb.conf which is a physical directory.

3.3 MATLAB Web Application Development

Matlab's Web applications compiled the key issue is the preparation for the Matlab Web server program, the key is to solve two problems: Matlab program to HTML page for input by Matlab program to generate a one parameter output data and images HTML documents. These two problems can be a template file to the following description of the design elements are as follows: Matlab's Web applications compiled the key issue is the preparation for the Matlab Web server program, the key is to solve two problems: Matlab program to HTML page for input by Matlab program to generate a one parameter output data and images HTML documents. These two problems can be a template file to the following description of the design elements are as follows: Treatment of output variables: Matlab program to process input variables, the calculated results, in its written structures outstruct. Such as the preservation of z calculated using the following syntax: outstruct.z=z; structure variables outstruct contains all the output variables, each variable with variable name as a member. Struct the Matlab's very flexible data structure, the grammar is suitable for any data type.

4 System Implementation

In the B / S mode, the system used passwords to restrict landing experimental system landing without legal status, to achieve the level of site security system. In the setting up of website's structure, the front desk adopted HTML webpage file to transfer relevant experiment processing procedures to realize, and ASP to achieve the system's programming. The first class website was a portal of the system, and it mainly realized prediction and information of the experiment project of correlated curriculum were released. It was the window that a discipline was exchanged at the same time; to carry on concrete one entry of experiment, obtained legal identity person could enter the fictitious experimental system in concrete entry of webpage this only, and carried on the relevant experiments of course arrangement. In addition, it had provided a large number of tool software download for user in website's homepage, and offered more convenience for users' use.

System ASP programming interface and the server Matlab web server components completed the test function through linking the CGI interface. Experimental interface frame structure had achieved the illustrated results. The experimental report module submitted had used the generic message board model. It had pictures and editing forms, and submitted the relevant experimental data for the students to lay a foundation. The system interface was designed experiment shows that provided the experimental description document, but also for students to preview provided experimental convenience. Teachers can reduce laboratory test instructions issued tedious work. It can achieve a reading test instructions from resource constraints, Students will be able to log off the system and download and read the test at any time guidance. Problems in the preview can give the teacher a message advice.

The traditional experiments to use a variety of test instruments, and different instruments used in the experiments are different, if the creation of the necessary equipment is comprehensive experiment more, and so many of the instruments is not only costly but also very troublesome to connect with each other, so the development Web-based virtual laboratory, using a computer system and the principle of virtual instrument can be a good simulation to solve these problems. Establishment of local area network using server mode can achieve virtual experimental platform of a machine terminal access to real-time experiment is completed, and can replace a number of pilot projects in the laboratory, saving the equipment investment, but also ease the shortage of school laboratories and equipment.In addition, the processing of experimental data on the use of graphics technology, such data results in favor of marking the teacher and students in the experiment can be easily observed when the results of the output curve, compared to its profound perceptual experiments.

Building web-based virtual circuit analysis laboratory for remote virtual laboratory. Teachers, students using the Internet release of the distant experimental requirements, the students by visiting the web-based virtual electrical and electronic test systems, into a virtual experimental system virtual experiments can be carried out by the input parameters, to observe and analyze the test results of simulation, easy to save the results, also facilitate the analysis of theoretical and experimental values between, and make test report. Teachers can collect and marking through the network of students lab reports, send the results to the students. This will help resolve the current experiment appear in distance learning difficulty[5].

5 Conclusions

The circuit analysis virtual laboratory made up the deficiencies of the traditional laboratory, and it made practical teaching more lively and more flexible design of the experiment. In the innovation reform process of experimental teaching, the virtual laboratory reduced the dependence on hardware devices, and it was also the direction of the development of experimental teaching reform. Through continuous exploration and improvement, on the campus network the experimenter used the virtual experiment platform remote client access pilot project, and completed a simulation of various circuit analysis experiments. The results also timely deposited into a computer, and experimental analysis was also done on the computer. Experimental platform and can be submitted to the system from the experimental. The virtual computer system for the majority of circuit analysis test provides teachers and students to more advanced and scientific experimental methods. The system was designed experiments report system, which has uploaded experimental report. For the majority of teachers and students, the virtual experiment system is a more advanced and scientific experimental methods.

References

1. Ding, S.C., Li, W.H., Yang, S.Z., Li, J.H., Yuan, G.C.: Design and Implement of the Electric and Electronic Virtual Experiment System. Journal of Networks 5(12) (2010)
2. Wang, S.L., Wu, Q.Y.: To carry out a virtual experiment system research and application. Computer Engineering and Science 22(2), 33–36 (2000)
3. Zhu, X.H., Fen, Y.T., Zhang, Y.J.: Component-based Virtual Instrument Development Method. Shanghai University (Natural Science Edition) (4), 357–360 (1999)
4. Gao, J.H., Yang, J.P., Qiu, A.Z.: Virtual instrument technology and build a virtual laboratory. Henan Normal University (Natural Science Edition) 36(2), 45–48 (2008)
5. Wang, T.L., Chen, D.S., Yuan, H.Y., Lei, F.: Electrical and Electronic Laboratory virtual network teaching platform. Research and Exploration in Laboratory 27(7), 50–53 (2008)

Identification and Evaluation Methods of Expert Knowledge Based on Social Network Analysis

Guoai Gu[1] and Wanjun Deng[2]

[1] School of Economics and Management,
Beijing University of Aeronautics and Astronautics, BUAA;
Institute of Labor Science and Law, Beijing Wuzi University
Beijing, China
jeff_9144@hotmail.com
[2] Institute of Comprehensive Development
Chinese Academy of Science and Technology for Development, CASTD
Beijing, China
dengwanjun1984@yahoo.com.cn

Abstract. By building expert knowledge network and social network analysis, this article proposes the identification and evaluation methods of expert knowledge. Based on the analysis on the relationship between the knowledge points, the expert knowledge network is built. Then through social network analysis, the article obtains expert knowledge fields at different levels and achieves to identify expert knowledge and establish the expert knowledge system. We also use the journal impact factor and network analysis indicators to realize the knowledge evaluation according to classification.

Keywords: expert knowledge, knowledge identification, social network analysis.

1 Introduction

Expert knowledge management is an important component of knowledge management. Experts play an important role in judging, undertaking and consultation of the projects of scientific research and economic development. In order to ensure to select the excellent experts whose knowledge fields are appropriate for project demands, it is needed to identify and evaluate expert knowledge.

The current common method of expert knowledge identification is first directly summarized by the man, and then is described in the forms of natural language, phrases, charts and so on [1, 2]. The manual method of obtaining expert knowledge is rather subjective and blind and thus the expert knowledge received is not necessarily accurate or complete. In addition, the method of expert knowledge is evaluation often the overall evaluation about expert knowledge [3, 4] and there are a few studies on expert knowledge evaluation according to the knowledge fields. Therefore, the article studies the objective identification and classifying evaluation methods of expert knowledge.

D. Jin and S. Lin (Eds.): Advances in MSEC Vol. 1, AISC 128, pp. 219–225.
springerlink.com

Social network is a set of the actors and the relations between actors, and the network structure is represented by the nodes and their connections. Social network analysis (SNA) is a method and instrument to analyze social network by analyzing the relationships of individuals within an organization. A big advantage of SNA is its theoretic perspective on relationships between actors. When relationship becomes a research object, the conventional statistics can not quantitatively analyze well [5]. Expert knowledge consists of a lot of different knowledge points which are related to the expert's research field. There may be the intrinsic relations between these knowledge points because they are associated with the some knowledge object [6, 7]. For example, the knowledge points "social interests" and "government interest" are not the same, but both they are related to the "interest", so the intrinsic relationship exists between them. This sort of relationship makes expert knowledge shown a set of knowledge points and their relations and expert knowledge forms a knowledge network. Thus SNA is an appropriate method to analyze expert knowledge network. The article will establish expert knowledge network, use SNA to identify and sort expert knowledge and mark the knowledge according to the fields.

2 Methodology and Data

In the field of management, SNA generally studies the relationship network in the research organization such as consulting, trust, friendship, intelligence, communication and workflow, to meet the actual demands within the organization such as decision-making, communication, personnel changes and organizational conflicts [8]. SNA promotes thinking and research patterns towards relationship, so it is more suitable for a variety of researches on social relations, including the relationship between expert knowledge points in the paper. This article will use the software UCINET 6.0, from the perspective of overall network analysis to study the network structure of expert knowledge. The following are the main network indicators used in the article [8].

Component: The correlations exist among the members in one part, but no correlations exist between the various parts. Those parts are called components.

Clique: It is the maximal complete sub-graph. That is to say in the nodes, there is a direct connection between any pair of nodes, and one clique can not be included in any other clique.

Network average distance: It is the average distance of all reachable pairs of nodes in the network. The distance between a reachable pair of nodes is the sum of the number of edges in the shortest path of the two nodes.

Cohesion index: It is an index based on the distance and it is to reflect cohesion of the whole network.

Network intensity: It is an index to reflect network cohesion and compactness. It is the ratio of the number of the practical edges and the number of the possible edges.

Network degree centralization: It is an index to assess the central tendency of the whole network by analyzing the point centrality of all the nodes.

This paper takes Chen Qingyun, a professor in Peking University, as an example, to carry out the expert knowledge network analysis. The data are from Prof. Chen's papers searched by China National Knowledge Internet during the latest 10 years (2002-2011) papers, a total of 15 papers.

3 The Establishment of Expert Knowledge Network

A Assumptions

Expert knowledge network takes the expert knowledge points as the nodes and the relations between knowledge points as the edges. According to the actual situation of expert knowledge, the following assumptions are proposed for establishing the expert knowledge network:

1) Assumption 1: Expert knowledge points are shown as the keywords or the substantives in the title and subtitles in the published papers and reports.

Experts express their knowledge by publishing the papers and reports and thus the expert knowledge points exist in the experts' papers and reports. If the paper summarizes the key words, the key words often reflect the important areas of the research and can become the knowledge points. If the paper does not sum up key words, the title and subtitles will reflected the important research areas in the article and paragraphs and knowledge points can come from the substantives of all the levels of titles. This method avoids text mining for the whole article to find knowledge points. It is relatively simple, effectively using keywords and titles, and it can involve all significant expert knowledge points.

2) Assumption 2: The scores of the expert knowledge points are different and the score is the value of impact factor of the journal which contains the expert knowledge point.

The expert has the different excellent degrees for all his knowledge points of one expert, so expert knowledge points should have the different evaluation scores. Many articles of the expert are published in the journal and the reports with strong reference values are tend to be reproduced in the journal. At the same time the level of the journal represents the level of the article. So the scores of the expert knowledge points can be calculated by the journal impact factor.

3) Assumption 3: Relationship matrix of the expert knowledge points is a symmetric matrix and the expert knowledge network is an undirected graph.

We assume that no direction exists in the relationship of the two knowledge points. For example, the relation of "knowledge management" and "knowledge" is the same with the relation of "knowledge" and "knowledge management". In fact the two relations are not exactly the same, but the difference has no obvious function for analyzing the knowledge structure. So relationship matrix of the expert knowledge points is a symmetric matrix and the expert knowledge network is an undirected graph.

B Expert Knowledge Network

Based on the above 3 assumptions, the expert knowledge network can be expressed as (1):

$$G= (K, R) \quad (1)$$

The set of nodes, knowledge points K={(k1,q(k1)), (k2,q(k2)),…, (kn,q(kn))}. Thereinto, ki, i= 1,2,…,n is the knowledge point and q(ki), i= 1,2,…,n is the score of ki, i= 1,2,…,n. The set of edges, the relations between the knowledge points R= {(rij)}, i,j= 1,2,…,n. Thereinto, rij is the relation between ki and kj.

Construction rules of the expert knowledge network include:

- Rule 1: The knowledge points k_i are the keywords or the substantives in the title and subtitles of the articles.
- Rule 2: The score $q(k_i)$ is the sum of the impact factors of the journals where the k_i is. If k_i appears in one journal, $q(k_i)$ is the value of the journal impact factor. If k_i appears in many journals, $q(k_i)$ is the sum of the impact factors of all those journals. For the journal impact factor in the paper, we use the impact factors of the top 1200 journals in the Beijing University core journals in 2009. The value range of the 1200 impact factors are [0.44, 8.619]. If the journal where the expert publishes his paper is not included in the 1200 journals, we view that the effect of this journal is too weak, the score of the corresponding knowledge point is 0, and the knowledge point is invalid.
- Rule 3: r_{ij} is the relation of the two knowledge points and can be represented the similar degree of the two knowledge points. r_{ij} is calculated by (2). In (2), $l(k_i, k_j)$ is the number of the same characters in the two knowledge points. min($l(k_i)$, $l(k_j)$) is the minimum of the numbers of the characters in the two knowledge points. For example, the relation of "social benefit" and "public benefit" is 2/min(4, 4)=0.5.

$$r_{ij} = \frac{l(k_i, k_j)}{\min\left(l(k_i), l(k_j)\right)} \quad (2)$$

4 Identification and Evaluation of Expert Knowledge

We will use SNA to identify and evaluate the expert knowledge.

1) The first step is to establish the Prof. Chen's knowledge network.

According to the Rule 1 and Rule 2, we collect 10 effective papers and 25 knowledge points. We grade the knowledge points according to the Rule 2 and calculate the relation between the two knowledge points according to the Rule 3. The Prof. Chen's knowledge network is obtained and seen in Figure 1. It can be seen that there is only a group of more concentrated nodes in the knowledge network and other nodes are very dispersive.

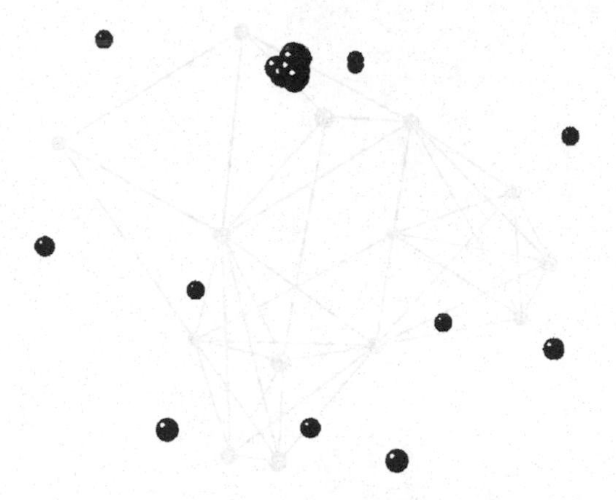

Fig. 1. Prof. Chen's knowledge network

2) The second step is component analysis of knowledge network to get the knowledge field in the first level.
The knowledge field in the first level can be obtained by component analysis and the score of each component is the sum of knowledge points which are included in the component.

Prof. Chen's 25 knowledge points are divided into 25 components. The largest one consists of 14 nodes and its score is 42.2. The second largest one only consists of 2 nodes and its score is 3.376. The other 9 simple points are 9 components respectively. Nodes in the largest one occupy 56% total nodes. These indicate the network has more single nodes. Again referring to the network indicators: the network intensity = 0.0720 and cohesion index = 0.222, the weak cohesion degree is further verified.

Thus, Prof. Chen's knowledge field is only concentrated in one component and Prof. Chen only has one knowledge field in the first level. According to the knowledge points included in the largest component, the knowledge field in the first level is public management and its score is 42.2. For other knowledge points, Prof. Chen's research has no system.

3) The third step is to assess the knowledge field in the first level, public management, by the network indicators.
We take 14 knowledge points included in the field of public management to establish a knowledge sub-network (seeing Figure 2). Figure 2 tells us that the network of public management is a knowledge system with close contact.

The specific network indicators are calculated and analyzed as follows. The Network intensity is 0.2320, the network average distance is 1.582, and the cohesion index is 0.720. The values indicate Prof. Chen has a systematic research on public management. In addition, the network degree centralization is 18.83%. It further shows the research on public management does not focus on some knowledge point, but it is an interconnected system.

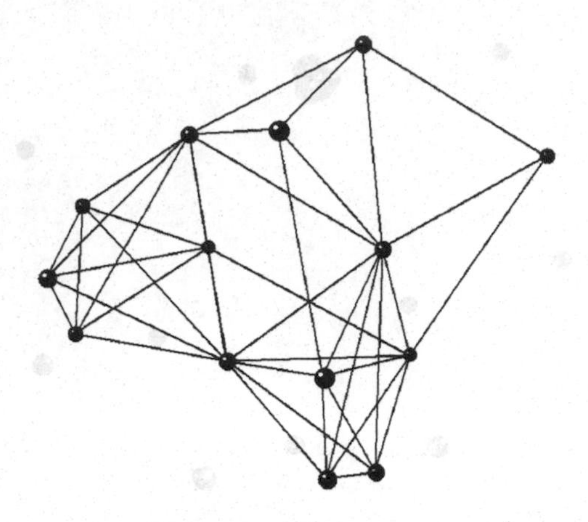

Fig. 2. Prof. Chen's knowledge sub-network of public management

4) The fourth step is to get the knowledge field in the second level by clique analysis on the public management.
The clique analysis is carried out on the knowledge sub-network of public management. We set the smallest scale as 4 and obtain 3 cliques. That is to say the knowledge field of public management involves 3 knowledge fields in the second level. They are interest research, social research and government research respectively and their scores are 16.880•11.816 and 6.752 respectively.

The identification and evaluation of Prof. Chen's knowledge comes to the end and Chen's knowledge system can be represented in Table 1.

Table 1. Prof. Chen's knowledge System and knowlsdge scores

The Knowledge Field in the First level (Score)	**The Knowledge Field in the Second level (Score)**	**The Knowledge Points in the Third level (Score)**		
Public management (42.2)	Interest research (16.880)	Social interest (3.376)	Public interest (5.064)	Interest analysis (3.376)
		Government interest (1.688)	Interest structure (1.688)	Interest comparison human (1.688)
	Social Research (11.816)	Pubic management socialization (1.688)	Government management socialization (1.688)	Social interest (3.376)
		Social participation (1.688)	Social standard (1.688)	Social governance (1.688)
	Government research (6.752)	Pubic management socialization (1.688)	Government interest (1.688)	Government standard (1.688)
		Government management (1.688)		

5 Conclusions

By building expert knowledge network and social network analysis, this article proposes the identification and evaluation methods of expert knowledge. This method can objectively find expert knowledge at different levels and establish expert knowledge system. In addition, this paper utilizes the impact factor of the journal where the knowledge point is, to realize the knowledge evaluation according to knowledge classification. The evaluation is quantitative and objective.

When the knowledge of many experts in one organization are identified and evaluated by this method, the experts can be ranked according to the knowledge field and it can help organizations find the most appropriate experts.

References

1. Loveland, D.W., Valtorta, M.: Detecting ambiguity: An example in knowledge evaluation. In: Proceeding of 8th International Joint Conference on Artificial Intelligence, Karlsruhe, Federal Republic of Germany, pp. 182–184 (1983)
2. Cirover, M.D.: A pragmatic knowledge acquisition methodology. In: Proceeding of the 8th International Joint Conference on Artificial Intelligence, Karlsruhe, Federal Republic of Germany, pp. 436–438 (1983)
3. Qiao, D., Yang, J., Li, Z.: Knowledge obtaining, analysis, and evaluation of internet-based consultation system for scientific and technological programming. Science & Technology Progress and Policy 24(7), 127–131 (2007)
4. Guo, F., Hou, C., Dai, Z., Wang, X.: Modeling and simulation of ANFIS to evaluate software quality based on expert knowledge. Systems Engineering and Electronics 28(2), 317–320 (2006)
5. Otte, E., Rousseau, R.: Social network analysis: a powerful strategy, also for the information sciences. Journal of Information Science 28(6), 441–453 (2002)
6. Xi, Y., Dang, Y.: The discovery and representation methods of expert domain knowledge based on knowledge network. System Engineering 23(8), 110–115 (2005)
7. Gong, J., Liu, L.: Representing and m easuring experts'knowledge based on knowledge networ. Studies in Science of Science 28(10), 1521–1530 (2010)
8. Liu, J.: Lectures on whole network approach: A practical guide to UCINET. Truth & Wisdom Press, Shanghai People's Publishing House, Shanghai (2009)

Effect of Ammonium and Nitrate Ratios on Growth and Yield of Flowering Chinese Cabbage

Shiwei Song, Lingyan Yi, Houcheng Liu, Guangwen Sun, and Riyuan Chen*

College of Horticulture, South China Agricultural University, Guangzhou 510642, China
rychen@scau.edu.cn

Abstract. The effect of different ammonium and nitrate ratios (NH_4^+-N : NO_3^--N = 0:100, 25:75, 50:50 and 75:25) on growth and yield of flowering Chinese cabbage (*Brassica campestris* L. ssp. *chinensis* var. *utilis* Tsen et Lee) with 3 cultivars were studied in hydroponics. The results indicated that, compared with the complete nitrate treatment, plant height, stem diameter and biomass of flowering Chinese cabbage were increased in the low enhancement of ammonium (25%) in nutrient solution, while plant growth and biomass were decreased in the medium (50%) and high (75%) enhancement of ammonium. Compared with the complete nitrate treatment, low enhancement of ammonium (25%) in nutrient solution increased root activity of flowering Chinese cabbage, while it was decreased by the medium (50%) and high (75%) enhancement of ammonium treatments. Nutrient solution with 25% ammonium enhancement maintained a high root absorption capacity and increased plant biomass, so it was appropriate to hydroponics for flowering Chinese cabbage.

Keywords: flowering Chinese cabbage, ammonium and nitrate ratios, growth, yield.

1 Introduction

Nitrogen is one of the most important nutrients for plant growth, and ammonium nitrogen and nitrate nitrogen are two primary inorganic nitrogen forms absorbed by plants. It is generally believed that both nitrate and ammonium can produce sufficient nitrogen for plant growth, but nitrate is more secure [1]. Usually vegetable crops tend to absorb nitrate. But for many vegetable crops, especially leafy vegetables, compared with the complete nitrate, the appropriate proportion of ammonium in the nutrient solution had advantage, not only for a high biomass, but also good quality of product, especially for significantly reduced nitrate content [2]. Single supply of ammonium in nutrient solution would have toxic effects for many plants [3], and plant growth was seriously inhibited [4]. Therefore, the concentration and proportion of ammonium enhancement in nutrient solution must be appropriate.

The flowering Chinese cabbage (*Brassica campestris* L. ssp. *chinensis* var. *utilis* Tsen et Lee) is one of famous special vegetable in south China, also it has the largest grown area and yield in local area. Flower stalk of flowering Chinese cabbage is the

* Corresponding author.

D. Jin and S. Lin (Eds.): Advances in MSEC Vol. 1, AISC 128, pp. 227–232.
springerlink.com

edible organ, which is crisp and full of nutrient. Flowering Chinese cabbage had a rich germplasm resources, and its maturity characteristics, yield and quality varied among different genotypes [5]. Previous experiment studied the effect of nitrogen, phosphorus and potassium fertilizer on flowering Chinese cabbage [6]. While the effect of different nitrogen forms on growth and development of it, was scarcely been reported.

In this experiment, the effect of different ratios of ammonium and nitrate on growth and yield in flowering Chinese cabbage were studied under hydroponics. The aim was to provide optimum nutrition formulation for its high yield and quality of the production.

2 Materials and Methods

Materials and Treatments. The experiment was carried out in plastic greenhouse on vegetable research base of South China Agricultural University. 3 cultivars of flowering Chinese cabbage were used in the experiment, which were "Lvbao 70", "Youlv 80" and "Chixin No.2". Plug seedlings was started on November 8, 2010, with the medium of perlite, and seedlings with 3 true leaves were transplanted in the nutrient solution (standard 1/2 dose of Hoagland formula). 11 seedlings transplanted in one plastic hydroponic box (filled with 15 L nutrient solution) were looked as a repeat, and each cultivar had 3 repeats with randomized block arrangement.

1/2 dose of Hoagland formula was used as basic nutrient solution, with the total N 7.5 mM, total P 0.5 mM, total K 3.0 mM, total Ca 2.5 mM and total Mg 1.4 mM. The treatments were 4 different ammonium and nitrate ratios (NH_4^+-N : NO_3^--N = 0:100, 25:75, 50:50 and 75:25) with the same amount of total N (shown detail in Table 1).

Nutrient solution was replaced every week and ventilated through pump every 30 minutes. pH value of the solution was adjusted to around 6.2 every day. Materials were taken when they reached marketable maturity.

Table 1. Nutrition formula of different ammonium and nitrate ratios ($mM \cdot L^{-1}$)

$NH_4^+:NO_3^-$	KNO_3	$Ca(NO_3)_2 \cdot 4H_2O$	KH_2PO_4	$MgSO_4 \cdot 7H_2O$	$(NH_4)_2SO_4$	K_2SO_4	$CaCl_2$
0:100	2.5	2.5	0.5	1.4	—	—	—
25:75	0.625	2.5	0.5	1.4	0.9375	0.9375	—
50:50	—	1.875	0.5	1.4	1.875	1.25	0.625
75:25	—	0.9375	0.5	1.4	2.8125	1.25	1.5625

Measurement. Random samplings of flowering Chinese cabbage plant were measured plant height and stem diameter (at 5-6 node of flower stalk). The plant divided by root, rootstock and flower stalk respectively, was weighted for the fresh weight and dry weight (after drying at 70 °C to constant weight). The fresh weight of product organ (flower stalk above the 4th node) was named yield. Root activity of flowering Chinese cabbage was determined by TTC method [7].

Data Analysis. Statistical analysis of the data was performed with Duncan's method at 5% level, using the Version 16.0 of the SPSS software package.

3 Results

Effect of Ammonium and Nitrate Ratios on Plant Height and Stem Diameter of Flowering Chinese Cabbage

Different ratios of ammonium and nitrate in nutrient solution significantly affected plant height of flowering Chinese cabbage (Figure 1). Plant height was the highest in low enhancement of ammonium (25%) among the 4 treatments for all 3 cultivars, and it was significantly higher than the other 3 treatments for "Lvbao 70" and "Chixin No.2" cultivars ($p < 0.05$). The plant height in medium (50%) and high (75%) enhancement of ammonium treatments was significantly lower than that of the complete nitrate treatment and low enhancement of ammonium treatment, and that of high (75%) enhancement of ammonium treatments was the lowest.

Different ratios of ammonium and nitrate significantly affect the stem diameter of flowering Chinese cabbage (Figure 1). Stem diameter was the highest in low enhancement of ammonium (25%) among the 4 treatments, and it was significantly higher than the other 3 treatments for "Lvbao 70" and "Youlv 80" cultivars ($p < 0.05$). The stem diameter in medium (50%) and high (75%) enhancement of ammonium treatments was significantly lower than that of the complete nitrate treatment and low enhancement of ammonium treatment.

This indicated that low (or called moderate) enhancement of ammonium (25%) in nutrient solution would increase plant height and stem diameter of flowering Chinese cabbage, but too high ratios of ammonium (50% and 75%) would inhibit plant growth.

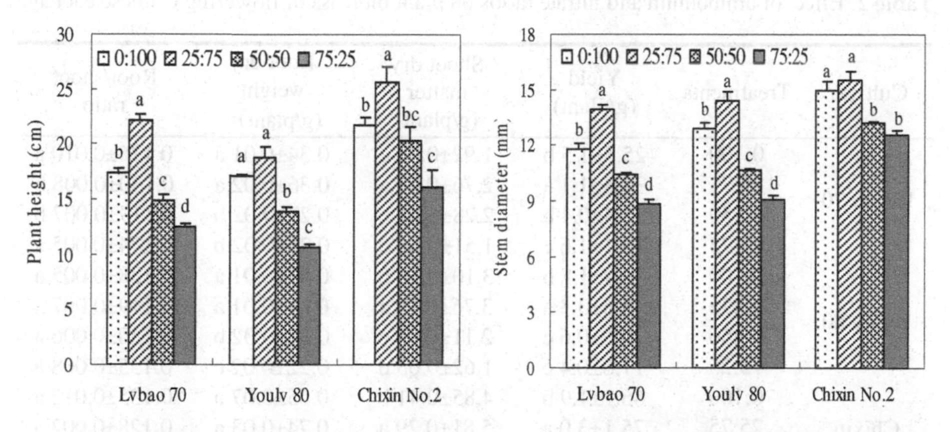

Fig. 1. Effect of ammonium and nitrate ratios on plant height and stem diameter of flowering Chinese cabbage

Note: The columns in the figure represented the average (n=3) and the vertical bars expressed as standard error. Different letters in the same cultivar indicated significant difference at 5 % level (Duncan method). The same as follows.

Effect of Ammonium and Nitrate Ratios on Plant Biomass of Flowering Chinese Cabbage

Yield of flowering Chinese cabbage was significantly affected by different nitrogen forms (Table 2), and the tendency was the same for 3 cultivars. The yield was the highest in low enhancement of ammonium (25%) treatment, followed by complete nitrate treatment and medium enhancement of ammonium (50%), and it was lowest in high enhancement of ammonium (75%) treatment. The shoot dry weight of plant also showed the same trend as yield. This indicated that low enhancement of ammonium (25%) in nutrient solution would increase plant biomass of flowering Chinese cabbage, but too high ratios of ammonium (50% and 75%) would decrease plant biomass. The yield and shoot dry weight had significant difference among cultivars: "Chixin No.2" was the highest, followed by "Youlv 80", and "Lvbao 70" was the lowest.

Root dry weight was clearly affected by different nitrogen forms. There was no significant difference between complete nitrate treatment and low enhancement of ammonium (25%) treatment in root dry weight for "Lvbao 70" and "Youlv 80", but it was significantly higher than in medium (50%) and high (75%) enhancement of ammonium treatments. While root dry weight for "Chixin No.2" cultivar was not affected by 4 different nitrogen forms.

Root/shoot ratio in "Lvbao 70" was significantly higher in low (25%) and medium (50%) enhancement of ammonium treatments than that in complete nitrate treatment and high enhancement of ammonium (75%) treatment. While there was no significant difference among 4 treatments in "Youlv 80" and "Chixin No.2" cultivars. This indicated that enhancement of ammonium in nutrient solution did not significantly affect the biomass distribution between aboveground and underground of flowering Chinese cabbage.

Table 2. Effect of ammonium and nitrate ratios on plant biomass of flowering Chinese cabbage

Cultivar	Treatments	Yield (g/plant)	Shoot dry matter (g/plant)	Root dry weight (g/plant)	Root/shoot ratio
Lvbao 70	0:100	25.5±1.5 b	1.92±0.04 c	0.34±0.01 a	0.176±0.010 a
	25:75	33.8±1.1 a	2.76±0.08 a	0.36±0.02 a	0.130±0.008 b
	50:50	25.4±0.4 b	2.28±0.01 b	0.27±0.02 b	0.119±0.007 b
	75:25	15.1±1.6 c	1.51±0.08 d	0.26±0.02 b	0.174±0.005 a
Youlv 80	0:100	32.9±1.8 b	3.10±0.16 b	0.43±0.01 a	0.138±0.005 a
	25:75	45.6±3.3 a	3.76±0.19 a	0.44±0.01 a	0.119±0.007 a
	50:50	20.6±0.6 c	2.11±0.05 c	0.27±0.02 b	0.130±0.006 a
	75:25	17.6±0.4 c	1.62±0.03 d	0.22±0.02 c	0.135±0.008 a
Chixin No.2	0:100	66.1±1.9 b	4.85±0.41 b	0.65±0.07 a	0.135±0.012 a
	25:75	75.1±3.0 a	5.81±0.29 a	0.74±0.03 a	0.128±0.002 a
	50:50	48.3±0.6 c	4.24±0.07 b	0.61±0.06 a	0.144±0.013 a
	75:25	43.7±4.1 c	4.00±0.11 b	0.61±0.05 a	0.153±0.012 a

Note: Data in the table were average ± standard error (n=3), and different letters for the same cultivar indicated significant difference at 5 % level (Duncan method).

Effect of Ammonium and Nitrate Ratios on Root Activity of Flowering Chinese Cabbage

Root activity of flowering Chinese cabbage was significantly affected by different nitrogen forms (Figure 2). Root activity of 3 cultivars had the same change tendency in 4 treatments. It was significantly higher in complete nitrate treatment and low enhancement of ammonium (25%) treatment than in medium (50%) and high (75%) enhancement of ammonium treatments, and the difference was significant for "Youlv 80" and "Chixin No.2" cultivars. The results indicated that root activity was decreased in medium (50%) and high (75%) enhancement of ammonium treatments, with the root absorption ability decline.

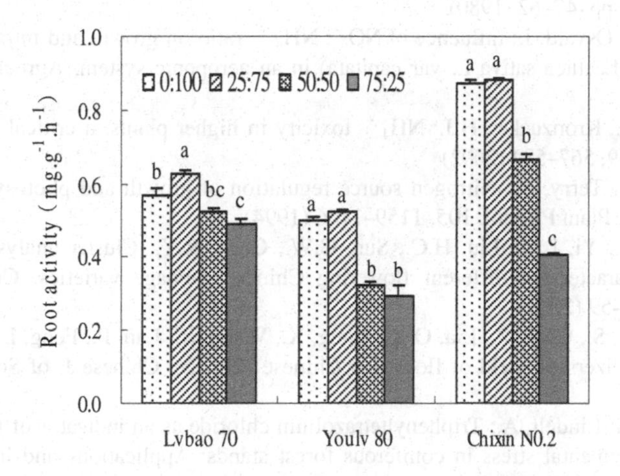

Fig. 2. Effect of ammonium and nitrate ratios on root activity of flowering Chinese cabbage

4 Discussion

Generally vegetable crops tend to absorb nitrate. However, studies had shown that compared with single nitrogen sources, most of the crops grown better and had higher nitrogen utilization rate in the nutrient solution with appropriate percentage of ammonium [8]. Usually, in greenhouse hydroponics, maximum growth and yield of tomato or pepper was obtained with an optimal ammonium concentration not exceeding 30% of total nitrogen [9-11].

Different nitrogen forms had significant effect on plant growth and yield of flowering Chinese cabbage. Plant height, stem diameter and yield were the biggest in low enhancement of ammonium (25%) treatment, indicating that nutrient solution with 25% ammonium promoted the growth of flowering Chinese cabbage. There were the same results in cabbage [12] and tomato [13] carried out by former research.

Medium (50%) and high (75%) enhancement of ammonium in nutrient solution reduced root activity of flowering Chinese cabbage, compared with complete nitrate treatment and low enhancement of ammonium (25%) treatment. Excess ammonium in nutrient solution was against root growth and activity. This may due to ammonium toxicity to plants [3] or rhizosphere acidification, where under high ammonium concentration conditions, plant root released H^+ during absorption of ammonium in order to maintain the body cation/anion balance [14].

Acknowledgment. This study was financially supported by China Agriculture Research System (CARS-25-C-04).

References

1. Hageman, R.H.: Effect of form of nitrogen on plant growth. In: Meisinger, J.J., Randall, G.W., Vitosh, M.L. (eds.) Nitrification Inhibitors-Potentials and Limitations, Madison, Wisconsin, pp. 47–62 (1980)
2. Demsar, J., Osvald, J.: Influence of NO_3^-: NH_4^+ ratio on growth and nitrate accumulation in lettuce (Lactuca sativa L. var capitata) in an aeroponic system. Agrochimica 47, 112–121 (2003)
3. Britto, D.T., Kronzucker, H.J.: NH_4^+ toxicity in higher plants: a critical review. J. Plant Physiol. 159, 567–584 (2002)
4. Raab, T.K., Terry, N.: Nitrogen source regulation of growth and photosynthesis in Beta vulgaris. L. Plant Physiol. 105, 1159–1166 (1994)
5. Song, S.W., Yi, L.Y., Liu, H.C., Sun, G.W., Chen, R.Y.: Cluster analysis on yield and quality characters of different flowering Chinese cabbage varieties. Guangdong Agri. Sci. 38, 56–59 (2011)
6. Cai, M., Li, S., Chen, Z., Lia, O.X., Kong, X., Wang, R., Lan, P., Feng, L.: Effect of N, P and K fertilizers on yield of flowering Chinese cabbage. Chinese J. of Soil Sci. 41, 126–132 (2010)
7. Clemensson-Lindell, A.: Triphenyltetrazolium chloride as an indicator of fine-root vitality and environmental stress in coniferous forest stands: Applications and limitations. Plant Soil 159, 297–300 (1994)
8. Gentry, L.E., Below, F.E.: Maize productibity as influenced by form and availability of nitogen. Crop Sci. 33, 491–497 (1993)
9. Sandoval-Villa, M., Guertal, E.A., Wood, C.W.: Greenhouse tomato response to low ammonium-nitrogen concentrations and duration of ammonium-nitrogen supply. J. Plant Nutr. 24, 1787–1798 (2001)
10. Claussen, W.: Growth, water use efficiency, and proline content of hydroponically grown tomato plants as affected by nitrogen source and nutrient concentration. Plant Soil 247, 199–209 (2002)
11. Xu, G.H., Wolf, S., Kafkafi, U.: Effect of varying nitrogen form and concentration during growing season on sweet pepper flowering and fruit yield. J. Plant Nutr. 24, 1099–1116 (2001)
12. Chen, W., Luo, J.K., Shen, Q.R.: Effect of NH_4^+-N/NO_3^--N Ratios on Growth and Some Physiological Parameters of Chinese Cabbage Cultivars. Pedosphere 15, 310–318 (2005)
13. Dong, C.X., Shen, Q.R., Wang, G.: Tomato Growth and Organic Acid Changes in Response to Partial Replacement of NO_3^--N by NH_4^+-N. Pedosphere 14, 159–164 (2004)
14. Gerendas, J., Zhu, Z., Bendixen, R., Ratcliffe, R.G., Sattelmacher, B.: Physiological and biochemical processes related to ammonium toxicity in higher plants. J. Plant Nutr. Soil Sc. 160, 239–251 (1997)

Research of Rapid Design Method for Munitions Based on Generalized Modularity

Chao Wang[a], Chunlan Jiang[b], Zaicheng Wang[c,*], and Ming Li[d]

State Key Laboratory of Explosion Science and Technology, Beijing Institute of Technology, Beijing, 100081, China
[a]xinxinfenghuo@163.com, [b]jiangchunwh@bit.edu.cn,
[c]wangskyshark@bit.edu.cn, [d]iseagle@bit.edu.cn

Abstract. According to personal salience of munitions, munitions 'modular division and interface definition means are researched based on generalized modularity. Both are introduced to munitions' design process and optimum design, and an example is given. Generalized modular design means ,which is adequate for different kinds of munitions design, is approved advance munitions' design speed, and get reference action to the munitions celerity design platform's foundation.

Keywords: generalized modularity, munitions design, flexible modular, optimum design.

1 Introduction

With the celerity change of modern science and war field environment, munitions variety tends to diversification and munitions design trends toward celerity. Most conventional munitions design is empirical design, in which actual test and scheme amendment are repeated. It went against munitions celerity Design, and couldn't take full advantage of acquired design knowledge. With the development of CAD, parameterization design, virtual prototype technology and knowledge engineering are applied comprehensively at the region of munitions design [1-4].

Based on generalized modular design means munitions design method is researched in this paper. It is the continuation and stretch of traditional modular design, combines kinds of design technique and as a result expedites munitions design speed.

2 Generalized Modularity

2.1 The Concept of Generalized Modularity

Generalized modular design is a rapid design method which has considered the salience of product at function, manufacture use and etc. products are actualized through the plot and combination of generalized modules. The kernel of generalized module is flexible module and virtual module[5].

* Corresponding author.

D. Jin and S. Lin (Eds.): Advances in MSEC Vol. 1, AISC 128, pp. 233–237.
springerlink.com

Flexible module is a parameterized model which has homology function structure and relatively constant interface. The value of modular parameter is not one fixed value, but a definite variation range. Giving specific value to the parameter of flexible module creates flexible modular example, the combination of which gets specific product. Virtual module is a compose cell which has relatively fixed subfunction and geometry topology. It is has its independent meaning just in CAD.

2.2 Generalized Modular Interface

The interface of flexible module is called flexible interface, it's a generalized interface, of which the topology shape is fixed and the parameters varies in a range. It's designed to connect each flexible module. The content of the flexibility interface includes topology shape, attended mode, connective principal and subordinate module, interface position etc.

3 The Design Cycle of the Munitions

The flow chart of munitions generalized modular design is like figure 1. In the flow chart, flexible modules are selected according to the range of tactical-technical indices. The best fit flexible modules are picked from the modular example library. The value of design parameter should be adjusted till it's up to the mark of tactical-technical indices.

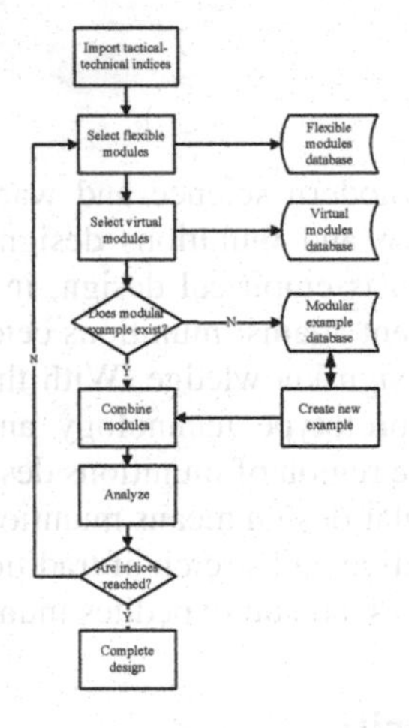

Fig. 1. Design cycle of the munitions

4 Plot Generalized Modules

4.1 Plot the Flexible Modules of Munitions

There are 2 steps while plotting flexible modules. First of all, different assemblies are sorted by different functions. Fig.2. is function modules of shrapnel.

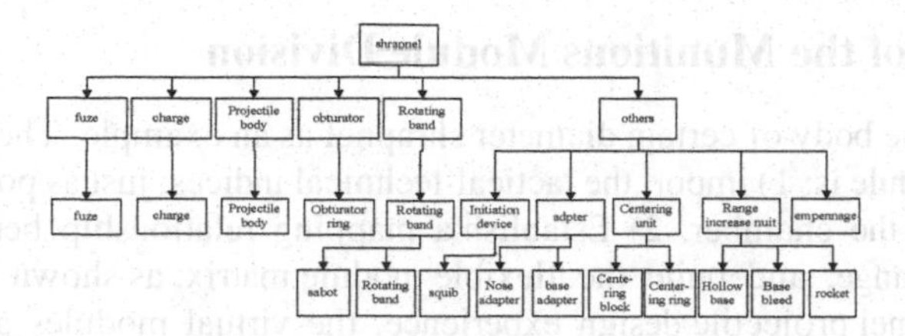

Fig. 2. Function modules of shrapnel

Munitions are designed according to the tactical-technical indices. Tactical-technical indices not only contain performance requirements, but also contain limitation of parameters just as quality of shrapnel. The structural features and layout of munitions is similar, and the flexible module is divided relatively simply. Basing on the structurally similar analysis, the existing munitions models are summed up. Then set the performance requirements as the vertical line, structure and layout as the horizontal line, and built the matrix of flexible modules, just as figure 3. The content of flexible module will play a role as optimize information in the design. The determination of the specific values in flexible module matrix should introduce of knowledge engineering.

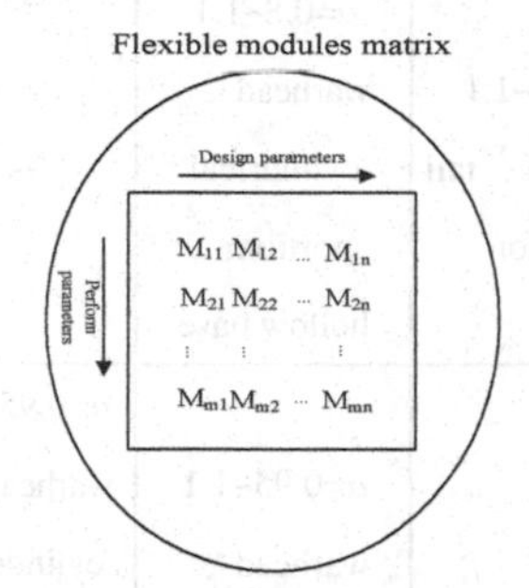

Fig. 3. Flexible modules matrix

4.2 Plot the Virtual Modules of Munitions

For Flexible modules, whose structure and design parameters are associated with a variety of tactical-technical indices, the plot of virtual modules is necessary. The division of the virtual module should ensure the minimum number of virtual modules which has a relatively independent features and fixed topology. The means of plot virtual modules is the same with flexible modules.

5 Optimization Based on Generalized Modular Design

In the design cycle, when the analysis results cannot achieve the intended targets, it is needed to optimize the design parameters. But a parameter's change may affect the performance of ammunition, and cause some difficulties to optimal design of ammunition. In the design of munitions, designers can change the value of the relative module to achieve the purpose of rapid adjustment.

6 Instance of the Munitions Module Division

The paper set the body of certain diameter shrapnel as an example. The steps of divide the flexible module is: 1) import the tactical-technical indices, just as power, range, and stability inside the chamber. 2) Establish a mapping relationship between the body structure and range, and build the flexible module matrix, as shown in Table 1. 2) Based on shrapnel projectile design experience, the virtual modules are plotted, and then the table of virtual modules is created. In the table, each module has a mapping relationship with a tactical-technical index. If one index is required to adjust, the related

Table 1. Matrix of flexible models

Structure / *Range L*	Regular shape	Long-distance shape	With hollow base	With Base bleed	With rocket device	With Base bleed-rocket device
27km	α=0.25~1.2 Normal	—	—	—	—	—
32km	—	α=0.8~1.1 warhead 、tail portion	α=0.8~1.1 warhead 、cylindrical portion、hollow base	—	—	—
37km	—	—	α=0.95~1.1 warhead 、hollow base	α=0.95~1.1 warhead 、cylindrical portion、base bleed	α=0.9~1.1 warhead 、cylindrical portion、rocket	—
42km	—	—	—	α=0.95~1.1 warhead 、base bleed	α=0.95~1.1 warhead 、rocket	α=0.95~1.1 warhead 、cylindrical portion、base bleed-rocket
47km	—	—	—	—	—	α=0.95~1.1 warhead 、base bleed-rocket

αL—the scope of range

Table 2. Virtual module table

Virtual module	Topology	The preferentially performing parameters	Primary parameter
Warhead		shape factor	length of warhead A
Cylindrical portion		power	length of cylindrical portion H
Centering portion		Inner trajectory stability	width of centering portion B
tail portion		range	length of tail portion E

module is considered as a priority. The table of the divided virtual modules is as table 2. The values in the table are only a reference.

7 Conclusion

In this paper, generalized modularization is introduced to the munitions design. The paper introduced the concept of generalize modularity, built the design cycle of munitions, did some research in the division of flexible modules and virtual modules, and analyzed the optimum design under generalized modularity. In addition, parameter design and knowledge engineering needs to be researched to cooperate with generalized modularity.

References

1. Wei, H., Zhu, H., Wang, D.: Munitions design theory (1985)
2. Yuan, Z., Wang, X., Hao, B.: The Formation for Three-Dimensional Parameter Drive Constraint Equation of High Explosive Shell. Journal of Shenyang Institute of Technology 23(2), 61–68 (2004)
3. Yang, Y., Qian, L., Lu, Y.: Study on virtual prototyping design method for conventional warhead. Manufacturing Automation 31(10), 114–118 (2009)
4. Yu, D., Qian, L., Ma, L., Deng, K., Peng, Y.: Research & Implementation of Conventional Warhead CAD System Based on KBE Technology. China Mechanical Engineering 13(14), 23–25 (2002)
5. Gao, W., Xu, Y., Chen, Y., Zhang, Q.: Theory and Methodology of Generalized Modular Design. Chinese Journal of Mechanical Engineering 43(6), 49–54 (2007)

Structured Evaluation Report of an Online Writing Resource Site*

Xing Zou

Department of Foreign Languages
Wuhan Polytechnic University
Wuhan, China
Chriskie2006@yahoo.com.cn

Abstract. Nowadays online writing has become a new way of developing one's writing ability. Through structured evaluation of an online writing resource site, this paper attempts to analyze the writing as a process, as a genre and its practicability in Chinese context. It turns out that writing is a social practice rather than a solitary activity. Learners will gain new insight into writing and it also has pedagogical significance for the teachers.

Keywords: online writing, evaluations, process, genre.

1 Introduction

The name and the URL of the online writing resource are Writing @CSU, Writing Resources (http://writing.colostate.edu/learn.cfm). I will choose the most significant part to analyze.

1.1 One New Insight into Writing

Online writing or writing online is a new way of learning how to write and developing one's writing ability. As for me, the unique sense of online writing is that writing is a social practice rather than a solitary activity---"Many of us think of writing as a solitary activity -- something done when we're alone in a quiet place. Yet most of our writing, like other forms of communication -- telephone conversations, classroom discussions, meetings, and presentations -- is an intensely social activity." (http://writing.colostate.edu/guides/processes/writingsituatio ns/)

1.2 My Comments

I agree to the above comments in terms of communicative writer-reader relationship. Any writing involves conveying ideas to the audience and the audience will respond to the writing, so writing becomes an effective way of communication like a face-to-face conversation, which let the writers bear the readers in mind, expect the readers'

* This research is supported by the Hubei Humanities and social science project " The Teaching of English Majors under the Guidance of the Implicit Knowledge in the Language Output" No:2011jytq139.

D. Jin and S. Lin (Eds.): Advances in MSEC Vol. 1, AISC 128, pp. 239–243.
springerlink.com

reactions and intend to share by making sense of their own writing so that they can find a sense of community and become more confident and less solitary. Actually, the online-writing websites provide readers and writers with helpful guide, lots of writing materials and chances of exchanging ideas, sharing what one has learned through writing experience and benefit from the wisdom of the others including some experts. They are supported by online-writing Guides; they can make use of Writing Tools; they can join in a Writing Activity, and get feedback on their writing, etc. All these writing activities make writing easier and interesting for the writers to share the pleasure of writing online, more important, they make writing socially significant.

If you have more than one surname, please make sure that the Volume Editor knows how you are to be listed in the author index.

2 Writing as a Process

Process approaches focus primarily on what writers do as they write rather than on textual features. As Figure 2.6 shows, the process approach includes different stages, which can be combined with other aspects of teaching writing (Coffin, C., Curry, M. J., Goodman, S., Hewings, A., Lillis, T. M., and Swann, J., 2003, P33-34). The iterative cycle of process approaches delineated in this Figure showed that this approach involves the linguistic skills of prewriting, planning, drafting, reflection, peer review, revision, editing and proofreading. This approach see writing primarily as the exercise of linguistic skills, and writing development as an unconscious process which happens when teachers facilitate the exercise of writing skills (Richard Badger and Goodith White, 2000, P155). In the On-line lab, I found one part that exhibits writing as a process:

Preparing to Write
Starting to Write
Conducting Research
Reading & Responding
Working with Sources
Planning, Drafting, & Organizing
Designing Documents
Working Together
Revising & Editing
Publishing
(http://writing.colostate.edu/guides/)

Each of the above steps has been illustrated in detail. In fact, the processes listed on the website are slightly different from what we learned from traditional course book (A Handbook of Writing)

1. To fix on a subject or theme
   To list an outline
   To write the first draft
   To revise the first draft: content, structure, sentences, diction, etc.
   To compose the final version
   To check the final version

Through comparison, one may find that the processes of online writing are of more mutual exchange, which indicates that writing is a social act. Actually, the non-linear process can be got access to in various orders at different points. One needn't follow it strictly and move through the stages one by one. Just like the on-line resource, you can click any button to get a clear idea about the stage and it gives you a hand-on guide for you to practice.

The printing area is 122 mm × 193 mm. The text should be justified to occupy the full line width, so that the right margin is not ragged, with words hyphenated as appropriate. Please fill pages so that the length of the text is no less than 180 mm, if possible.

3 Writing as a Genre

Genre refers to abstract, socially recognized ways of using language. And genre approaches see ways of writing as purposeful, socially situated responses to particular contexts and communities (Ken Hyland, 2003, P25). This approach see writing as essentially concerned with knowledge of language, and as being tied closely to a social purpose, while the development of writing is largely viewed as the analysis and imitation of input in the form of texts provided by the teacher (Richard Badger and Goodith White, 2000, P156). In the on-line lab, according to the social context of creation and use, texts with similar features are grouped together, therefore a linguistic community is established in each type, which involves interaction of the writer, reader and the text. Writing can be guided in different genres:

Composition & Academic Writing
Argument
Fiction
Poetry
Creative Non-Fiction
Writing about Literature
Scholarly Writing
Business Writing
Science Writing
Writing in Engineering
Writing for the Web
Speeches & Presentations
(http://writing.colostate.edu/guides/index.cfm?)

Major text types that student might frequently write or find difficult to write are listed here. Each one provides the learners with an explicit understanding of the text and a metalanguage by which to analyze them, which can help students understand that texts can be explicitly questioned, compared and deconstructured. Writing is not abstract activity, it becomes a social practice, varying from one community context to the next. It can be safely concluded that if a writer wants to create a certain composition, he must get himself familiar with the characteristics of such writing which appears in a certain context and appeals to certain group of readers. A successful writing depends on a desirable interaction within the community of the writer, reader and the text.

4 Overall Suitability as a Self-access Resource for Tertiary EFL Students in China

Students in China can benefit a lot from the writing lab. Since most of them don't form the habit of reading such on-line instructions outside the class, they rely too much on the in-class teaching, which is far from enough for them to improve their writing proficiency. This on-line lab will be of great help for them to raise their interest in and enhance their motivation for writing.

As for non-English majors, their writing competence has been prescribed in the book *College English Curriculum Requirements* (2007) which is the guideline for college English teaching in China. From its basic requirements to advanced requirements, writing ranges from daily message, letters, notes to thesis writing, presentation, paper report, etc. Students may refer to "writing as genres" and practice related activities.

And for English majors, they have more demands for linguistic skills. Sufficient in-class teaching time plus more motivation enable them to learn writing systematically. They can start with constructing a sentence, telling a story or with lots of exposure to reading. And this lab can provide a good platform for them to learn, to practice and to share. So they may combine these two approaches more. From this website, they can see the values of reading and writing and gain a lot of insights into them so that they can make sense of themselves. Apparently this lab will be of great significance to them both academically and socially.

5 Using the Online Resource for Teaching

On-line resources are also of pedagogical significance for the teachers, for it provides a detailed teaching guide for us. Different teachers may choose different way of teaching. During my preparation for the lesson, I'd like to use the Sample Lesson Format (http://writing.colostate.edu/guides/teaching/lesson_plans/pop2f.cfm) in the lab as a guide, considering my own students' needs. In class, this on-line resource will also serve as a reference, and I prefer to combine the process and genre approach, using Write To Learn (WTL) （http://writing.colostate.edu/guides/teaching/planning/wtl.cfm）. For example, when it comes to the topic "Pollution", before writing, I'll show them some pictures or short articles on pollution, carrying out a brain-storming activity using the "concept map" or "graphic organizer" to come up with the related category, cause, consequence and possible solutions. Then a model text will be deconstructed to raise students' consciousness of the structure and related social purposes. In the following session students will begin to plan a composition by joint construction with the teacher or partners. When their first drafts come out, they will start the peer review so that they can revise with the help of the teacher. After I give my final evaluation, students are required to write a reflection on what they've learnt from the writing, which I think they will benefit a lot from this experience. After class I will write my own reflection on the goal and result of the lesson so that I can adapt some part of my plan.

This on-line resource provides the teachers with new insight into both writing and teaching writing. Our Chinese teachers and students, in particular will find it a rewarding and meaningful experience in exploring this resource, though it will be a great challenge for us to adjust our traditional way of writing and teaching writing. So even if I don't teach writing in China, I really want to make an attempt to teach after I come back to China.

Acknowledgment. This work is supported by the Teaching and Researching Foundation of my university this year. Thanks to my learning experience in Singapore, I am grateful to Professor Ramona Tang in Nanyang Technological University and teachers of the university , who have helped me a lot in ushering me into the fascinating world of teaching methodology.

I am especially indebted to my husband, my colleagues and family members for their continuous support all the time. Without their encouragement, this paper could not be possible.

References

[1] Badger, R., White, G.: A process genre approach to teaching writing. ELT Journal 54(2), 155–156 (2000)

[2] Tomlinson, B.: Developing Materials for Language Teaching Continuum International Publishing Group Ltd. illustrated edition (2003)

[3] Coffin, C., Curry, M.J., Goodman, S., Hewings, A., Lillis, T.M., Swann, J.: Teaching Academic Writing: A Toolkit for Higher Education, pp. 33–34. Routledge, London (2003)

[4] Grabe, W.: Reading in a Second language: Moving from Theory to Practice. Cambridge University Press, New York (2009)

[5] Hyland, K.: Second Language Writing. Cambridge University press, Cambridge (2003)

[6] Zhang, L.: A Study into the Chinese English Learners Psychological Process in L2 Discursive Writing. Foreign Language Learning Theory and Practice 03 (2009)

[7] Liu, X.-Q.: "China English". Study and English Writing. Journal of Nanhua University 04 (2004)

[8] Lu, Z.H.: Discourse Analysis and the Teaching of English Writing. Shangdong Foreign Languages Journal 04 (2003)

[9] Muncie, J.: Finding a place for grammar in the EFL composition class. ELT Journal 56(2), 180–186 (2002)

[10] Wang, W., Wang, L.: L2 Writing Research in China: an Overview. Foreign Language World 03 (2004)

Inversion Calculation and Site Application for High-Resolution Dual Laterolog (HRDL) Tool

Zhenhua Liu and Jianhua Zhang

Xian Shiyou University, 710065, Xian, China
liuzhenhua@xsyu.edu.cn

Abstract. A fast and efficient inversion algorithm was suggested for the new developed high-resolution dual laterolog (HRLD) tool. The algorithm can yield reliable estimates for the depth of invasion zone and true-formation resistivity simultaneously. Both the behaviour of HRDL measurements and inversion were studied for various bed thickness from 0.4m to 4m. The present results indicated that the HRLD tool improved the vertical resolution obviously, but the HRLD measurements deviate the true value of formation due to the invasion and shoulder effect. These environment influences can be corrected from the inversion process, thus the original formation parameters can be obtained. A site application of the present inversion algorithm gave the distribution of the true-formation resistivity in a real formation scale. It is helpful for log analyists to evaluate the reservoirs.

Keywords: inversion, resistivity, invasion, logging.

1 Introduction

The determination of true-formation resistivity from inversion techniques or chart corrections is one of the major tasks of electrical-logging interpretation. Since well logging is affected by environment factors, such as borehole, invasion, shoulder beds, and so on; the logging data will deviate from the true-formation model. Inversion algorithm is introduced to recovery the original formation data from field measurements for various logging devices.

Formation resistivity is an important parameter for the determination of the oil/gas saturation of a reservoir. Laterolog, which consists of focused coils so that they have deep investigation depth [1], is a basis method to measure the formation resistivity. However, in many situations the traditional dual-laterolog (DLL) does not supply sufficient information in thinly bedded formations; hence its poor vertical resolution leads to the DLL measurement can not distinguish the thin reservoirs. A recent developed new high-resolution dual laterolog (HRDL) tool by China Petroleum Logging Co. Ltd. has higher vertical resolution and shorter coil system [2].

HRDL tool modified the behaviors of traditional dual laterolog device and has been applied in fields [3]. This paper suggested a fast and efficient inversion algorithm for HRDL tool and illustrated the results of both synthetic data inversion and case application. The present results are helpful for log analysts to determine the true formation resistivity and to evaluate an oil/gas or water reservoir using HRDL data.

D. Jin and S. Lin (Eds.): Advances in MSEC Vol. 1, AISC 128, pp. 245–250.
springerlink.com

2 Inversion Algorithm

In all environment factors, the invasion of drilling mud filtrate into formation make a strong influence on resistivity measurement. The HRDL tool yields two logging curves with shallow and deep detective depths respectively. Hence, we use the two curves to inverse the true-formation resistivity and the depth of invasion zone. For this purpose, we construct a two-dimensional axisymmetric formation model that included borehole (radius r_h and mud resistivity R_m), invaded zone (depth r_i and resistivity R_{xo}), formation (thickness H and true-formation resistivity R_t), and shoulder beds (resistivity R_{su} and R_{sd} for up shoulder and down shoulder respectively). The logging sonde is centered in the borehole.

During the inversion of HRDL data, the damping least square method [4] was used. Two logging curve data were fit from the forward model with two unknown parameters. The formula is

$$R_n = F_n(R_t, r_i, S_n) \qquad (n=1,\ 2) \tag{1}$$

Here, R_n is logging data, $R_1 = R_{HD}$ and $R_2 = R_{HS}$ mean deep and shallow logging of a HRDL tool respectively. R_t and r_i are inversed parameters. S_n means other parameter in the forward model.

Eq.(1) is a system of nonlinear equation. If the initial guesses of the model were given, it can be linearized according to Taylar progression:

$$R_n = R_n^0 + \frac{\partial F_n}{\partial R_t}\delta R_t + \frac{\partial F_n}{\partial r_i}\delta r_i \qquad (n=1,2)$$

This equation can be rewritten in a matrix

$$\vec{\varepsilon} = \vec{R} - \vec{R}^0 = J\delta\vec{P} \tag{2}$$

Where, $\vec{R} = (R_{HD}, R_{HS})^T$ is the vector of logging data. $\vec{R}^0 = (R'_{HD}, R'_{HS})^T$ is the vector of guessed data in forward model. ε is the vector of the difference between logging data and guessed data. $\vec{P} = (R_t, r_i)^T$ is the vector of inversed parameters and J is the Jacobi matrix. Eq.(2) is a system of linear equations related to modified step $\delta\vec{P}$. Its damping least square solution is

$$\delta\vec{P} = (J^T J + \eta I_0)^{-1} J^T \vec{\varepsilon} \tag{3}$$

Where η is damping factor and I_0 is a unit matrix. The forward model was linearized at initial value P_0, and then the increment $\delta\vec{P}$ was obtained by damping least square method. New model parameters $\vec{P}'$ are generated from the following formula

$$\vec{P}' = \vec{P}_0 + \delta\vec{P} \tag{4}$$

This new parameter was input into forward model and new guess data $\vec{R}'$ was generated. Thus, new difference vector

$$\vec{\varepsilon} = \overline{R} - \overline{R}' \tag{5}$$

and new Jacobi matrix were generated. Then new parameter vector increment $\delta\vec{P}$ was obtained from these results. Thus, a new iterations was informed and Eqs. (2) ~(5) were repeated until the expectant results that satisfy iteration error was reached through iteration process.

3 Synthetic Data Inversion

In order to check the characteristics of HRDL tool and the inversion algorithm, a synthetic formation model was constructed with true-formation resistivity R_t=5.0Ωm, mud resistivity R_m=1.0Ωm, borehole radius r_h=0.1016m, which corresponds to the diameter 8 inch for a borehole. The resistivity of invasion was set to be R_{xo}=15Ωm and the depth of invasion zone was 0.4m. The resistivities of shoulder beds R_{su}=R_{sd}=2.0. The reservoir thickness H was taken to be 4, 3, 2, 1.5, 1.0, 0.6 and 0.4m respectively in order to address the vertical resolution of a HRDL tool. Fig.1 shows the logging responses R_{HD} and R_{HS} for a HRDL tool. The vertical axis direction denotes the apparent resistivities and the horizontal direction denotes the depth from the surface in a formation. The deep and shallow data of HRDL measurements were draw in Fig.1 using heavy line and thin line respectively.

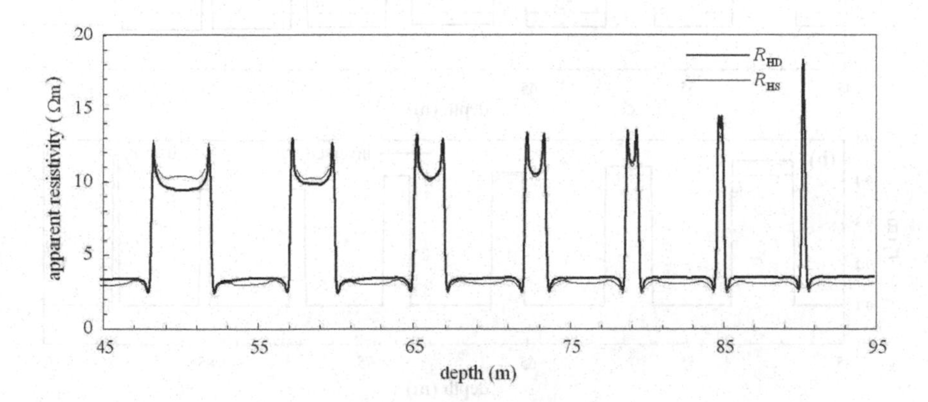

Fig. 1. The logging responses of HRDL tool for a synthetic formation model

Fig.1 showed that the HRDL tool can supply sufficient information in thinly bedded formations even for the thickness of 0.4m. In the present calculation for figure 1, the logging data deviated strong from the true-formation resistivity R_t=5.0Ωm due to the high resistivity of invasion zone, R_{xo}=15.0Ωm. Due to the high-resistivity invasion zone invading into the reservoir, the apparent resistivities of logging tool will increase usually. It illustrates that the invasion plays an important role in well logging and effects on the logging data seriously. In the boundary of formation and shoulder, the shoulder effect [5] influences the shape of logging curve for a HRDL tool.

Comparing with traditional dual laterolog device, the HRDL tool modified the vertical resolution.

The logging data illustrated in Fig.1 were input to the present inversion algorithm. The inversion iterations start with the initial guess values input. Fig.2(a) draw the inversion results of true-formation resistivity in heavy line using the logging curve of HRDL tool in figure 1. The true value of the formation resistivity was denoted by the thin line in Fig.2(a).

For a thick reservoir (H>2m), the inversed R_t agreed with the true value of formation resistivity. However, for a thin bed (H<1m), the errors between the inversed R_t and the true value were introduced because the low-resistivity shoulder effect become strongly for thin layers. Even though, the HRDL measurement can distinguish the thin layer such as 0.4m, whereas the vertical resolution of a traditional dual laterolog device is 0.6 ~ 0. 9 m.

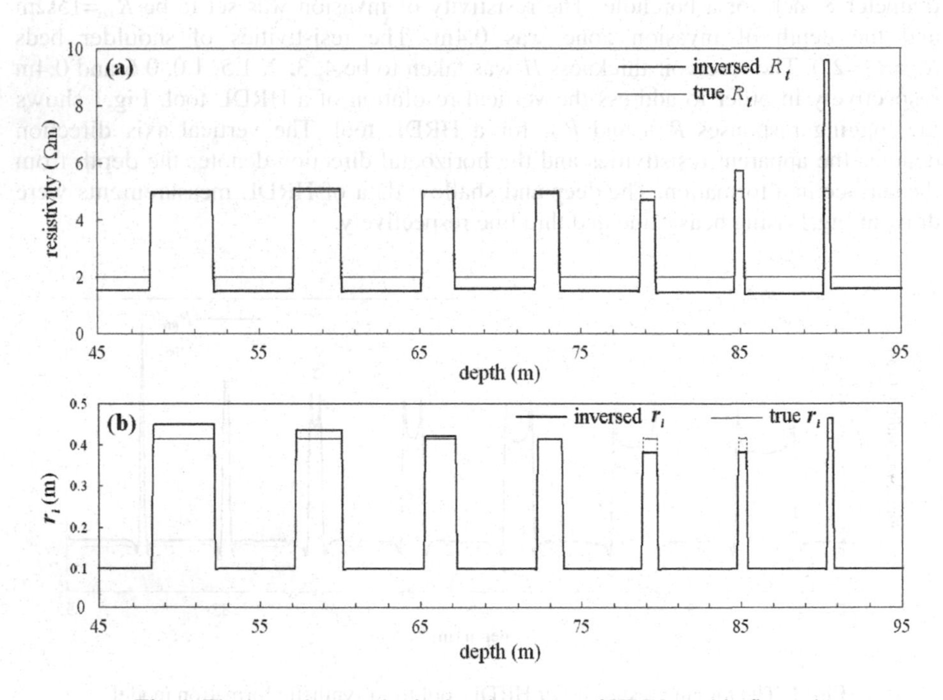

Fig. 2. The inversion results for logging curve of HRDL tool in figure 1

The true invasion depth (thin line as shown in Fig.2b) and the inversed value (heavy line as shown in Fig.2b) were drawn in Fig.2 as well. Good agreement of inversed results with true values was obtained. From the present inverted data, it is concluded that the HRDL inversion reaches the satisfied estimation, especially when the thickness of a reservoir H is greater than 1 meter. Usually, the inversion of HRDL tool enhances the reliability of inversion results.

4 Site Application

The present inversion algorithm was used for a well located at Western China. The well were logged using high-resolution dual laterolog tool and other resistivity logging device such as microspherically focused log (MSFL) device simultaneously. Figure 3a are the logging curves of HRDL and MSFL devices for a well, which located at Western China. The vertical axis direction denotes the apparent resistivities and the horizontal direction denotes the depth from 1910~1933m in the formation.

Usually, the true-formation resistivity R_t, invasion depth r_i , and the resistivity of invasion zone R_{xo} are unknown. Since a high-resolution dual laterolog tool provided two log curves only, so the logging data of microspherically focused log was regarded as the resistivity of invasion zone R_{xo} because it records the information in the vicinity of borehole [1]. For the present site example, it was drawn in long dashed line in Fig.3(b) after the formation was layered from the data of deep HRDL data.

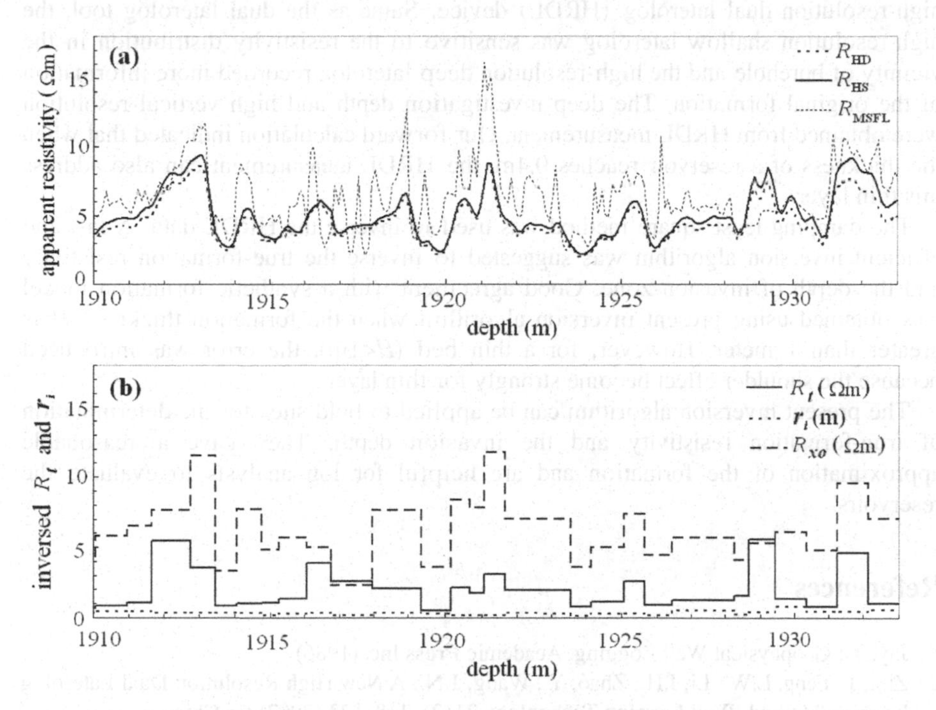

Fig. 3. (a) the logging curves of field HRDL measurement
(b) the inversion results for field HRDL logging curves

The HRDL measurement data at 1910~1933m in figure 3(a) were input into the present inversion iterations with other necessary input parameters such as mud resistivity R_m=0.8Ωm and borehole radius r_h=0.1016m. Before inversion iterations, the formation was layered from the deep logging curves of HRDL device, as shown in Fig.3(b). In figure 3(b), the solid line denotes the inversed true-formation resistivity; the dot line means the invaded depth of invasion zone. The present inversion results indicated that the apparent resistivity of HRDL measurement was higher than true-formation resistivity obviously due to high resistivity in the invasion zone. Also, this high resistivity in the invasion zone influences the response R_{HS} much more, which records the information of shallow detective depth device. Therefore, R_{HS} is greater than the high-resolution deep laterolog response R_{HD} , which records the deep depth information in a reservoir, as shown in Fig.3(a).

In logging interpretation, true-formation resistivity R_t is of importance in estimating original hydrocarbon saturation and evaluating a reservoir. It is related to saturation through hydrocarbon saturation formula [6]. If R_t had been obtained, hydrocarbon saturation can be estimated. Therefore the reservoir can be evaluated.

5 Conclusions

The shortages of a traditional dual laterolog (DLL) tool were improved using the high-resolution dual laterolog (HRDL) device. Same as the dual laterolog tool, the high-resolution shallow laterolog was sensitive to the resistivity distribution in the vicinity of borehole and the high-resolution deep laterolog recorded more information of the original formation. The deep investigation depth and high vertical resolution were obtained from HRDL measurement. Our forward calculation indicated that when the thickness of a reservoir reaches 0.4m, the HRDL measurement can also address this thin layer.

The damping least square method was used to inverse the HRDL data. A fast and efficient inversion algorithm was suggested to inverse the true-formation resistivity and the depth of invasion zone. Good agreement with a synthetic formation model was obtained using present inversion algorithm when the formation thickness H is greater than 1 meter. However, for a thin bed (H<1m), the error was introduced because the shoulder effect become strongly for thin layers.

The present inversion algorithm can be applied to field sites for the determination of true-formation resistivity and the invasion depth. They gave a reasonable approximation of the formation and are helpful for log analysts to evaluate the reservoirs.

References

1. Jay, T.: Geophysical Well Logging. Academic Press Inc. (1986)
2. Zhu, J., Feng, L.W., Li, J.H., Zhao, Y., Wang, J.N.: A New High Resolution Dual Laterolog Logging Method. Well Logging Technology 31(2), 118–123 (2007) (in Chinese)
3. Wang, H., Li, J.P., Wang, A.Y., Cheng, G., Xiong, X.Y., Li, J., Li, G.: Application of the High Resolution Bilateral Logging Tool. Petroleum Instruments 21(3), 27–28 (2007) (in Chinese)
4. Lawson, C.L., Hanson, R.J.: Solving Least-squares Problems. Prentice-Hall, Inc. (1974)
5. Tumer, K., Torres, D., Chemali, R.: A New Algorithm for Automatic Shoulder Bed Correction of Dual laterolog Tools. In: SPWLA 32nd Annual logging Symp., Midland, Texas (1991)
6. Schlumberger: Log Interpretation Principles / Applications. Houston, Schlumberger Educational Services (1987)

GPU Accelerated Target Tracking Method

Jian Cao, Xiao-fang Xie, Jie Liang, and De-dong Li

Department of Ordnance Science and Technology
Naval Aeronautical and Astronautical University
Yantai, China
ddcjd@163.com

Abstract. Unlike differential motion analysis and optical flow computation methods, target tracking method based on correspondence of robust feature descriptors can give vast improvements in the quality and speed of subsequent steps, but intensive computation is still required. With the release of general purpose parallel computing interfaces, opportunities for increases in performance arise. In this paper we present an accelerated target tracking method based on OpenCL to meet the real time requirement. First, it uses SURF algorithm to extract target feature and match them in following frames. Then, the bary method is used to compute the target displacement. Experimental results show that this method has strong robustness to little rotated, shielded, illumination changed of the target and decrease largely the computing time.

Keywords: target tracking, SURF, feature extraction, OpenCL, GPU.

1 Introduction

Video tracking is the most active topic in the fields of computer vision, image processing and pattern recognition. And it may become as complex as we wish by considering pan-tilt-zoorn (PTZ) cameras, and cameras with overlapping fields of view [1]. The commom methods include differential motion analysis, optical flow computation and interest points based mothod, etc.

This paper we implement a target tarcking methods includes two main steps. First, it uses features detection algorithm to extract target feature and match them in following frames. Then, the bary method is used to compute the target displacement.

During the whole tracking process, feature point detection and description generation is the most important step. Since features can be viewed from different angles, distances, and illumination, it is important that a feature descriptor be relatively invariant to changes in orientation, scale, brightness, and contrast, while remaining descriptive enough to guarantee match precision. We chose the Speeded-Up Robust Features (SURF) descriptor. Comparing those previous algorithms, such as the Scale-Invariant Feature Transform (SIFT)[2], SURF locates features adopt many approximations to increase the speed of subsequent matching operations with almost same matching precision, while themselves being less expensive to compute, as illustrated in Table 1[3]. However, SURF cannot yet achieve interactive frame rates on a traditional CPU.

D. Jin and S. Lin (Eds.): Advances in MSEC Vol. 1, AISC 128, pp. 251–257.
springerlink.com

Table 1. The Comparsion Result of SIFT, PCA-SIFT and SURF

Method	Time	Scale	Rotation	Blur	Affine
SIFT	common	best	best	common	good
PCA-SIFT	good	good	good	best	best
SURF	best	common	common	good	good

Recently, to meet the real time requirement, a multitude of work has emerged using the graphics processing unit (GPU) for scientific computing. The most notable in this context are NVIDIA's CUDA [4] and the Khronos group's OpenCL [5]. OpenCL is an open standard maintained by the Khronos group, and has received the backing of major graphics hardware vendors. OpenCL [5] have enabled relatively straightforward development of parallel applications that can run on commodity hardware across a number of platforms (CPU, GPU, Cell BE, etc). In this paper, we were interested in developing a very efficient parallelized (accelerated) version of the target tracking method in OpenCL.

The rest of this paper is organized as follows: Section 2 presents a short introduction to OpenCL working procedure. In section 3 we will present the SURF followed bary tracking algorithm briefly. Section 4 contains the GPU accelerated target tracking implementation by using OpenCL. In section5, the experiment results and anlysis will be exhibited. In section 6, conclusions are drawn and directions for future work are given.

2 OpenCL Working Procedure

An OpenCL program requires a number of steps to run. First a context must be created; the context describes a means to communicate with an OpenCL devices such as a single or multiple CPUs, GPUs, or FPGAs; however the focus of this discussion will only be on GPUs [7]· The context also retains and manages memory allocated on the device, and allows for reading and writing to this memory. Allocating memory for use by OpenCL is done through the use of a special malloc function clCreateBuffer. Other functions exist such as clCreateImage2D and clCreate-Image3D, which are useful for more specific applications. Reading memory is done through clEnqueueReadBuffer and clEnqueueWriteBuffer, both of these commands support synchronous and asynchronous operations. Reading and Writing do require the addition of a queue to perform their tasks. To submit commands to the device within a context we use a queue. The queue is specified for all kernels, reading and writing to device memory, and supporting functions such as barriers for synchronization. The OpenCL function, clCreateCommandQueue, creates the queue. Once an operation is enqueued it runs immediatly and asynchro-nously, however to enable synchronous communication OpenCL provides cl_events data types and the functions clWaitForEvents and clEnqueueWaitForEvents. Here cl_ events work as tokens allowing subsets of a group of operations to work together while preserving dependencies. To create a barrier over an entire queue though would be quite cumbersome using cl_events so OpenCL provides clEnqueueBarrier. And with the OpenCL function clCreate-ProgramWithSource the kernel will be compiled during runtime from strings containing the program or read in as pre-compiled binaries, the kernel arguments can be set by function clSetKernelArg. Once a program is compiled it can then be linked against its arguments to create a kernel. This kernel can be enqueued for immediate processing by the GPU. Once we receive the calculated result from the device memory back to CPU memory by clEnqueureadBuffer function, clReleaseKernel, ReleaseProgram, ReleaseCommandQueue, clReleaseContext and clReleaseMem-Object functions will follow to release the device memory and GPU resources [8].

3 The Target Tracking Algorithm

In this paper, the target tracking algorithm includes following 4 steps:.①Select target template form the image eliminated impulse noise by median filtering manually. ② Using SURF algorithm to extract target features and storges them into feature library.On the followed frame of image sequence, extract the bigger ROI around manual one(For example, if the manual region's size is about 100*100, the pending ROI's is about 200*200) and calculate it's SURF features. ③ Find matched SURF features. Due to the unpredictable distribution of matching feature points, the bary method is used to compute the target displacement. Take the frames displacement as the moving estimation of target to update ROI position and reduce the searching area for the following frame's SURF feature extraction.④Update target template by storging the new matched SURF features into feature library and execute step ② until all frames are handled.

In above processes, median filtering and SURF algorithm are intensive computation suit to execute on GPU with OpenCL, while the bary method and feature matching algorithm are suitable for CPU because of the low amount of processing data and high Latency of data exchange between CPU(HOST) and GPU(DEVICE). By the way, there's an excellent sample programme of median filtering in GPU computing SDK provided by Nvidia Corporation[8]. So in the rest of paper,we will mainly discuss the OpenCL SURF implementation.

3.1 The SURF Algorithm

This section reviews the original SURF algorithm. We defer some of the details to the next section, which discusses our OpenCL based GPU implementation, but we highlight the main points here.

SURF[9] locates features using an approximation to the determinant of the Hessian, chosen for its stability and repeatability, as well as its speed. An ideal filter would construct the Hessian by convolving the second-order derivatives of a Gaussian of a given scale$\sigma$with the input image. This is approximated by replacing the second order Gaussian filters with a box filter. Box filters are chosen because they can be evaluated extremely efficiently using the so-called integral mage *II*, defined in terms of an input image *I* as:

$$II(x, y) = \sum_{i=0}^{x}\sum_{j=0}^{y} I(i, j) \tag{1}$$

To achieve scale invariance, the filters are evaluated at a number of different scales, s, and the 3×3×3 local maxima in scale and position space form the set of detected features. Here $s = \sigma$, the scale of the Gaussians used to derive the box filters. A minimum threshold H_0 on the response values limits the total number of features. The location x0 of each feature is then refined to sub-pixel accuracy via

$$\hat{X} = X_0 - (\frac{\partial^2 H}{\partial X^2})^{-1} \frac{\partial H}{\partial X} \tag{2}$$

Where $X = (x, y, s)^T$ are scale-space coordinates and $H = |\det(H)|$ is the magnitude of the Hessian determi-nant. The derivatives of H are computed around x_0 via finite differences.

Rotation invariance is achieved by detecting the dominant orientation of the image around each feature using the high-pass coefficients of a Haar filter in both the x and y directions inside a circle of radius 6s. The size of the Haar filter kernel is scaled to be 4s×4s, and the sampling locations are also scaled by s, which is easily accomplished using the integral image.

The resulting 2D vectors are weighted by a Gaussian with $\sigma = 2.5s$ and then sorted by orientation. The vectors are summed in a sliding window of size $\pi / 3$, and the orientation is taken from the output of the window with the largest magnitude. Once position, scale, and orientation are determined, a feature descriptor is computed, which is used to match features across images. It is built from a set of Haar responses computed in a 4×4 grid of sub-regions of a square of size 20s around each feature point, oriented along the dominant orientation. Twenty-five 2D Haar responses (dx, dy) are computed using filters of size 2s × 2s on a 5 × 5 grid inside each sub-region and weighted by a Gaussian with $\sigma = 3.3s$ centered at the interest point [9].

The total algorithm runs in approximately 354 ms on a 3 GHz Pentium IV for an 800×640 image [9], or at just under 3 Hz.

4 OpenCL SURF Implementation

This section covers the major parts of our OpenCL SURF implementation. Our implementation has the following major steps:

Integral Image Computation

The primary workhorse of the SURF algorithm is the integral image, which is used to compute box filter and Haar filter responses at arbitrary scales in constant time per pixel. Since it must be computed over the entire image, it is one of the more expensive steps [10].

The computation of the integral image itself is a classic parallel prefix sum problem. It can be implemented as a prefix sum on each row, followed by a prefix sum on each column of the output. Consulting the sample 'oclScan' in OpenCL SDK, we designed the kernel called Integral-Img_kernel. This program operates strictly within a single workgroup and each work item in this workgroup performs a coalesced copy to a local array.

Feature Detection

Having constructed the integral image, we turn to the evaluation of the box filters used to locate interest points. The first step is to confirm the size of box filters. In Bay's paper, the values for each octave-interval pair dictate the size of the filter which will be convolved at that layer in the scale-space. The filter size is given by the formula:

$$\text{Filter Size} = 3(2^{\text{octave}} \times \text{interval} + 1) \tag{3}$$

Like Bay et al. [9], we use filters of size 9, 15, and 21 at the first three scales. Bay et al. derive their size 9 filters as the best box-filter approximation of the second-order derivatives of a Gaussian with scale $\sigma = 1.2$ and compute the scale associated with the rest of the filters based on the ratio of their size to that of the base filter. After reaching a filter size of 27, Bay et al. begin incrementing by 12, for several steps, then 24, etc., simultaneously doubling the sampling interval at which filter responses are computed each time the filter step size doubles. Like SIFT algorithm, SURF calculate the Hessian matrix, H, as function of both space x = (x; y) and scale σ

$$H(x,\sigma) = \begin{bmatrix} L_{xx}(x,\sigma) & L_{xy}(x,\sigma) \\ L_{xy}(x,\sigma) & L_{yy}(x,\sigma) \end{bmatrix} \tag{4}$$

Here $L_{xx}(x,\sigma)$ refers to the convolution of the second order Gaussian derivative $\partial^2 g(\sigma)/\partial x^2$ with the image at point $X = (x, y)$ and similarly for L_{xx} and L_{xy}. These derivatives are known as Laplacian of Gaussians.

Working from this we can calculate the determinant of the Hessian for each pixel in the image and use the value to find interest points. SURF performing conjunctions labeled D_{xx} D_{xy} D_{yy} between varying size box filters and integral image mentioned above to approximate Laplacian of Gaussians, and Hessian determinant using the approximated Gaussians:

$$\det(H_{approx}) = D_{xx}D_{yy} - (0.9D_{xy})^2 \tag{5}$$

The determinant here is referred to as the blob response at location $X = (x, y, \sigma)$. The search for local maxima of this function over both space and scale yields the interest points for an image.

Following detection ideas we set three OpenCL kernels named fasthessian_kernel nonmaxsuppression_Kernel and keypointinterpolation_kernel. fasthessian_kernel takes care about the conjunctions of varying size box filters and the IntegralImg_kernel output in device Global memory and establish the scale-space for further work. Nonmaxsuppression_Kernel search for local maxima of this function over both space and scale yields, check to see if we have a max (in its 26 neighbors) as described in Bay's paper. Finally use the algorithm in (2) keypointinterpolation _kernel output a vector of accurately localized sub-pixel interest points.

Orientation Detection

In order to achieve invariance to image rotation each detected interest point is assigned a reproducible orientation. Extraction of the descriptor components is performed relative to this direction so it is important that this direction is found to be repeatable under varying conditions. Like SIFT, conjunctions are made within a scale dependent neighborhood of each interest point detected by the Fast-Hessian.But in SURF algorithm integral images used in conjunction with filters known as Haar wavelets are used in order to increase robustness and decrease computation time.

In Bay's paper, to determine the orientation, Haar wavelet responses of size 4σ are calculated for a set pixels within a radius of 6σ of the detected point, where σ refers to the scale at which the point was detected. Within our kernel orientation_kernel the branching logic slow down efficiency because circle estimation.

The resulting 2D vectors are weighted by a Gaussian with $\sigma = 2.5s$ and then sorted by orientation. The vectors are summed in a sliding window of size $\pi / 3$, and the orientation is taken from the output of the window with the largest magnitude.

Feature Vector Calculation

To construct the feature vectors, axis-aligned Haar responses are computed on a 20s × 20s grid. The lattice points of the grid are aligned with the feature orientation, and the Haar response vector is rotated by this angle as well. Normalization is done on the GPU using another simple reduction to compute the vector magnitude. The kernel carry out above calculations named descriptors_kernel.

5 Experiment Results and Analysis

In this section, an experiment on the 800×600 continuous image sequences was carried out and those OpenCL kernels were programmed with OpenCL 1.1 in *.cl files separately. The test programme worked on the so common inexpensive computer (equipped with Intel Core2 Quad Q6600 2.4GHz and 4G RAM). The results shows as below.In figure 1, the arget template form the frame 1 image eliminated impulse noise is selected manually. Figure 2 shows the car tracking with the tracking method. Just because the SURF feature is not only a kind of robust feature which be invariant to luminance, translation, rotation and scale change, it also a kind of local feature for tracking partial occlusion target, as show in figure 3. And in figure 4, the tracking methods mentioned in this paper still track the target car partial appeared in view. Especially, with GPU accelerate the whole target tracking method took about 30 frames per second.

Fig. 1. Template selection

Fig. 2. Single frame image segment

Fig. 3. Partial occlusion target tracking

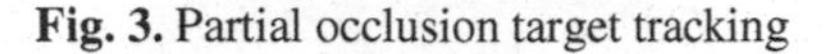

Fig. 4. Partial appeared target tracking

6 Conclusion

The method based on correspondence of Speeded up robust features (SURF) can give vast improvements in the quality and speed of subsequent steps, but intensive computation is still required[10]. In this paper we present an accelerated target tracking method based on OpenCL to meet the real time requirement. Experimental results show that this method has strong robustness to little rotated, shielded, illumination changed of the target and achieves interactive frame rates.

References

1. Sonka, M., Hlavac, V., Boyle, R.: Image Processing, Analysis, and Machine Vision, 3rd edn. Thomson, Canada (2008)
2. Lowe, D.G.: Distinctive image features from scale-invariant keypoints. International Journal of Computer Vision 60(2), 91–110 (2004)
3. Juan, L., Gwun, O.: A comparison of SIFT, PCA-SIFT and SURF. Journal of Image Processing (IJIP) 4(3), 143–152 (2008)
4. NVIDIA: CUDA web page (2009), `http://www.nvidia.com/object/cuda_learn.html`, KHRONOS: OpenCL overview web page, `http://www.khronos.org/opencl/`
5. Gohara, D.: OpenCL tutorials web page (2009), `http://www.macresearch.org/opencl`
6. NVIDIA Corporation. OpenCL Programming Guide for the CUDA Architecture Version 3.0. NVIDIA, Santa Clara (2010)
7. Wikipedia: OpenCL. Wikipedia (2011), `http://en.wikipedia.org/wiki/OpenCL` (July 05, 2011)
8. OpenCL API 1.0 Quick Reference Card. Khronos Group (2009)
9. Bay, H., Tuytelaars, T., Van Gool, L.: SURF: Speeded up Robust Features. In: Leonardis, A., Bischof, H., Pinz, A. (eds.) ECCV 2006, Part I. LNCS, vol. 3951, pp. 404–417. Springer, Heidelberg (2006)
10. Terriberry, T.B., French, L.M., Helmsen, J.: GPU Accelerating Speeded-Up Robust Features, DPVT (2008)

Web Database Access Technology Based on ASP.NET

Jin Wang

Guilin University of Electronic Technology Teaching Practice & Experiment Department, China
access_wj@163.net

Abstract. This article presents the principle about Web database access based on ASP.NET, expounds three kinds of connection methods and two accessing models that about ASP.NET page and database, and then analyzes and compares various means about using ADO.NET technology on database access. In the meantime, provides some utility codes. ASP. NET is currently an ideal choice for visiting Web database.

Keywords: ASP.NET, Web database, ADO.NET, data access.

1 Introduction

Data access is a key issue on web application development. General business application needs data-driven Web pages, developer must quickly access the data from different formats data source. Under the current windows environments, there are various technologies on web database access. Such as: CGI(Common Gateway Interface)、ADC(Advance Database Connector)、IDC (Internet Database Connector)、Java／JDBC、ASP and ASP．NET (Active Server Page) etc.

ASP.NET is a new generation technology of web application development that the Microsoft Company launches following the ASP, and provides a complete visual development environment for users. It closely integrates with the .NET Framework and provides a modular design method, supports for multiple programming languages, code and HTML design is separated. It is more clearly structure, more readable and high efficiency.

ADO.NET is an up to date database access technology at the Microsoft .NET platform. It has a brand new design concept, and has a significant innovations based on the original ADO. No matter which database the data sources come from, all of them can be effectively accessed. It is an important bridge between applications and databases.

During the implementation process of using ADO.NET technology to visit database, on account of unfamiliar with object model and basic development operation, usually lead to technical fault. In response to this fact, this article analyses and discusses the technology of database access in ASP.NET.

2 Access Principle

Through Server-side component ADO.NET, ASP.NET realizes web database visiting. ADO.NET provides optimized, facing web data access model for the .NET framework,

D. Jin and S. Lin (Eds.): Advances in MSEC Vol. 1, AISC 128, pp. 259–264.
springerlink.com

it has its own ADO. NET interface based on XML format and. NET architecture. ADO. NET provides two data access models for the application: connection mode and non-connection mode. Compared with the traditional, non-connection mode strengthens the reliability and stability. In this mode, once the application obtains the required data from the data source, it would disconnect the connection with the original data source, and store the obtaining data in the form of XML. After handling, it then made the connection with the original data and completes the data updating.

ADO. NET includes two core components: Dataset and .NET Framework data provider. The former is a core component of off-type structure, which realizes data access independently of any data source and manages local application data. The latter is a group component including objects about “Connection”, “Command”, “DataReader”, “DataAdapter”, and provides methods about data manipulation, data fast access, forward-only and read-only access. Among this, “Connection” provides connectivity with the data sources. “Command” provides database commands used for returning data, modifying data, running stored procedures and sending or retrieving parameter information. “DataReader” provides high-performance data flow from data source and it is a bridge that connects the data source with Dataset. “DataAdapter” uses “Command” objects to execute SQL command in the data source, and it would load the data into the Dataset, make the changes to the Dataset consistent with the data source.

3 Cnnection Method

Before an ASP.NET page file query, insert, update the database, it should establish a connection with the database first, and then do some corresponding operation process. The steps to establish database connection is:

1. Introduce associated ADO. NET namespace in the page file
2. Set database connection parameters
3. According to the connection parameters, create a database connection object
4. Execute database connection operation

3.1 Three Kinds of Connection Mode

There are three connection ways that ASP.NET visits database through ADO.NET, the first one, through ADO. NET Managed Provider could connect to any ODBC; the second, through ADO. NET Managed Provider could connect to any OLEDB data center; the third way through SQL Managed Provider could connect to MS SQL Server. Generally, in these three methods, SQL Managed Provider is most efficient, followed by is the ADO.NET Managed Provider + OLEDB, the worst one is ADO. NET Managed Provider + ODBC.

3.2 Two Access Model

ADO.NET provides two accessing models according to different forms of Web database. One is “DataReader” that read from data source based on stream. Another is through the Dataset to isolate heterogeneous data source. If only want to display data on a web page and seldom need to manipulate or change them, you can directly access the database by using ADO. NET. This model utilizes SqlDataReader object or OleDbDataReader object for fast reading. These classes are equivalent to ADO fast forward pointer (cursor). They keep an activity connection with data source, but can not make any changes. If want to get a row of data from sqlDataReader or OleDbDataReader, the "Read" method should be used.

If want to do a more complex interactive access, you should use DataSet object through ADO. NET. Compared with the traditional ADO technology, ADO.NET uses disconnect method. That is, in the ADO.NET, create a connection to the database without using cursors. Instead, fill data set (DataSet) with information copies extracted from a database. If the data set information has been changed, the corresponding information in database would not be changed. That is, you can easily change and manipulate data without worry, because the database connection is not the important one. If necessary, the Dataset can be connected back to the original data sources and apply all changes.

Schematic diagram of database access: Fig.1

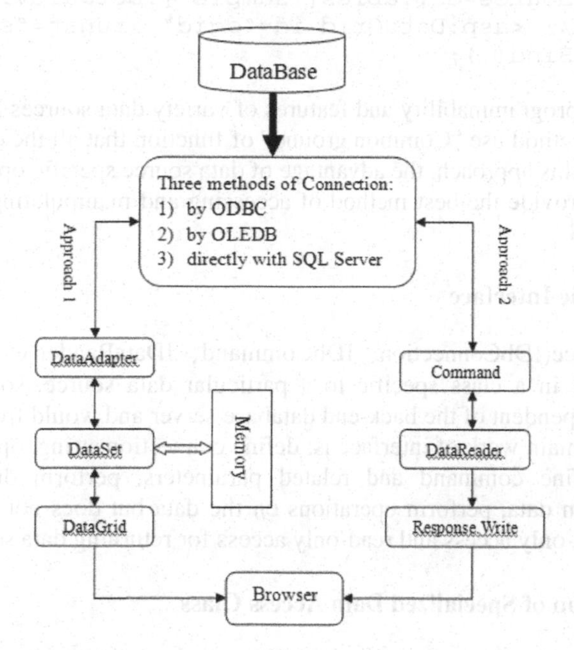

Fig. 1.

4 ASP.NET Database Access

Common database access way basically has the following kinds:

4.1 Using Common Data Access Methods

This is the standard procedure for ADO.NET. The method is: by using simple object model, such as "Connection", "Command" and "Recordset" etc, write out applications that could connect to all kinds of different data sources.

Sample code is as follows:

```
OleDbConnection Conn= new
OleDbConnection("Provider=Microsoft.Jet.OLEDB.4.0,"+
"Data Source= "+Server.MapPath("example.mdb"));
                                    // connection object
Conn.Open();               // open connection object
OleDbCommand Comm=new OleDbCommand ("select * from
sample", Conn) ;
                                    // query data from sample
OleDbDataReader dr= Comm.ExecuteReader ( );
                                          // Data read object
......
OleDbDataAdapter da=new OleDbDataAdapter ("select * from
sample", Conn);
DataSet  ds = new DataSet( );
Da.Fill(ds, "sample") ;
dgid.DataSource=ds.Tables["sample"].DefaultView;
         // <asp:DataGrid id="dgid"  runat="server"/>
dgid.DataBind( );
```

But in fact, the programmability and features of variety data sources are multifarious, and the above method use "Common ground" of function that all the data source have provided. So in this approach, the advantage of data source specific option will be lost, and could not provide the best method of accessing and manipulating information in various RDBMS.

4.2 Using Basic Interface

Through interface(IDbConnection、IDbCommand、IDataReader etc), the code will be encapsulated in a class specific to a particular data source, so the rest of the application independent of the back-end database server and would from its effects.

Usually, the main work of interface is: define connection string, open and close the connection, define command and related parameters, perform different type of command, return data, perform operations on the data but does not return anything, provide forward-only access and read-only access for returning data set.

4.3 Preparation of Specialized Data Access Class

The above two methods have some limitations, a good solution is to improve its level of abstraction, that is, by creating a specially class to encapsulate the using of specific data provider, and through data structures to exchange information with other level application which has nothing to do with such as: specific data source, typed "data set", objects collection etc.

We can create a special class for each supporting data sources in a particular program set, and in case of need, load them from the application according to the configuration files. So, if want to add new data sources to application, you would only need to achieve a new class aim at the defined rules for a group of general interface.

If want SQL Server as a data source and provide support, you could do the following definition :

```
namespace MYEXAMPIE{
public class Sample: IDbSample {
public DataTable GetSam( ){
String ConnStr =
ConfigurationSettings.AppSettings[ "ConnStr"];
using(SqlConnection conn = new SqlConnection(ConnStr)){
string cmdString = " SELECT  ID, name  FROM Sample " ;
SqlCommand cmd = new SqlCommand(cmdString, conn);
SqlDataAdapter da = new SqlDataAdapter(cmd) ;
DataTable dt = new DataTable("Sample ");
Da.Fill(dt);
return dt ; }}
public DataTable GetSamID(string ID) {......}
public bool InsertSam( ) {......}
......                      //  And other achieved method
}}
```

Sample class realize IDbSample interface. We can only create a new class that realize this interface when need to support new data sources.

This type of interface can be defined as following :

```
namespace Common{
public interface IDbSample{
DataTable GetSam();
DataTable GetSamID(string ID);
bool InsertSam(); }}
```

Other types of data sources are similar to the above code, and here, we would not give unnecessary details. The advantage of this method is: a high degree of decoupling, easy changing, suitable for team development. The benefits of decoupling are easy to test. When test, we can simulate a data access layer, test business logic independently, even can simulate various database errors, and also can test the data access layer separately.

5 Summary

In web site construction, there are some other ways to visit database by ASP.NET based on .NET technology, and generally, it is similar. In the actual development process, which way would be chosen? We should nimbly select according to the needs. In addition, through data control encapsulation functions, ASP. NET could be more flexible to control data, and reduce the required code when access and display the data. From traditional database to XML data storage, all kinds of data source can be connected to these controls, and all data sources can be handled in a similar format, greatly reduces the data-driven application development complexity. Currently, ASP.NET is an ideal choice for web database visiting.

References

1. Sun, Z.-J., Lu, L.: An Implement of Web Database Access Applications Based on Midware Technology. Computer Systems & Applications 1, 87–90 (2008)
2. Millett, S.: Professional ASP.NET Design Patterns. Wrox (2010)
3. Evjen, B., Hanselman, S., Rader, D.: Professional ASP.NET 3.5 SP1 edn. In C# and VB. Wrox (2009)
4. Zhou, S.-X.: Research on Data Access Technology Based on ADO.NET2.0. Computer Technology and Development 18, 144–146 (2008)

A Novel Approach for Software Quality Evaluation Based on Information Axiom

Houxing You[1,2]

[1] Department of Electronic Commerce, Jiangxi University of Finance and Economics, Nanchang, 330013, China
[2] School of Economics and Management, Tongji University, Shanghai, 200092, China
youhouxing@126.com

Abstract. Software quality evaluation is the focus problem of software product development, which is the lifeline of software industry. Toward the problem, in this paper, a software evaluation model is introduced, an evaluation method based on information axiom is proposed, and the evaluation procedure is given. Finally, an application case study is given, and the experimental result shows the feasibility and practicality of the proposed method. Compared with other methods, the evaluation process is conducted with non-weight information, and the proposed method has noticeable characteristic.

Keywords: software quality, evaluation model, evaluation method, information axiom.

1 Introduction

Software quality is the lifeline of software industry. It is a crucial factor of software industry development that whether software quality is evaluated objectively and fairly [1]. In the past thirty years, many scholars have paid close attention to the software quality evaluation, which basically reflects in the two respects: the first is how to construct software quality evaluation model, the second is how to develop software quality evaluation method.

In the first place, many software quality evaluation models have been proposed in the previous research. There were four famous models: CMM [2], Boehm [3], McCall [4] and ISO/IEC9126 [5]. The CMM model belongs to procedure-oriented quality evaluation, and the rest models belong to product-oriented quality evaluation. The ISO/IEC9126 model was widely adopted, which was expressed by a hierarchical structure, including quality factor, quality standard and quality measurement. The model was divided into six quality factors, including functionality, efficiency, reliability, maintainability, portability and usability [5]. Later, many evolution models derive from the ISO/IEC9126. In this paper, considering the software testing data provided by five software product developers, we introduce a software quality model on the basis of the ISO/IEC9126. The model consists of completeness, accuracy, efficiency, reliability, maintainability, portability and usability.

D. Jin and S. Lin (Eds.): Advances in MSEC Vol. 1, AISC 128, pp. 265–270.
springerlink.com

In the second place, many methods have been proposed. Otamendi et al. [6] used the AHP to evaluate simulation software for international airport management. Feng et al. [7] presented the grey quantitative evaluation model to evaluate software quality. Lu and Liao [8] proposed fuzzy sets theory based method to evaluate software quality. In addition, some integrated methods were introduced in the previous research [9]. However, the above methods must determine weights value. It is difficult for decision-makers to do this. For this, the information axiom [10] provides an effective approach for software quality evaluation, in which the priority of alternatives is determined by the size of information content, and the information content is measured via both design range and system range of alternative's attributes. Thus, the evaluation process can avoid determining the weights value. In view of the above mentioned characteristic, the information axiom is widely used in decision-making field [11]. Kulak [12] introduced the information axiom under fuzzy environment. The evaluation of the alternatives and the definition of functional requirements were defined by triangular fuzzy numbers. The proposed approach was applied to multi-attribute comparison of advanced manufacturing systems. Kahraman [13] applied the information axiom for teaching assistant selection problem. Coelho used axiomatic design principles as a decision making tool to determine one of the manufacturing technologies [14].

According to software quality evaluation model, the information axiom is introduced to evaluate software quality in this paper. The rest of this paper is organized as follows. In Section 2, the method of software quality evaluation is proposed, which is based on information axiom. An application case study is shown in Section 3. Finally, the concluding remarks are presented in Section 4.

2 The Proposed Method for Software Quality Evaluation

In this section, information axiom and the method for software quality evaluation are introduced, and the procedure is to be given at the end of this section.

2.1 Principles of the Information Axiom

The information axiom states that among those designs that satisfy the independence axiom, the design that has the smallest information content is the best design [10]. Information is defined in terms of the information content, I , that is related in its simplest form with the probability of satisfying the given FRs. I determines that the design with the highest probability of success is the best design. Information content I is defined as equation (1).

$$I = -lbP \tag{1}$$

Where P is the probability of achieving the functional requirement (FR). In practice, in any design situation, the probability of success is given by what designer wishes to achieve in terms of tolerance (design range) and what the system is capable of delivering (system range). The overlap between the designer-specified "design range" and the system capability range "system range" is the region where the acceptable solution exists. Therefore, in the case of uniform probability distribution function P may be expressed as equation (2).

$$I = lb\frac{S_r}{C_r} \tag{2}$$

Where S_r is system range, and C_r is common range. If FR is a continuous random variable, as shown in Fig.1, the probability of achieving FR in the design range may be expressed as equation (3).

$$P = \int_{dr}^{du} f(FR)d_{FR} \tag{3}$$

Where d_r and d_u are the lower bound of design range and the upper bound of design range respectively. $f(FR)$ is system probability density function for FR. So the information content is equal to equation (4).

$$I = -lbP = -lb\int_{dr}^{du} f(FR)d_{FR} \tag{4}$$

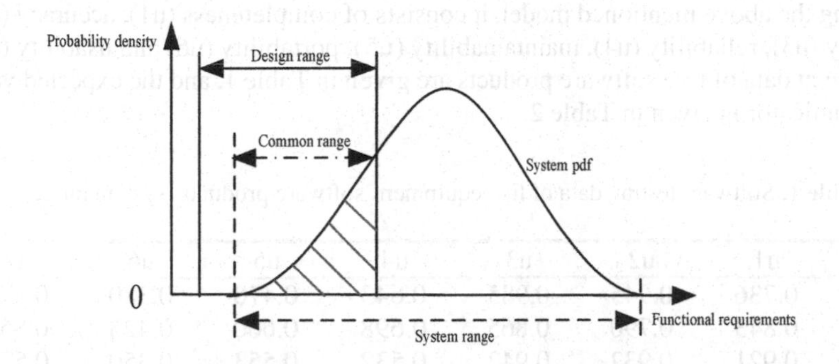

Fig. 1. Design range, system range, common range and probability density function

If we use single point value y_0 to express decision makers' expected value (design range) of certain attribute of scheme during calculation of information content, y_i is the attribute value of the ith scheme. According to statistical distribution, using exponential distribution density function, the information content may be expressed as equation (5).

$$I = lbe^{|y_i - y_0|} \tag{5}$$

2.2 The Method Based on the Information Axiom

According to the above mentioned theory, if we use the information axiom to evaluate software quality, we must obtain design range of each index of the model. Then the information content of each index is calculated by the calculation formula of the information content. In this paper, the design range value is obtained by the historical statistics data. After that, the total information content of each software product is summed, and the evaluation result is obtained by the information axiom. Namely, the software product has the smallest information content, which is the best alternative. The method based on the information axiom can be expressed as:

Step 1: According to the evaluation model, determine software product evaluation indicators.

Step 2: According to the relative importance of each indicator, determine the design range, which is expectations value scope of each indicator.

Step 3: According to the system range of software product, calculate information content of each indicator set.

Step 4: Calculate the total information content of each software product.

Step 5: Obtain the minimum value of the total information content.

Step 6: Obtain the ranking of software products by the information axiom.

3 Case Study

According the above mentioned model, it consists of completeness (u1), accuracy (u2), efficiency (u3), reliability (u4), maintainability (u5), portability (u6) and usability (u7). The relevant data of five software products are given in Table 1, and the expected value of each indicator is given in Table 2.

Table 1. Software testing data of five equipment software products (system range)

SP	u1	u2	u3	u4	u5	u6	u7
1	0.736	0.945	0.985	0.643	0.470	0.380	0.873
2	0.843	0.790	0.865	0.698	0.660	0.423	0.852
3	0.921	0.932	0.942	0.532	0.553	0.350	0.578
4	0.892	0.671	0.886	0.900	0.762	0.510	0.905
5	0.654	0.996	0.924	0.840	0.740	0.570	0.940

Table 2. The expected value of each indicator (design range)

SP	u1	u2	u3	u4	u5	u6	u7
EV	0.702	0.942	0.955	0.957	0.549	0.529	0.833

According to the proposed method, we can obtain information content of each software product, as shown in Table 3.

The ranking of SPS: SP5 $\succ$ SP4 $\succ$ SP1 $\succ$ SP2 $\succ$ SP3. SP5 is the best software product. Compared with other methods [15], the result has consistency.

Table 3. Information content of SPs

SP	u1	u2	u3	u4	u5	u6	u7	∑
1	0	0	0	0.453	0.114	0.215	0	0.782
2	0	0.219	0.130	0.374	0	0.153	0	0.876
3	0	0.014	0.019	0.614	0	0.258	0.367	1.272
4	0	0.390	0.099	0.083	0	0.027	0	0.599
5	0.069	0	0.044	0.169	0	0	0	0.282

4 Conclusions

In this paper, a software product evaluation model is introduced. On the basis of the model, an evaluation method based on information axiom is proposed, and the evaluation procedure is given. Compared with other methods, the evaluation process is conducted with non-weight information. The experimental result shows feasibility of the method. The method has stronger practicability. It will be popularized in other field, such as supplier selection, process selection, performance evaluation and so on.

Acknowledgments. This work was supported by Doctoral Fund of Ministry of Education of China (No. 200802470009) and Shanghai Leading Academic Discipline Project (No.B310).

References

1. Lan, Y.Q., Zhao, T., Gao, J.: Quality evaluation of foundational software platform. Journal of Software 20, 567–582 (2009)
2. CMU/SEI, CMM (2001)
3. Boehm, B.W., Brown, J.R., Lipow, M.: Quantitative evaluation of software quality. In: Proc. of the 2nd Int'l. Conf. on Software Engineering. IEEE Computer Society, Long Beach (1976)
4. McCall, J.A., Richards, P.K., Walters, G.F.: Factors in software quality, ol. I, II, III, Final Technical Report, RADC-TR-77-369, Rome Air Development Center, Air Force Systems Command, Griffiss Air Force Base (1977)
5. Software product evaluation quality characteristics and guideline for their Use, ISO/IEC Standard ISO-9126 (1991)
6. Otamendi, J., Pastor, J.M., Garciá, A.: Selection of the simulation software for the management of the operations at an international airport. Simulation Modelling Practice and Theory 16, 1103–1112 (2008)
7. Feng, J.X., Tang, R., Wang, S.W., Li, Z.B.: Study on software quality grey quantitative evaluation mode. Journal of Harbin Institute of Technology 37, 639–642 (2005)
8. Lu, X., Liao, J.M.: Study of software quality evaluation based on fuzzy sets theory. Journal of University of Electronic Science and Technology of China 36, 652–655 (2007)
9. Chen, C., Guo, J.W., Zhao, C.X.: A method of equipment software quality evaluation based on AHP-ELECTRE. Acta Armamentarii 31, 1481–1486 (2010)
10. Suh, N.P.: Axiomatic Design: Advances and Applications. Oxford University Press, New York (2001)

11. Kulak, O., Cebi, S., Kahraman, C.: Applications of axiomatic design principles: A literature review. Expert Systems with Applications 37, 6705–6717 (2010)
12. Kulak, O., Kahraman, C.: Fuzzy multi-attribute selection among transportation companies using axiomatic design and analytic hierarchy process. Information Sciences 170, 191–210 (2005)
13. Kahraman, C.: A new multi-attribute decision making method: Hierarchical fuzzy axiomatic design. Expert Systems with Applications 36, 4848–4861 (2009)
14. Coelho, A.M.G.: Axiomatic design as support for decision-making in a design for manufacturing context: A case study. International Journal of Production Economics 109, 81–89 (2007)
15. Saaty, T.L.: The Analytic Hierarchy Process. McGraw-Hill Book Company, New York (1980)

Research on Model of Ontology-Based Semantic Information Retrieval

Yu Cheng[1] and Ying Xiong[2]

School of Computer, Hubei University of Technology, Wuhan, China
[1] 958161176@qq.com, [2] 1654149247@qq.com

Abstract. This paper studies mainly how to apply the ontology into information retrieval system, so as to achieve semantic retrieval. Starting from the introduction of status quo of traditional information retrieval , and analyses of its main Problems, the paper describes the key technologies involved in the semantic retrieval, including ontology building a database, ontology reasoning, semantic search tools Jena and OWL language of ontology, which have been combined to complete the design of semantic retrieval model.

Keywords: Ontology, Information Retrieval, Jena, Semantic reasoning.

1 Introduction

With the development of information society, information rapid growth on the Internet has begun to show its complexity and diversity. The traditional keyword-based information retrieval technology can't meet people's needs of information search. The problem exits mainly in that the traditional information retrieval system can't understand the inner meaning of information resources and their relationship, i.e. the lack of semantic understanding. The information people want to obtain just can be matched to knowledge in the professional field. However, the results retuned by matching with keyword are usually only the literal information, getting far away from information people really want to get. The semantic search technology is the most promising methods to solve this problem, which has become one of the hot fields of information retrieval.

2 Key Technologies of Semantic Retrieval

A Ontology Description Language

Ontology description language is mainly used to describe the ontology which can be used to write a clear and formal concept description for a domain mode. Therefore, Ontology description language should meet the requirements of good semantic, supporting effective semantic reasoning and full expression. Nowadays, a variety of ontology description languages exist, that including RDF, RDFS, OWL, DAML, KIF, etc. The paper chooses OWL as the ontology constructing language. OWL is W3C

D. Jin and S. Lin (Eds.): Advances in MSEC Vol. 1, AISC 128, pp. 271–276.
springerlink.com

recommendation standard based on description logic, and absorbing Web resource as its describing object, XML language as described base. Description Logic (abbreviated DL) can reason conditionally ontology classes, instances, and its attributes to rich ontology semantics. Therefore, OWL ontology based on description logic is superior to other description languages in the concept expression and reasoning ability as well as the integration of Web resources' knowledge and logical testing when ontology library building. Relationship of ontology language and XML, RDF / RDFS, OWL is shown as the Figure 1.

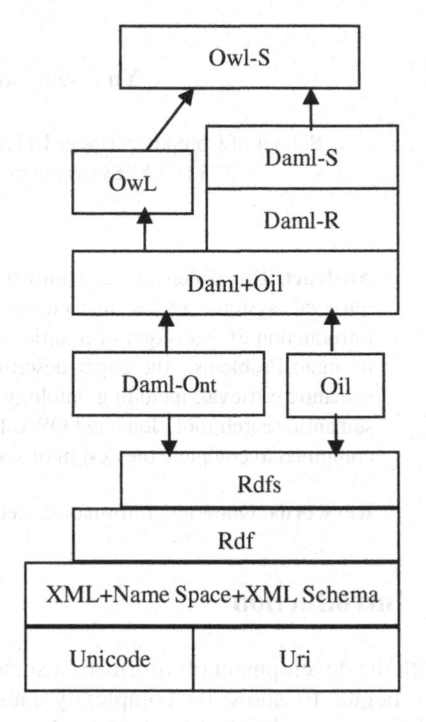

Fig. 1. Semantic Description Language Derivative Sketch Map based on XML

B Ontology Constructing Tool

There are many techniques and ways for building ontology, the paper selects Protégé as the ontology constructing tool. Protégé had been developed by the research team of School of Medicine, Stanford University. The thought of Protégé object is very similar to the principles of classification of ontology, which is easy to expand for developers to use a variety of plug-ins to increase the Protégé function. For example, supporting graphical ontology editing mode, Protégé can express the relationship among concepts through graphic representation; supporting database storage mode, built ontology can be recorded into the database for user-friendly operation; supporting group development, OWL files is easily extended, read and use the other ontology. Therefore, a large ontology can be put into several smaller parts, which is responsible by different developers for; supporting the logical testing, in order to avoid errors existing in persons' collaboration, the final merged ontology needs conceptual consistency testing and conflict testing, so as to detect and correct the contradictions in the concept of ontology and the error in case-attribute.

C Semantic Reasoning

Compared with traditional information retrieval, the biggest feature of semantic information retrieval is its introducing into the retrieval processing of the resource object. Data, information and knowledge are three concepts people daily exposed to, which is the object the user retrieve. The purpose of user's information retrieval is to obtain valuable knowledge. Domain ontology describes the relationship among the concepts, providing the logical rules semantic reasoning required, and the object of reasoning of the metadata information stored as XML / RDF or OWL, i.e. semantic reasoning allows computers to recognize and understand the structure of domain ontology and metadata information so that it can seek the closure of the existing information base in accordance with relevant logical rules.

D Semantic Search Tools

The paper makes currently more popular choice of the Jena Semantic Development Kit. Jena is toolkit of Java programmer development, which is open source and supporting RDF, OWL and semantic reasoning. Jena framework consists of three parts: the Graph, Enhanced-Graph and Mode layer. The Model layer is the entrance programmer operating RDF, OWL data. The Enhanced-Graph layer can provide a variety of different views for Graph layer which is responsible for data persistence, and display in the appropriate manner.

3 Ontology-Based Semantic Information Retrieval Model

3.1 Method of Semantic Retrieval

Semantic information retrieval is a combination of semantic retrieval, natural language, artificial intelligence technology. It analyzes search requests and information resource object from the perspective of the semantic understanding, which is a matching mechanism based on the relation of concepts. The key lies in representation of information resources and reasoning among concepts. Currently there are two type of semantic retrieval, which is based on ontology and concepts, the former achieve semantic search based on ontology constructing space, while the latter is based on relational database or conceptual dictionary to build the conceptual space. In summary, relationship of ontology, conceptual space and semantic retrieval can be shown as Figure 2:

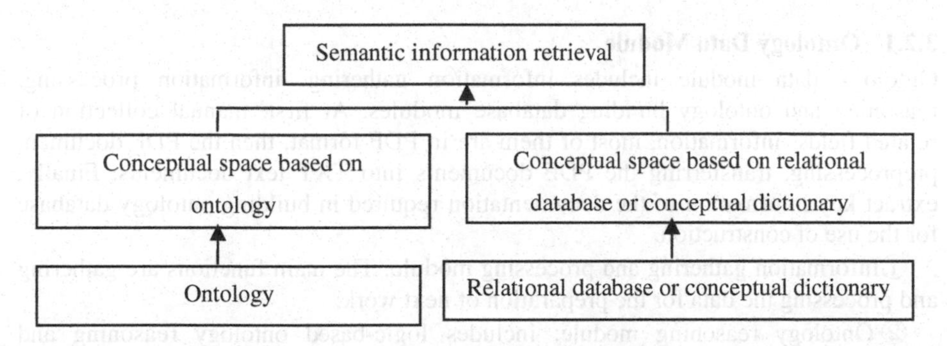

Fig. 2. The Method of Semantic Retrieval (the Relationship among Ontology, Space of Concept and Semantic Retrieval)

3.2 Design of Information Retrieval Model

Ontology-based semantic information retrieval model includes four modules of input and output of user, Lucene index, semantic query module and ontological data. Figure 3 shows the relationship of four modules.

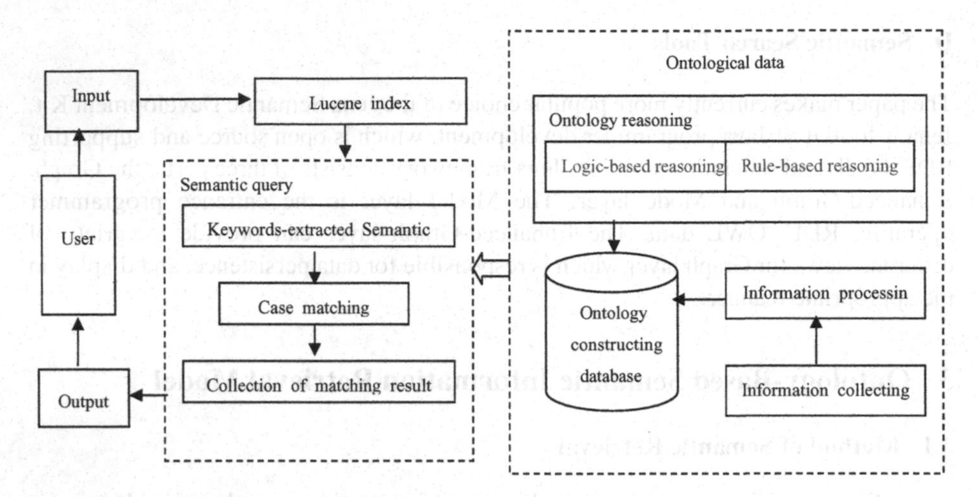

Fig. 3. Ontology-based semantic information retrieval model

The model carries on systematic semantic function, through the ontology database and ontology reasoning technology. System uses Protégé to build ontology structure, as well as the OWL ontology language to describe the ontology, and then fill in the content to the ontology database through Jena, open-source toolkit based on Java. At the same time, semantic reasoning has been carried on through Jena and the rule base for the ontology. Also Lucene, Java-based open-source toolkit, has been used to detach the word. Finally, information can be queried in the database through using the Jena toolkit.

3.2.1 Ontology Data Module

Ontology data module includes information gathering, information processing, reasoning and ontology building database modules. At first, manual collection of related fields' information, most of them are in PDF format, then the PDF document preprocessing, transferring the PDF documents into TXT text documents. Finally, extract key information of the documentation required in building ontology database for the use of construction.

①Information gathering and processing module: The main functions are gathering and processing the data for the preparation of next work.

② Ontology reasoning module: includes logic-based ontology reasoning and rule-based ontology reasoning. Logic-based ontology reasoning is mainly used for testing the logical correctness of concepts, cases and its attributes; rule-based ontology reasoning is for exploring the tacit knowledge and richening ontology database, which can be realized through the method that building the ontology database, calling Racer inference engine in Jena package, logic testing and modifying or deleting an entry by the use of the ontology inference engine based on Tableaux algorithm for the easy calls on ontology database, meanwhile, the preparing a reasoning rules, reading the OWL ontology file using Jena, then rule-based reasoning to ontology database based on the use of Jena embedded inference engine.

③Ontology building module: by using the ontology building language, OWL, to construct the ontology-database structure, then store information extracted from text documents into the corresponding classes and attributes of the ontology structure. Its implementing steps is to determine the domain ontology building process based on the skeleton rule and seven-step method according to the specific content of ontology, as shown in Figure 4.

3.2.2 Semantic Query Module

The key information of the user retrieving and extracted, should be matched with instances of ontology database to obtain the matching-results, which is assembly of instances combining with the critical information of the user. Any semantic query to one of the assembly of instances can obtain results, which is tri-assembly of instances. This can be realized through the method, that is to get natural language keywords processed through the systematic front-end module, to get the class instance corresponding to keywords by using listSubClasses way of ExtendedIterator class of Jena package, to traverse tri-assembly of instances corresponding to keywords through using listProperties methods of StmtIterator class, and the next Statement of Statement class, and the last to get the matching tri-assembly.

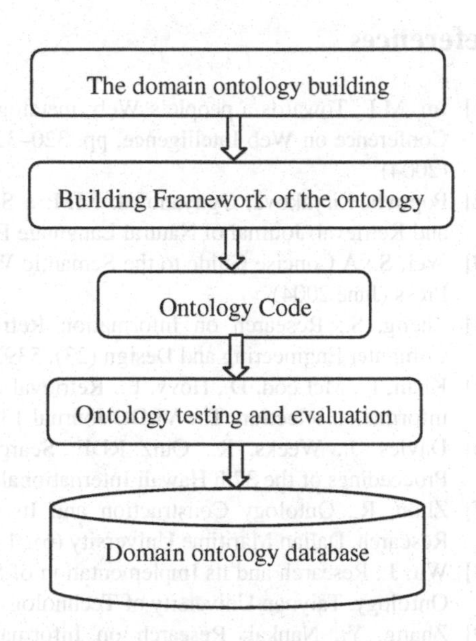

Fig. 4. Domain ontology building process

3.2.3 Lucene Indexing Module

Through Lucene indexing, the module, on the one hand, sends the results obtained by inputing information word to the semantic query module, for the semantic matching. On the other hand, it can meet the user's needs to traditional search through full-text index function by the use of Lucene.

3.2.4 User Input and Output Module

User input module is a simple search entrance. User output module is the graphical visualization of the result of semantic query module. Visualize graphic description of tri-assembly can be carried on by using owl2prefuse engineering package.

4 Inclusion

With the massive growth of information on the Internet, current information retrieval systems can't meet the user's information requirements. How to improve the current information retrieval systems to meet the growing information needs of users has become a very important issue. Ontology as a shared conceptual model has a better conceptual level and semantic description ability. The paper try to introduce the ontology into information retrieval systems, so as to semantic the traditional information retrieval systems, to realize clearly the real information needs of user, and understand real semantics of the information resources, to match information resources and user's information needs in the semantic level, to improve precision rate and recall rate of information retrieval system.

References

[1] Im, M.I.: Towards a people's Web: metalog. In: Proc. of IEEE/WIC/ACM International Conference on Web Intelligence, pp. 320–328. IEEE Computer Society, Washington DC (2004)
[2] Popovb, Kiryakova, Ognyanoffd: KIM: a Semantic Platform for Information Extraction and Retrieval. Journal of Natural Language Engineering 10(3), 375–392 (2004)
[3] Wei, S.: A Concise Guide to the Semantic Web, 1st edn., pp. 117–118. Higher Education Press (June 2004)
[4] Zheng, S.: Research on Information Retrieval System based on Ontology of Grid. Computer Engineering and Design (23), 5392–5399 (2009)
[5] Khan, L., McLeod, D., Hovy, E.: Retrieval effectiveness of an ontology-based model for information selection. The VLDB Journal 13(1), 71–85 (2004)
[6] Davies, J., Weeks, R.: Quiz RDF: Search Technology for the Semantic Web. In: Proceedings of the 37th Hawaii International Conference on System Sciences (2004)
[7] Zhou, R.: Ontology Construction and Its Application in Book Information Retrieval Research. Dalian Maritime University (6), 4–5 (2009)
[8] Wu, J.: Research and its Implementation of Semantic Retrieval System Based on Domain Ontology. Taiyuan University of Technology (May 2010)
[9] Zhang, Y., Nankai: Research on Information Retrieval Model Based on Ontology. Computer Application Research, 2240–2245 (August 2008)
[10] Tang, W.: Research and Its Application of Semantic Retrieval System Based on OWL. Wuhan University of Technology (June 2010)
[11] Shen, Y., Tian, A.: Research and Its Implementation of Information Retrieval System Based on Semantic Grid. Shandong University of Technology, 91–93 (January 2011)

Using Dynamic Fuzzy Neural Networks Approach to Predict Ice Formation

Qisheng Yan[1,2] and Muhua Ding[2]

[1] School of Digital Media, Jiangnan University, Wuxi 214122, China
[2] School of Science, East China Institute of Technology, Fuzhou 344000, China
yanqs93@126.com

Abstract. A prediction model of ice formation based on dynamic fuzzy neural network (D-FNN) combined with particle swarm optimization algorithm (PSO) are proposed. This method is applied to forecast the freeze-up date and break-up date of the Yellow River. The experimental results demonstrate that D-FNN can be used as a prediction system for the length of ice formation and the accuracy of forecasting is superior to those of support vector machine(SVM) and fuzzy optimization neural network(FONN). It is suggested that D-FNN is an effective and powerful tool for ice forecasting.

Keywords: dynamic fuzzy neural networks(D-FNN), PSO, ice forecasting, support vector machine(SVM), fuzzy optimization neural network(FONN).

1 Introduction

Ice run occurs under some meteorological, geographical and hydrological conditions. Ice flood occurs almost every year in Inner Mongolia reach of Yellow River, so ice flood prediction has become an important tool to support engineers to deal with ice flood problem. The traditional quantitative forecasting method includes multiple linear regression analysis and mathematical model established by Professor Shen based on heat exchange principle and the hydraulics for ice[1,2]. However, it is difficult to describe the nonlinear relationship between hydrological system characteristics and its influential factors by explicit mathematical model because of its complexity. In the recent years, artificial intelligence technologies technologies sucn as neural network and support vector machine are also used in the field and good effect is obtained[3-5]. Fuzzy logic and artificial neural networks are complementary tools that are used with each other to improve intelligent systems. fuzzy systems are high-level structures because of their usage of expert sights using linguistic variables. Two main advantages of fuzzy systems for the control and modeling applications are: (1) fuzzy systems are useful for uncertain or approximate reasoning, especially for the system with a mathematical model that is difficult to derive, and (2) fuzzy logic allows decision making with the estimated values under incomplete or uncertain information. Actually fuzzy systems do not have the ability to learn and they cannot be adapted to new environments, on the other hand, ANNs have learning ability but they do not comprehend with the user[6].In this paper, a prediction model of ice

D. Jin and S. Lin (Eds.): Advances in MSEC Vol. 1, AISC 128, pp. 277–283.
springerlink.com

formation based on dynamic fuzzy neural network (D-FNN)[7,8], and the particle swarm optimization algorithm (PSO) is applied for parametric optimization.

2 Dynamic Fuzzy Neural Networks

The new structure based on extended RBF neural networks to perform TSK model-based fuzzy system is shown in Fig.1. Comparing with standard RBF neural networks, the term "extended RBF neural networks" implies that: 1) there are more than three layers; 2) no bias is considered; and 3) the weights may be a function instead of a real constant.

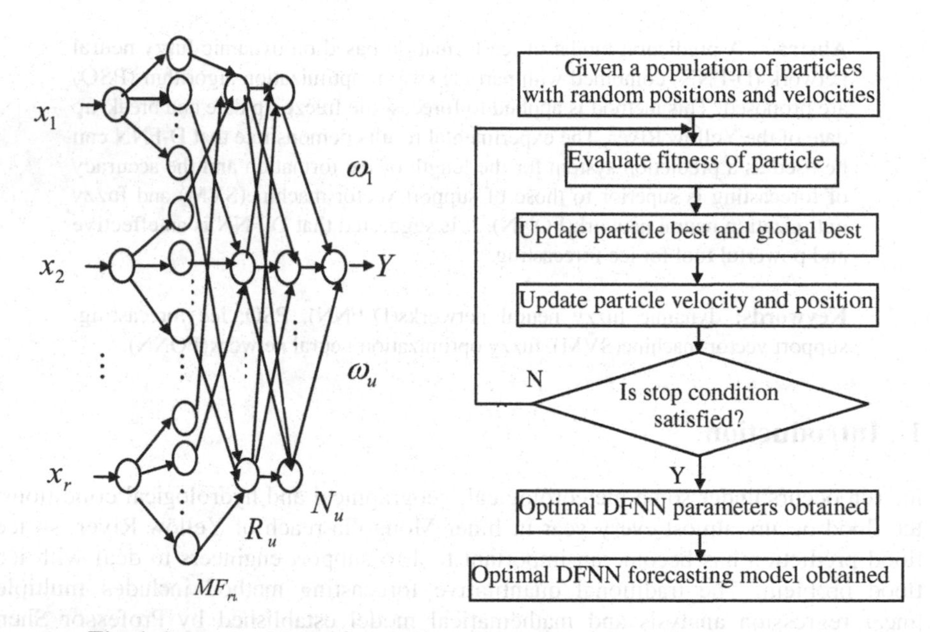

Fig. 1. Structure of D-FNN

Fig. 2. Optimizing the D-FNN parameters with PSO

Layer 1: Each node in layer 1 represents an input linguistic variable.

Layer 2: Each node in layer 2 represents a membership function (MF) which is in the form of Gaussian functions:

$$\mu_{ij}(x_i) = \exp[-\frac{(x_i - c_{ij})^2}{\sigma_j^2}], \quad i = 1,2,\cdots,r, \quad j = 1,2,\cdots,u \tag{1}$$

where μ_{ij} is the *j*th membership function of x_i, c_{ij} is the center of the *j*th Gaussian membership function of x_i, σ_j is the width of the *j*th Gaussian membership function of x_i, r is the number of input variables and u is the number of membership functions.

Layer 3: Each node in layer 3 represents a possible IF-part for fuzzy rules. For the *j*th rule R_j, its output is

$$\varphi_j = \exp\left[-\sum_{i=1}^{r}(x_i - c_{ij})^2 / \sigma_j^2\right] = \exp\left[-\left\|X - C_j\right\|^2 / \sigma_j^2\right], \quad j = 1, 2, \cdots, u \tag{2}$$

where $X = (x_1, x_2, \cdots, x_r) \in \Re^r$ and $C_j = (c_{1j}, c_{2j}, \cdots, c_{rj}) \in \Re^r$ is the center of the *j*th RBF unit. From Eq.(2) we can see that each node in this layer also represents an RBF unit.

Layer 4: We refer to these nodes as *N* (normalized) nodes. Obviously, the number of *N* nodes is equal to that of *R* nodes. The output of the N_j node is

$$\psi_j = \varphi_j / \sum_{k=1}^{u} \varphi_k, \qquad j = 1, 2, \cdots, u \tag{3}$$

Layer 5: Each node in this layer represents an output variable as the summation of incoming signals

$$y(X) = \frac{\sum_{i=1}^{u}\left[\omega_i \exp\left(-\frac{\left\|X - C_i\right\|^2}{\sigma_i^2}\right)\right]}{\sum_{i=1}^{u} \exp\left(-\frac{\left\|X - C_i\right\|^2}{\sigma_i^2}\right)} \tag{4}$$

where y is the value of an output variable and ω_k is the weight of each rule. For the TSK model

$$\omega_k = a_{k0} + a_{k1}x_1 + \cdots + a_{kr}x_r, k = 1, 2, \cdots, u \tag{5}$$

The learning algorithm of the DFNN comprises 4 parts: (1) criteria of rules generation; (2) allocation of premise parameters; (3) determination of consequent parameters and (4) pruning technology. For the *i*th observation (X_i, t_i),where t_i is the desired outpu, calculate the distance $d_i(j)$ between the observation X_i and the center C_j of the existing RBF units, i.e.,

$$d_i(j) = \left\|X_i - C_j\right\|, \quad j = 1, 2, \cdots, u \tag{6}$$

where u is the number of existing RBF units.

Find $$d_{\min} = \arg\min(d_i(j)) \tag{7}$$

If $$d_{\min} > k_d \tag{8}$$

an RBF unit should be considered.

On the other hand, define

$$\left\|e_i\right\| = \left\|t_i - y_i\right\| \tag{9}$$

If $$\|e_i\| > k_e \tag{10}$$

an RBF unit should be considered. Here, k_e, k_d were chosed as

$$k_e = \max[e_{\max} \times \beta^i, e_{\min}], \quad k_d = \max[d_{\max} \times \gamma^i, d_{\min}] \tag{11}$$

where $e_{\max}$ is the predefined maximum error, $e_{\min}$ is the desired accuracy of D-FNN, β is the convergence constant, $d_{\max}$ is the largest length of input space, $d_{\min}$ is the smallest length of interest, and γ is the decay constant. Besides, there are other four parameters to be tuned for setting an D-FNN, namely, k, k_{ω}, k_{err} and σ_0 .Readers are referred to Er *et al*[8] for details. The D-FNN performance is strongly dependent on these parameters. Specially, the preliminary experimental results demonstrate that the performance of DFNN are more sensitive for β, γ and σ_0, compared with other parameters. Thus particle swarm optimization algorithm (PSO) is used to determine the three parameters of dynamic fuzzy neural networks.

3 D-FNN Model for Forecasting Ice Formation

In this section, the application of D-FNN for predicting ice formation is illustrated. Two literature case studies are considered with data taken from[3,4,9]. The results obtained by D-FNN are compared with other intelligent methods, such as SVM and FONN.

3.1 Collection of the Data Set

The number of ice flood sample data at Sanhuhekou is 33 (Table 1). For the purpose of comparison, As in [3,4], the data corresponding to the first 28 are used to construct the training data set and the rest for testing.

3.2 Date Normalization

In order to eliminate dimension different, the following formula is used for data standardization and normalization, and then all input and output data were standardized and normalized to the range of [0,1]

$$x' = \frac{x - x_{\min}}{x_{\max} - x_{\min}} \tag{12}$$

where x' is the normalization data, $x_{\min}$ is the minimum data, $x_{\max}$ is the maximum data.

3.3 Constructing the D-FNN Forecasting Model

To predict break-up date a D-FNN structure with four inputs and one output is built. The cumulative positive temperature, water level, flow rate and the biggest ice thick freeze-up date period arc takcn as inputs. The break-up date is taken as output. In the training stage, firstly the parameters β, γ and σ_0 of D-FNN model are optimized by PSO, other parameters are decided by the empiric method. Fig.2 presents the process of optimizing the D-FNN parameters with PSO. The final parameters were determined: d_{max}=2, d_{min}=0.25, k=1.02, e_{max}=1.1, e_{min}=0.01, k_{ω}=1.01, k_{err}=0.00025, β=0.095, γ=0.959 and σ_0=0.46. Then, the forecasting model of break-up date based on PSO-DFNN model is constructed.

Table 1. Training sample data and forecasting value of ice flood at Sanhuhekou

no.	year	Cumulative positive temperature /°C	Water level /m	Flow rate /($m^3 \cdot s^{-1}$)	The biggest ice thick freeze-up date period /m	Break-up date /d
1	1968-1969	46.2	1019.55	1240	0.83	37
2	1969-1970	46.7	1019.4	1150	0.92	26
3	1970-1971	20.5	1019.44	713	0.75	37
4	1971-1972	61.0	1019.97	1150	0.92	44
5	1972-1973	8.6	1019.49	816	0.68	42
6	1973-1974	16.1	1019.81	893	0.70	34
7	1974-1975	19.5	1018.77	1150	0.78	40
8	1975-1976	8.0	1019.31	760	0.70	37
9	1976-1977	59.2	1019.44	849	0.82	32
10	1977-1978	8.4	1019.61	774	0.57	42
11	1978-1979	8.4	1018.97	692	0.59	42
12	1979-1980	140.5	1019.15	981	0.95	34
13	1980-1981	38.4	1019.53	851	0.83	44
14	1981-1982	15.7	1018.17	779	0.74	42
15	1982-1983	26.4	1018.82	1210	0.66	34
16	1983-1984	24.7	1019.45	1100	0.76	32
17	1984-1985	25.6	1018.95	745	0.80	35
18	1985-1986	28.2	1019.44	1080	0.72	35
19	1986-1987	16.4	1018.20	690	0.70	36
20	1987-1988	14.7	1018.70	680	0.65	34
21	1988-1989	18.4	1019.21	1170	0.55	39
22	1989-1990	14.5	1019.22	900	0.60	49
23	1990-1991	45.1	1018.43	820	0.50	39
24	1991-1992	18.6	1018.59	680	0.53	39
25	1992-1993	62.3	1020.20	1270	0.60	42
26	1993-1994	15.1	1019.14	590	0.53	41
27	1994-1995	2.5	1019.44	760	0.51	43
28	1995-1996	10.4	1020.25	1300	0.63	36
29	1996-1997	49.5	1019.43	1340	0.60	42
30	1997-1998	63.8	1019.96	1850	0.58	40
31	1998-1999	3.5	1019.18	820	0.48	44
32	1999-2000	37.6	1019.94	1020	0.52	40
33	2000-2001	12.8	1019.70	550	0.43	43

3.4 Results and Discussion

Fig.3 shows the comparison between the measured and predicted break-up date at Sanhuhekou by D-FNN for training data. In this study, averge time difference are used to evalute the performance of SVM, FONN and D-FNN models. Table 2 shows the forecasting results of D-FNN compared with SVM and FONN. It can be noted from Table 2 that D-FNN model has more better performance than SVM and FONN in forecasting break-up date at Sanhuhekou.

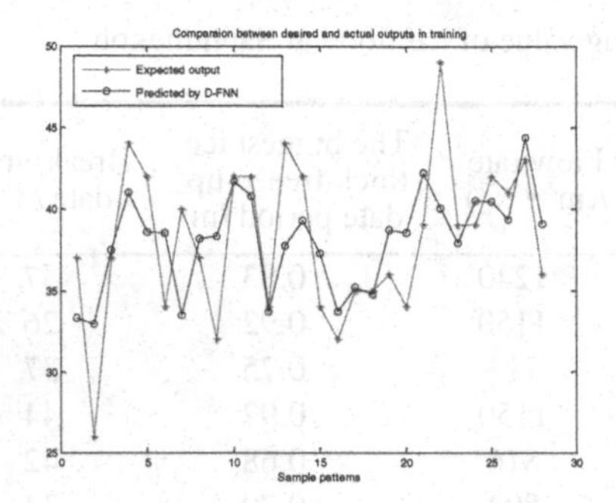

Fig. 3. Estimates of D-FNN in training period (at Sanhuhekou)

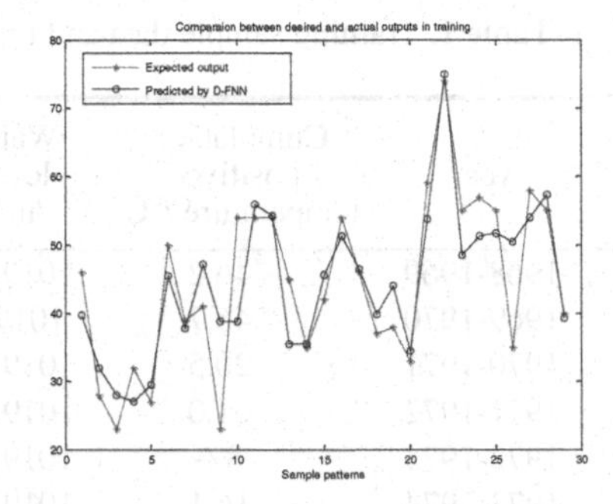

Fig. 4. Estimates of D-FNN in training period (at Bayangaole)

Table 2. Forecasting values of the test samples Sanhuhekou by three models

no.	year	observed value /(m^3s^{-1})	predicted value /(m^3s^{-1})			time difference /d		
			DFNN	SVM[4]	FONN[3]	DFNN	SVM[4]	FONN[3]
29	1996-1997	42	37	40	35	5	2	7
30	1997-1998	40	40	38	39	0	2	1
31	1998-1999	44	46	41	42	2	3	2
32	1999-2000	40	41	43	46	1	3	6
33	2000-2001	43	41	45	44	2	2	1
		averge time difference /d				2.0	2.4	3.4

And then, the method is used to predict freeze-up date at Bayangaole with data taken from[3,9]. For the purpose of comparison, the data corresponding to the first 29 are used to construct the training data set and the rest for testing in the same way. The comparison between the measured and predicted freeze-up date at Bayangaole by D-FNN can be seen in Fig.4 in training period. The performance statistics of different models in testing period are given in Table 3. As can be seen from table 3, the D-FNN has the best performance from the averge time difference viewpoints.

Table 3. Forecasting values of the test samplcs Bayangaole by three models

no.	year	observed value /(m^3s^{-1})	predicted value /(m^3s^{-1})			time difference /d		
			DFNN	SVM[4]	FONN[3]	DFNN	SVM[4]	FONN[3]
30	1997-1998	41	42	40	45	1	1	4
31	1998-1999	51	51	50	57	0	1	6
32	1999-2000	54	51	51	55	3	3	1
33	2000-2001	54	59	58	59	5	4	5
34	2001-2002	45	45	43	51	0	2	6
		averge time difference /d				1.8	2.2	4.4

4 Conclusion

To further improve prediction accuracy of ice formation, dynamic fuzzy neural networks(D-FNN) approach was used for forecasting break-up date and freeze-up date, where particle swarm optimization is used to select suitable parameters of D-FNN. The approach has some salient characteristics such as hierarchical on-line self-organizing learning, self-adaptive determination of structure, fast learning speed and real-time. The experimental results demonstrate that D-FNN is an effective and powerful tool for ice formation forecasting.

Acknowledgments. This study was supported by Foundation of Jiangxi Educational Committee(No.GJJ10500).

References

1. Shen, H.T.: Dynamic Transport of River Ice. J. Hydraul. Res. 28, 659–671 (1990)
2. Shen, H.T.: Under Cover Transport and Accumulation of Frazil Granules. J. Hydrol. Eng. 121(2), 184–195 (1995)
3. Chen, S.Y., Ji, H.L.: Fuzzy Optimization Neural Network BP Approach for Ice Forecasting. Shuili Xuebao (6), 114–118 (2004)
4. Li, Q.G., Chen, S.Y.: A SVM Regress Forecasting Method Based on the Fuzzy Recognition Theory. Advances in Water Science 16(9), 742–746 (2005)
5. Vapnik, V.N.: The Nature of Satistical Learning Theory. Springer, New York (1995)
6. Hocaoglu, F.O., Oysal, Y., Kurban, M.: Missing Wind Data Forecasting with Adaptive Neuro-Fuzzy Inference System. Neural Comput. & Applic. 18, 207–212 (2009)
7. Wu, S.Q., Er, M.J.: Dynamic Fuzzy Neural Networks: A Novel Approach to Function Approximation. IEEE Trans. Systems, Man, and Cybernetics. Part B 30, 358–364 (2000)
8. Er, M.J., Wu, S.Q.: A Fast Learning Algorithm for Parsimonious Fuzzy Neural Systems. Fuzzy Sets and Systems 126, 337–351 (2002)
9. Ji, H.L.: Factor Analysis for Ice Flood and Model Research for Freeze-up Time and Break-up Time in the Inner Mongolia Reach of the Yellow River. Inner Mongolia Agricultural University, Hohhot, P.R.China (2002)

The Design of Simple Experiment System for Automatic Control System

YiMing Li

Department of Computer,
Hunan Institute of Science and Technology.
Yueyang, China 414006
93210179@qq.com

Abstract. For the shortcomings high prices and big volume of traditional university laboratory experiment platform for automatic control principle, the paper designs the simple experiment system for automatic control principle based on double SCM system of AT89S52 and ATmega16 and virtual experimental software platform of LabVIEW. The system includes hardware and monitoring software interface of two parts. The system hardware part adopts the modular design concept, mainly including power module, signal generator module, basic experimental link module, control system link module and the data acquisition module. One signal generator and the data acquisition module contain lower level computer software programming. The upper level computer software is written by LabVIEW, and consists of serial parameters adjustment, display of data received, playback of data stored, and other functions. After the completion of the entire system design, the experimental result is consistent entirely with the simulation waveform of MATLAB through the step response for typical second-order oscillation link, and proves that the design of experiment system is feasible.

Keywords: automatic control principle, AT89S52, ATmega16, data collection, virtual oscilloscope.

1 Introduction

Automatic control principle is the most important professional course of automation for university. However, the theory of automatic control principle is very strong, and classroom instruction alone will not achieve the desired benefits. Thus, all universities have set up experiment of automatic control principle, through experimental teaching to enhance students' knowledge and understanding of automatic control theory, and classroom teaching knowledge into practical application. As the experimental instruments of automatic control principle are expensive, the experiment is not high degree of openness, and experimental results are not satisfactory. For this phenomenon, the paper designs a simple experimental platform for automatic control principle, the system includes power supply, signal generator module based on AT89S52, experimental aspects of basic modules, the control system part of modules, data acquisition modules and other modules. In addition, the experimental data collected by

D. Jin and S. Lin (Eds.): Advances in MSEC Vol. 1, AISC 128, pp. 285–290.
springerlink.com

the data acquisition module based on ATmega16 microcontroller will be sent to the PC for display and storage by the, virtual software platform of LabVIEW. The system has proved to operate easily and low price, and be suitable for use in university laboratory application.

2 The Introduction of Simple Experimental System for Automatic Control Principle

Simple experimental system for automatic control principle is shown as Figure1. Input voltage is 220V AC mains input, and the power module can provide the required voltage for the experiment system, including ±12V, +5V etc. The signal generator provides the required input signals, including step signal and sine wave, square wave, three-legged wave signal of frequency range from 0 to 100KHz. The analog link includes basic experimental link module and control system module. The data acquisition module sends the digital signals got by the experimental link to the PC through serial ports of RS232, which also includes conversion module of the signal level range form 0 to 5V.

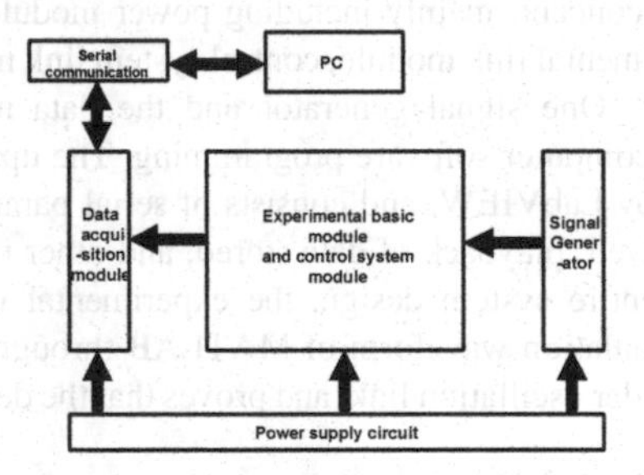

Fig. 1. System block diagram

The simple experimental platform of automatic control principle uses the method of module design. After the success of the design and debugging for separate module, then various modules make joint commissioning. Using this design method can make the design process simple and improve design efficiency.

2.1 The Design of Signal Generator Circuit

The diagram of signal generator is shown as Figure2, where the master chip AT89S52[1] receives the signal from the user's desired frequency and amplitude through the keyboard and LCD circuit. After the user sets the system, the main chip DDS of AD9833 produces the corresponding signals by pin control signals, and the signal is output by the output ports of the amplification circuit U7 composed by LM741[2].

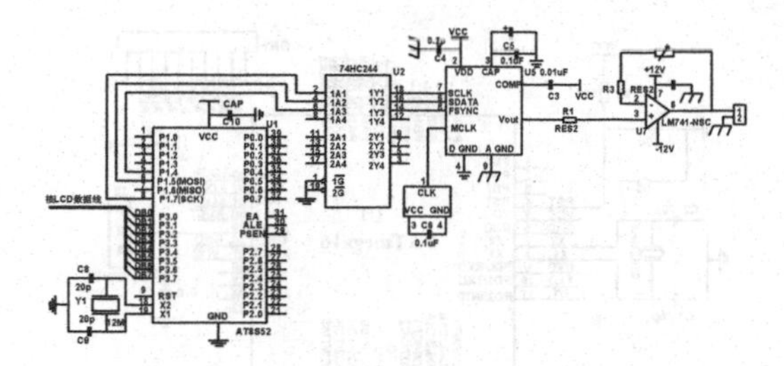

Fig. 2. The diagram of signal generator circuit

2.2 The Design of Multi-channel Data Acquisition Circuit

ATmega16[3] microcontroller chip in addition to other common feature of the same type, but also alone with 8 10-bit with optional differential input stage programmable gain ADC and programmable asynchronous serial interface of internal resources. System is to use the ATmega16 8 10-bit programmable gain of successive approximation type ADC and programmable asynchronous serial interface, internal resources, thus simplifying the circuit design and the difficulty of programming difficulty. ATmega16 microcontroller is a low-power CMOS 8-bit micro-controller[2]. ATmega16 MCU is used widely in industrial productions because the resources are very rich and cost-effective. ATmega16 microcontroller addition to other features common to the same type of microcontroller, has an optional 8-channel 10-bit differential input stage with programmable gain and programmable asynchronous serial interface ADC's internal resources. The system uses ATmega16 of 8-channel 10-bit programmable gain of successive approximation type ADC and programmable asynchronous serial interface, thus simplifying the circuit design and programming difficulty. The data acquisition circuit is shown in Figure2. ATmega16 can achieve technology design requirements easily with a simple conversion, crystal oscillator circuit and reset circuit. The analog signal can be input by any port through the eight analog input in Figure2, and can convert the analog signal into digital signal through the SCM internal process control. Then the SCM transfers the digital signal to PC with serial interface.

Since the internal ADC of ATmega16 is eight selecting one data channels, and in the realization of a channel data acquisition must be changed the values of select register ADMUX. In order to change channels at any time, the design uses a master-slave mode, and sends the value of ATmega16 to change the channel through the host computer. Meanwhile, after receiving the interrupt at the serial interface, the SCM changes ADC conversion delay between two adjacent values through the numerical code by receiving, so as to achieve the effect of changes in conversion rate. The system uses 0-7 for the channel option, and the system changes the sample rate value when the data is greater than 7. When the data change operation is completed, the procedure returns the main program immediately, and ADC conversion time is running by the new parameters.

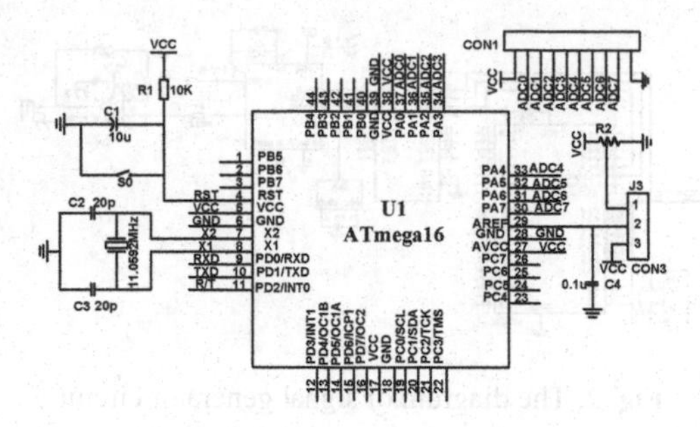

Fig. 3. The diagram of sampling circuit

Analog signal in Figure3 can enter any port through the eight analog inputs. The analog input signals becomes into digital through the internal process control of SCM. Since the internal ADC of ATmega16 is eight selecting one data channels, and in the realization of a channel data acquisition must be changed the values of select register ADMUX. In order to change channels at any time, the design uses a master-slave mode, and sends the value of ATmega16 to change the channel through the host computer. Meanwhile, after receiving the interrupt at the serial interface, the SCM changes ADC conversion delay between two adjacent values through the numerical code by receiving, so as to achieve the effect of changes in conversion rate. The system uses 0-7 for the channel option, and the system changes the sample rate value when the data is greater than 7. When the data change operation is completed, the procedure returns the main program immediately, and ADC conversion time is running by the new parameters[4][5].

3 Experimental Result and Analysis

After the simple experimental system of automatic control principle is complete, it can be test. The transfer function of typical second-order oscillation link is shown as type 1, and the analog circuit is shown as Figure4[6].

$$G(s)=\frac{U_0(s)}{U_1(s)}=\frac{10s+100}{s^2+10s+100} \tag{1}$$

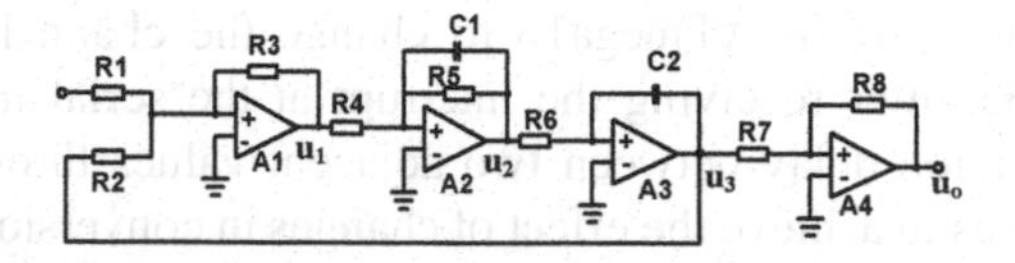

Fig. 4. The analog circuit of second-order oscillation link

The experimental waveform of output second-order oscillating link in the role of the step signal is shown as Figure5.

The output waveform of order oscillating link in the role of the step signal is shown as Figure6 for simple experimental system of automatic control principle. Compared with the simulation waveform of Figure5, it can be seen that experimental waveform is in accordance with the testing waveform in Figure6, indicating a good experimental system to achieve its experimental functions to meet the design objectives.

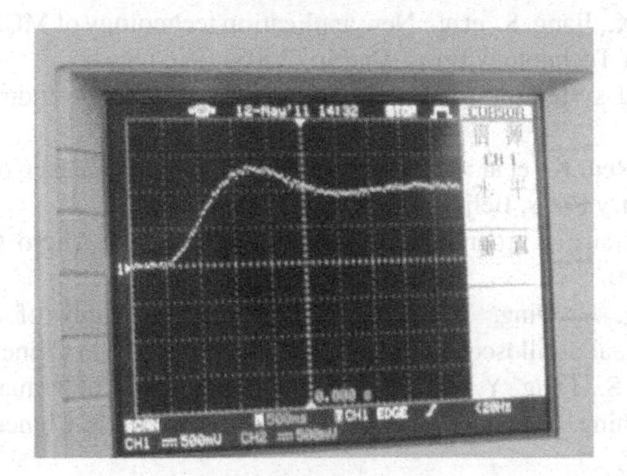

Fig. 5. The experimental waveform of output second-order oscillating link in the role of the step signal

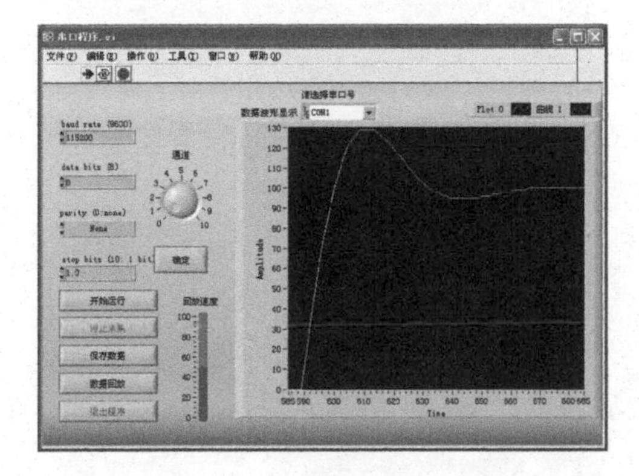

Fig. 6. The output waveform of order oscillating link in the role of the step signal

4 Conclusion

A simple experiment platform of automatic control principle is composed of hardware circuits and upper computer based on LabVIEW, including dual-chip of AT89S52 and ATmega16, LM741,etc. The system takes full advantage of the hardware and software resources of ATmega16 and AT89S52, quickly and easily implements the signal generation and data acquisition, and PC machine by LabVIEW to facilitate the data collection, monitoring, analysis, processing, storage and printout. The experimental system has a simple circuit, using the flexible and scalable features, and can be widely used in automatic control principle for university laboratories, to enable the students to carry out experiments to reduce experimental costs and save lab resources.

References

1. Zhang, Y., Peng, X., Jiang, S., et al.: New application technology of MCS-51 microcontroller. Harbin Institute of Technology Press, Harbin (2008) (in Chinese)
2. Peng, W.: Typical system design of microcontroller. Electronic Industry Press (2006) (in Chinese)
3. Chen, D., Du, J., Ren, K., et al.: Principles and development guidance of ATmega128 MCU. Mechanical Industry Press, Beijing (2005) (in Chinese)
4. Tan, L.: A data transfer programme of virtual oscilloscope. Micro Computer (1), 43–47 (2009) (in Chinese)
5. Huang, S., Rong, J., Ding, Y., et al.: The technology study of a novel and simple multi-channel virtual oscilloscope. Electronics, 25–26 (2010) (in Chinese)
6. Rong, J., Huang, S., Ding, Y., et al.: The application study of virtual oscilloscope in the experimental teaching. Computer Science 7, 100–101 (2010) (in Chinese)

Deformation Calculation for Uplift Piles Based on Generalized Load Transfer Model

Wenjuan Yao[1,*], Shangping Chen[1], and Shengqing Zhu[2]

[1] Department of Civil Engineering, Shanghai University, Shanghai 200072, China
wjyao@shu.edu.cn, shangping_chen@shu.edu.cn
[2] Division of Aerospace Engineering, Nanyang Technological University, 639798, Singapore
zsq-w@163.com

Abstract. In this paper, based on load transfer method, a deformation calculating method for uplift piles was presented, in which the nonlinear relation between the reaction force and the displacement of tension piles was simulated by the generalized load transfer model. The displacement on top of the uplift pile was adjusted by dichotomy, and at lower load levels, the deformation of uplift piles could be predicted satisfactorily. In order to facilitate analysis in engineering, a detailed analysis schedule for deformation of uplift piles were given. The analysis method for uplift piles established can be applied for the cases not only to calculate the settlement and bearing capacity of uplift piles, but also to analyze the load transfer law of uplift piles in layered ground. Comparisons between the calculated results showed good agreement with experimental data.

Keywords: uplift piles, nonlinear deformation, layered ground, method of deformation compatibility, load transfer function.

1 Introduction

With the development of urban construction, the development and utilization of underground space has been increasing attention of scholars home and abroad. The uplift piles are widely used to resist the uplift load, but theoretical study are far behind of engineering practice, and most studies focus on the ultimate bearing capacity, the deformation research is rather limited.

The deformation calculating for pressure piles has a mature theoretical study. However the deformation calculating theory for tension piles is relatively limited. Currently, the design of uplift piles always follows pressure piles by multiplying an uplift factor. Nevertheless, current research of design and calculation has remained mostly empirical, many field tests prove that this design is unreasonable, but sometimes even dangerous [1-3].

Studying the deformation of uplift piles would no doubt call for a systematic and sophisticated analysis. In this paper, based on load transfer method, a deformation calculating method for uplift piles was presented. This method presented can predict

* Wenjuan Yao, Ph.D, Professor, Department of Civil Engineering, Shanghai University, P. O. Box 47, No.149 Yanchang Road, Shanghai, China.

D. Jin and S. Lin (Eds.): Advances in MSEC Vol. 1, AISC 128, pp. 291–297.
springerlink.com

the displacement of the uplift piles satisfactorily. The efficient procedure corresponding to the method is given in order that the method can be applied in practice. The results can provide references to the design of uplift piles.

2 Basic Equation

With the static equilibrium of a tiny cell at depth Z along uplift pile, the skin friction resistance can be obtained by

$$\tau(z) = -\frac{1}{U}\frac{dQ(z)}{dz} \tag{1}$$

Compression $dS(z)$ of the tiny cell is

$$dS(z) = -\frac{Q(z)}{E_P A_P} dz \tag{2}$$

Differentiating Eq.(2) and substituting it into Eq. (1) gives the basic pile-soil load transfer differential function

$$\frac{dS^2(z)}{dz^2} = \frac{U\tau(z)}{E_P A_P} \tag{3}$$

where E_p, A_p, U are, respectively the concrete Young's modulus, cross-sectional area and cross-sectional perimeter of pile. The solution for pile shaft displacement $S(z)$ is depended on pile-soil load transfer differential function.

3 Pile-Side Load Transfer Model of Uplift Piles

Differentiating Eq.(2) and substituting it into Eq. (1) gives the basic pile-soil load transfer differential function.

The Load transfer method usually works well to simulate the load-settlement curves of pressure piles, and the accuracy of the results depends entirely on the accuracy of $\tau - z$ curves [4]. In addition to large-scale projects, most of the projects do not provide $\tau - z$ curves of test piles. In order to reasonably predict $\tau - z$ curves, domestic and foreign researchers have made a lot of load transfer function, in which the hyperbolic function used widely in engineering. The reason is that the hyperbolic function has few parameters, and the parameters have clear physical meaning.

In this paper, based on the traditional hyperbolic function, a generalized load transfer model which is in accord with working properties of pile-side soil, is established to simulate the working properties of uplift piles, as shown in Fig.1. The parameters for calculation are as follows: τ_u is the ultimate skin friction, resistance, β is the degraded coefficient, u_{s1} is the critical pile-soil settlement corresponding with τ_u , u_{s2} is the critical pile-soil settlement when the lateral friction reaches stability.

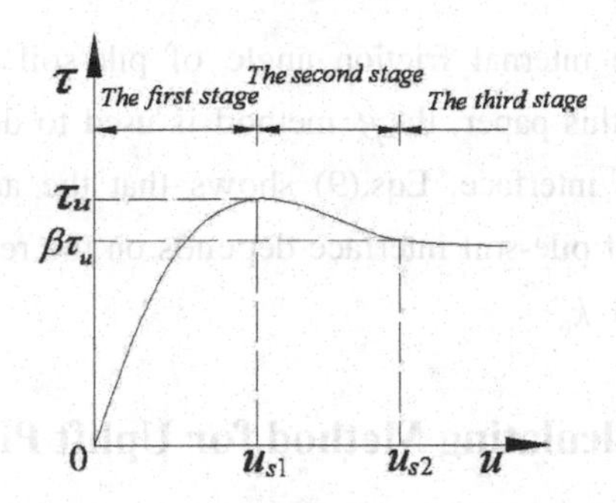

Fig. 1. Generalized hyperbolic load transfer model

The model above could properly explain the working properties of uplift piles, the expression of which is

$$\tau(z)=\frac{S(z)\left[a+bS(z)\right]}{\left[a+cS(z)\right]^2} \tag{4}$$

$$a=\frac{\beta-1-\sqrt{1-\beta}}{2\beta}\cdot\frac{u_{s1}}{\tau_u} \tag{5}$$

$$b=\frac{2-\beta+2\sqrt{1-\beta}}{4\beta\tau_u} \tag{6}$$

$$c=\frac{1+\sqrt{1-\beta}}{2\beta\tau_u} \tag{7}$$

where *a*, *b*, *c* are model parameters, their values are respectively given by Eqs.(5)-(7); $S(z)$, $\tau(z)$ are respectively settlement and skin friction resistance at the depth of *Z*.

4 Pile-Side Ultimate Friction Resistance

Neill [6] proposed α method and β method to determine the maximum shear stress of pile-soil interface:

$$\tau_u=\alpha c_u \tag{8}$$

$$\tau_u=\sigma_v' k_0 \tan\delta=\beta\sigma_v' \tag{9}$$

$$\sigma_v'=\gamma z \tag{10}$$

in which, c_u is undrained shear strength of soil around pile, σ_v' is in-situ vertical effective stress, γ is volume weight of soil, z is the calculating depth from the surface point, δ is the internal friction angle of pile-soil interface, k_0 is lateral pressure coefficient. In this paper, the β method is used to determine the maximum shear stress of pile-soil interface. Eqs.(9) shows that the accurately prediction of maximum shear stress of pile-soil interface depends on the reasonable determination of two parameters δ and k_0.

5 Deformation Calculating Method for Uplift Piles

According to geotechnical conditions the pile shaft is divided into N units, the number of which are n=1, 2, … , N, n=1 means pile head, while $n = N$ means pile tip. In calculating process, the pile-top displacement is adjusted by dichotomy. Then, the axial force and pile-side friction resistance for each segment is deduced according to the axial deformation coordination between pile and soil, until the pile-side total shear resistance is equal to the total load pile-top load. The analysis steps are as follows:

Step1: the uplift pile is divided into N units, as is shown in Fig.2.

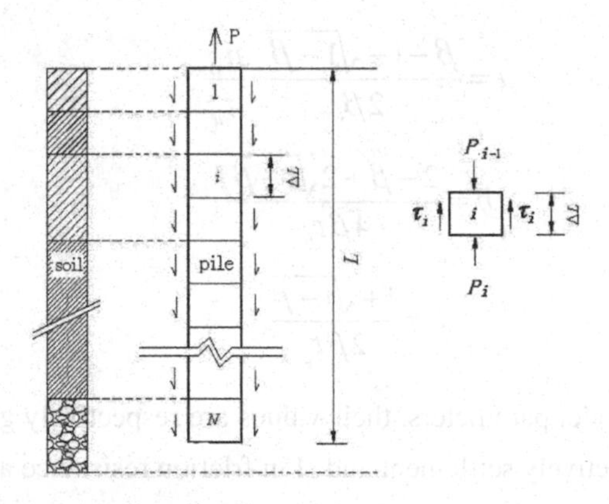

Fig. 2. Analytical model of uplift pile

Step2: the pile-head load is assumed to be P, and the pile-head displacement w_t, is assumed to be a small value. Hence, the top load for unit one:

$$P_{t1} = P \tag{11}$$

The top displacement for unit one is

$$w_{t1} = w_t \tag{12}$$

Step3: assume the average tension of unit one to be P_{t1}, the initial elastic deformation of unit one can be obtained by

$$e_1 = \frac{P_{t1}L_1}{A_pE_p} \tag{13}$$

where L_1 is the length of unit one.

Step4: the midpoint displacement of unit one is

$$\Delta w_1 = w_{t1} - e_1/2 \tag{14}$$

Step5: substitute the midpoint displacement of unit one into Eq.(4) to corresponding pile-side shear stress τ_1.

Step6: after the pile-side shear stress is determined, the total pile-side friction resistance of unit one is

$$T_1 = 2\pi R_0 L_1 \tau_1 \tag{15}$$

Step7: the axial load at the bottom of unit one is

$$P_{b1} = P_{t1} - T_1 \tag{16}$$

Step8: the average pulling force of unit one is

$$P_1 = (P_{t1} + P_{b1})/2 \tag{17}$$

Then, the modified elastic deformation of unit one is

$$e_1^{`} = \frac{P_1L_1}{E_pA_p} \tag{18}$$

*Step9:*comparing the elastic deformation e_1 and modified elastic deformation $e_1^{`}$ of unit one, and if the difference is greater than the limit value($10^{-6}$), assuming that $e_1 = e_1^{`}$, repeat step4 to step8, until the difference is less than the limit value.

Step10: the displacement at the bottom of unit one is

$$w_{b1} = w_{t1} - e_1 \tag{19}$$

Step11: the top load and displacement of unit two are equal to that of unit two, hence

$$P_{t2} = P_{b1} \tag{20}$$

$$w_{t2} = w_{b1} \tag{21}$$

Step12: repeating step3~step10, the displacement and load of unit two can be obtained. Calculating one by one until the unit *N*, the series of displacement and load of each unit can be obtained.

Step13: the total pile-side friction resistance is

$$T = \sum_{i=1}^{n} T_i \tag{22}$$

Step14: assume the a larger pile-top displacement $w_t^`$ (such as $w_t^` = d$) again, repeating step2 ~ step13, the another total pile-side friction resistance $T^`$ can be obtained.

Step15: the average displacement of pile top is

$$w_t^m = (w_t + w_t^`)/2 \tag{23}$$

repeating step2~step13, the total pile-side friction resistance T^m can be obtained.

Step16: if the difference between T^m and assumed pile-top load *P* is less than the given limit value, the pile-top displacement under pile-top load *P* is equal to w_t^m, and the iterative calculation finishes. But if the difference is greater than the given limit value and $(T^m - P)(T - P) < 0$, the $w_t^` = w_t^m$ is obtained. On the contrary, the $w_t = w_t^m$ is obtained. repeating step2 ~ step15, until the difference between T^m and assumed pile-top load P is less than the given limit value.

Step17: repeating step2 ~ step16 with diferent pile-top load, and the series of displacement of uplift piles can be obtained, then the load-settlement relationship of uplift piles is obtained.

6 Example

A bored cast-in-place pile of length 25 m is embedded in homogeneous clay. The concrete of pile is C35, the elastic modulus E_p is 3×10^4 *MPa*, and the radius of pile is 150*mm*. The designing value of single uplift pile is 150*kN*. According to the designing code, several uplift piles were carried out loading test on the vertical bearing capacity. As a result of the comparability of loading test, this paper just shows the results of one pile. The soil of this project site is mainly clay and sandy soil. The weighted average of physical and mechanical properties of soil is got by the thickness of the soil, the average volumetric weight γ is 18.1kN/m^3, the average inner friction angle φ is 16.2^0, and the soil around pile is mainly clay, therefore lateral pressure coefficient could use the coefficient of earth pressure at rest. After the weighted average of compression modulus of soil is obtained, then according to conversion relationship of compressed modulus and elastic modulus, the calculated elastic modulus E_s is 9.3*MPa*, Poisson's ratio υ_s is 0.4. When generalized load transfer model was used in the deformation calculation, the results calculated by the proposed method is in good agreement with the results measured, as shown in Fig.3.

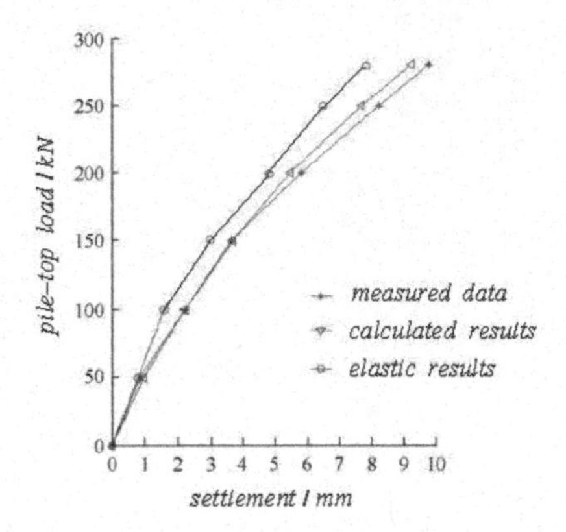

Fig. 3. Comparison between theory curves and the measured curves of uplift pile

7 Conclusions

The method of design and calculation of uplift piles still uses the theory of pressure piles. This theory cannot consider working properties of uplift piles. In this paper, after derivation of theory and comparison of example, the following conclusions were obtained:

1. A deformation calculating method for uplift piles was presented on base of load transfer method, which is in accord with working properties of pile-side soil.
2. In the method presented, the displacement on top of the uplift pile was adjusted by dichotomy, and at lower load levels, the deformation of uplift piles could be predicted satisfactorily. A detailed analysis steps for deformation of uplift piles were given to facilitate analysis in engineering.
3. The analysis method for uplift piles established can be applied for the engineering practice not only to calculate the settlement and bearing capacity of uplift piles, but also to analyze the load transfer law of uplift piles in layered ground.

References

1. Zhang, D.L., Yang, L.C., Wang, B.L.: Experimental analysis and calculation of the uplift resistance bearing capacity of bored cast-in-place pile with enlarged bottom. Journal of Underground Space and Engineering 2, 775–780 (2006)
2. Yang, M., Zhao, X.H.: An approach for a single pile in layered soil. Journal of Tongji University, 421–428 (1992)
3. Ilamparuthi, K., Muthukrisnaiah, K.: Experimental investigation of the uplift behaviour of circular plate anchors embedded in sand. Canadian Getechnical Journal 39, 648–664 (2002)
4. Coyle, H.M., Reese, L.C.: Load transfer for axially loaded piles in clay. ASCE, Journal of the Soil Illeclianics and Foundations Division 92, 1–26 (1966)
5. Kraft, L.M., Ray, R.P., Kagawa, T.: Theoretical t-z Curves. Journal of Geotechnical Engineering Division, SCE 107, 1543–1561 (1981)
6. O'Neill, M.W.: Side resistance in piles and drilled sliafts. J. of Geotechnical and Geoenvironmental Engineering 127, 3–16 (2001)

Teaching Linear Algebra: Multimedia, Strategies, Methods and Computing Technology Development

Jing Zhang

Basic Courses Department, Beijing Union University
Beijing, China
zhang1jing4@sina.com

Abstract. Teaching practice and the actual situation of students, to improve the quality of teaching linear algebra for a more in-depth discussion in the teaching process, and strengthen the quality of students in mathematics, mathematical thinking ability of students. Linear Algebra concepts and more abstract features of content, from the analysis of the conceptual level, to seize the key words, pay attention to the similarities and differences compared to the concept of teaching, with practical examples to introduce the concept in various different aspects, discusses how to improve the basic concepts of the course classroom teaching, to seize the key breakthroughs in difficult, improve teaching quality. Both theory and practice combined with the author's personal experience, for Linear Algebra teaching some of the problems discussed how classroom teaching in order to receive good teaching.

Keywords: Multimedia, computing technology, Linear Algebra, teaching.

1 Introduction

Mathematical modeling is to train students to use mathematics to solve practical problems courses to students in-depth research to understand the object information, to make simplifying assumptions, analysis of the inherent laws, etc., based on the use of mathematical symbols and language, the actual problem statement the mathematical model, using the results to solve practical problems, and accept the actual test. Linear Algebra and administered by the university science and engineering students in an important class of required courses, training its people's quality of mathematical thinking, and numerical computing power has an irreplaceable role. Many teachers are often encountered in teaching this course, students complain about dull, and even some students also have the psychological weariness.

However, due to linear algebra this course have a higher abstract and rigorous logic, you want to learn it has a certain degree of difficulty. Modern computer technology and the rapid development of mathematical software, but also for the mathematical modeling provide a very good platform for in-depth, quantitative analysis of the practical problems of the foundation. With the development of technology and the proliferation of computers, more and more disciplines and

D. Jin and S. Lin (Eds.): Advances in MSEC Vol. 1, AISC 128, pp. 299–304.
springerlink.com

practical problems involve linear algebra. For engineering students, linear algebra is so important, so we need to how to teach this course are discussed.

2 Multimedia

In actual fact, the key is the class skills. For example, a need to find the determinant of Example 5 to 6 slides, according to the order of words, then to the next slide does not remember the previous results, but the results of the previous step is not important, it is important to grasp the entire problem-solving ideas. Today in the modernization of teaching methods, mathematics, under the conditions permitting the use of multimedia teaching should be. Requirements of this determinant, we use what method? Analysis, the role of each small step is to validate our ideas, specific figures have been less important. Therefore, as can the traditional teaching methods and modern multimedia teaching methods combine multi-media technology as a means of teaching, will be able to achieve better teaching results. A lot of teachers oppose the math multimedia on the grounds that slide switch too fast, students accept the difficulties.

Teachers to use this rich emotional material that can stimulate students interest in learning. University mathematics classroom time is limited, but a lot of content involved, teachers can take advantage of new educational technology and teaching methods, broaden student's knowledge. Through multimedia courseware, teachers can easily teach this lesson to summarize the contents, but also on the knowledge systematic review. A lot of teachers oppose the math multimedia on the grounds that slide switch too fast, students accept the difficulties. Linear Algebra course fewer hours, more content, exercises a lot, but longer than that, the knowledge point of each class content was relatively large, significant savings in the use of multimedia will define a lot of writing on the blackboard, the subject of time, which can point to the knowledge better explanation. Meanwhile, the full use of modern teaching methods, some of the abstract can become clear, specific.

3 Computing Technology

In theory, an abstract space to continue to learn linear algebra basis; in applications, but also to solve practical engineering problems is an effective algorithm. Linear problem is widespread in all areas of science and technology, and programming to achieve by means of numerical calculation is inseparable from the linear method. Basic linear algebra problems, such as matrix inversion, solving linear equations, find the rank of the matrix and vector group, the largest independent group, the matrix of eigenvectors etc can be used to solve elementary transformations, elementary transformation to solve simple steps to a single, easy to compile program. As Mathematica, Maple and Matlab (Matrix Laboratory) and other mathematical software, widely used in teaching linear algebra can be the focus of teaching students of computing power from the transition to train students to apply mathematical knowledge to analyze and solve problems up. Therefore, in teaching linear algebra this course, emphasizing basic concepts, the importance of basic theories and methods of calculation, while it should allow students to learn to use mathematical software to solve the corresponding chapter exercises. In the United States and Europe currently use Matlab is very popular in college mathematics, engineering and scientific disciplines, Matlab is used as a teaching tool in many courses; in research institutions and industry Matlab is a high-quality new product research, development and analysis of the main tools.

Students learn mathematics through a positive initiative to mobilize the training, can improve the ability of students to analyze and solve problems, student's practical application ability. Teaching of linear algebra, at home and abroad have the same knowledge, but with a different foreign countries, the importance of foreign students apply skills and ability, they make appropriate arrangements before and after each chapter introduce the Matlab software knowledge, which theoretical knowledge can be boring together with computer applications, but also to keep up the pace, which is what we should learn from. In addition, the experimental class is to improve students' math proficiency platform for students to experience the use of mathematics in the process, recognizing the importance of mathematics and computer combination.

4 Practice Teaching

Through the above three major platforms, to enable students to achieve from the emotional to the rational, from theory to practice to the theory of cognitive processes, so as to stimulate their sense of innovation, exercise their creative thinking, to develop the final play the role of their ability to innovate. Matlab is currently the most widely used mathematical software, one of which is based on the matrix language evolved, for the vast majority of linear algebra content are involved, and its mathematical formulation and is very similar statements.

If the process of teaching linear algebra, introduction into the use of Matlab, and with case, can greatly enhance students' interest in learning and help students more deeply understand and master the basic concepts and methods. Of course, this should be part of the introduction of basic concepts, methods, after the introduction. University of mathematics with abstract features, students learn the characteristics of the building line, to "study - experiment - application" in the form of practical teaching system: to learn the basic theory of Mathematics, University of study and practice of knowledge-based platform; to use Mathematica, Maple, Matlab and other mathematical software, to solve simple math problems based platform for experimental practice; an integrated application of the theory of knowledge, mathematical modeling activities in a variety of innovative activities and extra-curricular math practice based application platform.

5 Applications and Modeling

Combination of theory and practice is the key to learning all kinds of knowledge, linear algebra is no exception. In teaching, I always combine different professions, given an instance of the application of teaching materials to introduce some linear algebra in economics, management, operations research, sociology, demography, genetics, biology and other aspects of the application example, let students truly feel is really necessary to learn linear algebra. For example, to the electrician profession students to explain the theory of linear equations, first introduced a circuit network and its application G R Kirchhoff Theorem to establish a system of linear equations, and linear equations from the abstract general concept, introducing line existence of solutions of equations and solution conditions; linear transformation in learning can be described as a linear network can be seen as part of the linear transformation of input into output; n a part of the series as a linear transformation of the n order effect (multiplied).

To focus on the application of linear algebra, linear algebra to try to solve various problems, you can also join in the process of teaching mathematical modeling. He spoke of the inverse matrix; inverse matrix can introduce students to the application of the password. Explaining the equations, you can trim the chemical formula of practical problems with the network flow example. Focus on linking theory with practice, starting from the actual study, discussion of theoretical issues and applied to practical problems, which introduced the theory development of actual production, the real significance of life, so that students interested in the theory that the learning of useful effective use.

Linear transformation is the study of electrical theory, a common tool. As a part of the linear network can be seen as the input into the output of a linear transformation, several links in series can be seen as a linear transformation followed by a few effects (multiplication). Introducing the vector space, the inter-city migration, farm animals of all ages a proportion of the distribution of the Markov chain mathematical model will give the students a great interest in learning.

6 Strategies

The introduction of new concepts, knowledge before and after convergence, focus and difficult to explain even the examples of the selection and arrangement, everywhere should be carefully considered and, where appropriate. Knowledge-point of convergence and transformation of knowledge into the network should strive to improve the overall analysis. It is precisely because the knowledge of linear algebra points are inextricably linked, integrated algebra problems and flexibility on the larger, should guide students to focus on finishing the series, convergence and transformation. At the same time, you can also consider linear algebra and students learn a combination of professional, so that students see the real application, increase the interest in learning.

Teachers should be teaching the details of more effort. Each lesson, the teacher can immediately to learn new content, combined with previously learned, reasonable cleverly arranged lectures. Later after class, teachers should also sum up, self test, good or do not find a place to improve future instruction. Crises-cross from the content point of view of linear algebra, after closely, interlocking, interwoven, so flexible problem-solving methods, and only keep the summary, to find out the internal relations, so that the knowledge mastery, interface and the entry point more familiar with, and ideas naturally broadened.

Pay attention to the blackboard in the lecture and multimedia teaching combination. I practice in the classroom teaching linear algebra found the task of writing on the blackboard too, such as the calculation of the determinant, solving linear equations such as writing on the blackboard will take a lot of classroom time. Teacher's writing on the blackboard has a strong demonstration of good writing on the blackboard for students are a beautiful visual experience; the students play a subtle role. Blackboard is a teacher on the blackboard to write the text, also known as mini-lesson plans, it is for teachers to improve the quality of teaching has a very important supporting role. Blackboard should reflect the study method for students to know how to acquire knowledge.

In view of this situation, the use of writing on the blackboard in the classroom and multimedia combination: the larger the matrix, determinant and other multi-media presentations, lively animation will not only leave a strong perceptions of students and increased the amount of information the classroom, sparking the enthusiasm of students, more importantly, saving teachers time writing on the blackboard, there is sufficient time to enable teachers to explain the concept. Good writing on the blackboard, help students understand and master the knowledge, help develop students' ability and non-intellectual factors, can effectively stimulate student interest in learning, inspire students thinking. Therefore, the traditional teaching and multimedia teaching combination, complementarily, so that the students learn easily, also easy to teach teachers can greatly improve the efficiency of classroom teaching. For more complex theory that can be derived on the blackboard so that students can keep up with the teacher's ideas.

7 Conclusions

In view of this situation, in the teaching process as much as possible to cite some typical examples of lives, discussion and analysis of practical problems, to guide students through the practical problems of its abstract nature. Enable students to independently explore the use of mathematical software and related applications textbook knowledge learned students apply skills and innovation. Linear algebra teaching, teachers should foster the concept of the times, the ongoing education reform, according to the actual situation of students, students learning linear algebra interest in improving teaching quality, as much as possible so that the teaching content and full of vitality times, effectively promote the construction of linear algebra course.

By mathematical experiments, to enable students to understand and apply mathematical knowledge and initial practice methods to solve practical problems of the whole process, and through computer experiments and mathematical software, the experimental results apply not only the derivation of theorems and formulas manually calculate conclusions, it reflect the principles of mathematics students, mathematical methods, modeling, computer operating software, and many other mastery of the content and application capabilities.

Acknowledgments. The work is supported by Funding Project for Academic Human Resources Development in Institutions of Higher Learning under the Jurisdiction of Beijing Municipality (PHR (IHLB)) (THR201108407). Thanks for the help.

References

1. Carlson, D.: Teaching Linear Algebra: Must the Fog Always Roll In? College Mathematics Journal 24(1), 29–40 (1993)
2. Carlson, D., Johnson, C.R., Lay, D.C., Porter, D.C., Duane Porter, A.: The Linear Algebra Curriculum StudyGroup Recommendations for the First Course in Linear Algebra. College Mathematics Journal 24(1), 41–46 (1993)
3. Kalman, D.: New Mathwright Library (Software Review). College Mathematics Journal 30(5), 398–405 (1999)
4. Zhang, J.: To realize the diathetic education of mathematical education and CAI by network, mathematical experiment course. In: ICETC 2010 - 2010 2nd International Conference on Education Technology and Computer, vol. 4, pp. V4467–V4471 (2010)
5. Lay, D.C.: Linear Algebra and Its Applications. Addison-Wesley (1997)
6. Wicks, J.R.: Linear Algebra: An Interactive Laboratory Approach With Mathematica. Addison-Wesley (1996)
7. Dorier, J.-L., Robert, A., Robinet, J., Rogalski, M.: On a research programme concerning the teaching and learning of linear algebra in the first-year of a French science university. International Journal of Mathematical Education in Science and Technology 31(1), 27–35 (2000)
8. Harel, G.: Learning and Teaching Linear Algebra: Difficulties and an Alternative Approach to Visualizing Concepts and Processes. Focus on Learning Problems in Mathematics 11(1-2), 139–148 (1989)
9. Gray, E., Tall, D.O.: Relationships between embodied objects and symbolic procepts: An explanatory theory of success and failure in mathematics. In: van den Heuvel-Panhuizen, M. (ed.) Proceedings of the 25th Conference of the International Group for the Psychology of Mathematics Education, Utrecht, Netherlands, vol. 3, pp. 65–72 (2001)
10. Stewart, S., Thomas, M.O.J.: Embodied, symbolic and formal thinking in linear algebra. International Journal of Mathematical Education in Science and Technology 38(7), 927–937 (2007c)

Research on Keyword Information Retrieve Based on Semantic

Xin Li[1] and Wanxin Dong[2]

[1] Computer Center, Liaoning University of Technology, Jinzhou, China 121001
[2] Human Resources and Social Security Bureau, Yantai, China 264003

Abstract. To deal with the problem of semantic missing during keyword retrieval, this paper proposes a semantic-based keyword information retrieval approach. Firstly, an improved word-concept relevance method, which measures the relevance between word and concept by considering both the internal and external correlations, is proposed. Next, by importing the improved word-concept relevance method into the classical statistical language model, a new statistical language model which is based on the relevance of word and concept is proposed. The experiments demonstrated good efficiency and performance of the approach proposed in this paper.

Keywords: Ontology, Keyword, Information retrieval, Semantic association.

1 Introduction

Information retrieval models can be broadly summarized into the following two categories: one is information retrieval model based on statistics [1], which including: Boolean model [2], extended Boolean [3], Bayesian model [3], vector space model [4].The other is ontology-based information retrieval model [5]. The feature of this model type is that, using certain keywords by users input or keyword weights as query conditions, according to calculate the similarity to extract the query results. The higher the similarity of the results, more front row will be, and the greater chance for the user to retrieve. Therefore, many researchers take this opportunity to put forward some algorithms, such as PageRank algorithm [2], Hilltop algorithm [6], etc. However, the improved algorithms are still based on keywords that do not involve the semantic understanding. In order to better understanding of the semantics, people made the second type of information retrieval model, which called ontology-based information retrieval model. This paper firstly proposed the improved algorithm (NKTCM method) for correlation between word and concept (NCR).

2 NCR

In the actual calculation, as similarity and NCR are having some connections, which can fully distinguish between them, so the impact to the whole system or theoretical studies is not great. We used similarity to instead of NCR. There are many methods to calculate the similarity, such as Cosine, Dice. But these methods are that the

D. Jin and S. Lin (Eds.): Advances in MSEC Vol. 1, AISC 128, pp. 305–310.
springerlink.com

presumption is completely independent to the next step, so in some certain, it limited its application.

The main method to calculate the NCR is, according to the degree of co-occurrence between word and concept in the document, the thinking of this method is relatively simple, and easily to understand. The basic idea is that the higher the level of co-occurrence between word and concept, the greater the NCR. Co-occurrence degree can be describe by particle size, then appeared the word size, concept particle size, and so on. The emergence of the concept of granularity, which made people to describe the co-occurrence level between word and concept more easily, and it also provides a more detailed ways and means.

3 NKTCM

The NKTCM can be described in two ways: the one is to consider the relationship between word and concept exists in document itself, called intrinsic. The other is to consider the relationship between documents, called external links.

3.1 Consider from Intrinsic

Let D_k be a vector space, it contains all of the words t_k, using space vector representation is $D_k = \{ d_j \mid d_j \in D \wedge t_k \in d_j \}$; Let D^i stand for the i-th word corresponding to the document under the concept of c_i, expressed by the formula is $D_k = \{ d_j \mid d_j \in D \wedge d_j \propto c_i \}$; Let c_j be the set d_j belongs to a collection of concepts, as $C_j = \{c_i \mid c_i \in C \wedge d_j \propto c_i\}$; The document Includes t_k and belong to c_j ： $D_k^i = D_k \cap D^i$; The collection of concept contains with t_k can be expressed as： $C^k = \{c_i \mid t_k \in d_j \wedge c_i\}$; $count(t_k; d_j)$ stands for the frequency of t_k appears in d_j; $tf_{t_k}^{c_i} = \sum_{d_j \in D_k^i} \frac{count(t_k; d_j)}{len(d_j)}$ stands for the frequency of t_k appears in c_i; $aw(t_k; c_i)$ expresses the NCR of c_j to t_k;

Based on the above conditions, we submit the formula (1), used to calculate the NCR of c_j to t_k.

$$aw(t_k; c_i)_{t_k \neq c_i} = \log(N / |C^k| + 1.0) \times \log(|D_k^i| / |D^i| + 1.0)(N / |C_k| + 1.0) \tag{1}$$

3.2 Consider from External Links

Here we consider the relationship between documents. All the documents can be as a whole, the smaller and the value of the document can be named as a valid document.

Suppose w_i is a document, which core is c_i, similar to the window; $t_k \mapsto w_{c_i}$ expresses t_k appear in w_i. $\underset{d_j \in D \wedge W_{c_i} \subseteq d_j \wedge t_k \mapsto W_{c_i}}{count}(d_j, D)$ is the number of documents that

t_k appear in w_i. $pf(t_k, c_i; d_j)$ is the frequency of t_k in d_j that appears in w_{c_i}. That is, $\sum_{W_{c_i}} count(t_k, W_{c_i})$. $Dist(t_k, c_i; d_j)$ is the distance of (t_k, c_i) in d_j. $Dist(t_k, c_i; d_j) = \frac{\sum_{Wc_i \subseteq d_j} \sum_{(t'_k \mapsto Wc_i) \wedge (t'_{k=t_k})} dis_{c_i}(t'_k, w_{c_i})}{pf(t_k, c_i; d_j)}$ $dis_{c_i}(t'_k, w_{c_i})$ is a special case, that is the distance between t_k in d_j and the center c_i. $\sum_{(t'_k \mapsto Wc_i) \wedge (t'_{k=t_k})} dis_{c_i}(t'_k, w_{c_i})$ is their distance pluses. In the same way, we can get:

$$cw(t_k; c_i)_{t_{k \neq c_i}} = \frac{1}{|\Gamma_i|} \sum_{r=1 \ldots \|\Gamma_i\|} \frac{\underset{d_j \in D \wedge Wc_{i,r} \subseteq d_j \wedge t_p \mapsto Wc_{i,r}}{count}(d_j, D) \times \log(\sum_{d_j \in D} pf(t_{p,c_{i,r}}; d_j) + 1.0)}{\sum Dist(t_{k,c_{i,r}}; d_j)} \quad (2)$$

NCR's final calculation: the method is to combine the formula (1) and formula (2), considering from the two different ways to get the NCR. As they descriptions are related, so they have the same natures, therefore, we can use linear thinking to combine them together. As formula (3) , α is their coefficient, where $0 =< \alpha <= 1$.

$$s(t_k; c_i)_{t_k \neq c_i} = \alpha \times aw(t_k; c_i) + (1 - \alpha) \times cw(t_k; c_i) \quad (3)$$

4 KCLSM-Improved Statistical Language Models

In represent of KL-difference model, we can get an improved statistical language model. A document can be divided into two parts; one is the semantic part, the other is non-semantic. Dealing with non-semantic part, we need to set the terms of independence, in order to facilitate the calculation. The calculation of the semantic part, we use the method of NKTCM. Using NCR to measure the correlation between word and concept, and introduces this correlation into statistical language model, in order to improve the utilization and usefulness of the model.

Thus, according to KL-difference model, $Score(Q, d')$ can be decomposed into formula (4). dy stands for the semantic part. dn stands for the non-semantic part. $d = d^y \cup d^n$ and $d^y \cap d^n \neq \varnothing$.

$$\begin{aligned} Score(Q, d) &= Score_{\theta_R}(Q, d^y) + Score_{\theta_{NR}}(Q, d^n) \\ &= \sum_{q_i \in Q} P(q_i \mid Q) \log P_{\theta_R}(q_i \mid d^y) + \sum_{q_i \in Q} P(q_i \mid Q) \log P_{\theta_{NR}}(q_i \mid d^n) \end{aligned} \quad (4)$$

In formula (4), θ_R is KL-difference model we use NCR. θ_{NR} is the original KL-differential model. Thus, the definition between the semantic part and the non-semantic are successful. Then combining formula (4) with the formula (1) and formula (2), we can get the formula (5) and formula (6).

$$Score(Q,d) = Score_{\theta_R}(Q,d^y) + Score_{\theta_{NR}}(Q,d^n) = \sum_{q_i \in Q} \frac{count(q_i,Q)}{|Q|} (\log(\lambda \sum_{c_j \in d^{y'}} s(q_i;c_j) P(c_j | d^{y'}) + (1-\lambda) \sum_{c_k \in d^{y''}} s(q_i;c_k) P(c_k | d^{y''}) + \log(1 + \frac{count(q_i,d^n)}{\mu \cdot P(q_i | D)})) + |Q| \log \frac{\mu}{\mu + |d^n|} \quad (5)$$

$$Score(Q,d) = Score_{\theta_R}(Q,d^y) + Score_{\theta_{NR}}(Q,d^n) = \sum_{q_i \in Q} \frac{count(q_i,Q)}{|Q|} (\log(\lambda \sum_{c_j \in d^{y'}} s(q_i;c_j) P(c_j | d^{y'}) + (1-\lambda) \sum_{c_k \in d^{y''}} s(q_i;c_k) P(c_k | d^{y''}) + \log(1 + \frac{count(q_i,d)}{\mu \cdot P(q_i | D)})) + |Q| \log \frac{\mu}{\mu + |d|} \quad (6)$$

The implication of Formula (5) is that, the entire document is divided into two independent parts: One is semantic, using the method of NKTCM based on statistical language model. Non-semantic part, using the original statistical language models, that is KL-difference model that have the Dirichlet prior method.

The implication of Formula (6) is that, the entire document is not divided into two independent parts: One is semantic part, also using the method of NKTCM based on statistical language model. But the non-semantic part, is not the simple non-semantic, we see the entire document as the non-semantic part. And also use the original statistical language models.

5 KIRBS

The principle of this system is to apply the Semantic Web technology, in particular, to apply the ontology technology and the RDF technology to the information retrieval system, so as to achieve discovery, retrieve and use information on the Internet.

The system is divided into the following four parts: semantic processing, information preprocessing, information organization and storage, and sort results. Semantic processing is mainly responsible for user to enter searching content. After the user submit searching request, whether the retrieval system can accurately express the user's search intention, which is a key to the ultimate success for retrieval. By semantic reasoning, we can find implied semantic relations between the documents, so as to improve the recall and precision of the information retrieval. With specific subject background, domain experts firstly construct the ontology of Semantic Web and form Web ontology base, in order to describe the general sharing concepts and terms, and the relationship, within the fields. This is a basic of the information organization for Semantic Web.

6 Experiments

6.1 Experiments for KCLSM

Processing. Reference model is formula (7). Dirichlet prior method has a shortage is, when the parameter μ is too small, it will be bad for the experiment, so we must give a large parameter, such as 250. To facilitate the description, we named the three models as following.

$$Score(Q,d)=\sum_{q_i\in T}P(q_i\mid Q)\times\log P(q_i\mid d)$$
$$=\sum_{q_i\in T}(P(q_i\mid D)\log(1+\frac{count(q_i,d)}{\mu\cdot P(q_i\mid D)})+\log\frac{\mu}{\mu+|d|})\tag{7}$$
$$=\sum_{q_i\in Q}(\frac{count(q_i,Q)}{|Q|})\log(1+\frac{count(q_i,d)}{\mu\cdot P(q_i\mid D)})+\log\frac{\mu}{\mu+|d|}$$

a) Formula (7) is a reference model, called Model C.
b) Formula (5), called Model 1. c) Formula (6), called Model 2.

During the experiment, we use Precision and Recall to compare the three models. When the parameters n, k is fixed at 20, λ has different values. With curve, we can express the effect of three models.

Result. First, we test Model 1 only. As the parameters n, k is fixed, so we fix λ's values at [0,1]. In the process of value, we found when λ's value is 0, Model 1 is the most primitive state, that is not consider the semantics of words contained in the concept of the impact of the document. As λ form 0 to 0.5, semantic gradually impact the Document, and there is directly proportional between them. The search curve of Model 1 is shown in Fig. 1. The search curve of Model 2 is shown in Fig. 2.

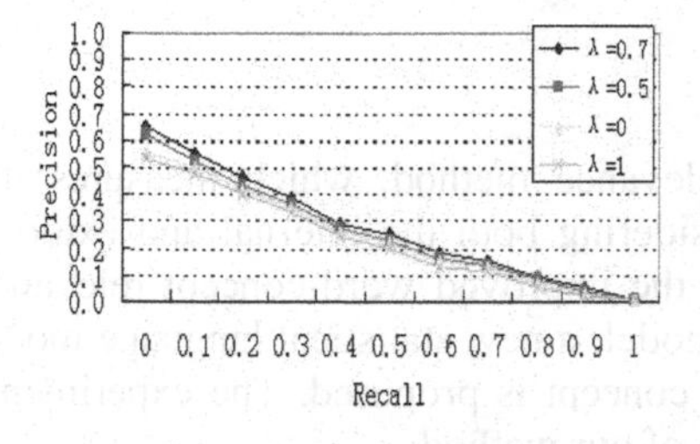

Fig. 1. The search curve of Model 1

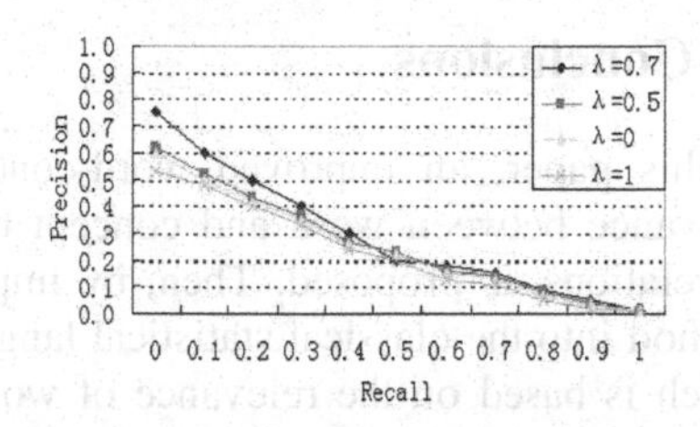

Fig. 2. The search curve of Model 2

As λ take 0.7 is the best state for both Model1 and Model2. So, λ take 0.7, we compare Model1, Model2 and Model C. And the results are as follows, shown in Fig. 3 and Fig. 4.

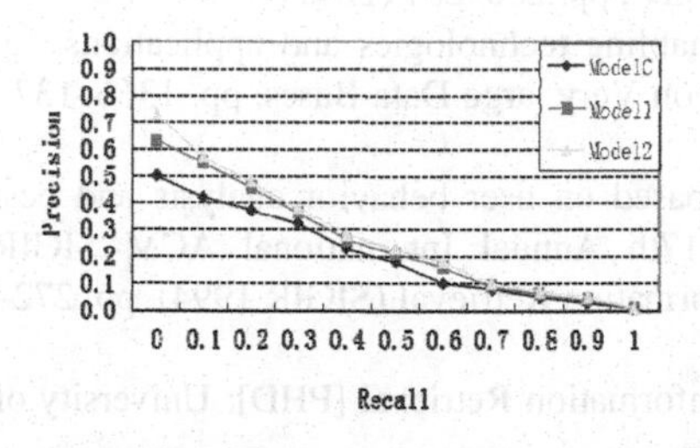

Fig. 3. The search curve of three models

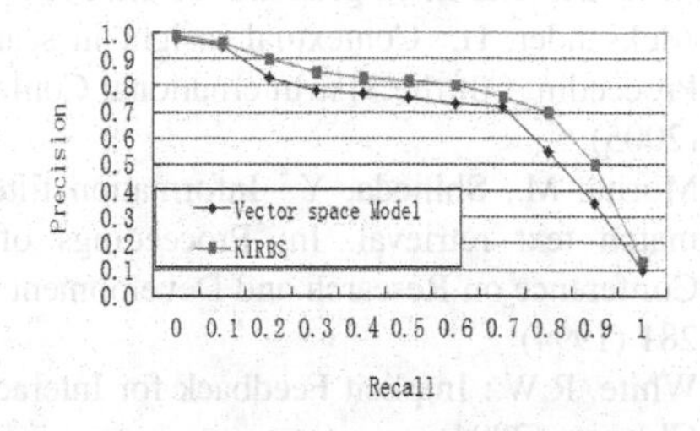

Fig. 4. Compare two system Models

It can be seen from the figure, model 1 and model 2 results are much higher than reference model (model C), and the effect of model 2 is the best.

6.2 Experiments for KIRBS

Processing. In the experiment, we select the top 50 documents as the reference documents, and the documents obtained from the semantic similarity or similar words, after re-calculated the weight, we can get the correlation. Then, cyclic approach, we can get the relationship between the Precision and Recall.

Result. Table 1 show the result of KIRBS compare with Vector space Model.

Table 1. Experiment Results

Retrieval Model	Precision	Recall
Vector space Model	72.56	73.23
KIRBS	76.62	75.54

The results can be seen, Precision and Recall is has a inverse proportional relationship. Compare with the vector space Model, KIRBS has more high efficiency, although both Precision and Recall are quite high.

7 Conclusions

In this paper, an improved word-concept relevance method, which measures the relevance between word and concept by considering both the internal and external correlations, is proposed. Then, by importing the improved word-concept relevance method into the classical statistical language model, a new statistical language model which is based on the relevance of word and concept is proposed. The experiments demonstrated good efficiency and performance of our method.

References

1. Mihalcea, R.F., Mihalcea, S.I.: Word semantics for information retrieval: Moving one step closer to the semantic web. In: Proceedings of the 13th International Conference on Tools with Artificial Intelligence, Southern Methodist Univ., pp. 280–287 (2001)
2. Aleksander, H.: Contextual insight in search: enabling technologies and applications. In: Proceedings of the 31st International Conference on Very large Data Bases, pp. 1366–1377 (2005)
3. Morita, M., Shinoda, Y.: Information filtering based on user behavior analysis and best match text retrieval. In: Proceedings of the 17th Annual International ACM SIGIR Conference on Research and Development in Information Retrieval (SIGIR 1994), pp. 272–281 (1994)
4. White, R.W.: Implicit Feedback for Interactive Information Retrieval [PHD]: University of Glasgow (2004)
5. Rocchio, J.: Relevance feedback in information retrieval. In: The SMART Retrieval System: Experiments in Automatic Document Processing, pp. 313–323 (1971)
6. Liu, R.L., Lin, W.J.: Incremental mining of information interest for personalized web scanning. Information Systems 30(8), 630–648 (2005)

The Implementation of a Specific Algorithm by Traversing the Graph Stored with Adjacency Matrix

Min Wang and YaoLong Li

College of Mathematics and Information Science, Weinan Teachers University,
Weinan, Shaanxi, China
wntcwm@126.com

Abstract. With analyzing the adjacency matrix storage structures of graph, the design and analysis on the specific algorithm of eliminating the redundant edges which forms ring routes in the undirected connected graph was introduced in detail in this paper based on the process of computing out the spanning tree of the graph through traversal process. The algorithm design ideas, implementation approaches and concrete steps, and the algorithm description in C were introduced in this paper, and then the algorithm was evaluated from the two aspects of time complexity and space complexity.

Keywords: Adjacency matrix, Graph traversal, Spanning tree of graph, Undirected connected graph, Ring route, Time complexity, Space complexity.

1 Introduction

Data structures and algorithms are important foundations for computer science disciplines, and they are an indivisible whole. To solve a specific problem, you need to analyze the problem, and combine organically the data structures and algorithms. In order to solve effectively the problem to be solved, the algorithm must be compatible with the data structure [1,2]. Graph is an important kind of nonlinear data structure, and its application has been infiltrated into linguistics, logic, physics, chemistry, telecommunications engineering, computer science and other branches of mathematics [3].

This paper will aim at the specific algorithm of eliminating the redundant edges which form the ring routes in an undirected connected graph. Based on the analysis on the adjacency matrix storage structure of the undirected graph, the design and analysis on the algorithm design ideas, implementation approaches and concrete steps will be introduced in detail based on the process of computing out the spanning tree of the graph through traversal process. The algorithm description in C will be introduced in this paper, and then it will be evaluated from the two aspects of time complexity and space complexity.

D. Jin and S. Lin (Eds.): Advances in MSEC Vol. 1, AISC 128, pp. 311–315.
springerlink.com

2 Problem Introduction

The description of the problem to be solved is as follows:

For any undirected connected graph *G*, there may contains some ring routes, this algorithm requests to remove some edges in *G* in order that *G* does not contain ring routes, and meanwhile the number of edges eliminated must be a minimum.

3 Problem Solving

The problem discussed here does not specify whether the graph *G* is a weighted graph, and the operations needed are just deleting the redundant edges, and without the weight correlation about the edges to be deleted. Therefore, you can regard the problem as a specific algorithm about a non-weighted undirected connected graph, and choose the popular adjacency matrix storage structure for it.

3.1 Adjacency Matrix Storage Structure of Graph

The adjacency matrix representation, also known as an array representation, uses a one dimension array to store the vertex information and a two dimension array, known as adjacency matrix, to store the relationship information between the vertex elements [3,5].

For a graph with *n* vertexs, its adjacency matrix *A* is a $n \times n$ square matrix. The formalized definition of *A* is as follows:

$$A[i,j] = \begin{cases} 1 & (v_i, v_j) \in VR \\ 0 & \text{otherwise}. \end{cases} \quad (1)$$

Where, *VR* is the set of relations between the vertexs. The direct edge, connecting two vertexs v_i and v_j, has the value of 1 at the corresponding intersection in the matrix.

Compressed Adjacency Matrix. Since the adjacency matrix of undirected graph *G* is a symmetric matrix, to delete or record an edge has been visited, you need to operate simultaneously at the lines of the two adjacent vertexs in the adjacency matrix. Obviously, there are duplicate operations, so we can consider a compressed adjacency matrix storage structure to store only its lower triangular matrix ($i \geqslant j$), thus the whole n^2 elements will be stored compressed into $n(n+1)/2$ spaces.

Assume that the one-dimensional array $SA[n(n+1)/2]$ is the adjacency matrix storage structure of *G*, then such one-to-one correspondence between $SA[k]$ and the matrix element $A[i,j]$ is as follows:

$$k = i\,(i+1) / 2 + j\,. \quad (2)$$

For any given pair of subscripts (i,j), the matrix element $A[i,j]$ can be found in *SA*; on the other hand, for all $k = 0,1,2,...,n(n+1)/2-1$, the position (i,j) of the $SA[k]$ in the matrix can also be determined [3].

Description in C of the Adjacency Matrix Storage Structure. The compressed adjacency matrix storage structure can be described in C as follows:

```
#define  MAX_VER  50          //Maximum number of vertexs
typedef struct ArcNode {
   int  adj;          //Value is 1 or 0
   InfoType  info;   //Other information of the edge
} ArcNode;
typedef struct{
   VerType vexs[MAX_VER];     //Array of vertexes
   ArcNode arcs[MAX_VER(MAX_VER+1)/2];
                          //Compressed adjacency matrix
   int vexnum,arcnum;
              //The current number of vertexes and edges
} MatrixGraph[3,5];
```

3.2 Algorithm Design and Analysis

This algorithm requires removing as little as possible redundant edges which form ring routes in the undirected connected graph *G*, the result will be a spanning tree of *G*, thus we can give the design ideas according to the ideas of obtaining the spanning tree of *G*.

Algorithm Design Ideas. The algorithm design idea is as follows:

First, you can record the edges visited during the process of depth-first search traversal or breadth-first search traversal of *G*, the spanning tree of *G* is composed of the edges visited, and then remove the edges not recorded (that is do not belong to the spanning tree of *G*).

Analysis on the Design Ideas. The proccess of recording the edges have been visited during the traversal of *G* can be implemented by the following two approaches:

Use the auxiliary array of edge node. Set an auxiliary one-dimensional array for the edges in G, and each edge node in it has the structure shown in Fig.1.

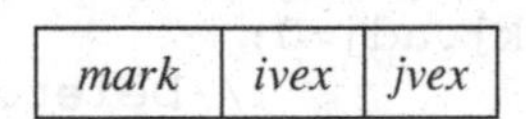

mark	*ivex*	*jvex*

Fig. 1. Edge node structure of the auxiliary array

Where, *mark* field records whether the edge has been visited, and its initial value is 0. Once an edge is visited, the *mark* field of the corresponding node is set to 1. *ivex* and *jvex* represent the positions of the two adjacent vertexs in *G*, and they are used to identify the edge.

The existing problem of this method is that the corresponding edge node in the auxiliary array can not be located simultaneously during the traversal process. Re-positioning the edge nodes in the auxiliary array reduces the time efficiency of algorithm. In addition, the introduction of the auxiliary array increases the space complexity of algorithm.

Effective Use of Adjacency Matrix Storage Structure. Let the info member of each ArcNode element in adjacency matrix record whether an edge was visited. The initial value of info is 0. Once an edge is visited during the traversal process, the corresponding info member is set to 1.

Obviously, for whatever consideration, the time efficiency or the space efficiency, the second method is more appropriate.

After obtaining the spanning tree of G, the process of removing from G the edges not marked needs only to set the *adj* members of the edges whose *info* members are 0 to be 0.

Algorithm Description in C. Suppose that the undirected connected graph G adopts the *MatrixGraph* storage structure aforementioned, G has n vertexes (v_1~v_n), which are stored respectively in *vexs[0..n-1]* units, and the elements of the *arcs* array are initialized. To ensure that all vertexes in G only be accessed once during the traversal process, you can set a global access indicating array *visited[n]*, whose initial value is 0. If vertex v_i is accessed, *visited[i]* will be set to 1 [3,5]. Without loss of generality, we start from the vertex v_i (*vexs[i], i=0~n-1*) to implement the depth-first search traversal on G, so this algorithm can be achieved using the depth-first search traversal algorithm. The algorithm description in C is as follows:

```
int visited[MAX_VER];          //Access indicating array
void DepthFirstSearch(MatrixGraph *G, int i)
{  visit(G->vexs[i]);  visited[i]=1;
   for(j=0;j<G->vexnum;j++)
      if(!visited[j]&&G->arcs[i(i+1)/2+j].adj==1)
      {   G->arcs[i(i+1)/2+j].info=1;
                               //Record the visited edge.
          DepthFirstSearch(G, j);    }
}
void DeleteRedundentEdge(MatrixGraph *G)
{  DepthFirstSearch(G, 0);
   for(i=0;i<G->vexnum;i++)
      for(j=0;j<i;j++)
      {   k= i(i+1)/2+j;
          if(G->arcs[k].adj==1&&G->arcs[k].info==0)
           G->arcs[k].adj=0;
                               //Delete the edge not
marked.
      }
}
```

3.3 Algorithm Evaluation

The algorithm of eliminating the redundent edges in undirected connected graph is achieved ultimately by function *DeleteRedundentEdge()*. This function includes two basic statements in its function body: the first statement is calling the recursive function *DepthFirstSearch()*, while the second is a double *for* loop. Recursive function *DepthFirstSearch()* executes the depth-first search traversal on G, and then obtains the spanning tree of G. Graph traversal process is essentially the process to locate the adjacent vertex for each vertex [3]. This algorithm adopts the adjacency matrix storage structure, and the time for locating the adjacent vertex for each vertex is nearly $O(n(n+1)/2)=O(n^2)$, where n is the number of vertexes. The number of times of the loop body (in double *for* loop) execution is nearly $O(n(n-1)/2)=O(n^2)$. Regarding the vertexes number n as the scale of the problem, and with the gradually increase of problem scale, the algorithm time complexity is nearly $T(n)=O(n^2)$.

In addition to the storage space used by the function itself and graph G, an auxiliary one dimension array *visited[MAX_VER]* was introduced, so the algorithm space complexity is linear order, that is $S(n)=O(n)$.

4 Conclusions

In this paper, the process of designing and analyzing the specific algorithm of eliminating the redundant edges, which form the ring routes in an undirected connected graph, was introduced in detail based on data structure. The algorithm description in C was also introduced in detail, and then it was evaluated from the two aspects of time complexity and space complexity. This work plays a guiding role in teaching the relevant chapters in "Data Structure" curriculum.

Acknowledgments. This work is supported by Scientific Research program funded by Shaanxi provincial education department (program No. 11JK0480), Research Fund of Weinan Teachers University (No. 11YKF011) and Research Fund of Weinan Teachers University (No. 11YKS014).

References

1. Wang, W.-D.: Data Structure Learning guidance. Xi'an Electronic Science and Technology University Press, Xi'an (2004)
2. Wang, M., Li, Y.: Designing on a Special Algorithm of Triple Tree Based on the Analysis of Data Structure. In: Lin, S., Huang, X. (eds.) CESM 2011. CCIS, vol. 175, pp. 423–427. Springer, Heidelberg (2011)
3. Yan, W.-M., Wu, W.-M.: Data Structures(C language edition). Tsinghua University Press, Beijing (2002)
4. Jiang, T.: Data Structures. Central Radio and TV University Press, Beijing (1995)
5. Geng, G.-H.: Data Structure—C Language description. Xi'an Electronic Science and Technology University Press, Xi'an (2005)

Research of Key Technologies in Development in Thesis Management System

Huadong Wang

Department of Computer Science
Zhoukou Normal University
Zhoukou, Henan
wanghuadong@zknu.edu.cn

Abstract. To achieve comprehensive and effective management of university thesis, using development of technology ASP.NET2.0, an university thesis management system based on B / S mode is designed and implemented. In this paper, several key technical in system development such as controlling of user access rights, using of master pages, production of Word format document, editing and publishing online information are focused on. These technologies have certain versatility in the Web application system. The results of the development and operation show that application of these key technologies improved the development efficiency and practicality of university thesis management system significantly.

Keywords: ASP.net, thesis management system, B/S, editing and publishing information online.

1 Introduction

In recent years, with the expansion of college enrollment, workload of thesis guidance and administrative increased exponentially, the traditional thesis management methods cannot meet the actual requirements, a new way to efficiently manage thesis process is urgent needed. With the popularity of computer networks in universities, the necessary hardware infrastructure and operating platform of office management by the campus network are provided. University graduation thesis management system achieve business logic of the graduate design process through the network, build a software platform between the teachers , students and managers that enables teachers and students can use the system to interact, to complete their course in the graduate design needs to be done , and then achieve the thesis network automation and management purposes.

In this paper, with design and development of university thesis management system, several key technical in system development such as controlling of user access rights, using of master pages, production of Word format document, editing and publishing online information are focused on.

D. Jin and S. Lin (Eds.): Advances in MSEC Vol. 1, AISC 128, pp. 317–322.
springerlink.com

2 Design of University Thesis Management System

2.1 System Architecture and Development Environment

System is based on ASP.NET three layers structure: presentation layer, business layer and data layer, as shown in fig 1. The application layer provides the user interface: the client IE browser; users access the system with IE browser. Business layer provide business functions, it is the core of the system. This layer provides function calls for the presentation layer, but it also calls functionality provides by the data layer to access the database. Data layer is in the bottom, using ADO.NET as the interface, dealing mainly with data request of the business layer, insert, modify and delete the data stored in the database.

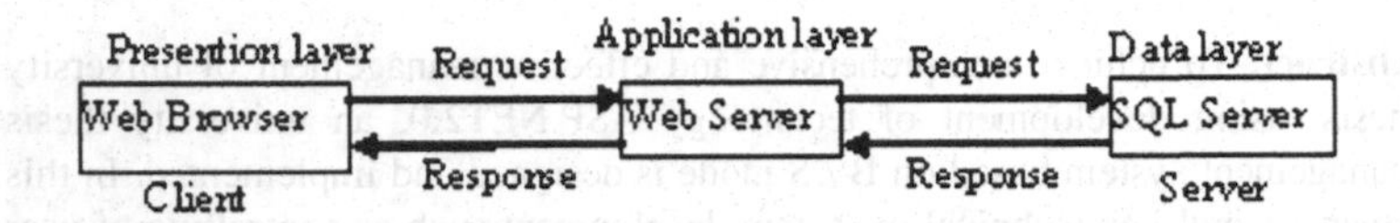

Fig. 1. Three layers structure of the system

System use Microsoft's Visual Studio2008 and Visual Developer2008 as a development platform, SQL Server 2005 as back-end database, based on B / S development model, build applications using ASP. NET technology, combined with C # and JavaScript.

2.2 System Function Design

System's main functions are: system management, teacher-tasking function, students-tasking function, defense management, forms printing, and similar paper detection. Systems management capabilities offer a simple system management to system administrators; teacher-tasking function help teachers to complete the subject of submission, submit the mission statement, opening report reviews, assessments, and other major work; students-tasking function help students to complete opening report submission, paper submission, etc.; respondent management features provide the respondent during the scoring function; form printing capabilities available to the teachers and students to print the relevant forms; identical paper detection can detect similar paper to reduce the phenomenon of student thesis.

According to the papers management processes and different user roles, establish four systems function "student management", "Teacher Management", "Department of Management" and "faculty management" and the corresponding sub-functions. Basic structure of the system is shown in Figure 2. Implementation of the system function, using some of the key technologies can not only improve system availability, but also has some versatility.

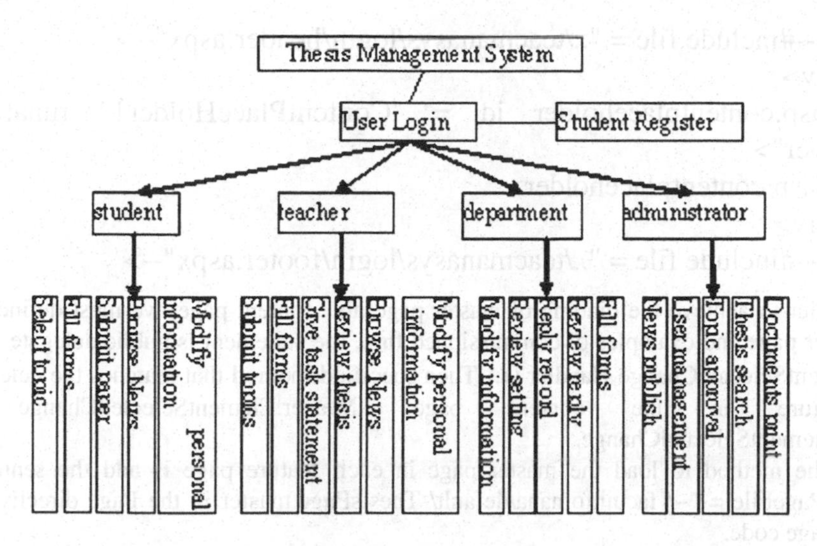

Fig. 2. System Functions

3 Usage of ASP. NET2.0 Master Pages

All pages of thesis management system have some common elements, such as copyright notices, navigation bar, Session judge and help Tips functions. We use ASP.NET2.0 master pages to create a unified user interface and styles, and to define the common elements of the system page.

Master pages are an ASP.NET file with extension of .Master, which can contain static text, HTML elements and ASP.NET server controls and other Web elements. Master page identificative by the special @ Master directive, the directive replaced the ordinary .Aspx page's @ Page directive, which written statements<%@Master Language = "C#"%> in master page code file. We design and define the master page file named Thesis Page. master. In this paper, use HTML table tags for page layout, set the system LOGO by an image element, display the copyright notice by a static text, and achieve navigation with the server control SiteMapPath. Add other controls in the master page needs to increase the common functions of system page is allowed.

In the master page, CSS cascading style sheets are used to achieve the reunification of the page style. the CSS cascading style sheets file set up by us named face.css, reference in the master page through <link> mark elements as follows:

```
<link href = "../css/face.css" rel = "stylesheet" type = "text/css">
```

Master page must be accessed by users with support by content page (ie, after binding). Content page is actually common .Aspx file. In college thesis management system, the common elements of the function code is written in two content pages, where the navigation bar, Session judge and other elements in the page header function code written in header.aspx file, and footer.aspx files are used to achieve function such as copyright notice, help tips and other elements in the bottom of the page. In the master page, ContentPlaceHolder controls of ASP. NET2.0 are called to identify common elements in the page display area, for example:

```
<!--#include file = "../teacmanasys/login/header.aspx"-->
<div>
<asp:contentplaceholder id = "ContentPlaceHolder1" runat = "server">
</asp:contentplaceholder>
</div>
<!--#include file = "../teacmanasys/login/footer.aspx"-->
```

In order to achieve the binding of master page and content page, we must define the master page and examples of commissioned first, the statement is public delegate void ElementSelectedChange Handler (); Then specified method that matches the delegate signature in the content page: Master.ElementSelectedChange = his.ElementSelectedChange.

The method to load the master page in each feature page is add the sentence asterPageFile = "~/ facuinfomana/default/ ThesisPage.master in the Page directive of the page code.

4 Classification Rights Management Based on RBAC

Many existing information management systems use a centralized access management, which all system maintenance and user permissions, etc. done by the system administrator to be responsible. However, in the actual information management work, often require the use of decentralized multi-level access control. For example, in the thesis management system, to meet the Department / Teaching And Research Section level management system requirements, business director can set the function operation range and data access permissions of the related Department Director, and the function operation range and data access permissions of other teachers access should set by the Department Director. Access control systems can effectively manage all information access requests and decide whether to allow users to access based on the system safe. Therefore, the study and design of the hierarchical rights management scheme can meet the management needs and solve the contradictions between hierarchical nature and the concentration of data management, became a key issues need to address.

In teachers integrated management system, based on role-based access control (RBAC) freedom customization of the role and operation rights are achieved. As required the appropriate role can be create for different jobs, and assign responsibilities based on user roles, user obtain the corresponding function and operating authority by assigned roles. An authorized user can have multiple roles, a role can form by multiple users; each role can perform different operations, each operation can also perform by different roles. Specifically, the allocations of user rights implement Teaching And Research Section and department two level management; centralized data storage, and data input / import implement the Department, Teaching And Research Section, teachers and students four level management, that all levels of legitimate users within the limits of their authority distributed data management system; the data submitted by next level of user should be reviewed by the superior users; access permissions of users of the system functions and data depends on their respective roles, to facilitate the implementation of role-based user security policies.

5 Editing and Publishing Information Online

Editing and publishing information online is a commonly used Web application component. Use university thesis management system as an example, online editing and publishing of related information of thesis are needed to achieve, so FCKeditor online text editor control is use to achieve the functionality. FCKeditor is compatible with most major browsers, including Internet Explorer, Mozilla Firefox, etc. In ASP. NET using FCKeditor control to achieve online text editor needs the following steps:

1) Download FCKeditor library archive from FCKeditor's official website (http://www.fckeditor. net) and copy FredCK.FCKeditorV2.dll to the bin directory of the current project;
2) Define FCKeditor setting item In ASP.NETWeb application configuration information Web.config:

```
<appSettings>
<add key = "FCKeditor:BasePath" value = "~/fckeditor/"/>
<add key = "FCKeditor:UserFilesPath" value = "/Files/"/>
</appSettings>
```

FCKeditor: BasePath is the default directory of the editor kernel, after set it can be used without specified BasePath by the FCKeditor instance attributes. In addition the control supports file upload, upload file type image, text, etc, FCKeditor: UserFilesPath is used to specify the directory where all uploaded files are.

3) Directive of add registered FCKeditor control in a Web Forms page used to publish information is: where TagPrefix attribute values determined the alias of only namespace in FCKeditor control; Namespace attribute defines a namespace associated; Assembly property defines the associated assembly, which is the dynamic link library files FredCK.FCKeditorV2.dll.

```
<%@Register TagPrefix = "FCKeditorV2" Namespace = "Fred-
CK.FCKeditorV2" Assembly = "FredCK.FCKeditorV2"%>,
```

4) the code for add the FCKeditor control to a Release information Web Forms page is <Fckeditorv2:fckeditor id = "FCKeditor1" runat = "server" DefaultLanguage= "zh-cn">

</Fckeditorv2:fckeditor>, Which Fckeditorv2 is the alias of the control Identified in the registration instructions; FCKeditor1 is designated as the control identifier, runat = "server" means that the control runs on the server side; DefaultLanguage= "zh-cn" indicates that the default language is Chinese.

5) In .cs file writing event code for FCKeditor control, define variable text_value. Users use FCKeditor control to edit the information to release and format processing them such as setting the text font, color, paragraph, etc. the information after editing is stored in the control object FCKeditor1.Value

properties in the form of a string, the stored information is the content and the HTML code to display the format of the page. In the page using the control achieve publishing information, click "Save" button, triggered event is assign the value of FCKeditor1.Value variable to text_value variable, and store the value of the variable in a database table.

6 Conclusion

The implementation method based on key technologies such as RBAC hierarchical rights management, Word format document production, editing and publishing online information describes in this article, has been in-depth application development in the university thesis management system, improved the development efficiency and practicality of system. These techniques have universal applicability in Web application development, and have reference value in the development of other Web application

References

1. Chen, X.: Design and Implementation of Journal Manuscript Management System Based on B/S. Journal of HangZhou Normal University(Natural Science) 5(1), 37–41 (2007)
2. Chen, W.: Solution of Manuscript Management System Based on Embedded Word Technology. Journal of NingBo Vocational and Technical College 12(5), 87–90 (2009)
3. Liao, W.F.: Thesis Management System Based on ASP.NET and XML Technology. Journal of HuNan Science and Technology College 29(8), 89–91 (2009)
4. Yu, W.F., Zhang, W.B., Liao, F.F.: Discussion of Three-tier Development Framework Application Based on ASP.NET. Software Guide 7(8) (2009)
5. Sun, B.X., He, Y.S., Wu, Z.X.: Implementation of Printing Industry's Online Quotation System Based on .NET Three-tier Architecture. Computer Knowledge and Technology 3(4), 599–600 (2009)
6. Xiong, Z.B., Chen, W.Y.: Design and Implementation of Universal Data Access Component Based on .NET. Software Guide 7(8), 95–97 (2009)

Evaluation Model of Business Continuity Management Environment on E-learning

Gang Chen

Shanghai University of Finance and Economic Shanghai P.R. China, 200433
School of Economic and Management, Tongji University Shanghai

Abstract. Business Continuity Management (BCM) becomes the most important assurance of E-learning. Environment of BCM is changing and affect implement of BCM. So we should frame evaluation model of Business Continuity Management Environment to improve quality of BCM. Opportunities and threats of business continuity management are derived from changes of environment. Organizations need to accurately identify characteristics and trends of environmental changes and adapt unpredictable changes of environment. Evaluation model of Business Continuity Management Environment has five parts: environmental observation, environmental exploration Scout, environmental analysis, environmental assessment and environmental prediction.

Keywords: Business Continuity Management, Evaluation Model, E-learning, Environment.

1 Introduction

The Environment of Business Continuity Management is a multi-level, multi-dimensional, interaction, causes and results of each system's collection. Each system is composed by its associated attributes, properties and multi-systems relationship. Attribute and relationship between systems attribute, and relationship between systems become composition of environment.

2 Dimensions of Business Continuity Management Environment

Environment of Business Continuity Management can be divided into three dimensions. The first dimension can be divided into two categories ------internal environment and external environment. The second dimension can be divided into controllable environment and uncontrollable environment. The third dimension can be divided into general environment and task environment, just as desired by figure 1.

D. Jin and S. Lin (Eds.): Advances in MSEC Vol. 1, AISC 128, pp. 323–327.
springerlink.com

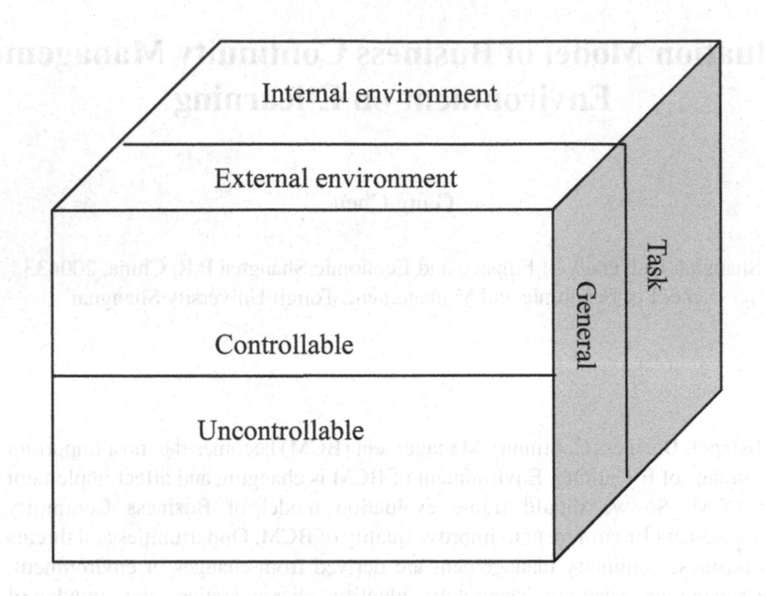

Fig. 1. Dimensions of Business Continuity Management Environment

2.1 Internal Environment and External Environment

Internal Environment of business continuity management refers to the organization's information architecture, application architecture and technical architecture. Information architecture includes Information flow, information and communication standards and other management standards, information sharing and information integration. Application architecture includes information systems which service E-learning such as teaching system, exam system, course resources system, students resign system and teacher service system. Technical architecture includes software, hardware, net work and database center.

External environment includes economic, political, cultural, geographical, weather, natural disasters, and other environments. They have a direct impact on organization's environmental factors.

2.2 Controllable Environment and Uncontrollable Environment

Environment can be divided into controllable and uncontrollable. Controllable environment should be affected by organization's ability, organization's resources and environmental characteristics. Uncontrollable environment affects business continuity services. It would be constantly adjust and adapted.

2.3 General Environment and Task Environment

General environment is universal significance of factors that might affect organization, but not very clear correlation between environmental factors, including economic factors, political and legal situation, social background, technical factors, and so on.

Task environment refers to a specific meaning and special significance of environment. It refers to achievement of organizational goals and has a direct impact on the environment.

3 Dynamic Factors of Business Continuity Management Environment

Dynamics factors of business continuity management environment can be judged from following aspects.

3.1 Objects of Environment Changes

Objects of environmental changes and numbers of environmental changing elements are complex features for judging dynamic environment. When environmental factors are changing more, organizations have to face more complex environment. According to numbers of elements and changing relationship between environmental elements factors we can determine degree of complexity of environment.

3.2 Rate of Environmental Changes

According to changing rate of environmental elements, we can find which environmental element is most dynamic. Then we can get rate of environmental changes.

3.3 Direction of Environmental Changes

Direction of environmental changing is main basis which we use to predict dynamic environment. Depending on result of predictable, direction of environmental can be divided into predictable and unpredictable. Predictability of direction of environmental changes depends on people's subjective cognitive ability and mastery of related information.

3.4 Frequency of Environmental Changes

As evolutionary algebra of evolutionary algorithms is determined by time, the best way to calculate changes in frequency is average of twice changes in evolution of algebra.

3.5 Complexity of Environmental Change

Intensity of environmental changes depends on distance of optimal solution between old environment and new environment. Some times optimal solution of new environment can be found from old optimal solution or just transform by simple mutation from old optimal solution.

3.6 Predictability of Environmental Changes

Environmental Changes are following certain trend or random. According to evolutionary algorithms we can forecast direction and complexity of next change and so on.

3.7 Cycle Length and Cycle Accuracy of Environmental Changes

Cycle length and cycle accuracy can be determined by optimal solution evolutionary algorithms. In addition, changes in environment are divided into fixed ----same changes occur for each cycle, static ----no change, cycle ----return to previous state, synchronous ----not necessarily at same time with different mobile search space compared, which can move entire search space---- and replacement changes ----optimal solution from a search space to another search space.

4 Dynamic Environment Parameters of Business Continuity Management

Dynamic environment parameters of business continuity management can be described with time and related to numbers of dynamic environmental systems.

If E is set as environment function, parameter is set as following. A1: internal environment, A2: external environment, B1: controllable environment, B2: uncontrollable environment, C1: general environment, C2: task environment.

R_{A_1} and R_{A_2} as the value of A-dimensional parameter, R_{B_1} and R_{B_2} are value of B-dimensional parameter, R_{C_1} and R_{C_2} are value of C-dimensional parameter. These values of parameters are obtained by using probability analysis---- take probability distribution in a cycle of environmental function. The environmental function is as following:

$$E = \begin{bmatrix} A_1, A_2 \\ B_1, B_2 \\ C_1, C_2 \end{bmatrix} \begin{bmatrix} R_{A_1}, R_{B_1}, R_{C_1} \\ R_{A_2}, R_{B_2}, R_{C_2} \end{bmatrix} \qquad \begin{aligned} & R_i = \left\{ \sum_{j=1}^{8} a_{ij} b_{ij} \right\}, \\ & i = A_1, A_2, B_1, B_2, C_1, C_2 \\ & j = 1, 2, 3, 4, 5, 6, 7, 8 \end{aligned}$$

Range of values of A, B and C dimension is {1, 2}. It represents environmental parameters as---- 1 means less weight, 2 means greater weight.

Ri indicates dimension dynamics of environmental parameters. The value of each Ri is a_{ij} which is one of factors determine extent of dynamic environment. a_{i1} is object of environmental change. a_{i2} is rate of environmental change. a_{i3} is direction of environmental change. a_{i4} is frequency of environmental change. a_{i5} is complexity of environmental change. a_{i6} is predictability of environmental change. a_{i7} is environmental change period length and a_{i8} is accuracy of environmental change cycle. b_{ij} is the weight of a_{ij}. Its' range is {1, 2}. 1 means less weight. 2 means greater weight.

5 Conclusion

Environmental observation, environmental patrol, and early detection of environmental assessment, environmental prediction information is basis for environmental analysis. Based on analysis in environment, external or internal auditors could conduct environmental assessment and predict changes in environment. Business continuity program responds to environmental changes.

Acknowledgment. This work is supported by Leading Academic Discipline Program 211 Project for Shanghai University of Finance and Economics (the 3rd phase).

References

1. Gang, C.: Risk Evaluation of Business Continuity Management by Using Green Technology. In: Zaman, M., Liang, Y., Siddiqui, S.M., Wang, T., Liu, V., Lu, C. (eds.) CETS 2010. CCIS, vol. 113, pp. 86–92. Springer, Heidelberg (2010)
2. Chen, G.: IT service system design and management of educational information system. In: 2009 1st International Conference on Information Science and Engineering, pp. 307–310 (2009)
3. Van, C.L., Morhaim, L., Dimaria, C.-H.: The discrete time version of the Romer model Economic Theory (2002)

Knowledge Discovering and Innovation Based on General Education Curriculum Reform around Classical Reading in Colleges

ShuQin Wu

Public Management College
Shandong Institute of Business & Technology
Yantai, P.R. China, 264005
wsq010@126.com

Abstract. What the urgent things of Chinese high education reform is to discover knowledge and innovate knowledge, instead of having knowledge solidification. Specific methods are reconstruct the curriculum system, which means that reading the original classic study of ancient and modern, accompanied with the transformation of the theory, principle including the spot mode of General history. Classic education is to cultivate people's aspirations for comprehensive development. It plays an important role in opening up the limits of professional education and general education, training education of college students' knowledge, and wisdom, and strengthening students' sense of classic, issue awareness, consciousness of responsibility and life consciousness.

Keywords: Knowledge discovering and innovation, General education, Classical reading, Higher education.

1 Introduction

What the urgent things of Chinese high education reform is to discover knowledge and innovate knowledge, instead of having knowledge solidification. Since the 1990s, China's higher education sectors, from various angles, have explored the mode reform of the talent training. Academician Yang Shu zi offered Culture quality education, which is thrust into the intersection of science and Humanities quality education. Prof. Gan Yang proposed general education, which centers on core curriculum and teaching assistant system. There is connection between "General education" and "cultural education", both of them reflect the aspiration for comprehensive development. Based on practice of many years in teaching reform of higher education and research on the outcome of the general education in universities at home and abroad, we considered curriculum system construction is the key to the reform of general education, and classical education is the core of the curriculum system construction.

D. Jin and S. Lin (Eds.): Advances in MSEC Vol. 1, AISC 128, pp. 329–335.
springerlink.com

2 Present Situation of General Education Curriculum in China

General education, as a practicality beyond the utilitarian and humanistic education, is relatively professional education and brought up, which stressed comprehensive development of individuals, comprehensive personality spirit rather than the tools' people. A student in the four years of study at the university can get not only the basic academic training, more important is to develop myself to be an "educated person ". As traditional discourse said, the purpose of university is to cultivate a person possessing not only "ability "and "cultured "but also the "road to ask "and "respecting for the moral ".

General education will eventually need to be implemented in practice in general education levels, more common practice is to learn to implement microcosmic level course from United States. Li Manli (1999) divided general education into four types: distribution of practice required, classic course type, core curriculum type and free elective type. Some famous universities in China go ahead in the general education curriculum system reform, such as Wuhan University, Peking University, Fudan University, Zhejiang University and other universities have made an attempt at the undergraduate curriculum of students. Peking University in math and science, social science, philosophy and psychology, history, language literature and art five areas open series of quality courses, as the general courses of quality education in school, meanwhile, centering on the professional course, it make the extension for the humanities and natural sciences related. For the implementation of the general education, in the part of the public elective courses, Wuhan University picks 51 more universal course as elective courses for guidance on general education, each two credits. In addition, there are about 200 any general education elective courses, each of them about two to three credits. As Wuhan university regulations said, every student should not get less than 12 credits in general courses, in the humanities and social science, math and science, and the Chinese civilization and foreign civilization, interdisciplinary field, each area at least take two credits, the students who get the degree in the humanities and social sciences should get at least four credits in math and science, the degree in math and science should complete at least four credits in the humanities and social sciences. Students take this professional repeat with or of close course, not included in the general credits. Interdisciplinary course to practice is accepted both credits.

General education reform in Chinese colleges and universities, mostly in the form of public elective courses allow students freedom of choice. Pervasive problem is that courses such as introduction to general theory are too much, original courses in classical East and west too little. General education course in College normally does not have any read request, even lists building reading material, there is no hard and fast requirement. General course is characterized by "culture" or "enlightenment", and is not in the pursuit of "knowledge" or "theory". It is a "lesson", not "taught", focusing on moist, cultivating, a gradual and probation. Just listening to the teacher in class to do some introduction to the knowledge, students can not exert active thinking, and lacking in problem consciousness, which is far from the targets that cultivate modern awareness and comprehensive personality of citizens.

3 Classical Education at the University of Chicago and Its Enlightenment

General education in foreign university contains three aspects the humanities, the social and natural. "After the history traditions" is the core and soul essence of the general education for the graduate in American, which is based on reading the classics of Western history as the main course. R. M. Hutchins, who is considered as the representative of the classical education, originally intended to reform the school education centering on the professional training.

American society and education is surrounded by pragmatism being popular by from the 1930 s and 1940 s, in which many people worship blindly material civilization, leading to a decrease in the level of social morality. In the area of education, most universities tend to be science, vocational and professional education, emphasizing the social experience the students receive, practice is more important than academic and the pursuit of knowledge becomes the vassal of practical education. At this point, Many scholars again review and affirm the value of liberal education, propose transformation of the classical liberal education, for example, opening comprehensive and integrity general education courses at the University, leading the first climax of the development of general education.

R. M. Hutchins, who entered the University of Chicago in 1929, when he was just thirty-years-old, witnessed the social conditions in the United States, deeply concerned, launched a comprehensive critique of the status and direction of American higher education at that time, criticized that American higher education has been completely astray, full of utilitarianism, pragmatism, professionalism, scientific doctrine, the technical doctrine, the vulgarization of market-oriented direction.. He thundered, if going on like this, universities will lose "the concept of University" or "University of the Road", meanwhile, only become a hodgepodge.

R. M. Hutchins stressed that, University of the Road firstly means the common spiritual culture foundation among different departments and professional, which require all students should accept a common education, what he has claimed is "general education".

On the basis of the eternal philosophy of education, Hutchins gets the general education theorization and systematization, and introduces the theory compatible with this classic curriculum. His greatest ambition and goal in practice is to build a new four-year undergraduate school within the University of Chicago, in which we implement general education centering on the classical read in four years.

But his ideas, at that time, become public criticism, the strongest criticism is first professors at the University of Chicago, especially the natural science and social science professors almost unanimously opposed it. This is because the University of Chicago has begun to represent the most new research University in United States since it was built in 1892. It is completely dominated by research institute; all faculties highly stress professional study, and emphasize the need to have students entering professional studies as soon as possible. Therefore, they think Hutchins ' general education program is to reverse back to the medieval School of modern University. As a result, the University of Chicago was in serious disagreement even abruption. The scheme was not passed until 1942, when University of Chicago established a most intensive general education system for undergraduate education in the modern research University, and since then it become famous for "Hutchins Institute". The core curriculum of general education in university of Chicago will not be contemporary pop scholars work, and what students study is already recognized as classics. For example, students in the University of Chicago can read all Plato, Shakespeare, Flaubert and Kafka and others work among any of the core courses of "Humanities". In the Harvard report of 1945, "Humanities" recommended compulsory reading materials: Homer and the Greece tragedy two; Platonic dialogues, the Bible, and Virgil, Dante; Shakespeare; Jay Hamilton, and Leo Tolstoy. In "Social science "school open the course called "Western thought and system ", reading material: Plato, and Aristotle, and Aquinas; Machiavelli, Luther, Badin, Montesquieu, Locke, Rousseau, Adam Smith, Bantam, and mill. Throughout the evolution of general education in the United States, like top general education courses in the United States', who's all basic reading materials, are around Western classics.

R. M. Hutchins shaped the modern general education with classical reading as the center, making university of Chicago become apotheosis as the general education model. At present, the University of Chicago is widely seen as a center of general education of University in American.

4 The Importance of Classical Education Courses

At present, many colleges and universities in general education in China has made many attempts to reform. As the prevalence of utilitarian of educational philosophy in the University, there is a clear tendency of knowledge-based general education in many schools which set general education curriculum, course content, form, and knowledge structure are unreasonable, the most prominent thing is the complete range of elective education courses, vast amount of courses, taking the core curriculum courses as the "interest " class which is outside professional course and selected at will, rather than the main courses and the basis of undergraduate academic training. This is needed to correct.

In ancient and modern time, there is communication between humanistic education model and the classical mode of education, classical literature should become the basis materials of general education courses for teaching. In order to achieve the goal of humanistic education, which is to cultivate the qualified, perfect "people ", What the core curriculum courses teach is the "metaphysical " rather than "physical " device, which means to teach the outstanding achievements of the civilization and the human spirit rather than specific operative technique. Therefore, the classics carrying the human spiritual naturally become materials of general education curriculum. Ancient and modern philosophy of life implied in the classic literature is the most fundamental teachings to achieve the purpose of educating people. This is why American universities list "Lao Zi ", "Confucius" as the materials of general education.

Since 2005, on the basis of general education patterns in the domestic and abroad, Shandong Institute of Business and Technology implements an education model combined with the introduction of Chinese and Foreign Culture Classics Tutorial Course (CFCCT) and Developing and Designing Oneself Course (DDO). Following the classical core status, we advocate the return to the mode of main body of student tutorials, fully mobilizing the initiative and the development and design of self-consciousness of the students, we have achieved certain results. Based on years of university education teaching reform of practice and on University pass general education results of research at home and abroad, we think, the key of general education reform is construction of courses system, and classic education is core of courses system construction, and also we argue, general education and professional education does not contradict, effective way of coordination is to rely on professional classic, within courses gradually penetrate general education concept.

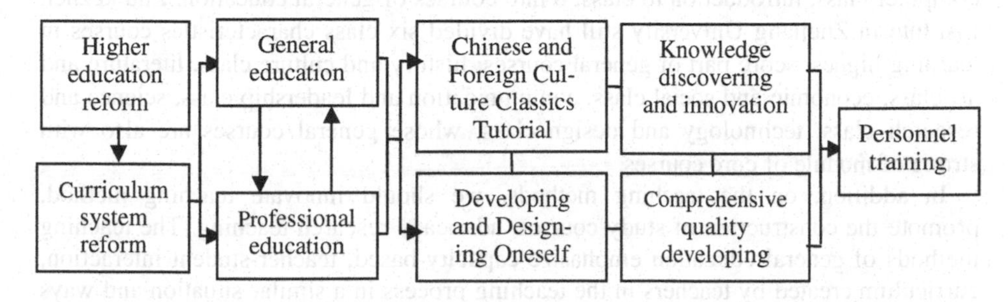

Fig. 1. Higher education reform models in Shandong Institute of Business and Technology

In courses teaching of general education, the core point is to increase the students' scientific reason and humanistic spirit. In traditional teaching, we have always stressed the "instructive of teaching ", that is in various courses teaching, we must consciously penetrate Humanities quality education. Before the class, teachers should recommend students one or two of the major classical works, layout tasks such as taking Notes, developing students' reading habits of professional classical in order to change the traditional way of teachers-introduction to knowledge. These attempts are of positive significance, contribute to dispelling tension between general education and professional education, which make people realize that they are not contraction, but can be symbiotic harmony. At the same time we should be aware that the effect of individual teachers alone, sporadic exploring is limited. It is not easy for general education to penetrate through the professional courses teaching, which has higher requirements for teachers' culture literacy, professional level, teaching ability. if in teaching practice everyone can really aware that, and make efforts to do it, the quality of teachers team must will obvious improved, which will led overall the benefit of general education to upgrading, helping us establish a opening teachers team of general education.

We should integrate the course system on the basis of the concept of general education. Core course is now the mainstream of curriculum reform of university general education in China; however, good implementation of a good core curriculum design of general education is very difficult. Construction of curriculum system of general education in China is basically built around a modular core curriculum, there is agreement on the core content of the course and the highlight on the value among the schools in china, which focus on the cultural heritage, humanism, critical awareness, exchange of Chinese and Western, the combination of Arts and Sciences, and six aspects of classical reading content. For example, "six large module" core courses in Fudan University, "four range " general courses in Hong Kong University (culture inheritance: the relationship of people and itself tradition; nature, and technology and environment: the relationship between people and substances environment; social and culture: the relationship between people and social culture; self and Humanities: relationship of people and itself and Humanities) ,which are quite obvious taking the way of general core courses system. Although putting general courses such as political class, army body class, foreign language class, computer class, introduction to class, o into courses of general education, Zhu kezhen Institute in Zhejiang University still have divided six class characteristics courses in learning highest score part of general courses: history and culture class, literature and art class, economic and social class, communication and leadership class, science and research class, technology and design class, whose general courses are also with strong of module of core courses.

In addition, on the teaching methods, we should innovate teaching method, promote the construction of study courses, advocacy research teaching. The teaching methods of general education emphasize capacity-based, teacher-student interaction, curriculum created by teachers in the teaching process in a similar situation and ways of scientific research, in which we make students through scientific research initiative on access to knowledge, application of knowledge to solve problem, so as to complete the related courses. Seminar teaching must have the following basic features: Firstly, it stresses cooperation between teachers and students. The teacher is not only the transfer of knowledge but also the facilitators of learning; Secondly, stressing that partnerships between students. It is necessary to develop students ' independent study of literacy, but also the ability to develop students' cooperation and exchange. What is more, the learning process was also stressed. Mr. Yang Gan raised the discussion about the way of teaching in small classes. After a discussion which is the course taught in small classes, students are divided into several small classes, less than 15 people, which is led by teaching assistants, to discuss the content of the text. The fundamental intention of class discussion is to help teachers' teaching and students' reading present more details in the text. Class discussion system originating from the United States of Colombia University, is not only the popular with in United States colleges and universities, and widely in Japan, and Korea and Hong Kong University in China. Discussion in small classes can cultivate habits of students for a discussion on the issue, learning how to discuss scientific issues. It is a research-oriented learning mode, by which we develop students' ability of independent learning and acquiring knowledge, maximize to inspire students' positive thinking, cultivate students ' innovative potential, improve the comprehensive quality of students and humanities..

Of course, implementation of other measures are needed to tie in with the construction of curriculum system of general education, management, and further expand on the discussion and practice, such as from the campus cultural construction, specifically highlighting the classic sense, effectively put general education into the model of talents training.

5 Conclusion

We should focus on building general education curriculum system around the classic education, which has great theoretical and practical significance for getting through boundaries between professional education and general education, enhancing students' humane strength and personal accomplishment, cultivating college students' knowledge, education and wisdom, strengthening the students' classic consciousness, problem consciousness and life consciousness, clarifying the essential connotation of general education, in order to solve the root problem that current educational mode does not adapt to the requirement of developing comprehensive quality of college students, which need more people to study.

References

1. Pang, H.: Predicament and hope of general education, p. 123. Beijing Institute of Technology Press, Beijing (2009) (Ch)
2. Li, N., Zhou, J.: The Comparison of the Construction of General Education Curriculum between Chinese and American University and its Revelation. Science & Technology Progress and Policy (14), 147–151 (2011) (Ch)
3. Yan, H.: An Analysis of Value Orientation of General Education under the Guidance of Holistic Education. Higher Education of Sciences (3), 4–7 (2011) (Ch)
4. Yuan, G., Zhou, Q.: Holistic Education and general education in University. Modern University Education (5), 6–10 (2008) (Ch)
5. Sun, H.: The core of general education: construction "classical reading" main course. Journal of Social Science of Hunan Medical University 3, 202–203 (2008) (Ch)
6. Zou, Y.-X.: Holistic Education, Great Education, Human Education——What kind of education is a successful education. Theory Research (14), 208–210 (2011) (Ch)
7. Forbes, S.H.: Holistic Education: An Analysis of Its Ideas and Nature, pp. 2–4. Foundation for Education Renewal, Brandon (2003)

Based on Embedded Image Processing Technology of Glass Inspection System*

Lian Pan[1,2,a], Xiaoming Liu[1,2,b], and Cheng Chen[1,2,c]

[1] Hubei Province Key Laboratory of Systems Science in Metallurgical Process (Wuhan University of Science and Technology), Wuhan 430081. China
[2] College of Information Science and Engineering, Wuhan University of Science and Technology, Wuhan 430081. China
[a] plwisco@163.com, [b] lilianliu1214@126.com, [c] 283408633@qq.com

Abstract. Some defects which are produced during the glass production will badly affect the quality of glass. It is necessary to detect the glass defects. The on-line glass defect inspection system is introduced in this article by using TMS320C6711-150 DSP and EPM1270GT144C5 CPLD as control parts and functional models as periphery to make up the hardware part. The software system is designed according to the nowadays-popular digital image processing technology. This system will be the important thing to improve the production level, intelligent grade of on-line glass defect inspection; increase the product quality, production efficiency and automation level; decrease the investment and cost of that.

Keywords: CCD, Embedded system, image processing technology, glass defect inspection.

1 Introduction

Some defects are produced during the actual glass production because of bubbles, stones, tin-point, inclusion, optical distortion[1] etc. Installing automated detecting equipments in glass production line is an important detection means to guarantee glass quality. It can be better to judge all kinds of problems existing in glass production process, to guide technicians to analyze and adjust them, and it can be more rapidly and accurately to classify and cut glass.

Aiming at float glass production process, this article introduces a kind of on-line glass defect inspection system by using high-speed line scanning CCD camera as system image sensor and embedded data processing module[2] based on CPLD and DSP as control part. It can discriminate various glass defects, marking and cutting optimally at the same time.

* Supported by Hubei Province Key Laboratory of Systems Science in Metallurgical Process(Wuhan University of Science and Technology).

D. Jin and S. Lin (Eds.): Advances in MSEC Vol. 1, AISC 128, pp. 337–342.
springerlink.com

2 System Composition and Working Principle

The whole glass defect inspection system preliminary designed consists of light source, optical system, high-speed CCD camera, CPLD acquisition module, DSP data processing module, defect mark circuit, glass mark models and optimal cutting models. As shown in figure 1.

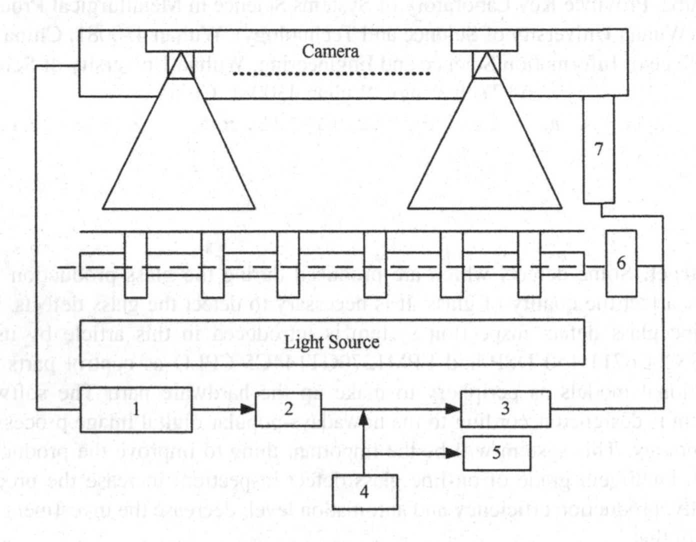

Fig. 1. Glass defect inspection system

1-Embedded data processing module, 2-Host computer, 3-Communication circuit
4-Keyboard, 5-Display, 6-Mark, 7-Cutting model

The inspection system captures images of glass and scan moving glass by high-speed CCD camera. Main functions of CPLD are to collect and obtain images, then to calculate grayscale for each pixel of images and judge the grayscale value of defects. Functions of DSP are to receive data of grayscale value of defects from CPLD, to complete identifying, analyzing and judging for defects data, according to data characteristic parameters of different defects, then to output processed results to host computer system. Basic glass image processing flow is shown in figure 2.

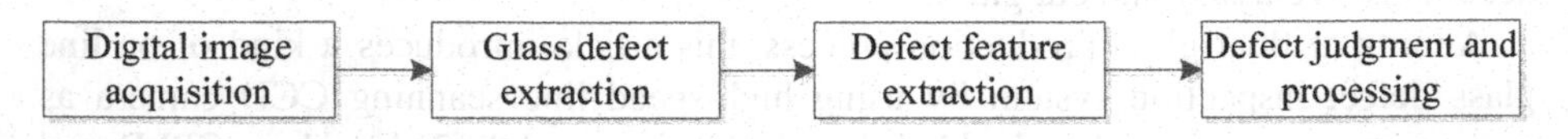

Fig. 2. Glass image processing flow

Digital Image Acquisition
We use the CCD (Charge Coupled Device) camera to capture images of detected glass and turn optical signal into electric signal., then converts it output to digital signals through embedded data processing module.

Glass Defect Extraction
Detecting object is glass defect. It is to extract the defects of glass images of appropriate methods and to prepare for the next process.

Defect Featurc Extraction
Area, diameter, location, the most value of grayscale, bubbles of stone image etc can form eigenvector[3] to be as the basic foundation for judging and processing of defects. And it can be used to judge and deal with various defects types.

Defect Judgment and Processing
Contrast the eigenvector of defecting glass with that of the known defects. Then according to the result, it can be judged about the defects' attribution category. This article mainly takes area, diameter, location, the most value of grayscale of bubbles and stone images as judging basis.

3 System Hardware Design

In order to achieve on-line processing requirements and realize mass data[4] processing, this article identifies the strategy of analyzing defect data only. In addition, it adopts two level processor consist of defect extraction and defect data analysis to achieve data processing of large amount of data structure.

3.1 CPLD Defect Extraction Processor

This system has specific requirements for high-speed real-time processing. With extracting and processing defect data, the output speed of front liner array CCD camera is up to 2x40 Mb/s. To judge the output value of the CCD camera, it needs to judge two output data within 25 ns, calculate relative coordinates and analyze whether they are defect pixels. If it is, storing its grayscale value and coordinates for further analysis. According to characteristics with high-speed calculation and mass data, we select the EPM1270 GT144C5 CPLD as defect extraction processor of the system.

3.2 DSP Data Analysis Processor

This system uses TMS320C6711-150 DSP chip of TI Company to read and analyze the output data of CPLD to obtain specific features of defects. That dual-core high-performance processor with CPLD+DSP makes costs be lower, volume be smaller. And it is with incomparable advantages than PC system in the ways of stability, energy consumption, installation, anti-jamming. The system overall design diagram is shown in figure 3.

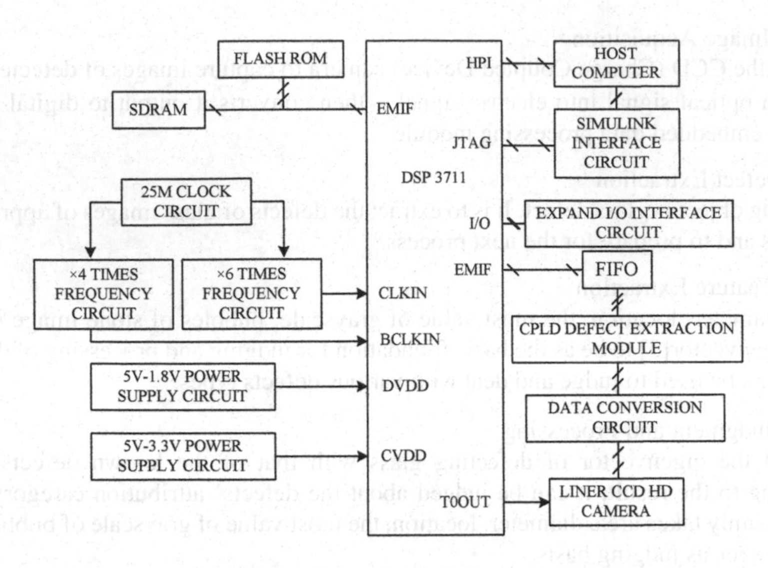

Fig. 3. System overall diagram

4 Image Processing Algorithm Design

When flat glass on the line gets through light source and CCD camera in average speed, its image signal is captured and converted into digital video signal, with work flowchart as shown in figure 4. The image preprocessing of glass defect detection is used to adjust the image contrast, highlight the important details and improve image visual quality. The methods mainly includes the contrast enhancement, edge detection[5] and smoothing image noise, etc.

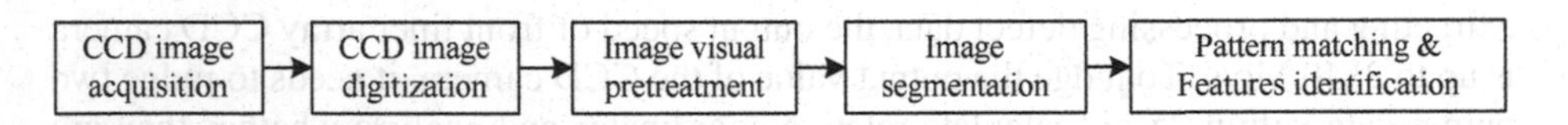

Fig. 4. Image processing work flowchart

We choose neighborhood average or median filter. As eliminating noise, it can still protect boundary information. It is a kind of nonlinear filter, which is effectively to remove interference of isolated point and line segment, especially for some irregular random noise. The image signal noises acquired by this system are in small range on the whole.

After detecting glass defects, it needs to make sure each defect size, location and shape to classify and record with defect recognition and threshold segmentation. Threshold segmentation is the technology and process to segment threshold into several specific regions. The algorithm has two main steps: first to identify the thresholds which need to be segmented; then to contrast segmented thresholds with pixel grayscale value to segment pixels of images. It is the key of segmentation to determine the threshold. This article chooses iteration threshold segmentation method aiming at managing specific application situation of thc scene. The threshold determined finally is the optimal threshold value. With the gray processing for true-color images collected by the system, grayscale of background and foreground changes smoothly, so we choose iterative method to determine the threshold.

5 Prototype Test

After hardware design, software design and algorithm optimization of the system, we test the main technical parameters of the function prototype, with specific data shown in table 1.

It can be found from the prototype test data, when increasing running speed of flat glass, it will increase detecting errors and the resolution that it is harder to detect tiny scratch of defects. When the running speed of glass is less than 2m/s, system functions meet the design requirements basically.

Table 1. Test for main technical parameters of the function prototype

Serial number	Testing project	Level operation speed of glass m/s(Test condition)		
		≤ 0.5m/s	0.5~2.0m/s	≥ 2.0m/s
I	System stability	Stable	Stable	Stable
II	Measurable flat glass width	3m	3m	2.5m
III	Horizontal longitudinal testing resolution	0.2mm×0.2mm	0.2mm×0.2mm	0.3mm×0.3mm
IV	Defect size detecting errors	Less than 1mm	Less than 1mm	1.5mm
V	Detecting defects types	Bubbles, stones, blob-shaped defects & small scratches	Bubbles, stones, blob-shaped defects & small scratches	Bubbles & stones

6 Conclusion

This system provides a set of advanced on-line glass defect automated detecting means, which combines high-speed CCD camera technology, CPLD+DSP dual-core two-level processor technology with digital image processing technology, to replace traditional artificial glass defect detecting means. It will be widely applied and extended in production fields of glass industry.

References

1. An, N., Lin, S., Liu, H., et al.: Study of image processing method and its application. Journal of Instrumentation S1, 309–310 (2006)
2. Wang, L., Lan, T., Wang, S.: Performance and practice of automated on-line glass defect detecting equipments. Journal of Glass (5), 46–48 (2003)
3. Cheng, H.: Design and implementation of automated glass defect inspection system. Xi'an Jiaotong University, Xi'an (1998)
4. Cheng, Z., Wu, H.: An improved median filtering algorithm. Journal of Huazhong University of Science 25(7), 40–42 (1997)
5. Qin, P., Ding, R.: Multilevel median filtering technology based on threshold value decomposition. Microcomputer and Application 22(12), 50–52 (2003)

Wavelet Weighted Multi-Modulus Blind Equalization Algorithm Based on Fractional Lower Order Statistics

Fang Xu[1], Yecai Guo[1,2,3], and Jun Guo[1]

[1] Nanjing University of Information Science and Technology, Nanjing 210044, Jiangsu, China
[2] Jiangsu Technology and Engineering Center of Meteorological Sensor Network, Nanjing 210044, Jiangsu, China
[3] Jiangsu Key Laboratory of Meteorological Observation and Information Processing, Nanjing 210044, Jiangsu, China
lisa.xf@163.com, guo-yecai@163.com, 659848231@qq.com

Abstract. When environmental noise keeps to fractional lower order α -stable distribution, the convergence performance of the traditional blind equalization algorithm is unstable. To overcome the deficiency, on the basis of combining weighted multi-modulus blind equalization algorithm (WMMA) with fractional lower order statistics (FLOS), a wavelet weighted multi-modulus blind equalization algorithm based on FLOS is proposed. This proposed algorithm uses FLOS to suppress the α -stable noise, utilizes WMMA to adjust the modulus in the cost function, and uses normalized orthogonal wavelet transform to reduce the autocorrelation of the input signals. The computer simulation indicates that the proposed algorithm is suitable for the equalization of higher order QAM signals. Compared with constant modulus algorithm based on FLOS and WMMA based on FLOS, the convergence performance of this proposed algorithm is superior.

Keywords: α -stable noise, Fraction lower order statistics, Blind equalization, Weighted multi-modulus.

1 Introduction

In the traditional blind equalization system, noise is assumed to be mainly Gaussian, but in some practical applications, the noise has a significant peak pulse characteristic, this kind of non-Gaussian noise has a long tail, such as water acoustic signal, this type of noise is usually described as α -stable distribution [1]. However, the equalization performance of constant modulus blind equalization algorithm (CMA) seriously descends. In view of the existence of fractional lower order statistics (FLOS) [2] for α -stable distribution noise, FLOS can be introduced into CMA. But this algorithm is not suitable for higher order Quadrature Amplitude Modulation (QAM) signals, in document [3], a weighted multi-modulus algorithm (WMMA) which is suitable for this kind of signals is proposed. This algorithm modifies modulus adaptively in the process of equalization, and it is good at convergence performance.

A wavelet weighted multi-modulus blind equalization algorithm based on FLOS is proposed in this paper. On the condition that the noise obeys α -stable distribution,

D. Jin and S. Lin (Eds.): Advances in MSEC Vol. 1, AISC 128, pp. 343–348.
springerlink.com

this proposed algorithm combines WMMA with wavelet transform theory, utilizes the characteristic of WMMA to make the output constellation clear and tight, and the autocorrelation of equalizer's input signals is reduced via wavelet transform [4], as a result, the convergence rate increases and mean square error reduces. The computer simulation results show that this proposed algorithm can suppress α -stable noise and has good convergence performance for higher order QAM signals.

2 α-Stable Noise

α -stable distribution doesn't have unified closed probability density function (PDF), it is usually represented in its fundamental function [5]

$$\varphi(u) = \exp\{ jau - \gamma |u|^{\alpha} [1 + j\beta \operatorname{sgn}(u) \omega(u,\alpha)]\} . \tag{1}$$

where, $\operatorname{sgn}(\cdot)$ is the sign function,

$$\omega(u,\alpha) = \begin{cases} \tan(\pi\alpha/2), & \alpha \neq 1 \\ (2/\pi)\lg|u|, & \alpha = 1 \end{cases} . \tag{2}$$

where, α is the characteristic index, γ is the dispersion coefficient, β is the symmetric parameter, a is the location parameter.

If the characteristic index satisfies the condition: $0 < \alpha < 2$, this distribution is called as fractional lower order α -stable distribution, and its higher order statistics and even second order statistics don't exist.

3 FLOS Based Constant Modulus Blind Equalization Algorithm

Considering that only statistics whose order is smaller than α are limited [6] in the fractional lower order α -stable noise, the fractional lower order statistics based constant modulus blind equalization algorithm (FLOSCMA) is proposed. The cost function of this algorithm is defined as

$$J = \mathrm{E}[|e(n)|^{p}](1 \le p < \alpha < 2) . \tag{3}$$

Error function ($e(n)$) is one form of the error functions of CMA and given by

$$e(n) = |z(n)| - \sqrt{R_{CM}} \, (R_{CM} = \mathrm{E}\{|\boldsymbol{a}(n)|^{4}\} / \mathrm{E}\{|\boldsymbol{a}(n)|^{2}\}) . \tag{4}$$

By the stochastic gradient method, the iterative formula of weight vector is obtained as follows

$$\boldsymbol{f}(n+1) = \boldsymbol{f}(n) - \mu |e(n)|^{(p-1)} \operatorname{sgn}(e(n)) z(n) \boldsymbol{y}^{*}(n) / |z(n)| . \tag{5}$$

where, $\operatorname{sgn}(\cdot)$ is the sign function, $()^{*}$ represents conjugation, $z(n)$ is the output signal of equalizer, $\boldsymbol{y}(n)$ is the input signal of equalizer, $\boldsymbol{f}(n)$ is weight vector of equalizer, $a(n)$ is the transmitted signal, μ is the iteration step size, n is the time sequence, the below is same.

4 FLOS Based Weighted Multi-Modulus Blind Equalization Algorithm

4.1 Weighted Multi-Modulus Blind Equalization Algorithm (WMMA)

The cost function of WMMA is defined as [3]

$$J_{WMMA} = \mathrm{E}[e_r^2(n) + e_i^2(n)] \ . \tag{6}$$

where,

$$e_r(n) = |z_r(n)| - \sqrt{|\hat{z}_r(n)|^{\lambda_r} R_r} \ , e_i(n) = |z_i(n)| - \sqrt{|\hat{z}_i(n)|^{\lambda_i} R_i} \ . \tag{7}$$

$$R_r = \mathrm{E}[a_r^4(n)] / \mathrm{E}[|a_r(n)|^{2+\lambda_r}] \ , R_i = \mathrm{E}[a_i^4(n)] / \mathrm{E}[|a_i(n)|^{2+\lambda_i}] \ . \tag{8}$$

where, $z_r(n)$ and $z_i(n)$ are respectively the real part and imaginary part of $z(n)$, $\hat{z}_r(n)$ and $\hat{z}_i(n)$ are respectively the real part and imaginary part of the decision value of $z(n)$, λ_r and λ_i are respectively the real part and imaginary part of the weighted factor, and $\lambda_r, \lambda_i \in [0,2]$, $a_r(n)$ and $a_i(n)$ are respectively the real part and imaginary part of $a(n)$.

The iterative formula of weight vector is given by

$$f(n+1) = f(n) - \mu(e_r(n) z_r(n) / |z_r(n)| + j e_i(n) z_i(n) / |z_i(n)|) y^*(n) \ . \tag{9}$$

Formula (9) indicates that WMMA modifies the modulus of signals dynamically in accordance with the output signals of equalizer. For the square constellations, $R_r = R_i$, and $\lambda_r = \lambda_i = \lambda$.

4.2 Fractional Lower Order Statistics Based WMMA

According to the advantages of WMMA, by introducing WMMA into FLOSCMA, fractional lower order statistics based WMMA (FLOSWMMA) is established, it can restrain the α-stable noise, and its cost function is written as

$$J_{FLOSWMMA} = \mathrm{E}[|e_r(n)|^p + |e_i(n)|^p] \quad (1 \le p < \alpha) \ . \tag{10}$$

The iterative formula of equalizer's weight vector is given by

$$\begin{aligned} f(n+1) = f(n) - \mu(&|e_r(n)|^{p-1} \mathrm{sgn}(e_r(n)) z_r(n) / |z_r(n)| \\ &+ j |e_i(n)|^{p-1} \mathrm{sgn}(e_i(n)) z_i(n) / |z_i(n)|) y^*(n) \ . \end{aligned} \tag{11}$$

FLOSCMA makes the output signals converge to a round in the statistical sense. Unlike FLOSCMA, in FLOSWMMA, the real part and the imaginary part of signal modulus are no longer constants, they are decided by the decision values of equalizer's output signals, and this method makes the output signals converge to multiple rectangles. So, FLOSWMMA can eliminate the phase ambiguity and improve the convergence performance in the environment of non-Gaussian noise.

5 Wavelet Transform-FLOSWMMA

Because orthogonal wavelet transform [4] can improve the convergence performance, when it is introduced to FLOSWMMA, we get Wavelet Transform-FLOSWMMA (WT-FLOSWMMA), its principle diagram is shown in figure 1.

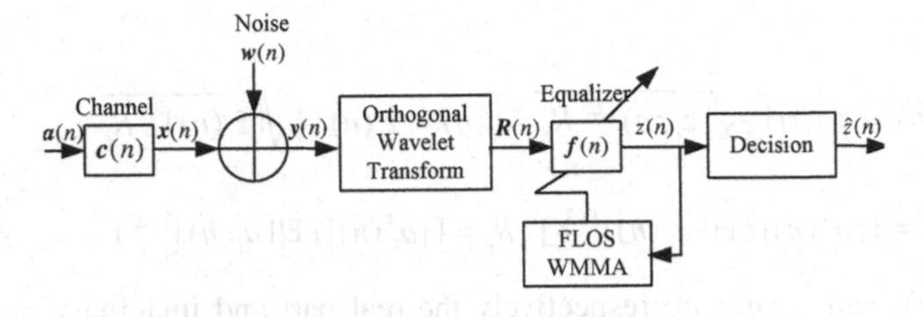

Fig. 1. The Principle Diagram of WT-FLOSWMMA

The equalizer's input signals which pass through orthogonal wavelet converter are turned into

$$\boldsymbol{R}(n)=\boldsymbol{Q}\boldsymbol{y}(n) \quad . \tag{12}$$

where, $\boldsymbol{Q}$ is orthogonal transform matrix, then the equalizer's output is written as

$$z(n)=\boldsymbol{f}^T(n)\boldsymbol{R}(n) \quad . \tag{13}$$

The iterative formula of weight vector is given by

$$\begin{aligned}\boldsymbol{f}(n+1)=\boldsymbol{f}(n)-\mu\hat{\boldsymbol{R}}^{-1}(n)(|e_r(n)|^{p-1}\operatorname{sgn}(e_r(n))z_r(n)/|z_r(n)|\\ +j|e_i(n)|^{p-1}\operatorname{sgn}(e_i(n))z_i(n)/|z_i(n)|)\boldsymbol{R}^*(n) \quad .\end{aligned} \tag{14}$$

where, $\hat{R}^{-1}(n)=\mathrm{diag}[\sigma_{j,0}^2(n),\sigma_{j,1}^2(n),\cdots,\sigma_{J,k_J}^2(n),\sigma_{J+1,0}^2(n),\cdots,\sigma_{J+1,k_J}^2(n)]$, $\mathrm{diag}[\cdot]$ represents a diagonal matrix, j is the scale, k is the translation, J is the maximum of scale, k_J is the maximal translation of wavelet function when its scale is J, $\sigma_{j,k}^2(n)$ and $\sigma_{J+1,k}^2(n)$ are respectively estimations of $r_{j,k}(n)$ and $s_{J,k}(n)$'s average power, they are deduced by the following formula

$$\begin{cases}\sigma_{j,k}^2(n+1)=\beta'\sigma_{j,k}^2(n)+(1-\beta')|r_{j,k}(n)|^2\\ \sigma_{J+1,k}^2(n+1)=\beta'\sigma_{J+1,k}^2(n)+(1-\beta')|s_{J,k}(n)|^2\end{cases} . \tag{15}$$

where, $r_{j,k}(n)$ is the wavelet transform coefficient when scale parameter is j, translation parameter is k, and time is n; $s_{J,k}(n)$ is the scale transform coefficient when scale parameter is J ,translation parameter is k, and time is n; β' is the smoothing factor, and $0<\beta'<1$. Document [4] indicates that orthogonal wavelet transform makes the signals' correlative matrix close to a diagonal line and the energy mainly concentrates around the diagonal line, that is to say, the correlation of signals are declined. Therefore, the convergence rate of WT-FLOSWMMA is increased.

Moreover, considering that α -stable noise has peak pulse, we use the method of soft limiting proposed in document [7] to weed out the large abnormal value of the equalizer's input signals.

6 Computer Simulations

The channel is $\boldsymbol{c}=[0.9656,-0.0906,0.0578,0.2368]$, the transmitted signals adopt 256 QAM. The environment noise is α -stable noise, the signal to noise ratio (SNR) is 30 dB, the parameters setting of the noise are as follows: $\alpha=1.7$, $\beta=a=0$, γ is decided by SNR, and $\gamma=\sigma^2/10^{SNR/10}$ (σ^2 is the variance of input sequence). The length of the equalizer is 16, and the weight coefficient is initially set to 0 except for the 8th tap which is set to 1. For FLOSCMA, the step size is 0.00001, for FLOSWMMA, the step size is 0.00002, the weight factor is 1.7, for WT-FLOSWMMA, the step size is 0.009, the weight factor is 1.8. 4000 times' Monte Carlo simulation result is shown in figure 2.

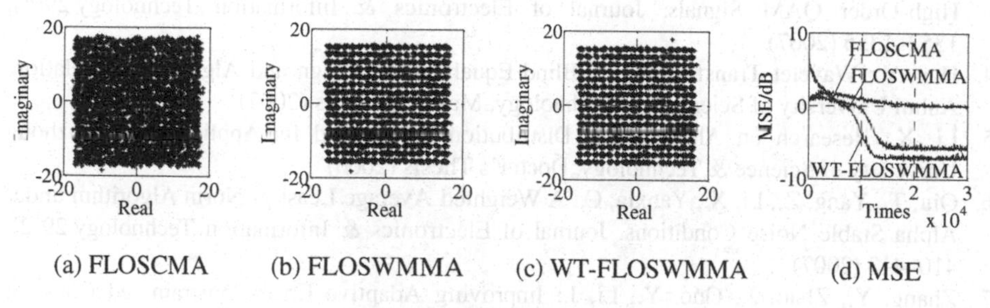

(a) FLOSCMA (b) FLOSWMMA (c) WT-FLOSWMMA (d) MSE

Fig. 2. Simulation Result

From Fig. 2(d), it can be concluded that compared with FLOSWMMA and FLOSCMA, WT-FLOSWMMA has a drop of about 1dB and 7 dB for convergence error and has an improvement of 2000 steps and 6000 steps for convergence rate, the output constellation of WT-FLOSWMMA is the most clear.

7 Conclusion

CMA is a relatively mature blind equalization algorithm, but it only uses the amplitude message of equalizer's output signals, has phase ambiguity, and while it is used to equalize higher order QAM signals, its convergence performance drops. For the environment of α-stable noise, WT-FLOSWMMA is proposed in this paper. This proposed algorithm utilizes FLOS to restrain α-stable noise, uses the prior knowledge to adaptively modify the modulus of signals in the process of iteration, and reduces the correlation of input signals by making the input signals pass through orthogonal wavelet converter, as a result, the equalization performance is improved. The simulation results validate that in equalizing higher order QAM, WT-FLOSWMMA has lower convergence error and faster convergence rate.

Acknowledgments. This paper is supported by Specialized Fund for the Author of National Excellent Doctoral Dissertation of China (200753), Natural Science Foundation of Higher Education Institution of Jiangsu Province (08KJB510010) and "the peak of six major talent" cultivate projects of Jiangsu Province (2008 026), Natural Science Foundation of Jiangsu Province (BK 2009410), Natural Science Foundation of Higher Education Institution of Anhui Province (KJ2010A096), Jiangsu Preponderant Discipline "Sensing Networks and Modern Meteorological Equipment".

References

1. Li, C., Yu, G.: A New Statistical Model for Rolling Element Bearing Fault Signals Based on Alpha-Stable Distribution. In: Second International Conference on Computer Modeling and Simulation. ICCMS 2010, vol. 4, pp. 386–390. IEEE (2010)
2. Zha, D., Qiu, T.: Adaptive Mixed-norm Filtering Algorithm based on Noise Model. Digital Signal Processing 17(2), 475–484 (2007)
3. Xu, X., Dai, X., Xu, P.: Weighted Multimodulus Blind Equalization Algorithm for High-Order QAM Signals. Journal of Electronics & Information Technology 29(6), 1352–1355 (2007)
4. Han, Y.: Wavelet Transform based Blind Equalization Design and Algorithm Simulation. Anhui University of Science and Technology, Master's Thesis (2007)
5. Li, X.: Research on Alpha Stable Distribution Model and Its Applications. Huazhong University of Science & Technology, Doctor's Thesis (2006)
6. Qiu, T., Yang, Z., Li, X., Yanxia, C.: A Weighted Average Least p- Norm Algorithm under Alpha Stable Noise Conditions. Journal of Electronics & Information Technology 29(2), 410–413 (2007)
7. Zhang, Y., Zhao, J., Guo, Y., Li, J.: Improving Adaptive Error-Constrain- ed Constant Modulus Algorithm for Blind Equalization to Make it Suitable in -Stable Noise. Journal of Northwestern Polytechnical University 28(2), 203–206 (2010)

Orthogonal Wavelet Transform Blind Equalization Algorithm Based on Chaos Optimization

Yecai Guo[a], Wencai Xu[b], Fang Xu[c], and Jun Guo

College of Electronic and Information Engineering,
Nanjing University of Information Science & Technology,
Nanjing, 210044, China
[a] guo-yecai@163.com, [b] xuwencai0556@126.com, [c] lisa.xf@163.com

Abstract. For greatly improving the convergence rate of constant modulus blind equalization algorithm and avoiding local minima, orthogonal wavelet transform blind equalization algorithm based on chaos optimization is proposed, on the basis of analyzing chaos optimization algorithm and orthogonal wavelet transform based blind equalization algorithm. The proposed algorithm utilizes a short initial data segment to optimize equalizer weight vectors, can make the weight vectors approach the global minimum through a hybrid algorithm fusing chaos optimization algorithm and steepest descent method. The equalizer's input signals are pretreated by using orthogonal wavelet transform to reduce the auto-correlation and improve the convergence speed. The simulation results in the underwater acoustic channel show that the proposed algorithm has faster convergence speed and higher accuracy and smaller residual mean square error.

Keywords: Blind equalization, Orthogonal wavelet transform, Chaos optimization, Global minimum.

1 Introduction

In modern underwater acoustic communication, Inter-Symbol Interference (ISI) caused by multi-path transmission channels can affect the communication quality. It is necessary to employ blind equalization technique without training sequences for greatly reducing the influence of the ISI on communication quality. In all sorts of equalization algorithms, Constant Modulus blind equalization Algorithm (CMA) with simple structure, small computational loads, and good stability, is widely applied in a variety of digital transmission system [1]. However, CMA has the disadvantages of slow convergence speed, larger mean square error (MSE), and different tap locations of the initial weight vectors can make the algorithm converge to different local minimum points [2].As shown in [3], wavelet transformation (WT) may be used to transform the input signals of blind equalizer to reduce the autocorrelation of the input signals, whereas wavelet transform constant modulus blind equalization algorithm (WT-CMA) can improve the convergence rate. However, the updating equation of its weight vectors is obtained via the stochastic gradient descent algorithm, so the WT-CMA easily falls into local convergence in the searching process. As shown in [4], chaotic motion with the feature of ergodicity, randomness , and regularity, can traverse every

D. Jin and S. Lin (Eds.): Advances in MSEC Vol. 1, AISC 128, pp. 349–354.
springerlink.com

state without repetition according to inherent laws within a certain range. Chaos optimization algorithm based on chaotic motion rule can traverse every state via chaotic variables and make chaotic variables jump out of the local minima [5][6].

In this paper, in order to greatly overcome the defects of CMA and make full use of advantages of wavelet transformation and chaos optimization algorithm, Chaos optimization based orthogonal Wavelet Transform Constant Modulus blind equalization Algorithm(CWTCMA) is proposed. In this proposed algorithm, normalized orthogonal wavelet transformation [3] is introduced into CMA to reduce the auto-correlation of equalizer input signals and improve the convergence rate, chaos optimization algorithm used to make the weight vectors of CMA escape from local minima to approach the global minima.

2 Hybrid Optimization of Weight Vectors

We introduce chaos optimization algorithm into CMA to avoid the local convergence, the weight vectors of CMA are regarded as the optimization variables. The modified Logistic map may be employed for the iteration formula of equalizer's weight vectors.

In mathematical sense, the modified Logistic map is written as

$$x(n+1) = 1 - 2x^2(n) \ . \tag{1}$$

where, $x(n)$ represents chaotic variables within (-1,1) and n donates the time series.

We combine chaos optimization algorithm with steepest descent method to obtain hybrid algorithm with the characteristics of global optimization and fast convergence, so it can be used to optimize weight vectors of equalizer. Firstly, the global optimal point is searched via using chaos optimization algorithm to make the weight vectors close to global optimal point; second, the steepest descent method is used to search local minima in the neighborhood of the global minimum point. Finally, the hybrid optimization algorithm, which is helpful for making weight vectors escape from local minima, is employed to make weight vectors approach global minimum. The procedures of optimizing weight vectors and calculating the minimum of the cost function by using the hybrid algorithm are described as follows:

Step1: Define M_1 as the iteration times of steepest descent method, M_2 as the times of chaos optimization iteration, and M_3 as the times of mixed searching algorithm. Initialize the counter $i = 0$ and assume that the weight vector $f = f_0$.

Step2: Take the f as initial points to make optimization iteration for M_1 times by using steepest descent method and to get the equalizer's optimization weight vector f^{*1} and the value of cost function J^{*1}.

Step3: f^{*1} is mapped into the scope of chaos variables by equation (2).

$$x_i = (f_i^{*1} - c_i) / d_i \ . \tag{2}$$

where x_i denotes the initial value of the i th weight vector of chaos optimization algorithm, c_i, d_i are assumed as constants. The initial f^{*1} is iterated via chaos optimization algorithm for M_2 times to get the equalizer's optimization weight vector f^{*2}. In this case, the value of cost function J^{*2} is obtained.

Step4: Do $i = i + 1$. If $i > M_3$, stop the searching process, else turn to Step5.

Step5: Compare J^{*1} with J^{*2}, if $J^{*2} \geq J^{*1}$, let $f = f^{*1}$; else let $f = f^{*2}$, then return to Step2.

3 CWTCMA

On the basis of integrating CMA with chaos optimization algorithm and wavelet transform, Chaos optimization orthogonal Wavelet Transform Constant modulus blind equalization Algorithm(CWTCMA) is proposed. The principle diagram of the proposed algorithm is shown in Fig .1.

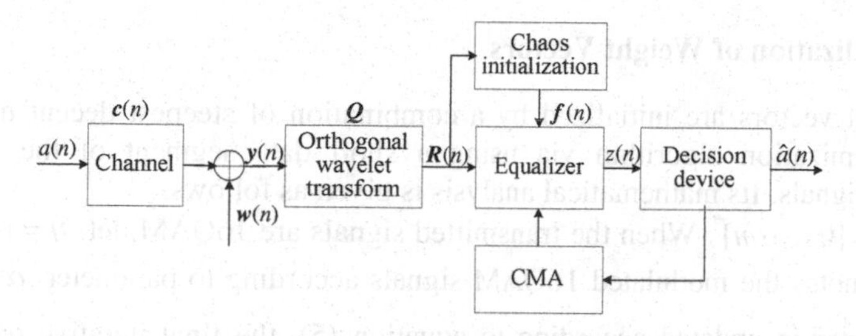

Fig. 1. Principle diagram of CWTCMA

where $a(n)$ denotes input signal sequences, $\boldsymbol{c}(n)$ is baseband channel's response vector, $\boldsymbol{w}(n)$ is noise vector, $\boldsymbol{Q}$ is an orthogonal wavelet transform matrix, $\boldsymbol{y}(n)$ is channel's output vector, $\boldsymbol{R}(n)$ is equalizer's input signals, $f(n)$ is weight vector of equalizer, $z(n)$ is equalizer's output signal, $\hat{a}(n)$ is the output of decision device.

3.1 Blind Equalization Algorithm Based on Orthogonal Wavelet Transform

As shown in figure 1, the input and output of the equalizer can be written as

$$\boldsymbol{R}(n) = \boldsymbol{Q}\boldsymbol{y}(n) \quad , \quad z(n) = \boldsymbol{f}^H(n)\boldsymbol{R}(n). \tag{3}$$

The cost function $J(n)$ is defined as

$$J(n) = E[e^2(n)]. \tag{4}$$

where, $e(n) = |z(n)| - \sqrt{R_{CM}}$ $(R_{CM} = \mathrm{E}\{|\boldsymbol{a}(n)|^4\} / \mathrm{E}\{|\boldsymbol{a}(n)|^2\})$.

According to the steepest decent method, the equalizer's weight vectors are updated as follows

$$\boldsymbol{f}(n+1)=\boldsymbol{f}(n)-\mu\hat{\boldsymbol{R}}^{-1}(n)z(n)\cdot[|z(n)|^2-R_{cma}]\boldsymbol{R}^*(n). \tag{5}$$

where, $R_{cma}=E\{|a(n)|^4\}/E\{|a(n)|^2\}$, μ is step-size factor, $\hat{\boldsymbol{R}}^{-1}(n)$ the orthogonal wavelet power normalization matrix and $\hat{\boldsymbol{R}}^{-1}(n)=\mathrm{diag}[\sigma_{j,0}^2(n),\sigma_{j,1}^2(n),\cdots\sigma_{j,k_j}^2(n),$ $\sigma_{J+1,0}^2(n),\cdots\sigma_{J+1,k_J}^2(n)]$. $\sigma_{j,k_j}^2(n)$ and $\sigma_{J+1,k_j}^2(n)$ represent the average power estimates of wavelet coefficient $r_{j,k}(n)$ and scale transform coefficient $x_{J,k}(n)$, respectively. They can be updated by following recursion equation.

$$\begin{cases}\sigma_{j,k}^2(n+1)=\beta\sigma_{j,k}^2(n)+(1-\beta)|r_{j,k}(n)|^2\\ \sigma_{J+1,k}^2(n+1)=\beta\sigma_{J,k}^2(n)+(1-\beta)|x_{J,k}(n)|^2\end{cases}. \tag{6}$$

where, $\beta\in(0,1)$ denotes smooth factor and its value is slightly less than 1. According to equations (3) ~ (6), the Wavelet Transform based CMA (WT-CMA) is obtained.

3.2 Initialization of Weight Vectors

The weight vectors are initialized by a combination of steepest decent method and chaos optimization algorithm via using a short data segment of the equalizer's receiving signals. Its mathematical analysis is given as follows.

Set $\alpha=[0,1,\cdots m]$. When the transmitted signals are 16QAM, let $\eta=16\mathrm{QAM}(\alpha)$ and η denotes the modulated 16QAM signals according to parameter α. After the weight vector is updated according to equation (5), the final iterative result $\boldsymbol{f}^{*1}$ is obtained. On the basis of obtaining final iterative result, we have

$$\boldsymbol{R}(n)=\boldsymbol{Q}*\boldsymbol{y}(n+L-1:-1:n),\quad z(n)=\boldsymbol{f}^H(n)\boldsymbol{R}(n). \tag{7}$$

where, L is the length of equalizer's weight vector and $n\in[1,N]$.

Modulation error $e(n)$ can be defined as

$$e(n)=\min(|z(n)-\eta|^2)\ . \tag{8}$$

$z(n)$ will change along with chaos iterations. In order to find the optimal chaos variable, we define the average modulation error (AME) as

$$\mathrm{AME}(k)=\frac{1}{N}\sum_{n=1}^{N}[\min(|z_k(n)-\eta|^2)]\ . \tag{9}$$

where $z_k(n)$ is the k th output signal of equalizer in the process of chaos optimization, N donates the number of $z(n)$. The real and imaginary part of $\boldsymbol{f}(n)$ are iterated for M_2 times , The minimum $\mathrm{AME}(k)$ is given to J^{*2}, and the corresponding weight vector is f^{*2}. The switching condition from CWTCMA to WT-CMA is given as

$$J^{*2}(i-1)-J^{*2}(i)<\zeta\ . \tag{10}$$

where ζ is a positive value, $J^{*2}(i)$ donates the value optimizing for M_2 times, i donates the i th mixed searching and $i=1,2\cdots M_3$. In optimizing process, if equation (10) is satisfied, the CWTCMA will switch to WT-CMA, otherwise, after the mixed searching for M_3 times, the CWTCMA will switch to WT-CMA.

4 Simulation Tests

In order to test the validity of the proposed CWTCMA algorithm, the simulation tests were be carried out and compared CWTCMA with WT-CMA and CMA. In our simulations, the channel $\mathbf{c}=[0.3132,-0.414,0.8908,0.3134]$, the signal-to-noise ratio (SNR) was set to 25dB, the length L of the equalizer was selected as 16, N was set to 20, For all algorithms, the fourth tap of weight vector f was set to 1 and the rests were set to 0. c_i was set to 0, d_i was chosen as 1,and $i=1,\cdots,L$. We utilized about 500 points within a data segment for initialization of weight vectors. ζ was selected as 10^{-5}. For 16QAM, the step-size μ in CWTCMA, WT-CMA , and CMA were chosen as 0.00001, 0.0002, and 0.0001,respectively. M_1, M_2, and M_3 were chosen as 500, 800, and 20, respectively. For 16PSK, the step-size μ in CWTCMA, WT-CMA, and CMA were chosen as 0.001, 0.002, and 0.001, respectively. M_1, M_2, and M_3 were chosen as 300,800,and 20. The simulation results were shown in Fig.2.

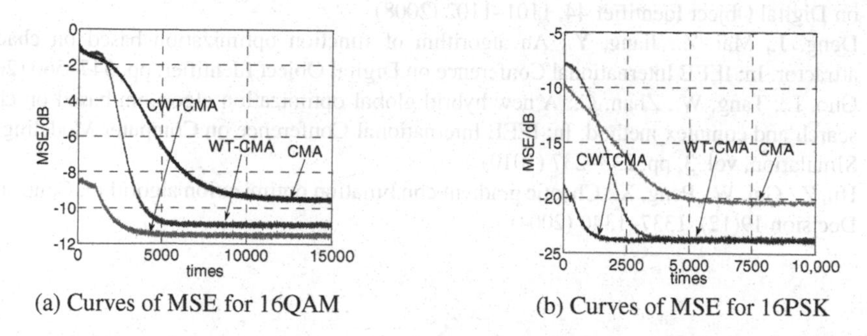

(a) Curves of MSE for 16QAM (b) Curves of MSE for 16PSK

Fig. 2. Simulation results

Fig.2(a) shows that the CWTCMA has a drop of about 2dB and 0.5dB for mean square error(MSE) comparison with WT-CMA and CMA. The CWTCMA has an improvement of about 1000 steps and 5000 steps for convergence rate comparison with WT-CMA and CMA. Fig.2(b) shows that the MSE of CWTCMA has a drop of about

3.5dB comparison with that of CMA and that its MSE is almost equal to that of WT-CMA. The CWTCMA has an improvement of about 2000 steps and 4000 steps for convergence rate comparison with WT-CMA and CMA.

5 Conclusions

The proposed CWTCMA integrating the chaos optimization algorithm with steepest descent method can make the weight vectors escape from local minimum and obtain the small mean square error and the fast convergence rate by wavelet transform and chaos optimization algorithm. Theoretical analyses and simulation results show the proposed CWTCMA algorithm can greatly improve the convergence rate and reduce the steady mean square error.

Acknowledgments. This paper is supported by Specialized Fund for the Author of National Excellent Doctoral Dissertation of China (200753), Natural Science Foundation of Higher Education Institution of Jiangsu Province (08KJB510010) and "the peak of six major talent" cultivate projects of Jiangsu Province (2008 026), Natural Science Foundation of Jiangsu Province (BK 2009410), Natural Science Foundation of Higher Education Institution of Anhui Province (KJ2010A096), Jiangsu Preponderant Discipline “Sensing Networks and Modern Meteorological Equipment”.

References

1. Abrar, S., Nandi, A.K.: An adaptive constant modulus blind equalization algorithm and its stochastic stability analysis. IEEE Transaction on Digital Object Identifier 17, 55–58 (2010)
2. Li, J., Zhao, J., Lu, J.: Simulation of constant modulus blind equalization algorithm initialized by support vector machines. Computer Simulation 25(1), 84–87 (2008)
3. Linfoot, S.L.: Wavelet families for orthogonal wavelet division multiplex. IEEE Transaction on Digital Object Identifier 44, 1101–1102 (2008)
4. Deng, J., Mai, Z., Jiang, Y.: An algorithm of function optimization based on chaostic attractor. In: IEEE International Conference on Digital Object Identifier, pp. 547–560 (2007)
5. Guo, L., Tang, W., Zhan, C.: A new hybrid global optimization algorithm based on chaos search and complex method. In: IEEE International Conference on Computer Modeling and Simulation, vol. 3, pp. 233–237 (2010)
6. Hu, Z., Gui, W., Peng, X.: Chaotic gradient combination optimization algorithm. Control and Decision 19(12), 1337–1340 (2004)

Research on Soil Heavy Metal Pollution Assessment System Based on WebGIS

He Shen[1], TingYan Xing[1], ShengXuan Zhou[2], YongHong Mi[1], and Xian Feng[1]

[1] School of Information Engineering, China University of Geosciences of Beijing, 100083 Beijing, China

[2] Geographic Information System Room, Surveying and Mapping Engineering Institute of Gansu, 730050 Lanzhou, Gansu, China

Abstract. In order to effectively evaluate heavy metal pollution in soil for the study area, based on the soil environment database and the map server of ArcGIS Server, the soil heavy metal pollution assessment system based on WebGIS was developed with ASP.NET(C#), JavaScript and HTML, System application of the sample soil pollution assessment model to evaluate the monitoring points and the index value derived from analysis of spatial interpolation. Finally, the results generate thematic maps based on the soil background values of Study area for visual display. The system achieved the soil quality of the environment online sharing, evaluation and consulting.

Keywords: WebGIS, Soil heavy metal element, Pollution assessment, Geostatistics.

1 Introduction

Since reform and opening, with the rapid development of China's industry and agriculture, soil heavy metal pollution has aroused more and more public concern. Heavy metal pollutants in the soil have features such as poor mobility, long residence time, difficult to be microbiologically degraded, highly toxic and accumulating effects, which have greater impacts on growth, yield and quality of crops, and impact on human health through the food chain. Therefore, soil heavy metal pollution problem has become an important content of today's environmental science study. As soil heavy metal pollution is influenced by natural conditions and human factors and has strong spatial correlation and dependence, we can use GIS technology to build soil environmental spatial database, study the spatial distribution characteristics of soil heavy metal using spatial analysis method provided by GIS, and conduct pollution assessment on them using a variety of pollution assessment methods.

2 WebGIS Technology

WebGIS is a computer information system that compacts, stores, processes, analyzes displays and applies geographic information under Internet environment, with the advantages of centralized management and distributed applications [1]. Geographic

D. Jin and S. Lin (Eds.): Advances in MSEC Vol. 1, AISC 128, pp. 355–360.
springerlink.com

information refers to the information that describes the spatial location and spatial relations of the earth's surface. By publishing and sharing spatial data on the Web via the Internet [2], users can browse the spatial data, produce special subject maps, and conduct various spatial retrieval and spatial analysis in the website from different places [3].

3 Overall System Design

3.1 Overall System Architecture

The system uses a common B/S network computing model, and is divided into 3 layers, namely data layer, application service layer and presentation layer. Its overall structure is as shown in Figure 1.

3.1.1 Data Layer

Data layer implements data logic, all data in the system is stored in this layer, so the data is separated from application logic, convenient for data maintenance. Specifically the implementation of database server that stores spatial data, attribute data, as well as knowledge data, and model data of the system.

3.1.2 Business Layer

Business layer implements application logic, which functions as a middle layer in the system. The server responds to the HTTP requests coming from the browser, application server communicates with ArcServer's application components via custom protocol, and with the data layer through the common interface, and loads data, invokes model, sets parameters, and conducts assessment.

3.1.3 Presentation Layer

Presentation layer implements client interface function, and is responsible for visual display of data and user interaction, specifically the implementation of client browser, through the browser in the user interface, users with different identities can send requests to the Web server, do personalized customization, explain the data returned by the Web server, and display the personalized service, thus implementing various system functions.

3.2 System Development and Runtime Environment

Software that supports system running includes WindowsXP operating system, SQLServer2005 database, map server, model component library, and other charts and report components. The entire system uses ASP.NET (C#) Web server-side for the development of language as well as DHTML, JavaScript client-side dynamic web technology

WebGIS platform uses ESRI's ArcGIS Server. ArcGIS Server is a server-based, powerful GIS product used for building enterprise-level GIS application service system that is centrally managed, supports multiple users, and has advanced GIS functionality, which provides spatial data management, 2D and 3D map visualization,

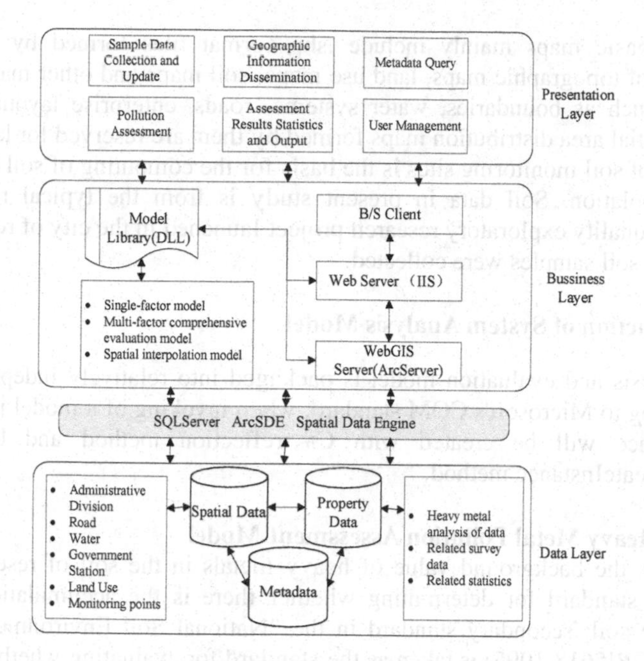

Fig. 1. Overall system diagram

data editing, spatial analysis and other ready-to-use applications and wide varieties of services [4].

Analysis and assessment model is packaged into relatively independent DLL files according to Microsoft's COM standard. When changes occur in sample monitoring data, model component is loaded on the server-side, according to model parameters input on the client-side, various analysis and assessment calculations on spatial data and attribute data read from the database are conducted, and the results are stored in the server-end database. SQLServer2005 is uniformly used for full relational storage of spatial data and attribute data, thus ensuring consistency in data maintenance.

3.3 System Functions

Investigation is carried out according to system requirements; the system mainly includes six functional modules: sample collection and update, pollution assessment, geographic information dissemination, assessment results statistics and output, metadata query, and user management.

4 Specific Implementation of System Functions

4.1 Data Acquisition and Processing

In this paper, spatial database is built based on the soil pollution status exploratory study monitoring site data of a city in south China, and through data layering, segment management, attribute coding and spatial index design.

Relevant basic maps mainly include .shp format files formed by scan vector quantization of topographic maps, land use maps, soil maps and other maps, different map layers such as boundaries, water systems, roads, enterprise layout maps, and urban residential area distribution maps formed by them are reserved for later use.

Selection of soil monitoring sites is the basis for the computing of soil heavy metal spatial interpolation. Soil data in present study is from the typical regional soil environment quality exploratory research project launched in the city of research area, a total of 114 soil samples were collected.

4.2 Construction of System Analysis Model

System analysis and evaluation model is packaged into relatively independent DLL files according to Microsoft's COM standard, when invoking of a model is needed, an object instance will be created with C# reflection method and by invoking Assembly.CreateInstance method.

4.2.1 Soil Heavy Metal Pollution Assessment Model

In this paper, the background value of heavy metals in the soil of research area is taken as the standard for determining whether there is the accumulation of heavy metals in the soil; secondary standard in the "National Soil Environmental Quality Standards" (GB15618-1995) is taken as the standard for evaluating whether the soil is contaminated, and whether the remediation of heavy metal pollution is needed. Evaluation methods include Nemerow index method, accumulation index method, and pollution load index method.

Nemerow index method is one of the most commonly used methods currently at home and abroad for comprehensive pollution index calculation. The formula is as follows:

$$P_i = \frac{C_i}{S_i} \tag{1}$$

Where Pi is pollution index of heavy metal;

Ci is measured value of heavy metal content;

Si is standard value of soil environmental quality (national secondary standard value [7])

4.2.2 Spatial Interpolation Model

After obtaining the pollution index of the sample sites, spatial interpolation should be performed to reveal the spatial variability status of the entire research area. The system comprehensively uses two local spatial interpolation methods, Kriging and IDW, so as to facilitate comparison of interpolation results.

Kriging method is also known as the spatial local interpolation method, where unbiased optimal estimation on regionalized variables is performed within a limited area based on variogram theory and structural analysis, and which is one of the main elements of geo-statistics. The application scope of Kriging method is the spatial correlation of regionalized variables, which can be expressed as:

$$Z(x_0) = \sum_{i=1}^{n} \omega Z(x_i) \tag{2}$$

Where *Z(x0)* is the value of unknown sample site, *Z(xi)* is the value of known sample site around unknown sample sites, *wi* is the weight of the *i-th* known sample site to unknown sample site, n is the number of known sample sites.

IDW method considers that several points which are closest in distance to the future sampling site have the largest contribution to the value of that sampling site, where the contribution is inversely proportional to distance. Its basic idea is to define the interpolation function *F(x, y)* as the weighted mean of each data point function *fk*. The weight of each data point is calculated as follows:

$$\lambda_k = \frac{d_k(x,y)^{-u}}{\sum_{k=1}^{n} d_k(x,y)^{-u}} \tag{3}$$

Where *dk(x, y)* —— value of u, distance from the *(x, y)* point to point k, is generally taken as being 1～3, and often 2.

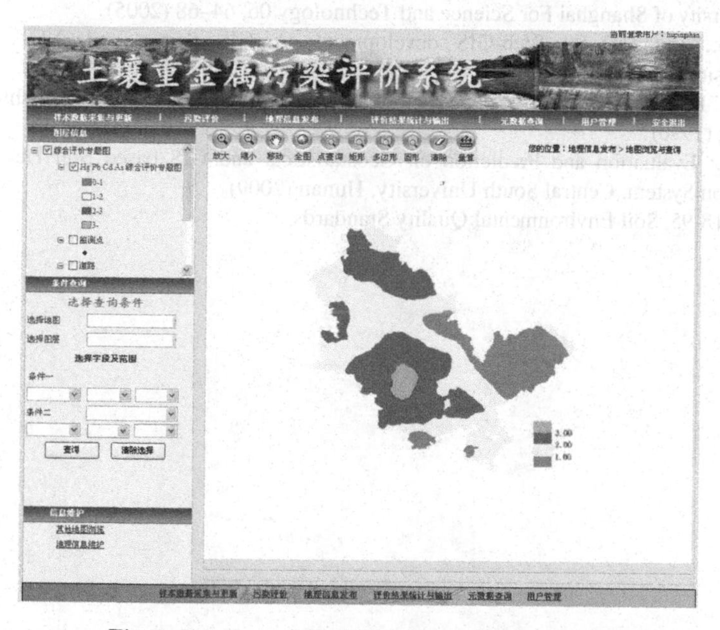

Fig. 2. Soil Heavy Metal Pollution System Interface

5 Conclusion and Outlook

In this paper, system is built based on WebGIS technology and geo-statistical methods, relevant spatial and attribute database construction as well as browse and query of map resources in the research area are achieved, and heavy metal pollution status of the region is assessed according to the relevant data of monitoring sites and soil evaluation criteria, the results are displayed in the form of thematic maps after spatial interpolation using geo-statistical methods. Through the development of this system, we can have a clear and intuitive understanding of the pollution status in the research area; meanwhile, networking characteristics of the system enables easier sharing and expansion.

As the spatial variability of soil environmental quality is influenced by a variety of complex environment, part of the assessment models and spatial interpolation methods used in this study still need constant improvement and expansion to make them more realistic.

References

1. Wang, S.: Design and Implementation on WebGIS-based Regional Traffic Information System in Beijing. Beijing Jiaotong University, Beijing (2009)
2. Wang, Y.: Integrated query of distributed spatial data sources. Computer Engineering and Design 06, 32–34 (2007)
3. Zhang, F., Cao, Q.-J.: Study on existing spatio-temporal data models. Journal of Uuniversity of Shanghai For Science and Technology 06, 64–68 (2005)
4. Liu, G., Tang, D.-S.: Web-GIS development: ArcGIS Server and .NET. Tsinghua University Press, Beijing (2009)
5. Hou, J., Huang, J.-X.: Theories and methods of geostatistics. Geological Publishing House, Beijing (1990)
6. Lin, Y.: Evaluation and Prediction on Geostatistical and GIS based Soil Heavy Metals Pollution System. Central South University, Hunan (2009)
7. GB15618-95, Soil Environmental Quality Standards

The Impact of Visual Transcoder on Self-similarity of Internet Flow

JingWen Zhu, Jun Steed Huang, and BoTao Zhu

School of Computer Science and Telecommunication Engineering
Jiangsu University, 212013, P.R. China
steedhuang@qq.com

Abstract. Due to the latest booming of the handheld devices, JPEG2000 is gaining momentum in medical image, military and surveillance, as the file size is small and easily fit on smart phones. This paper offers a systematical modeling method for characterizing JPEG2000. In this study, we holds two purposes: first, we wish to provide a measurement of the traditional JPEG streaming to get a feeling of how much self-similarity is in the JPEG wireless Internet flow; the second is to provide a practical model that the Internet Service Provider can use to design their optimum network load when JPEG2000 is used. For these reasons, we have also measured different Hurst parameters, under different compression qualities. The data considered in this paper comes directly from the GenieView camera project. Our result shows that the transcoding only slightly increases Hurst value by 1% or lower.

Keywords: Self-similarity, Hurst parameter, M/G/1 queue model.

1 Introduction

The phenomenon of self-similarity can be found in a various objects, such as the shape of the coastline, the trees, rolling hills and even the stock market movements. The study of self-similarity was initially applied to computer networks, successfully characterizing network traffic [8]. Later, many studies [5] have shown that the video traffic shows long-range dependencies, which indicates the self-similarity. But most of the researches have focused on the MPEG stream for its extensive application in video compression, while very few studies have done on M-JPEG stream.

It is well known that M-JPEG uses JPEG compression for each frame of the video sequence. Because frames are compressed independently, M-JPEG lacks interframe prediction, which limits its compression efficiency to 1:20 or lower [10]. Whereas modern interframe video formats, such as MPEG4 or H.264, achieve compression ratios of 1:50 or higher. However, when the smart phone screen is getting bigger, and also tablet PC market is picking up, so is e-book usage, researchers are either looking into H.265 or JPEG2000 as an alternative to H.264 or JPEG standard.

Therefore, based on previous studies, we make an assumption that the video stream such as JPEG and JPEG2000 show self-similarity on the time series and examine it via estimating the Hurst value. As a detector of self-similar phenomenon, the Hurst parameter indicates the degree of self-similarity, which provides an objective feeling of

D. Jin and S. Lin (Eds.): Advances in MSEC Vol. 1, AISC 128, pp. 361–366.
springerlink.com

how much the self-similarity is influenced under different transcoders. Finally, the M/G/1 queue model is introduced to estimate an explicit delay in JPEG2000 video traffic compare it with JPEG flow. Note that we just consider the wireless condition whose results may be different from wireless network, due to its sensitivity and instability.

2 An Overview of Self-similarity and Hurst Parameter

2.1 Intuitive Graphical Description of Self-similarity

Generally, a self-similar phenomenon means a process is exactly or approximately similar across all time-scales. Here, we present the M-JPEG video stream traces on different time-scales to give an intuitive feeling of self-similarity. In Fig. 1, plot (a) uses a time unit of 4 seconds; plot (b) uses 40 seconds. Observing the two plots, we find that they show a similar fluctuation on different time-scales. More detailed description can be found in [9].

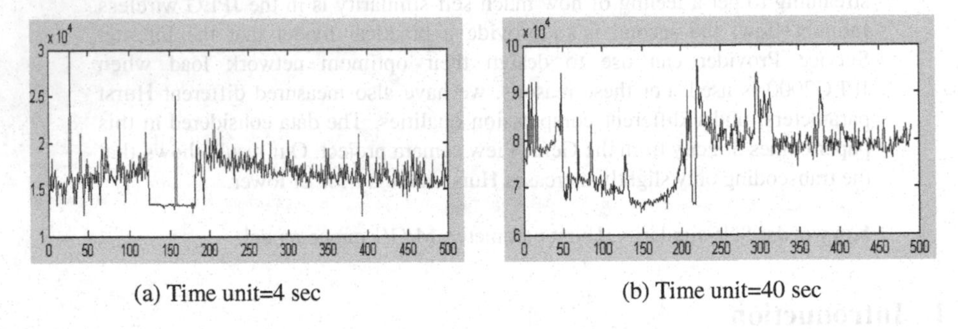

(a) Time unit=4 sec (b) Time unit=40 sec

Fig. 1. Pictorial demonstration of self-similarity

2.2 Estimation of Hurst Index for the Level of Self-similarity

We have just presented an intuitive graphical description of the self-similarity. In this section, we present a mathematical method to demonstrate it.

In current studies, the Hurst parameter is one of the most extensive measurements to describe the degree of self-similarity. From [4], we know that H takes value from 0.5 to 1. A value of 0.5 indicates the absence of self-similarity. Whereas, the closer H is to 1, the greater the degree of self-similarity is. By calculating the Hurst parameters in JPEG and JPEG2000 stream, we can examine their self-similar degree respectively.

In this paper, the Index of Dispersion for Counts (IDC) is introduced as an estimation of Hurst parameter. The method is studied in [2].

Consider a covariance stationary stochastic process $X=(X_1,X_2,\ldots)$ and $S_T=X_1+\ldots+X_2$, then IDC is defined as

$$IDC(T)=\frac{E[S_T-E[S_T]]^2}{E[S_T]}=\frac{E[(S_T)^2]-E^2[S_T]}{E[S_T]} \quad . \tag{1}$$

Under the pure fractal condition, IDC is given by the relation

$$IDC(T) - 1 = (T / T_0)^{\lambda} \ . \quad (2)$$

$$\lambda = 2H - 1 \ . \quad (3)$$

with $0 < \lambda < 1$. Taking the logarithm of both sides of the relation we get

$$\log(IDC(T) - 1) = \lambda \log(T / T_0) \ . \quad (4)$$

Thus we can obtain the slope λ from different point $(\log(F(T) - 1), \log(T / T_0))$ on the IDC curve. Using (3) we finally get the Hurst parameter.

$$H = \frac{1}{2}(1 + \lambda) \ . \quad (5)$$

3 Estimation of the Delay

In this section, we introduce M/G/1 queue model to estimate the explicit delay when JPEG2000 flow is used.

In [1], the author presents a detailed analysis of how the variability of job sizes affects the mean response time in an M/G/1/RR queue.

Let C^2 denote the squared coefficient of variation (SCV) of the packet size distribution:

$$C^2 = \frac{E[IDC]}{E[S_T]} \ . \quad (6)$$

The mean delay of the self-similar video stream is given by

$$E[T] = (1 + \rho * \frac{C^2 - 1}{2}) * \frac{1}{1 - \rho} \ . \quad (7)$$

where ρ denotes the workload of the system. Note that the value of delay is not the exact response time. It reflects the number of delayed packets. Data rate should be known in order to estimate the response time. According to [3], data rate varies in different scenarios, so packet number is more practical to calculate the delay.

4 Matlab Results

In our test and measurement for the GenieView camera project, a MJPEG stream is traced from a 400MHz wireless network and 4900 JPEG frames are recorded with time stamps for video only. Latter, we use a JPEG2000 encoder, Morgan JPEG2000, to convert the original wireless JPEG frames to JPEG2000 format with the convert quality as 50 and 150, targeted for BlackBerry 1.9GHz platforms.

4.1 Calculating Hurst Parameter

Using our traced the stream, we can plot its packet size distributions when different compression formats and qualities is used (see Fig.2). So we can analyze the impact of different transcoders on self-similar degree in video stream based on these flows. In this paper, we focus on Weibull distribution, due to its flexibility with different shape parameters [6].

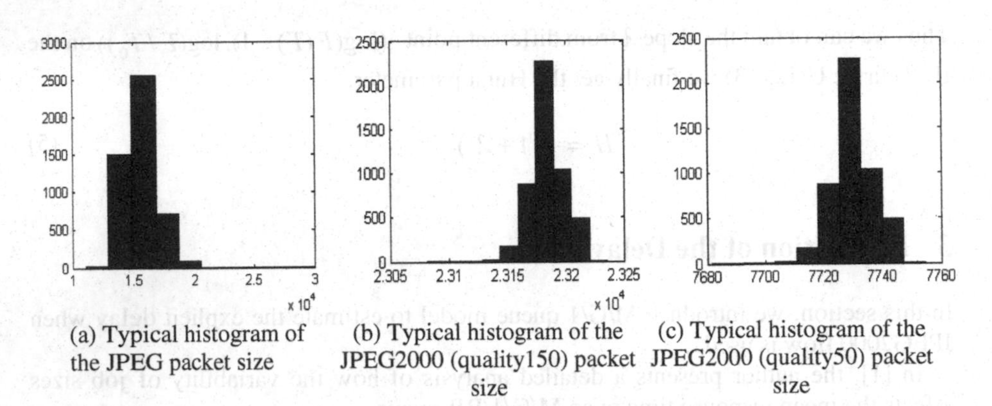

(a) Typical histogram of the JPEG packet size

(b) Typical histogram of the JPEG2000 (quality150) packet size

(c) Typical histogram of the JPEG2000 (quality50) packet size

Fig. 2. Visual compression of the packet size distribution between JPEG and JPEG2000 frames

IDC curves can also be plotted by MATLAB program, see in Fig. 3.

As the result shows, for both JPEG and JPEG2000 steam, their Hurst values are very close to 1, which demonstrates their strong self-similarity and abruptness [7]. Notably, the transcoding only increases the Hurst value by 1%. Further, we also calculate the Hurst value under different compression qualities when JPEG2000 is used. A striking consequence is that Hurst value neither increases nor decreases with the compression quality which means that self-similarity is an intrinsic property of the video stream, which won't be influenced by using different transcoders.

It needs to mention that since all the pictures are traced from the wireless network, we have no direct evidence that the wired network will have the same result. It is all due to the sensitivity and instability of wireless network, which is also an important factor of the high Hurst value.

4.2 Calculating the Delay

With obtained IDC curve, C^2 can be calculated from (6). We assume a set of ρ from 0.1 to 0.9, thus, users can design their network load according to the delay and the processing rate of their system. Table 1. gives our final result and a comparison of the delay between JPEG and JPEG2000 wireless Internet flows. We can also observe that JPEG2000 has less delay than JPEG.

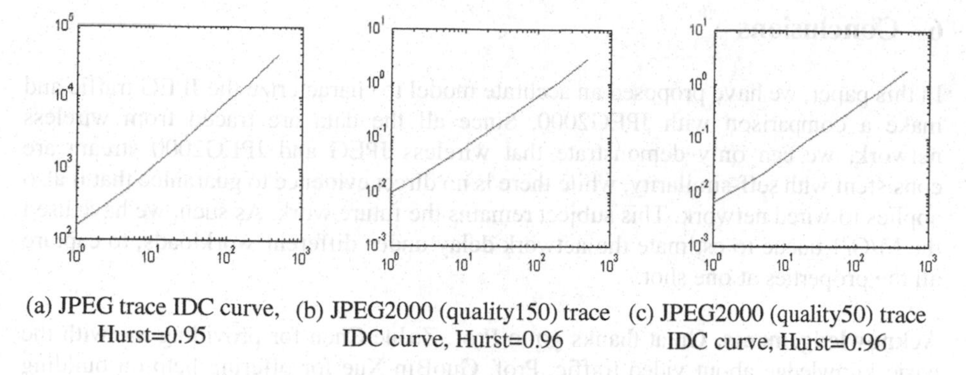

(a) JPEG trace IDC curve, Hurst=0.95 (b) JPEG2000 (quality150) trace IDC curve, Hurst=0.96 (c) JPEG2000 (quality50) trace IDC curve, Hurst=0.96

Fig. 3. The slope of IDC curve

Table 1. Explicit delay under different workloads

workload	Delay (JPEG)	Delay (JPEG2000)
0.1	1.0856	1.0556
0.2	1.1926	1.1250
0.3	1.3301	1.2143
0.4	1.5135	1.3334
0.5	1.7703	1.5000
0.6	2.1555	1.7500
0.7	2.7974	2.1667
0.8	4.0812	3.0001
0.9	7.9328	5.5002

5 The Future Research

The Internet of Things must incorporate traffic and congestion management. Traffic modeling and estimation can ensure efficient communication, load balance and end-to-end Quality of Service. This will sense and manage information flows, detect overflow conditions and implement resource reservation for time critical and life-critical data flows. Using above mentioned model to understand each applications.

The network management technologies will need depth visibility to the underlying seamless networks that serves the applications and services and check the processes that run on them, regardless of device, protocol, etc. This will require identifying sudden overloads in service response time and resolving solutions, monitoring IoT and web applications and identify any attacks by hackers, while getting connected remotely and managing all "smart things"/objects involved in specific applications from remote "emergency" centers, by identify their invariant signatures such as Hurst parameters.

6 Conclusions

In this paper, we have proposed an accurate model to characterize the JPEG traffic and make a comparison with JPEG2000. Since all the data are traced from wireless network, we can only demonstrate that wireless JPEG and JPEG2000 stream are consistent with self-similarity, while there is no direct evidence to guarantee that it also applies to wired network. This subject remains the future work. As such, we have used the M/G/1 queue to estimate the network delay under different workloads, to capture all the properties at one shot.

Acknowledgements. Great thanks go to Prof. ZuJue Chen for providing me with the basic knowledge about video traffic, Prof. GuoBin Xue for offering help on building MATLAB models, Mrs. Hong Qian for providing JPEG traces, and Mr. Mike Zhou for offering BlackBerry book.

References

1. Gupta, V.: Finding the optimal quantum size: Sensitivity analysis of the M/G/1 round-robin queue. In: SIGMETRIC Performance Evaluation Review, vol. 36, pp. 104–106. ACM, New York (2008)
2. Zhang, D., Ye, P.: Studies on Key Technologies of Optical Packet-Switch Network: Traffic Modeling, Traffic Characteristics Analysis and Optical Packet Assembly, pp. 51–57. Doctoral Dissertation Beijing University of Posts and Telecommunications (2006)
3. Lawal, F.: Sending Video Over WiMAX for Inter-Vehicle Communications. Master Thesis University of Ottawa (2011)
4. Park, C., Campos, F.H., Le, L., Marron, J.S., Park, J., Pipiras, V., Smith, F.D., Smith, R.L., Trovero, M., Zhu, Z.: Long Range Dependence Analysis of Internet Traffic. J. Applied Science 38(7), 1407–1433 (2011)
5. Yao, T., Tsang, D.H.K., Li, S.Y.: Cell Scheduling and bandwidth allocation for a class of VBR video connections. In: Global Telecommunications Conference, vol. 1, pp. 371–377. IEEE Press, Singapore (1995)
6. Duffy, K.R., Sapozhnikov, A.: The large deviation principle for the on/off weibull sojourn process. J. Applied Probability 45, 107–117 (2009)
7. Mehrvar, H.R., Le-Ngove, T., Huang, J.: Performance Evaluation of Bursty Traffic Using Neural Networks. In: Canadian Conference on Electrical and Computer Engineering, vol. 2, pp. 955–958. IEEE Press, Calgary (1996)
8. Barenblatt, G.I.: Scaling, Self-similarity and Intermediate Asymptotics. Cambridge University Press, Cambridge (1996)
9. Gomez, M.E., Santonja, V.: Analysis of self-similarity in I/O workload using structural modeling. In: 7th International Symposium on Modeling, Analysis and Simulation of Computer and Telecommunication System, pp. 234–242. IEEE press, College Park (1999)
10. What are MPEG and JPEG formats and What's better, `http://www.imakenews.com/kin2/e/article000195658.cfm`

The Research and Design of NSL-Oriented Automation Testing Framework

Chongwen Wang

School of Software, Beijing Institute of Technology, Beijing, China
wcwzzw@bit.edu.cn

Abstract. By analyzing the Selenium and other open source testing tool, the lack of Selenium and the design of testing scripts are given to discuss and try to improve to resolve problems of NLS. These improvements include the using of page elements, enhancement of the response of the heavyweight component, optimization of testing scripts for multi-language versions. The parallel execution strategy for multilingual test cases has been provided, through which the users can execute test cases of multi-language in a great number of test servers at the same time, greatly improving the overall testing efficiency. The testing framework proposed has been applied to the actual web product globalization testing, and achieved very good results.

Keywords: Automation testing, testing framework, NLS, Selenium.

1 Introduction

As Global software production and service to become an important development strategy for more international software companies, Software products need an international software design, development, testing and services to gain more international market share. So Testing for the NLS (National Language Support) becomes more and more important. But currently the existed automation test framework does not cover the content of global test. Therefore, we need to expand and improve the technology of global testing. Currently NLS-oriented test is used after function test and do not own a single automated test for itself. Many functional test scripts are reused in NLS-oriented test, so the problems found in NLS-oriented test will cause deep impact even need to refractor the whole software design. Therefore a special framework of automation testing is needed for NLS-oriented test. Based on the actual project requirements, we proposed a NLS-oriented automation testing system which based on the open source tools of Selenium, ANT and JUnit.

2 Choice of Automated Testing Framework

Software to be tested is collaboration office software based on B/S structure which uses WEB2.0 technology and provides all CRUD operation with basic data for the collaboration suite, with powerful data management capabilities. Testing work need to

D. Jin and S. Lin (Eds.): Advances in MSEC Vol. 1, AISC 128, pp. 367–373.
springerlink.com

support the globalization test with ten languages. The test script need to be reused and the operations which will be repeated need to cover all features. In common test automation frameworks, the test framework of keyword-driven/table-driven need to design a data sheet box of keyword and its development cycle is long. It is suitable for operation with large amounts of data and is not suitable for our project. The library test framework is not selected because the function libraries of our software are too complex. In this project, we select a hybrid testing framework with modularity test framework and data-driven testing framework for the script development.

Modular testing framework needs to create the scripts. The small scripts with tree structure can be constructed for a particular test case which provides modular features of program design. Despite it is difficult to be recorded; the code can be instead of script in the design of framework. It is simple and high controllable which makes the development more effective. Data-driven testing framework is a framework which reads input and output data from files and loads the data into the variables in capture or hand-coded scripts. It reads the testing data from testing script and stores in external files or database. In the testing process, it reads testing data from files dynamically. Extending test script by replacing the testing data with parameters, the test processes are not limited only with the testing data when test script is recorded. Through the way of parameters, we can read the test data from external data source or data generator. Then we can extend the test coverage and improve test flexibility.

3 The Choice of Automated Testing Solution

We intend to use the Selenium tool and use Java language to implement the automated test for our project. The test program is shown in Figure 1.

- Import Selenium and JUnit tools, we wrote the test scripts with Java language. By the Selenium tool we implemented the simulation of web operations. By the JUnit tool we organized the basic framework of scripts and verified the implementation checkpoints.
- Writing methods, using XML format file to parameter all data elements. Which data elements include basic configuration data, test record data, GUI element data and so on.
- Writing configuration file and using Ant tool to build the test project. Its main tasks include set environment parameters, start the Selenium Server, compile program, run the test cases and generate logs. Ant can be built automatically and JUnit can run the automation test. Using Ant and JUnit together, we can make the build and test process automatic. In order to ensure the labels of <JUnit> and <JUnitreport> can be identified by Ant tool, we need to copy three jar files which included by the JUnit to the directory of Ant library.

Compared to the general Selenium test solution, our program takes the following advantages. Firstly, it is easy to control with the Ant tool. If the test task changed, we can restructure the configuration files of Ant tool to implement the modified task. Secondly, by using the data-driven testing framework, all data is stored in the XML

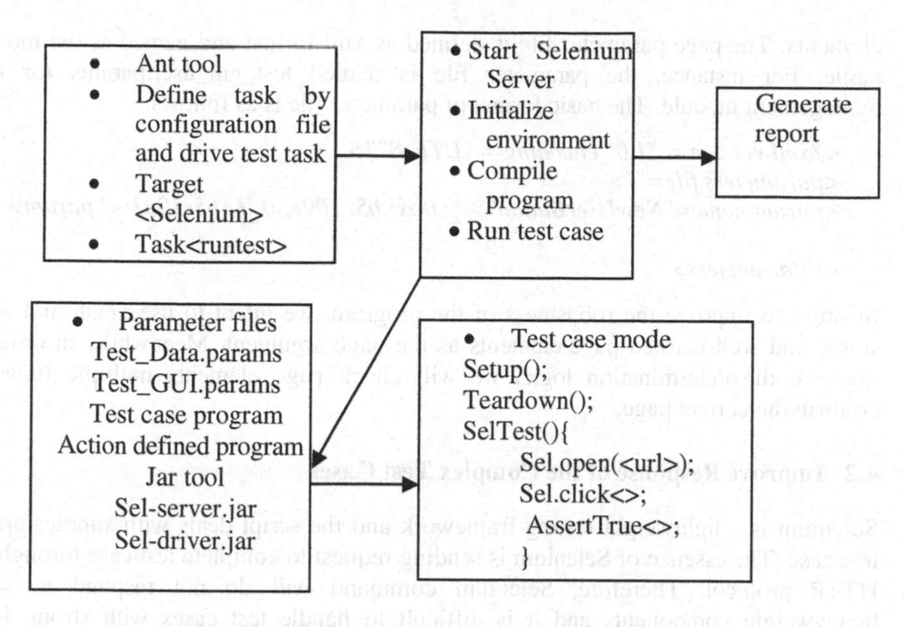

Fig. 1. Schematic diagram of the test program

format files and is easy for maintenance and modification. Finally, by saving the environment parameters, test data parameters, test result parameters and extracting test reports, it is convenient for tester to view and invoke test cases.

4 Improve Selenium Tool

Selenium tool has many advantages such as supporting multiple browsers, protocols and program languages and providing multiple ways to locate the elements of web applications. However, there are a lot of shortcomings for Selenium tool. Firstly, a lot of page elements such as name, id, xpath and so on were used in Selenium. Both the functional changes and UI reconstruction or interaction improvement will affect these elements which makes the test vulnerable. Secondly, it depends on the existence of specific data. The test will fail when individual data is no longer valid and such failure is not identified as the missing of function. Finally, Selenium is a lightweight testing framework and the script deals with simple-formed test case. In response to the above shortcomings, this paper proposed methods to improve Selenium tool and scripts.

4.1 Improvement for Page Elements

In Selenium toolkit, the operating functions use a lot of page elements as arguments directly and also depend on the existence of these data values. When the data in an individual page element is no longer valid, the test will fail. And the failure is not identified as the missing of the functions which makes the Selenium test fragile and the attributes in the same page element need to be modified repeatedly. In order to fix the above problem, we write a program to replace parameters by extracting the page elements. The page parameter file is defined as xml format and named as the module name. For instance, the parameter file is named test_gui_user.params for user management module. The basic format of parameter file is as follows.

```
<?xml version = "1.0" encoding = "UTF-8"?>
<parameters file="">
<param name="NewUserButton"> _sBox: b5c109e: 0_1: b5cl0al </ param>
... ...
</ Parameters>
```

In order to improve the robustness of the program, we intent to use xpath and other stable and well-marked page elements as the page argument. Meanwhile, in order to improve the determination logic, we will check page elements multiple times to confirm the correct page.

4.2 Improve Response of the Complex Test Cases

Selenium is a lightweight testing framework and the script deals with simple-formed test case. The essence of Selenium is sending request to complete test case through the HTTP protocol. Therefore, Selenium command will do not respond to some heavyweight components and it is difficult to handle test cases with strong logic business relationship

There are two solutions for the above situation. The first solution is to rewrite the JavaScript Event simulation code in Selenium toolkit. The disadvantage of this approach is that testers can not view the test process from the user perspective. The second solution is to use Java.awt.Robot to simulate the actual mouse movement. We need to rewrite the mouse movement code.

In Selenium toolkit, the HTML element can be located by Xpath, DOM API or other attributes value. By calling the Selenium function, the coordinates of HTML elements are returned. Originally, Selenium Server communicates with browser by AJAX (XmlHttpRequest) directly. Now it could invoke the method of Java.awt.Robot to simulate actual mouse movement after the coordinates of HTML elements were returned.

4.3 Support for Globalization Test

As the same with the majority automation testing tools, Selenium is also lack of globalization testing support. If the parameter file is not specified, Selenium Server will start a browser execution test with system default setting. The default language is the current browser version language. This test can not provide test data with multiple language and character set.

In order to improve the above drawback of Selenium toolkit and extend it for the globalization test, we design the structure of test scripts and add the environment setting and parameter files for the Selenium RC-packaged Java toolkit.

1) Make the elements of language-specific independent of the test script code and support script reuse of multiple languages

In order to reuse the whole test scripts for multiple languages, we make all elements which depend on the specific language independent of the test script code and store them in the parameter files. The structure design of the test script is as follows.

The source files of test cases are stored in the src directory and the parameter files are stored in the param directory. The parameter files used xml format which are divided into three categories according to the purpose of parameters and the relationship with languages.

a) The parameter files which stored fixed parameters.

In such files, we stored the parameter data which was not changed for different language versions. These fixed parameters can be the input string during test process or the property values of interface elements such as the test machine IP address, the id value of the button UI, the xpath value of table UI and so on.

b) The parameter files which stored the input random data according to the test language and character set.

According to the requirements of test language and character set, we analyze test cases which deal with string input/output directly or indirectly and extract the random data which can be replaced with any international input type. When the test is running, we can replace the input data to test the software capabilities of handling different languages and character sets. For instance, a new user name was needed when the test case of create new user is running. In this design, the user name can be replaced of data with any international input type.

2) Select the languages and character sets dynamically and support for multiple languages testing process

In order to meet the requirements of globalization test, we need initialize test environment such as add setting for test environment language and character set at first. The options of language and character set include not only the choice of current test software language versions, but also the choice of the current operating system language and character set. In our project, the major task is to achieve the options of the language and character set between the client operating system (Redhat4.0) and browser (Firefox3.0)

If the parameter file is not specified, Selenium Server will start a browser execution test with system default setting. The default language is the current browser version language. The setting of specific browser language and character encoding can be implemented through the following ways. Firstly, the test language and character set will be stored in the configuration file according to the choice of testers. Secondly, according to the browser language and character set of test requirements, we set the user language preference in Firefox parameter file which named prefs.js. Finally, we add the option of "firefoxProfileTemplate" and point it to the modified parameter file directory of Firefox when the Selenium Server is starting. For instance, we select the language Chinese and GB2312 character encoding as the user preference and add the following statements into the specified parameter file of Firefox which named prefs.js.

- export LC_ALL = zh_CN.gbk
- export LANG = zh_CN.gbk
- user_pref ("intl.accept_languages", "zh-cn")
- user_pref ("intl.charsetmenu.browser.cache", "GB2312")

In order to set the language and character set of client operating system (Redhat4.0), we can modify file in /etc/sysconfig/i18n or set the parameters of LC_ALL and LANG.

Meanwhile, the above method can be extended to the situation with multiple languages and character set environment. Firstly, all of the languages and character sets were stored in the configuration files. Then, the test was ran serially or parallel in the test machine. The serial execution means initialize test environment and loop test process step by step. And parallel execution means pass test information to multiple target machines through socket communication and initialize test environment parallel and run all of the test process.

3) Use regular expressions to deal with the translation unit contains special characters

In the above automated process of parameter file replacement, a large number of XML format translation file data need to be searched. However, some special characters written to the XML file will be replaced by entity references. Therefore, as shown in Table.1, the displayed string in target screen may be different with the translation string which stored in the XLIFF translation files.

Table 1. The characters replaced in the translation files.

Character shown in user interface	*Character in XLIFF format translation file*
&	&
‘	'
“	"
<	<
>	>

We can use regular expressions to solve the inconsistency displayed string between the original string and the translation string. Regular expressions provide a powerful, flexible and efficient way to handle text data. By the pattern matching method of regular expression, we can compare and match strings to determine the test results are correct or not.

5 Strategy for NLS-Oriented Test Task

When NLS-oriented test task is running, multi-language test cases tend to make the number of test cases exponentially growing which extend test time greatly. And the probability that cause some unexpected error in the test process will be higher. Therefore, our project uses parallel execution strategy for multi-language test. When the test languages in the instruction are over a number of test cases, we will split it into small tasks. Test controller program will find the idle test machine according to the status of test machine and the corresponding task amount. Then the test commands are sent to the test server through the socket communication. This parallel test platform with remote control will reduce the execution time of software test, raise the efficiency of execution and provide a valuable engineering reference for the large-scale software test. The test workflows of remote control are as follows.

The client proxy viewer and monitor were deployed in the test server which got the status of the test server by monitoring the relevant processes and updated the state property of test machines in the server database.

Testers accessed the front controlled interface based APEX through the WEB page, set up the task parameters and controlled the task execution.

In the control server, we find the idle test server according to the state information stored in the database, and then send the test command to the Agent program in test server through the socket communication.

The proxy program which was deployed in test server accepted the test task, controlled and driven the test script.

The client proxy viewer and monitor were deployed in the test server which got the status of the test server by monitoring the relevant processes. If the test end is detected, the test results will be sent to tester and the state property in controlled server will be updated.

6 Conclusion

In this paper through the production testing process of a WEB collaboration suite, we proposed an automation testing system of NLS-oriented which based on the Selenium, ant and JUnit tools. For the problems in NLS test, we improved the Selenium tool and test scripts. And we provided a parallel execution strategy for multi-language test which can run multiple test process in multiple servers and improved the overall test efficiency greatly. Of course, the test for the NLS, there are many issues involved need to continue in-depth study, such as multi-language environment deployment of test cases, test cases quickly build multi-language and how to automate the maintenance of test case library, and so on. However, there are many issues in NLS-oriented test to research for such as how to deploy test environment for multi-language's situation, how to build test cases for multi-language quickly, how to maintain a test case library automatically and so on.

References

[1] McMahon, C.: History of a large test automation project using selenium. In: Proceedings - 2009 Agile Conference, AGILE 2009, pp. 363–368 (2009)
[2] Bruns, A.: Web application tests with selenium. IEEE Software 26(5), 88–91 (2009)
[3] Xu, D.: A Tool for Automated Test Code Generation from High-Level Petri Nets. In: Kristensen, L.M., Petrucci, L. (eds.) PETRI NETS 2011. LNCS, vol. 6709, pp. 308–317. Springer, Heidelberg (2011)
[4] Patel, S.: TestDrive - A cost-effective way to create and maintain test scripts for web applications. In: SEKE 2010 - Proceedings of the 22nd International Conference on Software Engineering and Knowledge Engineering, pp. 474–477 (2010)
[5] Selenium, http://wiki.javascud.org/display/SEL/Home
[6] HtmlUnit, http://htmlunit.sourceforge.net/

Based on Self-manufactured Experimental Equipments, the Curriculums of Microcomputer Experiment Teaching Mode Reform and Innovation

Jiangchun Xu[1], Jiande Wu[1], Nan Lin[1], Ni Liu[2], Xiaojun Xue[1], and Yan Chen[1]

[1] Kunming University of Science and Technology, Kunming, Yunnan, China, 650051
[2] Sichuan University of Arts and Science, Dazhou Sichuan, China, 635000
jx19631018@163.com

Abstract. This paper describes the characteristics of university experimental teaching, and the necessity and urgency of reform. In the analysis of the current experimental teaching courses in computer problems, it gives experimental teaching measures and the reform of methods; combination of the KG89S 51-chip experiment platform which is independent researched for experimental teaching curriculum reform to improve the scale of innovative and comprehensive pilot project, focusing on capacity building of students, to stimulate students' interesting in learning, it had explored a new teaching experiment way for microcomputer courses curriculum reform initially.

Keywords: microcomputer courses, experimental teaching mode, self-manufactured aboratory equipment, innovative experiment.

1 Introduction

In the current higher education, experimental practice teaching is an important part of the teaching system in colleges and universities, got more and more people's high attention. Because experiments and practice teaching are the source of obtaining new knowledge, they are the key of knowledge and ability, the combination of theory and practice, they are the important mean of the training skills, cultivate the innovative consciousness and ability, it plays a very important position in the higher education system. Especially in history of tide of today's cultivate innovative talents, they have a very important practical significance to carry out the experiment practice teaching reform and innovation experimental project, and the research of training mode[1].

Microcomputer course mainly refers to "micro-computer design principles and procedures" and "Microcontroller Theory and Applications". The above courses are important professional basic courses in engineering which are related professional disciplines such as automation colleges, computer classes, and electrical and electronic and so on college; they are not only the basic courses of professional hardware computer applications, but also the application of highly professional and technical courses. Such courses are characterized by the teaching content of the abstract, difficult to learn, which requires a certain amount of programming ideas, and calls for microcomputer data transmission and processing of in-depth understanding, emphasizing the combination of

D. Jin and S. Lin (Eds.): Advances in MSEC Vol. 1, AISC 128, pp. 375–381.
springerlink.com

hardware and software, and its theoretical and practical are strong. At the same time, the object of teaching is wide, the teaching content informative, the educational content updates quickly, the students fell difficult to learn, and the teachers are difficult to teach[2]. Practice more, or it's hard to keep up with the experiment equipment performance and experimental project setting and open rate, so that it can't guarantee the education quality.

Ministry of Education issued a "Circular on Further Strengthening Undergraduate Education of the number of observations', requires colleges and universities to strengthen practical aspects such as experimental teaching, continue to improve experimental teaching methods, reform experimental teaching content, increase comprehensive and innovative experiment, to give students more hands-on opportunities to create, to improve the students ability of practice, ability of innovation and ability of independent research. Ministry of Education, "Undergraduate teaching level evaluation index system" (2004) documents, the comprehensive and designed experiments have clear standards, proposed specific ratio of experiments courses with comprehensive and design accounted for the total number of experimental program, and the effect requirements of design of experiments course. Therefore, it is necessary to depend on the characteristics of course and the current curriculum status of laboratory equipment, to analyze the experimental teaching model of microcomputer courses, so that we can reform experimental teaching ways, change the experimental teaching methods and training model that don't suite to the times. Independent research can meet experiment for experimental teaching and training the model of reform.

2 The Problems of Microcomputer Experiment Teaching

2.1 The Comprehensive Is Not Strong of the Basic Experiments

Previous experimental teaching is "according to a single dispensing," students without understanding, the completion of the experiment is mainly to verify the experiment. Namely: each of pilot project is given a experimental guidance whit too much detail, the steps are all written in a very specific, students do not need themselves to analysis, design, follow suit as long as you can. This experimental teaching to enable students to do a passive rather than active, to think problems independently, analyze problems and solve problems, stimulate students' interest and curiosity, ignoring to develop students' innovative and creative ability[3]. Meanwhile, some basic experiments lasted for decades, outdated content, and relatively simple, the lack of thinking and thought-provoking.

2.2 The Experiments with Innovative, Comprehensive, and Design Are Less

Currently, pilot projects in microcomputer courses, integrated, design and innovative pilot projects open a low rate. How to design some accord with students the knowledge structure of the comprehensive and simple design experiment, let the students in trials, in the experiment research, from the experiment in search for rule, improve students' innovative consciousness and ability is the present universities experiment teaching to be subjects of study[4].

2.3 Problems of the Equipments

At present the laboratory equipment or old, aging, many experiment content to stay in the 1990s level, serious behind the development of modern technology science, and experiment content also has a serious discrepancy with the reality of scientific and technological achievements and the production needs; although some experimental equipment for new purchased equipment, but also can't completely solve the existing problems above-mentioned in the experiment teaching project. Although the school in constantly increase of experimental equipment purchase investment, and made some achievements. But the experimental equipment construction not only requires a lot of money, while the new purchase the laboratory equipment on how to design innovation experiment projects, open the microcomputer course with comprehensive and designing, this involves the need to design teaching syllabus and other activities, and need to spend time waiting for construction and investment. This is to some extent, slowed the step of experimental teaching innovation and opening up. Our microcomputer courses for this experiment according to the characteristics of teaching, the first of a series of comprehensive and designing to innovative design of pilot projects for curriculum, combined with the research projects developed by teachers to meet the microcomputer class curriculum innovation pilot project out of the home-made equipment.

3 The Reform of Measures and Methods in the Microcomputer Class Experimental Teaching

3.1 Upgrading the Experimental Teaching Status

Over the years, experimental teaching has been in a subordinate position supporting the teaching of the theory, until now the situation has still not been completely changed. Embodied in the way of teaching, the traditional experimental teaching generally adopt a mode that "preview-classroom experiments- after-school teachers in Review", in this mode, there are often students do not attach importance to preview, they do not want to get involved in test, they do not carefully sum thinking after school, or even absent, plagiarism report and other undesirable phenomena. In the teaching content, the traditional experimental teaching of theory courses are generally of a teaching from the property, as in-depth understanding of theory learned auxiliary methods, too much emphasizing on theory as a means of verification, neglecting of developing student ability, lacking of students active participation and two-way communication between teachers and students, and having lost the opportunities of teachers to students to objectively examine practical ability ,the teaching methods do not largely reflect that the enhancing of the students' ability discovering and solving problems of scientific research and innovation, it also reduced the enthusiasm of students to participate in the experiment.

3.2 The Reform of Experiment Teaching Projects and Methods

Microcomputer courses experimental teaching reform should always focus on the themes "innovation", what is linked between experiment and innovation? And what difference between them? Experiment meaning of innovation itself, experiments do not experiment to experiments, experiments aimed at understanding the nature of things, grasp the inherent variation in order to use its laws, which is "innovation" means. How to recognize innovative experiment? We think that is from the experimental project function, content, technology, means and other aspects of to re-know and improve, improve the role of experiment and efficiency.

So we must re-development of the experimental curriculum, test plans and project redesign. Base on basic experiments in the curriculum reform project, increasing the design, comprehensive and innovative experiment proportion.

3.3 Innovation and Development Potential of Experimental Equipment

With the rapid development of science and technology, experimental teaching system, experimental teaching content are constantly changing and developing, so the experimental equipment must be configured with the experimental teaching system and experimental teaching content changing, the only way to meet the state innovative personnel training requirements. On the market some of the existing experimental equipment can not keep up the level of technological products development, and higher prices, to a certain extent, can not fully meet the growing range of tests, the need for deepening the depth experimental, training colleges and universities can not meet requirements. So tap the potential of existing laboratory equipment, based on the innovative applications, development and advanced self-made experimental equipment to meet the requirements of new curriculum pilot projects and experimental tools and model, which is especially important for colleges and universities.

4 Microcomputer Courses Development and Application of Laboratory Equipment

4.1 The Cause of Experimental Equipment Research and Development Process

Micro-computer technology is advancing rapidly, which determines the curriculum content to be constantly updated; the curriculum teaching must be subject to the latest development of the dynamic introduced to students. Experimental teaching system, experimental teaching content should be developed with innovative technology development, laboratory instruments and equipment must be configured with the experimental teaching system, experimental teaching content synchronization support to adapt to the evolving needs of society. The existing market available experimental equipment performance cannot completely meet the needs of the reform of the experimental teaching system and teaching experiment project, and limited funds decided not bought completely finished product equipment solutions to experiment instrument and equipment and supporting construction tasks. Experimental teaching system, experimental teaching content of the innovation determines the importance of home-made equipment, which can solve the major and difficult problem arisen from the experimental teaching system, experimental teaching content in the process of continuous innovation and development[5]. Experimental equipment hardware and software systems must be based on the experimental model to determine the content and design of pilot projects. Experiment equipment development flow chart is shown in Fig. 1.

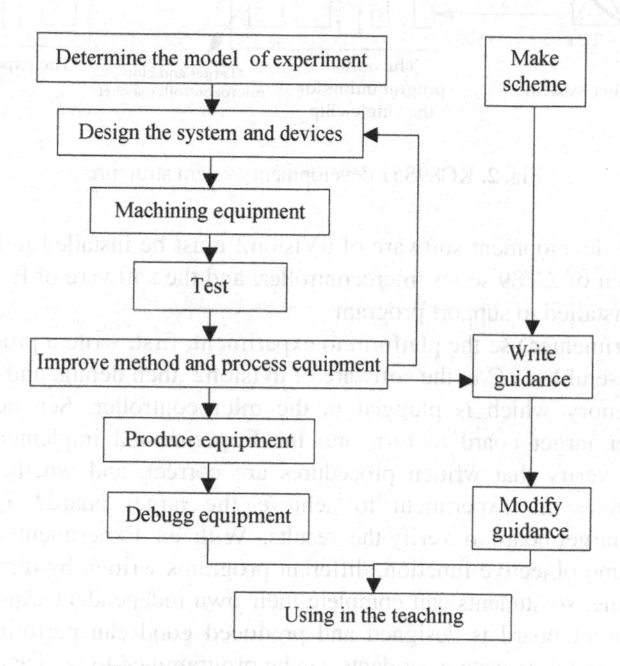

Fig. 1. The chart of the development flow

4.2 Experimental System Equipment Design

According to the characteristics of the microcomputer courses and the characteristics of the ISP, we use AT89S51 as the the main control chip to develop an experimental platform system[6]. The structure of the system is shown in Fig.2. This system with many functions contains a PC, an online serial programmer, a cable of ISP, an experiment board supports development of all microcomputers, and the software of programming. The online programmer is the core component.

A programming console as a single-chip microcomputer development tools, using of it can be programmed applications fast and correctly chip to or from single chip erased.Its main function is the development of the chip. Programmer can set the with one of two ways and PC computer communication, programmer receive the application through the debugging process in PC, and will be written into the burning of single chip, the goal programming console can USB interface, also can use the power supply + 5 V power supply or batteries; if you use ISP download cable programming, from time requires the COM interface, but requires LPT interface; if using a serial programmer for programming, there must be COM (RS-232) interface.

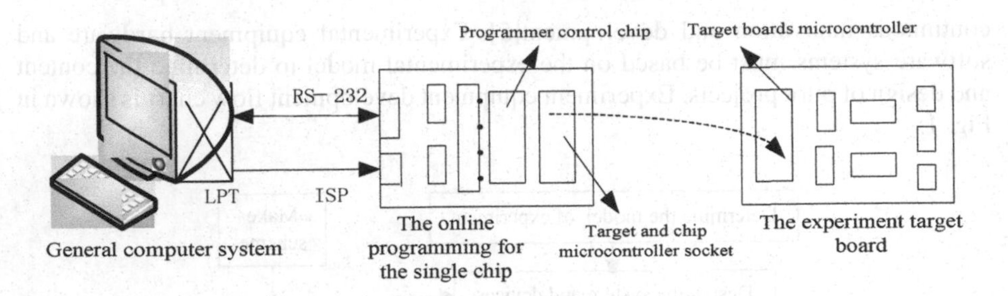

Fig. 2. KG89S51 development system structure

In Fig.2 the development software of µVision2 must be installed in PC to support the development of AT89 series microcontroller, and the software of Easy 51ProV5.2 also must be installed to support program.

When experimenter use the platform to experiment, first, write a program with the language of assemble or C in the software of µvision2 ,then debug, and last configure the Flash memory which is plugged in the microcontroller. Senond, unplug the microcontroller target board to turn into the Experimental implementation of the procedures to verify that written procedures are correct, and whether achieve the desired functions the experiment to achieve the target board? This can only be run by the target board to verify the results. With an Experimental on board to achieve the same objective function, different programs written by different students are not the same, so students can complete their own independent experimental. The teacher can target board is designed and produced good can perform a variety of functions of a circuit or system, students can be programmed to realize the function of the target board, training programming ability; also can be students want to achieve a certain function of their own design circuits or systems, and then through the programming achieve its function, training single-chip microcomputer system design and programming ability.

4.3 Application Results

Based on AT89S51 experiment platform system development success and put into use, after more than three years of use and operation, the use effect very good, equipment performance completely meet the functional requirements of curriculum of microcomputer experiment teaching system, experiment teaching content of reform. Of nearly 700 students, and evaluation of the effects of the survey results, see Table 1.

From the results of the survey questionnaire can be seen, the experimental teaching model after reform compares with the traditional teaching model, in the experimental system to improve the pilot project of proportion of innovative and comprehensive, the students of the experimental model of self-recognition and inspiring students to learn and achieved good results; also making experimental equipment system to ensure that the programming ability of students to get exercised and improved.

Table 1. The evaluation of the effect of reform and innovation survey results with teaching microcomputer experimental courses

Item / Mode	The verifiable experiment	The innovative and comprehensive experiment	The approval	The degree of the autonomous learning	The ability of programming	The satisfaction to the equipment
Traditional	85%	15%	60%	General	General	General
Reformational	15%	85%	98%	Well	Well	Well

5 Conclusion

From the characteristics of microcomputer class curriculum teaching, the reform experimental teaching system and teaching content ,through self-made equipment to achieve the experimental verification of pilot projects converse to innovative, design and integrated pilot projects; through the curriculum of microcomputer experiment teaching reform, has inspired the enthusiasm of students' autonomous learning, students of programming the ability to get the exercise; through the test of practice in recent years makes very good effect. The teaching mode of the reform and innovation practice can help similar course for reform.

References

1. Gao, Z.: Experimental innovation and innovation experimental analysis. Science and Technology Information Development and Economic 8, 144 (2004)
2. Han, Z.: Computer Principles course teaching reform and practice. Journal of Nanjing Engineering College (Social Science Edition) 12, 53 (2003)
3. Xie, Y., et al.: Set up a compreensive and designed experiments to improve students comprehensive quality. In: The fourth Session of National Universities Refrigeration and Air Conditioning Subject Development and Teaching Conference, vol. 4, pp. 1–29 (2006)
4. Zhou, H., Fang, Z., Liu, J., et al.: To strengthen laboratory construction management, reasonable use of teaching resources. Experimental Technology and Management 24(1), 141 (2007)
5. Meng, Y., Qin, G., Lu, F., et al.: To build a new experimental teaching system, create a mechanical engineering experimental teaching demonstration centers. Laboratory Research and Exploration 28(8), 23–28 (2009)
6. Chen, X., et al.: Kunming University of Science and technology KG89S 51-chip experiment platform manual 3(10), 2–5 (2008)

“Flexibility” of Software Development Method

Yu Gao and Yong-hua Yang

Department of Computer, Zhejiang Ocean University, Zhoushan, Zhejiang, P.R. China
gaoyu@zjou.edu.cn, yyh@zjou.edu.cn

Abstract. Contraposing the complex and fickle development tendency of software development process in recent years, “flexibility” of software development method is put forward. “Flexibility” of software development method is a new concept. Preliminary definition is given, to “flexibility” of software development method. The reason is expounded, for software development methods having “flexibility”. Manifestation of “flexibility” is expounded in software development method. The basic conditions supporting “flexibility” is discussed. Last, it is discussed that how to structure software development method possessing "flexibility".

Keywords: software development method, flexibility, network, tool, process.

1 Introduction

In the 1970s, "flexibility" concept began to appear in the machinery manufacturing industry. At the time, "flexibility" manufacturing system [1] was proposed. After 1990s, in order to enhance the ability to adapt the change for software products, and in order to meet the requirements of diversification and individuation, "flexibility" concept began to appear in the software field. The software system's "flexibility" becomes an important question to need people to study. Currently, around "flexibility" of software system, there have been many research articles [2-7].

We believe that the research field of "flexibility" should be expanded. People should not only study the "flexibility" of software system, should also study the "flexibility" of software development method.

So far, for the "flexibility" of software development method, people do not study it in depth, experts did not give a strict definition. In order to facilitate the research questions, we described the "flexibility" of software development methods so: In the software development process, with changes of objective circumstances, to allow adjusting the content of software development method, for this adjustable property of software development method, it is known as the "flexibility" of software development method.

In the past, when building and using software development methods, people do not consider the "flexibility", so a variety of traditional software development methods are the lack of "flexibility". Their contents are fixed for the traditional software development methods, development steps, methods and tools can not change. In the software development activities, the developer must implement these fixed contents.

D. Jin and S. Lin (Eds.): Advances in MSEC Vol. 1, AISC 128, pp. 383–387.
springerlink.com

In recent years, although some people have consider the problem adjusting contents of software development methods, but the "flexibility" concept of software development methods not been clearly put forward. For example, RUP method has attribute of "can cut" [8]. For example, XP method emphasizes "adapting the change" [9]. We think that "can cut" is a manifestation of "flexibility", and "adapting the change" is also a manifestation of "flexibility".

Round the "flexibility" of software development method, we have do some study. In the following we describe the study.

2 Software Development Method Should Possess "Flexibility"

Early software development activities have the following characteristics: In development activities, the changes of objective circumstances are very small, so the changes of development steps, methods and tools are also very small. That is to say, the contents of the software development methods do not require random adjustment in the development process. So, in the early activities of software development, considering "flexibility" of software development method is not necessary.

In recent years, user requirements and software development environments become increasingly complex, simple and fixed software requirements gradually transformed into complex and changing situation. Moreover, in the development process, it often is encountered that user make a request of changing in software function, performance or other content. The change of development environment and tools is growing gradually.

In order to adapt the complex and fickle new situation in development process, adjusting the content of traditional development method should be allowed. In this way, the content of the software development method is no longer fixed, and it possessed "flexibility". Its ability to cope with change obtains to enhance. Such, people have to consider "flexibility" of software development method.

First, if the software development methods possess "flexibility", then it can adapt to the fickle trend of development process. In recent years, software development process shows the complex and changeful trend. When there are the changes of situation, the contents of software development methods can change, so development tasks can complete successfully.

Second, if the software development methods possess "flexibility", then it can play developer's initiative and creativity. For traditional software development methods, its development steps, methods and tools are basic fixed, developers can only do in accordance with these fixed content. Traditional software development methods do not provide the opportunity to play a proactive and creative to developers. However, software development method of "flexibility" provides the opportunity for them. According to changing circumstances, developers can play to their initiative and creativity.

3 How to Embody "Flexibility" in Software Development Method

Software development methods include many contents, "flexibility" should embody in the different aspects.

3.1 To Embody "Flexibility" in Form of Development

Development form of traditional is that a number of developers of a development team work together in the same place. In order to facilitate the exchange of information, they are in close proximity to each other. They may be in the same office or adjacent offices.

Network closer to the distance between people, when exchanging information, geographic location can not restrict people. Members of development team both work at the same location, also can work in different location. Therefore, the traditional forms of development are clearly not applicable, and the new forms of development should embody the "flexibility".

"Diversity" should be results to embody "flexibility" in the development form. That is to say, developer can pre-designed a number of different forms of development, on the basis of changing circumstance, developer can free to decide form of development. For example, the forms of development can be concentrated form, at the same location, all members of the development team together work. Forms of development can also be dispersed form, members of the development team work in different location, using network to exchange information.

3.2 To Embody "Flexibility" in Tool of Development

In the early development activities, development tool is relatively small, ability of development tool is also relatively weak, and update of development tool is very slow. When designing content of a particular software development method, the range choosing development tools is relatively small. Once you've selected certain kinds of development tools, then it basically will not change.

In recent years, the types of development tools gradually increased. In order to complete a particular development work, according to the change of objective circumstances, people can choose different tools. This is manifestation of "flexibility" in the development tools. For example, in the coding phase, high-level language as a development tool, there are many kinds of high-level language. According to the specific circumstances of the coding phase, people can randomly choose in several high-level languages.

3.3 To Embody "Flexibility" in Process of Development

Software development process include: software analysis, design and implementation process. Traditional software development methods detailedly provided the specific methods for the analysis, design and implementation process, developers was asked to follow these specific methods. However, it is not to be considered that adjusting content of these methods on the basis of the change of circumstances.

In order to adapt the complex and changeful trends, development process should possess "flexibility". In the development process, the method should not be fixed in the analysis process or design process or realization process. The content of method should be allowed to adjust randomly, as the change of circumstances.

For example, to the analysis process, its difficulty is to obtain the user's needs. In formulating development plans, even if the method to get the user's needs is determined, but in the actual development, because the situation is constantly changing, pre-made plans may is not necessarily applicable. Adjusting the content of method is inevitable. At this point, manifestation of "flexibility" should be: based on possible scenarios, a variety of methods is prepared. In accordance with changing circumstances, developers select and use a method to complete the work obtaining the user's needs in development. Probably pre-designed a variety of methods are not applicable, then it is allowed that developer oneself design applicable methods.

4 Some Basic Conditions Supporting "Flexibility"

Network techniques are an important basic conditions supporting "flexibility" of software development method. As the development and application of network techniques, forms of development can not be restricted to geographic location. Developers can work in different locations, and select and use different forms of development. So that software development method possessed "flexibility".

Development tool are an important basic conditions supporting "flexibility" of software development method. The more types of development tools are, the greater range of the choice are. Now both in software analysis process, or in the software design process, or in the software implementation process, there are many different development tools. Through to choose development tools, developers can make the development various process with "flexibility".

Software component library are an important basic conditions supporting "flexibility" of software development method. People can reuse these materials stored in software component library. Various methods are all software components, they store in the software component library. For the analysis process or the design process or the implementation process, on the basis of the change of condition, people can randomly select in these methods.

5 How to Build Software Development Method Possessing "Flexibility"

According to the previous description, we know that software development method must have "flexibility". Current status quo is the lack "flexibility" to traditional software development method, complexity and polytrope is the change trend of software development process. Clearly, it is an important work to build software development method possessing "flexibility".

To build software development method possessing "flexibility", a relatively simple way is improving traditional software development methods. According to the previous description, the approach of improving is to add "flexibility" factors in the existing software development methods.

In order to determine the specific location adding "flexibility" factors, we should study from two aspects. One is to study the applicable field and development environment, from outside of software development method, we think the problem adding "flexibility" factors. We need to figure out which may be external situations of the frequently change. Then, according to these external situations we add "flexibility" factors. The other is to study the components of software development methods, from within of software development method, we think the problem adding "flexibility" factors. In the structure content of software development method, we find the part which possibility is relatively large to be change. For this part, we add "flexibility" factors.

Build software development method possessing "flexibility", innovation is another important way. That is to say, according to new ideas, based on the concept of "flexibility", we reconsider software development process.

6 Conclusion

In the above, we give an initial definition of "flexibility" of software development method, we discussed problems of "flexibility" of software development method. For example, how to embody "flexibility" in software development methods ? what are basic conditions to support "flexibility" ? how to build software development methods possessing "flexibility" ? We observed the change trend of software development process in recent years. We found that the software development method possessing "flexibility" is a promising software development method. Around the software development methods possessing "flexibility", there are many issues to study. We hope to make some contribution for these studies.

References

1. Liu, Y.-L.: Flexible Manufacturing Automation Introduction. Huazhong University of Science and Technology Press, Wuhan (2001)
2. Fang, S.-S., Shen, P.: Design and Development of Flexible MIS System. Nanjing: Journal of Nanjing University of Aeronautics and Astronautics 5, 597–599 (1993)
3. Cao, J.-W., Fan, Y.-S.: Concept, Methods and Mractice of flexibility software system. Computer Science 2, 74–77 (1999)
4. Shen, L.-M.: Flexibility software development technology. National defence industry press, Beijing (2003)
5. Fan, Y.-S., Wu, C.: Research on a Workflow Modeling Method to Improve System Flexibility. Journal of Software 4, 833–839 (2002)
6. Shen, L.-M., Mu, Y.-F.: Concept and measurement of software flexibility. Computer Integrated Manufacturing Systems 10, 1314–1320 (2004)
7. Li, A.-B., Huang, J.-Z., Bi, S.-B., Xie, X.-L.: Research and Application of Flexible Design in Software System. Application Research of Computers 1, 140–143 (2005)
8. Jacobson, I., Booch, G., Rumbaugh, J.: The Unified Software Development Process, 3rd edn. Addison Wesley, Boston (2004)
9. Beck, K.: Extreme Programming Explained. Addison Wesley, Boston (2000)

Direction of Arrival (DOA) Estimation Algorithm Based on the Radial Basis Function Neural Networks

Hong He[1], Tao Li[1], Tong Yang[2], and Lin He[2]

[1] Tianjin Key Laboratory for Control Theory and Application in Complicated Systems, Tianjin University of Technology, China
ahеho604300@126.com
[2] Tianjin Mobile Communications Co., Ltd

Abstract. According to the problem of the large calculated quantity and unavailable of multiple sources tracking in real time in the traditional DOA estimation algorithm which is disabled when locating sources that are greater than the number of array elements number, a new direction of arrival estimation algorithm based on the radial basis function neural networks in smart antenna is proposed in this paper to solve the problem. The proposed neural multiple-source tracking (N-MUST) algorithm is based on architecture of a family of radial basis function neural networks (RBFNN) to perform both detection and direction of arrival estimation. The model of neural network in direction of arrival estimation is created and trained in this paper. Simulation results which compared the traditional algorithm and the new one are indicated that the direction of arrival (DOA) estimation algorithm based on the radial basis function neural networks implement multiple-source tracking exactly and fast.

Keywords: Smart antenna, Probabilistic neural networks, General regression neural network, DOA estimation, Radial basis function neural networks.

1 Introduction

Smart Antenna Technology which combined Space Division Multiple Access (SDMA) with Frequency Division Multiple Access (FDMA), Time Division Multiple Access and Code Division Multiple Access, is one of the main key technologies of the third generation mobile communication. Smart Antenna makes full use of frequency resource. On the other hand, smart antenna can adaptive detect the number of signal and angles of arrival to track the desire sources, then produce maximum gain on the desire angles and produce zero trapped on the angles of the interference by the beamforming algorithm in the down-link. One of the main tasks of smart antenna is how to realize the direction of arrival (DOA) estimation in real time [1]. Only if work the task of DOA estimation can realize the beamforming algorithm in smart antenna. In the receiver of smart antenna, the received signals always are multipath signals and locating sources are greater than the number of array elements, so this case call higher request on the DOA estimation algorithm.

D. Jin and S. Lin (Eds.): Advances in MSEC Vol. 1, AISC 128, pp. 389–394.
springerlink.com

There are many shortcomings in traditional methods, such as large calculated quantity, slow convergence velocity, unavailable of multiple sources tracking in real time. In recent years, the algorithms of direction of arrival estimation and beamforming based on neural network are catching more attentions. [2~5] Radial Basis function have good characters such as fast convergence velocity, small calculated quantity and strong skills in nonlinear approximation[6], so Radial Basis Function Neural Network is used to estimate the directions of arrival signals in this paper.

2 Direction of Arrival Estimation Algorithm Based on Neural Network

The probabilistic neural network (PNN) and General Regression Neural Network (GRNN) are the transformations of Radial Basis Function Neural Network, and consist of input layer, hidden layer and output layer. We should create the model of DOA estimation based on neural network at first, and then generate input output pairs to train the network. Then the network can be employed to estimate the directions of arrival [7].

The structure diagram of RBFNN is shown in fig.1. Linear array composed of M equidistant elements which the separation distance is d. In between the blocks designated "sample data processing" and "post processing," as can be seen from Fig.1, the RBFNN consists of three layers of nodes: the input layer, the output layer, and the hidden layer. The block of sample data processing received signals, and the block of post processing produce outputs, and then we get estimation of direction of arrival signals.

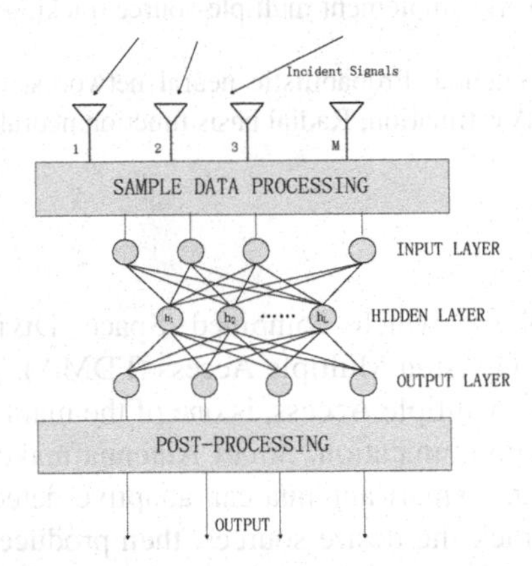

Fig. 1. The block diagram of Neural-network direction of arrival estimation system

In the block of sample data processing, signals which received by arrays are pre-processed by this block to input RBFNN. Correlation matrix contains sufficient information of incidentsignals, and R is Hermit matrix, so information contained by elements of R(i,j) and R(j,i) is identical, but elements in the diagonal line don't contain information of angles. So, in our design, the upper tri-angular half of is considered, An M M spatial correlation matrix can organized in a vector:

$$r = [R_{12}, R_{13}, \ldots, R_{1M}, R_{23}, R_{24}, \ldots, R_{(M-1)M}] \quad (1)$$

As known, signals which antenna arrays received are not ideal narrow-band signals. Elements in vector r need be separated into real and imaginary parts, and a new vector named z of real and imaginary parts denoted. The vector z is normalized by its norm prior to being applied at the input layer of the neural network:

$$z = \frac{b}{\|b\|} \quad (2)$$

Where $\|*\|$ Euclid Norm of b. Signals is input RBFNN after pre-processing, and they are estimated in two stages. [8]. The first stage is detection stage which consists of PNN.The entire angular spectrum (field of view of the antenna array) is divided in L sectors. The l th $(1 \le l \le L)$ RBFNN is trained to determine if one or more signals exist within the $[(p-1)\theta_\Delta, p\theta_\Delta)]$ sector. Then the signals which produced by the first stage are passed to the second stage, which estimates the DOA of these signals. In the second stage for DOA estimation, GRNN (General Regression Neural Network) is used in this paper. The reason for employing GRNN is that it has less human intervention parameters due to learning of GRNN rely on the sample data [9].

3 Simulation

To illustrate how a network working in the two stages and to understand the performance of this new algorithm in this paper, let us consider a case where linear array composed of 8 equidistant elements which the separation distance is d which $d = \frac{\lambda}{2}$ (λ is wavelength of electromagnetic wave in vacuum). The SNR of the signal sources used to train and test the network is 10dB. The correlation matrix was calculated from 200 snapshots of simulated array measurements.

3.1 Simulation in Detection Stage

In the stage of sources detection i.e. in PNN, Fig.2 is presenting the results obtained when PNN was designed for two users separated at 2°and with equal SNR 10dB. Then the PNN was tested for three users with same separation and values for SNR. The width of the sector is 20 from 80°to 100°. Similar experiment is presented in Fig.3, but for angles from 90°to 110°.

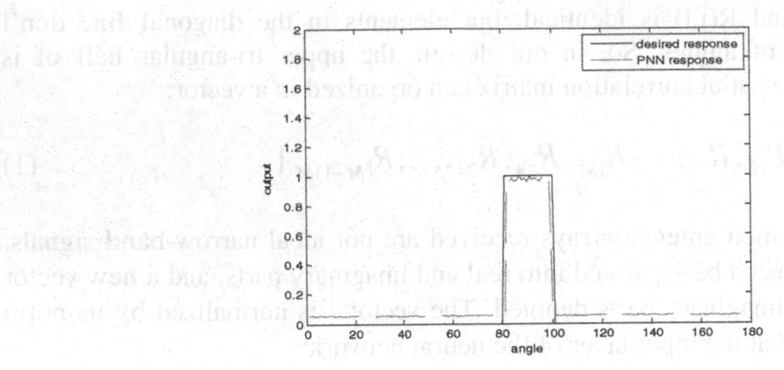

Fig. 2. Network trained for 2 users, separated at 2°with SNR 10dB,and tested for 3 users, separated at 2°with SNR=10dB,sector-(80°—100°)

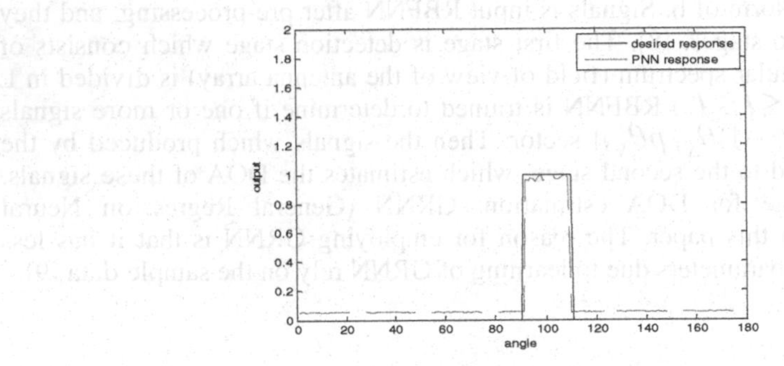

Fig. 3. Network trained for 2 users, separated at 2°with SNR=10dB, and tested for 4 users, separated at 2°with SNR=10dB, sector-(90°—110°)

The full line is presenting desired response and the dotted line is presenting PNN response. From the results it can be concluded that PNN is generalizing well except for some values and reach require of detecting sources in sectors in real time.

3.2 Simulation in DOA Estimation Stage

In the stage of sources detection i.e. in PNN, Fig.4 is presenting the results obtained when GRNN contrasted with MUSIC algorithm estimating direction of multiple sources which actual DOA of the sources is -60°, 20°and 60°. From the results it can be concluded that both two algorithms can estimate desired direction of arrival signals accurately to some extent. But GRNN's performance is closer to desired response than MUSIC. Similar experiment is presented in Fig.5, but for different direction of arrival sources which is -30°, 0°, 40°and70°. In the same way, we can see a better performance of GRNN than MUSIC.

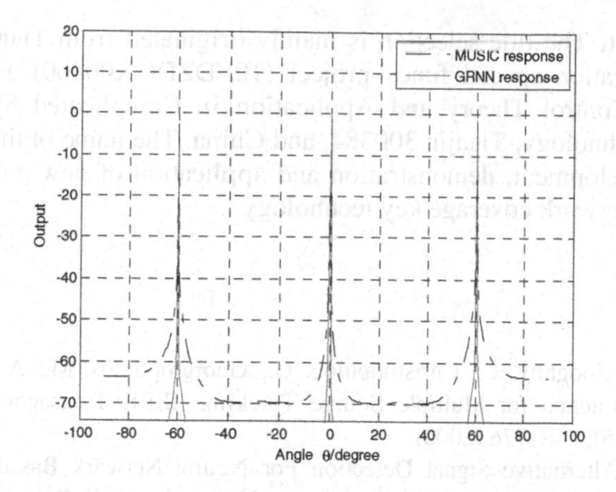

Fig. 4. Network tested for 3 users with SNR=10dB, contrasted with MUSIC algorithm output

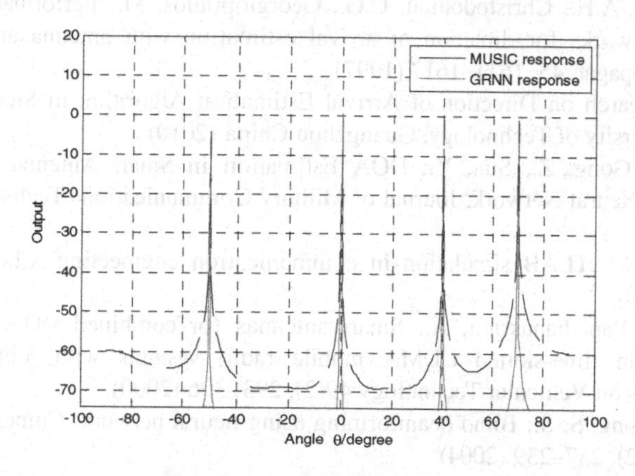

Fig. 5. Network tested for 4 users with SNR=10dB, contrasted with MUSIC algorithm output

On the other hand, simulations indicate that response time is shorter than MUSIC's when GRNN algorithm used in DOA estimation.

4 Conclusion

The DOA estimation algorithm used in this paper is divided into two stages. The first stage is to determine which sector sources in and the second stage is DOA estimation. Using this method can reduce the workload of the second stage. GRNN is used in the second stage of the DOA estimation stage. It's because of that GRNN has better approximation results ,on the other hand, GRNN is more dependent on the data sample and has less human intervention parameters than others when it being trained and then improve the reliability of the network. This allows convergence speed has been improved and the accuracy of identify corresponding increase in the MUST process.

Acknowledgment. The title selection is mainly originated from Tianjin science and technology innovation special funds project (10FDZDGX00400) and Tianjin Key Laboratory for Control Theory and Application in Complicated Systems, Tianjin University of Technology, Tianjin 300384, and China. The name of the project is "the research and development, demonstration and application of new generation mobile communication network coverage key technology".

References

1. Ahmed, H., El Zooghby, C., Christodoulou, G., Georgiopoulos, M.: A Neural Network-Based Smart Antenna for Multiple Source Tracking. IEEE Transactions Antennas and Propagation 48(5), 768–776 (2000)
2. Sarevska, M.: Alternative Signal Detection For Neural Network Based Smart Antenna. IEEE Faculty of Electrical Engineering, University of Belgrade. Serbia and Montenegro 36(6), 85–89 (2004)
3. El Zooghby, A.H., Christodoulou, C.G., Georgiopoulos, M.: Performance of radial basis function networks for direction of arrival estimation with antenna arrays. IEEE Trans. Antenna Propagat. 45, 1611–1617 (1997)
4. Le, Z.: Research on Direction of Arrival Estimation Algorithm in Smart Antenna. South China University of Technology, Guangzhou China (2010)
5. Zhang, Y., Gong, Z., Sun, Y.: DOA Estimation in Smart Antenna Based on General Regression Neural Network. Journal of Military Communications Technology 28(1), 23–25 (2007)
6. Zhang, D.: MATLAB simulation in communication engineering. China Machine Press, Beijing (2010)
7. Blanz, J.J., Papathanassiou, A.: Smart antennas for combined DOA and joint channel estimation in time-slotted CDMA mobile radio systems with joint detection. IEEE Transactions on Vehicular Technology 49(2), 293–306 (2000)
8. Sun, X., Zhong, S.-S.: Blind beamforming using neural network. Chinese Journal of Radio Science 19(2), 237–239 (2004)
9. Salvatore, B., Balanis, C.A., Jeffrey, F., et al.: Smart-antenna systems for mobile communication networks part 1. IEEE Antennas and Propagation Magazine 44(3), 145–154 (2002)

Interference Analysis of WCDMA Base from TD-SCDMA Terminals

Hong He[1], Xin Yin[2], Tong Yang[3], and Lin He[3]

[1] Tianjin Key Laboratory for Control Theory and Application in Complicated Systems, Tianjin University of Technology, China
heho604300@126.com
[2] School of Computer and Communication Engineering, Tianjin University of Technology, China
[3] Tianjin Mobile Communications Co., Ltd

Abstract. There are many factors to affect the 3G system performance, while adjacent channel interference is an important affecting factor to the coexistence network of WCDMA and TD-SCDMA. In this paper, the adjacent channel interference on WCDMA system which is caused by TD-SCDMA network on uplink is analyzed from uplink. The simulation results show that the number of interfering MS and the distance between BS and MS would affect the minimum allowed received power at WCDMA BS, and then it leading to degradation of the system capacity.

Keywords: TD-SCDMA, Adjacent channel interference, Uplink, Passes loss.

1 Introduction

Although 3G embraces a growing attention, for a long period of time it would not completely replace the 2G system on account of the same number of main format and the shared problems in the communication systems.

At present, various mobile base stations enjoy very high density covers, and the place suitable for base station is too much limited, so the third generation of the network construction inevitably reuses many radio station sites. Interference analysis and the necessary interference prevention measures must be considered to plan and design the third generation network in the case that base station is co-located and co-existent.

2 Principle of Frequency Interference

The co-existence interference of different frequency system is caused by the imperfection of the transmitter and receiver of the two systems [1]. The out-of-band radiation of interference system performs as transmitter's ACLR and spurious emission property, while the receiver selectivity in which system is interference performance for the receiver's ACS and blocking property. The result of these two factors' combined action can be measured by ACIR [2]. Transmitter's ACLR (emission property) and

D. Jin and S. Lin (Eds.): Advances in MSEC Vol. 1, AISC 128, pp. 395–400.
springerlink.com

receiver's ACS (receiving property) can be improved to develop system performance when two kinds of system co-exist in adjacent frequency.

3 The Types of Interference between WCDMA and TD-SCDMA Systems

WCDMA works in 1940～1955 MHz and 2130～2145 MHz while TD-SCDMA works in 1880～1900MHz and 2010～2025 MHz. In the 1920MHz, WCDMA and TD-SCDMA are in the adjacent frequency, therefore there are interference problems between WCDMA and TD-SCDMA. There are some possible situations as follows: WCDMA base interfere TD-SCDMA base; WCDMA terminal interfere TD-SCDMA base ; TD-SCDMA base interfere WCDMA base ; TD-SCDMA terminal interfere WCDMA base. This paper mainly studies the situation of TD-SCDMA terminal interfere WCDMA base.

4 Simulation Analysis

4.1 Process Analysis

The description of the scene Suppose that the two systems, WCDMA and TD-SCDMA, exist in the same cell, and the frequency of the two systems configured adjacent frequency band. A WCDMA base station and many TD-SCDMA terminals are here in this zone, and in this case, WCDMA system would inevitably suffer the influence of TD-SCDMA terminals in the adjacent frequency band. Simulation parameters are shown in table 1.

Table 1. Simulation parameters

Parameters	Symbol	Numerical value
Thermal noise density	N_{th}	-174dBm
Noise coefficient	F	5dB
Chip rate	W	3.84Mc/s
E_b/N_0	ρ	5dB
User data rate	R	12.2kb/s
Voice activity factor	α	0.5
Adjacent regions interference factors	β	0.55
TD-SCDMA terminal launch power	P_{MS_tx}	21dB
TD-SCDMA base launch power	P_{BS_tx}	43dB
Orthogonal factor	υ	0.4

4.2 Uplink Simulation

Uplink simulation means that WCDMA system suffers the interference of TD-SCDMA in the adjacent frequency band when the TD-SCDMA exists in the WCDMA base station service area. It is reflected by the minimum allowed received power of WCDMA base station and WCDMA system capacity loss [3].

4.2.1 The Minimum Allowed Received Power

The minimum allowed received power of WCDMA base station means traffic channel's minimum power input of receiver focusing controlling the calling quality.

For the first step ,it calculate the limiting capacity of WCDMA system, and the so-called limiting capacity equals to system capacity when signal of noise ratio from base station tends to be infinite [4]. The relevant formula (1) is shown below:

$$k = \frac{1}{1+\beta} + \frac{W/R}{\alpha \cdot \rho(1+\beta)} \tag{1}$$

WCDMA and TD-SCDMA are configuration of adjacent frequency on the base of their co-existence, and if service area contains WCDMA mobile station without TD-SCDMA mobile station, the minimum allowed received power of WCDMA base station is showed in formula (2).

$$P_1 = \frac{FN_{th}W \cdot \rho}{(W/R) - \alpha \cdot \rho[(1+\beta)k - 1]} \tag{2}$$

Considering the sole TD-SCDMA mobile station, the interference power can be calculated by the formula (3).

$$I_{Rx1} = P_{MS\_tx} - ACIR - PL \tag{3}$$

ACIR is the adjacent channel interference ratio. When the protection band width is 5MHz, ACIR is 32.8dB [5]. For convenience, we take 1940MHz as suitable carrier frequency in simulation propagation model of 3G and free space path loss. Assume that the distance between WCDMA base station and TD-SCDMA mobile station is d. So the PL (path loss) can be calculated by formula (4). (PL : dB, f : MHz, d : km)

$$PL = 32.45 + 20\log_{10} f + 20\log_{10} d \tag{4}$$

If interference power comes from each TD-SCDMA mobile station are equal, the total interference power is calculated by the following formula:

$I_{totle} = 10\log_{10} n + I_{Rx1}$, the n is the number of TD-SCDMA user.

When the two systems co-exist, the service area contains not only TD-SCDMA mobile stations, but adjacent channel interference the existing WCDMA mobile stations will produce. At this time the minimum allowed received power at WCDMA base station is calculated by formula (5).

$$P_2 = \frac{(FN_{th}W + I_{totle}) \cdot \rho}{(W/R) - \alpha\rho[(1+\beta)k - 1]} \tag{5}$$

Through the analysis of above, when the two systems co-exist in the same cell and the distance between WCDMA base station and TD-SCDMA mobile station is changing, it can get the minimum allowed received power at WCDMA base station whether or not adjacent channel interference existing in WCDMA.

Figure.1 is the simulation result of the minimum allowed received power at WCDMA base station when the distance between WCDMA base station and TD-SCDMA mobile station are 50m, 100m, 200m, 400m, and 800m. At the same time the adjacent protection bandwidth of the two systems is 5 MHz and the number of WCDMA user is 60.

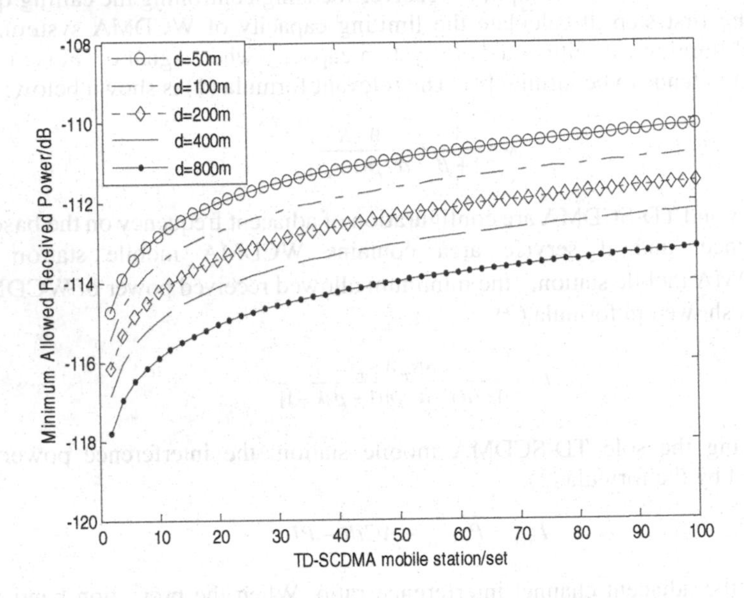

Fig. 1. The minimum allowed received power with protection bandwidth being 5MHz and the number of WCDMA mobile station being 60

From the simulation results we can see that the minimum allowed received power at WCDMA base station get bigger as the number of TD-SCDMA mobile station increasing while the distance between WCDMA base station and TD-SCDMA mobile station keep the same. Namely the interference is enlarging, which means the covered radius of the area reduced the same as capacity. While the number of TD-SCDMA mobile station keep the same, interference between systems increased as the distance between WCDMA base station and TD-SCDMA mobile station reduced. Namely the minimum allowed received power at WCDMA base station get bigger, which means the covered radius of the area reduced and the capacity descended.

4.2.2 Adjacent Channel Interference Bring Capacity Loss

When the two systems co-exist, WCDMA has the limited capacity as shown in formula (1) if WCDMA base station is not interfered by TD-SCDMA mobile stations. After WCDMA base station being interfered by TD-SCDMA mobile stations, WCDMA system's capacity change to k1. It can be showed by formula (6).

$$k_1 = \frac{1}{1+\beta} + \frac{W/R}{\alpha \cdot \rho(1+\beta)} - \frac{(FN_{th}W + I_{totle})}{\alpha \cdot (1+\beta)P_1} \tag{6}$$

So because of adjacent channel interference the capacity loss is $(1-k_1/k)\cdot 100\%$.

In order to directly reflecting the changing trends of the capacity loss, the adjacent protection bandwidth of the two systems is supposed to be 5 MHz ,the range of the distance between WCDMA base station and TD-SCDMA mobile station to be 100~800m and the number of TD-SCDMA mobile station to be 10 and 100.

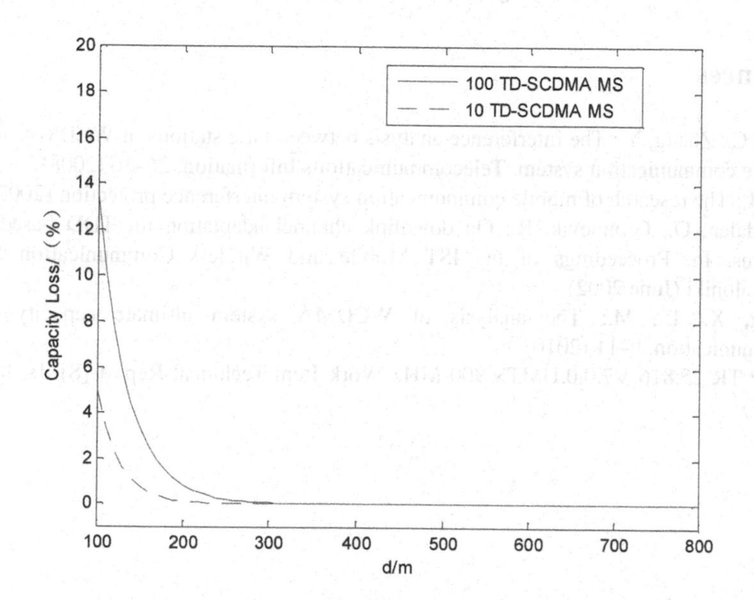

Fig. 2. Capacity loss with the protection bandwidth to be 5 MHz

Figure.2 is the simulation result of WCDMA uplink capacity loss which is in the situation of the protection bandwidth to be 5 MHz and the number of TD-SCDMA mobile station to be 10 and 100. It shows that as the distance between WCDMA base station and TD-SCDMA mobile station become short, the capacity loss become aggravated. Meanwhile the adjacent frequency interference enlarged with the increasing of TD-SCDMA mobile station which leads to the capacity loss aggravated.

5 Conclusions

At the time there are TD-SCDMA and WCDMA systems in the same cell. Through the analysis of the uplink, both the distance and the user number of TD-SCDMA mobile station that cause the minimum allowed received power at WCDMA base station are getting bigger, consequently leads to the reducing of covered area of WCDMA base station and to the aggravation of the capacity loss. Interference is affected by many factors. Therefore, if the mobile station number and the distance between base station and mobile station in the cell are effectively controlled, the interference would be and well reduced, together with the performance advancing.

Acknowledgment. The title selection is mainly originated from Tianjin science and technology innovation special funds project (10FDZDGX00400) and Tianjin Key Laboratory for Control Theory and Application in Complicated Systems, Tianjin University of Technology, Tianjin 300384, China. The name of the project is "the research and development, demonstration and application of new generation mobile communication network coverage key technology".

References

1. Fang, C., Zheng, Y.: The interference analysis between base stations of WCDMA and other mobile communication system. Telecommunications Information, 26–30 (2008)
2. Cao, J.: The research of mobile communication system interference protection (2007)
3. Grondalen, O., Gronnevik, R.: On downlink channel adaptation for TDD based LMDS systems. In: Proceedings of the IST Mobile and Wireless Communication Summit, Thessaloniki (June 2002)
4. Wang, X., Li, M.: The analysis of WCDMA system ultimate capacity. Mobile Communication, 9–11 (2010)
5. 3GPP TR 25.816 V7.0.0.UMTS 900 MHz Work Item Technical Reports[S]. [s. l.]: 3GPP (2005)

Research of Dispatching Method in Elevator Group Control System Based on Fuzzy Neural Network

Jun Wang[1], Airong Yu[2], Lei Cao[2], and Fei Yang[3]

[1] The 63rd Research Institute of PLA GSH, Nanjing Jiangsu, China
[2] Operational Software StaffRoom PLA University of Science and Technology
[3] Information Operations StaffRoom PLA University of Science and Technology
intraweb@163.com, yu_alice@163.com,
caolei.NJ@163.com, Faye.nj@163.com

Abstract. Elevator group control system (EGCS) with multi-objective, stochastic and nonlinear characteristics is a complex optimization system. It is hard to describe EGCS with exact mathematic model and to increase the capability of the system with traditional control method. After analyzing characteristic of typical traffic mode of elevator. This paper proposed a new simulation platform of an elevator group control system implemented in C# using the fuzzy-neural network technology. The comprehensive evaluation function of traffic signal is established and the right heavy of every evaluation factor (waiting time, riding time, energy consume, crowd degree) is studied by the neural network, so the elevator is dispatched optimally. The result of simulation shows that this method realizes reasonable elevator dispatching under various passenger traffic conditions and indicates the validity of this method.

Keywords: EGCS, Fuzzy Neural Network.

1 Introduction

Elevator group control system [1] are the same building many elevators as a whole to manage the control system, and its pursuit of the goals are based on different flow traffic conditions [2], select a reasonable scheduling program to coordinate the lift operation, so that ladder base to the most appropriate way of answering the call layer station staircase signal. However, because of elevator group control system control objectives of diversity, as well as elevator system inherent randomness and non-linear, it is difficult to establish a precise mathematical model, simply by traditional control methods difficult to improve the control performance of the system.

The use of expert knowledge fuzzy control rules to obtain a variety of control, elevator group control system can be a good deal of multi-objective nature of randomness and non-linear, but a simple lack of learning fuzzy control function, the runtime cannot amend the rules, and thus the system performance influenced by the expert knowledge. Neural network with non-linear, dynamic characteristics and strong learning capabilities for the establishment of the elevator group control system similar to a class of nonlinear dynamic systems, but because of the elevator group

D. Jin and S. Lin (Eds.): Advances in MSEC Vol. 1, AISC 128, pp. 401–407.
springerlink.com

control system is a multi-state, for the optimal input output mapping, the use of a simple neural network structure will make it very large, thereby increasing the network of offline and online learning.

Fuzzy logic and neural network constitutes a combination of fuzzy neural network Network (Fuzzy Neural Network, FNN), can effectively play their respective. Advantages and compensate each other enough to solve random, nonlinear problems, improve system performance.

2 Traffic Flow Model

Traffic flow has regularity and randomness, the regularity of the traffic flow is related to the people in the building; its randomness are different because working each the same time period the volume of traffic that is on each floor are random request for service a few passengers, passengers and the purpose of starting floor floors are random. A working mode of transport can be divided into up-peak traffic pattern, the down-peak traffic pattern, random interlayer traffic pattern and free traffic pattern.

(l) up-peak traffic pattern

Up-peak traffic patterns in general happen to work in the morning time, the passengers entered the elevator on the uplink to the various floors of the building work, and secondly, the strength of the uplink smaller peak occurred at the end of the afternoon rest time.

(2) down-peak traffic pattern

This happened at the evening after work, the peak after work is more strongly than the peak of morning.It exists the weaker down-peak at the beginning of mid-day rest.

(3) random interlayer traffic pattern

This traffic model is a normal traffic conditions, traffic demand between all floors of the basic balance. It exists in most of the time of day.

(4) The free traffic pattern

This traffic pattern usually occurs in the evening after work ,the next morning before work this time, as well as the rest of the afternoon time period. At rest day, the day will have different levels of the free traffic pattern exist.

3 Traffic Pattern Recognition Based on Fuzzy Neural Network Model of Elevator Control System

Fuzzy control technique using expert knowledge to obtain a variety of control rules can be a good deal of the elevator system of multi-objective, stochastic and nonlinear. The fuzzy control function lacks for the function of study; its rules cannot be modified on the runtime. In order to achieve a different transportation mode switching algorithm, it is necessary to identify the current transport mode, then switch to the corresponding algorithm. For Elevator Traffic Pattern Recognition using Fuzzy Neural Networks,

Neural network with non-linear, dynamic characteristics and a strong function of study, apply to set up similar to elevator group control system for a class of nonlinear dynamic systems, but because of the elevator group control system is a multi-state system, in order to get the optimal mapping, the use of a simple neural network structure will make it very large, thereby increasing the network offline and online study time.

Fuzzy logic and neural network constitutes a combination of fuzzy neural network [3]. It can effectively play their respective advantages and make up less than each other. Fuzzy neural network can resolve the stochastic, nonlinear and other issues in the Elevator group control system. Therefore, you can deal with fuzzy information by using neural network technology to resolve the automatic generation of fuzzy rules.

4 Formula of Elevator Group Control System

The goal of Elevator group control system is to shorten the average passenger waiting time and average time inside elevator; reduce energy consumption of elevators; improve operating efficiency and service quality. Taking all these factors, to the elevator i. The evaluation function of calls elevator signal can be integrated the following forms:

$$F_i = w_{ht} * F_{ht} + w_{ct} * F_{ct} + w_{nl} * F_{nl} + w_{yj} * F_{yj} \tag{1}$$

In the function, F_i is comprehensive evaluation value that called for the signal assigned to the elevator i, the elevator of the biggest comprehensive evaluation is the elevator of response signal. F_{ht}、F_{ct}、F_{nl}、F_{yj} separately are waiting time evaluation value, by the time the evaluation of the elevator, the evaluation of energy consumption values and staircase inside the Congestion evaluation value [4]. w_{ht}、w_{ct}、w_{nl}、w_{yj} is the corresponding weight coefficient separately to separate evaluation value, and $w_{ht} + w_{ct} + w_{nl} + w_{yj} = 1$. The different selection to weight coefficient reflects the traffic conditions at various traffic flow factors under different focus. Such as, this is the main purpose to reduce the waiting time and riding time on the peak of passengers,we can increased F_{ht}andF_{ct} the weight coefficient w_{ht} and w_{ct};

1) Waiting time wt.

When the new call elevator signal occurs, basing the floor under the call F_b , direction of the current D_b , the elevator floor F_c and direction D_c, We can calculate the arriving time of elevator by called for. Set up to run one floor time t_1, calling at one time t_2, elevators need to respond to the call to mission A, the elevator with the arrival of the furthest to the floor is F_{max}, Elevator reverse the farthest floor for the arrival F_{min}.

When D_c same as D_b ,and F_b ahead of F_c, the elevator can be called with the staircase signal to arrive:

$$wt = |F_b - F_0| * t_1 + A * t_2 \tag{2}$$

When D_c same as D_b ,and F_c ahead of F_b, the reverse operation of the elevator to arrive and then call the same staircase signal:

$$wt = (|F_{max} - F_0| + |F_{max} - F_{min}| + |F_b - F_{min}|) * t_1 + A * t_2 \tag{3}$$

When D_c with D_b the contrary, the elevator after the reverse-run to arrive the call for signal:

$$wt = (|F_{max} - F_0| + |F_{max} - F_b|) * t_1 + A * t_2 \quad (4)$$

2) Designate a short elevator time membership function $f_{wt}(wt)$.

$$f_{wt}(wt) = 1 - e^{-0.009wt^2} \quad (5)$$

3) Take the elevator time ct.
We can only know the direction of the passengers when the new called for signals occurred, we do not know the purpose of passengers layer, the paper in the assumption that passengers for the purpose of layer farthest layer, then the calculation of time taking elevator as follows:

$$ct = |F_{max} - F_b| * t_1 + A * t_2 \quad (6)$$

4) Taking time for a short elevator membership function $f_{ct}(ct)$.

$$f_{ct}(ct) = \begin{cases} 1, & ct > 20 \\ e^{-0.002(20-rt)}, & ct \le 20 \end{cases} \quad (7)$$

5) Energy consumption of elevator nl.
The power consumption of straight run can spend more lower power, so the energy consumption of elevator depends primarily on the number of start and stop, but taking into account the elevator unbalanced energy consumption between up elevator and down elevator, set the power energy consumption of down elevator for up elevator to 60%.We set up the calculation formula for nl:

$$nl = |F_{max} - F_b| * A + |F_b - F_{min}| * 0.65 * A \quad (8)$$

6) The energy consumption of the small membership function f_nl (nl).

$$f_{nl}(nl) = 1 - e^{-0.008en^2} \quad (9)$$

7) Elevator Congestion of the small membership function f_yj (yj).
Congestion of elevator can be used to describe the number, although the number of elevator cannot be accurately calculated, but through an internal elevator pressure sensor has been measured by estimating the value of passengers, yj express the number of passengers on the current elevator. We definition of a small elevator congestion membership function as follows:

$$f_{yj}(yj) = 1 - e^{-0.009yj^2} \quad (10)$$

5 Scheduling Method of Elevator Group Control System

The structure of fuzzy neural network in the paper is a multi-layer feed forward network structure[5], as shown in Figure 1.

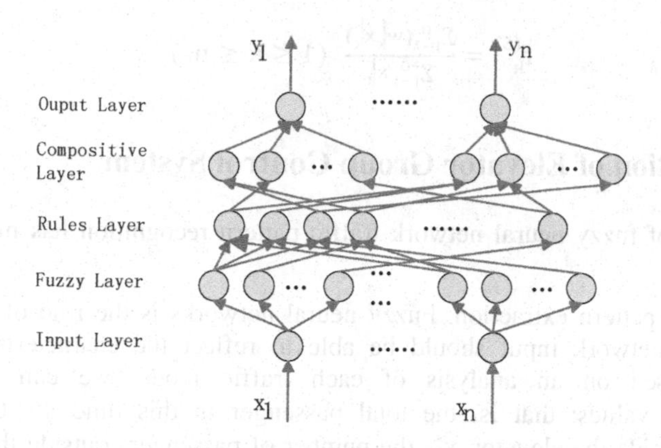

Fig. 1. The structure of Fuzzy neural network

The fuzzy neural network adopts a five-layer structure, the first layer is input layer, each node represents an input variable, the fifth layer on behalf of the output variables in the network node: the second layer and fourth layer nodes are fuzzy subset of nodes, separately for the express input and output variables membership functions: The third layer is the rules layer, each node on behalf of a rule, it is the second layer and fourth layer nodes connected on behalf of the specific composition of fuzzy rules.

Input layer send the input values to the next level directly, the number of neurons n_1 is same as the number of input variables n, that is, $n_1 = n$, for each neuron has:

$$f_k^{(1)} = x_k \ (1 \le k \le n_1) \tag{11}$$

The number of fuzzy neurons in the second layer n_2 relates to the input variables of upward layer n_1 and the number of each fuzzy subset.

So, $n_2 = \sum_{i=1}^{n_1} m_i$。

$$f_k^{(2)} = e^{-(\frac{x_i - m_i^j}{\sigma_i^j})^2} \ (1 \le k \le n_2) \tag{12}$$

Every neuron in the third layer on behalf of a fuzzy logic rules, fuzzy rules used to match conditions.

$$f_k^{(3)} = \min\left\{x_k^{j_1}, x_k^{j_2}, x_k^{j_3}, \dots, x_k^{j_k}\right\} \ (1 \le k \le n_3) \tag{13}$$

Among them,

$j_1 \in \{1,2,\dots,m_1\}$, $j_2 \in \{1,2,\dots,m_2\}$, $j_k \in \{1,2,\dots,m_k\}$,

Then the number of neurons in the layer is $n_3 = \prod_{i=1}^{n} m_i$.

The number of neurons in the fourth layer n_4 equals the number of all the fuzzy subset.

$$f_k^{(4)} = \min\{1 \textstyle\sum_{i=1}^{I_{4k}} x_k^i\} \ (1 \le k \le n_4) \tag{14}$$

I_{4k} is the number of input variables which connected with the kth neuron in current layer.

$$f_k^{(5)} = \frac{\sum_{j=1}^{so}(w_k^j x_k^j)}{\sum_{j=1}^{so} x_k^j} \quad (1 \le k \le n_5) \tag{15}$$

6 Simulation of Elevator Group Control System

Application of fuzzy neural network traffic pattern recognition has many necessary steps:

(l) Traffic pattern extraction. Fuzzy neural networks is the role of traffic pattern recognition, network input should be able to reflect the characteristics of traffic patterns. Based on an analysis of each traffic mode, we can identify three characteristic values: that is, the total passenger in this time x_1^*, the number of passengers inside the elevator x_2^*, the number of passengers outside the elevator x_3^*, the value of characteristic values can reflect the traffic characteristics in a time. It is suitable for identifying the traffic modes[6-9],

(2) Confirm the network structure. According to the total traffic flow, the number of all passengers, the number of passengers inside the elevator, the number of passengers outside the elevator can identified traffic mode.

(3) Training network through the sample. Select the samples: a sample of network traffic pattern recognition primarily on the basis of expertise to develop, experts can experience a more accurate reflection of traffic conditions and the relationship between the various modes. The input range of Network pattern recognition is [0,1], each sample of the output value is developed by the experience of experts; Training Network: First step, we achieve the function from samples, then we apply the SOM method to improve learning; the second step, we use neural network pattern recognition to training the traffic network, a sample of network traffic pattern recognition based on expert experience to develop[10-12].

Simulation environment parameters as follows: the level of building is 18,the number of elevator is 6, the elevator operation time for a layer is 2 second, the maximum load of elevator is 1500kg. the simulation results as shown in table 1.

Table 1. Simulation results

	50 peoples /5min	100 peoples /5min	150 peoples /5min	200 peoples /5min
Wt	7.8	8.9	8.8	9.6
Ct	14.6	16.3	17.3	18.4
Qt	61	72	83	84

We can find the average waiting time by elevator and the average time of inside the elevator is not an absolute increase with the traffic flow increases. The average number of start and stop the elevators is not a linear increase. It mainly because of fuzzy neural network control objectives in accordance with their corresponding weights was amended so as to achieve the elevator service requirements.

7 Conclusion

The paper advance a scheduling method of elevator group control system that can adapt to a variety of traffic conditions based on fuzzy neural network.

Through the simulation environment of the experimental data under the statistical analysis shows that this method of traffic flow in a variety of conditions can meet the requirements of passengers and achieve the purpose of energy saving systems.

References

[1] Li, D., Wang, W., Shao, C.: The intelligent Control technology of Elevator group control system. Control and Decision 16(9), 513–517 (2001)

[2] Zhang, J.: The Research of fuzzy neural network. The publishing company of industrial college in Haerbing, Haerbing (2004)

[3] Xu, Y., Luo, F.: The Research of Elevator group control system. Theory and Application of Control 22(6), 900–904 (2005)

[4] Goldfarb, D., Iyengar, G.: Robust portfolio selection problem. Mathematics of Operations Research 28(1), 1–38 (2003)

[5] Igarash, K., Take, I.S., Ishikawa, T.: Supervisory Control for Elevator Group with Fuzzy Expert System. In: 1994 Proceedings of the IEEE International Conference on Industrial Technology, pp. 133–137 (1994)

[6] Zong, Q., Tong, L., Xue, L.: The optimal scheduling method in Elevator group control system. Control and Decision 19(8), 939–942 (2004)

[7] Ben-Tal, A., Margalitt, Nemirovski, A.: Robust modeling of multi-stage portfolio problems. In: High Performance Optimization, pp. 303–328. Kluwer Academic Publisher, Dordrecht (2000)

[8] Malcolm, S.A., Zenios, S.A.: Robust optimization of power system capacity expansion under uncertainty. Journal of Operational Research Society 45(9), 1040–1049 (1994)

[9] Ben-Tal, A., Nemirovski, A.: Robust convex optimization. Mathematics of Operations Research 23(4), 769–805 (1998)

[10] Bertsimas, D., Thiele, A.: A Robust Optimization Approach to Supply Chain Management. In: Bienstock, D., Nemhauser, G.L. (eds.) IPCO 2004. LNCS, vol. 3064, pp. 86–100. Springer, Heidelberg (2004)

[11] Ben-Tal, A., Nemirovski, A.: Robust solutions of uncertain linear programs. Operation Research Letters 25(1), 1–13 (1999)

[12] Ben-Tal, A., Nemirovski, A.: Robust truss topology design via semi-definite programming. SIAM Journal on Optimization 7(4), 991–1016 (1997)

Modal of Dynamic Data Collection Based on SOA*

Airong Yu[1], Jun Wang[2], Lei Cao[1], and Yihui He[1]

[1] Operational Software StaffRoom PLA University of Science and Technology
[2] The 63rd Research Institute of PLA GSH, Nanjing Jiangsu, China
yu_alice@163.com, intraweb@163.com,
caolei.NJ@163.com, hyh.nj@163.com

Abstract. To solve the data integration problem in current business systems, a dynamic data integration model based on SOA is proposed. This model which is dynamic and application oriented extends data source to application oriented, component oriented or even service oriented new data source. The model provides a service oriented architecture model for the integration of heterogeneous and heterotypic data and data sharing. Thus data from each business system can easily be integrated and shared. Cross-platform integration of data resource is realized.

Keywords: SOA, XML, Web Service.

1 Introduction

Data integration is one of the most important parts of enterprise information systems.

Traditional methods usually build a temporary centralized base to integrate data[1]. However, they meet challenges when dealing with heterogeneous data sources. These raw data must be processed through business logic to become effective data. In fact, how to get the wanted and effective data is the key to data integration[3-5].

This paper proposes a dynamic data integration model which integrates data dynamically without building the temporary centralized base. Data integration is nearly real time and the data source which may be an application, a component, a service or even the result of data integration is not restricted to the common data storage base.

2 SOA and Related Technologies

The SOA implementations rely on a mesh of software services. Services comprise unassociated, loosely coupled units of functionality that have no calls to each other embedded in them. Each service implements one action.

Underlying and enabling all of this requires metadata in sufficient detail to describe not only the characteristics of these services, but also the data that drives them. Programmers have made extensive use of XML in SOA to structure data that they wrap in a nearly exhaustive description-container. Analogously, the Web Services

* Supported by Nature Science Foundation of Jiangsu Province under Grant, No.BK2010130.

D. Jin and S. Lin (Eds.): Advances in MSEC Vol. 1, AISC 128, pp. 409–414.
springerlink.com

Description Language (WSDL) typically describes the services themselves, while the SOAP protocol describes the communications protocols.

Web Service is independent, modulated application which is described, published, located and called through the Internet.the architecture of Web Service consists of three roles: service provider, service requester and service register. There are three operations on these roles: publish, search and binding.

SOA aims to allow users to string together fairly large chunks of functionality to form adhoc applications that are built almost entirely from existing software services. The larger the chunks, the fewer the interface points required to implement any given set of functionality; however, very large chunks of functionality may not prove sufficiently granular for easy reuse. Each interface brings with it some amount of processing overhead, so there is a performance consideration in choosing the granularity of services. The great promise of SOA suggests that the marginal cost of creating the n-th application is low, as all of the software required already exists to satisfy the requirements of other applications. Ideally, one requires only orchestration to produce a new application.

For this to operate, no interactions must exist between the chunks specified or within the chunks themselves. Instead, humans specify the interaction of services (all of them unassociated peers) in a relatively ad hoc way with the intent driven by newly emergent requirements. Thus the need for services as much larger units of functionality than traditional functions or classes, lest the sheer complexity of thousands of such granular objects overwhelm the application designer. Programmers develop the services themselves using traditional languages like Java, C, C++, C#, Visual Basic, COBOL, or PHP.

SOA services feature loose coupling, in contrast to the functions that a linker binds together to form an executable, to a dynamically linked library or to an assembly. SOA services also run in "safe" wrappers (such as Java or .NET) and in other programming languages that manage memory allocation and reclamation, allow ad hoc and late binding, and provide some degree of indeterminate data typing.

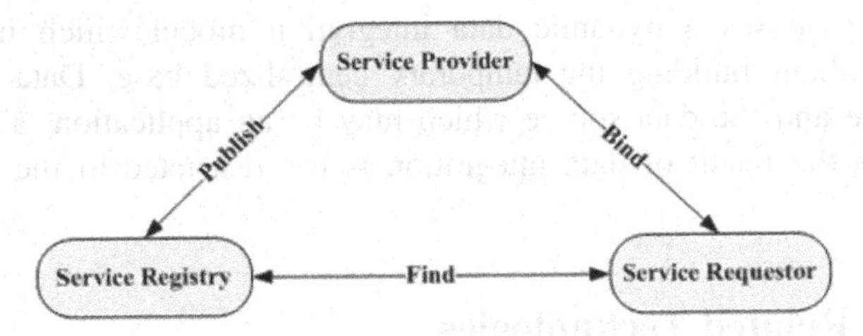

Fig. 1. Web Service architecture

3 Design of Dynamic Data Integration Model

Actually system properties critically depend on data integration. The following data integration issues need to be addressed:

Data Security: DoD (http://ia.gordon.army.mil/iaso/lesson01.htm) defined data security as the protection of data from unauthorized (accidental or intentional) modification, destruction, or disclosure. Kernochan [7] listed four impacts on security in SOA: 1) disguising data; 2) erasing data; 3) placing (old) data in secure facility; and 4) controlling data access.

Data Reliability: According to the United States General Accounting Office (GAO) [6], data reliability refers to the accuracy and completeness of computer processed data, given the intended purposes for use. Data source, routing, processing, and storage in the SOA system can be a source for reliability concern. Incorrect, inaccurate, outdated, and imprecise data can cause the system to be unreliable.

Data Integrity: Integrity is also a security concern because it also requires the assurance that data can only be accessed and altered by those authorized to do so[8]. In this sense, secure data means confidential and integral data. Two well-known integrity models are Biba Integrity model and Clark-Wilson Integrity model [2]. In an SOA system, each service can have an integrity level, and data produced by a high-integrity service can go into a low-integrity service, but not the other way around. This is opposite to the security management, where low-level data can go to a highlevel process, but high-level data cannot go to a lowlevel process[9]. The services and their composition structure determine the performance of SOA-based application. Thus, the data integrity of the SOA-based application depends on that of all constituent services and the composition structure.

Dynamic Composition: In an SOA system, a new application can be composed by specifying a workflow with reusable services, and services can be selected at runtime. As data are routed in an SOA system, they will be executed by various services selected at runtime. Thus, it may not be possible to know which service will be selected for execution of the data a priori, and thus the quality of processing by these services may not be known before execution. Dynamic composition affects almost all the issues in data integration[10].

Data Volume: An enterprise SOA system may process a large number of data at the data source, Furthermore; some of these data can be multimedia presentation data which have large storage requirements. The large volume of data puts the SOA system under stress and affects data integration and integrity significantly.

The architecture of dynamic data integration model based on SOA is illustrated in Fig.2. The technical hierarchy of the framework is constituted by 4 layers, namely uniform data accessor, XML view and SOA mapper, processing engine of XML data integration view and XML data integration view[14].

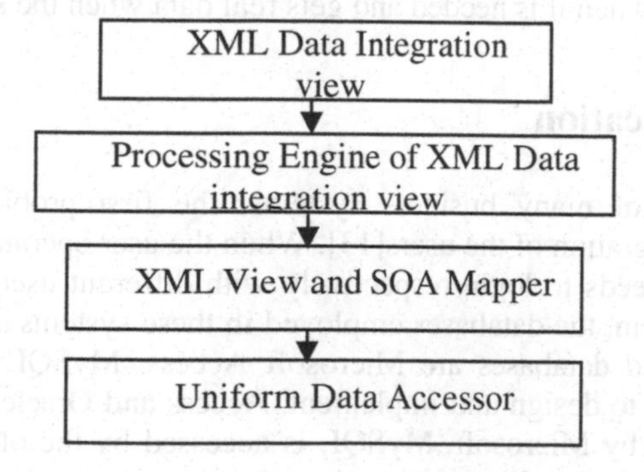

Fig. 2. Dynamic data integration model architecture

Uniform data accessor is the interface to various heterogeneous data sources. In our model, it is a group of components which can be equipped and loaded dynamically. Uniform data accessor must be programmed according to standard interfaces and take the design pattern and framework of data persistent layer for reference. There is no restriction on the type of data sources. So the data source may be database, common text file, XML file, or even an application system. Uniform data accessor is able to convert the original data stored in each data source into standard XML documents.

XML view and SOA mapper which contains view rule processor, flow processor, message processor and so forth encapsulates the uniform data accessor to the standard service of Web Service and then starts data integration through it. One of its most important functions is to convert original data from one or more data sources to more meaningful information through further logically processing. Cache is bought in to raise the efficiency and reliability of the whole system. Thus, even if the data source has something wrong, the system is able to go on. Business logic is preprocessed so as to reduce the complexity of the XML view and make it easy be understood.

Processing engine of XML data integration view offers three mechanisms.

- More than one heterogeneous data sources are presented as one virtual database. This virtual database comprises a series of realtime acquired XML files.
- The XML request from the user is analyzed and forwarded to the under layer. The final result is returned to the user with XML document.
- Metadata is presented with XML Schema which is the basis of XML view. XML Schema is like the table in relational database while XML view is just like the view corresponding to the table in relational database. Processing engine of XML data integration view itself is also provided as a web service. Thus the user can easily access it in a standard way. It can also be the data source of the other XML data integration views.

XML data integration view which expresses the integrated result is the metadata description of data integration. In fact, it is just the metadata representation of the integration result rather than the integration result. Just like the table view in relational database, the XML data integration view represents the constitution of practical data: where the data is from, which process is needed for table data, how to combine the data from different tables and so on. The user can see the scheme of data integration just with the XML view. Different XML view can be defined according to different requirement of data integration. This is dynamic data integration that defines the integration scheme when it is needed and gets real data when the system runs.

4 Model Application

In the integration of many business systems, the first problem is to realize the comprehensive integration of the users[13]. When the user operates on several systems, he or she usually needs to login respectively with different user name and password. To solve this problem, the databases employed in these systems are analyzed first. The frequently employed databases are Microsoft Access, MySQL and Oracle. We use Visual Studio 2005 to design and implement. Access and Oracle are accessed through ADO.Net provided by Microsoft. MySQL is accessed by the official driver provided by MySQL. So we need only to design the related classes and method for MySQL in the uniform data accessor. The related code is as follows:

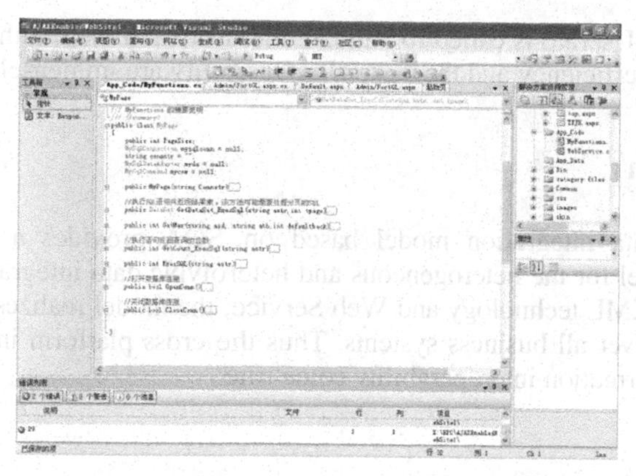

Fig. 3. Design interface of dynamic data integration model uniform data accessor

After finishing the programming work in the uniform data processing interface, we implement the XML data integration view as a web service. The interface of publishing information list is shown as follows:

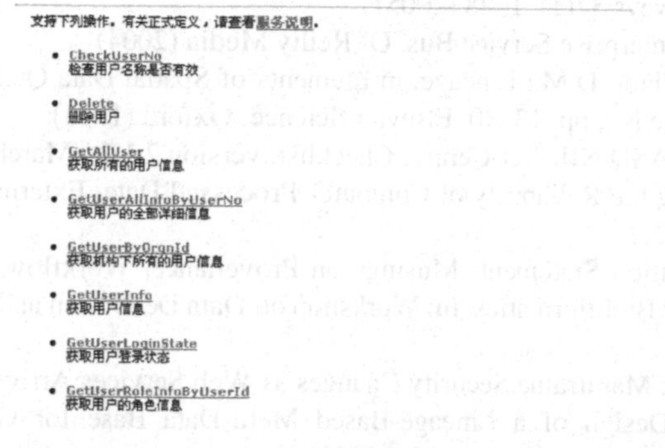

Fig. 4. Service list interface of Web Service in dynamic data integration model

Related service description information is shown in Fig.5.


```
<?xml version="1.0" encoding="utf-8" ?>
- <wsdl:definitions xmlns:soap="http://schemas.xmlsoap.org/wsdl/soap/"
    xmlns:tm="http://microsoft.com/wsdl/mime/textMatching/" xmlns:soapenc="http://schemas.xmlsoap.org/soap/encoding/"
    xmlns:mime="http://schemas.xmlsoap.org/wsdl/mime/" xmlns:tns="http://www.cysoft.dev/"
    xmlns:s="http://www.w3.org/2001/XMLSchema" xmlns:soap12="http://schemas.xmlsoap.org/wsdl/soap12/"
    xmlns:http="http://schemas.xmlsoap.org/wsdl/http/" targetNamespace="http://www.cysoft.dev/"
    xmlns:wsdl="http://schemas.xmlsoap.org/wsdl/">
  + <wsdl:types>
  + <wsdl:message name="SaveUserSoapIn">
  + <wsdl:message name="SaveUserSoapOut">
  + <wsdl:message name="SaveUserCYSoapHeader">
  + <wsdl:message name="SavePasswordsSoapIn">
  + <wsdl:message name="SavePasswordsSoapOut">
  + <wsdl:message name="SavePasswordsCYSoapHeader">
  + <wsdl:message name="DeleteSoapIn">
  + <wsdl:message name="DeleteSoapOut">
  + <wsdl:message name="DeleteCYSoapHeader">
  + <wsdl:message name="GetAllUserSoapIn">
  + <wsdl:message name="GetAllUserSoapOut">
  + <wsdl:message name="GetAllUserCYSoapHeader">
  + <wsdl:message name="GetUserInfoSoapIn">
  + <wsdl:message name="GetUserInfoSoapOut">
  + <wsdl:message name="GetUserInfoCYSoapHeader">
  + <wsdl:message name="GetUserAllInfoByUserNoSoapIn">
  + <wsdl:message name="GetUserAllInfoByUserNoSoapOut">
  + <wsdl:message name="GetUserAllInfoByUserNoCYSoapHeader">
  + <wsdl:message name="GetUserRoleInfoByUserIdSoapIn">
  + <wsdl:message name="GetUserRoleInfoByUserIdSoapOut">
  + <wsdl:message name="GetUserRoleInfoByUserIdCYSoapHeader">
  + <wsdl:message name="GetUserByOrgnIdSoapIn">
  - <wsdl:message name="GetUserByOrgnIdSoapOut">
      <wsdl:part name="parameters" element="tns:GetUserByOrgnIdResponse" />
    </wsdl:message>
```


Fig. 5. Service information description interface of Web Service in dynamic data integration model

Service CheckUserNo is called to validate the users and their authorized rights. So the development efficiency and the system extensibility are immensely improved.

5 Conclusion

The dynamic data integration model based on SOA provides a service oriented architecture model for the heterogeneous and heterotypic data integration and sharing. On the basis of XML technology and Web Service, the model realizes the data sharing and integration over all business systems. Thus the cross platform integration of data resource and information interoperability come true.

References

1. Bell, D., Padula La,L.: Secure Computer System: Unified Exposition and Multics Interpretation. Technical Report, MITRE Corporation (March 1976)
2. Bishop, M.: Computer Security: Art and Science (2002)
3. Bose, R., Frew, J.: Lineage Retrieval for Scientific Data Processing: A Survey. ACM Computing Surveys 37(1), 1–28 (2005)
4. Chappell, D.: Enterprise Service Bus. O' Reilly Media (2004)
5. Clarke, D.G., Clark, D.M.: Lineage, in Elements of Spatial Data Quality. In: Guptill, S.C., Morrison, J.L. (eds.), pp. 13–30. Elsevier Science, Oxford (1975)
6. DCIO, DOD OASD NII, Net-Centric Checklist, version 2.1.2, (March 31, 2004)
7. Gao.: Assessing the Reliability of Computer-Processed Data, External Version 1 (October 2002)
8. Goble, C.: Position Statement: Musings on Provenance, Workflow and (Semantic Web) Annotations for Bioinformatics. In: Workshop on Data Derivation and Provenance, Chicago (2002)
9. Kernochan, W.: Mainframe Security Changes as Web Services Arrive
10. Lanter, D.P.: Design of a Lineage-Based Meta-Data Base for GIS. Cartography and Geographic Information Systems 18 (1991)
11. Paul, R.: DoD Towards Software Services. In: Proc. of 10th IEEE International Workshop on Objectoriented Real-time Dependable Systems (WORDS 2005), pp. 3–6 (February 2005)
12. Portougal, V.: Business Processes: Operational Solutions for SAP Implementation. Idea Group Pub. (December 2005)
13. Ross, S.: Applied Probability Models with Optimizing Applications. Holden-Day, San Francisco (1970)
14. Singh, M.P., Huhns, M.N.: Service-Oriented Computing. John Wiley & Sons (2005)
15. Tsai, W.T.: Service-Oriented System Engineering: A New Paradigm. In: IEEE International Workshop on Service-Oriented System Engineering (SOSE), Beijing, pp. 3–8 (October 2005)

A Distributed Website Anti-tamper System Based on Filter Driver and Proxy

Jun Zhou[1], Qian He[2,*], and Linlin Yao[2]

[1] Wuhan Digital Engineering Institute, Wuhan, Hubei 430074, China
[2] Key Laboratory of Cognitive Radio and Information Processing of the Ministry of Education, Guilin University of Electronic Technology, Guangxi Guilin 541004, China
zhou_jun_123@263.net, treeqian@gmail.com, ikgg@hotmail.com

Abstract. The problem of Websites being attacked and has seriously affected the security of the Internet. A filter driver and proxy based website anti-tamper system (FPWAS) is proposed in this paper. FPWAS consists of five sub systems including web file monitor, content publish, web proxy, backup and recovery, and monitor center. The file filter driver is used to protect web program files are tampered illegally, and the protected web sites are run behind a web proxy which has intrusion detection function. Experiments show that FPWAS can work well and has acceptable concurrent performance.

Keywords: Website Protection, Anti Tamper, Filter Driver, Proxy, Intrusion Detection.

1 Introduction

In order to fully enjoy the use of information resources on the Internet, almost all government departments, companies, colleges, research institutes have established their own websites. However, some malicious intrusions easily exploit these vulnerabilities and flaws to attack the website, thereby undermining the normal operation of websites. The National Computer Network Emergency Response Technical Team Coordination Center of China has found that in the first half of 2008, the total number of sites in mainland China has been tampered with to 351, compared with the year 2007, an increase of nearly 23.7%. The China's Internet status white paper shows that China is facing serious network security threats in 2009, China's IP addresses are controlled by overseas more than 100 million, and over 42,000 websites are been tampered [1]. Thus, how to resist the invasion of illegal users, ensure the safety of websites have become a growing problem.

The reference 2 is based on a Web server component designed to achieve tamper-resistant pages, according to different needs of different built-in Web server technology. While adding a new Web server will need to re-design, it is difficult general. Currently a lot of well-known commercial systems such as InforGuard pages tamper system [3], Cylan webpage tamper system [4] are all based on kernel embedding or digital watermark. The windows I/O file filter can monitor the input/output file operations

* Corresponding author.

D. Jin and S. Lin (Eds.): Advances in MSEC Vol. 1, AISC 128, pp. 415–421.
springerlink.com

(create, read, write, rename, etc.) of each program, so filter driver is often used for anti-virus, security, backup, software snapshots and so on[6,7]. As an effective supplement of Firewall, Intrusion Detection System (IDS) check messages through the network and match them with attack signatures for safety testing. But both Firewall and IDS work at the network layer, can't fully understand the application layer protocols, so there must be much omissions and errors [5].

In this paper we design a filter driver and proxy based website anti-tamper system (FPWAS) that resists illegal website tampering and attacking. FPWAS uses a distributed system design, consists of five components including web file monitoring, content publish, web proxy, backup and recovery, and monitoring center. The windows I/O file filter driver is used to protect web program files, and the protected web sites are run behind a web proxy which has intrusion detection function. FPWAS can protect various language web programs, such as jsp, asp.net, php and so on.

The rest of this paper is organized as follows. Section 2 presents the architecture of FPWAS. Section 3 and 4 describe the designs of filter driver based web protection and proxy based web intrusion protection. Section 5 analyzes the functions and performance of FPWAS. Finally, we conclude the whole work in Section 6.

2 Architecture of FPWAS

A distribute website distribution anti-tamper system is designed for FPWAS to protect website comprehensively. FPWAS consists of five sub systems including web file monitor, content publish, web proxy, backup and recovery, and monitor center as shown in Fig 1.

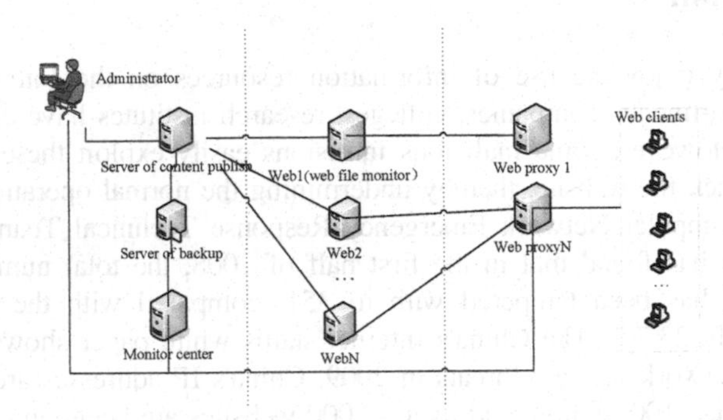

Fig. 1. Distribute architecture of FPWAS

The web file monitor subsystem runs on web servers including application and driver components. All the operations such as file creating, deleting and modifying about web directories are monitored by driver component which is based on filter driver programming. While illegal modifications are found, messages can be forbidden and sent to application components to warn.

The web proxy subsystem runs on independent servers. It not only achieve back-WWW server to forward requests and responses, but also checks semantics, protects dynamic pages, and resists SQL injection and cross site scripting attacks. Semantic check can filter illegal content efficiently, dynamic page protection compares Hash of key dynamic pages to find weather there is a unreasonable change or not, and prompt monitor center. All the HTTP contents of POST and GET are parsed based on web intrusion detection to prevent SQL injection and cross site scripting attacks.

The content publish subsystem is used to publish web program files to web server and backup server. Only these contents are legal through this subsystem, otherwise content are illegal. The processes of content publish use SSL and digital certificate technology.

The backup and recovery subsystem can backup program files of websites. When the websites files are lost or illegal modified, they can be recovered based on this subsystem.

The monitor center subsystem is a unified administrator platform for all these subsystems. The running information of web servers, webpage intrusion detection warning, interactive information between subsystems and other information are shown in this subsystem.

3 Web Files Protection Based on Filter Driver

All the Web files' activity is monitored based on filter driver. Firstly, the web directory, the exception directory and files in web directory that can be set freely by administrator. Only these contents are legal which are published through content publish subsystem. The filter processing flow is shown in Figure 2.

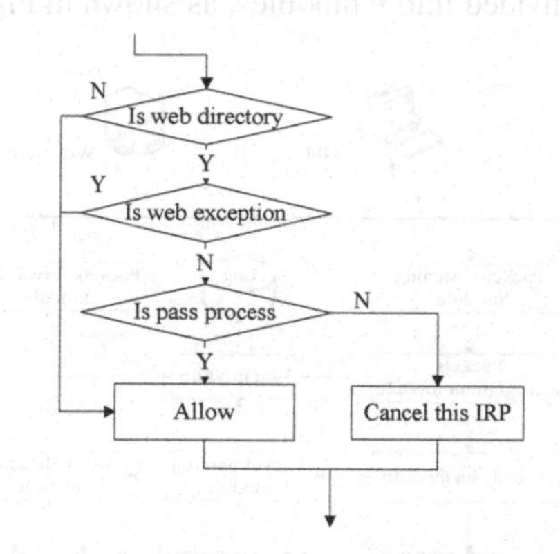

Fig. 2. Processing flow of web file operation filter

The kernel process of file protection is realized in the driver component of web file monitor subsystem. The main two steps are as follows:

Step 1: The dispatch function [IRP_MJ_CREATE] is set in DriverEntry. Modifying and overwriting file and directory may trigger this event, and the file location also can be got from this dispatch function [IRP_MJ_CREATE]. The main code is as follows:

```
DriverObject->MajorFunction [IRP_MJ_CREATE] = FsFilterDispatchCreate;
```

Step 2: Realize the function FsFilterDispatchCreate whose main code is as follows:

```
NTSTATUS FsFilterDispatchCreate(
__in PDEVICE_OBJECT DeviceObject, __in PIRP Irp)
{
    ...
CreateDisposition = (irpSp->Parameters. Create.Options >> 24) & 0x000000ff;
if(CreateDisposition==FILE_CREATE||CreateDisposition==FILE_OPEN_IF||CreateDispos
ition==FILE_OVERWRITE_IF)
 {BOOLEAN b=GetFullPathByFileObjectW (FileObject, fileName) ;
  if(b){
Forbidden = BlockFileOperation(fileName); // Decide whether a file can be operate or not
after matching. }}
   if(Forbidden)
{ return FsFilterDispatchForbidden(Irp);
return FsFilterDispatchPassThrough(DeviceObject, Irp);
}
```

4 Web Intrusion Protection Based on Proxy

In FAWAS, clients can't visit web servers protected, and they must be forward by the web proxy. There is an intrusion detection protection system (IDS) embedded in the proxy. The basic functional requirements of the web intrusion protection are to receive client requests and server responses, analyze the request and response messages, decrypt HTTPS message, forward the safe message and reject the dangerous ones, and at last log the client requests and detective results. The web proxy with intrusion detection protection is divided into 9 modules, as shown in Figure 3.

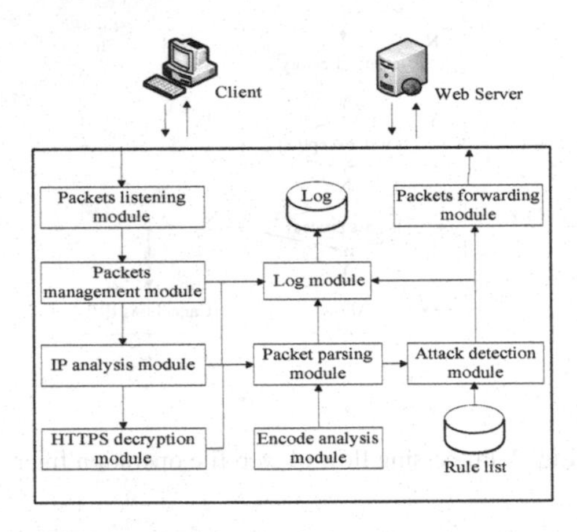

Fig. 3. Modules of Web proxy with intrusion detection protection

In the messages listening module, it listens to the ports of reserve proxy server in real time and receives client requests and server responses. If the number of client requests is over concurrency limit of system configuration concurrency limits, the redundant messages will be rejected and the client will get a no-response reply.

For each message entering the system, a single thread will be created from the thread pool thread and all the following operation to the message will run in the thread independently through the messages management module. If the message is forwarded or rejected, the thread will recycled by the thread pool.

There are both black list and white list in the IP analysis module. Firstly, the IP analysis parse module extracts the client IP address, then looks for the black list, If found, access is denied; Otherwise, looking for the white list, If found, the content is forwarded directly. And if client IP is neither in the black list nor the white one, request message will be forwarded to the next block. The lists can be build not only by the system automatically but also manually configured by the user.

The HTTPS decryption module can decrypt or encrypt the HTTPS messages between the client and the web servers. The Messages parsing module extracts various parts of the message which the next detection step needs, and all the parts are saved in an object list.

In order to cheat and cross the existing IDS, Many hackers use various methods allowed by the web server to encode part of HTTP header or payload, such as URL encoding. Through the encode analysis module, it's easy to identify the content of SQL injection and XSS attacks.

Attack detection module parse match the object list created at step 5 above with the attack rules list which is predefined by the experts. And if attack action is found, client request is denied. The module also detects the responses back from the web server, if privacy leaks, data theft and other danger acts are found, server responses will be prevented. The attack character set is constituted by a rules list, and a simple rule is as follow.

```
<rule>
No=No.001
Name=SQLInjection_Select_from
Where=RequestParam
Pattern=SELECT.*FROM
</rule>
```

This rule which is used to detect a class of "select * from" SQL injection attack matches the URI parameters of the HTTP header which and the variable "Pattern" is expressed by regular expression.

The messages forwarding module is responsible for forwarding the safe client request to the server and giving the server responses back to the client.

The log module will log the client access requests, discovered attacks, privacy leaks and other system actions. Logs can help the administrators to analyze system performance, track and position of the invaders.

5 Experiment Analysis

FPWAS is deployed in a local area network, the network topology of which is like Figure 4. Besides the publication and monitor center are deployed on the similar sever, the other subsystems are deployed on independent servers. The hardware of each server are same (CPU Xeon E5503 2G, 4G memory).

After monitoring settings control rules are set through the monitor center, the web server is protected by the web file monitor subsystem. All the operations such as file creating, deleting and modifying about web directories are monitored and controlled. When the illegal tampering taking place, the monitor center may warn and forbid it to happen. After embedding the file filter driver based monitor subsystem on web server, it is running very fast, almost no impact on access to the web site.

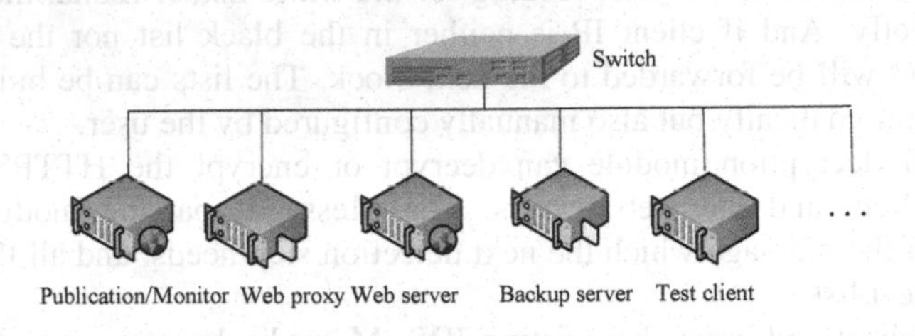

Fig. 4. Network topology of experiment

A simulating website is established using ASP programming, and a SQL vulnerability detection tool "Wis.exe"8 is used to scan that website. After scanning, there are nine possible SQL injection vulnerabilities. In the next, the simulating website is deployed on the FPWAS, and it is scanned by Wis.exe again. In this time, all the SQL injection vulnerabilities are not found, and the web proxy subsystem warn that there are attacks from test client which runs Wis.exe.

The concurrent performance of web proxy with IDS is tested using the tool "Web server stress tools"[8]. The stress tool is installed on the test client server which has a 1G bandwidth network card, the configuration is list in Table 1.

Table 1. Configuration of Webserver Stress Tool

Parameters	Values
Duration of single test	1min
Response timeout	120s
Analog network speed	1024Kbit/ s
Number of concurrent users	100,200…1000

With the growing number of concurrent users, the ratio of no response is nearly similar, but the delay of response is increasing. The rate of no response and the delay time with and without FPWAS are given in Figure 5 and 6.

When the concurrent users are 1000, the response time becomes 0.552 s, which can be accepted for normal customers, but there is an increase of 50% comparing with without FPWAS. Since there is always some no response according to Figure 7, the hardware performance of web and web proxy are all not good. So we think that the obvious increase of delay may be caused by the hardware of web proxy, the low process performance of which brings the delay. On the other hand, the delay should be decreased by optimizing the packets forwarding and attack detection modules of the web proxy in the next time.

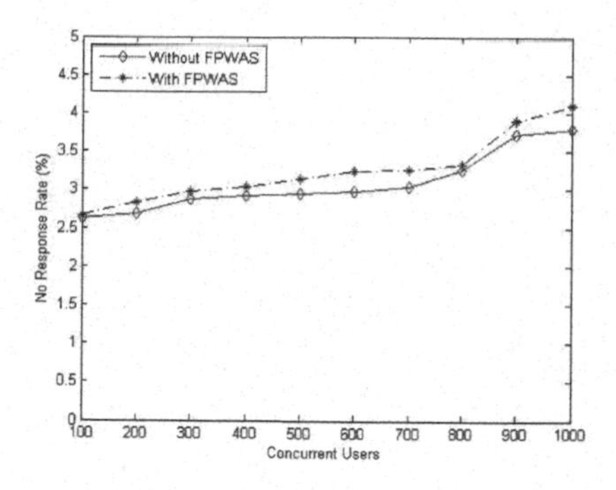

Fig. 5. The ratio of no response

Fig. 6. Delay time of website

6 Conclusion

In this paper we design a filter driver and proxy based website anti-tamper system that can avoid that the protected web program files are modified illegally, and that the protected websites are attacked by hackers. FPWAS consists of five sub systems including web file monitor, content publish, web proxy, backup and recovery, and monitor center. The file filter driver is used to protect web program files in the web file monitor subsystem. It is safe and fast for the web file monitor works in the low driver layer. The intrusion detection functions are embed in the web proxy, so the website written by various web program such as jsp, asp, php and so on can all be protected. We believe that FPWAS will be a valuable component for website protection. The concurrent performance of web proxy will be optimized in the future works.

Acknowledgment. This work was supported by the National Natural Science Foundation of China (No. 61163058) and Guangxi Natural Science Foundation of China (2011GXNSFD018023).

References

1. News Office of State, the White Paper of Status of Chinese Internet, `http://www.gov.cn/zwgk/2010-06/08/content/1622866.htm` (2010)
2. Zhang, L., Wang, L., Wang, D.: Model of Webpage Tamper-Proof System. Journal of Wuhan University (Science Edition) 55(1), 121–124 (2009)
3. `http://www.inforguard.com`
4. `http://www.jlsykj.com.cn`
5. Ryutov, T., Neuman, C., Kim, D., Zhou, L.: Integrated Access Control and Intrusion Detection for web servers. IEEE Transactions on Parallel and Distributed Systems 14(9), 841–850 (2003)
6. Oney, W.: Programming the Microsoft Windows Driver Mode, 2nd edn. Ebook. Microsoft Press, Washington (2003)
7. Zhang, F., Shi, C.-C.: Detailed Windows Driver Development. Electronic Industry Press, Beijing (2008)
8. Xiao, R.: WIS, `http://www.netxeyes.com/main.html`

The Application of TDL in the Windows Intelligent Control

WangNa and YanxiaPang

Computer and Information Department
Shanghai Second Polytechnic University
Shanghai, China
wnoffice@126.com

Abstract. TDL is an advanced programing language that can adapt to hard real-time control. This article presents the programming method of TDL and designs the timing program of the windows intelligent control system based on TDL= and describes how to use TDL in the actual application. The characteristic of separating the timing program from functional program in the process of TDL design makes TDL program independent from the running platform, improving the robustness, stability and coding reuse.

Keywords: TDL, timing program, windows intelligent control, hard real-time control.

1 Introduction

In recent years, with the development of computer technology, control technology and information technology, industrial production and management methods have stepped into production automation and control intelligence era. Especially in the applications of distributed embedded control system, we need standardized, real-time control module and I / O modules to implement system integration and build complex distributed systems that can adapt to the harsh environment. The main characteristic of distributed embedded control system is the interaction of software and hardware, such as production processes, intelligent traffic control network, distribution of the atmosphere, environment and mine safety monitoring. These applications all have hard real-time requirements.

TDL is a programming language applying to hard real-time control. The application of high-level TDL-based embedded control design is more suitable for hard real-time control application with constraints. A TDL program can clearly explain the integration of real-time interactive software and improve the reliability and reusability of control system. On a given platform, TDL compiler can automatically generate timing procedures to ensure the specified behavior of the object. TDL is especially suitable for embedded systems whose physical equipment involved process's behavior and time.

2 About TDL

Traditional real-time embedded software design is based on abstract mathematical models. Control engineers design the functionality and performance models by the

D. Jin and S. Lin (Eds.): Advances in MSEC Vol. 1, AISC 128, pp. 423–428.
springerlink.com

software tools solving the model based on the object behavior and environmental impact. Then the models are given to software engineers for coding, testing and optimizing on a given platform, until obtaining a satisfactory time behavior. In this process, the close correspondence between the code and model and the software reusability is lost and the complexity of software design increased. So it's difficult to apply the weakened software in different platforms.

TDL-based embedded control software separate functional program from specific platform and separate the time from the function. The platform independence makes the design has better real-time, reliability and reusability that is more suitable for embedded real-time distribution system. Firstly, control engineers and software engineers design a TDL program agreed to control design's functions and time. Then, the software engineers use the TDL compiler generates an executable code that can connect TDL run-time library to map the program to a given platform. TDL run-time library provides a middle layer for scheduling and communications which defines the interface between TDL executable program and platforms.

3 TDL Programming

3.1 TDL's Structure

TDL is designed for control applications requiring periodically reading the sensor, calling the task, updating implementation and conversing mode. TDL programming based on time-triggered and the communication from the sensor to the CPU and the CPU to the actuator are triggered by the global clock. The advantage is the design is platform-independent and compatible with any real-time operating systems, scheduling algorithms and real-time communication protocols.

TDL program is a multi-modal and multi-frequency system for long-running tasks. Its structure is hown in Figure 1 which including sensors, switches and actuators.

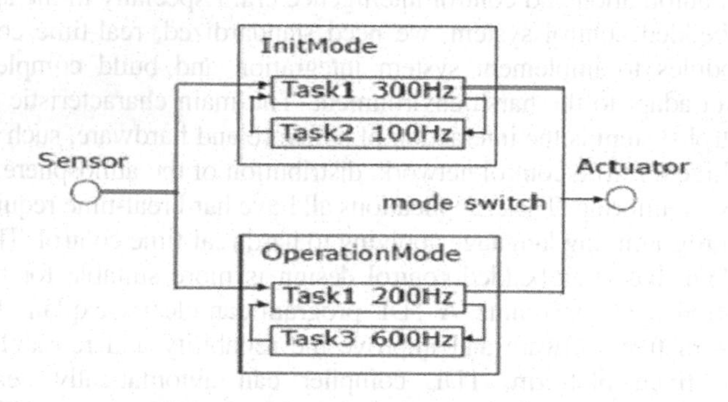

Fig. 1. TDL's structure

TDL's two core components are the periodical task calling and mode conversion. A TDL program specifies a series of models and each model also contains a series of tasks and mode conversion.

TDL introduced the concept of the module, re-organizing model statement. TDL's each mode declaration has a "start" at the beginning. In addition to the global output ports, TDL also provides task output ports. TDL hasn't a clear switch driver for the task and mode conversion which is merged into the model declaration.

3.2 TDL Program Flow

TDL program executes after switching to start mode. At logical zero time, there is neither the actuator updating nor mode conversion. At the execution point, TDL reads data from the sensor in real time in order to determine what to do next, even if there will be more than one module to access the same sensor at the same time. IT's shown as Figure 2:

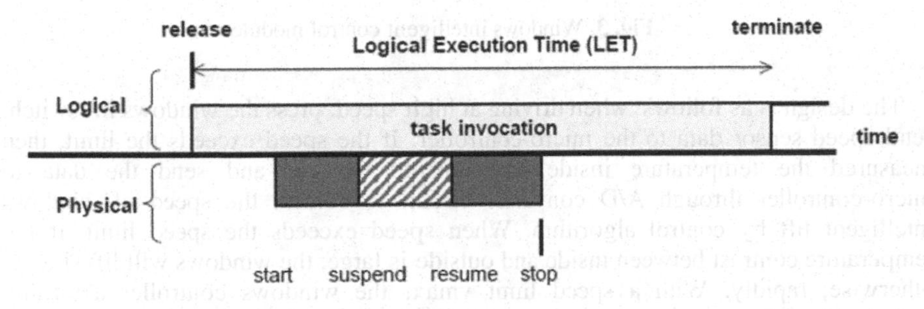

Fig. 2. TDL program model

TDL task is a periodic job, the updating of input and output ports respectively at the task cycle's start and end point. TDL task does not have to start at the beginning of the cycle, but should start and complete within the cycle. A TDL task can be considered as a unit of work. If mode switch occurred, the task does not continue to run.

TDL evaluates each monitor function in accordance with the order of top-down and implement the first mode switch which returns a true value, thus it provides a non-deterministic way with higher efficiency.

4 TDL-Based Windows Intelligent Control

4.1 The Composition and Design of Windows Intelligent Control

With the popularity of cars, the safety of cars has been more and more important.Considering traditional windows control system, when driving at a speed over the speed limit, if the temperature contrast between inside and outside is too large, the stability of vehicles will be impacted by the strong airflow, resulting in short-term discomfort to the driver, which may lead to traffic accidents. We will use the windows control system based on temperature and speed to give a brief description of TDL in the actual control design applications.

Windows intelligent control system consists of microcontroller,two sensors to read the current outside and inside temperature, a speed sensor and a motor drive module. It's shown in Figure 3.

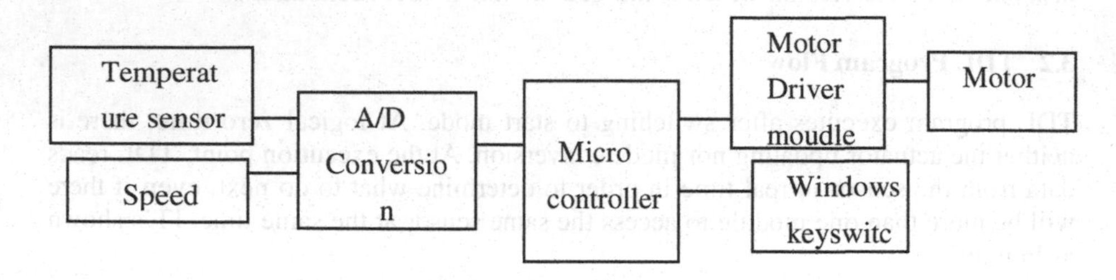

Fig. 3. Windows intelligent control module

The design is as follows: when driving at high speed, press the windows lift switch, send speed sensor data to the micro-controller, if the speed exceeds the limit, then measured the temperature inside and outside the car and send the data to micro-controller through A/D converter circuit to control the speed of windows intelligent lift by control algorithm. When speed exceeds the speed limit, if the temperature contrast between inside and outside is large, the windows will lift slowly, otherwise, rapidly. With a speed limit vmax, the windows controller algorithm includes the following steps:

(1) Obtaining a group of inside temperature X1, X2 ... Xn and a group of outside temperature Y1, Y2 ... Yn through temperature sensors.

(2) Obtaining the average of the two sets Xm and Ym. Temperature difference between inside and outside is n=|Xm-Ym|.

(3)Determining whether n is in the range of body's adapt, if not, the window motor forward rotates to control the windows up, on the contrary, the window motor reverse rotates to control the windows down.

4.2 The Structure of TDL Program in Windows Intelligent Control System

For TDL program, the first statement is all of the ports, then the tasks and mission modes. The componets in this case are as follows: port declaration consists of three sensor named tempinner,tempouter, carspeed, an Actuator window motor named wmachine.

Mission statement includes the task seeking the engine voltage named DCMotorController, measuring the temperature difference named TempSub, measuring the speed named SpeedTask, rising windows named WupTask, windows down named WdownTask, idle named IdleTaskk. TempSub detects the temperature difference according to the inside and outside temperature data from the sensors. WupTask and WdownTask control window up or down according to temperature difference and the car speed. Idle task initializes each port. It is noteworthy that all tasks are atomic execution unit that can not be interrupted in the implementation process, and has not set priority. Numbers 1 and 2 are the times of actuator updating and mode switch in a cycle.

Four different modes are declared: idle mode named idle, rising pattern named up, down pattern named down, closing the window mode named close. Each model has its own inherent pattern cycle, and specify the frequency of the actuator update, mode switching and task calls. As shown below, idle mode's period is 300ms, the other three mode's period are 100ms. TDL descripts this process with the following:

module ActuatorController {//This case describes the task of seeking the engine voltage and the remaining tasks in the module "carDynamics" which is imported here.

```
import carDynamics;
sensor double tempinner uses gettempinner;
sensor double tempouter uses gettempouter;
sensor double carspeed uses getcarspeed;
actuator int wmachine uses setwmachine;
public task DCMotorController {
input
double tempinner;
double tempouter;
double carspeed;
output
 double voltage;
uses
DCMotorControllerImpl(tempinner,tempouter,carspeed,voltage);
}
Start mode main [period=300ms] {
task
[freq=1] DCMotorController {
Tempinner:= carDynamics.dynamics.Tempinner;
Tempouter:= carDynamics.dynamics.Tempouter;
carspeed:= carDynamics.dynamics.carspeed;
actuator
[freq=1] voltage:= DCMotorController.voltage;
mode
[freq=1] if exitMain() then up;
}
```

```
mode up [period=100ms] {......}
mode down [period=100ms] {......}
mode close [period=100ms] {......}
}
```

4.3 The Execution of TDL Program in Windows Intelligent Control

In this control system, the start mode is idle, the program calls idletask with a period of 300ms to initialize ports of sensors and actuator, and calls tempsubtask with a period of 100ms to detect the temperature sensor, calls the speedtask with a period of 100ms to measured speed. Once the speed is greater than the maximum, and the temperature exceeds the maximum that human's body can withstand, it enters mode up during which wuptask with the period of 100ms is called. Once either the speed or the temperature does not meet the above, it enters mode down during which wdowntask with the period of 100ms is called. If the system detects a large rising resistance, it enters mode close which means the windows has safely shut down. At this point, one execution has finished and the system switch s to idle mode again.

5 Conclusion

In TDL-based embedded control system, TDL program is platform-independent due to the characteristics of separating timing program from functions program which improve the robustness, stability and code reusability of TDL program. Tasks calling and environment variables testing are triggered by the global clock, its time behavior has high degree of predictability that makes TDL be suitable for embedded control systems with hard real-time constraints. System in this article is only a partial application of TDL-based design in embedded auto. More intelligent control and optimization of timing issues are to be resolved in the embedded system that will lead to TDL's rapid devolpoment.

References

1. Christoph, M.K., Giotto, K.: A Time-Triggered Language for Embedded Programming. Proceedings of the IEEE (2003)
2. Kirsch, C.M.: Embedded Control Systems Development with Giotto. SIGPLAN Notices (2001)
3. Dr. (ETH) Josef Templ.: The Timing Definition Language
4. Pree, W., Temp, J.: lModeling with the Timing Definition Language (2006)

The System Design and Edge Detection on Island and Reef Based on Gabor Wavelet*

XiJun Zhu[1,2], QiuJu Bai[1], GuiFei Liu[1], and Ke Xu[1]

[1] College of Information Science & Technology, Qingdao University of Science and Technology, 266061, Qingdao, China

[2] Key Laboratory of Surveying and Mapping Technology on Island and Reef, State Bureau of Surveying and Mapping, China

{Zhuxj990,guifei_liu}@163.com

Abstract. Because of waves, the island and reef images have the characteristics such as large noise and big influence by light intensity changes, use Gabor wavelet's characteristics of multiple directions, multi-scale and can be able to determine the best sense of the texture frequency, this article proposed with Gabor wavelet edge detection method of the sea image edge by compared with a variety of edge detection operators. Matlab simulation results show that, by using this method the contour extraction more close to the real image, which verify the feasibility and effectiveness of the method. This paper using vc + + language design and realize the function of image preprocessing and reef edge detection. In addition, the system runs stably and basically achieved the requirement of the island and reef processing.

Keywords: island and reef, edge detection, Gabor wavelet, system design.

1 Introduction

In the 21st century, the whole world is on a large scale development of Marine resources, expanding ocean industries, developing the Marine economy. In order to maintain Marine rights and interests, and promote the economic development, China began to survey and map the island and reef. According to statistics, China's waters, the island area of 500 square meters up to 7372[1],but only part of the island have been roughly coordinated and mapped, and mapping information outdated, coordinate system is not uniform, also part of the island has yet to mapping information. So the study of mapping the island and reef is significant[2]. In order to be better on the island of surveying and mapping, we use Gabor wavelet to make edge detection of the given images.

Image edge is the most basic criterion of image detection. The traditional edge detection method, while having good real-time and easy implementation, but its positioning in edge and the anti-noise performance are often not very ideal. Use

* Supported By the Key Laboratory of Surveying and Mapping Technology on Island and Reef, State Bureau of Surveying and Mapping(2009B11), China.

D. Jin and S. Lin (Eds.): Advances in MSEC Vol. 1, AISC 128, pp. 429–434.
springerlink.com

Gabor function as filter on texture segmentation, can get multi-scale, multi-dimensional filtering results, comprehensive these results, we can capture most of the edge points in different directions.

Based on the discussion of the wavelet method algorithm used in island and reef images edge detection, We initially completed the reef image wavelet edge detection recognition system, This system aims to quickly identify and detect the island and reef' edge features in the given image, namely achieve faster or online identification function of island and reef edge.

2 The Traditional Edge Detection Operators

Edge of the image is in the form of not a continuous of the local features of image. Most of the traditional edge detection based on gradient maxima or zero-crossing second derivative to extract the image edges. The commonly used edge detection operators are the following[3]:

Sobel operator does edge detection from different directions. It is completed by using the neighborhood convolution of template in two directions and image in the image space. The algorithm is easy to achieve in space, and has a smoothing effect to noise. However, the algorithm also has the disadvantages of detecting pseudo-edge and a poor positioning accuracy.

Canny operator, using a Gaussian filter, and his first order differential, Mainly horizontal and vertical direction for processing, and then use linear interpolation to obtain the edge angle and gradient, and use the double-threshold algorithm to detect and connect the edge, Thus the boundary of obtained image is fine and positioning accuracy is higher. But it is very easy to produce double-edge, and sensitive to noise.

Robert operator using local differential operator to find the edge of the image, which is characterized by simple and intuitive, but the result is not very smooth. As Robert operator is usually generate a wide response near the edge of the image. Therefore, before using Robert operator should do the image thinning.

Prewitt operator use grayscale difference which between pixels from top to bottom, left and right neighbor reaching the extreme in the edge to do edge detection, It can remove the part of the pseudo-edge and has a smoothing effect to noise. But the edge detection accuracy is not high.

Log operator do gaussian low pass filtering to image by gaussian function, in order to eliminate the noise whose spatial scale much smaller than Gaussian space constant, then do high pass filtering by Laplacian and extract the zero crossing point. But, the effect of Log operator is better only in the case of obvious change in the gray.

3 2D Gabor Function

In recent years, Gabor wavelet transform are used widely in the information processing. Research shows that, Gabor wavelet can extract multi-direction frequency, multi-scale features of image in the particular area. 2D Gabor wavelet function defined as follows [4]:

$$\psi_{\mu,\nu}(z) = (\| k_{\mu,\nu} \|^2 / \sigma^2) \exp(-\| k_{\mu,\nu} \|^2 \| s \|^2 / 2\sigma^2)[e^{ik_{\mu,\nu}} - e^{-\sigma^2/2}] \tag{1}$$

This represented a sine wave after gaussian envelope modulated, $z = (x, y)$ shows a pixel in the image; $\| \bullet \|$ defines norm of vector; V represents the scale of the Gabor kernel function; μ represents the direction of the Gabor kernel function; Wavelength and the direction and the length of Gaussian window of the part of the shock are controlled by $k_{\mu,\nu}$, it is defined as $k_{\mu,\nu} = k_\nu e^{i\phi_\mu}$, Among them, $\phi_\mu = \pi\mu / 8$, represents the direction selectivity of the filter , change ϕ_μ can rotate the filter ; $k_\nu = k_{\max} / f^\nu$ is filter sampling frequency, $k_{\max}$ is the largest sampling frequency, f is sampling step in frequency domain , usually $f = \sqrt{2}$. Parameter determines the ratio of Gaussian window width to wavelength, the relationship between it and the bandwidth of the filter is expressed as:

$$\sigma = \sqrt{2 \ln 2} \left(\frac{2^\varphi + 1}{2^\varphi - 1} \right) \tag{2}$$

Among them, φ is half-peak bandwidth expressed by the octave. Gabor kernel is self-similar in formula (2), through appropriate scale changing and rotation of Gabor filter, that is, adjusting the scale parameter and the direction parameters, format a self-similar family of functions, namely, Gabor filters. Each Gabor kernel is a complex plane wave by envelope of gauss function. The first square brackets in formula (2) determines the shock part of the Gabor kernel, in order to eliminate the DC component of the image on the impact of two dimensional Gabor wavelet transform, the second in the formula (2) is used to compensate for the DC component, and this makes 2D Gabor function $\psi_{\mu,\nu}(z)$ as a complex function, the real part and the imaginary part are expressed respectively as:

$$R_{\mu,\nu}(z) = \frac{\| k_{\mu,\nu} \|^2}{\sigma^2} \exp(-\| k_{\mu,\nu} \|^2 \| s \|^2 / 2\sigma^2) \left[\cos(k_{\mu,\nu}) - \exp(-\frac{\sigma^2}{2}) \right] \tag{3}$$

$$I_{\mu,\nu}(z) = \frac{\| k_{\mu,\nu} \|^2}{\sigma^2} \exp(-\| k_{\mu,\nu} \|^2 \| s \|^2 / 2\sigma^2) \sin(k_{\mu,\nu}) \tag{4}$$

Thus, we can get image $I(x, y)$ through Gabor filter by the convolution to two-dimensional Gabor function and the image I.

We usually take a group of Gabor kernel function to obtain the local distinctive character with multi-scale and multi-direction, they have different scale parameters $\nu(\nu = 0, \cdots, V - 1)$ and different direction Parameters $\mu = (\mu = 0, \cdots, U - 1)$.

The selection of Gabor wavelet function parameter normally has:

Set $\sigma = 2\pi, k_{\max} = \pi / 2, f = \sqrt{2}$; U=8,so $\mu = 0,1,\cdots,7$

V=5, so $\nu = 0,1,\cdots,4$.

Of course, we can choose the most appropriate parameter values according to the actual situation. By further theoretical analysis [5] of 2D Gabor filter edge detection, we found that: for the step edge, in the case $\omega\sigma$ that is the product of Gabor filter space constant and frequency is close to 1.8 and the direction θ is $\pi/2$, edge detection has best results. Namely Gabor filter to step edge points has the largest response.

4 Simulation and Analysis of the Results

During the experiment, we use island and reef remote sensing images provided by open project funding of island and reef Key Laboratory, on the basis of the preprocessing, using Matlab to do simulation process of extracting and detection edge. First using the traditional edge detection operators to extract the edge of the island and reef, the result is shown as figure 1. And then based on Gabor wavelet doing image edge detection and extraction in different scales and direction, the procedure[6] to realization edge detection as follows:

①Choose the suitable Gabor wavelet function, do Gabor wavelet filter on the given island and reef image $I(x, y)$ in the same direction, different scales, transform scale as the case may be;

②In each different scales, get the convolution of each pixel and Gabor wavelet function, get local maximum value point, and get a possible edge;

③To set the threshold T on the edge of the image of different scales, to strike the edge point, and set the point greater than or equal to T as the edge point, set the point less than T to zero, to strike the edge of the image;

In this experiment, selected in a particular direction ($\mu = 3$), in multi-scale do Gabor filtering on island and reef image, and compared with the traditional edge detection operators treatment results.

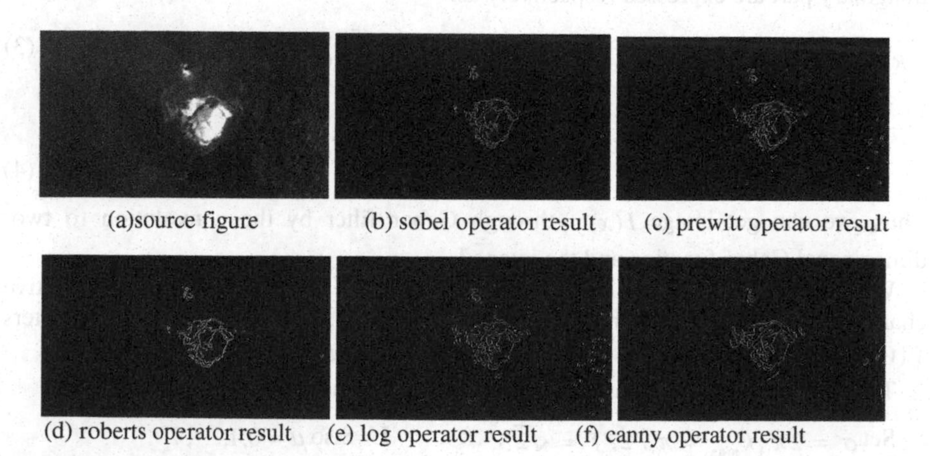

(a)source figure (b) sobel operator result (c) prewitt operator result

(d) roberts operator result (e) log operator result (f) canny operator result

Fig. 1. The traditional edge detection operators detection results

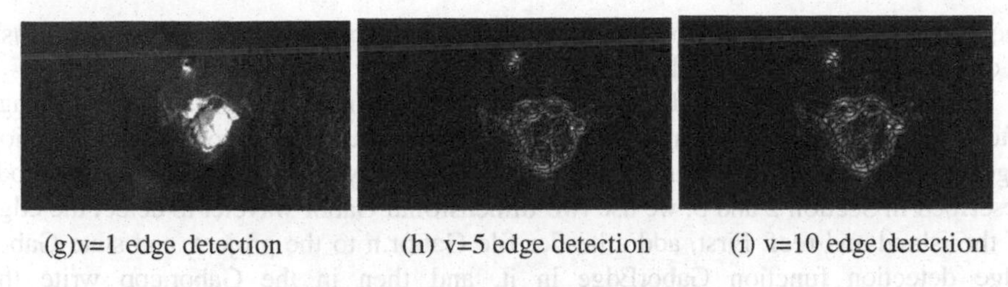

(g)v=1 edge detection (h) v=5 edge detection (i) v=10 edge detection

Fig. 2. Gabor wavelet detection results

In figure 1, figure 1 (a) is the island and reef source image , figure 1 (b), 1 (c), 1 (d), 1 (e), 1 (f) are respectively the island and reef outline extraction results by the use of the traditional edge detection operators. It can be seen from the simulation results, the edge detection effect of traditional edge detection operators on the island and reef of remote sensing images is not ideal, the image is detected more false edges, while the true edge of the target image is also lost a lot , anti-noise performance and the continuity of edge is poorer. Figure 2 (g), 2 (h), 2 (I), 2 (j) is the island and reef edge extraction results in the same direction ($\mu = 3$), different scales (v = 1, v = 5, v = 8, v = 10) by Gabor wavelet transformation. It can be seen from Figure 2, with the traditional edge detection operators detection results compared results detected by edge detection algorithm based on Gabor wavelet transform has good edge continuity, high detection accuracy , and has some anti-noise ability; In addition, it can also detect many details the classic operator cannot detect, there is also a certain resistance to the background, the outline extracted more closer to the real outline image. By comparison, can be seen in scale of v = 10 the detection effect is good.

5 System Design and Realization

The aim of this system is to quickly find the position and get the outline of the island and reef in the case of the pictures of island and reef are given, namely, it can achieve the goal of island and reef edge detection and recognition. By analyzing the whole system, it can be divided into 5 modules, the island and reef image denoising, image enhancement, the island and reef edge detection, feature extraction and recognition positioning module.

Using Microsoft Visual C + + as the development platform, the system is single document application based on MFC. In the menu bar, we add in the image denoising, image enhancement, island and reef edge detection, feature extraction and recognition positioning options.

The function of image denoising module is to denoise images of the island and reef, preparing for subsequent processing, which would reduce the pseudo-edge generated due to noise when extracting edge. The process is mainly realized in the view class. First, call command response function OnGray() to make island and reef image be gray-scale. Then respectively apply command response functions as OnAverage (), OnGaosi (), OnAlog () do 3 * 3 median filtering, gaussian smooth, 3 * 3 neighborhood average denoising to the island and reef image.

The image enhancement module is to enhance island and reef image after denoising, here, call the command response function OnDecorr () of menu item "contrast stretching" to deal with the island and reef image to stretch linear contrast, in order to extract edges of island and reef images.

The function of the island and reef edge detection module is to detect the image after denoising and enhancement edge, compared with the traditional edge detection algorithm simulation results and according to the theory and implementation methods described in Section 2 and 3, we use two-dimensional Gabor wavelet to detect the edge of the island and reef. First, add a header file Gabor.h to the project, and state Gabor edge detection function GaborEdge in it, and then in the Gabor.cpp write the realization of this function, finally do processing to island and reef image by calling the edge detection function of the command response function OnEdge(),set scale v to 10.

6 Conclusion

This paper presents a method of island and reef image edge detection based on Gabor wavelet transform theory, The results show that, based on Gabor wavelet transform can more accurately detect the edge information of the island and reef, the edge continuity is good, more accurate in positioning, and with some noise immunity, In addition, Gabor wavelet can be adjusted by changing the parameters for optimal detection. Initially completed the island and reef edge detection and recognition system, the system aims to achieve that in the case of the island and reef image is given, it can quickly identify and detect the image edge features , that is, realize the function of fast or online identification of island and reef.

Acknowledgment. This work was supported by the Key Laboratory of Surveying and Mapping Technology on Island and Reef, State Bureau of Surveying and Mapping, China (2009B11). We would like to thank Prof. Guo at the Key Laboratory of Surveying and Mapping Technology on Island and Reef for sharing their datasets.

References

1. Shi, Z., Cao, M.: A Study of Surveying and Mapping of Island and Tidal Flat Based on LiDAR. Bulletin of Surveying and Mapping 5, 49–53 (2007)
2. Zhang, Q., Shen, W.: Carry out national mapping of the island. In: China Surveying and Mapping Society of the Eighth National Congress and Comprehensive 2005 Annual Conference Proceedings, pp. 484–488. China Academic Journal Electronic Publishing House, Beijing (2005)
3. He, L.: MATLAB-based comparative analysis of remote sensing image edge detection. Computing Technology and Automation 29(2), 70–73 (2010)
4. Cao, L., Wang, D., Liu, X.: Face recognition algorithm Based on the 2D Gabor wavelet. Journal of Electronics and Information 28(3), 490–494 (2006)
5. Fu, Y.: Study of Edge Detection Theory, Fast Algorithms and Real-time Application Based On Gabor and Wavelet. Ph.D. Thesis of Zhejiang University, pp. 1–22 (2004)
6. Song, W., Wang, L., Cao, Y.: Studying the Edge Detection of Images of Plant Roots Based on Gabor Wavelet Theory. Process Automation Instrumentation 32(3), 24–28 (2011)

The Designs of the Function and Structure of the New Hydraulic Cylinder Test-Bed

DongHai Su and ZhengHui Qu

School of Mechanical Engineering, Shenyang University of Technology,
Shenyang, China
quzhenghui@163.com

Abstract. A new hydraulic test-bed of cylinder was designed and manufactured, which can perform hydraulic cylinder type test, factory test and the ultrahigh pressure test with 70Mpa. In addition, six hydraulic cylinders can be tested at the same time, and also the test-bed can simultaneously complete three types of test which are the routine test, high temperature test and ultrahigh pressure test. Besides, the new test-bed can achieve goals which are stepless and remote pressure control, multiple speed adjustment, proportional loading, automatic oil discharge, automatic control of oil temperature, noncontact measurement of hydraulic cylinder's stroke, and intelligent control of hydraulic system. In this paper, the performance and structure of the test-bed and the functions of hydraulic control system were described.

Keywords: hydraulic cylinder, test-bed, function, structure, automatic control.

1 Introduction

Cylinders are linear actuators whose output force or motion is in a straight line. Their function is to convert hydraulic power into linear mechanical power. Therefore, cylinder is an important part of the hydraulic system. And the performance of hydraulic cylinders often directly determines the performance of the machinery [1]. Moreover, the type test and the factory test of hydraulic cylinder are emphasis attached to testing the performance of the hydraulic cylinder, and it is of great significant to improve the performance and the quality of hydraulic cylinders. Also the accuracy of the measured data has a close bearing on the quality of the hydraulic cylinder. However, there are many problems of hydraulic cylinder test-bed in China, such as lacking of consistency in test methods, high labor intensity of operator, low accuracy of testing, poor degree of automation and inefficiency [2]. Meanwhile, the test items often cannot meet national standard. Therefore, we designed and manufactured a new test-bed of hydraulic cylinder to improve and overcome the above problems.

2 Functional Design of Test-Bed

2.1 Technical Parameters of the Hydraulic Cylinder Test-Bed

According to *Hydraulic Fluid Power- Test Method for the Cylinders* and *Hydraulic Fluid Power-Technical Specifications*, the main technical parameters of the new hydraulic cylinder test-bed are shown in the following.

D. Jin and S. Lin (Eds.): Advances in MSEC Vol. 1, AISC 128, pp. 435–440.
springerlink.com

Firstly, the accuracy grade of the test-bed is B. Secondly, the low pressure system can realize pressure of 8Mpa and flow of 270L/min. And the high pressure system can achieve pressure of 31.5Mpa and flow of 90L/min. Then the ultrahigh pressure can reach pressure of 70Mpa and flow of 6L/min. Thirdly, the maximum loading force is provided is 230T by loaded cylinder. And then, the test-bed can be used for different size cylinders for factory test and type test. And the largest size of the cylinder that can be tested is 320mm of cylinder bore and 12m of stroke. Finally, to meet production needs of enterprise, we also added the function of the ultrahigh pressure test.

2.2 Functional Design of the Hydraulic System

Considering the different requirements for temperature and pressure of the different tests along with the different functions of the system, the functions of test-bed are divided into four parts: routine test system, high temperature test system, ultrahigh pressure test system and oil discharge control system [3].

2.2.1 Routine Test System

The routine test system includes initial starting, the characteristics of actuating pressure, pressure test, durability test, leakage test, cushion test, load efficiency test and stroke test. And all these tests are included into the contents of national standard.

Currently, it is universally acknowledged that the problems of hydraulic cylinder test-bed are simple function, poor applicability and low efficiency. Therefore, to solve these problems, routine test system of the new test-bed has one high-pressure pump and three hydraulic pumps which the flow is 90L/min. Then the system can meet different size cylinders are tested under the different pressure, and can realize the complex speed control of the cylinder, steady reciprocating, fast response and high reliability. Moreover, there are two ways to control the pressure in the system and realize manual operation and automatic control to reduce labor intensity of operators. One is electrohydraulic pressure relief valve, the other is remote adjustment valve. Meanwhile, to reduce interference and error and to make sure the test result truthful and accurate, we set up an independent hydraulic reservoir at room temperature in the routine test system. The system can control the oil temperature at working by cooler, heater and temperature-sensing device. And with these designs, the system can work normally, and accomplish the automatic closed loop control of the oil temperature by industrial personal computer (IPC) and programmable logic controller (PLC) [4].

2.2.2 High Temperature Test System

The concept of the high temperature test is that the hydraulic cylinder reciprocating action one hour under the oil temperature of 90°C. There is no denying that the viscosity of the oil will decrease with temperature increasing. And decreasing in viscosity of the oil often results in increased leakage, poor lubrication and increased wear. Because the linear expansion coefficient and the conditions of heating or cooling are different in different components of system, the deformations of different components are not the same. So these factors often lead to the destroying of the original fit clearance, the increase of the frictional force, the locking of the valving element and the aging of components [5]. For these reasons, there is an independent oil reservoir at high temperature and an independent pump with some ancillary rigid tube in the high temperature test system.

These designs can protect other hydraulic components which are not involved in the test, enhance stability and reliability of system, extend the operating life of hydraulic components, and avoid interference or error between different types of test. Meanwhile, the system can also realize the automatic closed loop control of the temperature of high temperature reservoir by IPC and PLC.

2.2.3 Ultrahigh Pressure Test System

Generally speaking, the ultrahigh pressure can result in increasing of the leakage and reducing of the volumetric efficiency. It also can fast age the hydraulic components and damage the tube. Besides, ultrahigh pressure test system often requires sealing part with high strength to avoid breakdown. Furthermore, the increased pressure will lead to rising in temperature. So the system requires the sealing part has a good heat resistance. Because of the unique nature of ultrahigh-pressure test, test-bed has a specialized high pressure pump and ancillary rigid tube to provide a source of oil [6]. With these designs, it can separate the ultrahigh pressure test system from the other systems, protect other components from damaging and avoid sending the wrong signals by pressure transducer to interfere with the normal operation. In order to accelerate the speed of the oil into hydraulic cylinder and improve the system efficiency, the test-bed uses two three-way valves for reciprocating action of the hydraulic cylinder.

2.2.4 Oil Discharge Control System

In the past, hydraulic cylinder used to be drained oil offline. This approach increased the workload of the operator, caused waste and environmental pollution. However, the oil discharge control system of the new test-bed is composed of air pump, silencer, reversing valve, etc. And the system also has two pilot-controlled check valves which are controlled by hydraulic pressure, and two cartridge valves. By controlling the pressure of system with a special pump, it controls the check valves open or close, realizes two chambers of hydraulic cylinder draining oil online, and separates the oil with the gas when the oil is drained. Thereby, these designs increase the efficiency of draining oil, reduce the labor intensity, keep the test-bed clean and reduce production costs.

Moreover, the test-bed uses IPC as a host computer with PLC as a secondary computer to control the whole system. So that the whole control system has some functions, such as user-friendly, fault tolerance, help and guide. Meanwhile, all tests have two modes of operation which are manual operation and automatic control. Also there are other functions, such as real-time observation of tests, automatic data storage, automatic reciprocating, automatic timing and counting, maintaining the pressure and security [7]. The stroke of hydraulic cylinder is measured by laser displacement sensor. And the laser displacement sensor has the features of high accuracy, low interference and large stroke measurement. Therefore, the test-bed enables non-contact measurement of hydraulic cylinder's stroke.

3 Structural Design of the Test-Bed

According to functional design of test-bed and requirements of actual installation, overall structure of the test-bed consists of the power unit, the valve terminal and the control console. And the structure of the test-bed is shown in the Fig.1.

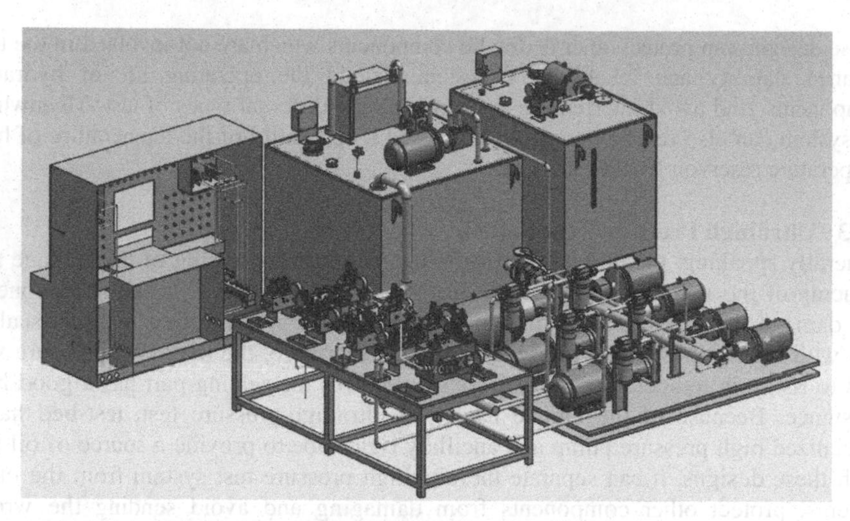

Fig. 1. The structure of the hydraulic cylinder test-bed

3.1 Power Unit

Taking into account that the requirement of oil temperature is different for different tests, the test-bed has two hydraulic reservoirs which are normal reservoir and high temperature reservoir to protect the hydraulic components and achieve synchronous test. Then the normal reservoir is needed to meet different tests to different cylinder bore, and performed the routine test by multiple hydraulic cylinders at the same time. Therefore, the test-bed needs much oil to ensure normal running of the system. And the reservoir fluid capacity is 3000L. On the contrary, the high temperature reservoir only runs the high temperature test, so the reservoir fluid capacity is 1500L. Furthermore, there are eight pumps in the routine test system and ultrahigh pressure test system of the test-bed. And they both use the normal reservoir to provide oil for the system. Due to its small size and light weight, the circulation pump adopts up-setting structure to play the role of filtering and cooling oil. The concept of the up-setting structure is that the pump is located on the cover of the reservoir to reduce the layout of rigid tube and to save space.

However, the quantity and volume of the remaining seven pumps are both larger, and the volume of the normal reservoir is also large. So these pumps adopt the side-setting structure. The meaning of the side-setting structure is that these pumps are installed in a special foundation to make the pumps absorb oil easier, improve efficiency of the system and to install and maintain more convenient. Moreover, there is only a pump in the high temperature system, so the up-setting structure is adopted [8]. And in order to prevent self-excited vibration and reduce vibration and noise, all electrical machines are equipped with vibration dampers and pallets. Besides, the location of the outlet port of reservoir is set lower. And the outlet port is connected to pumps by a butterfly valve. It leads to maintaining the power unit and replacing the oil more convenient and safer than before by opening and closing the butterfly valve. In a word, with these designs, the test-bed can alleviate cavitation and improve the ability of absorbing oil.

3.2 Valve Terminal

All control valves adopt the installing form of manifold block. And this installing form has many advantages, such as compact structure, convenient maintenance, low cost and relatively low pressure loss. As a result of the whole system is very complex, the number of test items and control valves is huge, the volume of manifold block is larger and pipe installation is more complex than other test-beds [9]. Therefore, there is an independent valve terminal next to the power unit, and it is provided a flow meter so that the whole system can be more easily observed and controlled. Among all manifold blocks, there are three interfluent manifold blocks to achieve all tests in the routine test system and adjust the speed of the hydraulic cylinder. Because the flow is large and diameter of rigid tube is thick, these manifold blocks should be placed close to the corresponding pumps to reduce cost and the losses of the power and pressure. For other manifold blocks, due to the flow is relatively small, they can be installed in the distal side of the valve terminal. Moreover, all tube is arranged on the base of electric machine and below the valve terminal by pipe clamps. And with these designs, it leads to installing and maintaining the tube easier and remaining the test-bed beautiful and clean appearance.

3.3 Control Console

There are some manual buttons, remote control valves, pressure gauges, numerical display devices and indicating lamps on the operation panel. And the interim manifold blocks, PLC, IPC and proportional amplifiers are installed in the control console. And with these designs, the test-bed can realize human-computer interaction, accurate test data, real-time date observation and recording, simple and safe operation, and beautiful appearance [10].

4 Conclusion

The new hydraulic test-bed of cylinder which is designed and manufactured can well achieve the desired goals, such as reliable system performance, fewer failures, high test precision, comprehensive test item, simple and safe operation, computer operation and control, compact structure, advanced technology performance and high degree of automation. It also can reduce the labor intensity, meet the different requirements of the different users for the hydraulic cylinder test-bed, save resource, reduce test and production costs and improve the product quality. Therefore, the new hydraulic test-bed of cylinder plays a major role to improve economic efficiency of enterprises.

Acknowledgments. Liaoning province science projects : Research on the key technology of the type test of hydraulic cylinder. Project number : 21010220031.

References

1. Chapple, J.P.: Principles of Hydraulic System Design. Information, Oxford (2003)
2. ChengZhi, W., YanPing, Z., Feng, R.: Improvement of the Experiment Platform for Cylinder Hydraulic System. J. Fluid Power Transmission and Control 4, 9–11 (2004)

3. ZhongXiu, P., SuoShan, H.: A New Test Stand for Vertical Hydraulic Cylinder. Building Machinery 288, 80–82 (2008)
4. LiJun, Z., KeMing, L.: The Application of Oil Temperature Control Method for Cylinder Stand. Chinese Hydraulics and Pneumatics 674, 56–59 (2008)
5. ZhiBin, T., YangJian, O.: The Design and Research of the Hydraulic Integration Testing Equipment in the High Temperature Condition. Chinese Hydraulics and Pneumatics 693, 21–23 (2009)
6. JianPing, J.: Application of Extra High Pressure Hydraulic Technology. Chinese Hydraulics and Pneumatics 680, 76–78 (2008)
7. XiaoQuan, S., XiLing, H.: Improvement in Control System of Hydraulic Experimental Stand with PLC. Applied Science and Technology 33, 23–25 (2006)
8. KunMin, Z.: The Type of YYKSDN Hydraulic Cylinder Test-bed Design. Hydraulics Pneumatics and Seals 68, 42–44 (1997)
9. FengTao, X., DanHong, W., YanMing, G., ChunQiang, J.: Design and Optimize Complex Hydraulic Manifold Block. Machine Tool and Hydraulics 36, 46–49 (2008)
10. Sha, L., JianZhong, W.: Design and Implementation of Cylinder Bench Autotest System. Machinery Design and Manufacture 12, 167–169 (2007)

A Positioning Algorithm for Node of WSN Based on Received Signal Strength

Xiang Yang[1,2], Wei Pan[3], and Yuanyi Zhang[4]

[1] School of Information Engineering, Wuhan University of Technology, WuHan, China
[2] School of Bowen Management,GuiLin University of Technology, GuiLin,541004, China
yangx@glite.edu.cn
[3] School of Mechanical and Control Engineering, GuiLin University of Technology, GinLin,350001, China
243546041@qq.com
[4] School of Architecture, Fuzhou University, Xueyuan Road No.2 Fuzhou, 350001, China
zyye@fzu.edu.cn

Abstract. Node localization plays a critical role in wireless sensor network. This paper emphatically analysised the signal strength based on ranging method(RSSI) of wireless location technology and the theory of three-point positioning method. On that basis, extended it and proposed a theoretical values combine experience values databaces based on the three-point positioning method. This positioning algorithm has a less relative errors by using different databases(experience value database and anchor node initial position distance database)according to different conditions. Finally, we verified the proposed method through MATLAB simulation. The results show that this positioning algorithm is effective. It can effectively solve the problem which may caused by environmental effect and the heavy actual measurement workload. What's more, it also can reduce the difficulties and workload of locating the random dropped wireless sensor network nodes.

Keywords: wireless sensor network, signal strength, three-point positioning, positioning algorithm.

1 The Nodes Localization Technique for WNS

At present, it has two different kinds of node positioning methods according to the location mechanism :range-based and range-free. The most commonly used method is RSSI(Received Signal Strength Indicator),which is especially applicable to the Large-scale wireless sensor network node localization. But with the influences of multi-radio, multipath propagation, non-line-of-sight and so on.. Some errors will be caused. Two most commonly wave propagation loss models used by researchers now is free space propagation model and logarithmic normal distribution model.

In the practical environment, the propagation path loss is different from theoretical value for various influences. It is more reasonable to adopts the logarithmic normal distribution model. The formula of calculating the path loss of the received anchor node information is as follows:

D. Jin and S. Lin (Eds.): Advances in MSEC Vol. 1, AISC 128, pp. 441–446.
springerlink.com

$$p(d)[dbm]=p(d_0)[dbm]-10n_p\lg(\frac{d}{d_0})+X_\sigma \quad (2)$$

In the formula(2), $p(d)$ is the strength of received signal when the distance between transceiver nodes is d; d_0 is the reference distance between transceiver nodes; $p(d_0)$ is the strength of received signal when the distance between transceiver nodes is d_0; n_p is the Path loss index which is determined by environment. X_σ is a random variable of Gaussian distribution which the mean value is 0 and the standard deviation is S. X_σ is mainly used to reduced the error of the signal strength, which reduced the positioning error.

2 Principle of Three-Point Fix Method

The principle of three-point fix method is showed in Fig. 1:

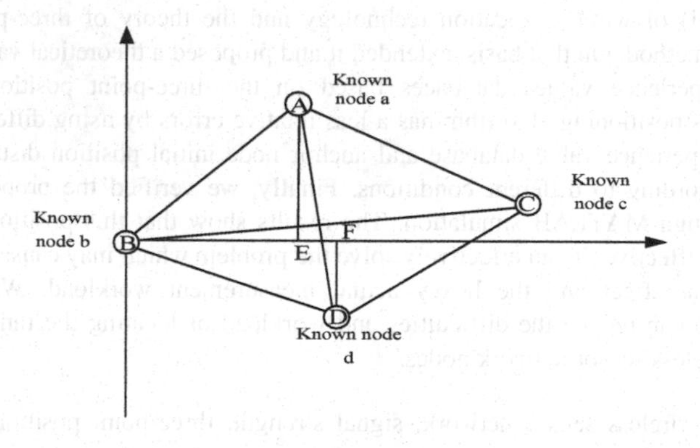

Fig. 1. *a,b,c* is the anchor nodes. Suppose the coordinate of *a,b,c* is given (x_a, y_a)、 (x_b, y_b)、(x_c, y_c),making the *A E* and *B F* respectively perpendicular to *x* axis, and then it is easy to calculate the coordinate of *D*: (x_b+L_{bf}, y_b-L_{df}), the existence of c is to assure the position of *D* relative to a and b which avoids the positioning influence caused by the symmetrical relation. So the core to use this arithmetic is to estimate the distance between AD、BD and CDL_{ad}, L_{bd}, L_{cd} accurately.

3 Theory of Signal Strength-Construction Empirical Value Database

In the literature[2] it forwarded a kind of positioning method which blends the signal strength's theoretical value and empiric value. The theorem is as the following picture shows:

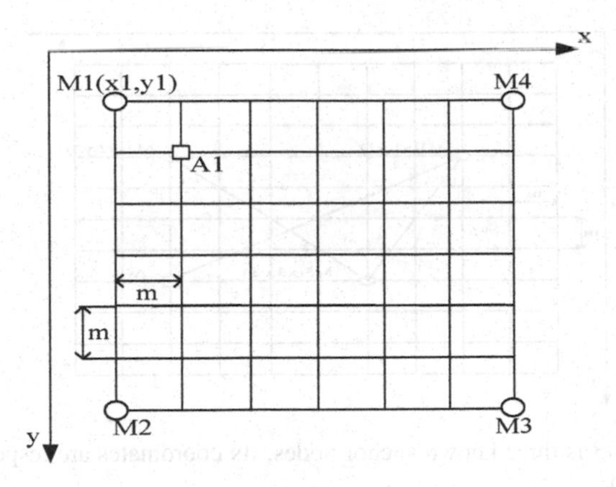

Fig. 2. In the simple and barrier free area of square ,set a anchor node in the four corners *M1,M2,M3,M4*. The coordinate is (x1,y1) , (x2,y2) , (x3,y3) , (x4,y4) respectively.

For the coordinate of A is (x1+m,y1+m) ,the distance from A to M1,M2,M3,M4 is respectively $\sqrt{2}$ m, $\sqrt{26}$ m, 5 $\sqrt{2}$ m, $\sqrt{26}$ m, so 2~4m is chosen as the step length as a matter of experience. Through the transformation formula: $p(d)[dbm] = p(d_0)[dbm] - 10n_p \lg(\frac{d}{d_0}) + X_\sigma$. (1)

The distance is transformed into the mean value of signal strength.

We can get the database of signal strength's theoretical value and constructe the experience value database, creating the congruent relationship of the collected signal strength between regional nodes with each anchor node:

$$(x_i, y_i) \sim \begin{pmatrix} SAVG_{i,1} & SDEV_{i,1} \\ SAVG_{i,2} & SDEV_{i,2} \\ SAVG_{i,3} & SDEV_{i,3} \\ SAVG_{i,4} & SDEV_{i,4} \end{pmatrix}$$

$SAVG_{i,1}$ is the mean valve of the node of number i to the first anchor node. $SDEV_{i,1}$ is the standard deviation.

5 Methods Improvement

This paper will put forward a method of based on the signal strength of three-point positioning theory database.

In a simple, clear square area and divide them into countless small square unit according to the step length m (m is between 2 ~ 4). At the intersection decorate three anchor nodes. Principle chart is as follows:

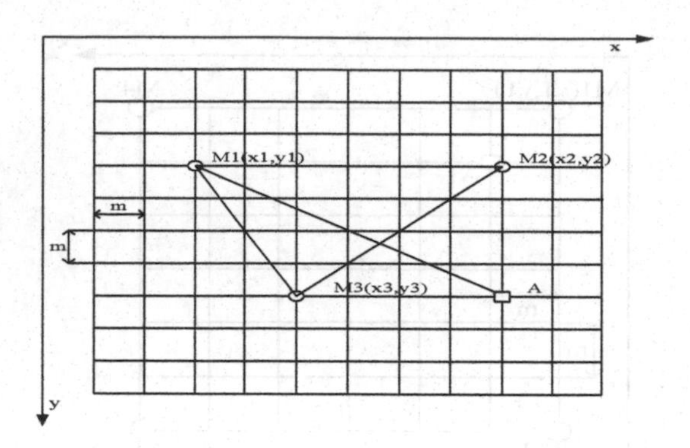

Fig. 3. *M1*, *M2*, *M3* is three known anchor nodes, its coordinates are respectively (x1,y1), (x2,y2), (x3,y3)

The distance from Point A to the M1 and M2, M3 are recorded as D1, D2, D3. The anchor point coordinate is a random position and known. D_i value and anchor node M_i are initial location related. By step length the distance of node A can be calculated. Therefore, we can establish an distance database based on the initial position of anchor nodes:

$$D_i \sim (x_i, y_i), (i = 1, 2, 3).$$

Then through the formula:

$$p(d)[dbm] = p(d_0)[dbm] - 10n_p \lg(\frac{d}{d_0}) + X_\sigma \quad (1)$$

transform the distance into signal strength mean value p1, p2, p3 and may get signal strength standard deviation through

$$\sigma_i = \sqrt{E\{[p - p_i]^2\}}, (i = 1, 2, 3). \quad (2)$$

Which can be used to build the corresponding relation between node A coordinates and the anchor point signal strength characteristics. Then we can establish signal strength theoretical database.

Through amounts of actual datas' acquisition and signal strength calculating mean value $\overline{RS} = (RS_1, RS_{2,....,} RS_n)$ (n is number of acquisition times) with noise signal $X_\sigma \sim N(0, S^2)$. Then get the signal strength mean value and standard deviation. Thus established the experience database:

$$(x_i, y_i) \sim \begin{pmatrix} SAVG_{i,1} & SDEV_{i,1} \\ SAVG_{i,2} & SDEV_{i,2} \\ SAVG_{i,3} & SDEV_{i,3} \end{pmatrix}$$

Fusion theory experience database and combining the database of anchor node initial position distance. We use different database in different environments.

6 MATLAB Simulation

On the basis of the created theoretical database and the actual positioning information of a wireless sensor node indoors. Following simulation diagram about signal strength and distance can be get from MATLAB simulation.

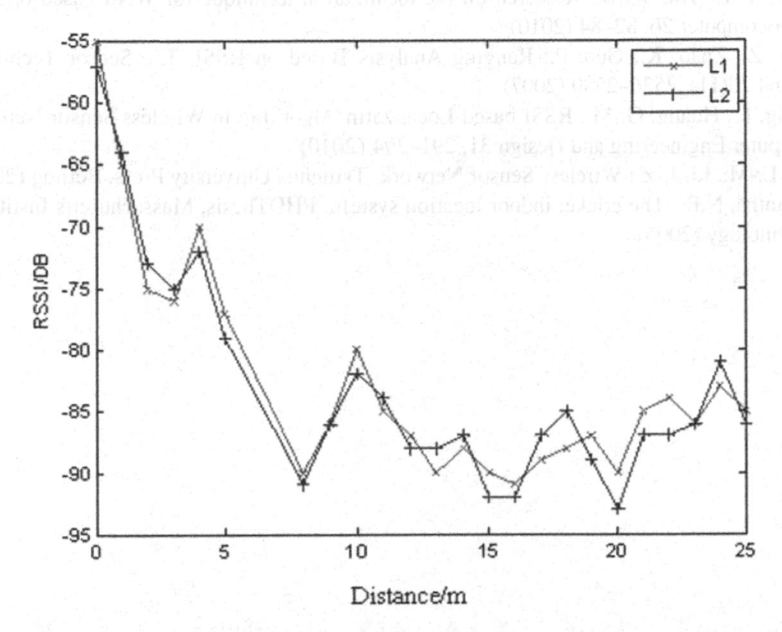

Fig. 4. In the diagram, the blue line shows the relationship between distance and signal strength which is get according to the theoretical database. The red line shows the relationship between distance and signal strength that was actually measured indoor.

The simulation shows clearly the result that using theoretical database is in close compliance with the actual measured value on the condition with little interference factors indoor.The changing trends of the curve is approximately in compliance with the formula $p(d)[dbm] = p(d_0)[dbm] - 10n_p \lg(\frac{d}{d_0}) + X_\sigma$.

7 Conclusion

This paper we adopts the three-point positioning method. It can effectively solve the problem which may caused by environmental effect and the heavy actual measurement workload in the process of using the signal strength orientation method by establishing database of anchor node initial position distance, theoretical database and experience value database. When the databank based on initial position distance of anchor nodes, the anchor nodes are fixed. As a result, the difficulties and workload of locating the random dropped wireless sensor network nodes are reduced.

References

1. Sun, L.-M., Li, J.-Z.: Wireless Sensor Network. Tsinghua University Press, Beijing (2005)
2. Chen, Y.-Z., Xu, W.-B.: Research on the localization technique for WNS based on RSSI. Microcomputer 26, 82–84 (2010)
3. Fang, Z., Zhao, K., Guo, P.: Ranging Analysis Based on RSSI. The Sensor Technology Joumal 20(11), 2526–2530 (2007)
4. Zhang, L., Huang, G.-M.: RSSI based Localizatin Algorithm in Wireless Sensor Networks. Computer Engineering and Design 31, 291–294 (2010)
5. Sun, L.-M., Li, J.-Z.: Wireless Sensor Network. Tsinghua University Press, Beijing (2005)
6. Priyantha, N.B.: The cricket indoor location system. PHDThesis, Massachusetts Institute of Tec-hnology (2005)

Linear Generalized Synchronization between Two Complex Networks

Qin Yao, Guoliang Cai, Xinghua Fan, and Juan Ding

Nonlinear Scientific Research Center, Jiangsu University,
Zhenjiang, Jiangsu 212013, China
glcai@ujs.edu.cn

Abstract. The main objective of this paper is further to investigate the generalized synchronization between two complex networks with time varying delays. Based on Lyapunov stability theory and Young inequality technique, new sufficient controllers are derived to guarantee the generalized synchronization between two complex networks with the different connection topologies, and numerical simulations are shown to illustrate the effectiveness of these controllers.

Keywords: Complex networks, Generalized synchronization, Time varying delays, Topology.

1 Introduction

Recently, complex networks' synchronization has been gaining increasing attention for its applications from communications networks to social networks, from the Internet to the World Wide Web. Synchronization has been studied various angles such as complete synchronization (CS) [1], projective synchronization (PS) [2-5], generalized synchronization (GS) [6, 7], and so on. More studies showed in [8-11]. The studies about synchronization between two coupled complex networks [12] are not so much as profound as others. The paper [7] considered the linear GS between two complex networks without time varying delays, and this paper investigates the GS between the two complex networks with time varying delays. In paper [7], a criterion for linear GS between two complex networks was attained by using the nonlinear control method, but in this paper, two modified systems will be characterized, and then a new method is introduced by using the technique of Young inequality to verify the GS.

2 Model Description and Preliminaries

In paper [7], the authors used the nonlinear control method to derive linear GS between the following drive and response systems.

The drive system was:

$$\dot{x}_i = f(x_i) + \sum_{j=1}^{N} c_{ij} \Gamma x_j \qquad i = 1,2,\dots,N. \tag{1}$$

D. Jin and S. Lin (Eds.): Advances in MSEC Vol. 1, AISC 128, pp. 447–452.
springerlink.com

The following controlled response system was :

$$\dot{y}_i = Ay_i + Bg(y_i) + \sum_{j=1}^{N} c_{ij}\Gamma y_j + u_i \qquad i = 1,2,..., N. \tag{2}$$

The parameters will define in (3) and (4). Now we consider the network with time varying delays. The drive system is:

$$\dot{x}_i(t) = f(x_i(t)) + \sum_{j=1}^{N} c_{ij}\Gamma x_j(t - \tau(t)) \qquad i = 1,2,..., N. \tag{3}$$

in which $\dot{x}_i(t) = (x_{i1}(t), x_{i2}(t),..., x_{in}(t))^{\mathrm{T}} \in R^n$ are the state variables of the i th node, $\tau(t)$ is the time-varying delay, $f : R \times R^n \to R^n$ is a smooth nonlinear vector valued function. For simplicity, we further assume that the inner connecting matrix Γ= diag $(\gamma_1,\gamma_2,\ldots,\gamma_n)$ and assume $\|\Gamma\| = \gamma > 0, C = (c_{ij})_{n\times n} \in R^{n\times n}$ is the coupling configuration matrix. The diagonal elements of matrix C are defined by $c_{ii} = -\sum_{j=1, j\neq i}^{N} c_{ij}, i = 1,2,..., N$.

The controlled response system is:

$$\dot{y}_i(t) = Ay_i(t) + Bg(y_i(t)) + \sum_{j=1}^{N} c_{ij}\Gamma y_j(t - \tau(t)) + u_i \qquad i = 1,2,...,N. \tag{4}$$

where $\dot{y}_i(t) = (y_{i1}(t), y_{i2}(t),..., y_{in}(t))^{\mathrm{T}} \in R^n$ are the state variables of the i th node, A and B are system matrices with proper dimensions satisfy $\|A\| \le \alpha, \|B\| \le \beta$, $g(\cdot)$ is a continuous function, u_i is the controller to be designed.

We now introduce the following definition, assumptions and lemma:

Definition 1. Give a variable mapping $\varphi : R^n \to R^n$, if systems (3) and (4) satisfy the following property: $\lim_{t\to\infty} \|y_i(t) - \varphi(x_i(t))\| = 0, i = 1,2,..., N$, thus we say that system (3) achieves GS with system (4).

For simplicity, in this paper, φ is selected as: $\varphi(x) = Px+Q$, in which P and Q are constant matrices. The GS is linear GS.

Assumption 1. The time delay $\tau(t)$ is a differential function with $0 \le \dot{\tau}(t) < 1$.

Assumption 2. For any different $x_1, x_2 \in R^n$, suppose there exists a constant L>0 such that

$$\|g(x_1) - g(x_2)\| \le L\|x_1 - x_2\|.$$

The norm $\|\cdot\|$ of a variable x is defined as $\|x\| = (x^{\mathrm{T}} x)^{1/2}$.

Lemma 1([12]). Suppose $e > 0, h > 0, r > 1, \frac{1}{r} + \frac{1}{q} = 1$, the next inequality holds:

$$eh \le \frac{1}{r}e^r + \frac{1}{q}h^q = \frac{1}{r}e^r + \frac{r-1}{r}h^{\frac{r}{r-1}}.$$

3 Generalized Synchronization

The synchronization error system for (3) and (4) is defined as:

$$e_i(t) = y_i(t) - Px_i(t) - Q, \qquad i = 1,2,\ldots,N. \tag{5}$$

Then the error system can be described as:

$$\begin{aligned} \dot{e}_i(t) &= \dot{y}_i(t) - P\dot{x}_i(t) \\ &= Ay_i(t) + Bg(y_i(t)) + \sum_{j=1}^{N} c_{ij}\Gamma y_j(t-\tau(t)) + u_i - Pf(x_i(t)) - P\sum_{j=1}^{N} c_{ij}\Gamma x_j(t-\tau(t)). \end{aligned} \tag{6}$$

Theorem 1. Suppose Assumptions 1 and 2 hold for the driving network (3) and the controlled response network (4), choose the controller u_i as:

$$u_i = Pf(x_i(t)) - M(y_i(t) - Px_i(t) - Q) - A(Px_i(t) + Q) + \sum_{j=1}^{N} c_{ij}\Gamma Q, \tag{7}$$

where M is a constant diagonal matrix with $M = \mathrm{diag}(\xi_1, \xi_2, \ldots, \xi_n)$, if M satisfies:

$$r\min(\xi_i) > r(\alpha + L\beta) + \gamma[(r-1)\left|c_{ij}\right|^{\frac{rv_{ij}}{r-1}} + \sum_{j=1}^{N}\frac{\alpha_j}{\alpha_i}\left|c_{ij}\right|^{r(1-v_{ij})}]. \tag{8}$$

where v_{ij} is a real constant, $r>1$, $\alpha_i>0$. Then networks (3) and (4) will achieve GS.

Proof. Substitute (7) to (6), and then we obtain:

$$\dot{e}_i(t) = Ae_i(t) - Me_i(t) + B(g(y_j(t)) - g(Px_i(t) + Q)) + \sum_{j=1}^{N} c_{ij}\Gamma e_j(t-\tau(t)). \tag{9}$$

Choose the following Lyapunov function:

$$V(t) = \sum_{i=1}^{N} \alpha_i\left|e_i(t)\right|^r + \Gamma\sum_{i=1}^{N}\alpha_i\sum_{j=1}^{N}(\left|c_{ij}\right|^{r(1-v_{ij})})\int_{t-\tau(t)}^{t}\left|e_j(s)\right|^r ds \tag{10}$$

Calculating the derivative of (10) along the trajectories of (9), we get:

$$\begin{aligned} \dot{V}(t) &= \sum_{i=1}^{N}\alpha_i r\left|e_i(t)\right|^{r-1}\mathrm{sign}(e_i(t))[Ae_i(t) - Me_i(t) + B(g(y_j(t)) - g(Px_i(t)+Q)) + \sum_{j=1}^{N}c_{ij}\Gamma e_j(t-\tau(t))] \\ &\quad + \Gamma\sum_{i=1}^{N}\alpha_i\sum_{j=1}^{N}\left|c_{ij}\right|^{r(1-v_{ij})}\left|e_j(t)\right|^r - \Gamma\sum_{i=1}^{N}\alpha_i\sum_{j=1}^{N}(1-\dot{\tau}(t))\left|c_{ij}\right|^{r(1-v_{ij})}\left|e_j(t-\tau(t))\right|^r \\ &\le \sum_{i=1}^{N}\alpha_i r[A\left|e_i(t)\right|^r - M\left|e_i(t)\right|^r + LB\left|e_i(t)\right|^r + \Gamma\sum_{i=1}^{N}\left|c_{ij}\right|\left|e_i(t)\right|^{r-1}\left|e_j(t-\tau(t))\right|] \\ &\quad + \Gamma\sum_{i=1}^{N}\alpha_i\sum_{j=1}^{N}\left|c_{ij}\right|^{r(1-v_{ij})}\left|e_j(t)\right|^r - \Gamma\sum_{i=1}^{N}\alpha_i\sum_{j=1}^{N}\left|c_{ij}\right|^{r(1-v_{ij})}\left|e_j(t-\tau(t))\right|^r \end{aligned}$$

Since

$$\left|c_{ij}\right|\left|e_i(t)\right|^{r-1}\left|e_j(t-\tau(t))\right| = \left|e_j(t-\tau(t))\right|\left|c_{ij}\right|^{1-v_{ij}}[\left|c_{ij}\right|^{\frac{v_{ij}}{r-1}}\left|e_i(t)\right|]^{r-1},$$

by the Lemma 1, we have:

$$\left|c_{ij}\right|\left|e_i(t)\right|^{r-1}\left|e_j(t-\tau(t))\right| \le \frac{1}{r}\left|c_{ij}\right|^{r(1-v_{ij})}\left|e_j(t-\tau(t))\right|^r + \frac{r-1}{r}\left|c_{ij}\right|^{\frac{rv_{ij}}{r-1}}\left|e_i(t)\right|^r,$$

because $\|A\| \le \alpha$, $\|B\| \le \beta$, $\|\Gamma\| = \gamma$, then we get:

$$\begin{aligned}\dot{V}(t) \le& \sum_{i=1}^{N}\alpha_i r[A|e_i(t)|^r - M|e_i(t)|^r + LB|e_i(t)|^r] + \Gamma\sum_{i=1}^{N}\alpha_i\sum_{j=1}^{N}|c_{ij}|^{r(1-v_{ij})}|e_j(t-\tau(t))|^r \\ &+ \Gamma\sum_{i=1}^{N}\alpha_i(r-1)|c_{ij}|^{\frac{rv_{ij}}{r-1}}|e_i(t)|^r + \Gamma\sum_{i=1}^{N}\alpha_i\sum_{j=1}^{N}|c_{ij}|^{r(1-v_{ij})}|e_j(t)|^r + \Gamma\sum_{i=1}^{N}\alpha_i\sum_{j=1}^{N}|c_{ij}|^{r(1-v_{ij})}|e_j(t-\tau(t))|^r \\ \le& \sum_{i=1}^{N}\alpha_i[r\alpha - r\min(\xi_i)M + L\beta + \gamma(r-1)|c_{ij}|^{\frac{rv_{ij}}{r-1}} + \gamma\sum_{j=1}^{N}\frac{\alpha_j}{\alpha_i}|c_{ij}|^{r(1-v_{ij})}]|e_i(t)^r|.\end{aligned}$$

where $e_i(t)=(e_1(t),e_2(t),\ldots,e_n(t))^{\mathrm{T}}$. Obviously, if condition (8) holds, we have: $\dot{V}(t) < 0$, namely, $\lim_{t\to\infty}\|e_i(t)\| = 0, i = 1,2,\ldots,N$, thus $\lim_{t\to\infty}\|y_i(t) - \varphi(x_i(t))\| = 0, i = 1,2,\ldots,N$. Therefore, the error system (9) is globally stable, so the drive network (3) with time varying delays can achieve GS with the response network (4).

4 Numerical Simulations

Let us consider the following Lü system:

$$\dot{x} = f(x) = \begin{pmatrix} a(x_2 - x_1) \\ cx_2 - x_1x_3 \\ -bx_3 + x_1x_2 \end{pmatrix}, \tag{11}$$

where a=36, b=3, c=20. The Rössler system is known as:

$$\dot{y} = Ay + Bg(y) = \begin{pmatrix} -y_2 - y_3 \\ y_1 + \omega y_2 \\ \lambda + y_3(y_1 - \theta) \end{pmatrix}, \tag{12}$$

where ω=0.2, λ=0.2, θ=5.7.

$$A = \begin{bmatrix} 0 & -1 & -1 \\ 1 & 0.2 & 0 \\ 0 & 0 & -5.7 \end{bmatrix}, \quad B = \begin{bmatrix} 0 & 0 & 0 \\ 0 & 0 & 0 \\ 0 & 0 & 1 \end{bmatrix}, \quad g(y) = \begin{pmatrix} 0 \\ 0 \\ y_1y_2 + 0.2 \end{pmatrix}. \tag{13}$$

Thus $\alpha = \|A\| = 5.7897$, $\beta = \|B\| = 1$.

We consider our driving system with 6-nodes:

$$\dot{x}_i(t) = f(x_i(t)) + \sum_{j=1}^{6} c_{ij} \Gamma x_j(t - \tau(t)) \qquad i = 1,2,\ldots,6.$$

where $f(x_i)$ is shown in Eq. (11). The response system is:

$$\dot{y}_i(t) = Ay_i(t) + Bg(y_i(t)) + \sum_{j=1}^{6} c_{ij} \Gamma y_j(t - \tau(t)) + u_i \qquad i = 1,2,\ldots,6.$$

in which $g(y_i)$, A and B are shown in Eq. (13), u_i is choosed in (7), $C=(c_{ij})_{6\times 6}$ is the configuration matrix and P is a non-diagonal matrix as follows.

In the numerical simulations, we select $r=2$, $L=1$, $Q=(0,0,0)^T$, $M=$ diag (30,30,40), $\Gamma=$ diag (1,1,1), $\|\Gamma\| = \gamma = 1$, $v_{ij}=0.5$, $\alpha_i=1$, $\alpha_j=0.1$, $\tau(t) = \frac{e^t}{1+e^t}$.

$$\begin{bmatrix} -4 & 3 & 0 & 0 & 1 & 0 \\ 1 & -6 & 2 & 0 & 0 & 3 \\ 2 & 1 & -3 & 0 & 0 & 0 \\ 0 & 3 & 0 & -7 & 4 & 0 \\ 0 & 0 & 0 & 4 & -4 & 0 \\ 1 & 0 & 1 & 0 & 0 & -2 \end{bmatrix}, \quad P = \begin{bmatrix} 2 & -1 & 0 \\ 0 & -1 & 0 \\ 0 & 0 & 1 \end{bmatrix}.$$

The initial values of the drive system and the response system are chosen as $x_i(0)=(0.3+0.1i, 0.3+0.1i, 0.3+0.1i)^T$, $y_i(0)=(2.0+0.7i, 2.0+0.7i, 2.0+0.7i)^T$. The numerical results show in Fig.1, respectively. One can see that the drive network (3) and the response network (4) achieve GS.

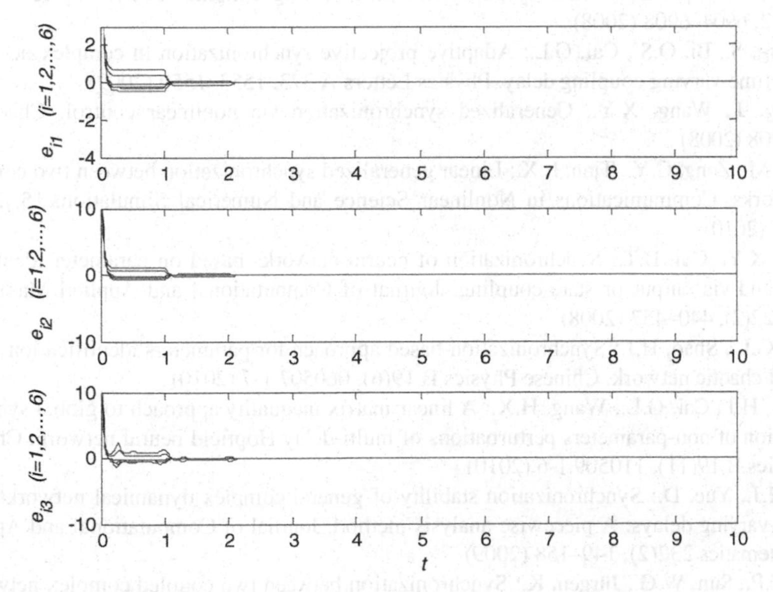

Fig. 1. The synchronization errors between networks (3) and (4)

5 Conclusions

In this paper, GS between two complex networks with time varying delays is studied by using Lyapunov stability theory and Young inequality technique. Although the GS is linear GS, the controller u_i is a nonlinear function and P is a non-diagonal matrix. A new sufficient condition is gained to guarantee the GS between two complex networks with different connection topologies, thus one may see that the result of paper [7] just was the special situations of this paper. Compared with some traditional papers, the approach presented in this paper is more comprehensive.

Acknowledgments. This work was supported by the National Nature Science foundation of China (Nos. 70571030, 90610031), the Society Science Foundation from Ministry of Education of China (Nos. 08JA790057), the Advanced Talents' Foundation of Jiangsu University (No. 07JDG054), and the Student Research Foundation of Jiangsu University (No. 10A147).

References

1. Xiao, Y.Z., Xu, W., Li, X.C.: Adaptive complete synchronization of chaotic dynamical network with unknown and mismatched parameters. Chaos 17, 033118 (2007)
2. Cai, G.L., Wang, H.X., Zheng, S.: Adaptive function projective synchronization of two different hyperchaotic systems with unknown parameters. Chinese Journal of Physics 47, 662–669 (2009)
3. Zheng, S., Dong, G.G., Bi, Q.S.: Adaptive modified function projective synchronization of hyperchaotic systems with unknown parameters. Communications in Nonlinear Science and Numerical Simulations 15, 3547–3556 (2010)
4. Sun, M., Zeng, C.Y., Tian, L.X.: Projective synchronization in drive–response dynamical networks of partially linear systems with time-varying coupling delay. Physics Letters A 372, 6904–6908 (2008)
5. Zheng, S., Bi, Q.S., Cai, G.L.: Adaptive projective synchronization in complex networks with time-varying coupling delay. Physics Letters A 373, 1553–1559 (2009)
6. Meng, J., Wang, X.Y.: Generalized synchronization via nonlinear control. Chaos 18, 023108 (2008)
7. Sun, M., Zeng, C.Y., Tian, L.X.: Linear generalized synchronization between two complex networks. Communications in Nonlinear Science and Numerical Simulations 15, 2162–2167 (2010)
8. Lou, X.Y., Cui, B.T.: Synchronization of neural networks based on parameter identification and via output or state coupling. Journal of Computational and Applied Mathematics 222(2), 440–457 (2008)
9. Cai, G.L., Shao, H.J.: Synchronization-based approach for parameters identification in delayed chaotic network. Chinese Physics B 19(6), 060507.1-7 (2010)
10. Shao, H.J., Cai, G.L., Wang, H.X.: A linear matrix inequality approach to global synchronization of non-parameters perturbations of multi-delay Hopfield neural network. Chinese Physics B 19(11), 110509.1-6 (2010)
11. Li, H.J., Yue, D.: Synchronization stability of general complex dynamical networks with time-varying delays: A piecewise analysis method. Journal of Computational and Applied Mathematics 232(2), 149–158 (2009)
12. Li, C.P., Sun, W.G., Jürgen, K.: Synchronization between two coupled complex networks. Physical Review E 76, 046204 (2007)

The Application of Digital Media Technology in Art Design

Qiang Liu[1], Lixin Diao[2], Guangcan Tu[3], and Linlin Lu[4]

[1] Art Academy, China Three Gorges University, Yichang, China
640690697@qq.com
[2] Yichang No.22 middle school, Yichang, China
276606771@qq.com
[3] Art department, Yangtze University, Jinzhou, China
29105800@qq.com
[4] Information Techonology Center, China Three Gorges University, Yichang, China
6269376@qq.com

Abstract. This paper research the application of digital media technology in art design. In the first place, this paper analyses the meaning of digital media technology. In the second place, this paper analyses the meaning of art design. In the third place, introduces the application of digital media technology in art design from three typical aspects. The three typical aspects are: 1) the application of digital media technology in animation design, 2) the application of digital media technology in exhibition design, 3) The application of digital media technology in costume design. Finally, is the conclusion and the prospect.

Keywords: digital media technology, art design, application.

1 Introduction

Every improvement of science and technology would have a tremendous impact on the field of art design. It is also a driving force for prosperity and development of art forms. Since the 20th century, the impact of technology on arts becomes more intense and rapid. Among them, the development of digital media technology created a broader performance space for the art. The art design based on the digital media technology became the modern sophisticated means and forms of visual expression. It is widely applied in the field of mass media and visual art design such as television, film, graphic art design, industrial design, exhibition art design, the architectural environment art design, etc. The digital media technology has become the necessary technique in the art design.

2 The Meaning of Digital Media Technology

The digital media technology has widely applications. But its meaning is still a lot of argument in academic circles. One wider spread defined is: the digital media

D. Jin and S. Lin (Eds.): Advances in MSEC Vol. 1, AISC 128, pp. 453–457.
springerlink.com

technology is a technique that can integrated treat text, sound, graphics, images and other information by the means of modern computing and communications, make the abstract information into perceived, manageable and interaction.

The digital media technology mainly research the theory, methods, techniques and systems about the acquisition, processing, storage, transmission, management, security, output of the digital media massage, and so on. The digital media technology is an integrated application of technology including various types of information technology such as computer technology, communication technology, information processing technology, etc. The key technology and content it involved mainly include digital information acquisition and output technology, digital information storage technology, digital information processing technology, digital communication technology, digital information management and security, etc.

The others digital media technologies include the integrated technology based on the key technology. For example, the streaming media technology based on the digital transmission technology and digital compressed technology is widely applied in the digital media network transmission. The computer animation technology base on the computer graphics technology is widely applied in the digital entertainment industry. The virtual reality technology based on the technology of human-computer interaction, computer graphics and display is widely applied in the fields of entertainment, broadcast, display and education.

3 The Meaning of Art Design

Art design has a very rich intension. It uses the form and aesthetic feeling of art, combined with social, cultural, economic, market, technology and many other factors, in the designs which is closely related to our lives. So that it not only has aesthetic function, also has a use function. Art design should be a certain perfect combination of physical features and mental function of human society. It is the inevitable product of the development of modern society. Art design by its nature is a creation activity of human.

Art design is an independent artistic discipline. Its research content and target are different of the traditional disciplines of the art. At the same time, art design is a highly comprehensive discipline. It involves social, cultural, economic, market, technology and many other aspects. It's aesthetic standards always changed with the change of these factors. On the one hand, art design itself has the objective characteristics of the natural sciences because it has close contact with the society's material, production and science and technology. On the other hand, art design has a special ideological color because it has close relationship with the certain socio-political, cultural, art.

Art design is divided into three categories: visual communication design, product design and environment design. Visual communication design include: lettering design, logo design, advertisement design, packaging design, CI (Corporate Identity) image design, format design, book design, exhibition design, multimedia design, etc [1]. Product design include: transportation design, daily design, furniture design, costume design, craft design, etc. Environment design include: urban planning and design, architectural environment design, interior environment design, garden landscape design, public art design, etc.

4 The Application of Digital Media Technology in Art Design

The digital media technology has penetrated deeply into the various aspects and links of art design. It has both the application of software technology and the application of hardware technology. Here is the specific introduction of some typical aspects.

4.1 The Application of Digital Media Technology in Animation Design

In the area of stop-motion animation, the development of digital media technology makes the production of stop-motion animation into a new era. The progress is mainly embodied in the shooting stage of the stop-motion animation. Software which is named "Stop Motion Studio" provides all the features of shooting stop-motion animation. It has the features of accurate position animation, real-time keying and equalizing the frame's brightness. It can monitor the shooting's effect real time so that it effectively avoids the post-production rework in the pre-shooting. It can real time compose the figure in the solid background into the scene by using foreground image keying and background image keying. It can regulate the relationship between the role and the scene so that it avoids a lot of adjustments in the composition. This new technology improves work efficiency greatly and also makes the picture more attractive, standardized.

In the area of 2D animation, the production of traditional 2D animation is 'paper animation' that the original painting, animation, color, background and all the work would be completed on paper. But by applying the digital media technology to the 2D animation, the paperless animation appeared. The so-called paperless animation actually refers to using computers and software to complete the whole animation process. At present, some software such as FLASH support the original painting, animation processing, background rendering, and post-production process of paperless animation. The creators use the computer to complete the character design, original painting, animation, background design, color designation, special effects, etc with digital media technology. It has many advantages such as low-cost, easy to master, high-quality, standard, simple output, etc. In particular, its storage mode can easily combined with 3D animation, digital post-technology, and the finished digital material can be reused. It reached the purposes of small investment, small risk and shorten production cycle.

The 3D animation is entirely the product of digital media technology. Animators use 3DS MAX, MAYA and other software first to create a virtual world in the computer and in accordance with the performance of object shape and size to create a role model and scene in this virtual 3D world. Then set the model trajectory, the virtual camera movements and other animation parameters. At last, assign the certain material and lights to the model. When all this is completed, the software operates automatically to generate the final picture. The 3D animation technology can simulate the real object movement pattern and the 3D space structure model. It has the attributes of preciseness, reality and infinite maneuverability. The 3D animation technology can also be used for television advertising and film special effects (such as explosions, smoke, rain, light effects, distortion, fantasy scene or role, etc.).

4.2 The Application of Digital Media Technology in Exhibition Design

The application of digital media technology has brought revolutionary changes to exhibition design. The most representative example is the virtual reality exhibition design. The virtual reality exhibition design is a means of information multimedia design based on the virtual reality technology. It simulates the 3D real environment and the information of the object to generate the digital virtual image through the special software such as VRML, Virtools, Cu1t3D, VR Platform, Quick Time VR, etc, integrates images, sound, animation and other multimedia information, and communicates those to people through the digital media [2]. It makes people to be personally on the virtual scene that is simulated by the computer through the sight, hearing, even skin.

The basic expression of the virtual reality exhibition design can be divided into three kinds: panorama exhibition, object exhibition and scenes exhibition. The panorama exhibition is that the object is relatively fixed while the lens position is relatively moving. The object exhibition is that the object is relatively moving while the lens position is relatively fixed. For example, the product rotate but the lens does not change when we show a product. The scenes exhibition is that the object and the lens position all can move. We can turn from one viewpoint to another viewpoint if we set some viewpoints. It is the integrated use of the panorama exhibition and the object exhibition.

The virtual reality exhibition design based on the 3D model is based on the 3D scene and 3D object model. It is usually realized through the professional modeling software such as 3DS MAX, MAYA, etc. This modeling method can represent the scenes and objects of the real world reality and expediently. At the same time, it can generate animation. At the present, the main 3D modeling software such as 3DS MAX, MAYA, Lightwave 3D, Softimage 3D, etc all can do the scene and object modeling, texturing and rendering. The effect of this kind of virtual reality exhibition is more abundant and the exhibition ability is more powerful.

4.3 The Application of Digital Media Technology in Costume Design

Digital media technology makes costume design develop in the directions of diversification, conceptualization, virtualization, popularity. With the maturing of digital media technology, costume design is more convenient and effective. For example, fashion illustration and effect drawing are usually made by the software named "CorelDraw" or the costume CAD system witch can meet requirements of business and customer accurately and efficiently. If designers prefer the stylized performance, they can choice the software named "Photoshop" and the software named "Painter". In particular, coupled with a digital camera, scanner, digital input board and pressure sensitive pen and other peripheral equipment, the habits of the designer's creative painting can be fully retained in these software. For example, there are dozens kinds of painting tools and paper with different styles in Painter. Through the rational use, designers can design such as gouache, watercolor, crayon, color of lead, oil and even painting style of clothing designs, and its lightness, saturation, save time and so much are better than traditional fashion painting work.

The virtual reality (VR) in digital media technology realized the three-dimensional clothing, animation and even virtual reality clothing show. Now, the bands which have the function of 3D virtual clothing include: the "System3D Sample" of PDA company in Canada, the "3D-Fashion Design System" of Computer Design Incorporation in America, the "Runway 3D" of PGM company in America, the "V-Stitcher" of PGM company in America, the "E-Design" of Lectra company in France, the "Investronica System" of Spanish, etc [3]. The 3D virtual clothing software can generate three-dimensional human model from the model library. Then draw the style line on the model by the "Three-dimensional cut" technology similar to the traditional clothing design, and generate clothing outline. Under the support of digital technology witch has analysis ability, with light and scene, the work it designed can represent realistic fabric, pattern and texture, and it can make designer "virtual reality" involve in the design creation.

5 Conclusion

Digital media technology promotes the development of art design. Art design has a unique vitality by the application of digital media technology. The relationship between design and technology are inseparable. Creative design, design ideas, design concepts by the aid of new technology have been further sublimation.

Digital media technology will rapidly develop. Its impact and participation on art design will be more in-depth. Art and science are together in our lives. So, the boundaries between art and science will become increasingly blurred. But, we can not have equated the art and science, and can not consider that new technology will make art into science or science into art. We must define the role of digital media technology in art design reasonably, so that we can really apply digital media technology in art design.

References

1. Debeve, P.E., et al.: Modelinga Rendering from Photographs: geometry. Computer Graphies, 11–20 (2006)
2. Barzel, R.: Procedure for computing. Based Modeling for Compute Graphies, 89-179 (2008)
3. Song, S.W.: Design and Implementation of the Image Interactive System Based on Human-Computer Interaction. Journal of Intelligent Information Management (IIM) 2(5) (May 2010)

A Metadata Architecture Based on Combined Index

Li Cai[1] and JianYing Su[2]

[1] School of Information Engineering, Chongqing City Management College, 401331 Chongqing, China
cailitg@163.com
[2] Library, Chongqing City Management College, 401331 Chongqing, China
tsgsujy@163.com

Abstract. Perfect metadata architecture can better support changes of user requirement and help users access to cared information quickly. Aiming at flaws of simple functions and poor flexibility and scalability of existing metadata architectures, metadata architecture based on combined index was brought out in the paper. The architecture introduced idea of P2P and hierarchy and stored metadata according to classifications of core elements, resource type core elements and individual element based on its basic structure. Basic performance analysis on the architecture indicates that it has advantages than MDS on query speed and improves extensibility and interoperability of metadata in heterogeneous system.

Keywords: data grid, metadata, combined index.

1 Introduction

In modern science research and applications, existing data management system, method and technologies can not meet requirements of high performance and large amount storage as well as distributed processing capability. Therefore, idea of data grid was presented. In the data grid, resources are distributed, and resource provider is also distributed. So it is extremely important for people to precisely seek for needed information. Under this background, the metadata came into being. Metadata is data about data, namely structured data that can accommodate any digital resources [1]. Metadata enable users can know history status of data, such as where the data come from, how long is the distribution, how much is update frequency and what is the meaning of data elements as well as which computation, transformation and filtering have been conducted on the data. In the instant changing network environment and data ever expanding days, perfect metadata architecture can better support changes of user requirement and help users access to cared information quickly [2-4]. In order to make up for some shortcomings of existing metadata architecture, the paper presented a kind of metadata architecture based on Combined Index (CI) and analyzed its performance. The paper is organized as follows: section 2 analyzes on existing metadata architecture; section 3 brings out architecture based on CI; section 4 analyzes on its performance and section 5 concludes our work.

D. Jin and S. Lin (Eds.): Advances in MSEC Vol. 1, AISC 128, pp. 459–464.
springerlink.com

2 Analysis on Existing Metadata Architecture

The existing metadata architecture includes Globus, GDMP and MCAT. Globus is a representative plan of parallel and distributed computing environment based on WAN and metacomputing directory service (MDS) is a information service module provided by Globus, which is the storage and manager of resource information [5]. It is accomplished by grid resource information service (GRIS) and grid index information service (GIIS) together. GRIS provides a common method to query current configuration, capability and resource information of computation grid. GIIS connects random GRIS into a continuous system image for exploration and query of grid applications. EDG mainly aims at European Nuclear Research Organization High Energy Physics applications to solve decomposition storage and processing problem of mass data. EDG system is built on the Globus data grid construction for the European data grid data management and image processing software (GDMP) [6]. Each site of EDG has a GDMP server and a file directory. Customer requests for data to GDMP and GDMP will query the directory service. If the directory exists, initiate data transmission process, or write pre-determined information to the directory. When a new file is generated, GDMP publish logical file name, meta-information and physical location information to the directory and notify user messages about new data. SRB is the storage resource agent in San Diego Supercomputing Center, which uses MCAT as a metadata pool system. It is a metadata register system achieved by relational database and used to store various metadata related to SRB [7]. SRB can query, establish and modify metadata in MCAT. Initially, MCAT uses centralized design, now it is being designed distributed to support data grid.

Although GIIS in MDS can connect some random GRIS into a continuous system image, it has not high query efficiency on metadata not concluded in GIIS. GDMP has simple functions and single target. It just achieves mete-information service simply. GDMP uses simple and centralized copy directory, so the flexibility and scalability are poor. Centralized MCAT supports complex attributes query. But for a distributed and scalable heterogeneous data system, it is not desirable. To make up for these shortcomings, the paper presents a kind of metadata architecture based on CI.

3 Metadata Architecture Based on CI

3.1 Architecture

Metadata architecture based on CI can be divided into application layer, data management layer and resource layer, as shown in Fig. 1. Metadata is stored in local metadata warehouse after processing. CI server is introduced in data management later to implement seamless connection cross-layer and cross-system on this layer. When the user access to metadata, it will automatically seek on local data and CI server. If there is needed data in local metadata warehouse, it will use from local. If there is not needed information local, it will seek for related information on the network through CI server.

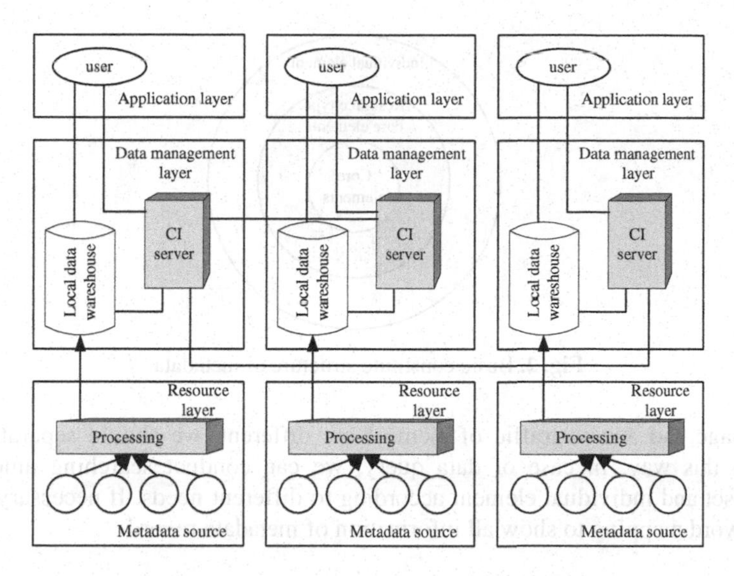

Fig. 1. Metadata management system based on CI

In case of metadata query through network, the idea of P2P is introduced. For example, when user at location *A* access to some metadata, it find that there is no related information in local metadata warehouse, it will send query request including keywords to it neighbor CI server through local CI server. Then, the neighbor CI server forwards it to its own neighbor CI server, and so on, till it find metadata meet requirement or TTL of query request is 0. If metadata meet conditions are found, query request response will return back to combined index server along back path. Afterwards, metadata warehouse at *A* will add the metadata.

In short, CI server can access information from other nodes in the network and can also provide information for other nodes, which avoids single node failure problem in centralized warehouse.

3.2 Storage of Local Data

The basic structure of meta includes three parts, namely core elements, resource type core elements and individual element, as shown in Fig. 2.

(1) Core element. It refers to generalized element in descriptions of various resource objects. In view of interoperability and inter-conversion of metadata, the element layer support metadata transformation among different system and different resources as well as generalized query tools.

(2) Resource type core elements. In the metadata specification development process of different types, we can design commonly needed element and qualifiers according to characteristics of resource objects, so as to ensure inter-conversion among metadata. Such kind of metadata is called as resource type core element.

(3) Individual element. It is developed according to attributes and characteristics of some specific resource objects, which only suitable to such kind of objects and can not be used for transmission.

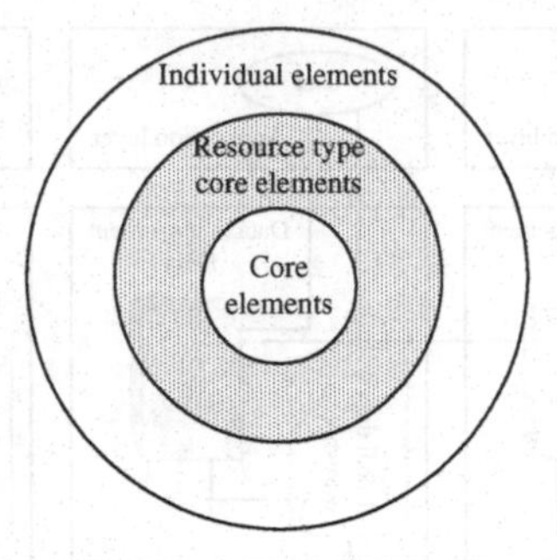

Fig. 2. Basic constitute structure of metadata

As usage and access traffic of element are different, we should separately store them, in this way, in case of data query, we can conduct searching among core element set and individual element according to different needs. If necessary, we can add keyword mapping to show all information of metadata record.

3.3 Combined Index

In the CI, interoperability among different metadata is achieved by constraints of semantic layer, syntax layer, structural layer and communication layer, as shown in Fig. 3.

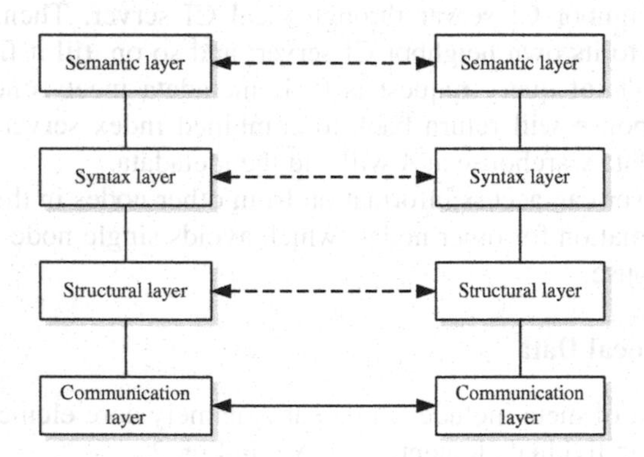

Fig. 3. Interoperability among different metadata in CI

(1) Structural layer. It is used to ensure inter-conversion between different metadata structures. We have mentioned above that classification storage of different elements has provided guarantee for interoperability in structural layer. Core element supports metadata conversion of different system and different types of resource as well as generalized query tools.

(2) Syntax layer. Syntax refers to that how the metadata be expressed and described. Different metadata specification has different syntax description method, such as MARC uses ISO2709 tape recording format as its standard for record description and exchange; metadata as DC does not regulate only one specific description can be used.

Here XML or other markup language in XML language family is recommended. Such kind of opening description method can ensure different metadata specific content can be processed and parsed by different heterogeneous systems.

(3) Semantic layer. It is used to ensure accurately analyze relationships between element semantics and to support semantic matching and transformation among elements with different metadata format. Generally speaking, a metadata specification can not achieve same conversion with that of other heterogeneous metadata specifications. Therefore, it can not ensure consistency if only close semantic principle be obeyed in case of semantic mapping.

4 Performance Analysis

4.1 Query Efficiency

Query searching of metadata object is basis for effective operation of metadata architecture. In DMS, for information that not be local, it should query other GIIS or GRIS level by level. The CI server proposed in the paper is based on P2P. For metadata not in local data warehouse, it can simultaneously link multiple adjacent CI servers and conduct query, while adjacent CI server also connects to its own CI server, so the architecture is superior to MDS in query speed.

In addition, according to principle of data locality, for metadata not in local warehouse but been query through CI server, it will store in local warehouse. In case of query again, it needs not to query other warehouse through CI server, which improve query efficiency.

4.2 Extensibility

The metadata architecture separately store different elements, while greatly improves flexibility of whole metadata architecture. Firstly, separation of core element from individual element enable the core element can ensure basic metadata will not be too long while reveal basic features of resources. As to individual element, it can randomly add new resource and new description according to resource status and user requirements. Secondly, P2P support combined index. It extends inter-query of metadata and compensate drawbacks of MDS and other centralized metadata query, which provide sufficient space for data storage and application in data grid.

4.3 Interoperability

Interoperability refers to capability of inter-use and information exchange among multiple systems. CI server in the system directly connects to other CI server in the network, which will not decrease its interoperability in case of extension. As core element and individual element are separately stored and four-layer mode in combined index, capability of understanding and transformation among different specifications can be improved. It can not only be operated by its host application system, but also be accepted and operated by other different heterogeneous applications operating systems.

5 Conclusion

The paper presented a metadata architecture based on CI and achieved classification storage in data grid. P2P layered combined index technology was introduced in the system. The local metadata was stored based on classification greatly improved extensibility and interoperability of metadata architecture.

References

1. Anderson, D.P.: Bonic: a system for public resource computing and storage. In: Proc of the 5th International Workshop of Grid Computing, pp. 4–10. Springer Press, Baltimore (2004)
2. Singh, G., Bharathi, S., Chervenak, A.: A metadata catalog service for data intensive applications. In: Proc of ACM/IEEE SC Conference, pp. 33–34. IEEE Computer Society Press, Phpenix (2003)
3. Liu, Z., Zhou, X.M.: A metadata management method based on directory path. Journal of Software 18, 236–245 (2007)
4. Wamg, Z.F., Zhang, W.S., Liu, Z.: The Meta-Data Management and Access Control in the Linux Cluster File System(LCFS). Computer Engineering and Science 27, 103–105 (2005)
5. Fitzgerald, S., Foster, I., Kesselman, C., von Laszewski, G., Smith, W., Tuecke, S.: A directory service for configuring high-performance distributed computations. In: Proceedings of The Sixth IEEE International Symposium on High Performance Distributed Computing, pp. 365–375 (August 1997)
6. Hoschek, W., Jaen-Martinez, J., Samar, A., Stockinger, H., Stockinger, K.: Data Management in an International Data Grid Project. In: Buyya, R., Baker, M. (eds.) GRID 2000. LNCS, vol. 1971, pp. 77–90. Springer, Heidelberg (2000)
7. Segars, W.P., Lalush, D.S., Tsui, B.M.W.: Modeling respiratory mechanics in the MCAT and spline-based MCAT phantoms. IEEE Transactions on Nuclear Science 48, 89–97 (2001)

Two-Stage Inventory Model with Price Discount under Stochastic Demand

Yanhong Qin[1] and Xinghong Qin[2]

[1] School of Management, Chongqing Jiaotong University, 400074, Chongqing, China
[2] School of Economics and Management, Tongji University, 200092, Shanghai, China
qinyanhong24@163.com, qinxinghong4515@sina.com

Abstract. The price discount is a very effective incentive strategy to improve the business interest of each node and even the overall interest of supply chain. By far, most researches in this field have been based on the certain demand, and there is no detailed verification of the theory before and after the implementation of price discount. So we will fetch up these shortages by studying a two-stage supply chain under stochastic demand, and obtain some important conclusions to provide some management insights.

Keywords: supply chain, price discount, stochastic demand, coordination.

1 Introduction

The researches on supply chain have been concerned by academia and business field together. It is difficult for enterprise to achieve their purpose of holding dominant position in the fierce competition only by the advantage of product quantity and quality. However, it is particularly important to improve the business interest of each node and even the overall interest of supply chain through effective optimization on the products of supply chain, and the price discount is a very effective incentive strategy to realize this. There were many researches on the price discount in the supply chain. Wu and Dan (2009) proposed the supply chain mode with price discount based on the traditional EOQ (Economic order quantity) model and set an optimal cost model under certain demand of distributor. He and Zeng (2007) maximized the total profit of the whole supply chain and increased the profit of both partners by an appropriate range of price made in the coordinate contract, under the assumption that the ratio of market demand is a constant. Hu and Wang (2008) improved the price discount contract to coordinate the supply chain under the emergency. Amy and Lau (2010) proved the effectiveness of using the wholesale-price discount contract for the manufacturer to coordinate the decentralized supply chain. Cai and Zhang (2009) proposed the price discount contracts can outperform the non-contract scenarios by Stackelberg model and Nash game theory in terms of supplier and retailer. Selcuk and Serpil (2008) discussed the efficiency of alternative effective sharing mechanisms, and proposed methods to design the coordinate discount schemes by taking buyer expectation into account, while he supplier was still able to coordinate the supply chain with high efficient level and obtained a significant

D. Jin and S. Lin (Eds.): Advances in MSEC Vol. 1, AISC 128, pp. 465–470.
springerlink.com

portion of the net savings. So in the research of supply chain coordination, the price discount is indeed a common and practical incentive mechanism and it is positive for improving the profits of each node partner in the supply chain and the entire supply chain. By far, most researches in this field have been based on the certain demand, and there is no detailed verification of the theory and case study before and after the implementation of price discount. So this article will give a detailed proof, and to achieve the combination of theory and practice, and we will simplify the two-stage supply chain as a distributor and many retailers and each retailer sells different products. In other words, the distributor supplies different kinds of products to every single retailer. The inventory model about the supply and demand sides before and after the implementation of price discount will be set on the assumption that the distributor will not stock out and both sides meet the optimal Pareto.

2 Notation

i : The index of retailers, and it is the product index, $i=1,2,3,\cdots n$. $F(x)$: The distribution function of demand, and $f(x)$ is its density function. d : The minimum demand, and D is the peak demand. w , w' : The wholesale price provided by distributor without discount and with price discount respectively; p_1 : The purchase price of distributor paying for its upstream suppliers; p_2 : The retail price at retailer; P_2' : The retail price at retailer after implementing price discount; l_i : The distance between the retailer and distributor; b_i , b_0 : The order cost at each time for the retailer i and distributor respectively; h_i , h_0 : The holding cost of unit product at retailer i and distributor respectively; k_i : The defect rate per product; $G(Q)$: The transportation price function of unit distance when retailer transports Q products. Let $G(Q)=a-b\ln(Q)$ (a, b>0) as many literatures do. T_i: The average turnover cycle without price discounts for the retailer; and $T_i'$is the average turnover cycle with price discounts. T_0 and T_0'have the same meaning for the distributor; v_i : The unit shortage cost per product for the retailer; β_0: The unit defect rate of product at distributor. ε : The pricing factor of the product in retail price, $p_i=w_i(1+\varepsilon)$ and α is pricing factor of the product for the distributor. C_0: The unit processing cost of defective product. θ: The discount rate provided by distributor.

3 The Inventory Model without Price Discounts

Before the collaboration between retailer and distributor, both of them optimize their own profit to make their order strategy, so we can model the profit separately.

The total profit of retailer i in the T th period is

$$\pi_r(Q_i,w)=\int_d^{Q_i} p_2 x f(x)dx+\int_{Q_i}^{D} p_2 Q_i f(x)dx-wQ_i-\int_0^{Q_i} h_i(Q_i-x)f(x)dx-$$
$$\int_{Q_i}^{D} v_i(x-Q_i)f(x)dx-Q_i k_i v_i-(a-b\ln(Q_i))l_i-b_i$$

If α_i ($\alpha_i \le \frac{Q_i}{\int_0^D xf(x)dx} \le 1$ as in Wu and Dan (2009)) denote the service level the probability without shortage, then The retailer's inventory model is:

$$\max_{\varepsilon,Q_i}\pi_r(Q_i,w)=\int_d^{Q_i} w(1+\varepsilon)xf(x)dx+\int_{Q_i}^{D} w(1+\varepsilon)Q_i f(x)dx-wQ_i-\int_0^{Q_i} h_i(Q_i-x)f(x)dx-$$
$$\int_{Q_i}^{D} v_i(x-Q_i)f(x)dx-Q_i k_i v_i-(a-b\ln(Q_i))l_i-b_i$$

$$\text{s.t}\quad \alpha_i \le \frac{Q_i}{\int_0^D xf(x)dx}\le 1, T_i\in I^+, \alpha_i, Q_i \ge 0$$

$T_i \in I^+$, is set of positive integer. We can compute the retailer's optimal order quantity Q_i^* by this programming problem.

From the above assumptions, we can see that the market demand faced by the retailer is continuous and stochastic. For all retailers, distributors adopt FCFS (first come, first served), and the previous order had no effect on the latter one. So the ordering time interval between the retailers observes the negative exponent distribution, and the density and distribution function is:

$$f_r(t)=\begin{cases}\mu e^{-\mu t}, t\ge 0\\ 0, t<0\end{cases} \qquad F_r(t)=\begin{cases}1-e^{-\mu t}, t\ge 0\\ 0, t<0\end{cases}$$

By the above assumptions, the supply capacity of distributor is much greater than the order requirement of the retailer, and there is no shortage at distributor. So we can assume that the order quantity of distributor Q_0 is integer multiple of the retailer's best order quantity n. Besides, once the order quantity of distributor is determined, it will not change for some operation constraints. Let $\pi_w(Q_0,T_0)$ be the distributor's profit function, and for $w=p_1(1+\alpha)$, so

$$\pi_w(Q_0,w)=Q_0 p_1\alpha-b_0-h_0\frac{(n-1)Q_i^*}{2}-C_0\beta Q_0$$

Suppose a service level α_0, then the distributor's order policy should meet the constraint: $\alpha_0 \le \frac{Q_0}{\sum_{j=1}^{n} Q_j} \le 1$. It can get the distributor's inventory model at a given service level α_0:

$$\max_{\alpha,Q_0} \pi_w(Q_0, w) = Q_0 p_1 \alpha - b_0 - h_0 \frac{(n-1)Q_i^*}{2} - C_0\beta_0 Q_0$$

$$\text{s.t}\ \alpha_0 \le \frac{Q_0}{\int_0^D xf(x)dx} \le 1, T_0 \in I^+, \alpha_0, Q_0 \ge 0$$

So We can get of the retailer's optimal order quantity Q_0^*.

4 The Inventory Model with Price Discounts

When the distributor and retailers achieve a long-term alliance relationship and the overall interests of the supply chain will increase through stimulating demand by price discounts. For the price elasticity, i.e. the demand is negative with product price, and then the order quantity of retailer and the market demand meet: $x = x_0 w^{-\lambda}$.

x_0 is a scale constant of demand, λ (λ>1) is the coefficient of demand-price elasticity, and the price discount rate is θ. It is easy to know that only $\frac{p_1}{w} \le \theta \le 1$, the distributor can profit, and $x' = x_0(w')^{-\lambda} = x_0(w\theta)^{-\lambda} = x_0 w^{-\lambda}\theta^{-\lambda} = x\theta^{-\lambda}$, i.e. $x' = x\theta^{-\lambda}$. For $0<\theta\le 1$ and $\lambda > 1$, $\theta^{-\lambda} > 1$, and $x' \ge x$. So the demand will increase after the implementation of price discounts and the optimal order quantity of retailer will also increase. After the implementation of price discounts, the retail price is p_2', the retailer's purchase price is w', and the total profit of retailer in the period T_i' is:

$$\pi_r'(Q_i', w') = \int_d^{Q_i'} p_1(1+\alpha)\theta(1+\varepsilon)xf(x)dx + \int_{Q_i'}^D p_1(1+\alpha)\theta(1+\varepsilon)Q_i' f(x)dx -$$
$$\int_0^{Q_i'} h_i(Q_i' - x)f(x)dx - \int_{Q_i'}^D v_i(x - Q_i')f(x)dx - Q_i' k_i v_i - (a - b\ln(Q_i'))l_i - b_i$$

Given the service level α_i, then the order policy of retailer should meet:

$$\alpha_i \le \frac{Q_i'}{\int_0^D xf(x)dx} \le 1 \quad (D > 0, \text{ is the peak demand})$$

At the same time, according to Pareto effectiveness, the order policy of retailer should meet: $\pi_r'(Q_i', w') \ge \pi_r(Q_i, w)$ (See Theorem 1). Therefore, after the implementation of the total price discounts, retailer's inventory model at a given service level α_i is:

$$\max_{\varepsilon,\theta,Q_i'} \pi_r'(Q_i', w') = \int_d^{Q_i'} p_1(1+\alpha)\theta(1+\varepsilon)xf(x)dx + \int_{Q_i'}^D p_1(1+\alpha)\theta(1+\varepsilon)Q_i' f(x)dx - p_1(1+\alpha)\theta Q_i'$$
$$- \int_0^{Q_i'} h_i(Q_i' - x)f(x)dx - \int_{Q_i'}^D v_i(x - Q_i')f(x)dx - Q_i' k_i v_i - (a - b\ln(Q_i'))l_i - b_i$$

We can get of the retailer's optimal order quantity $Q_i'^*$. For distributor with price discounts, the wholesale price is $w' = w\theta = p_1(1+\alpha)\theta$, and the total profit distributor in the period T_0' is: $\pi_w'(Q_0', w') = Q_0' w' - p_1 Q_0' - b_0 - h_0 \frac{(n-1)Q_i'^*}{2} - C_0\beta_0 Q_0'$

The key decision variable is the discount rate, which is determined by distributor:

$$\pi'_w(Q_0', w') = Q_0 p_1 (1+\alpha)\theta^{1-\lambda} - p_1 Q_0 \theta^{-\lambda} - b_0 - h_0 \frac{(n-1)Q_i'^*}{2} - C_0 \beta_0 Q_0 \theta^{-\lambda}$$

So $\dfrac{\partial(\pi_w'(Q_0', w'))}{\partial\theta} = Q_0 p_1 (1+\alpha)(1-\lambda)\theta^{-\lambda} + \lambda p_1 Q_0 \theta^{-\lambda-1} + \lambda c_0 \beta_0 \theta^{-\lambda-1}$

Since the best discount rate θ^* can't be obtained by using the second derivative, and in order to maximize the distributor's profits with price discounts, so we can search θ in [0, 1]. To search the θ^* as the distributor's optimal discount to maximize the formula (11), and the profits of distributor with discount should not be less than that without discount, so θ^* should subject to: $\pi_w'(Q_i', w') \geq \pi_w(Q_i, w)$

So the inventory model of distributor under given service level α_0 is:

$$\max_{\theta, Q_0', \alpha} \pi_w'(Q_0', w') = Q_0 p_1 (1+\alpha)\theta^{1-\lambda} - p_1 Q_0 \theta^{-\lambda} - b_0 - h_0 \frac{(n-1)Q_i'^*}{2} - C_0 \beta_0 Q_0'$$

In which, $T_0' \in I^+$,is set of positive integers. We can get of the optimal order quantity $Q_0'^*$ of distributor with discount.

5 Proof of the Relevant Theorem

Theorem1: For the retailers with price discounts, $\pi_r'(Q_i', w') \geq \pi_r(Q_i, w)$.

Proof. The total profit of retailer without price discount in the period T_i is:

$$\frac{\partial(\pi_r(Q_i, w))}{\partial Q_i} = -(p_2 + v_i + h_i)\int_0^{Q_i} f(x)dx + (p_2 + v_i)\int_0^{D} f(x)dx - k_i v_i - w = 0$$

And $F(Q_i) = \dfrac{(p_2 + v_i)F(D) - k_i v_i - w}{p_2 + v_i + h_i}$,

The optimal value of Q_i^* can be obtained by above equation.

Similarly $F(Q_i') = \dfrac{(p_2 + v_i)F(D) - k_i v_i - w'}{p_2 + v_i + h_i} = \dfrac{(p_2 + v_i)F(D) - k_i v_i - w\theta}{p_2 + v_i + h_i}$

We can draw the optimal order quantities $Q_i'^*$ after implementation of the price discounts from the above equation. Comparing (14) and (15), for $0 < \theta \leq 1$, and then $w\theta \leq w$. Obviously, $F(Q_i) \leq F(Q_i')$ and for $F(Q_i) = \int_0^{Q_i} f(x)dx$ is strictly increasing in Q_i, so $Q_i \leq Q_i'$. Combined with $x' > x$, i.e., market demand increase after the implementation of price discounts.

$$\pi_r'(Q_i', w')|_{Q_i'=d} - \pi_r(Q_i, w)|_{Q_i=d} = (1-\theta)\left[wd - p_2 d \int_d^D f(x)dx\right]$$

Theorem 2: For the distributor with price discounts, $\pi'_w(Q_0', w') \geq \pi_w(Q_0, w)$

Proof. Before the implementation of price discounts

$$\frac{\pi_w(Q_0,w)}{\pi'_w(Q_0',w')} = \frac{Q_0(w-p_1-C_0\beta_0)-b_0-h_0\frac{(n-1)Q_i^*}{2}}{Q_0(w\theta-p_1-c_0\beta_0)\theta^{-\lambda}-b_0-h_0\frac{(n-1)Q_i'^*}{2}} \leq \frac{Q_0(w-p_1-C_0\beta_0)-h_0\frac{(n-1)Q_i^*}{2}}{Q_0(w\theta-p_1-c_0\beta_0)\theta^{-\lambda}-h_0\frac{(n-1)Q_i'^*}{2}}$$

$$< \frac{w-p_1-c_0\beta_0}{(w\theta-p_1-c_0\beta_0)\theta^{-\lambda}-h_0\frac{n-1}{2}\frac{Q_i'^*}{Q_0}}$$

$$\frac{\pi_w(Q_0,w)}{\pi'_w(Q_0',w')} \leq 1, \text{ so } \pi'_w(Q'_0,w') \geq \pi_w(Q_0,w).$$

5 Conclusion

On the assumption of stochastic market demand and demand-price elasticity function, this paper has researched the price discount in the two-stage supply chain made up of single distributor and many retailers. And then, the profit model before and after implementation of price discount strategy was set for retailer, distributor and supply chain respectively, where the distributor supplied different products to each retailer. The price discount contract was proved to improve the profits of each partner and whole supply chain effectively, and when the profit of retailer in the stable tendency, the price discount provided by distributor became an effective method of coordinating the supply chain management.

References

1. Wu, Q., Dan, B.: Outsourcing logistics channel coordination with logistics service levels dependent market demand. International Journal of Services Technology and Management 11(2), 202–221 (2009)
2. He, Y.M., Zeng, J.P.: Supply Chain coordination under the price discount. Operations Research and Management Science 16(6), 38–41 (2007)
3. Hu, J.S., Wang, H.: The Price Discount Contract Analysis of Three-Level Supply Chain under Disruption. Chinese Journal of Management Science 15(3), 103–107 (2007)
4. Cai, G.S., Zhang, Z.G.: Game theoretical perspectives on dual-channel supply chain competition with price discounts and pricing schemes. International Journal of Production Economics 117(1), 80–96 (2009)
5. Selcuk, K., Serpil, S.Y.: Single-supplier/multiple-buyer supply chain coordination: Incorporating buyers' expectations under vertical information sharing. European Journal of Operational Research 187(3), 746–764 (2008)

Study a Text Classification Method Based on Neural Network Model

Jian Chen, Hailan Pan, and Qinyun Ao

School of Computer and Information, Shanghai Second Polytechnic University.
No.2360, Jinhai Road, Shanghai, China
{chenjian,hlpan,qyao}@it.sspu.cn

Abstract. Nowadays, data from all information channels expand rapidly, how to retrieve useful information becomes the critical issue. To find the readable and understandable information, classified data into different categories can help people deal with massive information within a short time. This paper provides a text classification method based on neural network model and delivers a reasonable performance.

Keywords: Text classification, Neural Network, Information Retrieval.

1 Introduction

With the rapid development of Internet, it contains a big amount of information, which constantly spreading onto the Internet. How to find the useful information with these randomly structured data is one of interests of current research area. Currently, automatic text classification approaches and models can automatically divide random information or text into different, pre-defined categories to help people understand data efficient.

1.1 Text Classification

Text classifications a process which classifies provided text into pre-defined categories or category based on the pre-defined categories. The most common categories would be sports, politics, economics, arts, etc. Text classification is a mapping process, it mapping a un-defined document into the known categories. This mapping process is one to one or one to many. It could be denoted as follow

For every

$$\langle d_i, c_i \rangle \in D \times C \tag{1}$$

(d_i denotes one document of documentation set D, C denotes classification set {c_1, c_2,… c_n}, if equation (1) is true, that means document d_i belongs to category c_i. or else if equation (1) is false., which can denote $D \times C \rightarrow \{T, F\}$.

D. Jin and S. Lin (Eds.): Advances in MSEC Vol. 1, AISC 128, pp. 471–475.
springerlink.com

But, there are some negative characteristics in text classification. One is high dimensional feature spaces. There are big amount of stand-by features when do feature extraction. A group of 100 of training texts may generate 10 times candidate features. Furthermore, it would generate more if the system adopts N-Gram. And the other is sparse distribution of characteristics. Feature space contains high dimension in text feature, because if a word were as one of text feature, it always would be the Moderated word in *corpus*. However, most of the feature words in feature space have zero frequency within a not document, it leads to the value of most characteristics represents document vectors also would be zero.

1.2 Neural Network

Neural network is a mathematic model which simulates structure and work process of human brain. There are several models of Neural network, back propagation (BP), counter propagation network (CPN), learning vector quantization (LVQ), Self-Organization Map (SOM) etc. most of them have same structure. A common neural network contains input layer, output layer and a few hidden layers, the amount of neurons equals to the amount of features of sample documents, output layers represents different categories and the amount of neurons represents the amount of categories. Figure 1 illustrates the typical overview of a neural network.

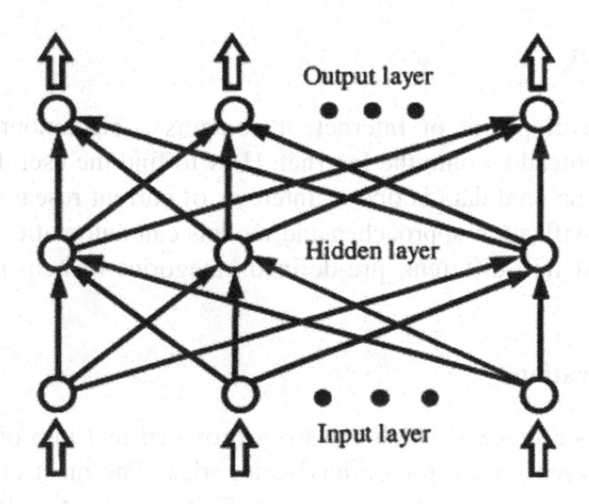

Fig. 1. Overview of Neural network

This paper represents a method uses BP neural network model to classify text can avoid such negative aspects. In this model, classification knowledge implicated stored on the weight, the value of weight could be calculated by iterative algorithm. According to the errors, the network could adjust the value of weight, which gradually approaches the errors.

A step of this algorithm would be follow:

1) Initialize the weight of each layers $W_{ij}$and threshold value θ_j, normally both of values would be a random and rather small value, like 0.0001.
2) Provides training sample documents, which can be denotes input vector I_n and output vector O_m. n,m = 1,...N;
3) Defines the relationship between input vectors and output vectors considering thresholds.

$$O_j = f(\sum_i w_{ij} I_{ij} - \theta_j) \tag{2}$$

O here denotes output node, w denotes weight of connection between nodes, and θ denotes threshold.

4) Calculate the errors:

$$e_i = O_j(1 - O_{j-1}) \sum_k e_{j-1} w_k \tag{3}$$

5) Adjust thresholds and weights

$$w_{ij}(n+1) = w_{ij}(n) + \alpha e_1 o_1 + \beta(w_{ij}(n) - w_{ij}(n-1)) \tag{4}$$

$$\theta_{ij}(n+1) = \theta_{ij}(n) + \alpha e_1 + \beta(\theta_{ij}(n) - \theta_{ij}(n-1)) \tag{5}$$

Here, α denotes learning rate, and β denotes bias. The loop ends until errors reach the minimum requirements.

2 Text Classification

This paper designed a system to divide text into seven categories as followed table, each category has about 200 training documents:

Fig.2 shows an overview structure of the text categorization task. The procedure for automatic text categorization can be divided into two sections, the training section and the test section, as shown in Fig.1. In the training section, documents which be listed in table 1 are training documents along with a category. Next, feature term via a feature selection process will be proceeded and generate an indices database, which is later used for the test phase. In the test phase, several new documents will be classified from the system, and also there are seven categories are allocated in these documents.

It is important to reduce the feature dimension before adopt neural network for text classification. There would be over thousands of dimensions for different groups of texts, it is very hard to analysis which dimension is important because most of them seem as spare. Hence, select proper model to reduce feature dimension and represent all of text is critical process. This paper select chi-square test (or X^2 test) to reduce feature dimensions.

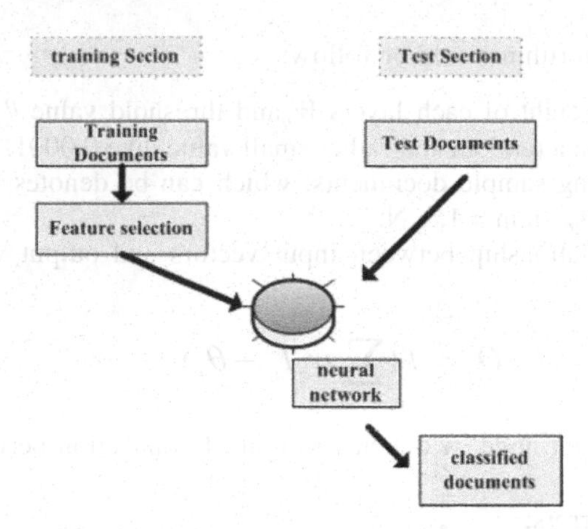

Fig. 2. Overview structure of text categorization

If C denotes a feature of category: then, with chi-square test can be denoted by following equation:

$$X^2(C) = \sum_t P(T_t \frac{N(a_2 a_4 - a_2 a_3)}{(a_2 + a_1)(a_3 + a_4)(a_1 + a_3)(a_2 + a_4)}) \qquad (6)$$

Here, a_1 denotes number of texts in category T which include feature C. a_2 denotes number of texts which not in category T but include feature C. a_3 denotes number of texts in category T which not include feature C. a_4 denotes number of texts which neither in category T nor include feature C. N denotes the total amount of training texts. With chi-square test, the dimensions of feature could be largely controlled under satisfied level.

3 Experiment and Results

It is very difficult to find standard benchmark sets for text classification, because each method would be tested under different circumstances performed differently. This paper use text within Reuter-21578 text collection, which selected 8821 documents for training and 1753 for testing.

Table 1. Shows the accurate of seven categories

Categories	Accurate rate
Sports	92.72%
Politics	90.34%
Culture	87.28%
Military	94.56%
I.T	96.13%
History	89.29%
Religion	88.75%

Table 1 shows the highest accurate rate belongs to I.T categories because this category contains more unique features for this category. On contrast, Culture & History obtains lowest rate because both categories share common features.

Since problems still bias performance of the text classification model, it is possible to adopt new feature diagram specified to text within History, Religion, and Culture topics.

5 Summary

Information retrieving is an important issue for understanding information, and text categorization is the key process to divide information into different topics for better understanding. This paper demonstrates a text classification method which can divided information into nine categories with Back-propagation neural network and with chi-square test, the dimensions of feature be controlled under satisfied level. For future research, it is possible to demonstrate different methodologies to illustrate the best to classified pre-defined categories.

References

1. Ou, G., Murphey, Y.L.: Multi-class pattern classification using neural networks. Pattern Recognition 40(1), 4–18 (2007)
2. Lindenbaum, M., Markovitch, S., Rusakov, D.: Selective Sampling for Nearest Neighbor Classifiers. Machine Learning 54(2), 125–152 (2004)
3. Thankachan, S.V., Hon, W.-K., Shah, R., Vitter, J.S.: String retrieval for multi pattern queries (2010)
4. Sadakane, K.: Succinct Data Structures for Flexible Text Retrieval Systems. Journal of Discrete Algorithms 5(1), 12–22 (2007)
5. Zobel, J., Moffat, A.: Inverted files for text search engines. ACM Computing Surveys 38(2) (2006)
6. Tam, A., Wu, E., Lam, T.W., Yiu, S.-M.: Succinct text indexing with wildcards. In: Proc. Intl. Symp. on String Processing Information Retrieval, pp. 39–50 (August 2009)
7. Hon, W.-K., Lam, T.-W., Shah, R., Tam, S.-L., Vitter, J.S.: Succinct Index for Dynamic Dictionary Matching. In: Dong, Y., Du, D.-Z., Ibarra, O. (eds.) ISAAC 2009. LNCS, vol. 5878, pp. 1034–1043. Springer, Heidelberg (2009)

Security Design and Implementation of Yangtze Gold Cruises Website

Huashan Tan, You Yang, and Ping Yu

School of Computer and Information Science
Chongqing Normal University
Chongqing, 400047, China
{6510388,40465742}@qq.com, youyung@yahoo.com

Abstract. Based on the analysis of common web security defects, some web design considerations about web data security and user identity authentication were pointed out. A substitute algorithm and its program used in web data were described. SQL injection attack (SIA) to web data couldn't operate due to the using of this algorithm. Values generated by MD5 algorithm was added to username and password, for the goal to ensure the user table storing security. To resist SIA while identity was authenticating, a method called program authentication was proposed. These safety measures or considerations may benefit to similar web design.

Keywords: Yangtze gold cruises, web security, SQL injection attack, substitute algorithm.

1 Introduction

To build international Yangtze gold travel stripe, Yangtze gold cruises appeared with Chongqing government improvement. Www.yangtzegoldcruises.com plays an important role. It not only plays as an enterprise information portal (EIP), but also is helpful to tickets selling.

Today, structure of dynamic web pages in front plus database behind was used in most website. Some parameters were obtained from the user request in web pages. Then parameters were constructed dynamically to SQL statements. SQL statements sent to database next. The results returned to the web pages from database finally. These steps finished the execution task of web pages. In the process, if hacker attack happened or Trojan horse invaded, the constructed SQL statements could be changed. The statements maybe accomplished the functions which were not the initial operations user wished. This is known as SQL injection attack (SIA).

Most websites nowadays used the rent space of ISP (Internet Service Provider). Program service and database service were managed by ISP. What can website builder do for web security? That is only the security of the application program and the web data. Current security standards for XML and web service were described by Nordbotton [1]. These standards plays great important role to the overall security of distributed system. Maguire pointed out: without proactive consideration for an application's security, attackers can bypass nearly all lower-layer security controls simply by using the

D. Jin and S. Lin (Eds.): Advances in MSEC Vol. 1, AISC 128, pp. 477–483.
springerlink.com

application in a way its developers didn't envision [2]. An application model called PWSSec(Process for Web Service Security) was proposed by Carlos and David [3]. Based on this model, it's easy to analysis the security between the models in information system. These research results are helpful to the website builder. For an instance, two considerations of security were implemented in www.yangtzegoldcruises.com. One is for web data, another for ID authentication. The web data was filtered by the proposed substitute algorithm so that the hidden Trojan horse could not be executed. The user ID was identified by the proposed program authentication so that the SIA could not be happened easily.

2 Web Data Submitting

Trojan horse may hide in the web data user submitted. It's very dangerous to web operation. Before the web data submitted to database, it's necessary to scan the data to find the possible signs or characters which could be used to attack the web.

The proposed substitute algorithm is such a method to filter the web data. The characters hiding in the web data were found by the algorithm and substituted to HTML codes. This substitution resulted in the web data not only was safety, but also was displayed normally. Before this substitute algorithm was applied, Trojan horse may embed in web data such as news table by hackers or viruses. In the worst situation, Trojan horse may cause the web disable. After this substitute algorithm was applied, even Trojan horse was embedded in web data, the key characters of Trojan horse program were filtered and changed. The substituted characters could recognize by HTML so that it did not affect the data display. However, the substituted characters changed the Trojan horse program so that the horse program could not run.

For more detail, the ASP program codes corresponding to above algorithm were described bellow also. In the codes bellow, the characters used for attack were numerable. Because of these special characters were defined in the program but not in database, so illegal users could not attack the web by database invading.

```
FUNCTION MYFILTRATE(S)
  DIM S1,I,X,N,Y
  S1=""
  N=LEN(S)
  FOR I=1 TO N
    X=MID(S,I,1)
    Y=ASC(X)
    SELECT CASE Y
      CASE 34,37,38,39,60,61,62,63
        S1=S1 & "&#" & Y & ";"
      CASE ELSE
        S1=S1 & X
    END SELECT
  NEXT
  MYFILTRATE=S1
END FUNCTION
```

Actually, the web data filtered by substitute algorithm cost more time than not filtered, but ensure the web security. How much time did this algorithm cost? The answer is low, very low, because of finite characters to substitute and the use of CASE statements in the algorithm. For example, running on a computer of Intel E6500 and 1GB RAM, if web data consisted 10 000 characters, the total submitting time of this web data is less than 1 second. Therefore, the time cost by substitute algorithm could ignore almost.

Another feature of this substitute algorithm was alternative. It means that the web data for general users should be filter, but was not necessary for administrator. First of all, the web data was security essentially submitted by administrator, because of the high level web management skill and good mind. Secondly, administrator could embed script program to web data just like hacker hide the Trojan horse in web data. The goal of this manner was to add web functions. For example, there was a group of Java script codes which used to play video. If these codes could embed to a web page, then a video from other web could be played in our web page. It's very clear, additional function added to the local web without any program changing. Furthermore, the data flow caused by this play manner is not counted to the local web. This decreased the money for these webs rent web space from other Internet service providers.

3 Identity Authenticating

3.1 Defects of Password Checking

Identity authentication is the foundation of the whole information safety system. It's the first defense line of network security, also the most important defense line. Identity authentication is such a technique to determine user's identity. Before communicating, identities of both communication sides should be identified. This is the basic requirements of identity authentication. Such technique solves the consistence problem between physical identity and digital identity. Furthermore, it supports other safety techniques to manage user's authorities. Data protection does not only let data store permanently and correctly, but also let data keep safe.

There are two type techniques of identity authentication over common network nowadays. One is password checking. Another is asymmetry encryption algorithm. Secret information of authentication user should transform over network if first identity authentication was used, and not if second was used.

Password checking is the popular method for identity authentication. A necessary condition should be satisfied in this method. Authenticator requests to identify should have an ID. This ID is unique in authenticator user's database. Generally, ID and password are included in this user database table. To ensure the authentication validity, some problems should be solved during the process of authentication. One problem is the password of requesting authenticator should be correct. The second problem is to keep the password security and not be substituted by others, during password transformation. The third problem is to identify the authenticator true identity before communication. Once identification was wrong, password may be delivered to a pseudo authenticator. Final problem is also the serious safety problem. That is all the users' password could be gotten and recognized by system administrator. These defects should be overcome to keep identity authentication right.

3.2 Improving of Password Checking

Adding Hash value to password could increase data security. Generally, identity information in a database is

username(ID) + password(psd).

Through Hash transformation, above information changed to

Hash(username(ID)) + Hash(password(psd)).

Before the transformation, username(ID) and password(psd) could be recognized from database or during information transmission. After the transformation, username(ID) and password(psd) could not be recognized even if the database table was opened.

MD5 (message-digest algorithm 5) is a popular Hash function. Any input character string with arbitrary string length could be transformed to an output integer with fixed length. The input string is meaningful. But the output integer is not. Fox example, if input password was "cqnucomputer", then the output of MD5 is "72A0D23C97F3C89CF4C40F73F6D5FBBD" over a 32 bits computer. It's easy to understand "cqnucomputer", but it's hard to interpret the later big integer. After MD5 transformation was used, identity information is security whether the information was transmitting or storing.

SHA is another Hash algorithm. SHA comes from MD5. 79 constants with 32 bits were used to calculate in this method. The final output of SHA is a completeness checksum with 160 bits. Due to the checksum length of SHA is longer than MD5, so SHA has higher security than MD5. But it's so sophisticated for SHA calculation. MD5 is used more widely than SHA.

4 SIA Resisting

4.1 Attack to Identity Authentication by SIA

SQL injection vulnerability results from the fact that most web application developers do not apply user input validation and they are not aware about the consequences of such practices. This inappropriate programming practice enables the attackers to trick the system by executing malicious SQL commands to manipulate the backend database [5-6]. One of the most important properties of SQL injection attack is that it is easy to be launched and difficult to be avoided. These factors make this kind of attack preferred by most cyber criminals, and it is getting more attention in the recent years.

Without loss of generality, two fields named username and password were included in user table. Suppose TID and TPSD represent the two fields of the user table. Suppose ID and PSD are two variables user submitted in the web login interface of application program. Every field and variable is the type of character. The traditional identity authentication is accomplished by constructing the following typical SQL statements:

```
SQLSTR="Select * from users where '" & ID & "'=TID and
'" & PSD & "'=TPSD"
SQLSTR="select count(*) from users where '" & ID &
"'=TID and '" & PSD & "'=TPSD"
```

Two SQL statements return record content and the number of record(s) satisfied. When the record(s) returned by the first statement or the number was greater than 0 by the second statement, it means success of identity authentication. Otherwise, it means failure.

Normally, ID and PSD user input in the web login program interface are not aggressive. For example, if the user typed “ABC” and “123” respectively, then the constructed SQL statements are:

```
SQLSTR="Select * from users where 'ABC'=TID and
'123'=TPSD"
SQLSTR="select count(*) from users where 'ABC'=TID and
'123'=TPSD"
```

Above two statements finished identity authentication normally.

Un-normally, if the user (may be a hacker) typed the following information: “1’=’1’ OR ‘ABC” and “123”, then the constructed SQL statements are:

```
SQLSTR="Select * from users where '1'='1' OR 'ABC'=TID
and '123'=TPSD"
SQLSTR="select count(*) from users where '1'='1' OR
'ABC'=TID and '123'=TPSD"
```

Under this situation, the conditions of two SQL statements are true forever. The hacker login successfully. This is known as SIA happened.

4.2 To Resist SIA by Program Verification

To resist SIA when user login in, methods such as limitation of input character, escape processing and prepared statement were used. Some characters user appreciated to use were not accepted by limitation method so that the user input freedom was restricted. Escape processing was restricted by its coding system, no universality existed in this method. Prepared statement method needed to construct a parameter procedure based related to database. Advanced permissions to database were necessary in this method. Additionally, prepared statement methods operations were complicated a little relatively. To discard these disadvantages, a naval method we called program authentication (PA) was applied to www.yangztegoldcruises.com. PA resists SIA outer database. The authentication procedure was accomplished by PA program.

PA did not submit any SQL querying to database. But the records of users table were read firstly, then the records were checked by program out the database. Because it’s unnecessary to construct SQL statement, therefore the problem of SAI to identity authentication was solved thoroughly.

The ASP codes of PA described bellow also. User could type any character as username and password in PA. Disadvantages of the three methods mentioned above disappeared.

```
ID=Trim(Request.Form("userid"))
PSD=Trim(Request.Form("userpassword"))
if ID<>"" and PSD<>"" then
  login=false
  sqlstr="select * from usertable"
  set rs=dbcon.execute(sqlstr)
  do while not rs.eof
  if ID=rs("TID") and PSD=rs("TPSD") then
      login=true
      exit do
    end if
      rs.movenext
  loop
  if login=true then
    'identity authentication OK.
  else
    'identity authentication fail.
  end if
end if
```

Relatively, there were two disadvantages of PA. One was that the computing task could not be accomplished by the database server which has high performance. This task can only finish by web server. Another was that the records of user table should be delivered to web server. This would cost time and occupy the network.

Application of the proposed program authentication depends on the size of user table. For the general webs of enterprise portal or news release, the size of its user table is around KB. The disadvantages brought by PA could be ignored. For the webs such as large forum, its logged users are very large. Reading the user table or transmitting the records may cost too much time. Therefore, PA is not suitable again to this case. Other methods [4] could be employed.

5 Conclusion

The proposed substitute algorithm could filter effectively the web data user submitted, so that the Trojan horse that hacker or virus embedded could be ill-functional. The users name and password were encrypted by MD5 Hash function. Even if the hacker could read the database, the data in database could not recognize easily. The proposed program authentication was used to identify the users ID. It could resist effectively the attack of SQL injection. The characters filtered by substitute algorithm were defined by web application program. Because of the characters were numbered, the time cost caused by the substitute algorithm could be ignore almost. The program authentication method did not care the user input character. Any character of computer could be typed. It was fit to the website with KB user table. No database operation and less time cost were the advantages of this method. Three methods presented in this paper were used in www.yangtzegoldcruises.com. Such websites like this could employee these methods also.

Acknowledgement. This work was supported by the Doctor Fund of Chongqing Normal University (10XLB006), and the Science Technology Researching Fund of Chongqing Education Committee (KJ100623, KJ110629).

References

1. Nordbotton, N.A.: XML and web service security standards. IEEE Communications Surveys and Tutorials 11(3), 4–21 (2009)
2. Maguire, J.R., Miller, H.G.: Web-application security: from reactive to proactive. IT Professional 12(4), 7–9 (2010)
3. Gutiérrez, C., Rosado, D.G., Fernández-Medina, E.: The practical application of a process for eliciting and designing security in web service systems Original Research Article. Information and Software Technology 51(12), 1712–1738 (2009)
4. Mitropoulos, D., Spinellis, D.: SDriver: Location-specific signatures prevent SQL injection attacks Original Research Article. Computers & Security 28(3-4), 121–129 (2009)
5. Ali, A.B.M., Shakhatreh, A.Y.I., Abdullah, M.S., et al.: SQL-injection vulnerability scanning tool for automatic creation of SQL-injection attacks. Procedia Computer Science 3, 453–458 (2011)
6. Kemalis, K., Tzouramanis, T.: SQL-IDS: A Specification-based Approach for SQL-Injection Detection. In: SAC 2008, Fertaleza, Ceara, Brazil, pp. 2153–2158 (2008)

The Effect of Multimedia CAI Courseware in the Modern Art Teaching

Guangcan Tu[1], Qiang Liu[2], and Linlin Lu[3]

[1] Art department, Yangtze University, Jinzhou, China
tuguangcan@126.com
[2] Art Academy, China Three Gorges University, Yichang, China
640690697@qq.com
[3] Information Techonology Center, China Three Gorges University, Yichang, China,
6269376@qq.com

Abstract. With the continuously introduction of modern teaching means, under the internet environment, use the multimedia technology to develop the CAI courseware of art course teaching could bring new effects of vision and auditory, abundant teaching content and information, change the traditional art teaching mode. The reasonable art design and arrangement made the teaching methods better than before.

Keywords: art teaching, CAI courseware , modern educational technology.

1 Introduction

In the traditional teaching of art, often speaking teachers, students painted the teaching methods, although this method has some effect, but a single comparison, change and the lack of attractive, long out of date in modern teaching concepts and methods.Education students from the target, due to the teaching of the characteristics of the object in the development of the times with the change, we can not use hard and fast to the old methods of teaching, monotonous inefficient teaching methods will only weaken interest in learning, on this point We reached the objective of teaching is very negative.On the other hand, as the popularity of computer technology and development, a growing number of modern media has been widely used in teaching art. Due to the involvement of modern media, has greatly enriched the classroom teaching methods and structures to modern with the traditional classroom teaching incomparable superiority.

Multi-media teaching is the teaching process, according to the teaching goals and teaching the characteristics of the object through the instructional design, rational choice and use of modern teaching media, and with the organic combination of traditional teaching methods, teaching the whole process of participation to a variety of media information On the role of students to form a rational structure of the teaching process to optimize the benefits.In the multi-media teaching, is the most common multi-media classroom instruction, which is the traditional media and modern media combine and complement each other, complementary, multi-level, multi-dimensional display the contents of teaching, both teachers in the Jingjiang inspired and modern The participation of the media in a timely manner, students learn to practice there are a

D. Jin and S. Lin (Eds.): Advances in MSEC Vol. 1, AISC 128, pp. 485–490.
springerlink.com

variety of alternate forms of teaching and effective to stimulate interest in learning and improve their attention so that students always take the initiative in learning, so as to achieve efficiency and improve teaching quality.Multi-media classroom teaching irreplaceable advantages. Known as multi-media teaching is the use of computer-specific and pre-production through the use of multimedia educational software to carry out the process of teaching and learning activities. It can be called computer-assisted teaching.

Computer-Aided Education (Computer Aided Instruction, referred to as CAI) is carried out under a variety of computer-assisted teaching activities, through dialogue and discussion of teaching students to arrange the process of teaching, training, teaching methods and technology. CAI multi-media to provide students with a personalized learning environment. Application of integrated multimedia, hypertext, artificial intelligence and knowledge base, and other computer technology to overcome the traditional teaching methods on a single, one-sided disadvantages. Its use can effectively shorten the learning time, improve the quality of teaching and teaching efficiency, optimize teaching objectives. With the advances in technology, multimedia technology because of its integration, control, interactive features such as the impact of multimedia teaching has become the hot spot in the field of teaching. CAI multimedia courseware for teaching as a media store a wealth of information, students can choose to study, students in the learning activities in a positive and active state of mind, so that a wide range of multimedia courseware for teaching CAI, compared with the conventional teaching methods Than, multimedia teaching from the intuitive, scientific, interactive, and so on, have shown great superiority.

With the continuous advance of the education reform, the quality of education in the process of reform and the status of the increasingly prominent role, the quality of education as an important part of the fine arts education --- more and more attention. We have said here is not the art of fine arts education institutions or professional art training techniques of pure abstract art learning theory, is designed to popularize knowledge of the arts, humanities improve the quality of the integrated arts education. Art education is a work full of creativity, art education, students can enhance human understanding of the history of civilization and culture they love nature, love life, love life, the true feelings to a more in-depth understanding of the law of the arts in order to broaden our horizons, mold, the wisdom of enlightenment, are effective in promoting the overall quality of students in the all-round development. For a long time, the main means of teaching art education to rely on slide shows, photo display and teachers to teach mainly verbal. But the teaching content in the form of small-han, a single tool can not effectively stimulate the enthusiasm of students and poor students in independent study. With modern means of teaching the constant introduction of Web-based environment, the use of multimedia technology development of fine arts curriculum CAI courseware for teaching students will bring new visual, auditory effect of a wealth of teaching content and information. This article will focus on art education and teaching of CAI courseware development and related issues to explore.

2 The Characteristics and Purpose of Art Teaching with Using Modern Educational Technology

The ultimate goal of art teaching is to improve the student's knowledge structure, promote the comprehensive development of students, students of health and aesthetic feeling of delight, the experience of the United States and appreciate the ability to establish the correct aesthetic concepts. However, the traditional fine arts education teaching model, teachers in a message issued by the position, and the students in a passive recipient of information, over-reliance on written materials and teachers to explain the lack of visual experience, making limited to classroom instruction book of knowledge Empty talk, teachers and students can not be carried out in a timely manner, the individual, in-depth exchange. Today, with the deepening of the reform of higher education, how to save more teaching hours to give way to the teaching profession, while at the same time maximize the use of the optional limited time to improve their overall quality of the humanities, is before us Problems. The use of multimedia network technology teaching is as good as most anticipated solving the problem of teaching means. Education and multi-media computer technology and network technology as the core, an integrated, interactive, resource sharing, and other features. Modern educational technology will be applied to art education, art education can significantly optimize the process of making art education can no longer rely solely on traditional print media and the conventional TV, listen to the media, students take the initiative, personalized learning provides free space , To the characteristics of hypertext, to provide rich and intuitive content to be fully demonstrated the effectiveness of teachers and students make the two-way communication has also been possible to achieve a harmonious co-operation between teachers and students so that the dynamic process of teaching , Inspired students and inspired the creation of innovative thinking. At the same time, the network environment, but also effectively control the cost of teaching, and maximize the goals of education. Futurist Nicholas Negroponte of the United States in the forecast was for many years before: we have entered a mode of artistic expression can be more lively and more participatory in a new era, we will have the opportunity to completely different way to spread And a wealth of experience sensory signals. This new approach is different from reading a book than on-site visit to the Louvre is more easy to achieve. The Internet will become the world's artists to display works of the world's largest art gallery, but also directly to the dissemination of works of art to the people's best tool.

3 The System Construction of Art Network CAI Courseware

(A) The Hardware System: the network of teaching art appreciation of the development of courseware for higher hardware requirements because of multimedia courseware contains a large number of visual and audio content, and its main production software such as the use of Photoshop, PowerPoint, flash, etc. Large graphics, audio production software, deal with them when computing and storage requirements are large, so the general requirements of computer hardware cpu to Pentium Ⅲ more than 128 megabytes of memory, with speakers, CD-ROM, scanners, digital cameras and so on.

(B) The Operating System: Windows98/2000/xp

(C) The Production of Software Commonly Used Software: Photoshop image-processing software plane, CorelDraw graphics software plane, 3DMAX animation software, Macromedia Flash animation software, AuthorWare the production of multi-media software, Power Point slide production software, FrontPage, Dream weaver and web design Web site management software, and so on.

(D) courseware design has been completed, but also through FTP upload to the campus network server release.

(E) landing students in the campus network registration can be carried out video-on-demand learning.

4 The Development of Art Teaching CAI Courseware

The development of art teaching CAI courseware generally include the following: the development of courseware for the preparation phase, the software script design, web page design software, network synthesis of the overall software and network software's release.

(A) Software Developed by the Preparatory Stage:
Teaching the art network CAI courseware development is an important characteristic of students --- and the autonomy of the individual. However, the design software should be in accordance with the requirements of the teaching content in the instructional design based on system design, software emphasized in the teaching content and teaching of a process of control. Although the network is teaching, teachers and students do not directly face-to-face exchange, but those Courseware must be of fine arts courses which have a more systematic classroom teaching ability and skill to master the ability to understand and fully familiar with the course of the emphasis and difficulty And according to the teaching system design plan will be difficult to re-analysis of the structure of the course to collect information on subjects related to the development of courseware to prepare audio-visual data collection phase of the image data collection.

To set up a three-dimensional nature of the multi-dimensional and multi-level knowledge of the structure of the courseware as a target point by the knowledge step by step into the deep to shallow as a whole, in addition to making each step has its own specific problems, but also Maintain links with the curriculum as a whole, and can be relatively independent in a position to become an independent unit. Art is not an isolated subject, said that if the fine arts in teaching aside the historical background, religious characteristics, aesthetic and psychological aside, only on the characteristics of art and art history of the development on, then surely the non-classroom teaching joy of life, students will be a Fog, and not knowing what. Therefore, art education to use an interdisciplinary approach to imparting knowledge, so that could be on the works of art, as well as on politics, religion, philosophy, economics and social structure of the knowledge combined. Research such as Leonardo da Vinci's "Last Supper" and "Mona Lisa" is not only study the works of visual style, these works reflect the life and the Renaissance era, it is a scientific discovery and the era of space exploration is Perspective of the times.

(B) Courseware Developed by the Script Design
CAI courseware for teaching art is usually based on the procedures to complete the preparation of the students through the computer in a picture frame and sound to

accept the teaching. As a result, CAI teaching art in the preparation of courseware, on the surface of the box for each frame of information must conduct a detailed design, a variety of side frame sequence is completed by a certain content as the main purpose of the study. According to the production flow chart included in the software system structure, knowledge element analysis, important information prompt manner, graphics, animation, sound, screen design, multimedia description of the relationship between the link script. As the fine arts education is the public face of arts education for all, the software in the design and production should take into account the media interface to be simple, easy to use, the media format must be affordable to the public, can be a wide range of players. Courseware in the preparation of the process, according to the course itself, there are characteristics of the purpose of the targeted selection of the design to be adopted by the media, means. Among them, a typical feature-sense approach of the use of analogy. For example, in music, art is bound to use the hearing means to lead students into the teaching curriculum, but the software designers can use the visual analogy for students to enhance the image of the music, color-awareness and understanding. In the visual arts, may be the introduction of hearing, touch the media to enhance the visual perception of ability and depth.

(C) Courseware Hyperlink Diagram of the Page

CAI courseware for teaching art network is a collection of web pages, CAI study is based on a page and start page design is a substantive CAI courseware design, including the following: ① text information processing. The main advantage of CAI courseware is informative and has a variety of convenient data presentation: enlarged rapid, rapid insertion and disappear, and so on, has expanded the amount of information so that students get the type of information has become very rich. ② page with the sub-page design: from the main branch of the learning process with a number of levels, with level-learning process. ③ the use of graphics software, animation presentation to guide what they have learned. Rational use of animated presentation can simplify the complexity of the study, making the students not just a passive recipient of the realization of interactive teaching.

(D) Of the Art network Teaching Evaluation Methods

Technology-based multimedia network teaching mode, significantly different from that in the past to memory as the center and a written examination as the main basis for the evaluation of a single method, it takes the multi-lateral, open, diverse teaching evaluation -- - Both teachers evaluate students as well as students of the evaluation of students, teachers, students, software for evaluation. Students can also study a unit, through its own software to bring the examination system for simulate random test papers. In such a network environment, the evaluation will be a process of evaluation, mainly mutual evaluation, supplemented by the examination and evaluation of the evaluation, and teachers can continue to guide the process of evaluation of students, correcting the direction of learning and methods of guidance, Students continue to stimulate learning and development; at the same time, teachers can also be based on student feedback information in a timely manner modify courseware to achieve the best teaching results. In addition, because of the special education network for teachers in the basic quality of the requirements have changed, he teaches students to evaluate not only the language of expression, organization, and other traditional teaching evaluation criteria, teachers also have strong computer Application of multimedia capabilities and technology capabilities, the ability of electronic design lesson plans and other modern educational technology.

References

1. Chang, R.-L.: Interdisciplinary art education. Capital Normal University Press (2000)
2. Yang, J.: Teaching Introduction to Art. Hubei Fine Arts Publishing House (February 2002)
3. Wu, H., et al.: Multimedia courseware for teaching of Mathematics and System Integration. Journal of Mathematics Education (2), 60–62 (2004)

A Virtual Laboratory Framework Based on Mobile Agents

Chao Yang and Gang Liu

Basic teaching and experimental center, Hefei University
Hefei, Anhui 230601, China
{c_yang5102,liugang}@hfuu.edu.cn

Abstract. This paper proposes a virtual laboratory framework based on mobile agents which can be applied to various experiment courses. We design two kinds of mobile agents including role agents and middleware services agents to simulate the behavior of human students and tutors and provide interactions and collaboration for students and tutors. The role agents provide the basic functionality for experiment courses: guiding, demonstrating, and explaining. And the middleware services agents enable the learning resources collaboration and the existing software application reuse to succeed. To demonstrate the feasibility of the proposed framework, Digital Circuit Virtual Laboratory (DCVL) has been successfully implemented applying it.

Keywords: virtual Laboratory, Intelligent Agent, E-learning, Experiment Courses.

1 Introduction

Due to the development of network technologies, is very prevalent and acceptable among people. In order to propose a rich distance education environment, a virtual laboratory should be provided which is a key on improving the quality of distance education since experiments are significant for most engineering and application courses [1]. A virtual laboratory is one of the new solutions offered by current technologies as an aid for teaching and researching, and it is provided for the learners to conduct course related experiments and simulations via network. [2].

The virtual laboratory can be classified into two categories according to its realization technologies. The first types of virtual laboratory adopt these popular WWW technologies including HTML, CGI, Java, Applet, Java Servlet and so forth to access the physical equipment used for experiments. Examples of these types of virtual laboratories are included in [3,4]. The second types of virtual laboratories merge intelligent agent approaches with virtual reality and artificial life. Typical examples include [5,6]. These virtual laboratories change the traditional lecturing method into a network enabled lecturing environment. But these virtual laboratories mostly based on client-server computing paradigm. When the virtual laboratory is getting larger, the environment will cause traffic jam. Thus these virtual laboratories are weak on environment adaptability and scalability. Moreover they can not share of many different resources between as many users as possible only if getting the source code and modifying existing standalone applications or developing new applications.

D. Jin and S. Lin (Eds.): Advances in MSEC Vol. 1, AISC 128, pp. 491–496.
springerlink.com

As we consider that the mobile agent (MA) approach is useful to solve these problems, we propose a framework based on mobile agent. We use design patterns to create various agents which can provide interactions and collaboration between teacher and learner or learner and learner in laboratory. The framework that reuses the existing software application or experimental platform will reduce the cost including time and money. Moreover, the proposed framework adapts java-based mobile agent system that can operate under different platforms and supports various multimedia. The different virtual laboratories can be designed and implemented based on the framework. The framework has been successfully applied to Digital Circuit virtual laboratory to demonstrate its feasibility.

2 System Architecture

In our proposed framework, we use the mobile agent techniques to construct the virtual laboratory. Figure 1 shows the mobile agent-based virtual laboratory framework. There are three major components in our virtual laboratory framework, including the mobile agent platform, mobile agents, and experimental platform.

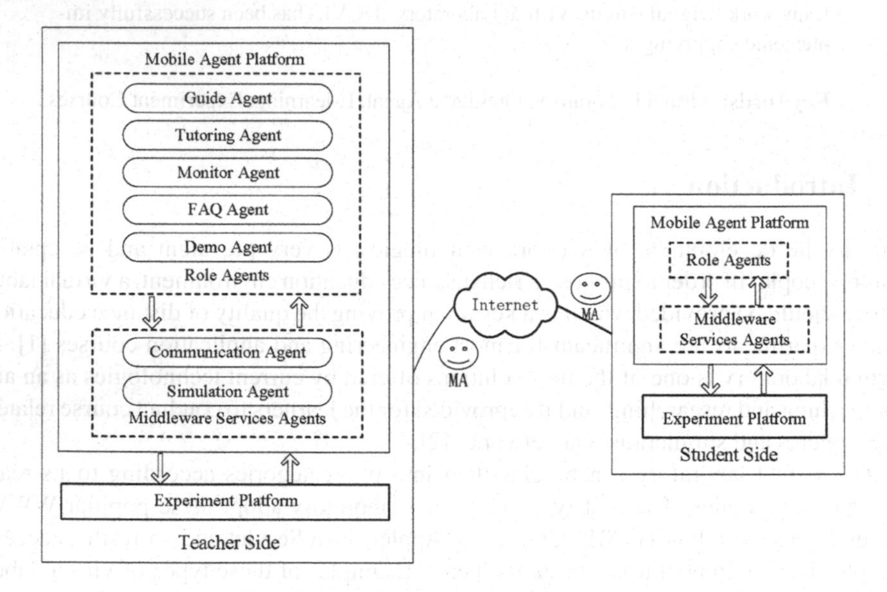

Fig. 1. Architecture of mobile agent-based virtual laboratory

2.1 Mobile Agent Platform

The mobile agent platform provides an execution environment for agents to execute [7]. The execution environment provides the resources required for agents to execute and communicate with other agents as well as with other resources and entities. Four important elements exist in the mobile agent execution environment, including agent management and control, agent communication services, Agent security services, and agent mobility service [8]. Agent management and control is responsible for managing all agents executing on the platform including monitoring and controlling access to resources as well as communication between agents executing on the local platform. Agent communication services provide communications facilities to agents executing within the environment. Agent security services include security services provided by the environment to executing agents. Agent mobility services enable agents to send themselves to other agents.

2.2 Mobile Agents

The mobile agent is a principal role in the virtual laboratory. Various mobile software agents provide interaction and collaboration between teacher, learners, and learning tools. There are two kinds of mobile agents for our framework, including middleware services agents and role agents. The role agents, such as Demo Agent, Guide Agent, Tutoring Agent, QA Agent, and Monitor Agent are designed for a learning environment. A teacher can use these role agents to assist learners to learn.

The middleware services agents are the key component to enable the learning resources collaboration to succeed. The Simulation Agent provides system function calls to control and gather experimental platform actions and information and then wrap around these resources in a software entity working. The Simulation Agent doesn't always need to be connected to some laboratory equipments or a specific real hardware resource. It can simply be a software simulation of some software application or some interface to data stored in a database. The Communication Agent supports interaction for the virtual laboratory to enable the message translation and communication in general with the Simulation Agent and the role agents. The role agents can command the experimental platform and monitor the student's operations via the Communication Agent.

2.3 Experimental Platform

The experimental platform consists of hardware and software environment to meet experimental requirements. In the software part, it is an existing application program, which could be a learning program as the CAI tool, design software like the CAD tool, simulation tool as Code Composer, Matlab, Altera, etc. or course management system for online learning as Blackboard, Moodle. The hardware environment in the experimental platform provides various hardware devices. The experimental platform could not require hardware support for some courses .For examples, experimental platform of Computer Principles course includes Code Composer, chip experimental controller and emulator. And the Matalab only contains the software platform and learners learn the experiments through simulation functions.

3 Applications of the Framework

Digital Circuit Virtual Laboratory (DCVL) was implemented by applying the framework we described above.

We adapt the IBM Aglet [9] as our mobile agent system for DCVL. Aglets were developed by the IBM Tokyo Research Laboratory. Aglets were the first Internet agent systems based on the Java classes. The aglets objects can move from one host on the network to another. That is, an Aglet that executes on a host can suddenly halt execution, dispatch to a remote host, and start executing again. The Aglet architecture consists of two main layers which are the Runtime Layer running on top of the Communication Layer [9]. The tight coordination between these two layers provides Aglets with their execution environment. The Aglets runtime layer defines the behavior of the API components, such as AgletProxy and AgletContext. It provides the fundamental functions for aglets to be created, managed, and dispatched to remote hosts. The Communication Layer, on the other hand, provides the basic mechanisms to allow Aglet's mobility and message passing through support of Agent Transfer Protocol (ATP) and Remote Method Invocation (RMI). It also supports agent-to-agent communication and facilities for agent management. The latest version of Aglet system is Aglets Software Developer Kit (ASDK) 2.0 which can be obtained in: http://www.trl.ibm.com/aglets/.

In DCVL all mobile agents are developed based on aglet design patterns [10] which can increase reuse and quality of code and at the same time reduces the effort of development of software system. In addition, Java Native Interface (JNI) technology was adapted for the middleware services agents. The aglet design patterns are classified to the three types including traveling, task and interaction. In DCVL, because the requirement that the network traffic should be low is given, the traveling patterns were applied to deal with various aspects of managing the movements of mobile agents, such as routing and quality of service. Also, these patterns allow us to enforce encapsulation of mobility management that enhances reuse and simplifies agent design. There are three abstractions for a role agent, GUI, Environment, and Middleware pattern. The GUI pattern provides friendly user interface for users to communicate with role agents. The Environment pattern provides role agents execution environments to allow different agents to move. The middleware pattern provides role agents a way to interact with the middleware services agents. Each type of role agent has different teaching and learning rules in a knowledge base which contains all data that has been perceived by the agent or produced as a result of its reasoning processes. Once the predicate of each rule is matched, the role agent will take predefined actions.

The DCVL experimental platform is composed of Altera's MAX+PLUS II and Altera's FLEX 10K devices which is FPGA FLEX EPF10K10LC84. MAX+PLUS II [11]provides a multi-platform, architecture-independent design environment that easily adapts to specific design needs. MAX+PLUS II also offers easy design entry, quick processing, and straightforward device programming. Altera's FLEX 10K devices are the industry's first embedded PLD (Programmable Logic Device)s. Based on reconfigurable CMOS SRAM elements, the FLEX architecture incorporates all features necessary to implement common gate array megafunctions. With up to 250,000 gates, the FLEX 10K family provides the density, speed, and features to integrate entire systems, including multiple 32-bit buses, into a single device.

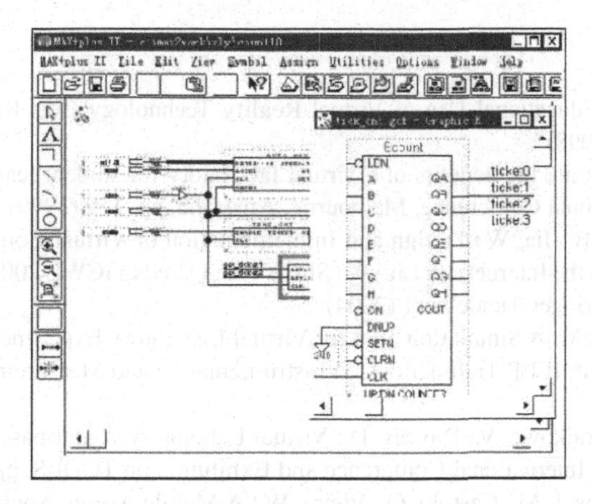

Fig. 2. A snapshot of MAX+PLUS II

The next three figures show the implemented DCVL in action. Figure 2 shows a snapshot of MAX+PLUS II. Figure 3 shows a guide agent running which will first initiate an experimental platform and then provide an agent list for the teacher. The teacher can dispatch various software agents providing to monitor, broadcast, demo, and cooperate with learners through network capability. For example, the teacher often needs to provide a demonstration of the experiment for students. The teacher can dispatch a demo agent to the students. The demo agent will carry a predefined demonstration example. Then the demonstration example will be presented step-by-step .Figures 4 show the creation of a demo agent and demo example selection for certain students.

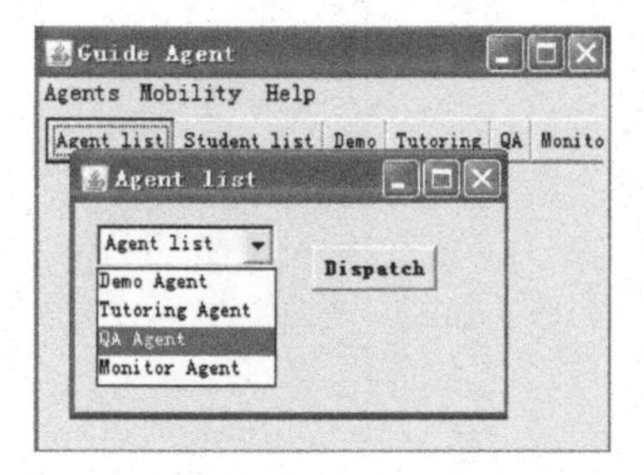

Fig. 3. A guide agent running

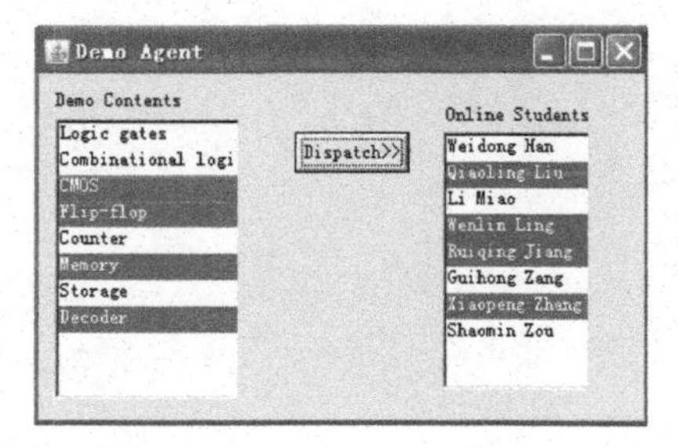

Fig. 4. The creation of a demo agent

4 Conclusions

In this paper a framework for virtual laboratory based on mobile agents was presented. By applying the proposed framework, the Digital Circuit Virtual Laboratory (DCVL) was implemented. In the framework the mobile agent techniques are used to provide teachers and students with various instructions and interactions including guiding, demonstrating, and explaining. Using design patterns approach in the development of the mobile agents is to increase reuse and quality of code and at the same time reduces the effort of development of virtual laboratory system.

References

1. Youngblut, C.: Educational Use of Virtual Reality Technology. Tec. Report. Inst. Defense Analyses, US (1998)
2. Chu, K.C.: What are the benefits of a virtual laboratory for student learning. In: HERDSA Annual International Conference, Melbourne, Australia, pp. 1–9 (1999)
3. Wang, J., Peng, B., Jia, W.: Design and Implementation of Virtual Computer Network Lab Based on NS2 In the Internet. In: Liu, W., Shi, Y., Li, Q. (eds.) ICWL 2004. LNCS, vol. 3143, pp. 346–353. Springer, Heidelberg (2004)
4. Ferreo, A., Piuri, V.: A Simulation Tool for Virtual Laboratory Experiments in a World Wide Web Environment. IEEE Transactions on Instrumentation and Measurement 48(3), 741–746 (1999)
5. Kimovski, G., Trajkovic, V., Davcev, D.: Virtual Laboratory-Agent-based Resource Sharing System. In: 39th International Conference and Exhibition on TOOLS, pp. 89–98 (2001)
6. Camarinha-Matos, L.M., Castolo, O., Vieira, W.: A Mobile Agents Approach to Virtual Laboratories and Remote Supervision. Journal of Intelligent & Robotic Systems 35(1), 1–22 (2002)
7. Lange, D.B., Oshima, M.: Mobile Agents with JAVA: The Aglet API. World Wide Web 1(3), 111–121 (1998)
8. Du, T.C., Li, E.Y., Chang, A.P.: Mobile agents in distributed network management. Communications of the ACM 46(7), 127–132 (2003)
9. IBM Japan Research Group. Aglets Workbench, http://aglets.trl.ibm.co.jp
10. Aridor, Y., Lange, D.: Agent design patterns: elements of agent application design. In: Proceedings of the Second International Conference on Autonomous Agents, pp. 108–115. ACM Press, New York (1998)
11. Heidergott, B., et al.: Max Plus at Work. Princeton University Press, Princeton (2006)

Research and Improvement of Resource Scheduling Mechanism in Alchemi Desktop Grid

CaiFeng Cao and DeDong Jiang

School of Computer Wuyi University, Jiangmen 529020, China
cfcao@126.com

Abstract. Alchemi provides concise and easy construction technology for central desktop grid system. But as the expansion of the scale of distributed system such as enterprise computing, pervasive computing and Internet computing, Alchemi needs to be expanded and improved in resource management , resource scheduling and other aspects .So Alchemi desktop grid middleware is studied throughly. The source codes are analyzed.The program of gaining resource information is repaired so to gaining the rich resource information. The algorithm of dynamic task mapping is designed,which realizes more accurate resources matching. Through tests, the improved system has good effect in running.

Keywords: Alchemi, desktop grid, grid resource scheduling, dynamic task mapping.

1 Introduction

Alchemi system, designed by the University of Melbourne in Australia, a .NET-based grid computing framework, provides the runtime machinery and programming environment required to construct desktop grids and develop grid applications. It allows flexible application composition by supporting an object-oriented grid application programming model in addition to a grid job model. Cross-platform support is provided via a web services interface and a flexible execution model supports dedicated and non-dedicated (voluntary) execution by grid nodes.

An Alchemi grid is constructed by some Manager nodes and more Executor nodes configured to connect to the Manager. Users can execute their applications on the cluster by connecting to the Manager. An optional component, the Cross Platform Manager provides a web service interface to custom grid middleware. The operation of the Manager, Executor, User and Cross Platform Manager nodes is described as Fig.1. [1].

Alchemi has conciseness, open-source, practical and other characteristics.It has already been used in many application projects. But with the development of application, the resources information gaining , resources scheduling and fault tolerance mechanism have been difficult to satisfy the demand of large-scale, multi-granularity grid task. For this, we study Alchemi kernel, modify some algorithms and procedures , and complete the tests.

D. Jin and S. Lin (Eds.): Advances in MSEC Vol. 1, AISC 128, pp. 497–502.
springerlink.com

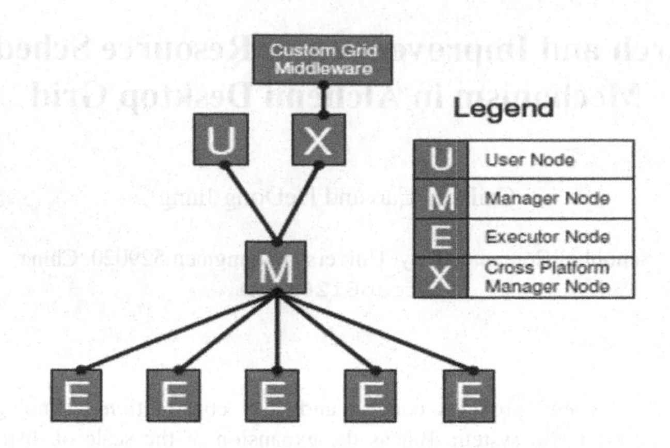

Fig. 1. The basic construction of Alchemi grid [1]

2 Alchemi Resource Management and Scheduling

Alchemi system's main resources are Executors.Resource discovery is finished by registration mechanism,in which every Executor registers its own information to one Manager. The resource information is recorded in the main database Alchemi.mdf, and refreshed at regular time. Resource scheduling is completed by Manager, which orders and chooses thread according to the priority of a thread and the principle of first come first served. All Executors in the cluster managed by one Manager are considered to be equal. Manager gains available Executor and distributes thread to it according to the state of Executor.

In the original resource information management system, only the Executor's CPU data is recorded, which can't fully describe the information of the resources. So it is necessary to acquire the status information of Executor's CPU, memory and hard disk.Then it is realizable to match the resources with tasks accurately.

3 New Program of Gaining Resource Information

The new program is created through modification of the original, which can access and refresh the status information of Executor's CPU, memory and hard disk.

3.1 Path Information of the Modified Program

Alchemi1.0.5 is installed in the folder Alchemic123.The common path is "\\\Alchemic123\Alchemi1.0.5\Alchemi-1.0.5-src-net-2.0\src\.Some files and lines that need to modify are shown below.

1)Alchemi.Core\Manager\IManager.cs"(248):void Executor_Heartbeat(SecurityCredentials sc,string executorId, HeartbeatInfo info); //add function definition

2)\Alchemi.Executor\GExecutor.cs"(812-852)://get HD and MEM information

3) \Alchemi.Executor\GExecutor.cs"(892):Manager.Executor_Heartbeat(Credentials, _Id, info);// add function definition

4)\Alchemi.Manager\GManager.cs"(719):public void Executor_Heartbeat(SecurityCredentials sc, string executorId, HeartbeatInfo info);//add function definition

5)\ Alchemi.Manager\MExecutor.cs"(182):public void HeartbeatUpdate(HeartbeatInfo info)//review function

6)\Alchemi.Manager\Storage\GenericManagerDatabaseStorage.cs"(625):protected void UpdateExecutorMemAndDisk(String executorId, float mem_max, float disk_mem, float mem_limit) //review function

7)\Alchemi.Manager\Storage\GenericManagerDatabaseStorage.cs"(661): public void UpdateExecutor(ExecutorStorageView executor) //review function

8)Alchemi.Manager\Storage\GenericManagerDatabaseStorage.cs"(676): UpdateExecutorMemAndDisk(executor.ExecutorId, executor.MaxMemory, executor.MaxDisk, executor.MemoryLimit); // add function definition

3.2 Example of the Program

```
Namespace Alchemi.Core.Executor
//HeartbeatInfo.cs
{ public struct HeartbeatInfo
  { public int Interval,PercentUsedCpuPower;
    public int PercentAvailCpuPower;
    public float mem_max,disk_max;
    public float mem_limit,disk_limit;
    ...
    public HeartbeatInfo(int interval,int used,int
    avail,float m_max,float d_max,float m_limit,float
    d_limit){…} }
  }
//Gexecutor.cs
{  ...
   ManagementClass mcHD=new ManagementClass(
   "Win32_DiskDrive"); //get HD information
   ManagementObjectCollection moHD=mcHD.GetInstances();
   Double diskmax=0;
   Float disklimit=0;
   Foreeach (ManagementObject tempob in moHD)
   { diskmax +=float.Parse(
      tempob.Properties["Size"].Value.Tostring()); }
    moHD.Dispose();
    mcHD.Dispose();

info.disk_max=float.Parse(diskmax.Tostring())/(1024*
    1024);
    long physicalMemory=0;
    ManagementClass mc=new ManagementClass(
   "Computer_System"); //get memory information
    ManagementObjectCollection moc=mc.GetInstances();
    Foreeach (ManagementObject mo in moc)
    { if (mo["TotalPhysicalMemory"]!=null)
      { PhysicalMemory+=long.Parse(
         mo["TotalPhysicalMemory"].Tostring());}
```

```
        }
        moc.Dispose();
        mc.Dispose();
        info.mem_max=float.Parse(PhysicalMemory.Tostring())
        /(1024*1024);
        ...
    }
```

4 Design of Grid Scheduler

Grid scheduler realizes machine selection and task scheduling. Grid scheduling includes four stages: resource discovery, resources selection, scheduling generation and job execution. Alchemi resource discovery is realized through Excutor,which registers its information to Manager automatically and refreshes at regular intervals.The procedure forms the table Alchemi. executor that is called resources list or resource pool. In resource selection and scheduling the resource characteristics and application properties must be considered. Alchemi desktop grid system is isomorphic and applicable to compute-intensive applications. With so little communication between tasks, the communication performance differences between hosts can be ignored.

Task scheduling is in two steps. The first step is to realize the mapping of application or task to one Maneger. The second step is to realize the mapping of thread to Excutor. This paper introduces dynamic task mapping algorithm.It not only ensures each task has least execution time, and avoid most tasks crowd in the best performance group,which will cause the load imbalance between machines.

4.1 Calculation of Resource Evaluation Value

For each Excutor, resource evaluation parameter includes CPU speed, memory capacity and hard disk capacity with the weight of 6, 3 and 1 respectively. Their computation formulas are as Fig. 2.

$$E\_cpu = W\_cpu * (1 - cpu\_usage) * \frac{cpu\_max}{cpu\_limit} \quad (1)$$

$$E\_mem = W\_mem * (1 - \frac{mem\_usage}{mem\_max}) * \frac{mem\_max}{mem\_limit} \quad (2)$$

$$E\_HD = W\_HD * (1 - \frac{HD\_usage}{HD\_max}) * \frac{HD\_max}{HD\_limit} \quad (3)$$

$$ER\_executor = E\_cpu + E\_mem + E\_HD \quad (4)$$

Fig. 2. Computation formulas of resource evaluation

In formula (1), E_cpu means Excutor's CPU evaluation value; W_cpu means CPU weight; cpu_usage, cpu_max, cpu_limit taken from Alchemi.excutor table, express respectively current load rate, actual rate and minimum rate of CPU. In formula (2),

E_mem means the evaluation value of Excutor memory; W_mem means its weight; mem_usage, mem_max and mem_limit taken from Alchemi.excutor table,express respectively the amount of current usage,total capacity and minimum limit of memory. Formula (3) expresses evaluation value of hard disk. In formula (4), ER_executor means resource evaluation value of one Excutor.

4.2 Dynamic Task Mapping Algorithm

The scheduled task set is called S, S = {S [0], S [1],... S [K-1]}, K is the total number of tasks. S [k] contains many threads,the set is {T [k, t] | t<=H, H is total thread number}. M is the collection of Managers participated in scheduling, M = {M [0], M[1],... , M [N-1]}, N is total Manager number. Each Manager manages a cluster.The set of resource evaluation values of all clusters is MR, MR = {MR [0], MR [1],... , MR [N-1]}, MR [i] represents M [i] evaluation value which is equal to the sum of its Executors value. For M [i] ,its Executor resource evaluation value set is {ER [i, j] | j<=P, P is Executor number in M [i]}.Then the algorithm description is as follows.

For each arrived task S[k] in Portal Server
  For each host M[j] in Managers set M
    searching out the greatest MR [i] and the corresponding M [i], connecting it.
   Endfor
  Serializing S [k] and related data files, sending them to M [i] through TCP channel.
Endfor
For arrived threads T[k,t] in M[i]
  Choosing the highest priority thread T[k,t], as well as the greatest ER [i,j], sending T [k,t] to ER [i,j] through TCP channel, then executing it. / * for same thread priority, according to the first come first served*/
Endfor

This algorithm firstly distributes tasks to Managers, which solves the problem of single point failure and visit bottleneck. Then realizes dynamic mapping for threads to Executors, which improves thread running efficiency. According to the algorithm, we rewrite the scheduler, and do corresponding test. One such application is "Pi Calculator" that calculates the value of Pi to 100 and 200 digits.Each thread calculates 10 digits. In a cluster with one Manager and 8 Executors, the original system running time is T1, new scheduler system running time is T2. When submitting 10 threads, 0<T2-T1 <3.5% * T1; When submitting 20 threads, 0<T2-T1 <4.8% * T1. The more threads submited, the more obvious the new algorithm has advantages.

5 Summary

Alchemi is an easy-to-use enterprise grid framework,which offers a simple way to set up and run a Windows-based compute grid,has a flexible programming model and simple tools for monitoring and administration.We are interested in it, deeply study it and do experiment on it. As an open-source project, the framework is in constant development. So we all do some works in its every aspects in order to make it better development. The next step we will introduce some intelligent scheduling algorithm□ do experiments further, and apply it to the practical projects.

References

1. Luther, A., Buyya, R., Ranjan, R., Venugopal, S.: Alchemi: A. NET-based Enterprise Grid Computing System. In: Proceedings of the 6th International Conference on Internet Computing (ICOMP 2005), Las Vegas, pp. 1–9 (2005)
2. Nadiminti, K., Luther, A., Buyya, R.: Alchemi: A. NET-based Enterprise Grid System and Framework User Guide for Alchemi 1.0. In: Grid Computing and Distributed Systems (GRIDS) Laboratory,Dept. of Computer Science and Software Engineering, The University of Melbourne, Australia (2005)
3. Zhang, H.L., Liao, X.Z., Zheng, H.T., Zhao, W., Leng, C.H.: Distributed Interactive Simulation Platform Based on Desktop Grid. Computer Integrated Manufacturing Sy stems 16(7), 1383–1389 (2010)
4. Li, M., Baker, M.: The Core Technology of Grid Computing. Translated by Wang Xianling, Zhang Shanqing and Wang Jingli, pp. 172–183. Tsinghua University Press, BeiJing (2006)
5. Wei, W.-D.: Decentralized Desktop Grid Platform Based on Unstructured Peer-to-Peer Networks. Computer Engineering and Design 32(1), 92–95 (2011)
6. Luther, A., Buyya, R., Ranjan, R., Venugopal, S.: Peer-to-Peer Grid Computing and a .NET-based Alchemi Framework. In: High Performance Computing: Paradigm and Infrastructure, Wiley Press, New Jersey (2005)
7. Md. Bikas, A.N., Hussain, A., Md. Shoeb, A.A., Md. Hasan, K., Md. Rabbi, F.: File Based GRID Thread Implementation in the. NET-based Alchemi Framework. In: Proceedings of the 12th IEEE International Multitopic Conference, pp. 468–472 (2008)

Customers Evaluation Effects of Brand Extension towards Brand Image of Chinese Internet Companies

Ming Zhou

School of Management, Zhejiang University, Hangzhou, China 310058
zmingzju@163.com

Abstract. The paper studied that the customers' evaluation of brand extension could affect the brand image of Chinese internet companies. Firstly, the paper did a comprehensive literature review about brand extension evaluation. Then, the paper proposed a synthetic model to testify our hypotheses, which identified the three aspects, perceived quality of parent brand, perceived fit between parent brand and extension products, difference of extension product, positively affected customer evaluation. And customer evaluation positively affected brand image. After data collected, the empirical analysis was carried through. In the final part, we analyzed the results and discussed the implications of the findings.

Keywords: Brand Extension, Customers Evaluation, Brand Image, Internet Industry.

1 Introduction

Brand is regarded as crucial intangible assets to enterprises. Nowadays, with the increasing significance of brand, there are quantities of companies using brand extension strategies for further development, especially for launching new products. Based on Aaker & Keller's opinion [1], brand extension is a marketing strategy using existed brand for the launch of new products in different product lines.

Up to 30 years, the research circles pour more interests on this field, to study the impact on the enterprises' development. It is a two-edged sword, Tauber [2] argues that organization can make full use of the brand association between parent brand and extension product to promote extension product. Consumers are willing to buy it based on the acquaintance of parent brand. So this strategy can reduce marketing costs and decrease the failure risk. At same time, it also can strengthen the image of key product of the brand. However, some other holds the different points. John & Loken [3] proved that extra brand extension can hurt the brand, as it obstruct the balance between extension product and parent brand. Ries & Trout [4] unreasonable brand extension can lead to psychological confusion, which will affect market position of parent brand. So the improper brand extension will affect the brand image so as to influence the long-term development of the company.

As internet industry growing giant and giant, the brand extension strategies also can be found in this area. Some companies provide a wide product lines from search engine to online C2C business. Based on the previous researches, it is common to

D. Jin and S. Lin (Eds.): Advances in MSEC Vol. 1, AISC 128, pp. 503–508.
springerlink.com

choose tradition industry as investigation samples, while it's seldom to find internet products. The most often used brands are NOKIA, McDonald's, Haier, Wahaha, etc. So what's the relation of brand extension, consumer evaluation, and brand image in internet industry? The paper is arranged as follows: Section 2 is the literature review on brand extension studies. At Section 3, we will introduce our methodology, conduct the hypotheses based on the literature review and our research objectives. After data collected, the empirical analysis is carried through. In the final part, we will analyze the results and discuss the implications of the findings.

2 Literature Review

There are plenty of papers about brand extension, especially in west world. From 1980's, the study on brand extension had never been stopped when Tauber [5] published the famous paper "Brand Franchise Extension: New Product Benefit from Existing Brand". There are two kind of brand extension based on direction: vertical extension and horizontal extension. Simon George [6] differentiated brand extension and product line extension. The change of size, quantity, price, technological process can be only seen as product line extension. The brand extension is the enterprises introduce a new product in other product line by using an existed brand, which had already built highly consumers recognition.

As to consumer evaluation, the epoch-making model is proposed by Aaker & Keller [7] in 1990. They proved that consumers attitude towards extension products can be influenced by following factors: perceived quality of parent brand, the perceived fit, or the similarity between parent brand and extension product, and the difference of producing extension products. Meanwhile, the similarity contains three dimensions: Complement, Substitute, and Transfer. Yang Ming [8] tested the model in Chinese contact, and put forward some adjustment about A&K model. He added brand association and delete the difference of producing extension products.

Alokparna Basu and Deborah Roedder [9] held that compared to analyzing thinkers, general flow thinkers are more willing to accept high-span brand extension of luxury products. It's better to use lower span extension of functional products. Fu, John & Qu [10] add two factors on Aaker &Keller's theory: the perceived fit based on brand image consistence, the competition condition of extension products. These two factors are critical for consumers evaluation, and the former factor are more important than the factors in A&K model.

JINWEI HOU [11] summarized previous literature and proposed the following five factors toward consumer evaluation. First factor is extension character, such as perceived fit and technology. Secondly, the characters of parent brand, which contains brand width and brand power. The third is enterprise character, for instance, enterprise size and reputation. The next is consumer character, such as product knowledge and culture difference. The last is market strategy. The market position, the sequence of extension, advertisement, pricing are can affect the evaluation. What's more the evaluation can change the attitude of parent brand in reverse.

There are also lots of papers related to brand image. Keller [12] raised brand image is a brand recognition in consumers memory, which is reflected by brand association. In other words, brand image can be regarded a set of feeling, expectation, attitude of a brand. Brand image consisted of perceived value, brand personality and organization image.

3 Hypothesis and Methodology

In this part, we carried out a group interview for the research samples in internet industry. Then, based on the literature review and research objective, we assumed the possible hypotheses. After that, we collected data for empirical analysis through website and spot collection.

Table 1. Brand extension-Customer evaluation-Brand image Model

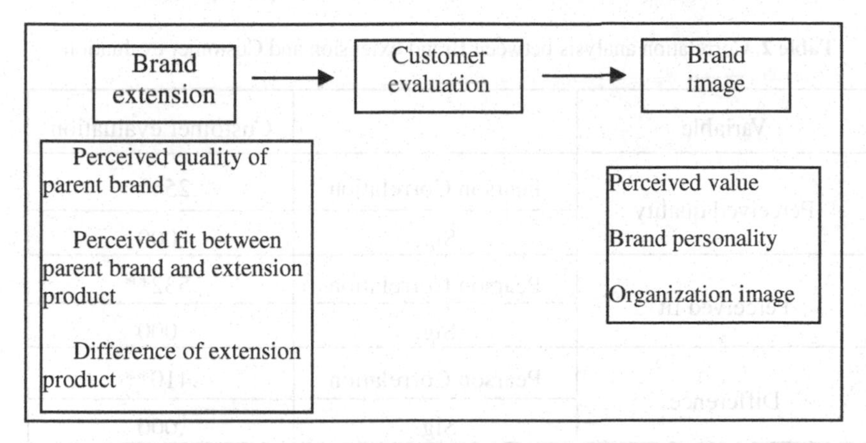

According to the literature review, we assumed the following hypotheses: (Table 1).

H1: Perceived quality of parent brand positively affects customer evaluation.

H2: Perceived fit between parent brand and extension product positively affects customer evaluation.

H3: Difference of extension product positively affects customer evaluation

H4: Customer evaluation positively affects brand image, which contains three dimensions, perceived value, brand personality and organization image.

For the group interview, we invited 10 master candidates, who have more than 5 year's internet age. In this interview, we got to know they are very familiar with brand of Baidu and Tencent. Baidu is a great search engine in China, enjoying 80 percent market share in mainland China. Tencent provide instant messaging software- QQ, whose quantity of registered users exceeds 670 million. Today, both of them develop lots of service and products through brand extension. For instance, Baidu and Tencent have browser, safety software, online market, social networks, etc. respectively. So we developed the questionnaire highly related to the extension products and parent brand of Baidu and Tencent.

The measuring scale is consulted by mature scale developed by Aaker & Keller. Based on the measuring scale, we design the questionnaire, using the frequent 7-point Likert scale to measuring the factors. We collected 206 valid consumer data, through which we conduct analyze the relationship between variables. And we used SPSS16 and AMOS 17 research software to testify our hypotheses.

4 Empirical Analysis

We made a correlation analysis about these variables. We use Correlation coefficient as a measure to the linear relationship strength between two variables. The results are as follows (Table 2/3): "Sig." means two-tailed tested P value. The "**" in the table stands for the significance when P<0.01. When the value is greater than 0, it means the two variables have positive correlation, while the value is less than 0, it means the two shows negative correlation.

Table 2. Correlation analysis between Brand extension and Customer evaluation

Variable		Customer evaluation
Perceived quality	Pearson Correlation	.254**
	Sig.	.000
Perceived fit	Pearson Correlation	.532**
	Sig.	.000
Difference	Pearson Correlation	.410**
	Sig.	.000

In this table, we can see the values of three Pearson Correlation values are greater than 0.01. The Sig. meets the test requirement. The second P value-Perceived fit is 0.532, which is the biggest one in the three variables of brand extension. And the other two are 0.254, 0.410 respectively. So the three data showed there is positive correlation between brand extension and customer evaluation. So the empirical analysis supports our hypotheses: Perceived quality of parent brand, Perceived fit between parent brand and extension product, Difference of extension product positively affects customer evaluation.

Table 3. Correlation analysis between Customer evaluation and Brand image

Variable		perceived value	brand personality	organization image
Customer evaluation	Pearson Correlation	.378**	.349**	.296**
	Sig.	.000	.000	.000

When it comes to the customer evaluation and brand image, the Pearson Correlation values are also greater than 0.01, The Sig. meets the test requirement. The largest value is P values between customer evaluation and perceived value. So it also means the H4 can be proved truly. In total, based on the results of correlation analysis, we can see that all of the variables have positive relationship.

5 Conclusion and Suggestion

Based on our empirical analysis, there comes our conclusion: Firstly, if the consumer perceived the quality of parent brand, they will make a high evaluation towards the brand extension strategies. Secondly, the perceived fit between parent brand and extension product is most important factor, which play the critical role in the evaluation to extension. Thirdly, the more difference the extension product has, the higher evaluation consumer make. On the second parts, we can see consumer evaluation also have positive effect on brand image. If the evaluation is positive, it will increase the parent brand image in consumer's reflection, while if negative, it damage the brand image.

From the conclusion, we can see even in internet industry, we also can get the most same results as in traditional industry. So when it comes to Chinese contact, the brand extension strategies should be considered the relation between parent brand and extension products. As Baidu are good at search engine, it is better to develop service or products related to retrieve and find effective results. So the map, resource database, tourism product may be more reliable than others. While comparing to Tencent, the QQ had such a high reputation in instant messaging software, the social networks, mobile phone service in connection may more easily succeed.

It's suggested that the internet companies should carried out an investigation about the reflection of consumers, as the failure of brand extension not only waste money, human capital, etc. it also hurt the company's image, which turned out a negative impact towards long-tern development.

References

1. Aaker, D.A., Keller, K.L.: The effects of Sequential Introduction of Brand Extensions. Journal of Marketing Research 29(1), 35–50 (1992)
2. Tauber, E.M.: Brand leverage: strategy for growth in a cost-control world. Journal of Advertising Research 28(4), 26–30 (1988)
3. John, D.R., Loken, B., Joiner: The Negative Impact of Extensions: Can Flagship Products Be Diluted. Journal of Marketing 62(1), 19–32 (1998)
4. Ries, A., Trout, J.: Positioning-The Battle for Your Mind. McGraw Hill, Illinois (2001)
5. Tauber, E.M.: Brand Franchise Extension: New Product Benefit from Existing Brand Name. Business Horizons 24(2), 36–41 (1981)
6. George, S.: Leveraging brand equity for developing appropriate brand extension strategies, Venice, T.A. Pai Management Institute, Manipal (2009)
7. Aaker, D.A., Keller, K.L.: Consumer Evaluation and Brand Extension. Journal of Marketing 54(1), 27–41 (1990)
8. Yang, M.: Brand extension effect evaluation model construction and empirical analysis. Business Times 20, 29–31 (2010)

9. Monga, A.B., John, D.R.: What Makes Brands Elastic? The Influence of Brand Concept and Styles of Thinking on Brand Extension Evaluation. Journal of Marketing 74(3), 80–92 (2010)
10. Fu, G., John, S., Qu, R.: Brand Extensions in Emerging Markets: Theory Development and Testing in China. Journal of Global Marketing 22(3), 217–228, 12p, 2 Charts (2009)
11. Hou, J.: Brand extensions: what do we know? Marketing Management Journal, 54–60 (2003)
12. Keller, K.L.: Conceptualizing, Measuring, and Managing Customer-based Brand Equity. Journal of Marketing 57(1), 1–22 (1993)

Research on the Fault Diagnosis of Excess Shaft Ran of Electric Submersible Pump

Fengyang Tao, Guangfu Liu, and Wenjing Xi

College of Information and Control Engineering, China university of Petroleum, DongYing, 257061, China

Abstract. The diagnosis strategy was studied about the excess shaft ran fault of floating Electric Submersible Pump (ESP) on the basis vibration signals. The acceleration signal acquisition system based on TMS320F2812 and MMA7260Q is designed and gathers the test wellhead vibration signals of normal unit and excess shaft ran unit. The vibration data were de-noised by wavelet and were analyzed and compared by using time-domain waveform, power spectrum, and wavelet decomposition method. The results show that the vibration signal of excess shaft ran unit was decomposed by 3 layers db3 wavelet, wavelet detail d3 of axial exists obvious frequency modulation phenomenon at about 100Hz, can be used as a basis for diagnosis. Application results show that the method can effectively diagnose the fault of excess shaft ran.

Keywords: Submersible pump units, Shaft ran, Vibration signals, Wavelet analysis, Fault diagnosis.

1 Introduction

Electric submersible pump (ESP) unit is the main device to exact oil in the world, which is an important technical measure to assure the high yield of the oil field in especially high water-cut stage. During application, many kinds of faults may occur, because the structure of ESP is relatively complex and running environment is bad [1]. One of the common faults is the excess shaft ran fault. Centrifugal pump mainly includes floating pump and pressured pump [2]. For floating ESP, due to the free movement between floating impellers and axis, the axial force is carried by protectors and spreads downward in shaft ran area. When the shaft ran is too little, axial force is difficult to spread download; when the shaft ran is too excessive, it is easy to lead to some problems, such as impellers “de-bonding”, which can effect ESP’s running life and production time efficiency and effectiveness, thus bring large economic losses to production.

The shaft ran of centrifugal pump is greatly related to design and manufacture. If the fault can be diagnosed before centrifugal pumps leave factory. Maintain and protect them in time to avoid unqualified centrifugal pump delivered. This can greatly last the average working life of centrifugal pump.

The excess shaft ran of ESP can bring vibration [3] [4]. This kind of vibration signal is non-stationary signal, while wavelet analysis is optimal to analyze the non-stationary

D. Jin and S. Lin (Eds.): Advances in MSEC Vol. 1, AISC 128, pp. 509–513.
springerlink.com

signal [5] [6]. It can construct eigenvalue needed by fault diagnosis and exact useful information directly.

2 Hardware Design

2.1 Vibration Signals Analysis of ESP

The excess shaft ran can lead to many serious problems, such as impellers wear and de-bonding, etc. These problems can cause shaft moving up and down, which increase the load of protector and thrust bearing and accelerate thrust bearing wear. Therefore, relative features of the excess shaft ran can be extracted by analyzing vibration signals, and this kind of fault can be diagnosed effectively.

When vibration propagates to wellhead through tube, its amplitude decreases while vibration frequency keeps a certain value. By measuring and analyzing vibration of wellhead, the running state of ESP units can be judged [7].

2.2 Design of the Acceleration Signal Acquisition System

In order to realize real-time diagnosis to faults of ESP units, a portable vibration signal acquisition system is designed. The DSP (Digital Signal Processor) TMS320F2812 is selected as CUP, which is suitable to high-speed and real-time processing; a capacitive three-dimensional acceleration sensor of Freescale--MMA7260Q is selected and these data are stored in common USB disk.

3 Processing the Vibration Signals with Wavelet De-Noising

Since multi-resolution analysis of wavelet transform can make signals begin in different frequency bands, signals can separate according to different frequency band. This characteristic is very useful to analyze complex vibration [8].

According to vibration feature, wavelet threshold was used to de-noise. The operating steps are:

(1) According to vibration features, decompose signals by 4layers with db5 wavelet, and get wavelet coefficients with noise.

(2) Adopting fixed threshold rule, firstly, estimate average value of each layer peak, and then based on experience, set each layer threshold by ratio.

(3) Based on every threshold, processing high frequency coefficients of each layer and reorganize them, then get the de-noising vibration signal.

4 Extracting Vibration Signal Features with the Combination of Power Spectrum

Since wavelet has good frequency analysis characteristic, it is widely applied in extracting characteristic parameters. Wavelet coefficient of each layer is the essence of a signal in time domain. It is difficult to judge the existence of faults by comparing the wavelet coefficients' difference directly. The slight difference of time domain signal is often obvious in frequency domain, and frequency-analysis has two methods: FFT and power spectrum. Because the resolution of FFT is low, the using effect is relatively poor. So power spectrum analysis is used to extract characteristic parameters by analyzing intra-layer wavelet coefficients. The method of signal spectral estimation includes classical and modern power spectral estimation. In classical power spectral estimation, assume that the data outside the time window is zero, so the resolution is poor and variance performance is poor. By using prior knowledge to assume the outside window data reasonably, the latter improves the quality of spectral estimation. In modern spectral estimation, the resolution of Burg and improved covariance method is much better. It is easy for Burg to occur many problems, such as spectral splitting, spectrum peak migration and false spectral peak, etc [9]. So improved covariance is used in spectral estimation and it is based on parameter model.

By using the acceleration signal acquisition system, multiple sets of tested ESP units from Shengli Oilfield were tested at test well. The data was analyzed and summarized from many aspects, such as time-domain waveform, power spectrum, wavelet decomposition waveform and each layer power spectrum, etc. Thus the diagnosis strategy is studied about the excess shaft ran fault. Analysis shows that the sample rate of vibration signal data is1kHz and the sample point of each group data are 2048.

The curve of vibration data in time-domain was analyzed by a large number of comparative analyses, but it is difficult to discover obvious fault feature information. The power spectrum of vibration curve was also analyzed, but the obvious and disciplinarian character still cannot be found.

And then, by using different wavelet basis, analyze many sets of units and find that the axial vibration data contains fault information. So analyze the axial data only in the following. In wavelet analysis, adopting db6 wavelet can get a better effect. When doing 3 layers db3 wavelet decomposition, wavelet detail d3 of the excess shaft ran units exists obvious frequency modulation phenomenon in axial, but all normal units do not exist this phenomenon. In the following, the examples given are typical normal unit CX018 and typical fault unit WG069.

The axial vibration data of two units were decomposed by 3 layers db6 wavelet, shown in Figure 1. For WG069, wavelet detail d3 of exists obvious frequency modulation phenomenon and amplitude changes remarkably. While normal unit don't exist this phenomenon and it's a common phenomenon with excess shaft ran units.

Wavelet detail d3 in axial of CX018 and WG069 was analyzed by power spectrum analysis, shown in Figure 2. By analyzing and comparing power spectrum, for WG069, the waveform of wavelet detail d3 in axial exists obvious frequency modulation phenomenon. There are two signal components at about 100Hz with strong energy and very close frequency--for WG069, they are 97.12Hz and 100.36Hz, and other units resemble, such as TD121, at 100.10Hz and 108.32Hz. Since the pole number of ESP is 1 and power-frequency is 50Hz, Motor Speed is close to 50r/s. The two frequencies are called "Close to Two Times Frequency". Through analyzing vibration data of many sets of excess shaft ran units and after decomposing 3 layers db6 wavelet of excess shaft ran units, it is found that wavelet detail d3 of axial exists obvious frequency modulation phenomenon in two times frequency (a few units in one times frequency) , but other normal units don't exist this phenomenon. So this vibration signal feather can act as the feather of shaft ran excess fault.

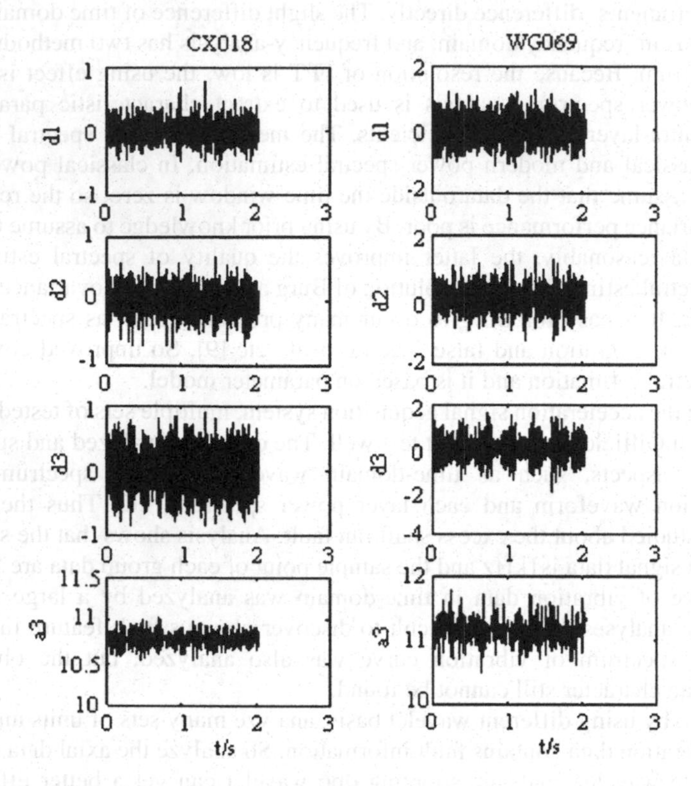

Fig. 1. Axial vibration signal wavelet decomposition results of test CX018 normal unit and WG069 fault unit

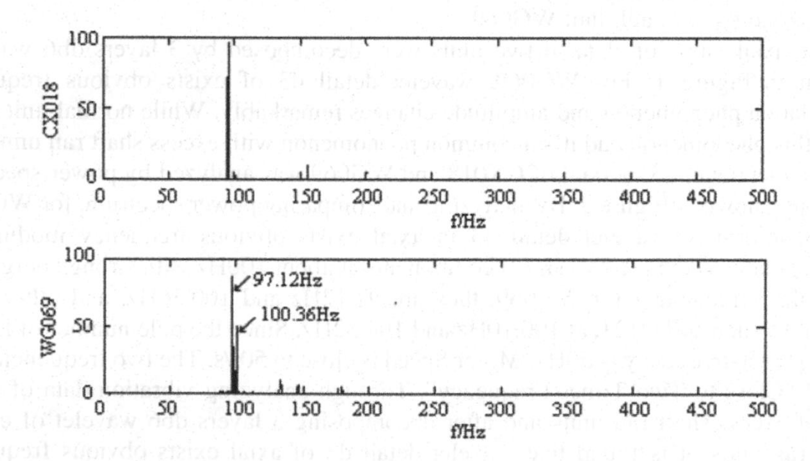

Fig. 2. Power spectrum of wavelet details d3 of CX018 normal unit's axial vibration and WG069 fault unit's axial vibration

5 Conclusion

Analyze the vibration signal of 12 set units from ShengLi oilfield by homemade portable vibration signal acquisition system, and take apart to test the ESP whose vibration features are various. Through testing four excess shaft ran ESP units, find that floating impeller ran is beyond requirements range in different degree. The test results show that this analysis method is effective to diagnose the excess shaft ran fault.

According to the vibration signal features and wavelet analysis advantages, the paper analyzes the gathered data. The analysis results indicatc that, for excess shaft ran units, after 3 layers db3 wavelet decomposition, when sampling frequency is 1 kHz, wavelet detail d3 of axial exists obvious frequency modulation phenomenon in two times frequency. Take these as vibration signal characters of the excess shaft ran fault to analyze and diagnose 12 sets of units' vibration signal. Practical diagnosis results show that excess shaft ran fault of ESP can be identified effectively by this method. The excess shaft ran fault diagnosis is realized in the paper. This avoids unqualified ESP leaving factory and assures their quality, thus establishes foundation for improving ESP's average working life.

Acknowledgments. Fund message: National Science And Technology Major Project (2011ZX05024-002-009).

References

1. Yu, J., Feng, D., Lu, Y.: Research on ESP Diagnosis Based on Neural Network. Machine 32(5), 54–56 (2005)
2. Mei, S., Shao, Y., Liu, J.: Technology of ESP (first volume), pp. 49–98. Petroleum Industry Press, Beijing (2004)
3. Sun, Y., Jia, X.: Research on System Safety Evaluation Method of ESP. Oil Field Equipment 32(5), 15–17 (2003)
4. Trevisan, F.: Exaltacao, Ph.D. Modeling and Visualization of Air and Viscous Liquid in Electrical Submersible Pump 11, 42-48 (2009)
5. Li, X.: Time-Frequency Analysis of Mechanical Vibrating Signal. Journal of Shandong Institute Technology 3 (1999)
6. Pang, P., Ding, G.: Vibration Diagnosis Method Based on Wavelet Analysis and Neural Network for Turbine-generator. In: Control and Decision Conference, pp. 5234–5237 (2009)
7. Xi, W.: Research on Fault Diagnosis of Electric Submersible Pumps Based on Vibration Detection. Dongying: China University of Petroleum (East China) 5 (2008)
8. Liu, B.: Selection of Wavelet Packet Basis for Rotating Machinery Fault Diagnosis. Journal of Sound and Vibration 258(3-5), 567–582 (2005)
9. Liu, Z., Li, L., Zhao, D.: Modern Spectra Estimation and its Application Efficiency. Oil Geophysical Prospecting (sup. 2), 5–11 (2009)

An Empirical Research of Effects of Realness on Microblogging Intention Model

Zhijie Zhang[1], Haitao Sun[1], and Huiying Du[2]

[1] School of Economics and Management, Beijing University of Posts and Telecommunications, Beijing, China
zhangzhijie143@163.com
[2] School of Information Management, Beijing Information Science & Technology University, Beijing, China

Abstract. A survey of 8 hypotheses to study effects of realness on microblogging intention was carried out based on 328 students from four universities and colleges in China. A TAM model was constructed to test the interior structure of microblogging intention. The results of this research show that realness has significant positive impacts on microblogging intention by influencing usefulness, enjoyment, trust and ease of use of microblogging. The implication for microblogging operators is that they should encourage the use of real personal information.

Keywords: microblogging, intention, SEM.

1 Introduction

Microblog is a Social Networking Services in the form of blogging. Akshay(2007) defines microblog as a new form of communication in which users can describe their current status in short posts distributed by instant messages, mobile phones, email or the Web[1]. Some microblogging services offer features such as privacy settings, which allow users to control who can read their microblogs, or alternative ways of publishing entries besides the web-based interface. These may include text messaging, instant messaging, E-mail, or digital audio. CNNIC (2011) published a report which estimated, first half of 2011, the number of China microblogging users growing rapidly, had increased from 63.11 million to 195 million, with an increase of 208.9% in the last six months.

Despite of the more rapid development of microblogs, the corresponding research still stay in an early stage, and the existing research can't fully explain the reasons for the prevalence of social networking sites. Here, An application of Davis' Technology Acceptance Model (TAM) is adopted. It's one of the most frequently used methodologies for studies on the adoption of new Information Systems (IS). In particular, we extend the basic TAM to study key features of SNS or personal perceptions affecting the use of microblogging services through Internet websites. As far as we know, there have been little studies conducted on identifying the psychological process of using social network service.

D. Jin and S. Lin (Eds.): Advances in MSEC Vol. 1, AISC 128, pp. 515–520.
springerlink.com

2 Literature Review

2.1 TAM

The technology acceptance model (TAM) (Davis, 1989) is grounded in both TRA and TPB. TAM was especially tailored for modeling user acceptance of an information system with the aim of explaining the behavioral intention to use the system. TAM proposes that perceived usefulness (PU) and perceived ease of use (PEU) are of prime relevance in explaining the behavioral intention to use IS. As noted by Venkatesh and Davis (2000), a better understanding of these would enable us to design effective organizational interventions that might lead to increased user acceptance and use of SNS.

2.2 Microblogging

Microblogging is a typical application of SNS, and online social networks represent a fast growing phenomenon and are emerging as the web's top application (Chiu et al., 2008). Microblogging is offered compared with blog and can be understood a micro form of blog. But a microblogging differs from a traditional blog in that its content is typically smaller in both actual and aggregate file size. Microblogs "allow users to exchange small elements of content such as short sentences, individual images, or video links"(Kaplan Andreas M.,Haenlein Michael, 2011)[2]. Recently, several studies have been made in microblog. For instance, past research by Hughes and Palen(2009) compares the behaviour of microblog users in mass convergence and emergency events from more general use[3]. Although the microblogging has become more and more popular, the specific research on microblogging perceived factors is very limited, needing further and deeper research.

3 Research Framework

Realness. One of the biggest features of Microblogging sites is that the authenticity of the registered users' information is higher. Basically social networking sites encourage the use of real personal information (for example: name, e-mail, graduate schools). Users on most SNS normally do not aim to make new friends. Instead, they link their social networks in real life online to make further contacts (Boyd & Ellison, 2008)[4]; hence, Mutual acquaintances in SNS helps to connect to more mutual friends (e.g., friend recommendation mechanism) and interaction and sharing between more friends creates a greater sense of pleasure (Powell, 2009; Tapscott, 2008). At the same time, new friends who knew through acquaintances can enhance mutual trust, thereby enhancing trust of the whole social networking sites. Thus in this work, we propose:

H1: Realness positively affects enjoyment of a social network service.
H2: Realness positively affects perceived usefulness of a social network service.
H3: Realness positively affects trust of a social network service.

Enjoyment. Moon and Kim (2001) defined enjoyment as “the pleasure the individual feels objectively when committing a particular behavior or carrying out a particular activity" and found in their study that enjoyment is a key factor for user's acceptance of the Internet. Many scholars (Kang & Lee, 2010) have considered SNS as a pleasure-oriented information system, where users continue use with stronger motivation if they have more intense perceived enjoyment from it[5]. Thus in this work, we propose:

H4: Enjoyment positively affects user intention to use of a social network service.

Ease of Use. Perceived ease of use is defined as "the degree to which a person believes that using a particular system would be free of effort" (Davis, 1989a, 1989b). It has been widely known that systems' perceived ease of use, as well as perceived usefulness, have a direct influence on usage of an information system, and perceived usefulness mediates the effect of perceived ease of use on usage (Agarwal & Prasad 1999; Davis et al., 1989; Jackson et al., 1997; Venkatesh, 1999)[6]. As a result, we propose:

H5. Perceived ease of use positively affects perceived usefulness of a social network service.

H6. Perceived ease of use positively affects user intention to use of a social network service.

Usefulness. Davis (1989) defined usefulness as "the degree to which a person believes that using a particular system would enhance his or her job performance," when the individual feels a system is useful, he or she thinks positively about it. Some scholars (Sledgianowski & Kulviwat, 2009) have discovered that users' perceived usefulness in SNS affects positive intention to use the SNS. Thus in this work, we propose:

H7. Perceived usefulness positively affects user intention to use of a social network service.

Trust. The impact of trust on attitudes is based on the credibility dimension of the trust. Recently, trust has taken center stage as a serious issue in SNS (Gambi, 2009). Thus in this work, we take Trust as the fourth intermediate variable, and propose:

H8: Trust positively affects user intention to use of a social network service.

Intention. In this study, Intention means users are willing to use social networking sites, such as register, and publishing their own articles and photographs, as well as browsing and sharing others' articles, photos and videos, etc.

The TAM model used in this study is shown in Fig.1.

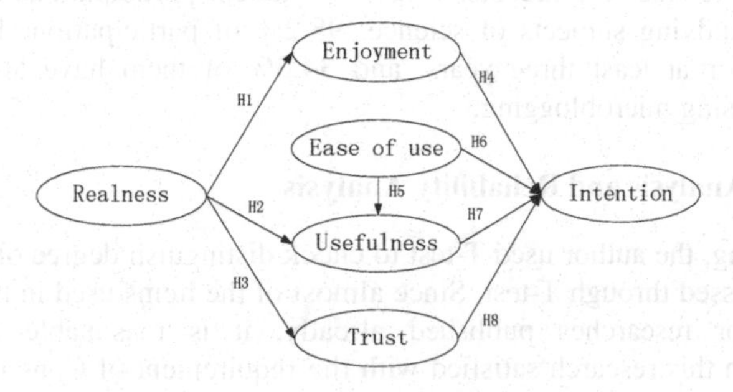

Fig. 1. TAM Model Used in This Study

4 Methods

4.1 Participations and Procedure

A customized questionnaire was used for our research purpose. All participations in this research are volunteers, and most of them are students from four universities and colleges in Beijing. Firstly, participants were asked to answer whether they had participated in social networking service websites. If so, after briefly elucidating our research purpose, they were invited to participate and complete the survey questionnaire.

Researchers numbered all the questionnaires collected, cleaned the data, and entered data into SPSS13.0 after initial examination. All the missing data and wrong data were disposed by using methods related. Then the researchers transferred data from SPSS13.0 into LISREL8.53, used SEM to analyze them.

4.2 Tools and Methods

Reliability analysis was finished by using SPSS13.0. Alpha (Cronbach α) coefficient, a reliability coefficient most frequently applied, is used to test the internal reliability of the measuring questionnaire. Validity analysis was finished by using LISREL 8.53 to do confirmatory factor analysis.

Structural equation modeling (SEM), which grows out of and serves purposes similar to multiple regression, but in a more powerful way which takes into account the modeling of interactions, nonlinearities, correlated independents, measurement error, is a method mostly used in the study of society science. SEM model was finished by using LISREL 8.53.

5 Results

5.1 Sample Description

Questionnaires were distributed to 340students at four universities and colleges in Beijing; 334 were returned. Of these returned questionnaires, six were only partially completed and therefore excluded from the data analysis, resulting in an effective response rate of 96.5%. The distribution of gender was quite balanced, with 158 (48.2%) of the female respondents. 79.2% of student participations are from cities, while 56.7% studying subjects of science. 98.2% of participations have connected with Internet for at least three years, and 34.3% of them have at least one-year experience of using microblogging.

5.2 Validity Analysis and Reliability Analysis

Before analyzing, the author used T-test to check distinguish degree of each items, all the 32 items passed through T-test. Since almost of the items used in this research are from papers or researches published already, it is reasonable to believe the questionnaire in this research satisfied with the requirement of Content Validity. The author used confirmatory factor analysis to check Structural Validity of this research.

4 items were removed due to necessary changes. Important goodness-of-fit indices of confirmatory factor analysis are: $\chi2(1032.45)$; RMSEA(0.076); NFI(0.94); NNFI(0.95); CFI(0.98); IFI(0.96); RFI(0.94); GFI(0.85).

Internal reliability analysis for each factor in confirmatory factor analysis is as follows: Realness (0.8253); Enjoyment (0.7423); Ease of use (0.7853); Usefulness (0.7952); Trust(0.8312); Intention (0.7651).

5.3 SEM

Figure 2 shows result of SEM.

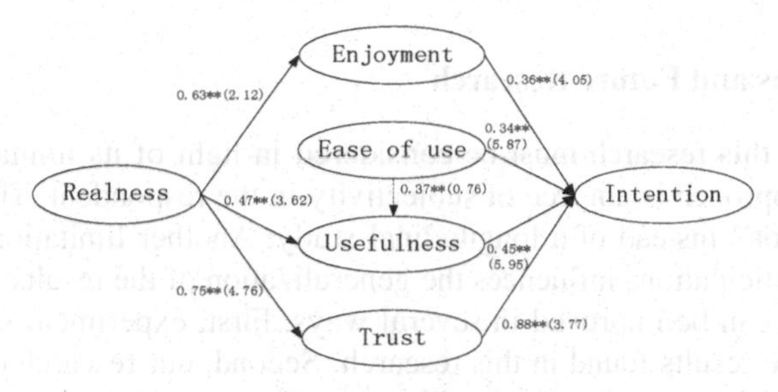

Fig. 2. Result of SEM

Important goodness-of-fit indices of SEM model are shown as below: $\chi2(1021.97)$; RMSEA(0.074); NFI(0.94); NNFI(0.95); CFI(0.96); IFI(0.95); RFI(0.94); GFI(0.85).

Among the 8 hypotheses, 10 (H1, H2, H3, H4, H6, H7, H8) are supported by the result, while 1 (H5) is rejected. Section Discussion discusses these results and their implications in greater detail.

6 Discussion

6.1 Findings and Managerial Implications

Through the empirical study, the authors have demonstrated the model and assumptions of factors that influence the intention to use microblogging sites. Main conclusions are as follows:

Realness of information of majority uses' has a positive impact on intention to use the microblogging site, of course, taking Enjoyment, Usefulness and Trust as intermediate variables. The websites are based on real information which will facilitate the users to find their acquaintances' ID. Due to the authenticity of the users' information, we can increase confidence in the whole network, and the usefulness of microblogging sites can be enhanced.

The perception of enjoyment has positive impact on the users' attitudes. In microblogging sites, users can publish and share other people's articles, photos and videos.

Trust, another variable that may affect the behavioral intention, is confirmed in this research. It's consistent with the reality. When the user are using social networking sites to communicate, good protection in privacy and personal information would encourage users to continue to use the site, and may recommend it to other people.

It's proved that perceived ease of use of microblogging sites has conclusive effect on intention. If the users feel it difficult to use the sites, they may give up using the social networking site.

The effect of usefulness of perception to the intention of using the microblogging sites can also be demonstrated, which is consistent with the former research.

According to analysis, the intention between consumer using microblogging and users' acceptance of new technologies is similar. Therefore TAM is used for reference to develop the influence factor model of microblogging using intention, which is verified in the end.

6.2 Limitations and Future Research

The findings of this research must be considered in light of its limitations. First, the questionnaire approach is not free of subjectivity in the respondent. The questionnaire was a ''snap-shot'' instead of a longitudinal study. Another limitation is that the use of student as participations influences the generalization of the results.

The research can be improved in several ways. First, experiment method could be used to retest the results found in this research. Second, our research could have been improved if a random sample of microblogging site users had been selected. Finally, time series study can be introduced to study how the preferences of microblogging user change over time.

References

1. Java, A., Finin, T.: Why we Twitter: understanding microblog usage and communities. Association for Computing Machinery, 56-65 (2007)
2. Kaplan Andreas, M., Michael, H.: The early bird catches the news: Nine things you should know about micro-blogging. Business Horizons 54(2) (2011)
3. Hughes, A.L., Palen, L.: Twitter adoption and use in mass convergence and emergency events. International Journal of Emergency Management 6(3/4), 248–260 (2009)
4. Boyd, D.M., Ellison, N.B.: Social network sites: definition, history, and scholarship. Journal of Computer-Mediated Communication 13, 210–230 (2008)
5. Kang, Y.S., Lee, H.: Understanding the role of an IT artifact in online service continuance: An extended perspective of user satisfaction. Computers in Human Behavior 26, 353–364 (2010)
6. Jackson, C., Chow, S., Leitch, R.: Toward an understanding of the behavioral intention to use an information system. Decision Sciences 28, 357–389 (1997)

A Study on the Relation between Moldingroom's Temperature and the Cock Issue in FDM Techniques

Hao Wu and Zheng Yang

Department of Design, School of Urban Design, Wuhan University,
430072 Wuhan, China

Abstract. The cock issue is one of the main problems arisen during the process of applying FDM facility into the Product Design field. Through analysis and experiments, this paper concludes that lower temperature in molding room can lead to the cock issue.

Keywords: FDM, temperature of moulding room, the cock issue.

1 Introduction

Compared to other rapid prototyping technologies, FDM techniques has a promising application in product development, due to its virtue of low equipment costs, rare environmental pollution. However, FDM is faced with many problems in this process, and the cock issue is a main one of them.

The cock issue is referring to the separation between layers in the FDM process, which is often illustrated by the phenomenon of one side cocking. The location and reason of cock issue is various, but the damage of the issue is so big. A FDM product is made of layers, which are accumulated one layer by one layer [1]. Therefore, if one layer was cocked, then the layer on it would be damaged. As a result, the FDM process will be interrupted. This result either leads to a product lack of precision, or the whole product ruined.

2 Aassumption

The cock issue is mainly caused by uneven material contraction among layers. The material of FDM product formed through three phases. In the first phase, the material is heated up from solid to liquid. After it comes out of the nozzle, the material is transformed into solid again. At this time, the temperature of the material is lowered to room temperature. In the process, the stress among layers is accumulated, which will cause the layer to shrink until layers are cocked. The formula to calculate the shrinkage is as follows:

$$\Delta L=\delta\times l\times\Delta t \text{ [2]}$$

Where is coefficient of linear expansion of certain material is dimension of the product, and it is difference in temperature.

D. Jin and S. Lin (Eds.): Advances in MSEC Vol. 1, AISC 128, pp. 521–524.
springerlink.com

Through this formula, one can conclude that it will be up if it is up, and the stress is up, which will lead to cock issue. According to the analysis, when temperature of the nozzle and other conditions are given, the lower the temperature of moulding room, the easier the cock issue will be caused. The argument of the assumption is as follows.

3 Argument of Assumption

3.1 Procedure of the Experiment

Take temperature of molding room as a variable, and other parameters as constants, then see how different temperatures affect the cock issue, and make a conclusion about the relation between molding room's temperature and the cock issue in FDM.

3.2 Selection for a Benchmark

In order to observe the situation of FDM product in different temperature of moulding room, we must select the must include the typical characters of product, and it may be a prism, column, taper, hemisphere or cirque, which is basic to a compound entity. The benchmark of this experiment is illustrated by Fig.1[3]:

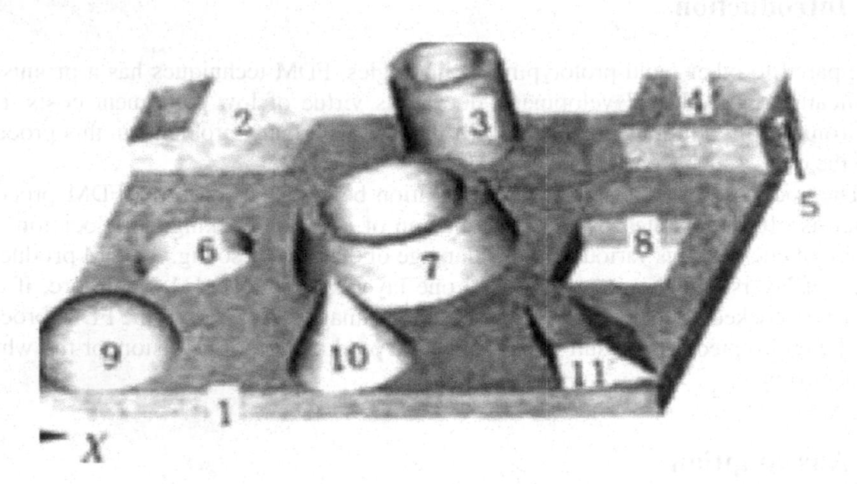

Fig. 1. Benchmark for the experiment

3.3 Condition of Experiment

3.3.1 Equipment and Material: MEM-450, ABS

MEM-450, based on FDM, is one kind of RP equipments. ABS is a material used by MEM-450.

3.3.2 Main Constants

In order to make sure other factors invariable, we must preset some parameters, the requirements of the parameters are as follows:

According to character of the benchmark, we must select suitable parameters to smooth the process of the experiment.

Once given, the parameters can't be reset any more.

Table 1. Main constants for the experiment

Parameters		**Value**
Thickness of layer		0.2
Scan speed	Profile	60
	Fill	65
	Support	75
Nozzle speed	Profile	0.90
	Fill	0.85
	Support	0.70
Fill density	Fill space	2
	Support space	4
Nozzle temperature		235°C

3.3.3 Variables

According to some research [4], the suitable scale of temperature of moulding room for average products is 40-50°C when the material is ABS. We should lengthen the scale until the limits of MEM-450 and pick some typical temperature as the value of the experiment's variable. 30°C is near room temperature and can be the lower limit, while 55°C can be chosen as the upper limit of the experiment, since it is the upper limit of MEM-450. At the end, we set the temperature scale as [30°C, 35°C, 40°C, 45°C, 50°C, 55°C].

3.4 Results and Discussions

The results of the experiment is showed by following Table 2:

Table 2. Different situation of benchmark at different moulding room's temperature

Moulding room's temperature (℃)	Situation of benchmark
30	Obvious cock phenomenon lead to interruption
35	Finished but slight cock phenomenon
40	No cock phenomenon but split at the bottom
45	No cock phenomenon
50	Compact profile
55	Compact profile

The note of Tab.2 is as follows:

In certain temperature, the benchmark was split, which mean there was split on certain location of the benchmark but not worse enough to lead to cock issue. We can say that, split is a early stage phenomenon of cock issue.

Certain situation of benchmark in Tab.2 is caused by certain moulding room's temperature and certain benchmark causes certain situations. If the value of moulding room's temperature in Tab.2 and the benchmark changed, the result would change as well.

Selection of certain temperature of moulding room is used to illustrate the relation between moulding room's temperature and the cock issue, we can't conclude the most suitable temperature to process the benchmark in this experiment.

Through the results, we can see that: at 30°C, the cock issue is so severe that the process is interrupted; at 35°C or 40°C, the benchmark is finished but cock issue shows up; 45°C is a watershed, after which there is no cock phenomenon. Therefore, we can conclude that cock phenomenon can disappear when moulding room's temperature is increasing.

4 Conclusions and Prospects

Though analysis and related experiment, this article concludes that, if other factors are given, the lower the moulding room's temperature is, the more obvious the cock phenomenon is. When the temperature is up, the cock issue will mitigate. Due to this conclusion, users of FDM can avoid cock issue by increasing moulding room's temperature. However, the suitable temperature to certain equipment and product is not reached. We should pay some attention to this problem in the future.

References

[1] Huang, S., Shen, Y., Huang, J.: Prospects of Rapid Prototyping Technology. China Mechanical Engineering 11(1-2), 195–196 (2000)

[2] He, X.: Research of the Control System and Technology of the Fused Deposition Modeling. Huazhong University of Science and Technology, 37–38 (2005)

[3] Zhou, G., Guo, D., Jia, Z.: Research on process parameter optimization of fused deposition modelling. Journal of Dalian University of Technology (4) (2002)

[4] Peng, A., Zhang, J.: Research on the Interlayer Stress and Warpage Deformation in FDM. Journal of Huaihai Institute of Technology(Natural Sciences Edition) (2), 16–19 (2007)

The Gracefulness of a Kind of Unconnected Graphs

Yan-Hua Yu, Wen-Xiang Wang, and Li-xia Song

North China Institute of Science and Technology,
Sanhe 065201, Hebei, China

Abstract. The paper demonstrates that for positive integer n, m, where then unconnected graph. $K_{n,m}\bigcup(\overline{K_2}\vee\overline{K_n})$, $K_{n,m}\bigcup(P_2\vee\overline{K_n})$, $K_{n,m}\bigcup(P_3\vee\overline{K_n})$, $K_{n,m}\bigcup(P_1\vee P_{2n+2})$ and $K_{n,m}\bigcup St(2n)$ are graceful graphs.

Keywords: Graph, graceful graph, unconnected graph.

1 Introduction

Labeling different kinds of graph has been a hot issue in graph theory all along. It not only has theoretical significance in respect of mathematics such as solving the question that a complete graph decomposes into isomorphic sub-graph through graceful label, but also has widely applications in military and science fields such as error correcting code design, communication network, measuring atomic position in a crystal structure, radar pulse, missile guidance code design, etc. So far, most studies work on the gracefulness of graphs is related to the gracefulness of connected graphs [1-6] rather than unconnected ones [7-9]. This paper particularly focuses on the gracefulness of unconnected graphs of completely bipartite $K_{n,m}$ by proving the gracefulness of unconnected graph: $K_{n,m}\bigcup(\overline{K_2}\vee\overline{K_n})$, $K_{n,m}\bigcup(P_2\vee\overline{K_n})$, $K_{n,m}\bigcup(P_3\vee\overline{K_n})$, $K_{n,m}\bigcup(P_1\vee P_{2n+2})$, $K_{n,m}\bigcup St(2n)$. All the graphs involved in the following discussion are simple undirected graphs labeled by $G(V,E)$, let $V=V(G)$ is the set of the vertices of graph G, $E=E(G)$ is the set of the edges of graph G, $|E|$ is the number of the edges of graph G, $K_{n,m}$ is completely bipartite, P_n is the path of n (a number) vertices, $G_1\vee G_2$ is the union set of G_1 and G_2, $\overline{G}$ is complementary graph of graph G.

Definition 1. Graph $G(V,E)$,and let k be a positive integer. If there is a injection $f:V\rightarrow\{0,1,\cdots,|E|+k-1\}$, such that for edges that satisfy $uv\in E$, $f'(uv)=|f(u)-f(v)|$ induces a bijection $f':E\rightarrow\{k,k+1,\cdots,|E|+k-1\}$,

D. Jin and S. Lin (Eds.): Advances in MSEC Vol. 1, AISC 128, pp. 525–531.
springerlink.com

then we call G a k-graceful graph, call f a k-graceful label of graph G, 1-graceful graph is also called graceful graph, 1-graceful label is also called graceful label.

2 Graceful Graphs

2.1 The Graph $K_{n,m} \cup (\overline{K_2} \vee \overline{K_n})$

Theorem 1. For positive integer m, n, where $1 \le n \le m$, then $K_{n,m} \cup (\overline{K_2} \vee \overline{K_n})$ is a graceful graph.

Proof. Since $\overline{K_2} \vee \overline{K_n} = K_{2,n}$, as a result, the gracefulness of $K_{n,m} \cup (\overline{K_2} \vee \overline{K_n})$, can be proved by demonstrating that $K_{n,m} \cup K_{2,n}$ is a graceful graph.

Let $V(K_{n,m}) = \{x_1, x_2, \cdots, x_n; x_{n+1}, x_{n+2}, \cdots, x_{n+m}\}$, $V(K_{2,n}) = \{y_1, y_2, \cdots, y_n; y_{n+1}, y_{n+2}\}$, $E = E(K_{n,m} \cup K_{2,n})$, $|E| = mn + 2n$.

Define that the label of vertices of graph $K_{n,m} \cup K_{2,n}$ f is:

$$f(x_i) = i - 1, (i = 1, 2, \cdots, n),$$

$$f(x_i) = 3n + n(i - n - 1), (i = n+1, n+2, \cdots, n+m),$$

$$f(y_j) = mn + j - 2, (i = 1, 2, \cdots, n), \; f(y_{n+1}) = mn + n - 1,$$

$$f(y_{n+2}) = mn + 2n - 1.$$

To prove that label f is a graceful label of graph $K_{n,m} \cup K_{2,n}$.

(i) On account of

$$0 = f(x_1) < f(x_2) < f(x_3) < \cdots f(x_{n-1}) < f(x_n)$$
$$< f(x_{n+1}) < f(x_{n+2}) < \cdots < f(x_{n+m-4}) < f(x_{n+m-3}) < f(y_1) < f(y_2)$$
$$= f(x_{n+m-2}) < f(y_3) < f(y_4) < \cdots < f(y_{n-1}) < f(y_n)$$
$$< f(y_{n+1}) < f(x_{n+m-1}) < f(y_{n+2}) < f(x_{n+m}) = mn + 2n.$$

Therefore map $f : V(K_{n,m} \cup K_{2,n}) \to \{0, 1, 2, \cdots, mn + 2n\}$ is an injection.

(ii) For edges that satisfy $uv \in E$, let $f'(uv) = |f(u) - f(v)|$ then,

$$f'(x_i x_j) = (j - n + 2) - i + 1, (i = 1, 2, \cdots, n; j = n+1, n+2, \cdots, n+m),$$

$$f'(y_{n+1} y_j) = n - j + 1, (j = 1, 2, \cdots, n),$$

$$f'(y_{n+2} y_j) = 2n - j + 1, (j = 1, 2, \cdots, n).$$

Hence $1 = f'(y_{n+1}y_n) < f'(y_{n+1p}y_{n-1}) < f'(y_{n+1}y_{n-2}) < \cdots$

$$< f'(y_{n+1}y_2) < f'(y_{n+1}y_1)$$
$$< f'(y_{n+2}y_n) < f'(y_{n+2}y_{n-1}) < f'(y_{n+2}y_{n-2}) < \cdots$$
$$< f'(y_{n+2}y_2) < f'(y_{n+2}y_1)$$
$$< f'(x_n x_{n+1}) < f'(x_{n-1}x_{n+1}) < f'(x_{n-2}x_{n+1}) < \cdots$$
$$< f'(x_2 x_{n+1}) < f'(x_1 x_{n+1})$$
$$< f'(x_n x_{n+m-1}) < f'(x_{n-1}x_{n+m-1}) < f'(x_{n-2}x_{n+m-1}) < \cdots$$
$$< f'(x_n x_{n+m}) < f'(x_{n-1}x_{n+m}) < f'(x_{n-2}x_{n+m}) < \cdots$$
$$< f'(x_2 x_{n+m}) < f'(x_1 x_{n+m}) = mn + 2n$$

Therefore, map $f' : E(K_{n,m} \cup K_{2,n}) \to \{1, 2, \cdots, mn + 2n\}$ is a bijection.

Based on demonstration (i) and (ii), for positive integer m, n, where $1 \le n \le m$, graph $K_{n,m} \cup K_{2,n}$ is a graceful graph, in other words graph $K_{n,m} \cup (\overline{K_2} \vee \overline{K_n})$ is graceful graph.

Example 1. The graceful label of unconnected graph $K_{3,4} \cup (\overline{K_2} \vee \overline{K_3})$ is shown below

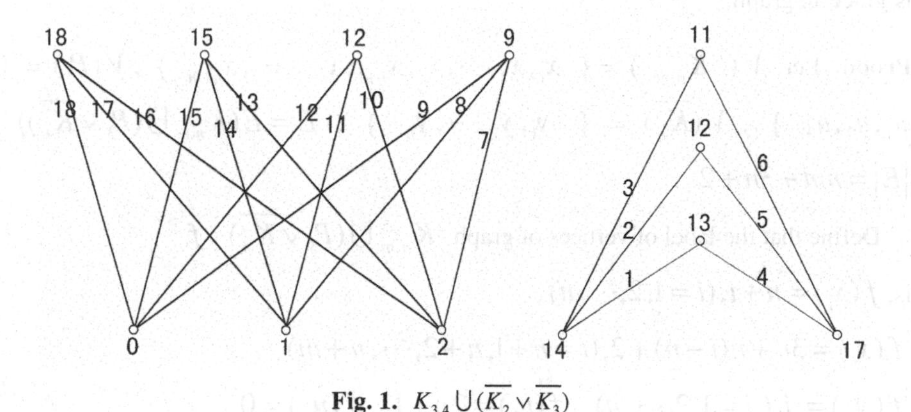

Fig. 1. $K_{3,4} \cup (\overline{K_2} \vee \overline{K_3})$

2.2 The Graph $K_{n,m} \cup (P_2 \vee \overline{K_n})$

Theorem 2. For positive integer m, n, where $1 \le n \le m$, then $K_{n,m} \cup (P_2 \vee \overline{K_n})$ is a graceful graph.

Example 2. The graceful label of unconnected graph $K_{3,4} \bigcup (P_2 \vee \overline{K_3})$ is shown below

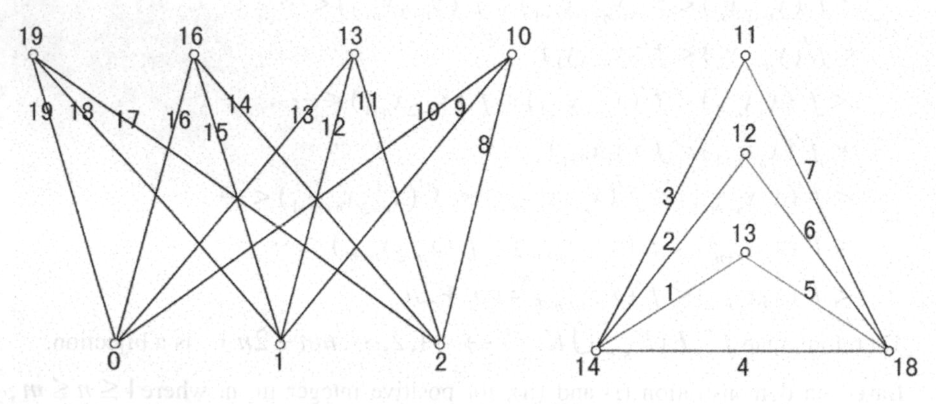

Fig. 2. $K_{3,4} \bigcup (P_2 \vee \overline{K_3})$

2.3 The Graph $K_{n,m} \bigcup (P_3 \vee \overline{K_n})$

Theorem 3. For positive integer m, n where $1 \le n \le m$, then $K_{n,m} \bigcup (P_3 \vee \overline{K_n})$ is graceful graph.

Proof. Let $V(K_{n,m}) = \{x_1, x_2, \cdots, x_n; x_{n+1}, x_{n+2}, \cdots, x_{n+m}\}$, $V(P_3) = \{u_1, u_2, u_3\}$, $V(\overline{K_n}) = \{y_1, y_2, \cdots, y_n\}$, $E = E(K_{n,m} \bigcup (P_3 \vee \overline{K_n}))$, $|E| = mn + 3n + 2$.

Define that the label of vertices of graph $K_{n,m} \bigcup (P_3 \vee \overline{K_n})$ f is, $f(x_i) = n + i, (i = 1, 2, \cdots, n)$,

$f(x_i) = 3n + n(i - n) + 2, (i = n+1, n+2, \cdots, n+m)$,

$f(y_j) = j, (j = 1, 2, \cdots, n)$, $f(u_1) = 2n + 1$, $f(u_2) = 0$,

$f(u_3) = mn + 3n + 2$.

To prove that label f is the graceful label of graph $K_{n,m} \bigcup (P_3 \vee \overline{K_n})$.

(i) On account of

$$0 = f(u_2) < f(y_1) < f(y_2) < \cdots < f(y_n) < f(x_1) < \cdots < f(x_{n-1})$$

$$< f(x_n) < f(u_1) < f(x_{n+1}) < f(x_{n+2}) < \cdots < f(x_{n+m-1})$$
$$< f(x_{n+m}) = f(u_3) = mn+3n+2$$

Therefore, map $f : V(K_{n,m} \cup (P_3 \vee \overline{K_n})) \to \{0,1,2,\cdots,mn+3n+2\}$ is injection

(ii) For all edges that satisfy $uv \in E$, let $f'(uv) = |f(u) - f(v)|$, then:

$$f'(u_2 y_j) = j, (j = 1,2,\cdots,n), \quad f'(u_1 y_j) = 2n+1-j, (j = 1,2,\cdots,n),$$
$$f'(u_3 y_j) = mn+3n+2-j, (j = 1,2,\cdots,n),$$
$$f'(u_2 u_1) = 2n+1, f'(u_2 u_3) = nm+3n+2$$
$$f'(x_i x_j) = 2n+n(j-n)-i+2, (i = 1,2,\cdots,n; j = n+1,n+2,\cdots,n+m)$$

Hence

$$1 = f'(u_2 y_1) < f'(u_2 y_2) < \cdots < f'(u_2 y_{n-1}) < f'(u_2 y_n)$$
$$< f'(u_1 y_n) < f'(u_1 y_{n-1}) < \cdots < f'(u_1 y_2) < f'(u_1 y_1) < f'(u_1 u_2)$$
$$< f'(x_n x_{n+1}) < f'(x_{n-1} x_{n+1}) < \cdots < f'(x_2 x_{n+1}) < f'(x_1 x_{n+1})$$
$$< f'(x_n x_{n+2}) < f'(x_{n-1} x_{n+2}) < \cdots < f'(x_2 x_{n+2}) < f'(x_1 x_{n+2}) < \cdots$$
$$< f'(x_n x_{n+m-1}) < f'(x_{n-1} x_{n+m-1}) < \cdots < f'(x_2 x_{n+m-1}) < f'(x_1 x_{n+m-1})$$
$$< f'(x_n x_{n+m}) < f'(x_{n-1} x_{n+m}) < \cdots < f'(x_2 x_{n+m}) < f'(x_1 x_{n+m})$$
$$< f'(u_3 y_n) < f'(u_3 y_{n-1}) < \cdots < f'(u_3 y_2) < f'(u_3 y_1) < f'(u_3 u_2)$$
$$= mn+3n+2$$

Therefore, map $f' : E(K_{n,m} \cup (P_3 \vee \overline{K_n})) \to \{ 1,2,3,\cdots,mn+3n+2 \}$ is a bijection. Based on demonstration (i) and (ii), for positive integer m, n, where $1 \le n \le m$, $K_{n,m} \cup (P_3 \vee \overline{K_n})$ is a graceful graph.

Example 3. The graceful label of unconnected graph $K_{3,4} \cup (P_3 \vee \overline{K_3})$ is shown below.

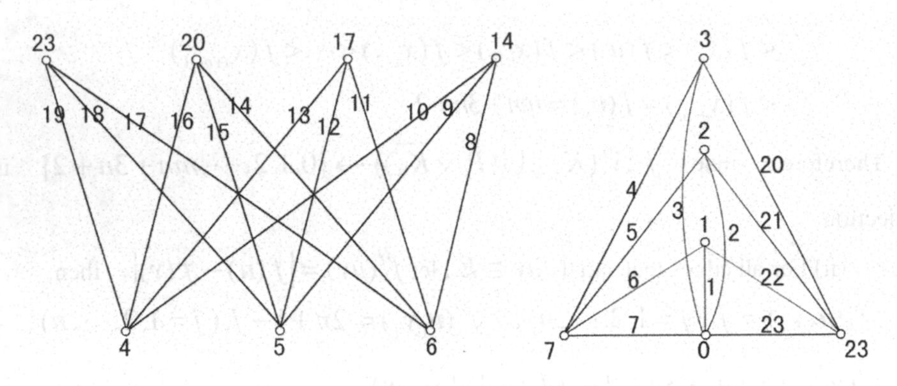

Fig. 3. $K_{3,4} \cup (P_3 \vee \overline{K_3})$

2.4 The Graphs $K_{n,m} \cup (P_3 \vee \overline{K_n})$ and $K_{n,m} \cup St(2n)$

Theorem 4. For positive integer m,n, where $1 \leq n \leq m$, then $K_{n,m} \cup (P_1 \vee P_{2n+2})$ and $K_{n,m} \cup St(2n)$ are graceful graphs.

3 Conclusions

We proved that for positive integer n, m, where then unconnected graph. $K_{n,m} \cup (\overline{K_2} \vee \overline{K_n})$, $K_{n,m} \cup (P_2 \vee \overline{K_n})$, $K_{n,m} \cup (P_3 \vee \overline{K_n})$, $K_{n,m} \cup (P_3 \vee \overline{K_n})$ and $K_{n,m} \cup St(2n)$ are graceful graphs.

Acknowledgement. This work described here is partially supported by the grants from the National Natural Science Foundation of North China Institute of Science and Technology (No. 2005A-13, B009018).

References

1. Cheng, H., Yao, B., Chen, X., Zhang, Z.: On graceful generalized spiders and caterpillars. Ars. Combin. 87, 181–191 (2008)
2. Yang, Y.S., Rong, Q., Xu, X.R.: A class of graceful graphs. J. Math. Research and Exposition 24, 520–524 (2004)
3. Yang, Y., Xu, X., Xi, Y., Li, H., Haque, K.: The graphs $C_9^{(t)}$ are graceful for t=0, 1mod(4). Ars. Combin. 79, 295–301 (2006)
4. Yang, Y., Xu, X., Xi, Y., Huijun, H.: The graphs $C_9^{(t)}$ are graceful for t=0, 3mod(4). Ars. Combin. 85, 361–368 (2007)
5. Youssef, M.Z.: On Skolem-graceful and cordial graphs. Ars. Combin. 78, 167–177 (2006)

6. Wei, L.-X., et al.: The researches on gracefilness of two kinds of unconnected graphs. Journal of Shandong University (Natural Science) 43, 90–96 (2008)
7. Wei, L.-X., Zhang, K.-L.: The Graceful Label of Several Kinds of join graph. Journal of Zhongshan University (Natural Science) 42(3), 10–13 (2008)
8. Bi, S.-Y., Li, X.-F., Lu, X.: k-Gracefulness and the nature of Algebra of $C_4 \cup St(m)$. Journal of Jilin University (Natural Science) (2), 19–22 (1999)
9. Lu, X., Li, X.-F.: k-Gracefulness and the nature of Algebra of $C_3 \cup St(m)$. Journal of Changchun College of Post and Telecommunication 18(4), 17–20 (2000)

Six-Order Symplectic Integration in Quasi-classical Trajectory Computation

Xian-Fang Yue*

Department of Physics and Information Engineering, Jining University,
Qufu 273155, China
xfyuejnu@gmail.com

Abstract. Six-order symplectic integration was used in the quasi-classical trajectory method to carry out stereodynamics computation. In the case of Li + HF (v=0, j=0) → LiF + H reaction, we found that the product rotational angular momentum is not only aligned perpendicular to the reagent relative velocity vector ***k***, but also orientated along the negative *y* axis. Collision energy effect on the reaction was also investigated, which demonstrated that the product rotational alignment and orientation are monotonously enhanced by the increasing collision energy.

Keywords: Six-order symplectic integration, Quasi-classical trajectory, Alignment, Orientation.

1 Introduction

The Runge-Kutta method, a numerical integrator, was used to solve a partitioned Hamiltonian of a Hamiltonian system in the past studies. Nevertheless, the Runge-Kutta integration destroys the symplectic structure of the Hamiltonian system because it is not a symplectic method, which leads to non-conservation of the total energy [1]. Even though the energy conservation is within the required range, the nonsymplectic structure may also result in significant trajectory errors. Especially for the long-lived complex collisions or unimolecular decay, the evolution of trajectories requires a necessarily long time to reach the ends, in which the error of total energy will accumulate. Therefore, the trajectory evolved from the integrator is qualitatively susceptible due to the effects of accumulated integration error. A technique known as "back-integration" combines a fourth-order Runge-Kutta initialized fourth-order Admas–Moulton–Hamming predictor-corrector integrator (RK4-AMH4), which can ensure the total energy conservation within a guaranteed range. This technique has been used in the quasi-classical trajectory calculations in many years [2]. Although there is not enough evidence to show that the conservation of total energy can guarantee the correctness of trajectories, it is undoubtedly the necessary condition of a correct end of trajectory. The symplectic integration method has generally been accepted as an alternative numerical solution in "molecular dynamics" as the second-order symplectic Verlet "leapfrog" integrator. Indirect reactions with a long-lived

* Corresponding author.

D. Jin and S. Lin (Eds.): Advances in MSEC Vol. 1, AISC 128, pp. 533–538.
springerlink.com

complex require longer integration times, which have a greater demand for numerical methods with improved speed and accuracy. However, the use of higher-order simulation methods in the quasi-classical trajectory computation is very rare [3]. In previous reports, Schlier and Seiter [4-5] introduced tests of some new symplectic integration sixth-order and eighth-order routines applied to the solution of classical trajectories for a triatomic model molecule and its molecular vibrations. They demonstrated that, among a great number of integrator approaches, the symplectic ones are the most efficient and need the smallest computational expense at a prescribed accuracy level.

In the present work, we have applied six-order symplectic integration routines to study the stereodynamics (Alignment and Orientation) of a typical reaction of Li + HF (v=0, j=0) → LiF + H with a quasi-classical trajectory method.

2 Methodology and Theory

In the symplectic integrator, the classical Hamilton's equations are integrated numerically for motion in three dimensions. In a given Hamiltonian system whose Hamiltonian can be partitioned, i.e., written as H = T(*p*) + U(*q*), we define the derivative terms in the Hamiltonian movement equations as $hq_i = \partial H / \partial p_i$ for the generalized momentum, and $hp_i = \partial H / \partial q_i$ for the generalized coordinates; let *dt* be the full time step. The algorithm can be expressed as [6]

```
do i = 0, 1, 2, . . . , n - 2, n - 1
   p = p + dta(2i)hp
   q = q + dta(2i + 1)hq
enddo
   p = p + dta(2n)hp.                                          (1)
```

The last step can be concatenated with the first step in a continuing calculation. Then there are 2n substeps in every step, where n is 8 for six-order symplectic integrator. Thus, the number of the calling to the potential derivatives is also n. The coefficients are defined by 2n + 1 values of a(*i*), which is taken directly from Ref. 4-5 without revision. Further details of the sixth-order symplectic integrator may be found in the series of articles by Schlier and Seiter [4-5].

A most popular and accurate ground 1^2A' state PES is used in the present study for the Li + HF(v=0, j=0) → LiF + H reaction, which is developed by Aguado et al. [7]. In the center of mass (CM) frame, the product rotational polarization can be depicted through three angular distributions $P(\theta_r)$, $P(\phi_r)$, and $P(\theta_r, \phi_r)$,. Here, the *z*-axis of the CM frame is parallel to the reagent relative velocity vector ***k***, while the *zx*-plane (also called the scattering plane) contains ***k*** and ***k'*** (the product relative velocity vector) with ***k'*** on the $x \geq 0$ half plane. The *y*-axis is perpendicular to the scattering plane. θ_r is the angle between ***k*** and ***j'***. ϕ_r is the dihedral angle between the scattering plane and the plane containing ***k*** and ***j'***. Batches of 5×10^5 trajectories are run for the reactants. The trajectories are started from 20 Å initial CM separation and then running on the ground 1^2A' electronic state at the collision energies of 97, 136, and 363 meV. The Hamilton's motion equation is solved by the symplectic integration method with the integration step of 0.1 fs, which was found enough for the total energy and angular momentum conservation. The truncated number used in the following expansions of $P(\theta_r)$, $P(\phi_r)$, and $P(\theta_r, \phi_r)$ is 18, 24 and 7, respectively [8].

$$P(\theta_r) = \frac{1}{2}\sum_k [k] a_0^k P_k(\cos\theta_r), \tag{2}$$

where $[k]$=2k+1, $a_0^k = \langle P_k(\cos\theta_r)\rangle$.

$$P(\phi_r) = \frac{1}{2\pi}(1 + \sum_{even, n\geq 2} a_n \cos n\phi_r + \sum_{odd, n\geq 1} b_n \sin n\phi_r), \tag{3}$$

where $a_n = 2\langle\cos n\phi_r\rangle$, $b_n = 2\langle\sin n\phi_r\rangle$.

$$P(\theta_r, \phi_r) = \frac{1}{4\pi}\sum_k \sum_{q\geq 0} [a_{q\pm}^k \cos q\phi_r - a_{q\mp}^k i \sin q\phi_r] C_{kq}(\theta_r, 0), \tag{4}$$

where $a_{q\pm}^k = 2\langle C_{k|q|}(\theta_r, 0)\cos q\phi_r\rangle$, k is even,

$a_{q\pm}^k = 2i\langle C_{k|q|}(\theta_r, 0)\sin q\phi_r\rangle$, k is odd.

Here, C_{kq} are the modified spherical harmonics, and the angular bracket means an average over all reactive trajectories and/or over all scattering angles.

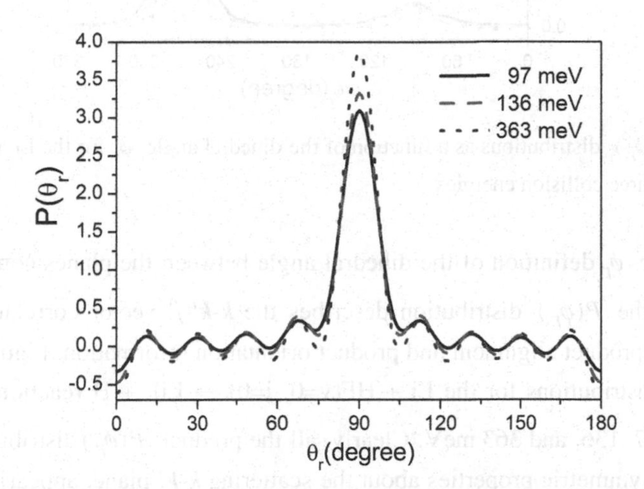

Fig. 1. The $P(\theta_r)$ distributions as a function of the polar angle θ_r for the Li + HF → LiF + H reaction at three collision energies

3 Results and Discussion

The product $P(\theta_r)$ distribution describes the ***k-j'*** vector correlation with ***k·j'***=cos(θ_r). Figure 1 displays the calculated product $P(\theta_r)$ distribution of the Li + HF(v=0, j=0) → LiF + H reaction at collision energies of 97, 136 and 363 meV, respectively. Obviously, the $P(\theta_r)$ distribution is symmetric about θ_r =90°, and illustrates a distinct peak at θ_r =90° for each of the collision energy. This indicates that the product rotational angular momentum vector ***j'*** is aligned perpendicular to the relative velocity direction ***k***. A prominent trend is easily observed from Fig. 1 that the increasing collision energy enhances the product rotational alignment, which means that the product rotational alignment becomes stronger with the increase of the collision energy.

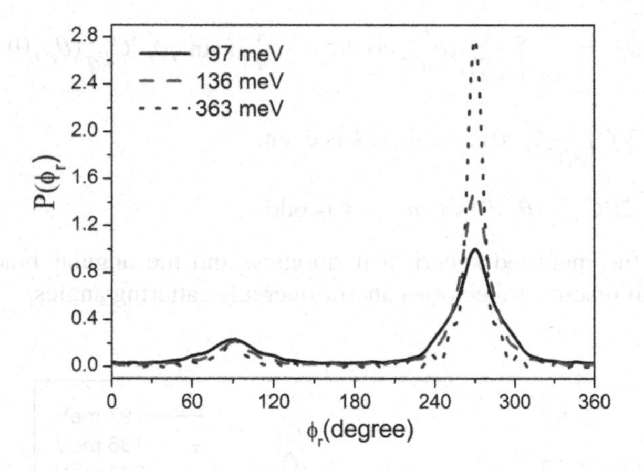

Fig. 2. The $P(\phi_r)$ distributions as a function of the dihedral angle ϕ_r for the Li + HF → LiF + H reaction at three collision energies

Under the ϕ_r definition of the dihedral angle between the planes consisting of ***k-k'*** and ***k-j'***, the $P(\phi_r)$ distribution describes the ***k-k'-j'*** vector correlation and can provide both product alignment and product orientation information. Figure 2 displays the $P(\phi_r)$ distributions for the Li + HF(v=0, j=0) → LiF + H reaction at collision energies of 97, 136, and 363 meV. Clearly, all the product $P(\phi_r)$ distributions in Fig. 2 illustrate asymmetric properties about the scattering ***k-k'*** plane, appearing one large peak at about ϕ_r =270° and one small peak at about ϕ_r =90°, respectively. This feature conveys the information to audience that most product rotational angular momentum tend to align along the direction of *y* axis which is perpendicular to the scattering ***k-k'*** plane, and orientate along the negative directions of *y* axis. That is to say, the product molecules prefer a counterclockwise rotation (see from the negative direction of y axis) in the plane parallel to the scattering plane. As shown in Fig. 2, the peak at about ϕ_r =270° gradually and monotonously increase with the collision energy increasing, while the peak at about ϕ_r =90° has no distinct change with the collision energy increasing. This indicates that the product rotational alignment becomes stronger along with the increasing collision energy.

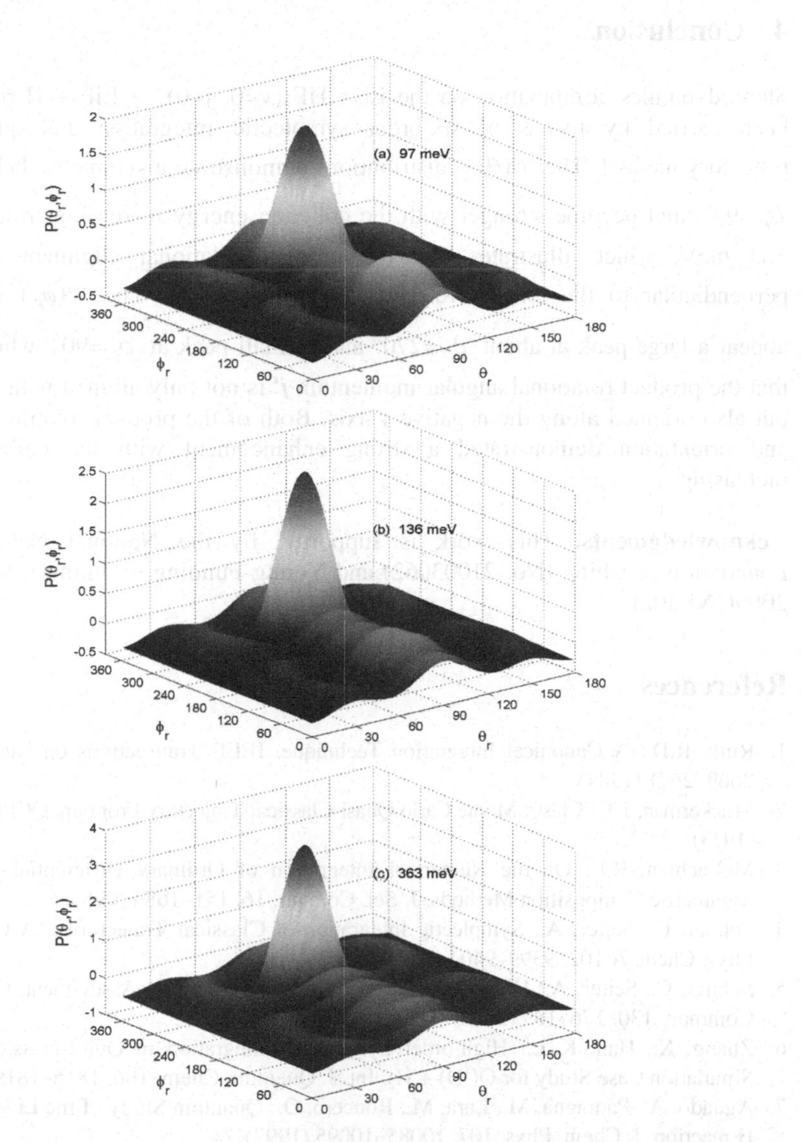

Fig. 3. The $P(\theta_r, \phi_r)$ distributions as a function of both the polar θ_r and dihedral angle ϕ_r for the Li + HF → LiF + H reaction at collision energies of (a) 97 meV, (b) 136 meV, and (c) 363 meV, respectively.

Figure 3a, b, and c show the joint $P(\theta_r, \phi_r)$ distributions for the Li + HF → LiF + H reaction at collision energies of 97, 136, and 363 meV, respectively. The tomographical features are in good consistence with the separate $P(\theta_r)$ and $P(\phi_r)$ distributions.

4 Conclusion

Stereodynamics computation on the Li + HF (v=0, j=0) → LiF + H reaction have been carried by means of six-order symplectic integration and quasi-classical trajectory method. The $P(\theta_r)$ distributions demonstrate a symmetric behavior about θ_r=90°, and become stronger with the collision energy increasing from 97, 136 to 363 meV, which illustrates that the product rotational alignment tend to be perpendicular to the reagent relative velocity vector ***k***. The $P(\phi_r)$ distributions appear a large peak at about ϕ_r=270° and a small peak at ϕ_r=90°, which indicates that the product rotational angular momentum ***j'*** is not only aligned with respect to ***k***, but also oriented along the negative *y* axis. Both of the product rotational alignment and orientation demonstrated a strong enhancement with the collision energy increasing.

Acknowledgments. This work is supported by the National Natural Science Foundation of China (No. 21003062) and Young Funding of Jining University (No. 2009QNKJ02).

References

1. Ruth, R.D.: A Canonical Integration Technique. IEEE Transactions on Nuclear Science, 2669–2671 (1983)
2. Muckerman, J.T.: Clastr: Monte Carlo Quasi-Classical Trajectory Program. QCPE 11, 229–236 (1973)
3. McLachlan, R.I.: On the Numerical Integration of Ordinary Differential Equations by Symmetric Composition Methods. J. Sci. Comput. 16, 151–168 (1995)
4. Schlier, C., Seiter, A.: Symplectic Integration of Classical Trajectories: A Case Study. J. Phys. Chem. A 102, 9399–9404 (1998)
5. Schlier, C., Seiter, A.: High-Order Symplectic Integration: An Assessment. Comput. Phys. Commun. 130, 176–189 (2000)
6. Zhang, X., Han, K.-L.: High-order Symplectic Integration in Quasi-classical Trajectory Simulation Case Study for $O(^1D) + H_2$. Int. J. Quantum Chem. 106, 1815–1819 (2006)
7. Aguado, A., Paniagua, M., Lara, M., Roncero, O.: Quantum Study of the Li + HF → LiF + H reaction. J. Chem. Phys. 107, 10085–10095 (1997)
8. Yue, X.-F., Cheng, J., Feng, H.-R., Li, H., Emilia, L.W.: Stereodynamics Study of Reactions $N(^2D)$ + HD → NH + D and ND + H. Chin. Phys. B 4, 43401 (2010)

Identity for Sums of Five Squares

Gaowen Xi

College of Mathematics and Physics,
Chongqing University of Science and Technology
Chongqing, 401331, P.R. China
xigaowen@163.com

Abstract. In this paper, we prove a q-expansion formula by means of the Liu's expansion formula. The new q-expansion formula leads to obtain the identities for sums of five squares and sums of five triangular numbers.

Keywords: q-series, q-differential operator, expansionformula, identities for sums of five squares, identities for sums of five triangular numbers.

1 Introduction

In [4], Z. G. Liu utilizes Leibniz formula for the q-difference operator to obtain the following q-expansion formula:

$$f(b)=\sum_{n=0}^{\infty}\frac{(1-aq^{2n})(aq/b;q)_n b^n}{(q,b;q)_n}\left[D_{q,x}^{n}\{f(x)(x;q)_{n-1}\}\right]_{x=aq}, \tag{1}$$

where $f(b)$ is a formal series in b. This expansion formula leads to the new proofs of the Rogers-Fine identity, the nonterminating ${}_6\varphi_5$ summation formula, Watson's q-analog of Whipple's theorem, Andrew's identities for sums ofthree squares and suns of three triangular numbers are also derived, etc.

In this paper, we shall derive a q-expansion formula using (1). By the new q-expansion formula, we will obtain the following identities for sums of five squares and sums of five triangular numbers.

2 Some Basic Facts

Before the proof of the theorems, we recall some definitions, notations and known results which will be used in the proof. We shall follow the notation and terminology in [3]. For two complex q and x, the shifted factorial of order n with base q is defined by

$$(x;q)_n=\begin{cases}1, & n=0,\\ (1-x)(1-xq)(1-xq^2)\cdots(1-xq^{n-1}), & n=1,2,\cdots\end{cases} \tag{2}$$

$$(x;q)_\infty=\lim_{n\to\infty}(x;q)_n=\prod_{k=0}^{\infty}(1-xq^k),\quad |q|<1. \tag{3}$$

D. Jin and S. Lin (Eds.): Advances in MSEC Vol. 1, AISC 128, pp. 539–544.
springerlink.com

Where $|q|<1$ in order for the infinite products to be convergent. It is easy to check that the shifted factorial with negative integer order is give by

$$(x;q)_{-n}=\frac{1}{(xq^{-n};q)_n}=\frac{(-q/x)^n q^{n(n-1)/2}}{(q/x;q)_n}. \tag{4}$$

The product and fractional forms of shifted factorial are abbreviated compactly to

$$(\alpha,\beta,\cdots,\gamma;q)_n=(\alpha;q)_n(\beta;q)_n\cdots(\gamma;q)_n.$$

Following Bailey [2] and Slater [6], The basic hypergeometric series ${}_{r+1}\phi_s$ is defined by

$${}_{r+1}\phi_s\left[\begin{matrix}a_0, & a_1, & \cdots, & a_r\\ & b_1, & \cdots, & b_s\end{matrix}\middle| q;\ x\right]=\left[\begin{matrix}a_0, & a_1, & \cdots, & a_r\\ q, & b_1, & \cdots, & b_s\end{matrix}\middle| q\right]_n \times_n [(-1)^n q^{n(n-1)/2}]^{r-s} x^n. \tag{5}$$

The q-differential operator about x is defined by

$$D_{q,x}\{f(x)\}=\frac{f(x)-f(qx)}{x}. \tag{6}$$

By convention, $D_{q,x}^0$ is understood as the identity.

The Leibniz rule for $D_{q,x}$ is the following identity [5, P.233]

$$D_{q,x}^n\{g(x)h(x)\}=\sum_{k=0}^{n} q^{k(k-n)}\begin{bmatrix}n\\k\end{bmatrix} D_{q,x}^n\{g(x)\}\, D_{q,x}^{n-k}\{h(q^k x)\}. \tag{7}$$

It is easy verify the following property of $D_{q,x}$:

$$D_{q,x}^n\left\{\frac{(tx;q)_\infty}{(sx;q)_\infty}\right\}=s^n(t/s;q)_n\frac{(tq^n x;q)_\infty}{(sx;q)_\infty}. \tag{8}$$

On stetting $t=0$, we have

$$D_{q,x}^n\left\{\frac{1}{(sx;q)_\infty}\right\}=\frac{s^n}{(sx;q)_\infty}. \tag{9}$$

3 Identity for Sums of Five Squares

In this section, we will obtain identities for sums of five squares and sums of five triangular number.

Theorem 3.1. We have

$$\frac{(bt,bu,aq,asq,avq;q)_\infty}{(b,bs,bv,atq,auq;q)_\infty}=\sum_{n=0}^{\infty}\frac{(1-aq^{2n})(a,aq/b;q)_n b^n}{(1-a)(q,b,auq;q)_n}$$
$$\times\sum_{k=0}^{n}\begin{bmatrix} n \\ k \end{bmatrix}\frac{(avq;q)_k(u/v;q)_{n-k}v^{n-k}}{(atq;q)_k}\sum_{j=0}^{k}\begin{bmatrix} k \\ j \end{bmatrix}q^{(k-j)(n-1)}$$
$$\times s^j(asq;q)_{k-j}(t/s;q)_j. \tag{10}$$

Proof. Now, we apply (1) to the function

$$f(b)=\frac{(tb,ub;q)_\infty}{(b,sb,vb;q)_\infty}. \tag{11}$$

Taking

$$g(x)=\frac{(x;q)_{n-1}(tx;q)_\infty}{(x,sx;q)_\infty}=\frac{(tx;q)_\infty}{(xq^{n-1},sx;q)_\infty}=,$$
$$h(x)=\frac{(ux;q)_\infty}{(vx;q)_\infty}=,$$

in the Leibniz formula and using (8) and (9), we find that

$$\left[D_{q,x}^n\{f(x)(x;q)_{n-1}\}\right]_{x=aq}=\left[\sum_{k=0}^{n}\begin{bmatrix} n \\ k \end{bmatrix}q^{k(k-n)}D_{q,x}^k\left\{\frac{(tx;q)_\infty}{(xq^{n-1},sx;q)_\infty}\right\}\right.$$
$$\left.\times D_{q,x}^{n-k}\left\{\frac{(uq^kx;q)_\infty}{(vq^kx;q)_\infty}\right\}\right]_{x=aq},$$

$$\left[D_{q,x}^k\left\{\frac{(tx;q)_\infty}{(xq^{n-1},sx;q)_\infty}\right\}\right]_{x=aq}=\left[\sum_{j=0}^{k}\begin{bmatrix} k \\ j \end{bmatrix}q^{j(j-k)}D_{q,x}^j\left\{\frac{1}{(xq^{n-1};q)_\infty}\right\}\right.$$
$$\left.\times D_{q,x}^{k-j}\left\{\frac{(tq^jx;q)_\infty}{(sq^{j1x},sx;q)_\infty}\right\}\right]_{x=aq}$$
$$=\frac{(atq;q)_\infty(aq;q)_{n-1}}{(aq,asq;q)_\infty(atq;q)_k}\sum_{j=0}^{k}\begin{bmatrix} k \\ j \end{bmatrix}$$
$$\times q^{(k-j)(n-1)}s^j(asq;q)_{k-j}(t/s;q)_j.$$

$$\left[D_{q,x}^{n-k}\left\{\frac{(uq^kx;q)_\infty}{(vq^kx;q)_\infty}\right\}\right]_{x=aq}=\left[v^{n-k}q^{k(n-k)}(u/v;q)_{n-k}\frac{(uq^nx;q)_\infty}{(vq^kx;q)_\infty}\right]_{x=aq}$$
$$=\frac{(auq;q)_\infty(avq;q)_k(u/v;q)_{n-k}}{(avq;q)_\infty(auq;q)_n}v^{n-k}q^{k(n-k)}, \tag{12}$$

So by (12), we have

$$\left[D_{q,x}^{n}\{f(x)(x;q)_{n-1}\}\right]_{x=aq}=\left[\sum_{k=0}^{n}\begin{bmatrix}n\\k\end{bmatrix}q^{k(k-n)}D_{q,x}^{k}\left\{\frac{(tx;q)_\infty}{(xq^{n-1},sx;q)_\infty}\right\}\right.$$

$$\left.\times D_{q,x}^{n-k}\left\{\frac{(uq^{k}x;q)_\infty}{(vq^{k}x;q)_\infty}\right\}\right]_{x=aq}\qquad q^{(k-j)(n-1)}s^{j}(asq;q)_{k-j}(t/s;q)_{j}$$

$$=\sum_{k=0}^{n}\begin{bmatrix}n\\k\end{bmatrix}\frac{(auq;q)_\infty(atq;q)_\infty(avq;q)_k(u/v;q)_{n-k}(aq;q)_{n-1}}{(aq,asq;q)_\infty(avq;q)_\infty(auq;q)_n(atq;q)_k}$$

$$\times\sum_{j=0}^{k}\begin{bmatrix}k\\j\end{bmatrix}q^{(k-j)(n-1)}s^{j}(asq;q)_{k-j}(t/s;q)_{j}v^{n-k}$$

$$=\frac{(btq,buq;q)_\infty}{(aq,asq,avq;q)_\infty}\sum_{k=0}^{n}\begin{bmatrix}n\\k\end{bmatrix}\frac{(aq;q)_{n-1}(avq;q)_k(u/v;q)_{n-k}v^{n-k}}{(auq;q)_n(atq;q)_k}$$

$$\times\sum_{j=0}^{k}\begin{bmatrix}k\\j\end{bmatrix}q^{(k-j)(n-1)}s^{j}(asq;q)_{k-j}(t/s;q)_{j}.$$

On substituting this into (1), we obtain (10).

In (10), setting $u=v=1$, we have

$$\frac{(bt,aq,asq;q)_\infty}{(b,bs,atq;q)_\infty}=\sum_{n=0}^{\infty}\frac{(1-aq^{2n})(a,aq/b;q)_n b^n}{(1-a)(q,b,atq;q)_n}$$

$$\times\sum_{j=0}^{n}\begin{bmatrix}n\\j\end{bmatrix}q^{(n-j)(n-1)}\ s^{j}(asq;q)_{n-j}(t/s;q)_{j}. \tag{13}$$

The (13) is the theorem 6 of [4]. So the special case of theorem 3.1 is the theorem 6 of [4]. By theorem 6 , we can obtain some important classical identities.

Theorem 3.2 (Identity for sums of five squares). We have

$$\sum_{n=0}^{\infty}(-1)^n r_5(n)q^n=\left[\sum_{n=-\infty}^{\infty}(-1)^n q^{n^2}\right]^5$$

$$=\frac{(q;q)_\infty^5}{(-q;q)_\infty^5}$$

$$=1+8\sum_{n=1}^{\infty}\sum_{k=0}^{n}\frac{(-1)^{n+k}(q;q)_k}{(1+q^{n-k})(-q;q)_k}\sum_{j=0}^{k}\frac{(-1)^j(-q,q^{n+1-k};q)_{j-1}}{(q,-q^{n+1-k};q)_j}$$

$$\times q^{-j(n+j-k)+k(n-1)+n}. \tag{14}$$

Proof. In (10), setting $a=s=v=1$, $t=u=-1$, $b=-q$, we have

$$\frac{(q;q)_\infty^5}{(-q;q)_\infty^5}=1+\sum_{n=1}^{\infty}\frac{(1-q^{2n})(q;q)_{n-1}(-1;q)_n(-q)^n}{(q,-q,-q;q)_n}\sum_{k=0}^{n}\begin{bmatrix}n\\k\end{bmatrix}\frac{(q;q)_k(-1;q)_{n-k}}{(-q;q)_k}$$

$$\times\sum_{j=0}^{k}\begin{bmatrix}k\\j\end{bmatrix}q^{(k-j)(n-1)}\ (q;q)_{k-j}(-1;q)_{j}$$

$$=1+4\sum_{n=1}^{\infty}\sum_{k=0}^{n}\frac{(-1)^n(q,q^{n+1-k};q)_k(-q^{n+1};q)_{-k-1}}{(q^{n+1};q)_{-k1}(-q;q)_k}\sum_{j=0}^{k}\frac{(-1;q)_j q^{(k-j)(n-1)+n}}{(q;q)_j}$$

$$=1+4\sum_{n=1}^{\infty}\sum_{k=0}^{n}\frac{(-1)^n(q,q^{n+1-k};q)_k q^{k(n-1)+n}}{(-q^{n-k};q)_{k+1}(-q;q)_k}\sum_{j=0}^{k}\frac{(-1;q)_j q^{-j(n-1)}}{(q;q)_j}. \tag{15}$$

In (4.2) and (5.5) 0f [1], taking $b=1,\ a=q^{n-k}$, we find that

$$\sum_{j=0}^{k}\frac{(-1;q)_j q^{-j(n-1)}}{(q;q)_j}=\frac{2(-1)^k(-q^{n+1-k};q)_k}{(q^{n+1-k};q)_k}$$
$$\times\sum_{j=0}^{k}\frac{(-q,q^{n+1-k};q)_{j-1}}{(q,-q^{n+1-k};q)_j}(-1)^j q^{-j(n+j-k)}. \tag{16}$$

Substituting (16) into (15), we obtain (14). The theorem 3.2 is proved.

Theorem 3.3 (Identity for sums of five triangular numbers). We have

$$\frac{(q^2;q^2)_\infty^5}{(q;q^2)_\infty^5}=\sum_{n=0}^{\infty}\frac{1+q^{2n+1}}{1-q^{2n+1}}\sum_{k=0}^{n}\frac{(q^2,q^{2(n+1-k)};q^2)_k}{(q^{2n+1-2k};q^2)_k(q;q^2)_{k+1}}$$
$$\times\sum_{j=0}^{k}\frac{(q;q^2)_j}{(q^2;q^2)_j}q^{n(2k-2j+1)}. \tag{17}$$

Proof. In (10), by replacing q by q^2, we have

$$\frac{(bt,bu,aq^2,asq^2,avq^2;q^2)_\infty}{(b,bs,bv,atq^2,auq^2;q^2)_\infty}=\sum_{n=0}^{\infty}\frac{(1-aq^{4n})(a,aq^2/b;q^2)_n b^n}{(1-a)(q^2,b,auq^2;q^2)_n}$$
$$\times\sum_{k=0}^{n}\frac{(q^2;q^2)_n(avq^2;q^2)_k(u/v;q^2)_{n-k}v^{n-k}}{(q^2;q^2)_{n-k}(q^2;q^2)_k(atq^2;q^2)_k}$$
$$\times\sum_{j=0}^{k}\frac{(q^2;q^2)_k}{(q^2;q^2)_{k-j}(q^2;q^2)_j}q^{2(k-j)(n-1)}$$
$$\times s^j(asq^2;q^2)_{k-j}(t/s;q^2)_j.$$

By taking $b=q^3,\ a=q^2,\ s=v=q^{-2},\ t=u=q^{-1}$, we obtain

$$\frac{(1-q)^3(q^2;q^2)_\infty^5}{(1-q^2)(q;q^2)_\infty^5}=\sum_{n=0}^{\infty}\frac{(1-q^{4n+2})(q^2,q;q^2)_n q^{3n}}{(1-q^2)(q^2,q^3,q^3;q^2)_n}$$
$$\times\sum_{k=0}^{n}\frac{(q^2;q^2)_n(q^2;q^2)_k(q;q^2)_{n-k}q^{-2(n-k)}}{(q^2;q^2)_{n-k}(q^2;q^2)_k(q^3;q^2)_k}$$
$$\times\sum_{j=0}^{k}\frac{(q^2;q^2)_k}{(q^2;q^2)_{k-j}(q^2;q^2)_j}q^{2(k-j)(n-1)}$$
$$\times q^{-2j}(q^2;q^2)_{k-j}(q;q^2)_j$$

$$
\begin{aligned}
&=\sum_{n=0}^{\infty}\frac{(1+q^{2n+1})(1-q)^2q^n}{(1-q^2)(q;q^2)_{n+1}}\sum_{k=0}^{n}\frac{(q^2;q^2)_k(q;q^2)_{n-k}q^{2nk}}{(q^{2n+2};q^2)_{-k}(q^3;q^2)_k}\\
&\quad\times\sum_{j=0}^{k}\frac{(q;q^2)_j}{(q^2;q^2)_j}q^{-2jn}\\
&=\sum_{n=0}^{\infty}\frac{(1-q)^3(1+q^{2n+1})q^n}{(1-q^2)(1-q^{2n+1})}\sum_{k=0}^{n}\frac{(q^2,q^{2(n+1-k)};q^2)_kq^{2nk}}{(q^{2n+1-2k};q^2)_k(q;q^2)_{k+1}}\\
&\quad\times\sum_{j=0}^{k}\frac{(q;q^2)_j}{(q^2;q^2)_j}q^{-2jn}.
\end{aligned}
$$

Hence the theorem 3.3 is proved.

Acknowledgements. This research is funded by Research Foundation of Chongqing University of Science and Technology, the project No. is CK2010B03.

References

1. Andrews, G.E.: The fifth and seventh order mock theta function. Tran. Amer. Math. Soc. 293, 113–134 (1986)
2. Bailey, W.N.: Generalized hypergeometric series. Cambridge University Press, Cambridge (1935)
3. Gasper, G., Rahman, M.: Basic hypergeometric series, Encyclopedia of Mathematics and its Applications. Cambridge University Press, Cambridge (1990)
4. Liu, Z.G.: An expansion foumula for q-series and applications. The Ramanujan Jounal 6, 429–447 (2002)
5. Roman, S.: More on q-umbral calculus. J. Math. Anal. Appl. 107, 222–252 (1985)
6. Slater, L.J.: Generalized hypergeometric functions. Cambridge University Press, Cambridge (1966)

The Optimal Traits of Semiorthogonal Trivariate Matrix-Valued Small-Wave Wraps and Trivariate Framelets

Delin Hua

School of Education, Nanyang Institute of Technology, Nanyang 473000, China
jhnsx123@126.com

Abstract. Wavelet analysis has been a powerful tool for solving many complicated problems in natural science and engineering computation. In this work, the notion of orthogonal matrix-valued trivariate small-wave wraps and wavelet frame wraps, which are generalization of univ-ariate small-wave wraps, is introduced. A new procedure for construct-ing these vector-valued trivariate small-wave wraps is presented. Their characteristics are studied by using time-frequency analysis method, Banach space theory and finite group theory. Orthogonal formulas concerning the wavelet packs are established. The biorthogonality formulas concerning these small-wave wraps are established. Moreover, it is shown how to draw new Riesz bases of space $L^2(R^3, C^v)$ from these small-wave wraps. An approach for designing a sort of affine triariate dual frames in three-dimensional space is presented.

Keywords: Convergence theorem, vector-valued trivariate small-wave wraps, Riesz sequence, time-frequency analysis approach.

1 Introduction and Notations

The wavelet theory has been one of powerful tools for researching into wavelets. Although the Fourier transform has been a major tool in analysis for over a century, it has a serious laking for signal analysis in that it hides in its phases information concerning the moment of emission and duration of a signal. Since 1986, wavelet analysis has become a developing branch of mathematics. Wavelet packets have been applied to signal processing [1], image compression [2] so on. Coifman and Meyer are those who firstly introduced the notion of univariate orthogonal wavelet packets. Shen[3] constructed the multivariate orthogonal wavelet packets. Wavelet packets include multiple orthonormal basis , which means that a signal could be represented in many different ways by using wavelet packets. But the performances in presenting the specified signal are different using different bases. The one which could provide the best performance according to some criterion will be the best basis. Nowadays, most of the related studies use the algorithm proposed by Coisman and Wickerhanser to select the best basis. Vector-valued wavelets are a class of generalized multiwavelets. Xia [4] introduced the notion of orthogonal vector-valued wavelets. Moreover, he studied the existence and construction of vector-valued wavelets. However, vector-valued wavelets [5] and multiwavelets are different. Hence, studying vector-valued wavelets is useful in multiwavelet theory and representations of signals.

D. Jin and S. Lin (Eds.): Advances in MSEC Vol. 1, AISC 128, pp. 545–551.
springerlink.com

It is known that the majority of information is multi-dimensional information. Thus, it is significant and necessary to generalize the concept of multivariate wavelet packets to the case of multiple vector-valued multivariate wavelets. Based on some ideas from [3], the goal of this paper is to discuss the properties of multiple vector-valued multivariate wavelet packets. Let R and C be all real and all complex numbers, respectively. Z and N denote, respectively, all integers and all positive integers. Set $Z_+ = \{0\} \cup N, a, s \in N$ as well as $a \geq 2$ By algebra theory, it is obviously follows that there are a^2 elements $d_0, d_1, \cdots, d_{a^2-1}$ in $Z_+^3 = \{(z_1, z_2, z_3) : z_1, z_2, z_3 \in Z_+\}$ such that $Z^3 = \bigcup_{d \in \Omega_0} (d + aZ^3)$; $(d_1 + aZ^3) \cap (d_2 + aZ^3) = \phi$, where $\Omega_0 = \{d_0, d_1, \cdots, d_{a^2-1}\}$ denotes the aggregate of all the different representative elements in the quotient group $Z^2 / (aZ^2)$ and order $d_0 = \{\underline{0}\}$ where $\{\underline{0}\}$ is the null element of Z_+^2 and d_1, d_2 denote two arbitrary distinct elements in Ω_0 .Let $\Omega = \Omega_0 - \{\underline{0}\}$ and Ω, Ω_0 to be two index sets. Define, By $L^2(R^3, C^s)$, we denote the set of all vector-valued functions $L^2(R^3, C^s) := \{\hbar(y) = (h_1(y)), h_2(y), \cdots, h_u(y))^T : h_l(y) \in L^2(R^3), l = 1, 2, \cdots, s\}$, where T means the transpose of a vector. For any $\hbar \in L^2(R^3, C^s)$ its integration is defined as $\int_{R^3} \hbar(y) dy = (\int_{R^3} h_1(y) dy, \int_{R^3} h_2(y) dy, \cdots, \int_{R^3} h_s(y) dy)$ and the Fourier transform of $\hbar(y)$ is defined by

$$\hat{\hbar}(\omega) := \int_{R^3} \hbar(y) \cdot \exp\{-i\langle y, \omega \rangle\} dy, \tag{1}$$

where $\langle y, \omega \rangle$ denotes the inner product of y and ω. For multiple vector-valued functions $< \hbar, \Lambda >$ denotes their symbol inner product, i.e.,

$$\langle \hbar, \Lambda \rangle := \int_{R^3} \hbar(y) \Lambda(y)^* dy, \tag{2}$$

where * means the transpose and the complex conjugate.

Definition 1. A sequenc $\{\hbar_n(y)_{n \in Z^3} \subset L^2(R^3, C^s)\}$ is called an orthogonal set, if

$$\langle \hbar_n, \hbar_v \rangle = \delta_{n,v} I_s, \quad n, v \in Z^3, \tag{3}$$

where I_s stands for the $s \times s$ identity matrix and $\delta_{n,v}$, is generalized Kronecker symbol, i.e., $\delta_{n,v} = 1$ as $n = v$ and $\delta_{n,v} = 0$, otherwise.

Definition 2. A sequence of vector-valued functions $\{\hbar_n(y)\}_{n \in Z^3} \subset W \subset L^2(R^3, C^s)$ is called a Riesz basis in W , if it satisfies (i) For any $\Lambda(y) \in W$, there exists a unique $s \times s$ matrix sequence $\{P_n\}_{n \in Z^3}$ such that

$$\Lambda(y) = \sum_{n \in Z^3} P_n \hbar_n(y), \quad y \in R^3. \tag{4}$$

(ii) There exist constants $0 < C_1 \le C_2 < \infty$ such that, for any constant matrix sequcnce $\{P_n\}_{n\in Z^3}$, we have,

$$C_1 \|\{P_n\}\|_\dagger \le \left\| \sum_{n\in Z^3} P_n(y)\hbar_n(y) \right\|_2 \le C_2 \|\{P_n\}\|_\dagger \tag{5}$$

Where $\|\{P_n\}\|_\dagger$ denotes the norm of sequence $\{P_n\}_{n\in Z^3}$

We consider any functions $\hbar(x), \lambdabar(x)$ in space $L^2(R^3)$ with the inner product:

$$\langle \lambdabar, \hbar \rangle = \int_{R^2} \lambdabar(x)\, \overline{\hbar(x)}\, dx\,, \quad \hat{\hbar}(\omega) = \int_{R^2} \hbar(x) e^{-x\cdot\omega} dx\,.$$

As usual, $\hat{\hbar}(\omega)$ denotes the Fourier transform of any function $\hbar(x) \in L^2(R^3)$.

2 Frames and Vector-Valued Multiresolution Analysis

Definition 3. Let Ω be a separable Hilbert space and Λ be an index set. We recall that a sequence $\{\Gamma_\iota : \iota \in \Lambda\} \subseteq \Omega$ is a frame for Ω if there exist two positive constants A_1, A_2 such that

$$\forall\, \xi \in \Omega, \quad A_1 \|\xi\|^2 \le \sum_{\iota\in\Lambda} |\langle \xi, \Gamma_\iota \rangle|^2 \le A_2 \|\xi\|^2, \tag{6}$$

where $\|\xi\|^2 = \langle \xi, \xi \rangle$, and A_1, A_2 are called frame bounds. A sequence $\{\hbar_\iota : \iota \in \Lambda\} \subseteq \Omega$ is a tight one if we can choose $A_1 = A_2$. A frame $\{\hbar_\iota : \iota \in \Lambda\}$ is an exact frame if it ceases to be a frame when any one of its elements is removed. If $A_1 = A_2 = 1$, then it follows from (1) that

$$\forall\, \xi \in \Omega, \quad \xi = \sum_{\iota\in\Lambda} \langle \xi, \Gamma_\iota \rangle \Gamma_\iota$$

A sequence $\{\Gamma_\iota : \iota \in \Lambda\} \subseteq \Omega$ is a Bessel sequence if (only) the upper inequality of (1) follows. If only for all $\hbar \in U \subset \Omega$, the upper inequality of (1) holds, the sequence $\{\Gamma_\iota : \iota \in \Lambda\} \subseteq \Omega$ is a Bessel sequence with respect to (w.r.t.) U. Moreover, we assume that $U \subset \Omega$ is a closed subspace, if for all $\lambdabar \in U$, there exist two real numbers $A_1, A_2 > 0$ such that $A_1 \|\lambdabar\|^2 \le \sum_{\iota\in\Lambda} |\langle \lambdabar, \Gamma_\iota \rangle|^2 \le A_2 \|\lambdabar\|^2$, the sequence $\{\Gamma_\iota : \iota \in \Lambda\} \subseteq \Omega$ is called an $U-$ subspace frame. In this section, we introduce the notion of vector-valued multiresolution analysis.

Definition 4. A vector-valued multiresolution analysis of $L^2(R^3, C^s)$ is a nested sequence of closed subspaces $\{X_j\}_{j\in Z}$ such that (i) $X_j \subset X_{j+1}$, $j \in Z$; (ii) $\bigcap_{j\in Z} X_j = \{O\}$ and $\bigcup_{j\in Z} X_j$ is dense in $L^2(R^3, C^s)$, where O denotes $s \times s$ zero matrix; (iii) $\hbar(y) \in X_j \Leftrightarrow \hbar(ay) \in X_{j+1}, \forall j \in Z$; (iv) there is a vector-valued function $G(y)\in X_0$ such that $\{G_n(y):=G(y-n):n\in Z^3\}$ form a Riesz basis for subspace $G(y)\in X_0$ .Since $G(y)\in X_0\subset X_1$, by definition and (4) there exist a finite supported constant $s\times s$ matrix sequence $\{M_n\}_{n\in Z^3}$ such that

$$G(y)=\sum_{n\in Z^3} M_n G(ay-n), \tag{7}$$

$$\widehat{G}(a\omega)=\mathrm{M}(\omega)\widehat{G}(\omega), \omega\in R^3, \tag{8}$$

where $\mathrm{M}(\omega) = \frac{1}{a^2}\sum_{n\in Z^2} M_n\cdot e^{-i\langle n,\omega\rangle}$. Equation (7) is called a refinement equation and $G(y)$ a vector-valued scaling function. Let $U_j, j\in Z$ be the direct complementary subspace of subspace X_j in subspace X_{j+1} and there exist a^3-1 vector-valued fun -ction $\Phi_\rho(y)\in L^2(R^3,C^s), \rho\in\Omega$, such that the translates and dilations of $\Phi_\rho(y)$ form a Riesz basis of U_j, i.e.

$$U_j = clos_{L^2(R^3,C^s)}\langle\Phi_\rho(a^j\cdot - n): n\in Z^3, \rho\in\Omega\rangle. \tag{9}$$

Since $\Phi_\rho(y)\in U_0\subset V_1, \rho\in\Omega$ there exist a^2-1 finite supported constant matrix sequences $\{B_n^\rho\}_{n\in Z^3}$ such that

$$\Phi_\rho(y)=\sum\nolimits_{n\in Z^3} B_n^\rho f(ay-n),\ \ \rho\in\Omega, \tag{10}$$

The vector-valued functions $f(x)\in L^2(R^2,C^s)$ is said to be an orthogonal one, if

$$\langle G(\cdot), G(\cdot-n)\rangle = \delta_{0,n} I_s, n\in Z^3, \tag{11}$$

We say that $\Phi_\rho(y)\in L^2(R^3,C^s), \rho\in\Omega$ are orthogonal vector-valued wavelets associated with an orthogonal vector-valued scaling functions $\Phi_\rho(y)$ if $\{\Phi_\rho(y-n), n\in Z^3, \rho\in\Omega\}$ is a Riesz basis of U_0 , and

$$\langle G(\cdot), \Phi_\rho(\cdot-n)\rangle = 0, \rho\in\Omega, n\in Z^3. \tag{12}$$

$$\langle\Phi_\rho(\cdot), \Phi_\rho(\cdot-n)\rangle = \delta_{\rho,\mu}\delta_{0,n} I_s, \rho,\mu\in\Omega. \tag{13}$$

3 The Orthogonality Traits of Vector-Valuedwavelet Wraps

To introduce the vector-valued wavelet packets, we set

$\Psi_0(y)=G(y), \Psi_\rho(y)=\Phi_\rho(y)$, $P_n^{(0)}=M_n, P_n^{(\rho)}=B_n^{(\rho)}, \rho\in\Omega, n\in Z^3.$

Then, the equations (7) and (10), which $G(y)$ and $\Phi_\rho(y)$ satisfy can be written as

$$\Psi_\rho(y) = \sum\nolimits_{n\in Z^3} P_n^{(\rho)} \Psi_{\underline{0}}(ay-n), \rho \in \Omega_0. \quad (14)$$

For any α and the given vector-valued orthogonal scaling function $G(y)$, let

$$\Psi_\alpha(y) = \Psi_{a\beta+\lambda}(y) = \sum_{n\in Z^3} P_n^{(\lambda)} \Psi_\beta(ay-n), \quad \lambda \in \Omega_0.\ \lambda \in \Omega_0 \quad (15)$$

Definition 5. The family of vector-valued functions $\{\Psi_{a\beta+\lambda}(y), \beta \in Z_+^2, \lambda \in \Omega_0\}$ is called vector-valued wavelet packs with respect to the vector-valued scaling function $f(y)$ where $\Psi_{a\beta+\lambda}(y)$ is given by (16).

$$\Psi_\alpha(x) = \Psi_{a\sigma+\mu}(x) = \sum_{k\in Z^2} Q_k^{(\mu)} \Psi_\sigma(ax-k), \quad \mu \in \Gamma_0, \quad (16)$$

where $\sigma \in Z_+^2$ is the unique element such that $\alpha = a\sigma + \mu$, $\mu \in \Gamma_0$ follows.

Lemma 1[4]. Let $F(x), \tilde{F}(x) \in L^2(R^3, C^v)$. Then they are biorthogonal if and only if

$$\sum_{k\in Z^3} \hat{F}(\gamma + 2k\pi)\hat{\tilde{F}}(\gamma + 2k\pi)^* = I_s. \quad (17)$$

Lemma 2[6]. Assume that $\Psi_\mu(x) \in L^2(R^3, C^s)$, $\mu \in \Gamma$ an orthogonal vector-valued wavelets associated with orthogonal scaling functions $\Psi_{\underline{0}}(x)$. Then, for $\mu, \nu \in \Omega_0$, we have

$$\sum_{\rho\in\Omega_0} \mathcal{P}^{(\mu)}((\gamma + 2\rho\pi)/a)\mathcal{P}^{(\nu)}((\gamma + 2\rho\pi)/a)^* = \delta_{\mu,\nu} I_s. \quad (18)$$

Lemma 3[6]. Suppose $\{\Psi_\alpha(x),\ \alpha \in Z_+^2\}$ are wavelet packets with respect to orthogonal vector-valued functions $\Psi_{\underline{0}}(x)$. Then, for $\alpha \in Z_+^3$, we have

$$[\Psi_\alpha(\cdot), \Psi_\alpha(\cdot - k)] = \delta_{\underline{0},k} I_s,\ k \in Z^3. \quad (19)$$

Theorem 1[8] Assume that $\{\Psi_\alpha(x),\ \alpha \in Z_+^3\}$ are wavelet packets with respect to orthogonal vector-valued functions $\Psi_{\underline{0}}(x)$. Then, for $\beta \in Z_+^3, \mu, \nu \in \Omega_0$, we have

$$\langle \Psi_{a\beta+\mu}(\cdot), \Psi_{a\beta+\nu}(\cdot - m)\rangle = \delta_{\underline{0},k}\delta_{\mu,\nu} I_s,\ m \in Z^3. \quad (20)$$

Theorem 2. If $\{G_\beta(x), \beta \in Z_+^3\}$ is vector-valued wavelet wraps with respect to a pair of biorthogonal vector scaling functions $G_0(x)$, then for any $\alpha, \sigma \in Z_+^3$, we have

$$[G_\alpha(\cdot), \tilde{G}_\sigma(\cdot - k)] = \delta_{\alpha,\sigma}\delta_{\underline{0},k} I_v,\ k \in Z^3. \quad (21)$$

Proof. When $\alpha = \sigma$, (21) follows by Lemma 3. as $\alpha \neq \sigma$ and $\alpha, \sigma \in \Gamma_0$, it follows from Lemma 4 that (21) holds, too. Assuming that α is not equal to β, as well as at least one of $\{\alpha,\sigma\}$ doesn't belong to Γ_0, we rewrite α,σ as $\alpha = a\alpha_1 + \rho_1$, $\sigma = a\sigma_1 + \mu_1$, where $\rho_1, \mu_1 \in \Gamma_0$.

Case 1. If $\alpha_1 = \sigma_1$, then $\rho_1 \neq \mu_1$. (21) follows by virtue of (19), (20) as well as Lemma 1 and Lemma 2, i.e.,

$$[G_\alpha(\cdot),\tilde{G}_\sigma(\cdot-k)] = \frac{1}{(2\pi)^3}\int_{R^3}\hat{G}_{a\alpha_1+\rho_1}(\gamma)\hat{\tilde{G}}_{a\sigma_1+\mu_1}(\gamma)^* \cdot \exp\{ik\cdot\gamma\}d\gamma$$

$$= \frac{1}{(2\pi)^3}\int_{[0,2\pi]^3}\delta_{\rho_1,\mu_1} I_s \cdot \exp\{ik\cdot\gamma\}d\gamma = O.$$

Case 2. If $\alpha_1 \neq \sigma_1$, order $\alpha_1 = a\alpha_2 + \rho_2$, $\sigma_1 = a\sigma_2 + \mu_2$, where $\alpha_2, \sigma_2 \in Z_+^3$, and $\rho_2, \mu_2 \in \Gamma_0$. If $\alpha_2 = \sigma_2$, then $\rho_2 \neq \mu_2$. Similar to Case 1, (21) follows. As $\alpha_2 \neq \sigma_2$, order $\alpha_2 = a\alpha_3 + \rho_3$, $\sigma_2 = a\sigma_3 + \mu_3$, where $\alpha_3, \sigma_3 \in Z_+^3$, $\rho_3, \mu_3 \in \Omega_0$. Thus, taking finite steps (denoted by κ), we obtain $\alpha_\kappa \in \Gamma_0$, and $\rho_\kappa, \mu_\kappa \in \Gamma_0$.

$$8\pi^2[G_\alpha(\cdot),\tilde{G}_\sigma(\cdot-k)] = \int_{R^3}\hat{G}_\alpha(\gamma)\hat{\tilde{G}}_{\sigma_1}(\gamma)^* \cdot e^{ik\cdot\gamma}d\gamma$$

$$= \int_{R^3}\hat{G}_{a\alpha_1+\lambda_1}(\gamma)\hat{\tilde{G}}_{a\beta_1+\mu_1}(\gamma)^* \cdot \exp\{ik\cdot\gamma\}d\gamma = \cdots\cdots\cdots\cdots$$

$$= \int_{[0,2\cdot a^\kappa\pi]^2}\{\prod_{l=1}^{\kappa}\mathcal{Q}^{(\rho_l)}(\gamma/a^l)\}\cdot O\cdot\{\prod_{l=1}^{\kappa}\tilde{\mathcal{Q}}^{(\mu_l)}(\gamma/a^l)\}^* \cdot \exp\{-ik\cdot\gamma\}d\gamma = O.$$

Therefore, for any $\alpha,\sigma \in Z_+^3$, result (21) is established.

For any $v = (v_1, v_2, v_3) \in Z^3$, the translation operator S is defined to be $(S_{va}\lambda)(x) = \lambda(x - va)$, where a is a pasitive constant real number.

Theorem 3[7]. Let $\phi(x), \tilde{\phi}(x), \hbar_\iota(x)$ and $\tilde{\hbar}_\iota(x)$, $\iota \in J$ be functions in $L^2(R^3)$. Assume that conditions in Theorem 1 are satisfied. Then, for any function $f(x) \in L^2(R^3)$, and any integer n, we have

$$\sum_{u\in Z^3}\langle f,\tilde{\phi}_{n,u}\rangle\phi_{n,u}(x) = \sum_{\iota=1}^{15}\sum_{s=-\infty}^{n-1}\sum_{u\in Z^3}\langle f,\tilde{\hbar}_{\iota:s,u}\rangle\hbar_{\iota:s,u}(x). \tag{22}$$

References

1. Telesca, L., et al.: Multiresolution wavelet analysis of earthquakes. Chaos, Solitons & Fractals 22(3), 741–748 (2004)
2. Iovane, G., Giordano, P.: Wavelet and multiresolution analysis: Nature of ε^∞ Cantorian space-time. Chaos, Solitons & Fractals 32(4), 896–910 (2007)
3. Zhang, N., Wu, X.: Lossless Compression of Color Mosaic Images. IEEE Trans. Image Processing 15(16), 1379–1388 (2006)
4. Chen, Q., et al.: Biorthogonal multiple vector-valued multivariate wavelet packets associated with a dilation matrix. Chaos, Solitons & Fractals 35(2), 323–332 (2008)

5. Shen, Z.: Nontensor product wavelet packets in $L_2(R^2)$. SIAM Math. Anal. 26(4), 1061–1074 (1995)
6. Chen, Q., Qu, X.: Characteristics of a class of vector-valued nonseparable higher-dimensional wavelet packet bases. Chaos, Solitons & Fractals 41(4), 1676–1683 (2009)
7. Chen, Q., Shi, Z.: Construction and properties of orthogonal matrix-valued wavelets and wavelet packets. Chaos, Solitons & Fractals 37(1), 75–86 (2008)
8. Chen, Q., Huo, A.: The research of a class of biorthogonal compactly supported vector-valued wavelets. Chaos, Solitons & Fractals 41(2), 951–961 (2009)
9. Chen, Q., Wei, Z.: The characteristics of orthogonal trivariate wavelet packets. Information Technology Journal 8(8), 1275–1280 (2009)

The Nice Features of Two-Direction Poly-scale Trivariate Small-Wave Packages with Finite Support

Bingqing Lv[1] and Jing Huang[2]

Department of Basic Course, Dongguan Polytechnic, Dongguan 523808, China
[1]zas123qwe@126.com, [2]nysslt88@126.com

Abstract. The rise of wavelet analysis in applied mathematics is due to its applications and the flexibility. In this article, the notion of biorthogonal two-directional compactly supported trivariate wavelet wraps with poly-scale is developed. Their properties are investigated by algebra theory, means of time-frequency analysis method and, operator theory. An approach for designing a sort of affine triariate dual frames in three-dimensional space is presented. The direct decomposition relationship is provided. In the final, new Riesz bases of space $L^2(R^3)$ are constructed from these wavelet packets. Moreover, it is shown how to draw new Riesz bases of space $L^2(R^3,C^v)$ from these wavelet wraps.

Keywords: two-direction, trivariate, small-wave packages, Riesz bases, iterative approach, operator frame, Riesz representation thorem.

1 Introduction

At present, image interpolation algorithms based on wavelet transform are mainly based on multiresolution analysis of the wavelet. Raditionally, short-time Fourier Transform and Gabor Transform were used for harmonic studies of nonstationary power system waveforms which are basically Fourier Transform-based methods. To overcome the limitations in these existing methods, wavelet transform based algorithm has been developed to estimate the limitations the frequency and time information of a harmonic signal. Multiwavelets can simultaneously possess many desired properties such as short support, orthogonality, symmetry, and vanishing moments, which a sin-gle wavelet cannot possess simultaneously. Already they have led to exciting applications in signal analysis [1], fractals [2] and image processing [3],and so on. Vector-valued wavelets are a sort of special multiwavelets Chen [4] introduced the notion of orthogonal vector-valued wavelets. However, vector-valued wavelets and multiwavelets are different in the following sense. Pre-filtering is usually required for discrete multiwavelet transforms [5] but not necessary for discrete vector-valued transforms. Wavelet wraps, owing to their nice characteristics, have been widely applied to signal processing [6],code theory, image compression, solving integralequation and so on. Coifman and Meyer firstly introduced the notion of univariate orthogonal wavelet wraps. Yang [7] constructed a-scale orthogonal multiwavelet wraps that were more flexible in applications. It is known that the majority of information is multidimensional information. Shen [8] introduced multivariate orthogonal wavelets which may be used in a wider field. Thus, it is

D. Jin and S. Lin (Eds.): Advances in MSEC Vol. 1, AISC 128, pp. 553–559.
springerlink.com

necessary to generalize the concept of multivariate wavelet wraps to the case of multivariate vector-valued wavelets. The goal of this paper is to give the definition and the construction of biorthogonal vector-va valued wavelet wraps and designt new Riesz bases of $L^2(R^3,C^v)$.

2 The Preliminaries on Vector-Valued Function Space

Let Z and Z+ stand for all integers and nonnegative integers, respectively. The multiresolution analysis is one of the main approaches in the construction of wavelets. Let us introduce two-direction multiresolution analysis and two-direction wavelets.

Definition 1. We say that $f(x)\in L^2(R^3)$ is a two-direction refinable function if $f(x)$ satisfies the following two-direction refinable equation:

$$f(x)=\sum_{v\in Z^3} b_v^+ f(4x-v)+\sum_{u\in Z^3} b_v^- f(u-4x), \tag{1}$$

where the sequences $\{b_v^+\}_{v\in Z^3}\in l^2(Z^3)$ and $\{b_v^-\}_{v\in Z^3}\in l^2(Z^3)$ are also called positive-direction mask and negative-direction mask, respectively. If all negative-direction mask are equal to 0, then two-direction refinable equation (2) become two-scale refinable equation (1). By taking the Fourier transform for the both sides of (2), we have

$$\widehat{f}(\gamma)=b^+\left(e^{-i\gamma/4}\right)\widehat{f}(\gamma/4)+b^-\left(e^{-i\gamma/4}\right)\overline{\widehat{f}(\gamma/4)}, \tag{2}$$

where $b^+(z)=(1/64)\sum_{v\in Z^3} b_v^+ z^u$, $z=e^{-i\omega/4}$ is called positive-direction mask symbol,and $64b^-(z)=\sum_{v\in Z^3} b_v^- z^u$ is called negative-direction mask symbol. In order to investigate the existence of solutions of the two-direction refinable equation (2), we rewrite the two-direction refinable equation (2) as

$$f(-x)=\sum_{v\in Z^3} b_v^+ f(-4x-v)+\sum_{v\in Z^3} b_v^- f(v+4x), \tag{3}$$

By taking the Fourier transform for the both sides of (4), we have

$$\overline{f(\gamma)}=\overline{b^+(e^{-i\gamma/4})\widehat{f}(\gamma/4)}+\overline{b^-(e^{-i\gamma/4})}\widehat{f}(\gamma/4), \tag{4}$$

Form the refinement equation (3) and the refinement equation (5), we get that

$$\widehat{F}(\gamma)=\begin{bmatrix}\hat{\phi}(\gamma)\\ \hat{\phi}(\gamma)\end{bmatrix}=\begin{bmatrix}b^+\left(e^{-i\gamma/4}\right) & \overline{b^-\left(e^{-i\gamma/4}\right)}\\ b^+\left(e^{-i\gamma/4}\right) & b^-\left(e^{-i\gamma/4}\right)\end{bmatrix}\begin{bmatrix}\widehat{f}\left(e^{-i\gamma/4}\right)\\ \widehat{f}\left(e^{-i\gamma/4}\right)\end{bmatrix} \tag{5}$$

By virtue of the positive-direction mask $\{b_v^+\}_{v\in Z^3}$ and the negative-direction mask $\{b_v^-\}_{v\in Z^3}$, we construct the following matrix equation:

$$F(x)=\begin{bmatrix}f(-x)\\ f(x)\end{bmatrix}=\sum_{u\in Z^3}\begin{bmatrix}b_{-u}^- & b_{-u}^+\\ b_u^+ & b_u^-\end{bmatrix}F(4x-u) \tag{6}$$

Definition 2. A pair of two-direction function $f(x),\ \widetilde{f}(x) \in L^2(R^3)$ are biorthogonal ones, if their translate satisfy

$$< f(x),\ \widetilde{f}(x-k) > = \delta_{0,k},\quad k \in Z^3, \tag{7}$$

$$< f(x),\ \widetilde{f}(n-x) > = O,\quad n \in Z^3, \tag{8}$$

where $\delta_{0,k}$ is the Kronecker symbol. Define a sequence $V_j \in L^2(R^3)$ by

$$S_j = clos_{L^2(R^3)} \left\langle 2^j f(4^j \cdot -k), 2^j f(l - 4^j \cdot) : k, l \in Z \right\rangle,\quad j \in Z.$$

where "clos" denote the closure of a space by a function $f(x)$.

A two-direction multiresolution analysis is a nested sequence $\{S_j\}_{j \in Z}$ genera-ted by $f(t)$, if it satisfies: (*i*) $\cdots \subset S_{-1} \subset S_0 \subset S_1 \subset \cdots$; (ii) $\bigcap_{j \in Z} S_j = \{0\}$ $\bigcup_{j \in Z} S_j$ is dense in $L^2(R^3)$, where 0 is the zero vector of $L^2(R^3)$; (*iii*) $h(x) \in S_0$ if and only if $h(4^j x) \in V_j$, $j \in Z$; (*iv*)There exists $f(x) \in V_0$ Such that the sequence $\{f(x-k), f(n-x) : k, n \in Z\}$ is a Riesz basis of S_0.

Four two-directional functions $\psi_\iota(x)$ $(\iota \in \Lambda \triangleq \{1,2,3\})$ are called two-directional wavelets with scale four associated with the scaling function $f(t)$, if the family $\{\psi_\iota(x-k),\ \psi_\iota(n-x) :\ k, n \in Z^3, \iota = 1,2,3\}$ forms a Riesz basis of W_j, where $V_{j+1} = V_j \oplus W_j$, $j \in Z$, where $\oplus$ denotes the direct sum. Then $\psi_\iota(x)$ satisfies the following equation:

$$\psi_\iota(x) = \sum_{v \in Z^3} q_{v,\iota}^{+} f(4x - v) + \sum_{v \in Z^3} q_{v,\iota}^{-} f(v - 4x),\quad \iota \in \Lambda \tag{9}$$

Implementing the Fourier transform for the both sides of (10) gives

$$\widehat{f}_\iota(\gamma) = q_\iota^{+}(e^{-i\gamma/4}) \widehat{f}(\gamma/4) + q_\iota^{-}(e^{-i\gamma/4}) \overline{\widehat{f}(\gamma/4)},\ \iota \in \Lambda. \tag{10}$$

Similarly, there exist two seq.s $\{\tilde{b}_u^{+}\}_{u \in Z^3}, \{\tilde{b}_u^{-}\}_{u \in Z^3} \in l^2(Z^3)$, such that

$$\widetilde{f}(x) = \sum_{u \in Z^3} \tilde{b}_u^{+} \widetilde{f}(4x - u) + \sum_{u \in Z^3} \tilde{b}_u^{-} \widetilde{f}(u - 4x). \tag{11}$$

The Fourier transforms of refinable equation (12) becomes

$$\widehat{\widetilde{f}}(\gamma) = \widetilde{b}_k^{+}(e^{-i\gamma/4}) \widehat{\widetilde{f}}(\gamma/4) + \widetilde{b}_k^{-}(e^{-i\gamma/4}) \overline{\widehat{\widetilde{f}}(\gamma/4)}\cdot \tag{12}$$

Similarly, there also exist two sequences $\{\widetilde{q}_{u,\iota}^{+}\}, \{\widetilde{q}_{u,\iota}^{-}\} \in l^2(Z)$ so that

$$\widetilde{\psi}_{\iota}(x)=\sum_{u\in Z^3}\widetilde{q}^{+}_{u,\iota}\widetilde{\phi}(4x-u)+\sum_{u\in Z^3}\widetilde{q}^{-}_{u,\iota}\widetilde{\phi}(u-4x). \tag{13}$$

By taking the Fourier transforms for (14), for $\iota\in\Lambda$, we have

$$\widehat{\widetilde{\psi}}_{\iota}(4\gamma)=\widetilde{q}^{+}_{\iota}(e^{-i\gamma})\widehat{\widetilde{f}}(\gamma)+q^{-}_{\iota}(e^{-i\gamma})\overline{\widehat{\widetilde{f}}(\gamma)}. \tag{14}$$

We say that $\psi_{\iota}(t),\widetilde{\psi}_{\iota}(t)\in L^2(R)$ are pairs of biorthogonal two-direction wave-lets associated with a pair of biorthogonal two-direction scaling functions $f(x)$, $\widetilde{f}(x)\in L^2(R^3)$, if the family $\{\psi_{\iota}(x-n),\psi_{\iota}(n-x):n\in Z^3\}$ is a Riesz basis of subspace W_0, and they satisfy the following equations:

$$\left\langle f(x),\widetilde{\psi}_{\iota}(x-u)\right\rangle=\left\langle f(x),\widetilde{\psi}_{\iota}(u-x)\right\rangle=0,\ \iota\in\Lambda,u\in Z^3, \tag{15}$$

$$\left\langle \widetilde{f}(x),\psi_{\iota}(x-u)\right\rangle=\left\langle \widetilde{f}(x),\psi_{\iota}(u-x)\right\rangle=0,\quad \iota\in\Lambda,\ u\in Z^3, \tag{16}$$

$$\left\langle \widetilde{\psi}_{\iota}(x),\psi_{\nu}(x-u)\right\rangle=\delta_{u,0}\delta_{\nu,\iota},\ u\in Z^3,\ \iota,\nu\in\Lambda, \tag{17}$$

$$\left\langle \psi_{\iota}(x),\widetilde{\psi}_{\nu}(u-x)\right\rangle=\delta_{u,0}\delta_{\iota,\nu},\ u\in Z^3,\ \iota,\nu\in\Lambda. \tag{18}$$

By replace t by $-t$ in the refinement equation (10), we get

$$\psi_{\iota}(-x)=\sum_{v\in Z^3}q^{+}_{v,\iota}f(-4x-u)+\sum_{v\in Z^3}q^{-}_{v,\iota}f(u+4x). \tag{19}$$

The refinement equ. (10) and (20) lead to the following relation formula

$$\Psi_{\iota}(x)=\begin{bmatrix}\psi_{\iota}(-x)\\ \psi_{\iota}(x)\end{bmatrix}=\sum_{k\in Z^3}\begin{bmatrix}q^{-}_{-k,\iota} & q^{+}_{-k,\iota}\\ q^{+}_{k,\iota} & q^{-}_{k,\iota}\end{bmatrix}F(4x-k),\qquad \iota\in\Lambda. \tag{20}$$

The frequency fiele form of the relation formula (18) is

$$\widehat{\Psi}_{\iota}(\omega)=\begin{bmatrix}\widehat{\overline{\psi}}_{\iota}(\omega)\\ \widehat{\psi}_{\iota}(\omega)\end{bmatrix}=\begin{bmatrix}\overline{q^{-}_{\iota}(e^{-i\omega/5})} & \overline{q^{+}_{\iota}(e^{-i\omega/5})}\\ q^{+}_{\iota}(e^{-i\omega/5}) & q^{-}_{\iota}(e^{-i\omega/5})\end{bmatrix}\begin{bmatrix}\overline{\widehat{f}(\omega/5)}\\ \widehat{f}(\omega/5)\end{bmatrix},\ \iota\in\Lambda. \tag{21}$$

Similarly for $\widetilde{\psi}_{\iota}(t)$ $(\iota\in\Lambda)$ and two mask symbols $\{\widetilde{q}^{+}_{v,\iota}\},\{\widetilde{q}^{-}_{v,\iota}\}$, $\in l^2(Z^3)$ we have

$$\widetilde{\Psi}_{\iota}(x)=\begin{bmatrix}\widetilde{\psi}_{\iota}(-x)\\ \widetilde{\psi}_{\iota}(x)\end{bmatrix}=\sum_{v\in Z^3}\begin{bmatrix}\widetilde{q}^{-}_{-v,\iota} & \widetilde{q}^{+}_{-v,\iota}\\ \widetilde{q}^{+}_{v,\iota} & \widetilde{q}^{-}_{v,\iota}\end{bmatrix}F(4x-v),\qquad \iota\in\Lambda. \tag{22}$$

By taking the fourier transforms for the both sides (23), it follows that

$$\widehat{\Psi}_\iota(\omega) = \begin{bmatrix} \overline{\widehat{\psi}_\iota(\omega)} \\ \widehat{\psi}_\iota(\omega) \end{bmatrix} = \begin{bmatrix} \overline{\tilde{q}_\iota^-(e^{-i\omega/5}) \quad \tilde{q}_\iota^+(e^{-i\omega/5})} \\ \tilde{q}_\iota^+(e^{-i\omega/5}) \quad \ddot{q}_\iota^-(e^{-i\omega/5}) \end{bmatrix} \begin{bmatrix} \overline{\widehat{f}(\omega/5)} \\ \widehat{f}(\omega/5) \end{bmatrix}, \quad \iota \in \Lambda. \tag{23}$$

By $L^2(R^3, C^v)$, we denote the aggregate of all vector- valued functions $H(x)$, i.e., $L^2(R^3, C^v) := \{\hbar(x) = (h_1(x), \ h_2(x), \cdots, h_v(x))^T : \ h_l(x) \in L^2(R^3), \ l = 1, 2, \cdots, \ v\}$, where T means the transpose of a vector. Video images and digital films are examples of vector-valued functions where $h_l(x)$ in the above $\hbar(x)$ denotes the pixel on the l th column at the point x. For $\hbar(x) \in L^2(R^3, C^v)$, $\|H\|$ denotes the norm of vector -valued function $\hbar(x)$, i.e., $\|\hbar\| := (\sum_{l=1}^{v} \int_{R^3} | h_l(x) |^2 \, dx)^{1/2}$ .In the below * means the transpose and the complex conjugate, and its integration is defined to be

$$\int_{R^3} \hbar(x) dx = (\int_{R^3} h_1(x) dx, \int_{R^3} h_2(x) \, dx, \cdots\cdots, \int_{R^3} h_v(x) dx)^T.$$

The Fourier transform of $\hbar(x)$ is defined as $\hat{\hbar}(\gamma) := \int_{R^3} \hbar(x) \cdot e^{-ix\cdot\gamma} dx$, where $x \cdot \gamma$ denotes the inner product of real vectors x and γ. For $F, H \in L^2 \ (R^3, C^v)$, their *symbol in ner product* is defined by

$$[F(\cdot), H(\cdot)] := \int_{R^2} F(x) H(x)^* dx, \tag{24}$$

Since $F(x) \in Y_0 \subset Y_1$, by Definition 3 and (4) there exists a finitely supported sequence of constant $v \times v$ matrice $\{\Omega_n\}_{n \in Z^3} \in \ell^2(Z^3)^{v \times v}$ such that

$$F(x) = \sum_{n \in Z^3} \Omega_n F(4x - n). \tag{25}$$

Equation (6) is called a refinement equation. Define

$$64 \cdot \Omega(\gamma) = \sum\nolimits_{n \in Z^3} \Omega_n \cdot \exp\{-in \cdot \gamma\}, \quad \gamma \in R^3. \tag{26}$$

where $\Omega(\gamma)$, which is $2\pi Z^2$ fun., is called a symbol of $F(x)$. Thus, (26) becomes

$$\hat{F}(4\gamma) = \Omega(\gamma) \hat{F}(\gamma), \qquad \gamma \in R^3. \tag{27}$$

Let $X_j, j \in Z$ be the direct complementary subspace of Y_j in Y_{j+1}. Assume that there exist 63 vector-valued functions $\psi_\mu(x) \in L^2(R^2, C^v), \mu \in \Gamma = \{1, 2, \cdots, 64\}$ such that their translations and dilations form a Riesz basis of X_j, i.e.,

$$X_j = \overline{(span\{\Psi_\mu(4^j \cdot - n) : n \in Z^3, \mu \in \Gamma\})}, \qquad j \in Z. \tag{28}$$

Since $\Psi_\mu(x) \in X_0 \subset Y_1, \quad \mu \in \Gamma$, there exist 63 finitely supported sequences of constant $v \times v$ matrice $\{B_n^{(\mu)}\}_{n \in Z^4}$ such that

$$\Psi_\mu(x) = \sum_{n \in Z^3} B_n^{(\mu)} F(4x - n), \ \ \mu \in \Gamma. \tag{29}$$

We say that $\Psi_\mu(x), \tilde{\Psi}_\mu(x) \in L^2(R^3, C^v), \mu \in \Gamma$ are pairs of biorthogonal vector wavelets associated with a pairof biorthogonal vector scaling functions $F(x)$ and $\tilde{F}(x)$, if the family $\{\Psi_\mu(x-n), n \in Z^3, \mu \in \Gamma\}$ is a Riesz basis of subspace X_0, and

$$[F(\cdot), \tilde{\Psi}_\mu(\cdot - n)] = [\tilde{F}(\cdot), \Psi_\mu(\cdot - n)] = O, \quad \mu \in \Gamma, \ n \in Z^3, \tag{30}$$

$$[\tilde{\Psi}_\lambda(\cdot), \Psi_\mu(\cdot - n)] = \delta_{\lambda,\mu}\delta_{0,n}, \quad \lambda, \mu \in \Gamma, \ n \in Z^3. \tag{31}$$

$$X_j^{(\mu)} = \overline{Span\{\Psi_\mu(4^j \cdot - n) : n \in Z^3\}}, \mu \in \Gamma, j \in Z. \tag{32}$$

Similar to (5) and (9), there exist 64 finitely supported sequences of $v \times v$ complex constant matrice $\{\tilde{\Omega}_k\}_{k \in Z^3}$ and $\{\tilde{B}_k^{(\mu)}\}_{k \in Z^3}$, $\mu \in \Gamma$ such that $\tilde{F}(x)$ and $\tilde{\Psi}_\mu(x)$ satisfy the refinement equations:

$$\tilde{F}(x) = \sum\nolimits_{k \in Z^3} \tilde{\Omega}_k \tilde{F}(4x - k) \tag{33}$$

$$\tilde{\Psi}_\mu(x) = \sum_{n \in Z^3} \tilde{B}_n^{(\mu)} \tilde{F}(4x - n), \quad \mu \in \Gamma. \tag{34}$$

3 The Biorthogonality Features of a Sort of Wavelet Wraps

Denoting by $G_{\underline{0}}(x) = F(x), G_\mu(x) = \tilde{\Psi}_\mu(x), \tilde{G}_{\underline{0}}(x) = F(x), \ \tilde{G}_\mu(x) = \tilde{\Psi}_\mu(x), \ Q_k^{(\underline{0})} = \Omega_k$, $Q_k^{(\mu)} = B_k^{(\mu)}$, $\tilde{Q}_k^{(\underline{0})} = \tilde{\Omega}_k$, $\tilde{Q}_k^{(\mu)} = \tilde{B}_k^{(\mu)}, \mu \in \Gamma, k \in Z^3$,. For any $\alpha \in Z_+^3$ and the given vector biorthogonal scaling functions $G_{\underline{0}}(x)$ and $\tilde{G}_{\underline{0}}(x)$, iteratitively define,

$$G_\alpha(x) = G_{4\sigma+\mu}(x) = \sum_{k \in Z^3} Q_k^{(\mu)} G_\sigma(4x - k), \qquad \mu \in \Gamma_0, \tag{35}$$

$$\tilde{G}_\alpha(x) = \tilde{G}_{4\sigma+\mu}(x) = \sum_{k \in Z^3} \tilde{Q}_k^{(\mu)} \tilde{G}_\sigma(4x - k), \qquad \mu \in \Gamma_0. \tag{36}$$

where $\sigma \in Z_+^3$ is the unique element such that $\alpha = 4\sigma + \mu, \ \mu \in \Gamma_0$ follows.

Definition 3. We say that two families of vector-valued functions $\{G_{4\sigma+\mu}(x), \sigma \in Z_+^3, \mu \in \Gamma_0\}$ and $\{\tilde{G}_{4\sigma+\mu}(x), \ \sigma \in Z_+^3, \ \mu \in \Gamma_0\}$ are vector-valued wavelet packets with respect to a pair of biorthogonal vector-valued scaling functions $G_{\underline{0}}(x)$ and $\tilde{G}_{\underline{0}}(x)$, resp., where $G_{4\sigma+\mu}(x)$ and $\tilde{G}_{4\sigma+\mu}(x)$ are given by (35) and (36), respectively.

Theorem 1[8]. Assume that $\{G_\beta(x), \beta \in Z_+^3\}$ and $\{\tilde{G}_\beta(x), \beta \in Z_+^3\}$ are vector-valued wavelet packets with respect to a pair ofbiorthogonal vector-valued functions $G_{\underline{0}}(x)$ and $\tilde{G}_{\underline{0}}(x)$, respectively. Then, for $\beta \in Z_+^3, \mu, v \in \Gamma_0$, we have

$$[G_{4\beta+\mu}(\cdot), \tilde{G}_{4\beta+v}(\cdot - k)] = \delta_{\underline{0},k}\delta_{\mu,v} I_v, \ k \in Z^3. \tag{37}$$

Theorem 2[8]. If $\{G_\mu(x), \mu \in Z_+^3\}$ and $\{\tilde{G}_\mu(x), \mu \in Z_+^3\}$ are vector-valued wavelet wraps with respect to a pair of biorthogonal vector scaling functions $G_0(x)$ and $\tilde{G}_0(x)$, then for any $\alpha, \sigma \in Z_+^3$, we have

$$[G_\alpha(\cdot), \tilde{G}_\mu(\cdot - n)] = \delta_{\alpha,\mu}\delta_{\underline{0},n} I_v, \; n \in Z^3. \tag{38}$$

References

1. Telesca, L., et al.: Multiresolution wavelet analysis of earthquakes. Chaos, Solitons & Fractals 22(3), 741–748 (2004)
2. Iovane, G., Giordano, P.: Wavelet and multiresolution analysis:Nature of ε^∞ Cantorian space-time. Chaos. Solitons & Fractals 32(4), 896–910 (2007)
3. Zhang, N., Wu, X.: Lossless Compression of Color Mosaic Images. IEEE Trans. Image Processing 15(16), 1379–1388 (2006)
4. Chen, Q., Wei, Z.: The characteristics of orthogonal trivariate wavelet packets. Information Technology Journal 8(8), 1275–1280 (2009)
5. Chen, Q., Huo, A.: The research of a class of biorthogonal compactly supported vector valued wavelets. Chaos, Solitons & Fractals 41(2), 951–961 (2009)
6. Chen, Q., Cheng, Z., Feng, X.: Multivariate Biorthogonal Multiwavelet packets. Mathematica Applicata. 18(3), 358–364 (2005) (in Chinese)
7. Chen, Q., Shi, Z.: Construction and properties of orthogonal matrix-valued wavelets and wavelet packets. Chaos, Solitons & Fractals 37(1), 75–86 (2008)
8. Shen, Z.: Nontensor product wavelet packets in $L_2(R^2)$. SIAM Math. Anal. 26(4), 1061–1074 (1995)
9. Chen, Q., Shi, Z.: Biorthogonal multiple vector-valued multivariate wavelet packets associated with a dilation matrix. Chaos, Solitons & Fractals 35(2), 323–332 (2008)

The Excellent Traits of Multiple Dual-Frames and Applications

Kezhong Han

Information and Statistics Vocational College, Zhengzhou Institute of Aeronautical Industry Management, Zhengzhou 450011, P.R. China
hjk123zas@126.com

Abstract. The notion of a generalized multiresolution structure and the concept of subspace bivariate dual pseudoframes are introduced. It is proved that generalized frame operators for a Hilbert space $L^2(R^2)$ is algebra homomorphism. The pyramid decomposition scheme of a generalized multiresolution structure(GMRS) is proposed. By using the norm in $L^2(R^2)$ as a measure tool, pertubation of generalized frames and generalized Riesz bases are established. Moreover, affine bivariate dual-frame expansions of space $L^2(R^2)$ are constructed by virtue of the pyramid decomposition scheme.

Keywords: Bivariate dual-frames, orthogonal pseudoframes, Banach frames, pyramid decomposition scheme, Bessel sequence, frame operator, iteration.

1 Introduction

The nice advantage of wavelets is their time-frequency localization property. Recently, wavelet tight frames have attracted more and more attention, just because they have nice time-frequency localization property, shift-invariances, and more design freedom. Wavelet tight frames have been widely used in denoising and image processing. Tight frames generalize orthonormal systems. They preserve the unitary property or the relevant analysis ans synthesis operator. Frames are intermingled with exciting applications to physics, to engineering and to science in general. Frames didn't start in isolation, and even now in its maturity, surprising and deep connectionsto other areas continue to enrich the subject. The subjects are well explained, and they are all amenable to the kind of numerical methods where wavelet algorithms excel. Wavelet analysis is a particular time-or space-scale representation of signals which has found a wide range of applications in physics, signal processing and applied Mathematics in the last few years. In 1946 D. Gabor [1] filled this gap and formulated a fundamental app-roach to signal decomposiion in terms of elementary signals. Gabor's approach quickly become a paradigm for the spectral analysis associated with time-frequency methods. The frame theory has been one of powerful tools for researching into wavelets. In 1952, Duffin and Schaeffer [2] introduced the notion of frames for a separable Hilbert space. Later, Daubechies, Grossmann, Meyer, Benedetto, Ron revived the study of frames in [3], and since then, frames have become the focus of active research, both in theory and in applications,

D. Jin and S. Lin (Eds.): Advances in MSEC Vol. 1, AISC 128, pp. 561–567.
springerlink.com

such as signal processing, image processing and sampling theory. Every frame(or Bessel sequence) determines an analysis operator, the range of which is important for a lumber of applications. The notion of Frame multiresolution analysis as described by [4] generalizes the notion of multiresolution structure by all-owing non-exact affine frames. The rise of frame theory in applied mathematics is due to the flexibility and redundancy of frames, where robustness, error tolerance and noise suppression play a vital role [4,8]. The concept of frame multiresolution analysis (FMRA) as described in [8] generalizes the notion of MRA by allowing non-exact affine frames. It is well known that the majority of information is multi-dimensional information. Thus, it is significant generalize the concept of multivariate wavelet packets to the case of multivariate wavelets. Based on on some ideas from Shen [6], the goal of this paper is to discuss the properties of vector-valued multivariate wavelet packets. Inspired by [2] and [5], we introduce the notion of an 2-band generalized multiresolution structure(GMS) of space $L^2(R^2)$, which has a pyramid decomposition scheeme. It also lead to new constructions of affine frames of $L^2(R^2)$. Let Ω be a separable Hilbert space. We recall that a sequence is a frame for H if there exist positive real numbers L, B such that

$$\forall \lambda \in \Omega, \qquad L\|\lambda\|^2 \le \sum_\iota |\langle \lambda, \eta_\iota \rangle|^2 \le B\|\lambda\|^2 \tag{1}$$

A sequence $\{\eta_\iota\} \subseteq \Omega$ is a Bessel sequence if (only) the upper inequality of (1) holds. If only forall $\lambda \in X \subset U$, the upper inequality of (1) holds, the sequence $\{\eta_\iota\} \subseteq \Omega$ is a Bessel sequencewith respect to (w.r.t.) Ω. If $\{\eta_\iota\}$ is a frame, there exist a dual frame $\{\eta_\iota^*\}$ such that

$$\forall\, g \in \Omega, \quad g = \sum_\iota \langle g, \eta_\iota \rangle \eta_\iota^* = \sum_\iota \langle g, \eta_\iota^* \rangle \eta_\iota . \tag{2}$$

The Fourier transform of an integrable function $\hbar(x) \in L^1(R^2) \bigcap L^2(R^2)$ is defined by

$$F\hbar(\omega) = \hat{\hbar}(\omega) = \int_{R^2} \hbar(x)\, e^{-2\pi i x\omega} dx, \quad \omega \in R^2, \tag{3}$$

which, as usual, can be naturally extended to functions in the space $L^2(R^2)$. For a sequence $c = \{c(v)\} \in \ell^2(Z^2)$, we define its discrete-time Fourier transform as the function in $L^2(0,1)^2$ by

$$Fc(\omega) = C(\omega) = \sum_{v \in Z^2} c(v)\, e^{-2\pi i x\omega} dx\,, \tag{4}$$

Note that the discrete-time Fourier transform is 1-periodic. Let $T_v f(x)$ denote integer translates of a function $f(x) \in L^2(R^2)$, i.e $(T_v f)(x) = f(x - v)$, $f_{n,v}(x) = 2^n f(2^n x - v)$. Let $f(x) \in L^2(R^2)$ and and let $V_0 = span\{T_v f : v \in Z^2, f(x) \in L^2(R^2)\}$ be a closed subspace of space $L^2(R^2)$. Assume that $\mathrm{H}(\omega) := \sum_v |\hat{\hbar}(\omega + v)|^2 \in L^\infty[0,1]^2$.

In $[5]$, the sequence $\{T_v \hbar(x)\}_v$ is a frame for V_0 if and only if there exist positive constants L_1 and L_2 such that $L_1 \le \mathrm{H}(\omega) \le L_2$, a.e.,

$$\omega \in [0,1]^2 \setminus \mathbb{N}, \quad \mathbb{N} = \{\omega \in [0,1]^2 : \mathrm{H}(\omega) = 0\}. \tag{5}$$

2 Bivariate Frame Multiresolution Analysis and Dual-Frames

Let r be a positive integer, and $\Lambda = \{1, 2, \cdots, r\}$ be a finite index set. We consider the case of multiple generators, which yield multiple pseudoframes for subspaces of $L^2(R^2)$. In what follows, we consider the case of generators, which yield affine pseudoframes of integer grid translates for subspaces of $L^2(R^2)$. Let $\{T_v \phi_\iota\}$ and $\{T_v \widetilde{\phi}_\iota\}$ $(\iota \in \Lambda,\ v \in Z^2)$ be two sequences in $L^2(R^2)$. Let Ω be a closed subspace of $L^2(R^2)$. We say $\{T_v \phi_\iota\}$ for-ms a pseudoframe for the subspace Ω with respect to (w.r.t.) $\{T_v \widetilde{\phi}_\iota\}$ $(\iota \in \Lambda,\ v \in Z^2)$ if

$$\forall\, \hbar(x) \in \Omega, \qquad \hbar(x) = \sum_{\iota \in \Lambda} \sum_{v \in Z^2} \langle \hbar, T_v \phi_\iota \rangle T_v \tilde{\phi}_\iota(x). \tag{6}$$

It is important to note that ϕ_ι and $\tilde{\phi}_\iota$ need not be contained in Ω.Consequently,the positions of $T_v \phi_\iota$ and $T_v \tilde{\phi}_\iota$ are not generally commutable $[5]$, i.e., there exists $\Gamma(x) \in \Omega$ such that

$$\sum_{\iota \in \Lambda} \sum_{v \in Z^2} \langle \hbar, T_v \phi_\iota \rangle T_v \tilde{\phi}_\iota(x) \neq \sum_{\iota \in \Lambda} \sum_{v \in Z^2} \langle \hbar, T_v \tilde{\phi}_\iota \rangle T_v \phi_\iota(x) = \hbar(x). \tag{7}$$

Definition 1. We say that a bivariate generalized multiresolution structure (BGMS) of form $\{V_n, \phi_\iota(x), \tilde{\phi}_\iota(x)\}_{n \in Z, \iota \in \Lambda}$ of $L^2(R^2)$ is a sequence of closed linear subspaces $\{V_n\}_{n \in Z}$ of $L^2(R^2)$ and 2r elements $\phi_\iota(x), \tilde{\phi}_\iota(x) \in L^2(R^2)$ such that (i) $V_n \subset V_{n+1}$, $n \in Z$; (ii) $\bigcap_{n \in Z} V_n = \{0\}$; $\bigcup_{n \in Z} V_n$ is dense in $L^2(R^2)$; (iii) $g(x) \in V_n$ if and only f $g(2x) \in V_{n+1}$ $\forall n \in Z$ implies $T_v g(x) \in V_0$, for $v \in Z$; (v) $\{T_v \phi_\iota(x), \iota \in \Lambda, v \in Z^2\}$ forms an affine pseudoframe for V_0 with respect to $\{T_v \tilde{\phi}_\iota(x), \iota \in \Lambda, v \in Z^2\}$.A necessary and sufficient condition for the construction of an affine pseudoframe for Paley-Wiener subspaces is prese-nted as follows.

Proposition 1[6]. Let $f(x) \in L^2(R^2)$ satisfy $|\widehat{f}|$ a.e. on a connected neighbourhood of 0 in $\left[-\frac{1}{2}, \frac{1}{2}\right)^2$, and $|\widehat{f}| = 0$ a.e. otherwise. Define $\Xi \equiv \{\omega \in R^2 : |\hat{f}(\omega)| \geq C > 0\}$, and $V_0 = PW_\Xi = \{\phi \in L^2(R) : \mathrm{supp}(\hat{\phi}) \subseteq \Lambda\}$. Then for $\tilde{f} \in L^2(R^2)$, $\{T_v f, v \in Z^2\}$ is an affine pseudoframe for V_0 with respect to $\left\{T_v \widetilde{f}, v \in Z^2\right\}$ if and only if

$$\overline{\hat{f}(\omega)} \hat{\tilde{f}}(\omega) \chi_\Xi(\omega) = \chi_\Xi(\omega) \quad \text{a. e.}, \tag{8}$$

where χ_Λ is the characteristic function on Λ. Moreover, if $\widetilde{f}(\omega)$ is the above condi-tions then $\{T_v\phi, v\in Z^2\}$ and $\{T_v\widetilde{\phi}, v\in Z^2\}$ are a pair of commutative affine pseudoframes for V_0, i.e., $\forall\, \hbar\in V_0$,

$$\hbar(x)=\sum_{k\in Z^2}\langle \hbar, T_k\widetilde{f}\rangle T_k f(x)=\sum_{k\in Z^2}\langle \hbar, T_k f\rangle T_k \widetilde{f}(x). \qquad (9)$$

The filter banks associated with a GMS are presented as follows. Define filter functions $B_0(\omega)$ and $\widetilde{B}_0(\omega)$ by the relaton $B_0(\omega)=\sum_{v\in Z^2} b_0(v)e^{-2\pi i\omega}$ and $\widetilde{B}_0(\omega)=\sum_{v\in Z^2}\widetilde{b}_0(v)e^{-2\pi i\omega}$ of the sequences $b_0=\{b_0(v)\}$ and $\widetilde{b}_0=\{\widetilde{b}_0(v)\}$, respectively, wherever the sum is defined. Let $\{b_0(v)\}$ be such that $B_0(0)=2$ and $B_0(\omega)\neq 0$ in a neighborhoood of 0. Assume also that $|B_0(\omega)|\leq 2$. Then there exists $f(x)\in L^2(R^2)$ (see ref.[8]) such that

$$f(x)=2\sum_{v\in Z^2} b_0(v)f(2x-v). \qquad (10)$$

Similarly, there exists a scaling relationship for $\widetilde{f}(x)$ under the same conditions as that of b_0 for a sequence $\widetilde{b}_0=\{\widetilde{b}_0(v)\}$, i.e.,

$$\widetilde{f}(x)=2\sum_{v\in Z^2}\widetilde{b}_0(v)\widetilde{f}(2x-v) \qquad (11)$$

3 The Bivariate Affine Multiple Dual-Frames of Translate

We begin with introducing the concept of dual-frames of translates.

Definition 2. Let $\{T_{va}\Upsilon,\ v\in Z^2\}$ and $\{T_{va}\widetilde{\Upsilon},\ v\in Z^2\}$ be two sequences in $L^2(R^2)$. Let U be a closed subspace of $L^2(R^2)$. We say $\{T_{va}\Upsilon,\ v\in Z^2\}$ forms an affine pseudoframe for U with respect to $\{T_{va}\widetilde{\Upsilon},\ v\in Z^2\}$ if

$$\forall\, f(x)\in U,\ f(x)=\sum_{v\in Z^2}\langle f, T_{va}\widetilde{\Upsilon}\rangle T_{va}\Upsilon(x) \qquad (12)$$

Define an operator $K: U\to \ell^2(Z^2)$ by

$$\forall\, f(x)\in U,\quad Kf=\{\langle f, T_{va}\Upsilon\rangle\}, \qquad (13)$$

and define another operator $S:\ell^2(Z^2)\to W$ such that

$$\forall c=\{c(k)\}\in \ell^2(Z^2),\ Sc=\sum_{v\in Z^2} c(v)T_{va}\widetilde{\Upsilon}. \qquad (14)$$

Theorem 1. Let $\{T_{va}\Upsilon\}_{v\in Z^2} \subset L^2(R^2)$ be a Bessel sequence with respect to the subspace $U \subset L^2(R^2)$, and $\{T_{va}\widetilde{\Upsilon}\}_{v\in Z^2}$ is a Bessel sequence in $L^2(R^2)$. Assume that K be defined by (7), and S be defined by(8). Assume P is a projection from $L^2(R^2)$ onto U. Then $\{T_{va}\Upsilon\}_{v\in Z^2}$ is pseudoframes of translates for U with respect to $\{T_{va}\widetilde{\Upsilon}\}_{v\in Z^2}$ if and only if

$$KSP = P. \tag{15}$$

Proof. The convergence of all summations of (7) and (8) follows from the assumptions that the family $\{T_{va}\Upsilon\}_{v\in Z^2}$ is a Bessel sequence with respect to the subspace Ω, and he family $\{T_{va}\widetilde{\Upsilon}\}_{v\in Z^2}$ is a Bessel sequence in $L^2(R^2)$ with which the proof of the theorem is direct forward.

Proposition 2[5]. Let $\{T_{va}f\}_{v\in Z^2}$ be pseudoframes of translates for V_0 with respect to $\{T_{va}\widetilde{f}\}_{v\in Z^2}$. Define V_n by

$$V_n \equiv \{\Upsilon(x) \in L^2(R^2) : \Upsilon(x/4^n) \in V\}, \quad n \in Z, \tag{16}$$

Then, $\{f_{n,va}\}_{v\in Z^2}$ is an affine pseudoframe for V_n with respect to $\{\widetilde{f}_{n,va}\}_{v\in Z^2}$.

Definition 3. Let $\{V_\ell, f, \widetilde{f}\}$ be a given GMS. We say that the GMS has a pyramid decom-position scheme if there are band-pass functions $\Upsilon_l, \widetilde{\Upsilon}_l \in L^2(R)$, $l \in I$ such that

$$\forall \Phi(x) \in L^2(R^2), \sum_{v\in Z^2} \langle \Phi, \widetilde{f}_{1,va} \rangle f_{1,va} = \sum_{v\in Z^2} \langle \Phi, T_{va}\widetilde{f} \rangle T_v f + \sum_{l\in\Lambda}\sum_{v\in Z^2} \langle \Phi, T_{va}\widetilde{\Upsilon}_l \rangle T_{va}\Upsilon_l .$$

Theorem 2[6]. Let $\{V_n, f, \widetilde{f}\}$ be a given BGMS. Assume that integer grid translates of each, , $f, \widetilde{f}, \Upsilon_l, \widetilde{\Upsilon}_l$, are all Bessel seq. in $L^2(R)$, denoting $\underline{H}(\omega) = \sum_{v\in Z} |f(\omega+v)|^2$. Then (25) holds if and only if

$$\sum_{l=0}^{3} B_l(\omega)\overline{\widetilde{B}_l(\omega)\underline{H}(\omega)} = 4\underline{H}(\omega), \text{ a.e.} \tag{17-1}$$

$$\sum_{l=0}^{3} B_l(\omega+1/4)\overline{\widetilde{B}_l(\omega+1/4)\underline{H}(\omega)} = 0 \text{ a.e.} \tag{17-2}$$

$$\sum_{l=0}^{3} B_l(\omega+1/2)\overline{\widetilde{B}_l(\omega+1/2)\underline{H}(\omega)} = 0 \text{ a.e.} \tag{17-3}$$

$$\sum_{l=0}^{3} B_l(\omega+3/4)\overline{\widetilde{B}_l(\omega+3/4)\underline{H}(\omega)} = 0 \text{ a.e.} \tag{17-4}$$

Corollary 1. Let $\{V_n, f, \tilde{f}\}$ be a given GMS. Assume that functions $f, \tilde{f}, \Upsilon_l$ and $\tilde{\Upsilon}_l, l \in L$ in $L^2(R^3)$ are such that (26) holds. Then for any integers n and d with $n < d$, we have

$$\forall \Phi \in V_d, \Phi = \sum_v \langle \Phi, \tilde{f}_{n,va} \rangle f_{n,va} + \sum_{l \in I} \sum_{\sigma=n}^{d-1} \sum_v \langle \Phi, \tilde{\Upsilon}_{l:\sigma,va} \rangle \Upsilon_{l:\sigma,va} \tag{18}$$

Theorem 3. Let $f, \tilde{f}, \Upsilon_l, \tilde{\Upsilon}_l,\ l \in I$ are functions in $L^2(R)$. Assume that conditions in Theorem 4 are satisfied. Then, for all functions $\Phi \in L^2(R)$, and any integer σ,

$$\sum_v \langle \Phi, \tilde{f}_{\sigma,v} \rangle f_{\sigma,v} = \sum_{l=1}^{2} \sum_{n=-\infty}^{\sigma-1} \sum_{v \in Z} \langle \Phi, \tilde{\Upsilon}_{l:n,v} \rangle \Upsilon_{l:n,v}. \quad \text{in } L^2(R)$$

$$\forall \Phi \in L^2(R).\ \Phi = \sum_{l=1}^{2} \sum_{n=-\infty}^{\infty} \sum_{v \in Z} \langle \Phi, \tilde{\Upsilon}_{l:n,v} \rangle \Upsilon_{l:n,v} \quad \text{in } L^2(R)$$

Especially, if $\{\Upsilon_{l:n,v}\}_{v \in Z^2}$ and $\{\tilde{\Upsilon}_{l:n,v}\}_{v \in Z^2}$ are Bessel sequences, then they are a pair of affine frames.

Example 1. Let $H(t) \in L^2(R^3, C^v)$ be 4-coefficient orthogonal vector-valued scaling function satisfy the following equation:

$$H(t) = A_0 H(2t - \mu_0) + A_1 H(2t - \mu_1) + \cdots + A_3 H(2t - \mu_7), \quad \mu_0, \cdots \mu_7 \in Z^3$$

where $A_3 = O,\ A_0(A_3)^* = O,\quad A_0(A_0)^* + A_1(A_1)^* + A_2(A_2)^* + A_3(A_3)^* = 8I_3.$

$$A_0 = \begin{pmatrix} \frac{\sqrt{2}}{2} & \frac{\sqrt{2}}{2} & 0 \\ -\frac{1}{2} & \frac{\sqrt{2}}{3} & 1 \\ 0 & 0 & \frac{2\sqrt{3}}{3} \end{pmatrix},\ A_1 = \begin{pmatrix} 1 & 0 & 0 \\ 0 & \frac{\sqrt{2}}{6} & 0 \\ 0 & 0 & \frac{\sqrt{3}}{3} \end{pmatrix}, A_2 = \begin{pmatrix} \frac{\sqrt{2}}{2} & -\frac{\sqrt{2}}{2} & 0 \\ \frac{1}{2} & \frac{\sqrt{2}}{3} & -1 \\ 0 & 0 & \frac{2\sqrt{3}}{3} \end{pmatrix},$$

$$B_1^{(1)} = -diag(\sqrt{2}, \sqrt{106}/6, 2\sqrt{6}/3),\ B_1^{(2)} = diag(-\sqrt{2}, -\sqrt{106}/6, 2\sqrt{6}/3).$$

Applying Theorem 2, we obtain that $G_t(t) = B_0^{(t)} H(2t - \mu_0) + B_1^{(t)} H(2t - \mu_1) + \cdots + B_4^{(t)} H(2t - \mu_3), t = 1, 2, 3$ are orthogonal vector-valued wavelet functions associated with the orthogonal vector-valued scaling function.

4 Conclusion

A necessary and sufficient condition on the existence of a sort of orthogonal vector-valued trivariate wavelets is presented. An optimal algorithm is provided.

References

[1] Zhang, N., Wu, X.: Lossless of color masaic images. IEEE Trans. Image Delivery 15(6), 1379–1388 (2006)
[2] Shen, Z.: Nontensor product wavelet packets in $L_2(R^2)$. SIAM Math Anal. 26(4), 1061–1074 (1995)
[3] Xia, X.G., Suter, B.W.: Vector-valued wavelets and vector filter banks. IEEE Trans. Signal Processing 44(3), 508–518 (1996)
[4] Chen, Q., Qu, G.: Characteristics of a class of vector-valued nonseparable higher-dimensional wavelet packet bases. Chaos, Solitons & Fractals 41(4), 1676–1683 (2009)
[5] Chen, Q., Huo, A.: The research of a class of biorthogonal compactly supported vector-valued wavelets'. Chaos, Solitons & Fractals 41(2), 951–961 (2009)
[6] Chen, Q., Wei, Z.: The characteristics of orthogonal trivariate wavelet packets. Information Technology Journal 8(8), 1275–1280 (2009)
[7] Chen, Q., Shi, Z., Cao, H.: The characterization of a class of subspace pseudoframes with arbitrary real number translations. Chaos, Solitons & Fractals 42(5), 2696–2706 (2009)
[8] Chen, Q., Cao, H.: Construction and decomposition of biorthogonal vector-valued wavelets with compact support. Chaos, Solitons & Fractals 41(4), 2765–2778 (2009)

Finite-Time Control of Linear Discrete Singular Systems with Disturbances

Yaning Lin[1] and Fengjie An[2]

[1] School of Science, Shandong University of Technology, Zibo, 255049, People's Republic of China
yaning_lin@163.com

[2] The No. 4 Middle School of Jiaozhou, Qingdao, 266300, People's Republic of China
jzszafj@163.com

Abstract. The concept of finite-time boundedness for linear discrete singular system is induced in this paper. Finite-time control problem is considered for linear discrete singular systems with exogenous disturbances, of which the disturbance satisfies a dynamical system. By Lyapunov functional method and linear matrix inequality (LMI) technique, the sufficient conditions of finite -time boundedness (FTB) via state feedback controller for linear discrete singular systems are provided. Then, the conditions are translated to feasibility problems involving restricted linear matrix inequalities (LMIs). Finally, an example is given showing the effectiveness of the proposed method.

Keywords: finite-time boundedness (FTB), linear discrete singular systems, linear matrix inequality (LMI), exogenous disturbances.

1 Introduction

In the last decade, big effort has been spent on studying the robust stability problem for linear systems, the work of many control scientists and engineers has mainly focused on robust Lyapunov stability. In practice, a system could be stable but completely useless because it possesses undesirable transient performances. So, one should not only be interested in system stability (e.g. in the sense of Lyapunov), but also in transient performances, such as bounds of system trajectories. To study the transient performances of system, the concept of finite-time stability was proposed by Dorato[1]. Amato et al. [2] have extended the definition of finite-time stability (FTS) to the definition of finite-time boundedness (FTB) in order to take into account the presence of external disturbances. Some work has been done on the finite-time control of linear systems, such as [3, 4, 5]. Amato et al.[3] discussed the finite-time control problem of linear systems subject to time-varying parametric uncertainties and steady exogenous disturbances. Amato et al.[4] discussed the finite-time control problem of discrete time linear system. Amato et al.[5] provided a design method for a dynamic output feedback controller which makes the closed-loop system finite-time stable. [6~8] extended the concept of finite-time stability (FTS) and finite-time boundedness (FTB) for linear systems to linear singular systems, and solved the finite-time control problem of linear

D. Jin and S. Lin (Eds.): Advances in MSEC Vol. 1, AISC 128, pp. 569–573.
springerlink.com

singular systems. To the best of our knowledge, the finite-time control of linear discrete singular systems has not been investigated in the literature.

In this paper, based on [6], we first generalize the concept of finite-time boundedness (FTB) and finite-time stability (FTS) to linear discrete singular systems. By Lyapunov functional method and linear matrix inequality (LMI) technique, the sufficient conditions of finite-time boundedness (FTB) via state feedback controller for linear discrete singular systems are provided. Then, the conditions are translated to feasibility problems involving restricted linear matrix inequalities (LMIs). At last, an example shows the effectiveness of the proposed method.

2 Description of Problem and Preliminaries

Consider the following linear discrete singular systems

$$\begin{cases} Ex(k+1) = Ax(k) + Bu(k) + G\omega(k), \\ \omega(k+1) = F\omega(k), \end{cases} \tag{1}$$

where, $x(k) \in \mathbb{R}^n$ is the state, $u(k) \in \mathbb{R}^m$ is control input, $\omega(k) \in \mathbb{R}^l$ is the disturbance; E is a singular matrix with $rankE = r < n$, other matrices A, B, G, F are of appropriate dimensions. In this paper, we assume that the initial value $\omega(0)$ satisfies the constraint: $\omega^T(0)\omega(0) \le d$, $d \ge 0$.

The following definitions and lemmas are used in this paper.

Definition 1.[9] System $Ex(k+1) = Ax(k)$ is said to be causal, if $\deg ree(\det(zE - A)) = rankE$.

Definition 2. (FTB) The causal system

$$\begin{cases} Ex(k+1) = Ax(k) + G\omega(k), \\ \omega(k+1) = F\omega(k), \ \omega^T(0)\omega(0) \le d. \end{cases} \tag{2}$$

is said to be finite-time boundedness with respect to $(\delta_x, \varepsilon, R, d, N)$ with $R > 0$, $\varepsilon > \delta_x > 0$, $d > 0$, $N \in \mathbb{N}$. If $x^T(0)E^T REx(0) \le \delta_x^2$, then $x^T(k)E^T REx(k) \le \varepsilon^2$, for all $k \in \{1, 2, \ldots, N\}$.

Lemma 1. Linear discrete singular system $Ex(k+1) = Ax(k) + G\omega(k)$ is causal, if there exists a symmetric matrix $P \in \mathbb{R}^{n \times n}$ such that

$$E^T PE \ge 0, \quad A^T PA - E^T PE < 0. \tag{3,4}$$

The paper's aim is to find the state feedback controller $u(k) = Kx(k)$ such that the closed-loop system is causal and FTB with respect to $(\delta_x, \varepsilon, R, d, N)$.

3 Main Results

Theorem 1. The linear discrete singular system (2) is causal and FTB with respect to $(\delta_x, \varepsilon, R, d, N)$, if there exist positive matrices $Q_1 > 0,\ Q_2 > 0$ and a scalar $\gamma > 1$ satisfying

$$\begin{bmatrix} -\gamma E^T PE & 0 & A^T P \\ 0 & F^T Q_2 F - \gamma Q_2 & G^T P \\ PA & PG & -P \end{bmatrix} \le 0, \tag{5a}$$

$$\gamma^N (\lambda_{\max}(Q_1)\delta_x^2 + \lambda_{\max}(Q_2)d) < \lambda_{\min}(Q_1)\varepsilon^2, \tag{5b}$$

where $P = R^{\frac{1}{2}} Q_1 R^{\frac{1}{2}}$.

Proof. Assume $x^T(0)E^T REx(0) \le \delta_x^2$, $\omega^T(0)\omega(0) \le d$. Construct Lyapunov function as $V(x(k),\omega(k)) = x^T(k)E^T PEx(k) + \omega^T(k)Q_2\omega(k)$. By Schur complement, (5a) is equivalent to

$$\begin{bmatrix} A^T PA - \gamma E^T PE & A^T PG \\ G^T PA & G^T PG + F^T Q_2 F - \gamma Q_2 \end{bmatrix} \le 0. \tag{6}$$

(6) means that $A^T PA - \gamma E^T PE \le 0$. Considering $\gamma > 1$, $A^T PA - E^T PE < 0$. For P is a positive matrix, so $E^T PE \ge 0$. By Lemma 1, system (2) is causal. On the other hand, (6) means that

$$V(x(k+1), \omega(k+1)) \le \gamma V(x(k), \omega(k)).$$

Repeatedly,

$$V(x(k), \omega(k)) \le \gamma^k V(x(0), \omega(0)).$$

Noticed

$$V(x(k), \omega(k)) \ge \lambda_{\min}(Q_1) x^T(k)E^T REx(k)$$

and

$$V(x(0), \omega(0)) \le \lambda_{\max}(Q_1)\delta_x^2 + \lambda_{\max}(Q_2)d.$$

So, $x^T(k)E^T REx(k) \le \dfrac{\gamma^N(\lambda_{\max}(Q_1)\delta_x^2 + \lambda_{\max}(Q_2)d)}{\lambda_{\min}(Q_1)} \le \varepsilon^2$, for all $k \in \{1, 2, \cdots, N\}$. So, system (2) is FTB with respect to $(\delta_x, \varepsilon, R, d, N)$.

Using Theorem 1, we get our main conclusion.

Theorem 2. There exists a controller for system (1) such that the closed-loop is causal and FTB with respect to $(\delta_x, \varepsilon, R, d, N)$, if there exist positive matrices

$Q_1>0,\ Q_2>0$, nonsingular matrix L, matrix $\bar{K}$ and a scalar $\gamma>1$ satisfying (5b) and the following conditions

$$PB=BL,\begin{bmatrix} -\gamma E^T PE & 0 & A^T P+\bar{K}^T B^T \\ 0 & F^T Q_2 F-\gamma Q_2 & G^T P \\ PA+B\bar{K} & PG & -P \end{bmatrix}\le 0, \qquad (7a,7b)$$

where $P=R^{\frac{1}{2}}Q_1R^{\frac{1}{2}}$. And the feedback controller is $u(k)=L^{-1}\bar{K}x(k)$.

To solve (5b) and (7b), we need translate them to restrict linear matrix inequalities. For condition (5b), it is easy to check that condition (5b) is guaranteed by imposing the conditions

$$\lambda_1 I<Q_1<I,\ 0<Q_2<\lambda_2 I,\ \begin{bmatrix} \gamma^{-N}\lambda_1\varepsilon^2-d\lambda_2 & \delta_x \\ \delta_x & 1 \end{bmatrix}>0. \qquad (8a,8b,8c)$$

For condition (7b), since $rankE=r$, there exist nonsingular matrices C,D such that $CED=\begin{bmatrix} I_r & 0 \\ 0 & 0 \end{bmatrix}$. Letting $C^{-T}PC^{-1}=\begin{bmatrix} P_{11} & P_{12} \\ P_{12}^T & P_{22} \end{bmatrix}$, then

$$D^T E^T PED=D^T E^T C^T C^{-T} PC^{-1}CED=\begin{bmatrix} I_r & 0 \\ 0 & 0 \end{bmatrix}\begin{bmatrix} P_{11} & P_{12} \\ P_{12}^T & P_{22} \end{bmatrix}\begin{bmatrix} I_r & 0 \\ 0 & 0 \end{bmatrix}=\begin{bmatrix} P_{11} & 0 \\ 0 & 0 \end{bmatrix}.$$

Setting $E_1=D^{-T}\begin{bmatrix} 0 & 0 \\ 0 & -\alpha I_{n-r} \end{bmatrix}D^{-1}$, where α is some positive number, then,

$$-\gamma E^T PE+E_1=D^{-T}\begin{bmatrix} -\gamma P_{11} & 0 \\ 0 & -\alpha I_{n-r} \end{bmatrix}D^{-1}.$$

Obviously, if

$$\begin{bmatrix} -\gamma E^T PE+E_1 & 0 & A^T P+\bar{K}^T B^T \\ 0 & F^T Q_2 F-\gamma Q_2 & G^T P \\ PA+B\bar{K} & PG & -P \end{bmatrix}<0, \qquad (9)$$

then condition (7b) is satisfied. Therefore we have

Corollary 1. For system (1), there exists a controller such that the closed-loop sys -tem is causal and FTB with respect to $(\delta_x,\varepsilon,R,d,N)$, if there exist positive matrices $Q_1>0,\ Q_2>0$, nonsingular matrix L, matrix $\bar{K}$ and scalars $\gamma>1$, λ_i, $i=1,2$, satisfying condition (7a) and LMIs (8,9), where $P=R^{\frac{1}{2}}Q_1R^{\frac{1}{2}}$, matrices C, D

satisfying $CED = diag\{I_r, 0\}$, $E_1 = D^{-T}\begin{bmatrix} 0 & 0 \\ 0 & -\alpha I_{n-r} \end{bmatrix} D^{-1}$, α is some positive number. Then the feedback controller is $u(k) = L^{-1}\overline{K}x(k)$.

4 Example

Consider system (1) with

$$E = \begin{bmatrix} 1 & 0 \\ 0 & 0 \end{bmatrix},\ E_1 = \begin{bmatrix} 0 & 0 \\ 0 & -0.1 \end{bmatrix},\ A = \begin{bmatrix} 1 & 3 \\ -1 & -1 \end{bmatrix},$$

$$B = \begin{bmatrix} -1 & 1 \\ 0 & 1 \end{bmatrix},\ G = \begin{bmatrix} 6 & 0 \\ 0 & 4 \end{bmatrix},\ F = \begin{bmatrix} 0.1 & -0.5 \\ 0.02 & 0.4 \end{bmatrix}.$$

When $\delta_x = 1,\ \varepsilon = 10,\ R = I_2,\ d = 0.2,\ N = 3,\ \gamma = 2,$ we get

$$\overline{K} = \begin{bmatrix} 0.7387 & 1.3585 \\ 0.4228 & 0.3692 \end{bmatrix},\ L = \begin{bmatrix} 0.3099 & 0.1189 \\ -0.0298 & 0.4885 \end{bmatrix},\ \lambda_1 = 0.3,\ \lambda_2 = 10.5.$$

Then the finite-time state feedback controller for system (1) is $u(k) = \begin{bmatrix} 2 & 4 \\ 1 & 1 \end{bmatrix} x(k)$.

References

1. Dorato, P.: Short time stability in linear time-varying systems. In: Schrank, H.E. (ed.) IRE International Convention Record, Part 4., pp. 83–87. The Institute of Radio Engineers, Incopcrated, New York (1961)
2. Amato, F., Ariola, M., Abdallah, C.T., et al.: Dynamic output feedback finite-time control of LTI systems subject to parametric uncertainties and disturbances. In: Proc. European Control Conference, pp. 1176–1180. Springer, Berlin (1999)
3. Amato, F., Ariola, M., Dorato, P.: Finite-time control of linear systems subject to parameteric uncertainties and disturbances. Automatica 37, 1459–1463 (2001)
4. Amato, F., Ariola, M.: Finite-time control of discrete-time linear system. IEEE Trans on Automatic Control 50(5), 724–729 (2005)
5. Amato, F., Ariola, M., Dorato, P.: Finite-time stabilization via dynamic output feedback. Automatic 42, 337–342 (2006)
6. Feng, J.-E., Wu, Z., Sun, J.: Finite-time control of linear singular systems with parametric uncertainties and disturbances. Acta Automatica Sinica 31(4), 634–637 (2005)
7. Sun, J., Cheng, Z.: Finite-Time Control for One Kind of Uncertain Linear Singular Systems Subject to Norm Bounded Uncertainties. In: Proceedings of the 5th World Congress on Intelligent Control and Automation, vol. 6, pp. 980–984. Zhejiang University Press, Hangzhou (2004)
8. Sun, J., Cheng, Z.: Finite-time control for one kind of uncertain linear singular systems. Journal of Shandong University: Natural Science 39(2), 1–6 (2004)
9. Yang, D., Zhang, Q., Yao, B.: Descriptor Systems. Science Press, Beijing (2004)

Construction of Gray Wave for Dynamic Fabric Image Based on Visual Attention

Zhe Liu and Xiuchen Wang

Zhongyuan University of Technology, Zhongyuan Road 41,
Zhengzhou City, Henan Province, P.R. China
xyliuzhe@163.com

Abstract. According to the sense characteristic of image with human eyes, a new method is proposed in this paper. By this method, an image can be transfer to a sensory image which can produce rapid attention to visual. Firstly, a sensory threshold is set, the image is selected and segmented, and a sensory imge with only a few key points, which can show the characteristic change of image, is produced. Secondly, a function to fuse sensory image is given, and a sensory gray wave is formed. Finially, a trend coefficient is used to decrease the error produced by gray wave extraction. Experiments prove that this new method can transfer image better and more successfully.

Keywords: Vision, Fabric, Sense, Image, Sensory gray wave.

1 Introduction

The online detection of fabric defect is an important science problem to realieze the intelligent detection of fabric quality and has huge significance to improve enterprise efficiency. Up to now, there are not a mature online detection algorithm for fabric defect. Existing researches are forcuse on construction of static image feature of fabric. These methods are classified into two categories: spatial domain and frequency domain. In space domain, the image gray is extracted to show image feature. This analysis method includes Geometric method[1], Co-occurrence matrix [2], Neural network[3], Probabilistic statistic method[4], Fuzzy clustering[5], Local entropy[6], and Markov random field[7]. In frequency domain, fabric image is transfered to various spectrum to discribe fabric characteristic. This method includes Fourier transform[8], Gabor algorithm[9], and Wavelet transform[10].

Above methods have received some achievement of image analysis for static fabric. But there are many problems else, such as large calculation, high requirement for image quality, less classifications of recognition defects. Some algorithms, such as Neural network, have more steps like pretrainning step. In addition, the defect kinds of fabric are various and the image quality in dynamic state is worse than that in static state. Therefore, above algorithms are not suitable for online fabric detection. The main problems are as following: 1) how to decrease the calculation of algorithm to fit for the speed of online detection; 2) how to increase the classification of fabric defect of algorithm recognition to fit for defects change; 3) how to reduce the requirement of image quality to keep better accuracy in dynamic image recognition.

D. Jin and S. Lin (Eds.): Advances in MSEC Vol. 1, AISC 128, pp. 575–579.
springerlink.com

Therefore, a new method to construct a sensory gray wave of dynamic fabric image based on visual attention is proposed in this paper. This method can effectively produce a wave impacting vision in defect region when the fabric is moving at rapid speed. All this can provide a basis for further defect position and detection.

2 Model Establishment

2.1 Our Idea

In the research of recognition principle with human eyes, we summarized earlier results of characteristic study[11] and deeply knew appearence effect and principle with different characteristics. So we proposed the super fuzzy characteristic[12] to combine characteristics and explore the distribution rule of gray in defect region. In these researches, after a gray wave by gray value extraction of pixel points was constructed[13], we found that almost an abnormal wave mutation was procduced in corresponding position of defect. These mutations are not related to the defect classifications. As long as the defect procuced by some reason, these gray waves can correctly find the defect position. These gray waves are constructed by reading directly the gray value of pixel points of dynamic fabric image, and digital video are carry out simultaneously. The calculation is little, the speed is conform to the requirment of online detection. Therefore, our idea is that a sensory image is established on gray wave to construct an attracting model of dynamic fabric defect based on rapid sense.

2.2 Concrete Model

Let the size of pixel point in image window is $M \times N$, where ϖ is the mean of gray extreme. If σ exists and let

$$MAX_{\sigma} = MAX_{\varpi} \tag{1}$$

Then σ is considered as the threshold of sensory wave. MAX_{σ} denotes the number of extremum when the gray value of pixel point is large than the value of σ .The purpose of this step is to segment the image and decrease the pixel points under the condition of the result is not change.

Suppose μ_{σ} is the interval number of gray over the value of σ , the first sensory gray vave is extracted by the function as follow:

$$f = \frac{\sum_{j=1}^{M_{\sigma}} \frac{\sum_{i=1}^{N_{\sigma}} H_{i(j \times \mu_{\sigma}+1)}}{\frac{M}{\mu_{\sigma}} - 1}}{M_{\sigma} \times N_{\sigma}} \tag{2}$$

Where, the meaning of M_{σ}, N_{σ} and μ_{σ} are same, denoting the number of lateral low and vertical colum of pixel point when the gray value is greater than the value of σ .

To discribe the sensory gray wave, a gray trend coefficient k_g is introduced. This coefficient is to measure the gray value is increased or decreased with one side. It is calculated by

$$k_g = \frac{\sum_{i=1}^{\mathrm{int}(v/2)} \frac{T_i}{\mathrm{int}(v/2)}}{\sum_{i=\mathrm{int}(v/2)+1}^{v} \frac{T_i}{v-\mathrm{int}(v/2)}} \tag{3}$$

Where, T_i denotes the i th extremum in gray distribution wave, v denotes the number of the extreme, function int() is a rounding function. Taking value of the trend coefficient is [0,1], the bigger the value is, the more abnormal the image is.

3 Results and Analysis

3.1 Experiments

A Canon photography instrument is used to digitize fabric image, 20 kinds of fabric are used to test the method. Fabrics have different weave, they are plain, twill, and stain and so on. The resolution is set at 300 ppi, the size of image is made as $10cm \times 10cm$. The MATLAB7.0 is used to compile computer program. the sample of fabric is shown in Figure1(a), the sensory wave is shown in Figure1 (b), and the image of the sensory wave is shown in Figure1 (c).

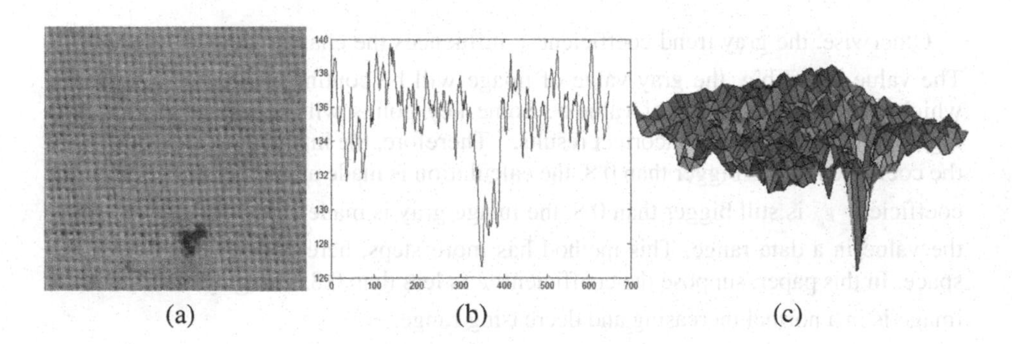

Fig. 1. Effect diagram of sensory gray wave

3.2 Analysis

It is important to determine the threshold value of sensory wave. if the value is too small, the pixel point will be more in processing image and the speed of calculation will decrease. If the value is big, many pixel points are filtered out and the result is not correct enough. Therefore, to ensure the value of σ in famula (1), we can use the formula as follow:

$$|MAX\_H_{\sigma} - MAX\_E_{\varpi}| < \varepsilon \tag{4}$$

Where, MAX_H_{σ} is the total number of pixel point when its gray value is greater than the value of σ, MAX_E_{ϖ} denotes the average of pixel gray in all coating region with gray extreme. Their values can be determined by the gray histogram shown in Figure 2, and it can be expressed by

$$\varepsilon = \frac{\sum_{i=1}^{3} num_{\min\_i}}{3} \tag{5}$$

Where, $num_{\min\_i}$ denotes the number of pixel point at the i th gray cluster.

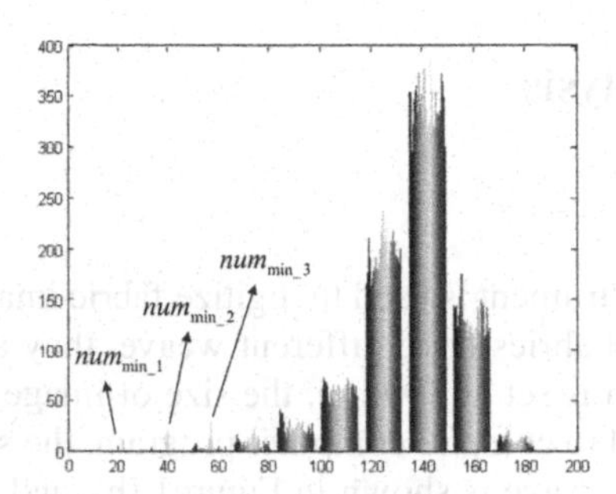

Fig. 2. Schematic diagram of value ε determination

Otherwise, the gray trend coefficient k_g influences the change trend of image gray. The value is too big, the gray value of image will be continue increase or decrease, which will cause the error of value σ. Some key points will be missed in low gray region, and will cause the incorrect results. Therefore, the image must be modified. If the coefficient k_g is bigger than 0.8, the calculation is made at another direction. If the coefficient k_g is still bigger than 0.8, the image gray is made along each row to made the value in a data range. This method has more steps, here is not listed for limited space. In this paper, suppose the coefficient k_g is less than 0.8. The gray distribution of image is in a normal increasing and decreasing range.

4 Results

In this paper, the sensory threshold is set to select and segment image with only a few key points according to the characteristic of sense image with human eyes. The sensory imge, which can show the characteristic change of image, is produced.The function to fuse sensory image,which can influence the image mutation better, is given. The trend coefficient is used to decrease the error produced by gray wave extraction. Experiments prove that this new method can transfer image to sensory gray image with obvious characteristic, it can extract the gray wave successfully and provide the basis for further all kinds of detection researches.

References

1. Wang, F., Jiao, G., Du, H.: Method of Fabric Defects Detection Based on Mathematical Morphology. Journal of Test and Measurement Technology 6, 515–518 (2007)
2. Zhou, C., Zhu, D., Xiao, L.: Textural defect detection based on label co-occurrence matrix. Journal of Huazhong University of Science and Technology 6, 25–28 (2006)
3. Liu, J., Zuo, B.: Application of BP nenral network on the idendification of fabric defects. Journal of Textile Research 9, 43–46 (2008)
4. Arunkumar, G.H.S., Eric, F.: Statistical approach to unsupervised defect detection and multiscale localization in two-texture images. Optical Engineering 2, 20–27 (2008)
5. Kuo, C.-F.J., Shih, C.-Y., et al.: Automatic Recognition of Fabric Weave Patterns by a Fuzzy C-Means Clustering Method. Textile Research Journal 2, 107–111 (2004)
6. Shen, Y., Liu, C., Wang, Z.: Defect detection of CMOS camera based on local entropy. Computer Engineering and Application 20, 71–73 (2009)
7. Kuo, C.-F.J., Chang, C.-D., et al.: Intelligence Control of On-line Dynamic Gray Cloth Inspecting Machine System Module Design. Fibers and Polymers 3, 394–402 (2009)
8. Tsai, D.-M., Kuo, C.-C.: Defect detection in inhomogeneously textured sputtered surfaces using 3D Fourier image reconstruction. Machine Vision and Applications 6, 383–400 (2007)
9. Mak, K.L., Peng, P.: An automated inspection system for textile fabrics based on Gabor filters. Robotics and Computer-Integrate Manufacturing 6, 359–369 (2008)
10. Li, L., Huang, X.: Realization of Orthogonal Wavelets Adapted to Fabric Texture for Defect Detection. Journal of Donghua University 2, 77–81 (2002)
11. Liu, Z.: Feature model of fabrics irregular defect. Key Engineering Materials 1, 566–571 (2001)
12. Liu, Z., Wang, X.: Image defect recognition based on super fuzzy characteristic. Journal of Multimedia 2, 181–188 (2010)
13. Wang, X., Liu, Z.: Pre-judgment model of image based on gray distribution wave. Applied Mechanics and Materials 6, 155–160 (2010)

Theoretical Framework of Responsible Leadership in China

XiaoLin Zhang and YangHua Lu

School of Management, Zhejiang University, 310000 Hangzhou, China
luyanghua@zju.edu.cn

Abstract. While the concept of responsible leadership has attracted scholarly interests for many years, there is no universally recognized definition. The current study represents the first attempt to understand the concept of responsible leadership, and forms a theoretical framework of responsible leadership for managers in the Chinese organizational context, from the perspective of Corporate Social Responsibility (CSR). Through a review of the literature, we constructed a theoretical framework of responsible leadership from the perspective of corporate social responsibility. The study creates a good start for the research on responsible leadership in Chinese organizational context, and has some important practical implication for management selection, assessment and training in China today.

Keywords: responsible leadership, corporate social responsibility, Chinese organizational context.

1 Introduction

The corporate scandals have triggered a broad discussion on the role of business in society. As a result, business leaders are increasingly held accountable for their actions and non-actions related to all stakeholders and society at large [1]. The enterprises face increasing pressure on corporate social responsibility and need to be more aware of the interests for the whole society [2]. Nowadays corporate social responsibility has drawn greater public attention, and banks or other institutional investors have reported social considerations to be a factor in their investment decisions [3]. Managers are responsible for formulating business strategies and are often deeply involved in promoting the image of their respective organizations through social responsibility practice. Furthermore, they may dramatically change the strategic direction of the organizations, including decisions pertaining to CSR [4]. Therefore the levels of managers' responsible leadership have great influence on their respective organizations.

Numerous researches have highlighted the importance of values-based leadership theory [5]. However, to the best of our knowledge, none of these studies specifically address the role of responsible leadership. The responsibility element is at the heart of what effective leadership is all about, and managers can not be effective as a leader without responsibility [6]. However, available leadership theory neither explicitly nor adequately address the nature and challenges of leadership that is both responsible and focused on performance [7].

D. Jin and S. Lin (Eds.): Advances in MSEC Vol. 1, AISC 128, pp. 581–585.
springerlink.com

2 The Concept of Responsible Leadership

Theoretically, responsible leadership draws from findings in leadership ethics, developmental psychology, psychoanalysis, stakeholder theory and systems theories [8]. Maak and Pless suggest that leaders need “relational intelligence” in order to connect and interact effectively and respectfully with people and stakeholders from various backgrounds, diverse culture and with different interests, inside and outside the organization [9], which means that responsible leaders need both emotional and ethical qualities to guide their action and behavior in interaction. They define responsible leadership as the art and ability of building, developing and sustaining trust relationships with all relevant stakeholders, coordinating responsible behavior to achieve a meaningful, shared corporate vision [10]. Maak also introduces "social capital" into the study of responsible leadership concept [11]. Pless further explores the relationship between responsible leadership behavior and the underlying motivational systems, with clinical and normative lenses [8].

The research on responsible leadership for managers in China is still at the embryonic stage. In order to study the concept of responsible leadership in China, we must be more aware of the unique organizational context in China. Chinese companies often use "emotional investment" to motivate their staff [12], and many managers pay more attention to the relationship benefit from their subordinates when assessing their subordinates' performance [13]. This phenomenon frequently occurs under the Chinese organizational context, and is closely related to specific organizational culture in China. Therefore, this study explores the interaction mechanism of responsible leadership, proposing a model of relationship among constructs, requisite inputs, and corresponding outcomes for responsible leadership in China, which could be applied to advance theory, research, education, and implementation.

3 Theoretical Framework of Responsible Leadership

Based on the previous research, we first started to develop a theoretical framework of responsible leadership. We aimed at depicting a proposed system of relationships between responsible leadership and input/output variables, and provided the rationale for the inclusion of critical inputs and propositions about their relationships to responsible leadership.

We had further excavation into the antecedent variables of responsible leadership from the perspective of leaders' characteristics, explored the factors that leads to different level of responsible leadership. Also, we went into the consequences of responsible leadership from the perspective of corporate social responsibility, to identify the impact of responsible leadership on organizations. Therefore this study analyzed and integrated existing literatures related to responsible leadership, and put forward a theoretical framework of responsible leadership especially from the perspective of corporate social responsibility.

3.1 Characteristics of Leaders

Researchers have cited this proposed input as a prerequisite to become a responsible leader. Top management's character and behavior will affect the extent of the corporate social responsibility practice [6]. Their selfless and integrity personality create favorable conditions for good relationships, and have positive impact on their subordinates [14].

On the other hand, the ethical values of business leaders determine their companies' ethical level. Leaders take corporate social responsibility behavior because of their own ethical values [15]. The effective leadership model increasingly emphasis on the values of leaders and leadership qualities that influence the strategic decision-making and execution, including corporate social responsibility-related decisions and actions [16].

3.2 Organizational Performance

The level of responsible leadership will influence the final organizational performance [7]. Ethical leadership has a positive predictive effect on organizational performance [17]. The visionary leaders who strongly hold the concept of stakeholders will lead the company to obtain better financial performance, and their subordinates are more willing to pay extra efforts to benefit their company [6].

3.3 Corporate Social Responsibility

Different to the social responsibility dimension of responsible leadership which is at the individual level, the concept of corporate social responsibility refers to the socially responsible strategy and activities of the company as a whole. Business leaders play a key role in the formulation and implementation of corporate social responsibility activities. If we neglect their influence on corporate social responsibility in the study of leadership, we may get incorrect conclusions for the causes and consequences of corporate social responsibility activities [4]. Senior leaders must recognize the sustainability objective, ensuring that this sustainable vision is consistent with the company's strategies, policies, and culture, and convey this information to their employees [18]. That is to say, responsible leadership motivates business managers to lead social responsibility activities in their respective companies, and results in a good corporate image of social responsibility.

3.4 The Mediating Effect of Corporate Social Responsibility

Some research found that companies in which responsible leaders implement corporate social responsibility showed much better performance than others which do not achieve the same social standards [19]. This is also an important driving force for managers to continue maintaining social responsibility. However other studies indicated that corporate social responsibility activities of responsible leaders can not only increase the financial performance, but are counter-productive [19]. Although the research findings have drawn different conclusions about the correlation between corporate social responsibility and organizational performance, at least it can be inferred that corporate social responsibility and organizational performance do have certain interaction in between. Managers' responsible leadership plays a leading role in corporate social responsibility practice, and corporate social responsibility may affect the company's operating performance to some extent. Therefore responsible leadership would possibly indirectly affect organizational performance through the mediating effect of corporate social responsibility.

Based on the above analysis, the theory framework of responsible leadership is formed. This theoretical framework of responsible leadership is an integrative constructs of leadership that addresses ethical characteristics of leaders, and challenges of leadership that are both responsible and focused on performance.

4 Conclusions

In conclusion, this study explores the characteristics of responsible leadership, which provides reference and basis to assess the managers-job fit. According to managers' different levels of responsible leadership, companies can design corresponding appropriate training plans and measures, so as to comprehensively improve managers' awareness of corporate social responsibility. Therefore this study has important practical implications for managers' selection, assessment and training in China.

References

1. Pless, N., Maak, T.: Responsible Leaders as Agents of World Benefit: Learnings from Project Ulysses. Journal of Business Ethics 85, 59–71 (2009)
2. Sun, L.P., Ling, W.Q., Fang, L.L.: The Analysis on Socio-cultural Source of Moral Leadership (in Chinese). Theorists 5, 166–168 (2009)
3. McGuire, J.B., Sundgren, A., Schneeweis, T.: Corporate social responsibility and financial performance. Academy of Management Journal 31, 354–372 (1988)
4. Waldman, D., Siegel, D., Javidan, M.: Components of CEO Transformational Lleadership and Corporate Social Responsibility. Journal of Management Studies 43, 1703–1722 (2006)
5. Clark, K.E., Clark, M.B.: Choosing to lead. Leadership Press, Richmond (1996)
6. Waldman, D.A., Galvin, B.M.: Alternative Perspectives of Responsible Leadership. Organizational Dynamics 37, 327–341 (2008)
7. Lynham, S.A., Chermack, T.J.: Responsible Leadership for Performance: A Theoretical Model and Hypotheses. Journal of Leadership and Organizational Studies 12, 73–88 (2006)
8. Pless, N.: Understanding Responsible Leadership: Role Identity and Motivational Drivers. Journal of Business Ethics 74, 437–456 (2007)
9. Pless, N., Maak, T.: Relational Intelligence for Leading Responsibly in a Connected World. In: 65th Annual Meeting of the Academy of Management, Honolulu (2005)
10. Maak, T., Pless, N.M.: Responsible Leadership in a Stakeholder Society: A Relational Perspective. Journal of Business Ethics 66, 99–115 (2006)
11. Maak, T.: Responsible Leadership, Stakeholder Engagement and the Emergence of Social Capital. Journal of Business Ethics 74, 329–343 (2007)
12. Ling, W.Q., Fang, L.L.: Study on Chinese Culture and Management of Sino-Japanese Joint Venture (in Chinese). Chinese Journal of Management Science 1, 47–56 (1994)
13. Chen, Y.Y., Zhang, Y.C.: Political Considerations in Performance Evaluation: Context-Based Analysis of Chinese Content (in Chinese). Management Review 20, 39–49 (2008)

14. Song, Y.: The Role of Personality in Leadership Activities and Leaders' Personality Cultivation. Leadership Science 52, 48–49 (2002)
15. Jones, T.M.: Instrumental Stakeholder Theory: A Synthesis of Ethics and Economics. Academy of Management Review 20, 404–437 (1995)
16. House, R.J., Aditya, R.: The social scientific study of leadership: Quo vadis? Journal of Management Studies 23, 409–474 (1997)
17. Brown, M.E., Trevino, L.K., Harrison, D.A.: Ethical leadership: A Ssocial Learning Perspective for Construct Ddevelopment and Testing. Organizational Behavior and Human Decision Process 97, 117–134 (2005)
18. Szekely, F., Knirsch, M.: Responsible Leadership and Corporate Social Responsibility: Metrics for Sustainable Performance. European Management Journal 23, 628–647 (2005)
19. Waddock, S., Graves, S.: The Corporate Ssocial Performance-Financial Performance Link. Strategic Management Journal 18, 303–319 (1997)

Research on Dual-Ring Network Based TDM Ring and CSMA Ring

Haiyan Chen[1], Li Hua[2], and Donghua Lu[3]

[1] Northeastern University at Qinhuangdao, Institute of Internet of Things and Information Security, Hebei, China 066004
[2] School of Information Science and Technology, Zhanjiang Normal College, Zhanjiang, China 524048
[3] Department of Mechanic and Electric, Qinhuangdao Institute of Technology, HeBei, China 066100

Abstract. In this paper, we introduce the TDM ring and CSMA ring, put forward dual-ring based on TDM ring and CSMA ring. Using Event-driven simulation strategy of discrete time, we design the simulation experiment. Some performance parameters are obtained. The results show that the scheme of dual-ring has advantage over single TDM ring or CSMA ring in some indices such as the average number of data packets waiting in buffer queue and average delay time of data packet.

Keywords: Dual-Ring, TDM Ring, CSMA Ring, Event-driven simulation strategy.

1 Introduction

Time-division multiplexing (TDM)[1] is a process in which all nodes to achieve the purpose of sharing transmission medium their using different timeslots sending data packet.TDM ring is several nodes using TDM ways in ring to transmit data. Its transmission rate is higher, but its timeslots length is fixed no suitable for variable-length data packets. When data packet is too long, timeslots cannot hold and when data packet is too short, the timeslot is waster.

Carrier Sense Multiple Access (CSMA) [2] is a probabilistic Media Access Control (MAC) protocol in which a node verifies the absence of other traffic before transmitting on a shared transmission medium. The realization of CSMA is relatively easy, and performance is modest.

In order to make use of the advantages of TDM and CSMA, we put forward the dual-ring network based on TDM and CSMA to meet the requirements of integrated data service.

2 The Working Mode

The dual-ring network is made up of N nodes. The transmission direction of two rings called outer ring and inner ring is same. The access mode of inner ring is CSMA and

D. Jin and S. Lin (Eds.): Advances in MSEC Vol. 1, AISC 128, pp. 587–591.
springerlink.com

the access mode of outer ring is TDM. The data Packets reach every node randomly, each node can buffer the waiting data packet. The two rings work independent, the buffer data packets in each node can be send using TDM mode or dual-ring mode. In TDM ring, each node allocates the special timeslot. It is shorter and appears periodically. Its arriving is fast. Because the mode sending data of the nodes in CSMA ring is random, in order to avoid the nodes sending data packet meantime, when data packet of each node in the buffer reach a certain value and the node sensing transmission medium is free, the node in CSMA ring will send data packet in its buffer. Therefore, when a node sends data packet in CSMA ring, it will send data packet in the TDM ring simultaneously. The schematic diagram of dual-ring network based on TDM and CSMA scheme is shown in Figure1. The arriving data packets can send through single TDM ring or dual-ring, but their sending rate is different. The sending rate of dual-ring is the sum of TDM ring and CSMA ring.

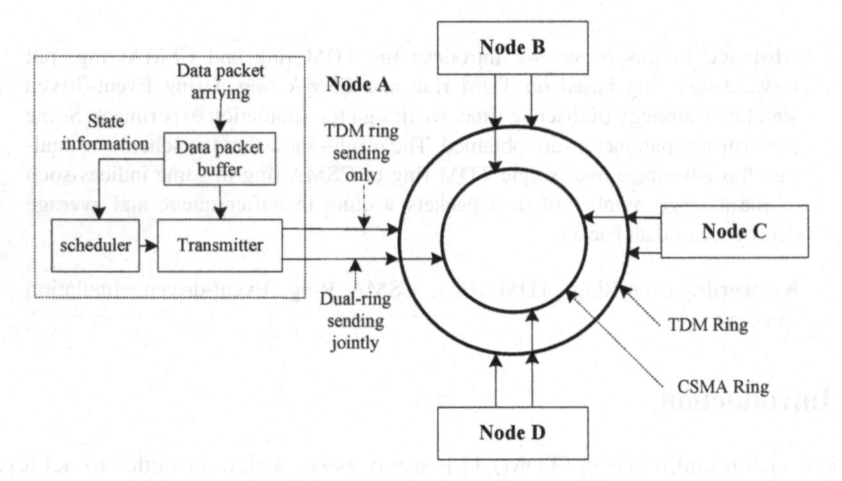

Fig. 1. Schematic diagram of dual-ring network based on TDM and CSMA scheme

3 Simulation Design

The simulation programs adopt the event scheduling strategy thoughts of discrete-time. According to the Event-driven simulation strategy[3-5], when the simulation clock pushes to a moment of the certain event to occur, the corresponding subroutines will be called. This subroutine will handle to some parameters and compute the time of the next event to happen, so reciprocating.

The simulation programs include main program, sending data packet subroutine, data packet arriving subroutines, data packet leaving subroutines, state scheduling subroutine of node and dual-ring interface.

In order to simplify the analysis, only analyse operation process of a node.

(1) Main program
Firstly, to define some variable, to initialize some queuing parameter, then run the arriving data packets subroutines. According to the Event-driven simulation strategy,

along with call other subroutines. Finally, increase arriving rate and drew the associated graphics.

(2) Sending data packet subroutines
The program declares variables firstly and set the parameter of the first arrived data packet. Then judge "the statue of sending data packet=dual-ring jointly or only TDM ring", If the statue of sending data packet is the dual-ring jointly, the program is the period of dual-ring sending data.

By means of comparing the next arriving time of arriving data packet and next leaving time of leaving data packet, the program determines what kind of events occur in the next time point and will call different subroutine:

```
If "Arriving time of i-th data packet" < "Leaving time of j-th data packet"
        Call the event of data packet arriving
    else
        Call the event of data packet leaving
    end
```

Similarly, if the statue of sending data packet is the only TDM ring, the program will be the period of only TDM sending data. Meanwhile the program will judge that the next event is the arriving event of data packet or the leaving event of data packet and decide whether change the sending rate according to whether the TDM ring sending time is terminate. Then the program will call the arriving event of data packet, the leaving event of data packet or the sending event of data packet in only TDM:

```
If "Arriving time of i-th data packet" < "Leaving time of j-th data packet"
      If "the sending time of data packet in only TDM ring" < "Arriving time of
               i-th data packet "
          Call the event of sending data packet in only TDM ring
      else
          Call the event of data packet arriving
      end
else
      If "the sending time of data packet in only TDM ring" < "Leaving time of
               j-th data packet"
          Call the event of sending data packet in only TDM ring
     else
          Call the event of data packet leaving
     end
end
```

According to the Event-driven simulation strategy, when the setting simulation total time in coming, the simulation program no longer generate the next arriving time of data packet.

(3) Data packet arriving subroutines
The program declares variables and generates the parameter of the arriving data packet. Then the simulation time to the next arriving time is carried forward. The program will calculates number of arriving data packet, queue length, etc.

(4) Data packet leaving subroutines
The program declares variables and generates the parameter of the leaving data packet. Then the simulation time to the next leaving time is carried forward. The program will calculates number of leaving data packet, queue length, etc.
(5) State scheduling subroutines of the interface of node and dual-ring
The program is designed to judgment whether the state will change in the period TDM ring sending data alone. If this occurs, then need to change the corresponding packets sent rate. The condition of the state change: in the period TDM ring sending data alone, when data packets of each node in the buffer reach a certain value and the node sensing transmission medium is free.

4 Simulation Results

Experimental conditions: suppose data packets arrival time interval follows the geometric distribution, Sending data time follows geometric distribution and the sending rate is 0.3 during the period of TDM only, the sending rate is 0.5 during the period of dual-ring, as shown in figure 2and figure 3.

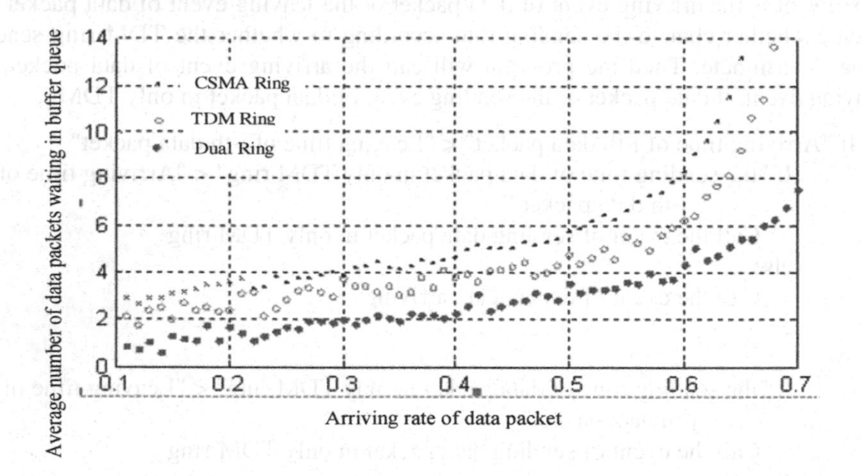

Fig. 2. The comparison of the average numbers of data package waiting in the buffer

(1) Average number of data packets waiting in buffer queue
With the increase of arriving data packet, result in the average number of data packets waiting in buffer queue increase in the mode of the TDM ring, CSMA ring and dual-ring. But the increase quantity of the dual-ring is less than other two rings independently, this reflect the sending capacity of dual-ring exceed other two rings.

(2) Average delay time of data packet
With the increase of arriving data packet, result in Average delay time of data packet increase in the mode of the TDM ring, CSMA ring and dual-ring. But the average delay time of dual-ring is obviously less than other two rings independently, as shown in figure 3.

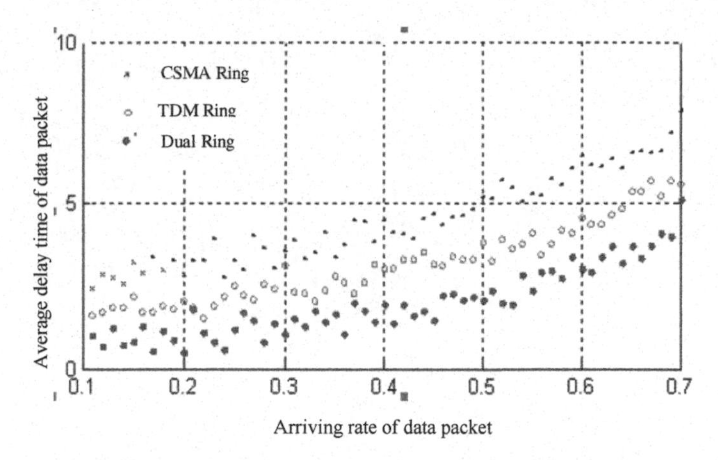

Fig. 3. The comparison of the average time delays of data package waiting in the buffer

5 Conclusion

In the paper, we introduce the dual-ring network based on TDM ring and CSMA ring. According to the Event-driven simulation strategy of discrete time, the simulation experiment program is carried out. Through comparing the performance of dual-ring, TDM ring and CSMA ring, we verify that the dual-ring based on TDM ring and CSMA ring jointly exceed the only TDM ring and only CSMA ring in the average number of data packets waiting in buffer queue and average delay time of data packet.

Acknowledgment. The authors would like to thank the support from Natural Science Foundation of China #50775198.

References

1. Keyur, P., Junius, K.: TDM services over IP networks. In: Military Communications Conference, MILCOM (2007)
2. Wang, X., Peter, C., Yie, H.J., Wai, L.: Performance comparison of CSMA/CD, CSMA/CA, CSMA/RI, CSMA/PRI and CSMA/PR with BEB. In: 5th IEEE Conference on Industrial Electronics and Applications, ICIEA 2010, pp. 1843–1848 (2010)
3. Chen, H., Shi, W., Yang, M.: Research on working vacations strategy of WLAN MAC layer. In: Advanced Materials Research, vol. 225-226, pp. 536–539 (2011)
4. Mani Chandy, K.: Event-Driven Applications: Costs, Benefits and Design Approaches. Presentation at Gartner Application Integration and Web Service Summit, California (2006)
5. Michelson, B.M.: Event-Driven Architecture Overview: Event Driven SOA Is Just Part of the EDA Story (2006), `http://dx.doi.org/JOI57I/bda2-2-06cc`

Improved Authentication Model Based on Kerberos Protocol

Xine You and Lingfeng Zhang

Loudi Vocational and Technical College, Loudi 417000 China

Abstract. An improved authentication model was suggested in this paper. Based on the analysis of the basic principle and security of Kerberos protocol, the lightweight level ticket was used in the improved model and the disadvantages of Kerberos protocol were well settled by using mixed key cryptosystem and USBKey two-factor authentication. Furthermore , the improved authentication protocol model was used to realize uniform authentication of multiple application system and the better security and easier feasibility of improved Kerberos protocol was proved。

Keywords: Kerberos protocol, ticket, USBKey two-factor authentication, Uniform authentication.

1 Introduction

Kerberos protocol is an authentication protocol based on broker, Certification center like a broker centralized carries out user authentication and distribution of electronic identity. It provides authentication method for the open network, authentication entity can be a user or customer service, this certification does not rely on the host operating system or the host's IP address, all hosts on the network do not need to be guaranteed for the physical security[1]. So it is widely used in authentication. However, the protocol has shortcomings of password-guessing attacks, replay attacks, the clock synchronization problems, key storage problems and other. To address these shortcomings, this paper on the basis of in-depth analysis of the Kerberos protocol system and the certification process put forward an improved model of the Kerberos authentication protocol against its shortcomings, and the model was applied to the unified authentication system, in security, stability and easy feasibility achieved good results.

2 Kerberos Protocol and Its Disadvantage Analysis

2.1 Basic Principle of Kerberos Protocol

Reality of Kerberos includes key distribution center(KDC) and a callable function library[2]. The KDC includes authentication server(AS) and ticket granting server(TGS). Client requests a ticket from the AS as a license service ticket, the ticket is encrypted by the client's secret key and sent to the client. In order to communicate with

D. Jin and S. Lin (Eds.): Advances in MSEC Vol. 1, AISC 128, pp. 593–599.
springerlink.com

the specified server, clients need to request an authorization ticket from the TGS. The ticket will be sent back to the client by TGS through a secure channel , the client presented this ticket to the target server. if the identity is no problem, the server allows customers to access its resource.

2.2 Disadvantages of Kerberos Protocol

From the viewpoint of security and the certification process, Kerberos protocol has the security flaws as follows:

(1) Password guessing attacks
When Users are to be online, they can share a key with the authentication server by typing the password on the users' platform. Password is vulnerable to eavesdropping and interception, the intruder can record the online session, which is analyzed by calculating the session key for password guessing attacks.
(2) Replay attack the problem with clock synchronization
In the Kerberos protocol, the timestamp is introduced in order to prevent replay attacks, which requires the entire network to achieve accurate clock synchronization in a distributed network. But it is difficult to achieve.
(3) Management and maintenance of key
KDC save a lot of shared keys, whose management and distribution are very complex. Special and careful security management measures are required. Whether KDC is broken or not, the system will pay a terrible price[3].
(4) Conversation eavesdropping
Conversation between the client and the server is generated by Kerberos key, so Kerberos can eavesdrop on the conversation with no evidence [4].

3 Improved Kerberos Protocol Authentication Model

Based on the above analysis on the Kerberos protocol security flaws, this paper improves the tickets, introduce the lightweight level tickets, and combines public key cryptography, digital certificates and USBKey two-factor authentication and other to improve the Kerberos' security.

3.1 Lightweight Level Ticket Model

The lightweight level ticket model is divided into two parts, which are ticket and ticket reference[5]. When the user authentication is successful by the AS, TGT and TGT REFERENCE are created for the users. TGT records the user's login information: user ID, user IP, user digital certificate, tickets' identification, time when tickets are created, time when tickets become invalid, and so on. TGT REFERENCE is a lightweight level user's certificate, which sign tickets' identification with the AS private key, and encrypt tickets' identification with the TGS public key. Its forms are: PUK_{TGS} {SessionID, PRK_{AS} {SessionID}}, in which PUK_{TGS} is the TGS's public key, and PRK_{AS} is the AS's private key, and SessionID is the TGT ticket's identity.

Similarly, the forms of ST tickets reference are: PUK_{S} {SessionID, PRK_{TGS} {SessionID}}, in which PUK_{S} is the S's public key, and PRK_{TGS} is the TGS's private key, and SessionID is the ST ticket's identity.

3.2 Improvement of Kerberos Protocol

The improved authentication model introduced the CA to issue and manage the certificate. certificates are stored in the LDAP directory server. The improved model specific certification process is shown in Figure 1.

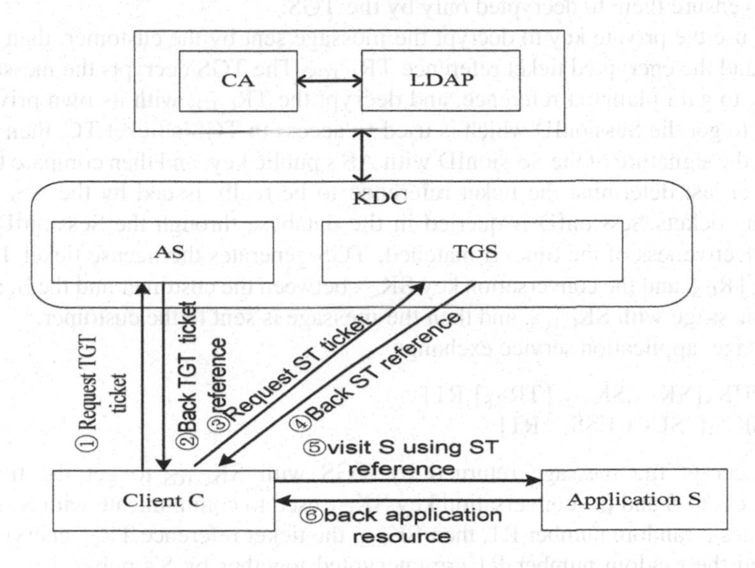

Fig. 1. Improved Kerberos authentication process model

The 1st stage: Authentication Service Exchange

1)$C \rightarrow AS: PUK_{AS}\{ ID_C , IP_C , ID_{TGS} , Cert_C \}$

2)$AS \rightarrow C: PUK_C\{ SK_{C,TGS} , SK_{C,TGS} \{TR_{C,TGS}\}\}$

$TR_{C,TGS} = PUK_{TGS}\{SessionID, PRK_{AS}\{ SessionID \}\}$

Client sends a requisition and the message about establishing connection with TGS to the AS. the message includes the client ID, the client IP, the TGS's ID, the client digital certificate $Cert_C$, in which the $Cert_C$ issued by the CA and stored in the USBKey. the USBKey encrypts the message with the AS's public key PUK_{AS}.

The AS decrypts the message, then obtains the ID_C and the $Cert_C$, then through the ID_C query the LDAP server to get a copy of the corresponding customer's digital certificate, then the copy is compared with the $Cert_C$, if matched, at last, a ticket $T_{C,TGS}$ and a ticket reference $TR_{C,TGS}$ are generated, and a conversation key is generated which is used for the secret communication between client C and TGS. The tickets are stored in a central database to prepare for verification later. the conversation key and the encrypted ticket reference are encrypted again with the client's public key, then are sent to the client again.

The 2nd stage: Ticket authorization exchange

3)$C \rightarrow TGS: PUK_{TGS}\{ SK_{C,TGS}, SK_{C,TGS}\{ TR_{C,TGS}\}, ID_S\}$

4)$TGS \rightarrow C: SK_{C,TGS} \{TR_{C,S}, SK_{C,S}\}$

$TR_{C,S} = PUK_S\{SessionID, PRK_{TGS}\{ SessionID \}\}$

Customers use the private key to decrypt the message delivered by the AS, then get the conversation key $SK_{C,TGS}$ and the ciphertext ticket reference $TR_{C,TGS}$ with the TGS, then request to communicate with the TGS to gain access to the application server S. The information which the customer delivers to the TGS includes the ID_S, the $SK_{C,TGS}$ and the $TR_{C,TGS}$ encrypted by $SK_{C,TGS}$. these information is encrypted by the TGS's public key to ensure them to decrypted only by the TGS.

The TGS use the private key to decrypt the message sent by the customer, then get the $SK_{C,TGS}$ and the encrypted ticket reference $TR_{C,TGS}$. The TGS decrypts the message with $SK_{C,TGS}$ to gain plaintext reference, and decrypt the $TR_{C\ TGS}$ with its own private key PRK_{TGS} to get the SessionID which is used to access to TGS's ticket TC, then the AS decrypts the signature of the SessionID with AS's public key, and then compare two SessionIDs, at last determine the ticket reference to be really issued by the AS. the corresponding tickets SessionID is queried in the database through the SessionID to verify the effectiveness of the time. If matched, TGS generates the license ticket $T_{C,S}$, its reference $TR_{C,S}$, and the conversation key $SK_{C,S}$ between the customer and the S, and encrypt the message with $SK_{C,TGS}$, and then the message is sent to the customer.

The 3rd stage: application service exchange

5)C→S:$PUK_S\{SK_{C,S}, SK_{C,S}\{TR_{C,S}\}, R1\}$
6)S→C:$SK_{C,S}\{$"SUCCESS ", R1$\}$

Customers decrypt the message returned by TGS with $SK_{C,TGS}$ to get the ticket reference access to S and the conversation key $SK_{C,S}$ used to communicate with S. The client generates a random number R1, then $SK_{C,S}$, the ticket reference $TR_{C,S}$ encrypted by $SK_{C,S}$, and the random number R1 are encrypted together by S's public key, and then all is sent to S.

The application server S receives the message, then the message is decrypted by its own private key to get the conversation key and the encrypted ticket reference $TR_{C,S}$. S unlocks $TR_{C,S}$ with $SK_{C,S}$, and then unlocks $TR_{C,S}$ with the private key again to get the SessionID, and decrypts the signature by TGS of SessionID with TGS's public key, and then compares two SessionIDs to determine the ticket reference to be indeed issued by TGS. If the verification succeeds, the SessionID in the database queries the corresponding tickets, and verify the effectiveness of time. If matched, with $SK_{C,S}$, S returns a message that includes a confirmation symbol and the random numbers R1 used to verify the identity of S for C. When the verification of the identity of S is passed, C can obtain service resources of S, and thus the process of the authentication protocol exchange ends.

4 Security Analysis of Improved Model

(1) Password guessing attacks avoided
In the improved certification model, encryption is achieved by using the client's public key, and two-factor authentication is achieved by using USBKey for the client, and thus password guessing attacks are avoided effectively.

(2) Dependence on clock synchronization reduced and replay attacks prevented
In the improved certification model, the adoption of random numbers instead of timestamps reduces dependence on clock synchronization; Meanwhile, the

communication between the client and AS is encrypted by the public key. Only when the customer use his own private key to decrypt the communication, he can get the conversation key with TGS, and the private key is stored in customer's USBKey, thus, replay attacks because of attackers' intercepting the conversation key are avoided effectively.

(3) The burden on the KDC key storage management reduced
In the improved certification model, the public key cryptography is adopted, the CA is responsible for the generation of public / private key and the issuance of digital certificate. Customer's private key is stored in USBKey, and only the public key information can be stored in KDC. So, the task of management and distribution of key is greatly reduced, at the same time even if attackers break the KDC's database to get only the public key, this could not form effective attacks, thus, the security of the system is greatly improved.

(4) The security of notes improved
In the improved certification model, the lightweight level ticket is introduced, and in the certification process the ticket reference rather than the ticket is transmitted, thus, on the one hand the security of ticket to ensure the safety of ; on the other hand, the network load is reduced.

5 The Application of the Improved Kerberos Authentication Model

With the rapid development of network technology, Digital Campus construction has also been extended, and a variety of application system based on the campus network are followed, and the service is increasing day by day. When users access different applications, they often require frequent logins and authentication operations, the flaws of independent certification on application systems are gradually emerging [6]. Therefore, this group based on the improved Kerberos authentication model establishes a unified identity authentication system, and network users are uniformly managed, authenticated and authorized.

5.1 Structure of Uniform Authentication System

The structure of unified identity authentication system is shown in Figure 2, in which central authentication and authorization server is based on the improved Kerberos authentication model.

5.2 The Achievement of Central Authentication and Authorization Server

The central authentication and authorization server is the core of entire unified authentication system, which stores the user identity information, access information and business application system information, and uniformly authenticates all users, delivers tickets and tickets reference to legal users, and generates digital certificates, public / private key pairs and the conversation keys and so on.

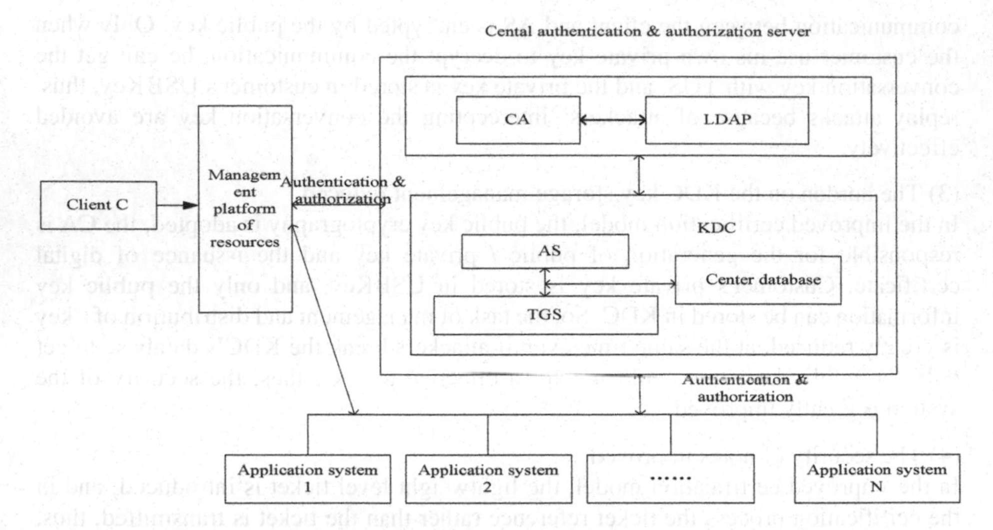

Fig. 2. Structure of Uniform Authentication System

Encryption and decryption, key generation, certificate operation, and signing operation of the system are all implemented by using the JAVA language to achieve the system's platform independence. The following focuses on the implementation of AS module and TGS module.

5.2.1 Implementation of AS Module

AS module's main function is to authenticate the user identity, deliver TGS's ticket TGT and TGT's references for legitimate users, and generate the conversation key for users to communicate with TGS. The workflow of AS is shown in Figure 3.

In the system the TGT is not sent to users but stored in the central database, and the TGT's reference is sent to the user.

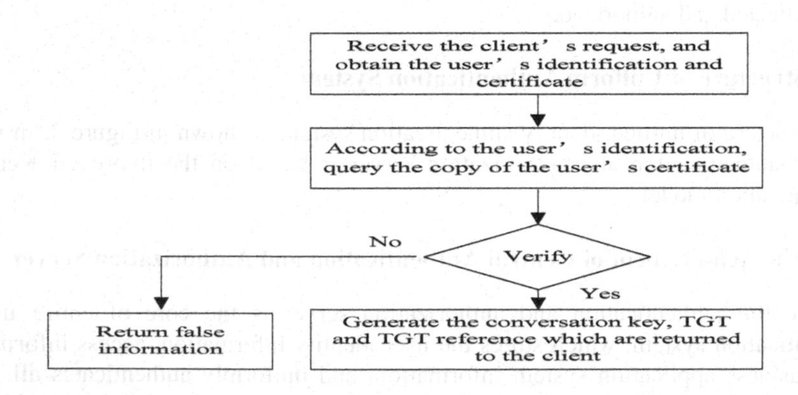

Fig. 3. Workflow of AS

5.2.2 Implementation of TGS Module

In the system, TGS's main task is to deliver ST ticket and ST ticket reference to the user whose identity has been authenticated by AS, and generate the conversation key for the client to communicate with the application system. The workflow of TGS is shown in Figure 4.

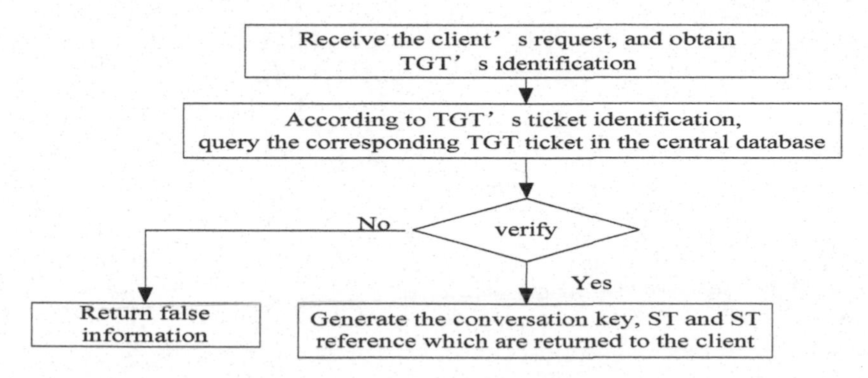

Fig. 4. Workflow of TGS

6 Conclusion

On the basis of full analysis on the Kerberos protocol, the shortcomings of the agreement were indicated in this paper, and an improved authentication model was put forward in which the lightweight level ticket, hybrid cryptosystem and USBKey two-factor authentication are introduced. Password guessing attacks and replay attacks are avoided better, and the security of the protocol is effectively improved, and the burden on the KDC key storage is reduced. The correct Kerberos protocol authentication model is applied to the unified authentication system, which possesses better security and stability.

References

1. Hu, Y., Wang, S.: Research on Kerberos identity authentication protocol based on hybrid system. Journal of Computer Application 29(6), 1659–1661 (2009)
2. Zhang, L., Yang, X.: Scheme of enhanced security for Kerberos protocol. Computer Engineering and Desig. 30(9), 2124–2126 (2009)
3. Hu, Z., Zeng, Q.: Improved Kerberos Protocol Based on Visual Cryptography. Computer Engineering 35(18), 159–163 (2009)
4. Wu, C., Liu, L.: Improving Kerberos system based on the key agreement protocol of identity based cryptograph. J. University of Shanghai for Science and Technology 32(4), 305–308 (2010)
5. Butle, F., Cervesato, I., Jaggard, A.D., et al.: Formal Analysis of Kerberos 5. Theoretical Computer Science 36(7), 57–87 (2006)
6. Law, L., Menezes, A., Qu, M., et al.: An efficient protocol for authenticated key agreement. Designs, Codes and Cryptography 28(2), 119–134 (2003)

Predicting Arsenic Concentration in Rice Plants from Hyperspectral Data Using Random Forests

Jie Lv and Xiangnan Liu*

School of Information Engineering, China University of Geosciences (Beijing), Beijing 100083, China
liuxncugb@163.com

Abstract. Accurate prediction of Arsenic concentration is important for food safety and precision farming. We explore the feasibility of predicting Arsenic concentration in rice plants from hyperspectral data using random forests (RF) in the Arsenic polluted farm lands. Canopy spectral measurements from rice plants were collected using ASD field spectrometer in Suzhou, Jiangsu Province. Rice plants were collected for chemical analysis of Arsenic concentration. Prediction of Arsenic concentration was achieved by a random forests approach. The results show that the random forests approach achieved an R^2 value of 0.84 and an MSE value of 3.97. The results indicate that it is possible to predict concentration of Arsenic in rice plants from hyperspectral data using random forests.

Keywords: Arsenic, hyperspectral, random forests, rice, stress.

1 Introduction

The Arsenic contamination by heavy metals has caused great concern worldwide, it pose a threat to human health through food chain [1]. Rice is the staple food for more that half of the world's population, especially in China. Therefore, it is imperative to predict Arsenic concentration in rice plants for sate food production.

When plants are under stressed by heavy metals, their reflectance spectra increase in the visible region due to inhibiton of photosynthesis and changes in the content or structure of proteins in the photosynthetic apparatus, and decrease in the near infrared region due to damage to leaf cell walls and mesophyll tissue [2]. Therefore, it is feasible to predict Arsenic concentration in plants with the help of hyperspectral remote sensing and random forests.

Hyperspectral remote sensing has been widely applied in vegetation science, and agricultural crop management, most of these studies have concentrated on linking biophysical variables [3, 4], leaf biochemical contents [5–7], water contents [8, 9], plant stress [10–13] to spectral response of plants.

Some machine learning methods, such as artificial neural networks (ANN), support vector machines (SVM), and random forests (RF) could potentially deal better with nonlinear spectral responses of plants under heavy metal stress. Random forests was

* Corresponding author.

D. Jin and S. Lin (Eds.): Advances in MSEC Vol. 1, AISC 128, pp. 601–606.
springerlink.com

proposed by Breiman in 2001, which are a recent extension of decision tree learning [14]. Random forests have shown superior performance over other machine learning techniques for both prediction and classification in a variety of applications. In this paper, we apply random forests to predict Arsenic concentration in Arsenic stressed rice plants from hyperspectral data.

2 Material and Methods

2.1 Hyperspectral Reflectance Measurement

Canopy reflectance spectra were collected from rice using a Analytical Spectral Devices (ASD) FieldSpec 3 spectrometer, which has a spectral range between 350–2500 nm with a sampling interval of 1.4 nm in the 350–1000 nm range and 2 nm in the 1000–2500 nm range.

The ASD field of view was set to 25° with a 1 m sensor height above rice canopies. All measurements were performed on clear days between 10:30 am and 1:30 pm local time. A white Spectralon reference panel was used under the same illumination conditions to calculate relative reflectance spectra by dividing leaf radiance by reference radiance for each wavelength. Ten scans were measured at each point, and averaged to produce a single spectrum. 60 spectral measurements were made altogether.

2.2 Rice Plants Sampling Collection and Chemical Analyses

A total of 60 surface rice samples and 60 soil samples were collected from the paddy fields. Eight rice leaves sampling was performed from 3-5 rice plants at every spectral measurement site.Heavy metals in rice samples were extracted according to the basic methods in China [15]. For determination of total Arsenic in rice plants, the subsamples were digested with a mixed acid (1:3:4 HNO_3:HCL:H_2O).Arsenic concentration was determined by atomic fluorometry. The determination of Arsenic in soil was the same as that of rice plants.

2.3 Random Forests

Random Forest is a general term for ensemble methods using tree type classifiers $\{h(x, \Theta_k), k=1,\ldots\}$ where the $\{\Theta_k\}$ are independent identically distributed random vectors and x is an input pattern [14]. It does not overfit, runs fast and efficiently on large datasets such as hyperspectral remote sensed data. It does not require assumptions on the distribution of the data, which is interesting when different types or scales of input features are used. Random forest can handle thousands of variables of different types with many missing values, as a result, random forests can deal better with high dimensional data and use a large number of trees in the ensemble. These outstanding advantages make it suitable for remote sensing prediction.

Fig 1 shows the general process of random forests, it can be seen that random forests uses the best split of a random subset of input features or predictive variables in the division of every node, instead of using the best split variables, which redues

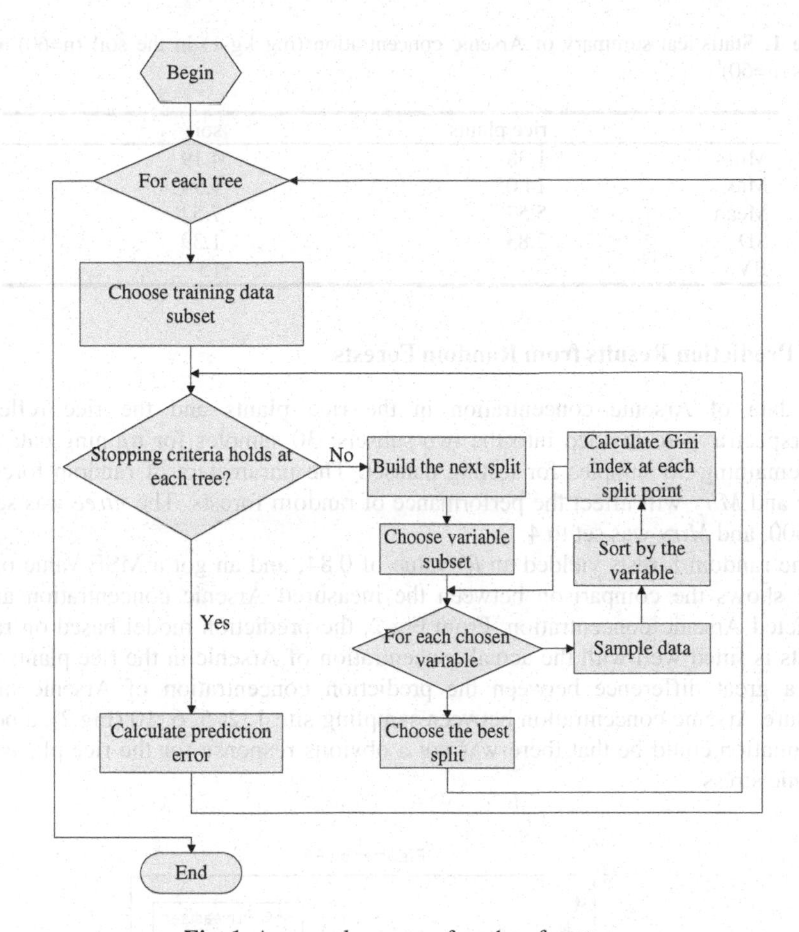

Fig. 1. A general process of random forests

the generalization error. Additionally, random forests adopts bagging or bootstrap aggregating to make them grow from different training data subsets to increase the diversity of the trees.

3 Results

3.1 Arsenic Concentration in the Soil and Rice Plants

The concentration of Arsenic in the rice plants and soil are presented in Table 1. The mean Arsenic concentration in the soil was 7.53 mg kg-1, although mean concentration of Arsenic in the soil was below national background value, some soils samples exceeded national background values. A possible explanation for the contamination of Arsenic in the soil is that there are some cadmium stearate factories in the city. The mean concentration of Arsenic in the rice was above the Chinese maximum allowable concentration in foods (GB2762–2005). It indicates that the rice plants were under Arsenic stress in the farm lands.

Table 1. Statistical summary of Arsenic concentration (mg kg-1) in the soil (n=60) and rice plants (n=60)

	rice plants	soil
Min	1.38	4.19
Max	14.05	9.94
Mean	5.57	7.53
SD	2.83	1.39
BV		15

3.2 Prediction Results from Random Forests

The data of Arsenic concentration in the rice plants and the rice reflectance hyperspectra were divided into the two subsets: 30 samples for training dataset and the remaining 30 samples for testing dataset. The parameters of random forests are *ntree* and *Mtry* will affect the performance of random forests. The *ntree* was selected as 2000, and *Mtry* was set to 4.

The random forests yielded an R^2 value of 0.84, and an got a MSE value of 3.97. Fig.2 shows the comparison between the measured Arsenic concentration and the predicted Arsenic concentration. From Fig 2, the prediction model based on random forests is fitted well with the actual concentration of Arsenic in the rice plants. There was a great difference between the prediction concentration of Arsenic and the measure Arsenic concentration between sampling site 1, 2, 5, 6, 10 (Fig.2), a possible explaniation could be that there was not a obvious response for the rice plants under Arsenic stress.

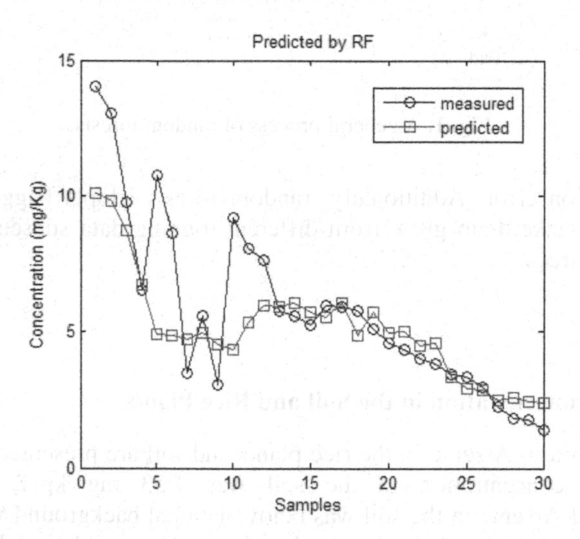

Fig. 2. The comparison between the predicted Arsenic concentration and the measured Arsenic concentration

4 Conclusion

The study showed that the canopy reflectance of heavy metal stressed rice can be used to predict Arsenic concentration in rice plants by random forests. Hyperspectral remote sensing and random forests are shown to be a very promising method for rapid estimation of concentration of Arsenic in rice plants. A satisfactory relationship was found for rice canopy reflectance spectra, with R^2 values of 0.84 and MSE value of 3.97.

To monitor plant health in a large scale, spaceborne hyperspectral imaging spectrometer such as Hyperion would provide a coverage of 7.7 km×42 km with 242 bands of 10 nm. The possibilities of applying random forests in spaceborne hyperspectral imaging spectrometer have to be explored.

Acknowlegements. This research was supported by the National Natural Science Foundation of China (No. 40771155) and National High-tech Research and Development Program of China (No. 2007AA12Z174). We are grateful to Dr. Li Yuzhong and Dr. Xu Chunying from Chinese Academy of Agricultural Sciences for technical assistance with sampling and chemical analysis.

References

1. Huang, S.S., Liao, Q.L., Hua, M., et al.: Survey of heavy metal pollution and assessment of agricultural soil in Yangzhong district, Jiangsu Province, China. J. Chemosphere. 67, 2148–2155 (2007)
2. Mysliwa, K.P., Strzalka, K.: Influence of metals on biosynthesis of photosynthetic pigments. In: Prasad, M.N.V., Strzalka, K. (eds.) Physiology and Biochemistry of Metal Toxicity and Tolerance in Plants, pp. 201–302. Kluwer Academic Publishers, Dordrecht (2002)
3. Clevers, J.G.P.W.: Application of the WDVI in estimating LAI at the generative stage of barley. J. ISPRS. J. Photogramm. 46, 37–41 (1991)
4. Schaepman, M.E., Koetz, B., Strub, G.S., Itten, K.I.: Spectrodirectional remote sensing for the improved estimation of biophysical and -chemical variables: two case studies. J. Int. J. Appl. Earth. Obs. 6, 271–282 (2005)
5. Imanishi, J., Nakayama, A., Suzuki, Y., Imanishi, A., Ueda, N., Morimoto, Y., Yoneda, M.: Nondestructive determination of leaf chlorophyll content in two flowering cherries using reflectance and absorptance spectra. J. Landscape. Ecol. Eng. 6, 219–234 (2010)
6. Tian, Y.C., Yao, X., Yang, J., Cao, W.X., Hannaway, Y., Zhu, Y.: Assessing newly developed and published vegetation indices for estimating rice leaf nitrogen concentration with ground- and space-based hyperspectral reflectance. J. Field. Crop. Res. 120, 299–310 (2011)
7. Bannari, A., Khurshid, K.S., Staenz, K., Schwarz, J.W.: A Comparison of Hyperspectral Chlorophyll Indices for Wheat Crop Chlorophyll Content Estimation Using Laboratory Reflectance Measurements. J. IEEE. T. Geosci.Remotes. 41, 6770–6775 (2007)
8. Cheng, T., Rivard, B.W., Azofeifa, A.S.: Spectroscopic determination of leaf water content using continuous wavelet analysis. J. Remote. Sens. Environ. 2, 659–670 (2011)
9. Danson, F.M., Steven, M.D., Malthus, T.J., Clark, J.A.: High-spectral resolution data for determining leaf water content. J. Int. J. Remote. Sens. 13, 461–467 (1992)

10. Dunagan, S.C., Gilmore, M.S., Varekamp, J.C.: Effects of mercury on visible/near-infrared reflectance spectra of mustard spinach plants (Brassica rapa P.). J. Environ. Pollut. 148, 301–311 (2007)
11. Franke, J., Mewes, T., Menz, G.: In Requirements on spectral resolution of remote sensing data for crop stress detection. In: Proceedings of the IEEE International Geoscience & Remote Sensing Symposium, Cape Town, South Africa, Jul 13-17, pp. I–184–I–187 (2009)
12. Naumann, J.C., Anderson, J.E., Young, D.R.: Remote detection of plant physiological responses to TNT soil contamination. J. Plant. Soil. 329, 239–248 (2010)
13. Carter, G.A.: Responses of leaf spectral reflectance to plant stress. J. Am. J. Bot. 80, 239–243 (1993)
14. Breiman, L.: Random forests. J. Machine Learning 45, 5–32 (2002)
15. Agricultural Chemistry Committee of China, Conventional Methods of Soil and Agricultural Chemistry Analysis, Science Press, Beijing (in Chinese) (1983)

Design and Implementation of Online Experiment System Based on Multistorey Architectures*

Wei Li, WenLong Hu, and YiWen Zhang

School of Computer Science and Technology, Anhui University, Hefei, Anhui, China

Abstract. The Online Experiment System, taking advantage of the mature network technology and database technology, can make up for the scarcity of traditional teaching model. Online Experiment System consummates the experiment aspect in teaching model with improving the quality of teaching and relieving the pressure of teacher. It is an important part of realizing the teaching informatization. This paper has analysed the process of design and implementation of Online Experiment System from the perspective of multistorey architectures and achieved this system drastically.

Keywords: Online Experiment System, Multistorey Architecture, DataBase.

1 Introduction

As the development of science technology at high speed, the computer is playing a more and more important role in our modern life. The omnipresent application of network platform has clearly explained that the technology of computer and network has made a huge step to maturation. At the same time, it has forced the traditional teaching procedure of higher education to face to the innovation with popularization of higher education and advent of lifelong education[1]. Unfortunately, the method of higher education has been still limited to the parochial traditional teaching model.

In order to meliorate this situation and break through this localization of traditional model, we expect to design and exploit a special online experiment system. Through this system, we can really do apply the advanced computer and network technology into the experiment tache of higher education so as to relieve the teaching pressure and stimulate and realize the transformation of current teaching model's informatizaon fundamentally.

2 Background of Technology

2.1 .NET

Essentially, .NET is usually explained as one kind of Development Platform which has defined the Common Language Subset (CLS). As we know, this is a sort of mixed

* Fund Project:Reform Project of Anhui Provincial Department of Education (Project Grand No:2008jyxm274).

language that provides the canonical seamless integration between the Language and class libraries. In addition, .NET has unified the programme class libraries and the perfect support for Extensible Markup Language (XML), the communication standard of next generation network, makes the process of exploitation easier and more quickly[2]. The most important point is as the base of Microsoft .NET platform, the general .NET framework and the simplified .NET framework which aims to some special facilities have provided an efficient and secure development environment for the Wed Service of XML and other applications[3].

2.2 SQL Server

SQL(Structured Query Language), is a kind of powerful standard language of Relational database. Currently, nearly all Relational databases have supported SQL, and many software companies have started to make further expansion and modification for the basic command sets of SQL[4]. Besides, being built on operating system of Microsoft, which provides the powerful Client-Server platform, drives SQL Server being perfectly competent for multiple parallel Relational database system simultaneously[5]. In the meantime, SQL Server has inherited the excellent graphic interface of Microsoft, bringing SQL the easy and convenient operation.

3 System Design

3.1 Summarization of System

Basically, the main users of Online Experiment System are students, teachers and system managers. After logging on the system, the users of system just can attain the relevant functions according to their identities. For example, as a student, you can choose your favorite courses after checking the correlative information of teachers or courses. when enter the system with the identity of teacher, users can manage their own courses in the system, such as add or modify the information of courses and one course's experiment project, upload the questions of experiment. Another identity ,allowed to enter the system, is administrator. The duty of administrator mainly focuses on managing the users of system centrally and supervising the application of system. Besides, the Online Experiment System has included the judgement tool installed on the server part, which is mainly used to monitor whether there are messages from client part and judge the answers submitted and return the results automatically.

3.2 Analysis of System's Framework

To divide the system scientifically, the principle of high cohere and low coupling should be followed by the designers all the time. High cohere can be comprehended as one kind of design mind which emphasizes the concentration of inner function of module and good Package so that there is not necessary to know the inside details clearly while calling the functions. Low coupling means degree of coupling should be as low as possible to reduce the mutual effect between each other[6].

As mentioned, web part and the judgement tool installed on the server both have adopted Multistorey Architectures that has divided the system into three layers, which are interface layer, business layer and data-access layer. Each layer has its own duty, and gets the support of data from its underlayer.

For the interface layer, we can regard it as the core role in the proscenium interface of system. Commonly its content is stored in the aspx files and we can add codes to the relative cs files which can deal with the events of proscenium pages. When it is necessary, the interface layer can submit the business applications to its underlayer; business layer provides business logic method for interface layer. That means when the interface demands some operation, logic process or data, the business will accept the application, and accomplish the disposal. When business layer needs to read the data from database, it will request its underlay-data-access layer; the data-access layer has provided the method of data access so as to response the data application of business layer. At the same time, the server part also disposals the familiar Multistorey Architectures. The system's framework is described as the figure 1:

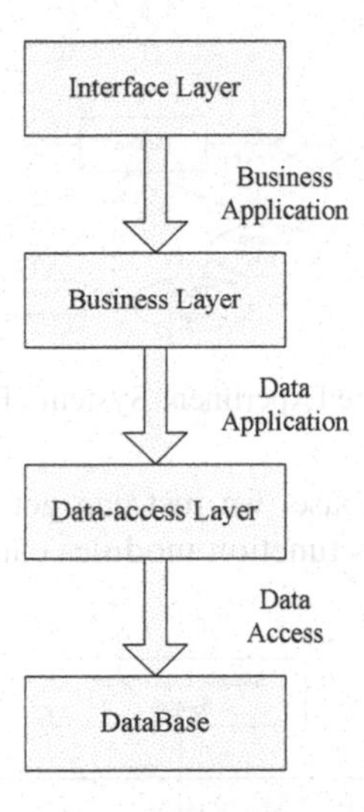

Fig. 1. Online Experiment System's framework

4 Design of Database and System Module

The Online Experiment System has adopted the convenient and high security SQL Server as its database. Because the system needs to interact with database all the time, whether design of database is suitable to the appliance of system will determine difficulty of development and system's efficiency. We finally ascertain the structure of database is like this:

Administrator information table: record some basic information about administrator, such as administrator's ID, user name, user password. we set the administrator's ID as the key of table.Teacher and student information tables' structure is similar to administration information table. So we won't describe them here.

Course information table: record some basic information about course, such as course's ID, teacher of course, information of course. We set the course's ID as the key of table.Experiment and experiment question information tables' structure is similar to course information table. So we won't describe them here.

System event table: record some relevant event information in the process of using the system, such as event's ID, content of event, page of event, operator's IP. We set the event's ID as the key of table.

The system's E-R figure is like figure 2:

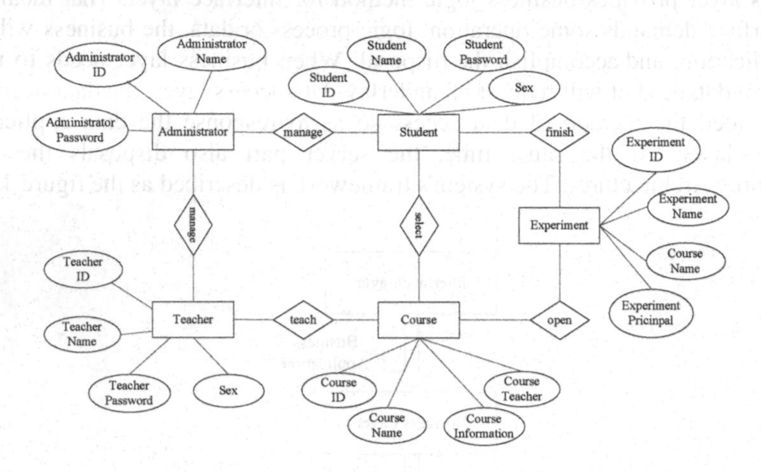

Fig. 2. Online Experiment System's E-R figure

After finishing design of database, we just can get the system's function modules by requirement analysis. System's function modules can be described like figure 3:

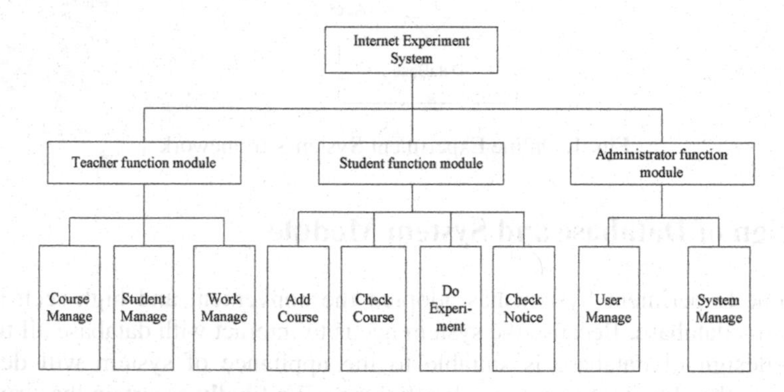

Fig. 3. Internet Experiment System's function module figure

Of course, these modules also include some submodules.

5 System Principal and Implementation of Key Technology

5.1 Logging on Module of System

Before using the system, user has to pass through the identify validation. After providing the correct user name, corresponding password and user type, user could be allowed to enter or employ the system.The background of Interface layer will take advantage of parameters formed from message user has filled in to call the VerifyUser function to examine whether user name is legal. The main code is like this:

```
if (ds_user.Tables[0].Rows.Count == 0)
        {
            status.Value = false;
            return status;
        }
        string pwd =
ds_user.Tables[0].Rows[0]["userpwd"].ToString();
        if (pwd != userpwd)
        {
            status.Value = false;
            return status;
        }
        status.Value = true;
```

Namely, the function will search the record from database whose user name is equal to what user has written in proscenium, and compare the record information with message written in proscenium. If the message is validated successfully, the page will jump to the relevant user page.

5.2 Teacher's Editing Question Module

After enter the system, teachers can check the information about their own courses, including experiment message of course. On the page of experiment message, system allows teacher to update existed experiment message or upload experiment questions, and so on. The code to upload experiment questions is like this:

```
string datainfilename = Session["userid"].ToString() +
BussinessHelper.getFilenameFromDatetime();
string dataoutfilename = datainfilename;
string datainfilepathname = BussinessHelper.getWorkPath()
+ "\\dataInOutFile\\" + datainfilename + ".in";
```

```
string dataoutfilepathname = BussinessHelper.getWorkPath()
+ "\\dataInOutFile\\"+ dataoutfilename  +   ".out";
```

When teacher adds some experiment question, system will call the getFileNameFromDatetime function in business layer to name experiment's input file and output file, and store the content of textbox in the proscenium into named files automatically. Finally, input files would be named .in files and output files would be named .out files which will be stored under an unified file directory together.

5.3 Server's Automatic Judgement Module

After student submits his answers, system should ensure to return the results of answers to student right now. In order to accomplish this function, we have deployed an judgement tool on the server to supervise messages from server at any moment and judge the submitted answer immediately. Here, we take programme question judgement as an example to explain the process of system's judgement module. The main code is like this:

```
switch (resultinfo.Result)
   {
      case CodeCompileRunResult.CompileError:
      message = string.Format("-1|compile error! \n{0}",
resultinfo.Message);
      break;
      case CodeCompileRunResult.RuntimeError:
      message = string.Format("-1|running error! \n{0}",
resultinfo.Message);
      break;
      case CodeCompileRunResult.OK:
      string str_output = resultinfo.Message;
      if (File.Exists(this._outputfile)){
         string str_standardoutput =
File.ReadAllText(this._outputfile);
if (str_output == str_standardoutput){
               message = "1|AC";
            }
         else{
               message = "-1|Wrong Answer!";
            }
         }
```

```
            break;
            default:
            message = "-1|system error! ";
            break;
    }
```

After student submits the code of the question, the judgement tool will make use of its own compiler to compile the submitted code and return the information. If the code is compiled successfully, system will return compared result with correct answer. If not, system will return error message of compiler.

6 Conclusion

How to apply the technology of computer and Internet into today education, to release teaching pressure and improve teaching quality, is the hot issue all the time. By introducing the design and implementation of Online Experiment System, this paper has proven feasibility from the aspect of technology. Of course, there are still some disadvantages in the system, we need to improve it gradually in the learning process. We hope the implementation of this kind of system can offer some help to similar teaching information in the future.

References

1. Liu, C.: Changes and Operation Mechanism in Higher Education Teaching Process at Information Age (in Chinese). Xiamen University, Xiamen (2003)
2. Hilt, D.Z.: Complete Guide to C# (in Chinese). Electronic Industry Press, Beijing (2008)
3. Ma, Y., Xiong, Q.: NET Application Development Overview (in Chinese). In: 6th National Joint Computer Conference Proceedings, pp. 259–263. Beijing University of Posts and Telecommunications Press, Beijing (2002)
4. Wang, S., Sa, S.: Introduction to Database System (in Chinese). Higher Education Press, Beijing (2006)
5. Zhu, X.: Tuning Strategies and Research based on SQL Server Database (in Chinese). Harbin Institute of Technology, Harbin (2009)
6. Lu, W.: Maintainability for the software system architecture analysis and design (in Chinese). Shanghai Jiao Tong University, Shanghai (2004)

CT-PCA Algorithm Based on Contourlet Transform and KPCA for Image Enhancement

Qisong Chen*, Maonian Wu, Xiaowei Chen, and Yanlong Yang

College of Computer Science& Information, Guizhou University, 550025 Guiyang, China
ytgcqs1@yahoo.com.cn, maonian_wu@hotmail.com,
gzu@vip.sina.com.cn, yyl1980@163.com

Abstract. Removing noises and increasing articulation are two important aims in digital image processing. But they are inconsistent in many methods. To enhance the definition and vision effect of digital images, a novel image enhancement algorithm named CT-PCA is proposed based on many methods. Firstly, on the basis of low-pass filter, adaptive gray mathematical morphology has been adopted in contourlet domain coefficients. On the other hand, image de-noising is executed on different frequency sub-images by kernel principal component analysis method. Finally, the clearest image is acquired with inverse contourlet transform. So contourlet transform, kernel principal component analysis and gray mathematical morphology are harmonious in this method. The simulation experimental results on images show that the proposed algorithm not only decreases image noise effectively, but also improves image PSNR.

Keywords: Gray mathematical morphology, Contourlet transform, Kernel principal component analysis, Image enhancement.

1 Introduction

Image processing typically relies on accurate, simple, and tractable model. Usually, digital images acquired from optical or electronic cameras tend to suffer interference. These spots degrade the articulation of images. There are many methods to remove noise in space domain and frequency domain, such as average filter, median filter, Wiener filter, discrete Fourier transform, discrete cosine transform, wavelet transform and so on. A block in image de-noising is that particular pixels in images would be clear up as well as noisy pixels. So, an integrated algorithm is raised in this project.

The obvious shortcoming for wavelets in two dimensions is the limited ability in capturing directional information. To overcome this deficiency, many researchers have recently considered multiscale and directional representations that can capture the intrinsic geometrical structures such as smooth contours in natural images. Some examples include the steerable pyramid [1], brushlets [2], complex wavelets [3], and the curvelet transform [4]. In particular, the curvelet transform, pioneered by Candès and Donoho, was shown to be optimal in a certain sense for functions in the continuous domain with curved singularities.

* Correspondence author.

D. Jin and S. Lin (Eds.): Advances in MSEC Vol. 1, AISC 128, pp. 615–621.
springerlink.com

2 Related Theories Analysis

In general image de-noising theories and method, some filters are quite effective for noise removal and have been widely used for removing impulse noise. However, they tend to modify both undisturbed good pixels and noisy pixels. Other types of algorithms are good at image edge preservation, but have litter effects in de-noising.

2.1 Mathematical Morphology

Mathematical morphology is a very powerful tool used for the analysis and representation of binary and grayscale images.

In mathematical morphology theory, sets are adopted as theory descriptions of digital signals and digital images, and functions represent gray-level signals and images [5-7]. The extension of translation-invariant binary morphology to the gray-level case was first derived based on the set representation of functions.

In early form, mathematical morphology approach handles binary images as sets and probes them with a structuring element (SE). Usually the SE is a set smaller than the original image on which it is translated.

Erosion and dilation are two basic morphological operations. All the other morphological transforms can be composed from them. Mathematical morphological transforms can separate intricate image from undesirable parts and decompose complex shapes into more meaningful forms [8]. Mathematical morphology has provided solutions to many computer vision problems, such as noise suppression, feature extraction, edge detection etc [9].

2.2 Contourlet Transform

The Fourier transform is an excellent theory suited for studying stationary signal. But it fails to exactly determine when or where a particular frequency component occurred. Gabor proposed a kind of short time Fourier transform with a fixed sized window to overcome the problems of Fourier transform. However it does not describe local changes in frequency content which prompted the use of Wavelet transform for the purpose of signal analysis. However, wavelet transform fails to characterize efficiently edges along different directions as the implemented wavelet transform is a separable transform constructed only along the horizontal and vertical directions [10].

The major drawback of wavelets in two dimensions is their limited ability in capturing directional information. To overcome the major drawback of wavelets in two dimensions, recently many researchers have considered multiscale and directional representations. These methods can capture the intrinsic geometrical structures such as smooth contours in natural images.

The contourlet transform is constructed by combining the Laplacian pyramid with the Directional Filter Bank (DFB). Compaired with wavelet transform, the contourlet transform is a multi-directional and multi-scale transform. But the pyramidal filter bank structure is unsuited for a denoising applications because the contourlet transform has very little redundancy.

Therefore, da Chuna et al. [11] introduced the nonsubsampled contourlet transform which is a fully shift-invariant, multi-scale and multi-direction expansion of the contourlet transform.

2.3 Kernel Principal Component Analysis

As a linear transformation technique, principal component analysis (PCA) is widely used in spectral extracting of the multidimensional original data. In another coordinate system the data are transformed so the first coordinate represents the largest variance, and the second coordinate to the second largest variance, and so forth.

Kernel principal component analysis can be derived using the known fact that principal component analysis can be carried out on the dot product matrix instead of the covariance matrix [12-13].

Let X be a p-dimensional random column vector having a zero empirical mean. The PCA transformation tries to find an orthonormal projection $p\times p$ matrix W such that $s = W^T X$, where the elements of W are the eigenvectors of the covariance matrix of X .

Suppose one is given a set of sample data $X = \{x_1, x_2, ... x_n\}, x_i \in R^N$. The KPCA algorithm maps the data set into a high dimensional feature space F via a function $\Phi(\cdot)$. The covariance matrix of the mapped data $\Phi(x_i)$ is

$$\overline{C} = \frac{1}{n} \sum_{i=1}^{n} \Phi(x_i)\Phi^T(x_i) \tag{1}$$

Suppose the data was centered in F , as $\frac{1}{n}\sum_{i=1}^{n}\Phi(x_i) = 0$, The principal components are then computed by solving the eigenvalue problem

$$\overline{C}V = \overline{\lambda}V \tag{2}$$

where eigenvalue $\overline{\lambda} > 0$, and V is corresponding eigenvector V . Furthermore, as seen from equation (4), all nonzero eigenvalue must be in the span of the mapped date, i.e., $V \in span\{\Phi(x_1),...,\Phi(x_n)\}$, there exist coefficients $a_j (j = 1,2,...,n)$:

$$V = \sum_{j=1}^{n} a_j \Phi(x_j) \tag{3}$$

Defining a $n\times n$ matrix K as

$$K_{ij} = (\Phi(x_i), \Phi(x_j)) \tag{4}$$

3 CT-PCA Algorithm for Images

3.1 Mathematical Morphology in Noise Image

As mentioned before, a smart method is proposed in digital images denoise and enhance. Firstly, simple median filter is applied on source image to smooth image. Secondly, mathematical morphology is applied on smoothing image to detect and enhance the edge, so that the original edge details are preserved. Thirdly, contourlet transformation is used to image, and on different frequency sub-images, KPCA method is adopted to denoise. Finally, distinct image is attained with inverse contourlet transform.

Where $f(x,y)$ and $g(x,y)$ corresponding to the object and image function, mathematics model of digital imaging system could be indicated as:

$$g(x,y)=h(x,y)*f(x,Y)+n(x,y) \tag{5}$$

the symbol * represents convolution algorithm. With Fourier transform, PSF (point spread function) is equivalent to a low-pass system. Most parts of useful information concentrate in the center of frequency domain.

Mathematical morphology. Nowadays, mathematical morphology offers many theoretical and algorithmic tools to many research areas and inspires new directions from the fields of signal processing, image processing, pattern recognition and machine vision, and so on.

According to equations 6 and 7, binary erosion and dilation are defined as follow:

$$A\Theta B = \bigcap_{x\in B} (A)_{-x} \tag{6}$$

$$A \oplus B = \bigcup_{x\in B} (A)_x \tag{7}$$

where A, B are sets of Z2; $(A)_x$ is the translation of A by x, and -B is the reflection of B. These are defined as

$$(A)_x = \{Z^2 \mid c = a + x \text{ for some } A \in A\} \tag{8}$$

$$-B = \{x \mid \text{for some } b \in B, x = -b\} \tag{9}$$

Set A is the image under process and set B is the structuring element.

The definitions of binary opening and closing are, respectively, $A \circ B = (A\Theta B) \oplus B$, $A \bullet B = (A \oplus B)\Theta B$. In terms of grey-scale morphology, set f is the input image and b is the structure element then the dilation of b to f is defined as

$$f \oplus b(s) = \max\{f(s-x)+b(x) \mid (s-x) \in D_f, x \in D_f\} \tag{10}$$

and the erosion of b to f is defined as

$$f\Theta b(s) = \min\{f(s+x)-b(x) \mid (s+x) \in D_f, x \in D_b\} \tag{11}$$

D_f and D_b are the domain of f and b, s and x were space vector of integer Z2.

Opening operation can eliminate the convex domains which were not according with the SE of the image. Closing operation can fill the concave domains which were not corresponding with the SE of the image, while retaining those who match the structural elements.

3.2 KPCA in De-noise Image

Substituting equation (1), equation (3) and equation (4) into equation(2), the following equation can be obtained

$$Ka = N\overline{\lambda}a \tag{12}$$

Eigenvalue $\lambda_1 \geq \lambda_2 \geq \cdots \geq \lambda_n$ and corresponding eigenvector $a^1, a^2, \cdots a^n$ can be obtained by solving equation (12). If the data was not centered in F, this can be done by substituting the kernel-matrix K with $\overline{K} = K - I_n K - K I_n + I_n K I_n$, where I_n is a $n \times n$ matrix with $(I_n) = 1/n$. Suppose x is a sample data and its map in F is $\Phi(x)$, to extract the feature information with KPCA, we can compute the projection of $\Phi(x)$ onto V^k by $(V^k \cdot \Phi(x)) = \sum_{i=1}^{n} a_i^k \Phi^T(x_i)\Phi(x) = \sum_{i=1}^{n} a_i^k K(x_i, x)$, where a_i^k is the coefficient of corresponding eigenvector of the i-th eigenvalue. Using the rule of $\sum_{i=1}^{p} \lambda_i / \sum_{i=1}^{n} c_{ij} > 0.9$, we can determine the number of important principal components. The first p principal components can be employed to construct the data when the sum of variances of the first p principal components 90% of the sum of whole variance.

There are 3 common kernel functions: polynomial function, RBF function and Sigmoid function. RBF function is adopted in this research.

Polynomial function: $K(x, y) = [\gamma(x \cdot y) + 1]^d$; RBF function: $K(x, y) = \exp(-\| x - y \|^2 / 2\sigma^2)$; Sigmoid function: $K(x, y) = \tanh[a(x \cdot y) + b]$.

3.3 Contourlet Transform in Image with KPCA

The contourlet transform consists of two filter banks, the one is Laplacian pyramid, the other is directional filter bank. The first stage is adopted to generate a multiscale representation of the 2-D data. The second stage processes the band pass subband image to capture the directional information on discontinuities. The output of the second filter bank is named contourlet coefficients. The contourlet transformation provides a multidirectional and multiscale representation of the image. It has been show that redundant and shift-invariant transformation produce better results in denoising.

Contourlet transform of image with KPCA is consisted of these steps:

Step1: applied Contourlet transformation to the image, carry out multiscale and multidirectional decompose.

Step2: let X be the coefficient matrix of subbands in the finest scale, calculate the X's covariance matrix: C_x and eigenvector ξ_x, eigenvalue λ_x.

Step3: let the cost fuction $y = \| X - \xi_n^T \xi_n X \|_2^2$, find the value n while y is the minimum.

Step4: for the first n eigenvector ξ_n, reconstruct subband coefficient $y = \xi_n^T X \xi_n$.

Step5: the denoising image is get after reverse contourlet transform using y and other leveles' coefficient.

4 Experiment Result and Discuss

Matlab 2008 is employed for computation and analysis in this simulation experimental. The input is 256×256 pixels standard test image Lina. Using contourlet decomposing, in the coarse-to-fine order, Laplacian pyramid stage is 3, the directional number is 2,4 and 8.

Table 1. PSNR value of different methods in denoising different noise

	Average filter	Median filter	Wiener filter	CT-PCA methods
Salt and Pepper Noise	24.8730	25.6422	24.4559	30.3647
Random noise	25.8964	25.1577	26.8546	28.7853
Gaussian	25.0657	24.1840	25.9765	28.0154

Table 2. MSE value of different methods in denoising different noise

	Average filter	Median filter	Wiener filter	CT-PCA methods
Salt and Pepper Noise	0.0042	0.0030	0.0047	2.8678e-004
Random noise	0.0025	0.0035	0.0021	0.0021
Gaussian	0.0028	0.0045	0.0028	0.0027

Table 3. NAE value of different methods in denoising different noise

	Average filter	Median filter	Wiener filter	CT-PCA methods
Salt and Pepper Noise	0.0768	0.0543	0.0758	0.0071
Random noise	0.0674	0.0697	0.0683	0.0653
Gaussian	0.0786	0.0703	0.0704	0.0689

With contourlet hard threshold denoising and wavelet hard threshold denoising as compare, in this experiment, apply multilayer threshold as the hard threshold of above mentioned, at the certain level, threshold is $\lambda_s = \gamma\sqrt{2\log N} \cdot MED[d_{j,k}]/0.6745$.

5 Conclusion

A novel image enhancement algorithm named CT-PCA is proposed to enhance the definition and vision effect of digital images. This method is based on gray mathematical morphology method, contourlet domain transform, and kernel principal component analysis method.. The simulation experimental results on images show that the proposed algorithm not only decreases image noise effectively, but also improves image PSNR, MSE and NAE data.

Acknowledgments. This work is work is supported by the project of Province Natural Science Fund of Guizhou (QKHJ [2009]2112) and Project of Science and Technology of Guiyang (No.ZG [2010] 1-57).

References

1. Simoncelli, E.P., Freeman, W.T., Adelson, E.H., Heeger, D.J.: Shiftable multiscale transforms. IEEE Trans. Inf. Theory 38(2), 587–607 (1992)
2. Meyer, F.G., Coifman, R.R.: Brushlets: a tool for directional image analysis and image compression. J. Appl. Comput. Harmon. Anal. 5, 147–187 (1997)
3. Kingsbury, N.: Complex wavelets for shift invariant analysis and filtering of signals. J. Appl. Compu. Harmon. Anal. 10, 234–253 (2001)
4. Candès, E.J., Donoho, D.L.: Curvelets—a surprisingly effective nonadaptive representation for objects with edges. In: Cohen, A., Rabut, C., Schumaker, L.L. (eds.) Curve and Surface Fitting. Vanderbilt Univ. Press, Nashville (2000)
5. Matheron, G.: Random Sets and Integral Geometry. J. Wiley & Sons (1975)
6. Serra, J.: Introduction to Mathematical Morphology. Computer Vision, Graphics, and Image Processing 35, 283–305 (1986)
7. Maragos, P.A.: A Representation Theory for Morphological Image and Signal Processing. IEEE Trans. Pattern Analysis and Machine Intelligence 11(6), 586–599 (1989)
8. Haralick, R.M., Sternberg, R., Zhuang, X.: Image analysis using mathematical morphology. IEEE Trans., PAMI 9(4), 532–550 (1987)
9. Gonzalez, R.C., Woods, R.E.: 'Digital image processing. Addison-Wesley, New York (1992)
10. Mallat, S., Zhong, S.: Characterization of signals from multiscale edges. IEEE Trans. Pattern Anal. Machine Intell. 14, 2207–2232 (1992)
11. da Cunha, A.L., Zhou, J., Do, M.N.: The Nonsubsampled Contourlet Transform:Theory, Design, and Applications. IEEE Trans. on Image Processing 15(10), 3089–3101 (2006)
12. Kirby, M., Sirovich, L.: Application of the Karhunen-Loeve procedure for the characterization of human faces. IEEE Transactions on Pattern Analysis and Machine Intelligence 12(1), 103–108 (1990)
13. Scholkopf, B., Smola, A.J., Muller, K.-R.: Nonlinear component analysis as a kernel eigenvalue Problem. Neural Computation 10, 1299–1319 (1998)

A Robust STBC-OFDM Signal Detection with High Move Speed Environments

Yih-Haw Jan

Dept. of Computer and Communication Engineering
Nan Kai University of Technology
Caotun, Taiwan 542, ROC.
how@nkut.edu.tw

Abstract. In this paper, we propose a robust STBC-OFDM signal detection to combat intercarrier interference (ICI) over high move speed conditions. Successive interference cancellation (SIC) is an effective detection technique for STBC-OFDM systems. In this paper, a modified SIC detection is applied to the STBC-OFDM systems. Simulation results show that when the move speed is up to 250 km/hr and the SNR is 30dB, the modified SIC can achieve the performance of less than 10^{-5}. Moreover, the performance with some conventional methods are also evaluated.

Keywords: MIMO-OFDM, STBC-OFDM, signal detection, SIC.

1 Introduction

Multiple-input multiple-output (MIMO) systems offer an increase in diversity. The technique of diversity is aopted to cope with the effects caused by fast time-varying channels. Diversity not only improves the quality of wireless communication, but also increases efficiency of transmission power and spectral bandwidth. Space-time block codes (STBC) is a spectral efficient transmit diversity techniques which have significant performance and low complexity[1],[2]. Besides, Orthogonal frequency-division multiplexing (OFDM) systems offer high receiving quality in frequency-selective fading channels [3]. In wireless transmission standards, the combination of OFDM system and MIMO framework is set as basic requirement (e.g., worldwide interperability for microwave access (Mobile WiMAX))[4].

In the mobile communications, the movement of receiver will raise Doppler effect to cause more acute channel variations. Time variations of the channel lead to loss of subcarrier orthogonality and result in intercarrier interference (ICI) in OFDM systems. The problem is especially acute when the mobile is high. It needs ICI mitigation to eliminate subcarrier correlation.

This paper puts emphasis on the signal detection, by ICI mitigation, to apply in STBC-OFDM system. Four detections, full detection, diagonal detection, LS-ZF detection and the modified SIC, are utilized to analyze the performance of STBC-OFDM system.

The rest of this paper is organized as follows. In Section 2, the system model of STBC-based MIMO-OFDM is introduced. In section 3, some detection methods for

D. Jin and S. Lin (Eds.): Advances in MSEC Vol. 1, AISC 128, pp. 623–628.
springerlink.com

STBC-OFDM in multi-path fading channels are reviewed. In Section 4, we present the modified SIC algorithm. The simulation results are shown in Section 5. And Section 6 gives the conclusions.

2 System Model

STBC-based MIMO-OFDM transmission system in 2ISO is indicated in Fig. 1. In two consecutive OFDM blocks, X_{2n+0} and X_{2n+1} are transmitted from two antennas. Equivalent space-time transfer matrix is

$$G_{2,STBC} = \begin{bmatrix} X_{2n+0} & X_{2n+1} \\ -X^*_{2n+1} & X^*_{2n+0} \end{bmatrix} \quad (1)$$

in which

$$X_{2n+p} = \begin{bmatrix} x_{2n+p,0} & \cdots & x_{2n+p,N-1} \end{bmatrix}^T, \quad p\text{=0 and 1} \quad (2)$$

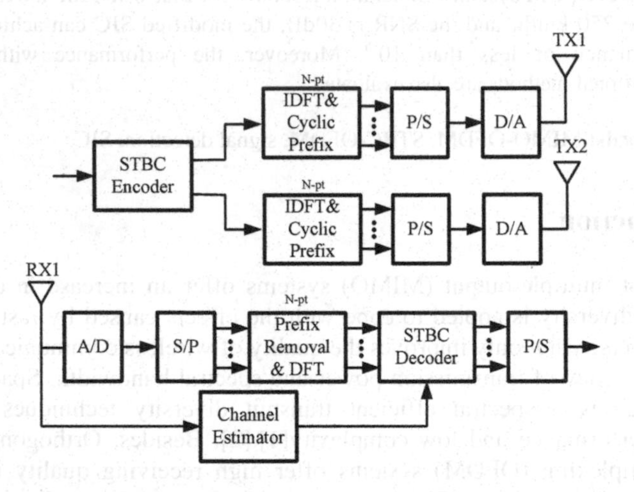

Fig. 1. STBC-OFDM transmit diversity system with one 2ISO

where $x_{2n+p,k}$ is the symbol transmitted by the k-th subcarrier in block $(2n+p)$. N is the number of subcarriers in OFDM. After demodulation, the system model corresponding time can be presented as:

$$\begin{bmatrix} Y_{2n+0} \\ Y^*_{2n+1} \end{bmatrix} = \begin{bmatrix} H^{(1)}_{2n+0} & H^{(2)}_{2n+0} \\ H^{(2)*}_{2n+1} & -H^{(1)*}_{2n+1} \end{bmatrix} \begin{bmatrix} X_{2n+0} \\ X_{2n+1} \end{bmatrix} + \begin{bmatrix} W_{2n+0} \\ W^*_{2n+1} \end{bmatrix} \quad (3)$$

in which, $H^{(i)}_{2n+p}$ is the channel frequency response (CFR) from the i-th transmit antenna to the receiving antenna in block $2n+p$. For simplification, (3) can be represented as

$$Y_{ST} = H_{ST} X_{ST} + W_{ST} \quad (4)$$

3 Signal Detection Method

Some STBC-OFDM techniques will be reviewed in this section. Those are Alamouti full-matrix detection, Alamouti diagonal-matrix detection, and LS-ZF detection.

Using Alamouti full-matrix detection means detecting signals by taking the full channel response matrix into consideration. To be simply put, following Alamouti full-matrix detection will be represented as full detection. By (4), the full detection of 2ISO STBC-OFDM system can be represented as:

$$\hat{\boldsymbol{X}}_{ST,full} = \boldsymbol{H}_{ST}^{H}\boldsymbol{Y}_{ST} \tag{5}$$

in which, $(\cdot)^H$ represents that the matrix does Hermitian transpose operation.

Alamouti diagonal-matrix detection is a special case of full detection. It only takes the diagonal value of frequency domain response into consideration to lower the computational complexity. By (4), the diagonal detection of 2ISO STBC-OFDM system can be represented as:

$$\hat{\boldsymbol{X}}_{ST,diag} = \boldsymbol{H}_{ST,diag}^{H}\boldsymbol{Y}_{ST} \tag{6}$$

where $\boldsymbol{H}_{ST,diag}$ is a diagonal matrix with its diagonal elements given as the diagonal values in $\boldsymbol{H}_{ST}$.

By (4), the LS-ZF detection of 2ISO STBC-OFDM system can be represented as:

$$\hat{\boldsymbol{X}}_{ST,LS-ZF} = (\boldsymbol{H}_{ST}^{H}\boldsymbol{H}_{ST})^{-1}\boldsymbol{H}_{ST}^{H}\boldsymbol{Y}_{ST} \tag{7}$$

LS-ZF is applied to eliminate jamming caused by other subcarriers; however, it also leads to the amplification of interference.

4 Proposed Modified SIC Detection

The modified SIC considers whole channel response to detect signals and accomplish it with matrix inversion, i.e. LS-ZF. If the receiving end can accurately estimate channel information, the algorithm of modified SIC in 2ISO STBC-OFDM system is:

0) Initial: $i \leftarrow 0$, $\boldsymbol{H}_0=\boldsymbol{H}_{ST}$, $\boldsymbol{Y}_0=\boldsymbol{Y}_{ST}$, $\boldsymbol{K} \equiv \{k_0\ k_1 \ldots k_{2N-1}\}$
1) $j=\arg_j \max\|(\boldsymbol{H}_i)_j\|^2$
2) $\boldsymbol{Z}_i=(\boldsymbol{H}_i^H\boldsymbol{H}_i)^{-1}\boldsymbol{H}_i^H=\boldsymbol{H}_i^+$
3) $\boldsymbol{U}(k_i)=\boldsymbol{Z}_i(j)\boldsymbol{Y}_i$
4) $\boldsymbol{X}_{STest}(k_i)=\mathrm{Q}(\boldsymbol{U}(k_i))$
5) $\boldsymbol{Y}_{i+1}=\boldsymbol{Y}_i - \boldsymbol{X}_{STest}(k_i)\boldsymbol{H}_i(j)$
6) $\boldsymbol{H}_{i+1}=[\boldsymbol{H}(0)\ldots\boldsymbol{H}(j-1)\ \boldsymbol{H}(j+1)\ldots\ \boldsymbol{H}(2N-1)]$ $\quad i \leftarrow i+1$; go to 1)

In this procedure, first, we substitute frequency domain channel matrix and receiving signal for the initialization of the modified SIC and obtain a collection

$\boldsymbol{K}$ of the best detection order in step 0). The j-th subcarrier with the maximum norm value obtained in step 1) undergoes the detection first.

Step 2) to 4) calculates the corresponding ZF-nulling vector and decisive value and defines constellation point, in which $\boldsymbol{Z}_i$ stands for the i-th recursive ZF-nulling vector, the dimension is $1\times 2N$; $\boldsymbol{Y}_i$ is the i-th recursive receiving signal after demodulation, the dimension is $2N\times 1$; $\boldsymbol{X}_{STest}$ stands for the estimated value of the transmit signal $\boldsymbol{X}_{ST}$ of STBC-OFDM.

Step 5) is the elimination of the detected vector in receiving vector, $\boldsymbol{H}_i(j)$ means the j-th column in $\boldsymbol{H}_i$. Step 6) is elimination of the estimated frequency domain channel response of subcarrier signals. Therefore, the vector of $\boldsymbol{H}_i$ will get smaller and smaller until signals on every subcarrier are processed.

5 Simulation

Simulation results will be implemented to verify the performance of each detection method to STBC-OFDM system. In this paper, the carrier frequency is 3.5 GHz, the number of subcarriers is 256, the bandwidth of the system is 1.25MHz, and the modulation adopted is QPSK. The transmission channel considered is COST207 Model, bad urban (BU) selected [5]. Move speeds considered are 75 km/hr, 125km/hr and 250 km/hr and their corresponding normalized Doppler frequency are 0.0497, 0.0829, and 0.1659 respectively.

Fig. 2 indicates 2ISO STBC-OFDM system's bit error rate (BER) performance when using full detection, diagonal detection, and LS-ZF in the BU. At 75 km/hr move speed, full detection and diagonal detection present error floor. Observe that LS-ZF can achieve 10^{-4} at signal-to-noise ratio (SNR) 20 dB. In addition, at move speed of 125 km/hr and 250 km/hr, because orthogonality is destroyed, full detection and diagonal detection's performance is almost the same. We can see that LS-ZF can achieve 10^{-4} and 10^{-3} BER at SNR 30 dB. Although full detection takes the whole matrix into consideration, other subcarrier will still influence accuracy of signal detection.

We see that using the LS-ZF can gain superior performance than full detection and diagonal detection; therefore, we will have performance analysis of the modified SIC and LS-ZF.

Fig. 3 is the performance of the modified SIC of 2ISO STBC-OFDM system in the environment of BU. The move speed is observed to rise from 125 km/hr to 250 km/hr. The performance of the modified SIC is significantly better than LS-ZF. The modified SIC deducts the interference of the detected signals from the correlation between subcarriers. Until subcarrier with smaller energy is detected, the interference from other subcarriers is gradually excluded. When the SNR is 30 dB, the modified SIC can achieve the BER of less than 10^{-5}, while the LS-ZF can only achieve the BER of about 10^{-3}. Therefore, the higher the SNR is, the better performance of the modified SIC will show.

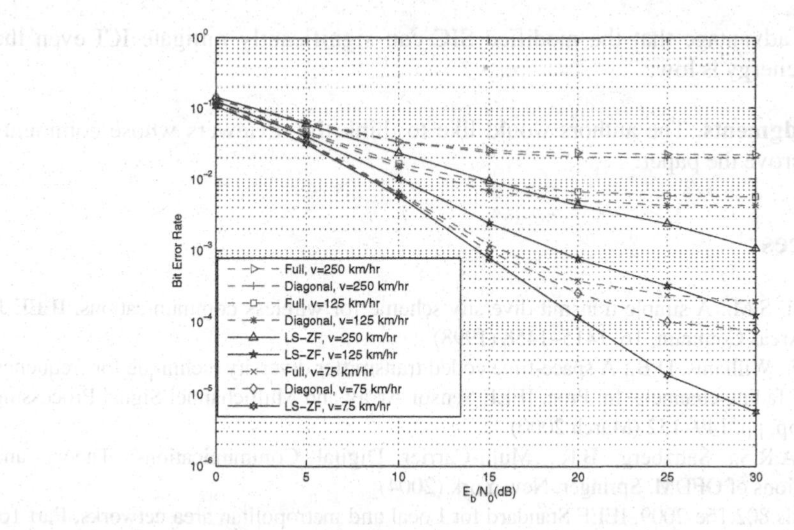

Fig. 2. Performance of 2ISO STBC-OFDM system when using full, Diagonal, and LS-ZF detections in the environment of COST207 BU

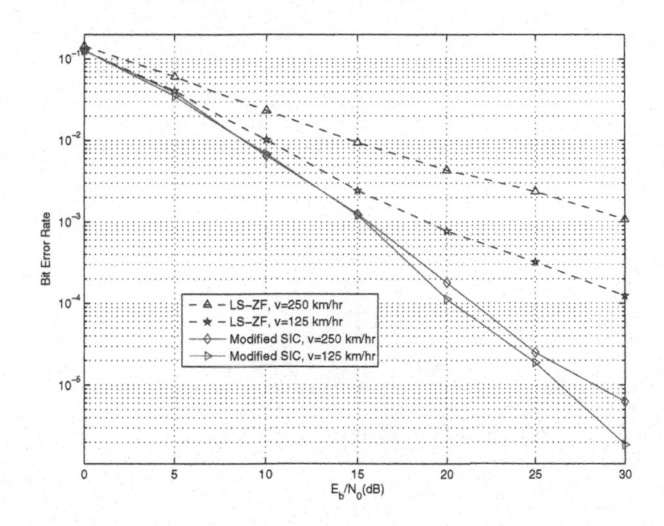

Fig. 3. Performance of 2ISO STBC-OFDM system when using modified SIC and LS-ZF detections in the environment of COST207 BU

6 Conclusion

This paper puts emphasis on the robustness of STBC-OFDM signal detection. With high move speed, full detection's and diagonal detection's performance will deteriorate. To further enhance the performance of the system, modified SIC detection is implemented. STBC-OFDM system boosts its performance by adopting the modified SIC to eliminate correlation between subcarriers. When with a high move speed, the advantage that the modified SIC can significantly mitigate ICI even the subcarrier energy is low.

Acknowledgments. The authors would like to think the reviewers whose comments helped improve the paper.

References

1. Alamouti, S.M.: A simple transmit diversity scheme for wireless communications. IEEE J. Select. Areas Commun. 16, 1415–1458 (1998)
2. Lee, K.F., Williams, D.B.: A space-time coded transmitter diversity technique for frequency selective fading channels. In: Proc. IEEE Sensor Array and Multichannel Signal Processing Workshop, pp. 149–152 (March 2000)
3. Bahai, A.R.S., Saltzberg, B.R.: Multi-Carrier Digital Communications: Theory and Applications of OFDM. Springer, New York (2004)
4. IEEE Stds 802.16e-2009, IEEE Standard for Local and metropolitan area networks, Part 16: Air Interface for Fixed and Mobile Broadband Wireless Access Systems, IEEE, May 29 (2009)
5. Patzold, M.: Mobile Fading Channels. John Wiley & Sons, Chichester (2002)

Application in Evaluating Driving Fatigue Influence Factors by Grey Interval Theory

Jixuan Yuan, Zhumei Song, and Shiqiong Zhou

Shenzhen Institute of Information Technology,
Shenzhen, Guangdong Province, China
{Yuanjx,Songzm,Zhousq}@sziit.com.cn

Abstract. Many researches focused on the formation mechanism of driving fatigue were made at home and abroad, but the importance ranking of various factors which leads to driving fatigue is seldom concerned. Based on previous researches, it arranges different factors causing driving fatigue in order by using the theory of grey interval. And the result shows that this method is more scientific and effective.

Keywords: Grey Interval Theory, driving fatigue, evaluation method.

1 Introduction

Cars are the most popular traffic vehicles in our daily life. As the car number arising, more and more traffic accidents occurred. In all the reasons, driving fatigue is an important one. Many researchers make thorough studies and give a lot of models to prevent driving fatigue. Many researches focused on the formation mechanism of driving fatigue were made at home and abroad, but the importance ranking of various factors which leads to driving fatigue is seldom concerned.

In reference [1], the author established comparison matrix to analyze the factors which affected the Model U Effect Stable Period (MUESP). But only sleep quality, living environment, car indoor environment, car outdoor environment, driving conditions and driver's self-conditions are taken into account as influence factors, all the factors has been paired compared and the paired comparison matrix has been given directly without any explains for how to calculate the comparison weight. It is not suitable for accurate evaluation of the influence factors index for MUESP.

This paper adopts grey interval theory to analyze all the influence factors for MUESP, and after five experts giving their evaluations, the accurate evaluation of the influence factors index for MUESP has been obtained by the analysis of grey interval theory.

2 Grey Interval Theory

The Interval grey number is not a exact value but a value scope. That is to say, the interval grey number is variable; its range varies with the interval grey number. Let's

D. Jin and S. Lin (Eds.): Advances in MSEC Vol. 1, AISC 128, pp. 629–633.
springerlink.com

suppose a evaluator give a scope $[\alpha, \beta]$ to some influence factor g. If there are n evaluators give their results to influence factor g, the results will become a Interval grey number serial $[\alpha_1, \beta_1]$, $[\alpha_2, \beta_2]$, …, $[\alpha_n, \beta_n]$. The comprehensive evaluation value of this serial is as follow:

$$E(g) = \frac{1}{2} \frac{\sum_{i=1}^{n} (\beta_i^2 - \alpha_i^2)}{\sum_{i=1}^{n} (\beta_i - \alpha_i)}. \tag{1}$$

The variance of this serial is as follow:

$$D(g) = \frac{1}{3} \frac{\sum_{i=1}^{n} [(\beta_i - E(g))^3 - (\alpha_i - E(g))^3]}{\sum_{i=1}^{n} (\beta_i - \alpha_i)}. \tag{2}$$

Therefore, the Standard deviation of this serial is $S = \sqrt{D(g)}$. The possible range of g can be calculate as $[E(g) - S, E(g) + S]$, the effective range of this evaluation serial is as follows:

$$[\xi_i, \eta_i] = [\alpha_i, \beta_i] \cap [E(g) - S, E(g) + S].$$

If there are k influence factors, k interval grey number serials can be drawn as $[\alpha_{1j}, \beta_{1j}]$, $[\alpha_{2j}, \beta_{2j}]$, …, $[\alpha_{nj}, \beta_{nj}]$. The comprehensive evaluation value of one serial is as follows:

$$E_j(g) = \frac{1}{2} \frac{\sum_{i=1}^{n} (\beta_{ij}^2 - \alpha_{ij}^2)}{\sum_{i=1}^{n} (\beta_{ij} - \alpha_{ij})}. \tag{3}$$

The variance of this one serial is as follow:

$$D_j(g) = \frac{1}{3} \frac{\sum_{i=1}^{n} [(\beta_{ij} - E_j(g))^3 - (\alpha_{ij} - E_j(g))^3]}{\sum_{i=1}^{n} (\beta_{ij} - \alpha_{ij})}. \tag{4}$$

Therefore, the effective range of every evaluation serial is as follows:

$$[\xi_{ij},\eta_{ij}]=[\alpha_{ij},\beta_{ij}]\cap[E_j(g)-S_j,E_j(g)+S_j]$$

$$(i=1,2,\cdots,n)\ (j=1,2,\cdots,k).$$

Then the effective level of the comprehensive evaluation value to the j_{th} influence factor by the i_{th} evaluator is as follows:

$$Q_{ij}=\frac{(\eta_{ij}-\xi_{ij})}{(\beta_{ij}-\alpha_{ij})}. \tag{5}$$

Among all the evaluators, the effective level of the i_{th} evaluator is as follows:

$$W_i=\frac{1}{k}\sum_{j=1}^{k}Q_{ij}. \tag{6}$$

In fact, if the effective level of the evaluator $W_i<0.1$, the evaluation result is not correct. It should be reject his evaluation result, and re-sorting the evaluation result again according to formula (7).

At last, the ranking result value drawn from comprehensive evaluation of every influence fact by each evaluator is as follows:

$$e_j=\sum_{i=1}^{n}\frac{\xi_{ij}+\eta_{ij}}{2}W_i\quad(j=1,2,\cdots,k)\ . \tag{7}$$

Formula (7) is the method for sort every influence factor, in next part it will show you how to get the final ranking result.

3 Example

According to the grey interval evaluation method, influence factors to MUESP such as sleep quality, living environment, car indoor environment, car outdoor environment, driving conditions and driver's self-conditions would be evaluated by five evaluators. Remarks to the pair comparison are not simply '1', '3', '5', '1/3', '1/5' but 'more important', 'important', 'the same level' 'unimportant' and 'more unimportant'. One of the evaluation indexes are assigned as an Anchor. In this sample 'sleep quality' are assigned as standard comparison item, and all the comparison results are shown in paragraph 1.

Para 1. Comparison results by experts

Influence items	Sleep quality (Anchor)	living environment	car indoor environment	car outdoor environment	driving conditions	driver's self-conditions
Expert 1	the same	unimportant	more unimportant	the same	more unimportant	unimportant
Expert 2	the same	more unimportant	more unimportant	important	the same	the same
Expert 3	the same	unimportant	the same	important	the same	important
Expert 4	the same	more unimportant	unimportant	important	important	the same
Expert 5	the same	unimportant	unimportant	the same	the same	more important

According to the grey interval theory introduced last part, the effective level of every expert is calculated as paragraph 2:

Para 2. The effective level of every expert

Expert No.	expert1	expert2	expert3	expert4	expert5
effective level	0.476	0.844	0.604	0.608	0.744

According to the result of Para 2, each expert's evaluation result is available, and final evaluation result of every influence factor can be drawn as paragraph 3.

Para 3. Final evaluation result of every influence factor

	Sleep quality	living environment	car indoor environment	car outdoor environment	driving conditions	driver's self-conditions
Rank index	7.4	3.55	4.26	10.23	5.12	6.65
Rank index after normalization	0.72	0.35	0.42	1	0.5	0.65

The result of Para 3 shows the most important influence factor to driving fatigue is 'car outdoor environment', the second important influence factor is 'sleep quality', the third is 'driver's self-conditions'. This rank index result matches our daily driving experience, driving fatigue happens the highest probably on expressway.

In reference 1, it is not correct from neither scientific nature nor preciseness that the author gave the same weight to 'sleep quality' and 'car outdoor environment'.

And his ranking index result only expresses his thought himself. Using grey interval method, we can both save the trouble to give the exact value such as '3', '1/3' and collect many experts' evaluations. And those advantages ensure the evaluation result more correct.

4 Conclusion

Many researches focused on the formation mechanism of driving fatigue were made at home and abroad, but the importance ranking of various factors which leads to driving fatigue is seldom concerned. Based on previous researches, this paper arranges different factors causing driving fatigue in order by using grey interval method. And the result shows that this method is more scientific and effective.

Acknowledgement. This works is supported by the Natural Science Foundation of Guangdong Province (Grant No. 9151503102000014).

References

1. Jian, J.: Research on Driving Fatigue Mechanism and Feedback and Selection Mode. Southwest Jiaotong University (2002)
2. Kuwano, S., Namba, S., Fastl, H., Schick, A.: Evaluation of the Impression of Danger Signals Comparison between Japanese and German Subjects. In: Schick, A., Klate, M. (eds.) 7. Oldenburger Symposium, BIS Oldenburg
3. Otto, N., Amman, S., Eaton, C., Lake, S.: Guidelines for Jury evaluations of Automotive Sounds. SAE TECHNICAL Paper 1999-01-1822
4. Solomon, L.N.: Semantic reactions to systematically varied sounds. J. Acoust. Soc. Am. 31, 986–990 (1959)
5. Zhou, K., Feng, S.: Evaluating the conceptual design schemes of complex products based on FAHP using fuzzy number. Journal of Systems Engineering and Electronics (03) (2005)

An Improved Semantic Differential Method and Its Application on Sound Quality

Jixuan Yuan, Zhumei Song, and Shiqiong Zhou

Shenzhen Institute of Information Technology,
Shenzhen, Guangdong Province, China
{Yuanjx,Songzm,Zhousq}@sziit.com.cn

Abstract. The evaluators frequently couldn't give a definition value but an interval to the evaluation objects during their evaluation course, and then Grey Interval Method(GIM) was proposed. Semantic Differential Method(SDM) was fit for those evaluators without experience and training. Therefore, a method that introduced GIM into SDM is proposed in this paper. It can allow the evaluators give their options more favoringly. At last, a sound quality example was given to prove the method.

Keywords: Grey Interval Theory, semantic differential method, evaluation method, sound quality, loudness.

1 Introduction

There are some differences on evaluation method of sound quality because of different understanding, and many factors will affect the evaluation results such as physical acoustics and psychoacoustics. The popular subject evaluation methods are including paired comparison method and semantic differential method etc. This article made some improvement to semantic differential method based on grey interval method, and then an air-conditioner under different conditions is subject evaluated, the result is satisfied.

The air-conditioner samples in are listed in Table. 1, including Loudness and A-weighted Sound Pressure Level (SPL).

Table 1. The Result of Noise Samples

Sample No.	A-weighted Sound Pressure Level (dB)	Loudness (sone)	Sample No.	A-weighted Sound Pressure Level (dB)	Loudness (sone)
1	42.4	4.88	6	42.2	4.82
2	40.1	4	7	46	7.25
3	38.8	3.34	8	44.4	6.67
4	44.1	5.91	9	44.7	6.48
5	42.8	5.27			

D. Jin and S. Lin (Eds.): Advances in MSEC Vol. 1, AISC 128, pp. 635–639.
springerlink.com

In this experiment, the jury evaluated the nine samples with words like 'quiet' or 'loud'. Corresponding to the jury's feeling, each grey value was introduced to represent the loudness degree: '0-1' represent 'the quietest', '1-2' represent 'quieter', '2-3' represent 'quiet', '3-4' represent 'the same', '4-5' represent 'loud', '5-6'represent 'louder', '6-7' represent 'the loudest'. For example, the jury gave the result as '3-4', it means the jury's option is 'the same'.

Among the nine samples, we appoint the 5th sample as 'Anchor'. That is to say, the evaluation result of the 5th sample are all 'the same'.

2 Improved Semantic Differential Method

The Interval grey number is not a exact value but a value scope. That is to say, the interval grey number is variable; its range varies with the interval grey number. Let's suppose a evaluator give a scope $[\alpha,\beta]$ to some influence factor g . If there are n evaluators give their results to influence factor g , the results will become a Interval grey number serial $[\alpha_1,\beta_1]$, $[\alpha_2,\beta_2]$, ... , $[\alpha_n,\beta_n]$. The comprehensive evaluation value of this serial is as follow:

$$E(g)=\frac{1}{2}\frac{\sum_{i=1}^{n}(\beta_i^2-\alpha_i^2)}{\sum_{i=1}^{n}(\beta_i-\alpha_i)}. \tag{1}$$

The variance of this serial is as follow:

$$D(g)=\frac{1}{3}\frac{\sum_{i=1}^{n}[(\beta_i-E(g))^3-(\alpha_i-E(g))^3]}{\sum_{i=1}^{n}(\beta_i-\alpha_i)}. \tag{2}$$

Therefore, the Standard deviation of this serial is $S=\sqrt{D(g)}$. The possible range of g can be calculate as $[E(g)-S,E(g)+S]$, the effective range of this evaluation serial is as follows:

$$[\xi_i,\eta_i]=[\alpha_i,\beta_i]\cap[E(g)-S,E(g)+S].$$

If there are k influence factors, k interval grey number serials can be drawn as $[\alpha_{1j},\beta_{1j}]$, $[\alpha_{2j},\beta_{2j}]$, ..., $[\alpha_{nj},\beta_{nj}]$. The comprehensive evaluation value of one serial is as follows:

$$E_j(g) = \frac{1}{2}\frac{\sum_{i=1}^{n}(\beta_{ij}^2 - \alpha_{ij}^2)}{\sum_{i=1}^{n}(\beta_{ij} - \alpha_{ij})}. \tag{3}$$

The variance of this one serial is as follow:

$$D_j(g) = \frac{1}{3}\frac{\sum_{i=1}^{n}[(\beta_{ij} - E_j(g))^3 - (\alpha_{ij} - E_j(g))^3]}{\sum_{i=1}^{n}(\beta_{ij} - \alpha_{ij})}. \tag{4}$$

Therefore, the effective range of every evaluation serial is as follows:

$$[\xi_{ij}, \eta_{ij}] = [\alpha_{ij}, \beta_{ij}] \cap [E_j(g) - S_j, E_j(g) + S_j]$$
$$(i = 1, 2, \cdots, n)\ (j = 1, 2, \cdots, k).$$

Then the effective level of the comprehensive evaluation value to the j_{th} influence factor by the i_{th} evaluator is as follows:

$$Q_{ij} = \frac{(\eta_{ij} - \xi_{ij})}{(\beta_{ij} - \alpha_{ij})}. \tag{5}$$

Among all the evaluators, the effective level of the i_{th} evaluator is as follows:

$$W_i = \frac{1}{k}\sum_{j=1}^{k} Q_{ij}. \tag{6}$$

In fact, if the effective level of the evaluator $W_i < 0.1$, the evaluation result is not correct. It should be reject his evaluation result, and re-sorting the evaluation result again according to formula (7).

At last, the ranking result value drawn from comprehensive evaluation of every influence fact by each evaluator is as follows:

$$e_j = \sum_{i=1}^{n}\frac{\xi_{ij} + \eta_{ij}}{2}W_i \quad (j = 1, 2, \cdots, k). \tag{7}$$

Formula (7) is the method for sort every influence factor, in next part it will show you how to get the final ranking result.

3 Example

The subjective evaluation results are shown in Table. 2.

Table 2. Evaluation results of loudness

samples / Evaluator	1	2	3	4	5 (Anchor)	6	7	8	9
1	quieter	quieter	the quietest	louder	the same	quieter	the loudest	louder	louder
2	quiet	quieter	quieter	the same	the same	the same	the loudest	louder	louder
3	quiet	quieter	quieter	louder	the same	quieter	louder	loud	loud
4	quieter	the quietest	the quietest	the same	the same	quieter	louder	louder	louder
5	the same	quiet	quieter	quiet	the same	loud	the loudest	loud	loud
6	quiet	quieter	quieter	louder	the same	quieter	louder	louder	louder
7	quieter	quiet	quieter	louder	the same	the same	louder	louder	loud
8	quiet	quieter	the quietest	louder	the same	quieter	the loudest	louder	loud
9	quiet	quieter	quieter	louder	the same	the same	louder	louder	louder
10	quiet	quieter	quieter	the same	the same	quieter	louder	loud	louder

According to the improved semantic differential method, we can get the effective level of 10 evaluators as follows:

Table 3. Effective levels of 10 evaluators

Evaluator No.	1	2	3	4	5	6	7	8	9	10
Effective level	0.62	0.74	0.69	0.54	0.31	0.75	0.67	0.58	0.75	0.71

In Table 3 most evaluators' effective level are above 0.5 except that the 5th evaluator's effective level is 0.31. In order to keep the accuracy of evaluation, after rejecting the 5th evaluation, a new effective level result was recalculating as follows:

Table 4. Effective levels of 9 residual evaluators

Evaluator No.	1	2	3	4	6	7	8	9	10
Effective level	0.63	0.63	0.64	0.52	0.74	0.63	0.53	0.71	0.64

In Table 4 each residual evaluator's effective level changed more balance, which shows that the evaluation result is more credible. Then the effective level was made a substitution in Equation 7, and a sorting from loud to quiet is as follows:

Table 5. Sorting result of the samples from loud to quiet

Sample No.	7	8	9	4	5	6	1	2	3
Sorting value	33	29.96	29.43	23.76	19.87	16.08	12.41	8.52	6.79
Loudness (sone)	7.25	6.67	6.48	5.91	5.27	4.82	4.88	4	3.34

In Table 5 we can find that the loudness sorting list was similar to their A-weighted Sound Pressure Level. Only the 6th sample was reversed from the 1st sample. But the two samples is nearly the same in loudness, it belongs to the acceptable range of the error.

4 Conclusion

The evaluators frequently couldn't give a definition value but an interval to the evaluation objects during their evaluation course, and then Grey Interval Method(GIM) was proposed. Semantic Differential Method(SDM) was fit for those evaluators without experience and training. Therefore, a method that introduced GIM into SDM is proposed in this paper. It can allow the evaluators give their options more favoring. At last, a sound quality example was given to prove the method.

Acknowledgement. This works is supported by the Natural Science Foundation of Guangdong Province (Grant No. 9151503102000014).

References

1. Zhou, K., Feng, S.: Evaluating the conceptual design schemes of complex products based on FAHP using fuzzy number. JSEE (03) (2005)
2. Kuwano, S., Namba, S., Fastl, H., Schick, A.: Evaluation of the Impression of Danger Signals Comparison between Japanese and German Subjects. In: Schick, A., Klate, M. (eds.) 7. Oldenburger Symposium, BIS Oldenburg
3. Otto, N., Amman, S., Eaton, C., Lake, S.: Guidelines for Jury evaluations of Automotive Sounds. Sae Technical Paper 1999-01-1822
4. Solomon, L.N.: Semantic reactions to systematically varied sounds. J. Acoust. Soc. Am. 31, 986–990 (1959)

Design and Application of Linux-Based Embedded Systems

Chunling Sun

School of Information Science and Technology, Heilongjiang University
chunling6666@163.com

Abstract. Linux-based systems have become a key support in the global electronic market. The latest data Based on third-party shows that. The Applications of embedded systems has occupied more than 40% of the entire computer applications around the world. This paper will introduce a key design of a Linux-based embedded systems, and the application in the teaching course.

Keywords: Embedded systems, Linux-based embedded systems, Design of hardware, Design of software.

Modern computer systems and computing power according to their different function, can be divided into the desktop for personal information processing systems, image processing, advanced graphics workstation, a host of scientific computing systems, as well as being present in a variety of mechanical and electrical equipment, embedded system. Currently, the embedded technology and embedded products have penetrated into industrial control systems, information appliances, communication equipment, instrumentation, military technology, and all areas of daily life.

1 Linux-Based Embedded System

1.1 What Is a Linux-Based Embedded System

Embedded systems based on application-centric, integrated computer software and hardware technology, communication technology and microelectronics technology, "tailored" approach to the required the embedded into the application system equipment for the application system functionality, reliability, cost, size, power demanding a dedicated computer system. Embedded system consists of an embedded processor and related support hardware, embedded operating system and application software, etc., is available independently of the "device."

With the advances in electronic technology Linux as a free software, get a great deal of development, embedded systems and Linux combination is increasingly being optimistic. Linux has its own set of tool chain, easy to create your own embedded system development environment and cross-operating environment, and embedded system development across the simulation tools obstacles. Linux kernel has a small, high efficiency, open source and so on.

D. Jin and S. Lin (Eds.): Advances in MSEC Vol. 1, AISC 128, pp. 641–645.
springerlink.com

Embedded Linux is increasingly popular Linux operating system modified so that it can be cut in the embedded computer system running an embedded Linux operating system on the Internet not only inherited the unlimited resources but also has embedded the open source operating system characteristics.

1.2 Advantages of Linux as an Embedded Operating System

1.2.1 Can be applied to a variety of hardware platforms. Linux uses a unified framework to manage the hardware, from one hardware platform to another hardware platform changes and the upper application-independent. Linux can be freely configured, does not require any license or business relationship, the source code is freely available. This makes the use of Linux as the operating system does not have any copyright disputes, which will save a lot of development costs.

1.2.2 Linux is similar to kernel-based, with full memory access control, supports a large number of hardware and other characteristics of a common operating system. All of its open source program, anyone can modify and GUN General Public License issued under. In this way, developers can customize the operating system to meet their special needs.

1.2.3 Linux users are familiar with the complete development tools, almost all Unix systems applications have been ported to Linux. Linux also provides a powerful networking feature, a variety of optional window manager (X Windows). Its powerful language compiler GCC, C++, etc. can also be very easy to get, not only sophisticated, but easy to use.

2 Linux-Based Embedded System Design

A typical embedded system usually consists of an embedded microprocessor (or microcontroller), various control sensor interfaces, communication network interface and application-specific or control software and other parts. As IC manufacturing technology development, more integrated chips. In modern embedded systems, microprocessor cores and peripheral interfaces are often integrated into an embedded chip. We can follow the system would be embedded chip is divided into several categories, including a microcontroller for control systems for hand-held communications devices and embedded chips for multimedia data processing DSP chips.

2.1 Hardware Design

The choice of hardware platform is an embedded processor of choice, the choice of what kind of processor cores depending on the application area, the needs of users, cost, ease of development and other factors. In determining the embedded processor core, the peripherals must also consider the needs of the situation, choose a suitable processor. the demand for bus, there is no Universal Serial Interface UART, the need for the USB bus, there is no Ethernet interface within the system need to SPI bus peripheral interface, the need for A / D and D / A converter. Figure 1 is a typical embedded Linux development environment that includes the host workstation or PC support GDB debugging tools BDI2000.

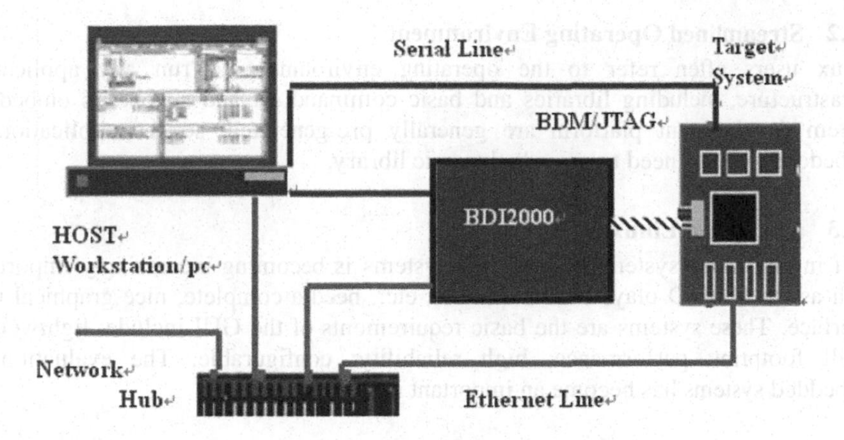

Fig. 1. Embedded Linux Development Environment

2.2 Software Design

Complete embedded Linux solutions include embedded Linux operating system kernel, operating environment, graphical interface and application software. As the special requirements of embedded devices, embedded Linux solutions in the core, the environment, GUI and so very different from the standard Linux, the main challenge is how to narrow FLASH, ROM and memory to achieve high-quality real-time tasks scheduling, graphical display, network communications and other functions.

2.2.1 Streamline Core

Linux kernel has its own architecture, in which process management, memory management and file system is the most basic of three sub-systems. Figure 2 simply expressed its framework. User process can directly or through the library system calls to access kernel resources.

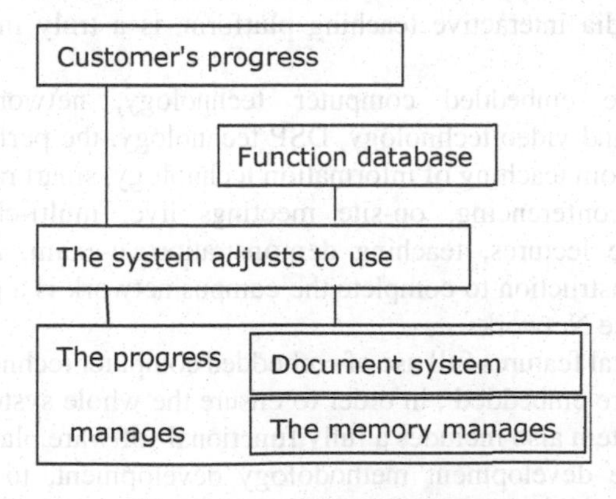

Fig. 2. Linux Kernel Structure

2.2.2 Streamlined Operating Environment

Linux users often refer to the operating environment to run any application infrastructure, including libraries and basic command set and so on. As embedded system development platform are generally pre-generated, so the application of embedded systems need to provide dynamic library.

2.2.3 Embedded Linux GUI

GUI in embedded systems or real-time systems is becoming increasingly important, such as PDA, DVD player, WAP phones, etc., need a complete, nice graphical user interface. These systems are the basic requirements of the GUI include: lightweight, small footprint; performance; high reliability; configurable. The evaluation of embedded systems has become an important indicator.

3 Linux-Based Embedded System Applications in Teaching

Educational information in the education process is more comprehensive use of computer-based multimedia and network communication of modern information technology to promote a comprehensive reform of education, to adapt the information society is approaching the new requirements for the development of education. Embedded system as the most cutting-edge of contemporary science and technology is applied to the education sector is one of them, are adapted to the requirements of information technology in education in China, which will help to achieve leap-forward development of China's education.

Linux-based embedded system in the education sector is widely used, the following examples--Embedded Broadband campus.

System built on the standard Fast / Gigabit campus network, campus network, based on fully integrated with the campus network, but also with education MAN, seamless network of public community. The system will broadcast networks, teaching evaluation network, multimedia network, cable television, and so all the traditional analog network into the campus network, to achieve "multi-play", and continue to strengthen multi-media, video, interactive features that make the campus network to become a multimedia interactive teaching platform, is a truly multi-media campus network.

System will be embedded computer technology, network communication technology, audio and video technology, DSP technology, the perfect combination. It supports the classroom teaching of information technology, smart radio, live television broadcast, video conferencing, on-site meetings live, multi-class discussion of classroom, distance lectures, teaching demonstration, e-exam, and so on all the activities of daily instruction to complete the campus network is a professional school of Integrated Service Networks.

A serious technical features full use of embedded computer technology, all computer hardware devices are embedded , in order to ensure the whole system is very reliable, stable and safe. System also includes a fully functional software platform, Linux-based embedded software development methodology development, to ensure the whole system stable and reliable. System integration of existing school facilities, such as a

VCR, VCD, DVD, tape recorders, cable TV analog audio and video programs, etc., which inherit the existing mode of teaching.

To the client, for example, embedded hardware platform consists of CPU, MPEG-2 decoding, TV and monitor interface, MPEG-2 encoding, digital module analog AV, 10M/100M adaptive Ethernet interfaces, infrared interfaces, seven major module is the core of the whole system. Software platform consists of embedded Linux operating system, Ethernet driver, infrared interface driver, MPEG-2 decoder driver, TV and monitor interface drivers, MPEG-2 encoding driver, analog AV digital module driver Teachers terminal control module, classroom information upload module, receiver module lesson plans, lesson plans display module, the total control module and other accessories.

4 The Future of Linux-Based Embedded System

Linux is a national policy of China's software, the use of large groups, open system and the vast expanse of rich resources, making Linux the future must be the focus of our popularity and promotion, Linux has gained superiority. Now there are many network technologies, servers, network devices are based on the Linux operating system. Linux-based embedded systems in many fashion phones, PDA, media players and other consumer electronic products has been widely used.

References

1. Wang, C., Zhang, L., Dong, M.: Development of Linux-Based Embedded Systems. Computer Applications (July 22, 2002)
2. Wang, G., Zhan, G., Gui, W.: Design and Implement of a Virtual Client/Server Based Embedded Web Browser (THEWB). Application Research of Computers (April 19, 2002)
3. Zhang, S.: Design and Implementation of an Embedded Browser Based on Linux. Journal of Xinyang Normal University (Natural Science Edition) (April 20, 2007)
4. Gao, Z., Yuan, P., Wang, X.: A Preliminary Study on Embedded Linux Based Embedded GIS. Geomatics World (February 5, 2004)
5. Chen, J., Chen, W.: One Video Player Which Runs on Embedded Linux Browser. Application Research of Computers (June 18, 2001)

Detection of Human Movement Behavior Rules Using Three-Axis Acceleration Sensor

Hui Xu, Lina Zhang, and Wenting Zhai

School of Information Science & Engineering,
Shenyang University of Technology, Shenyang, P.R. China
xhimage@163.com

Abstract. Human movement behavior rules are important parameters for identifying human individual. In this paper, three-axis acceleration sensor is used for sensing body movement behavior, and a microcontroller is used as the core to complete information collecting, recording, storing, and communicating with PC. The velocity and displacement curves are derived from integral operation of the acceleration value. The correctness of hardware systems and processes methods are verified by using planar movements and exported trajectories. Step and squats movement experiments are proceeded and the data are collected and analyzed. The movement frequency is extracted which represents the basic parameter of movement behavior.

Keywords: human movement, behavior rules, three-axis acceleration sensor, movement trajectories, movement frequency.

1 Introduction

Detection of human movement behavior rules is a new biometric identification technology, using the collected acceleration data to achieve the classification of human movement behavior and identify the athletic stance [1].

At present, there are two basic methods of detection of human behavior [2]: (1) The analysis is based on video images. Objects of real-time movement are monitored by the camera, venues are limited, and image processing algorithms have a high degree of complexity. (2) Detection is based on wearable device, compact design of the device is adaptable. These two methods are all needed for classification of motor behavior and establishment of the appropriate feature library.

Accelerometers are small, low cost and suitable as a detection device sensor. It can be used to measure the acceleration of gravity and body movement, suitable for measuring position and orientation of the movement of the body [3].

In this paper, using three-axis digital accelerometer has designed human movement behavior detection device, which gives the simple movement of the plane movement trajectory and frequency of human movement analysis.

2 Hardware Circuit Design

The hardware system consists of three-axis acceleration sensor module, microcontroller, memory, power supply, state switch, real-time clock, LCD module and serial

D. Jin and S. Lin (Eds.): Advances in MSEC Vol. 1, AISC 128, pp. 647–652.
springerlink.com

communication module [4]. Sensor measures the acceleration vector; real-time clock is used to record time for each test; LCD module is used to describe the trajectory of pointing; serial communication module is used for microcontroller and PC communications. The specific implementations design of the architecture is shown in Figure 1. Acceleration sensor data is transmitted to the PC machine through the serial port, and then MATLAB is use to process the data.

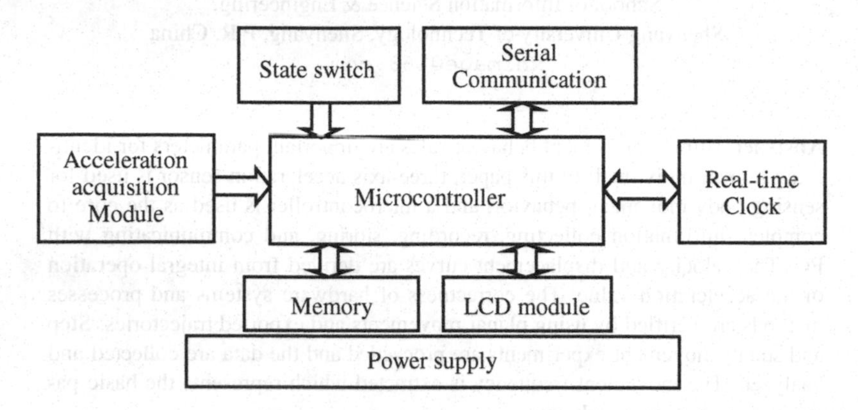

Fig. 1. Block diagram of system

Freescale MMA7455L is used as three axis detection module which is a MEMS accelerometer. Data transmission uses I^2C or SPI interface [5]. In this paper, 8-bit data output mode of the MMA7455L is used, and otherwise 10-bit mode can be used.

The design of microcontroller uses STC89C52, the data are stored in the AT24C256, the DS1302 real-time clock records test time. The liquid crystal display module OCMJ4X8C is used to show the movement trajectory.

3 Velocity and Displacement Calculation Method

Physical quantities which are measured by an acceleration sensor are only discrete acceleration values. In order to obtain displacement value, acceleration values must be integrated for two times [6].

Discrete velocity formula obtained by the discrete acceleration integral is shown as formula (1).

$$V(k) - V(m) = \sum_{i=m}^{k} a(i)T_s \tag{1}$$

Among them, m is the starting point of integral time; k is the follow-up time point of m, k=m, m+1, …… , n; $a(i)$ is the acceleration value detected at the i-th point; T_s is time interval between the two adjacent acquisition point; $V(k)$ and $V(m)$ are the velocity values corresponding to the k-th and m-th points.

Suppose the initial velocity is zero at the movement beginning, i.e., the velocity formula is shown as formula (2) when m=0.

$$V(m) = 0 \tag{2}$$

The velocity formula is as formula (3) by combining the formula (1) and (2).

$$V(k) = \sum_{i=m}^{k} a(i) T_s \tag{3}$$

Movement displacement formula obtained by the discrete velocity integral based on formula (3) is shown as formula (4).

$$S(k) = \sum_{i=m}^{k} V(i) T_s \tag{4}$$

Among them, m=0, k=m, m+1, … , n; $S(k)$ is displacement; $V(i)$ is the velocity corresponding to the i-th point.

4 Planar Movement Experiment and Result Analysis

Firstly, using the testing system did straight back and forth, square, oval, round, line and other planar movements. Figure 2 shows the X axis and Y axis acceleration, velocity and displacement curves which are drawn based on data of square movement on the desktop by MATLAB software. Movement trajectory is shown in figure 3.

According to figure 3 shows, square movement trajectory obtained by integral operation accurately reflects the X, Y axis of movement sequence, and through the movement locus can judge the laws of movement. Experimental results curves are ideal and the deviations are from the drift of sensor.

5 Human Movement Behavior Rules Detection

5.1 Squat Movement Detection

The test system is worn on the rear of the belt and needs to be guaranteed in the test position, X axis positive direction pointing in the direction of the human body is just above, Y axis positive direction pointing in the direction of the human body is left, and Z axis negative direction points the way forward people. Then collect the movement data. To human squats as an example, the human body first squats, stands up again, and repeats 5 times. Three-axis of movement acceleration curves are shown in figure 4.

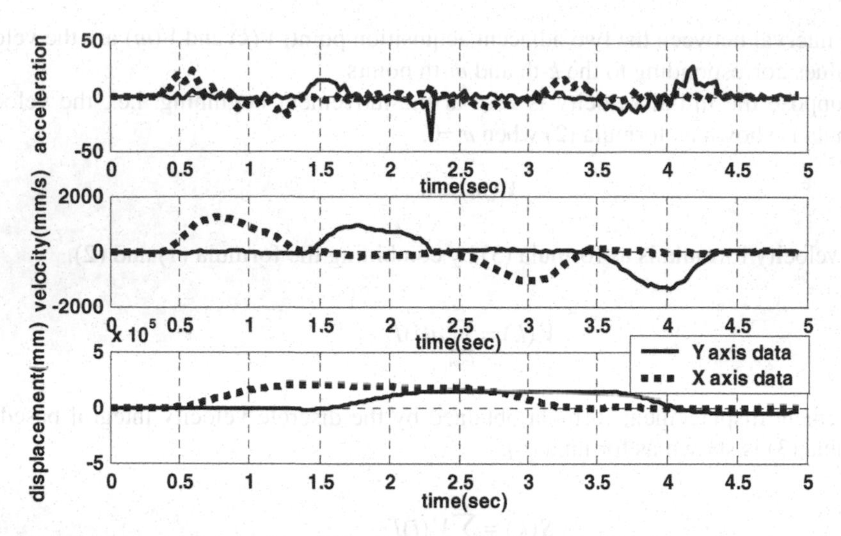

Fig. 2. Square movement X and Y axes of acceleration, velocity, displacement curves

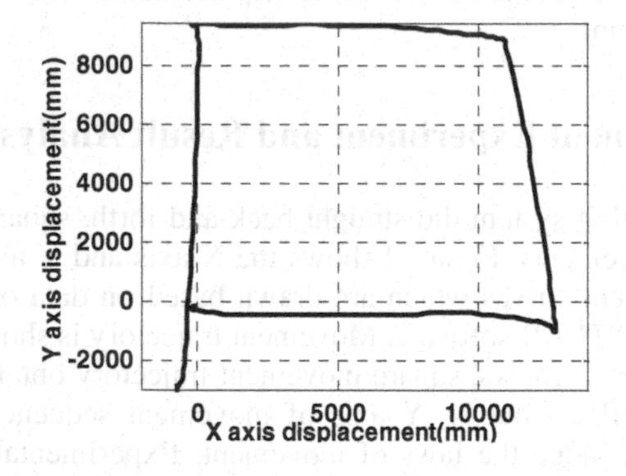

Fig. 3. Square movement trajectory

The figure shows, X axis and Z axis acceleration curves are obvious and both contain five complete cycles of movement. Because when the body squats, the body is downward and forward, and the body in horizontal direction is not changed, so the Y axis acceleration curve characteristic is not obvious. The X and Z axis acceleration curves can reflect the characteristics of the squats, that is, the peak stands for squat minimum. In order to analyze the Z axis spectrum as an example, periodic movement amplitude spectrum is obtained by using Fourier transform, take t=1s to t=8s point for analysis, contains 4 periodic movement. The Z axis amplitude spectrum of partial enlargement is shown in figure 5.

It can be seen the maximum amplitude Az of the Z axis acceleration amplitude spectrum is at the frequency 0.5787Hz by FFT.

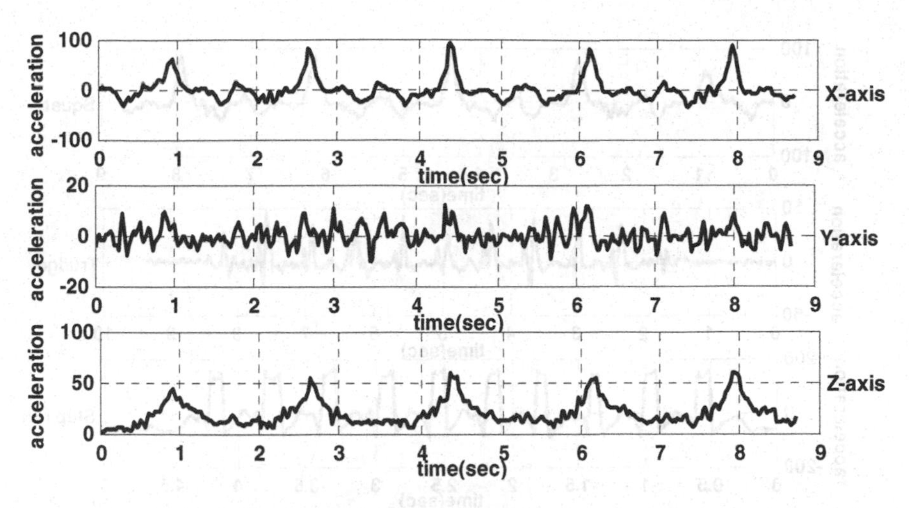

Fig. 4. Squats three axis acceleration curves

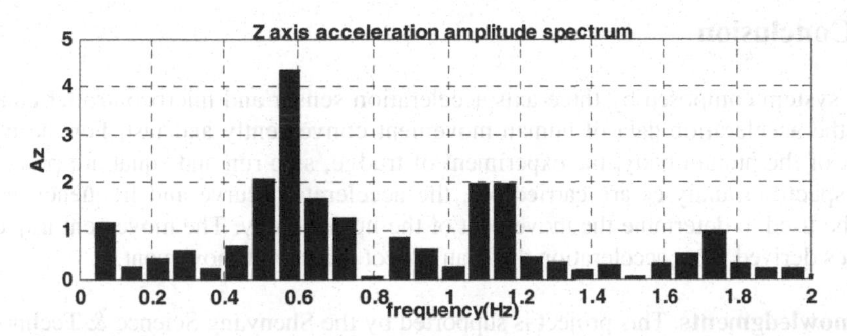

Fig. 5. Squats Z axis acceleration amplitude spectrum of partial enlargement

5.2 Kinds of Movement Detection

Three kinds of body movements of the X axis acceleration curve contrast diagrams are shown in Figure 6. Frequencies obtained by the program were 0.5787Hz, 1.5191Hz and 2.5318Hz. Combination of acceleration curve and the calculated frequency can more accurately determine the movement of the human body.

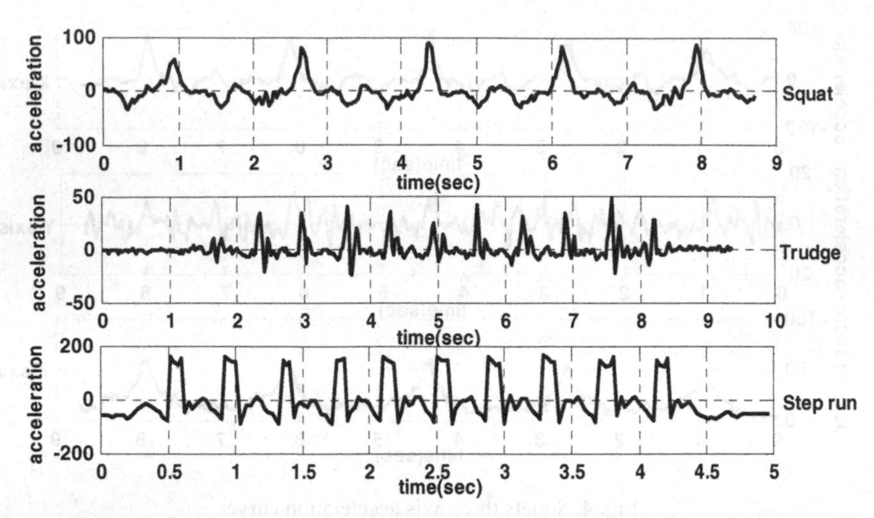

Fig. 6. Three kinds of human movement of the X axis acceleration curves

6 Conclusion

Test system composed by three-axis acceleration sensor and microcontroller can collect the acceleration data of human movement conveniently and fast. For the movement of the human body, the experiment of trudge, step run and squat are proceeded and spectrum analyses are carried out, the acceleration curve and frequency values can be used to determine the movement of the human body. The movement trajectory curves derived from acceleration data can reflect the law of movement.

Acknowledgments. This project is supported by the Shenyang Science & Technology Planning Item No. F10-213-1-00.

References

1. Tian, G.-J., Zhao, R.-C.: Survey of Gait Recognition. Computer Science (5), 20–22 (2005) (in Chinese)
2. Li, D.: Design and Implementation of Fall Detection Prototype System for Elderly People (Master Dissertation). Chongqing: Chongqing University (2008) (in Chinese)
3. Mathie, M.J., Celler, B.G., Lovell, N.H., et al.: Classification of Basic Daily Movements Using a Triaxial Accelerometer. Med. Biol. Eng. Comput. 42(5), 679–687 (2004)
4. Cao, Y.-Z., Cai, W.-C., Cheng, Y.: Body Posture Detection Technique Based on MEMS aAcceleration Sensor. Nanotechnology and Precision Engineering 8(1), 37–41 (2010) (in Chinese)
5. Gu, Y.-G., Shi, Y.-J., Zhou, X.-J., et al.: Study on Attitude Control System of Robot Based MMA7455. Manufacturing Automation 32(8), 15–17 (2010)
6. Xing, B.-W.: Sensors in the Model Rocket Set High Separation. Science and Technology Innovation Herald (6), 14–16 (2010) (in Chinese)

Research on Forecasting Models for Material Consumption

Qingtian Han, Wenqiang Li, and Wenjing Cao

Naval Aeronautical and Astronautical University, Yantai, China
hbluesky@163.com

Abstract. Forecasting the accurate assumes quantity of the material is very important for equipment management. In the paper, the different method of forecasting model were analyzed, such as simple and weighted moving average methods. Then the Exponential smooth method was given out. Finally an example was presented for the forecasting of the consume quantity. The result shows that the exponential smooth method is suitable and applicable.

Keywords: forecasting model, material management, moving average method.

1 Introduction

Time series forecasting has been paid attention to by the forecasting circles from 1969 when J. M. Bates and C. W. J. Granger put forward its theory and method[1]. The theory and methods of forecasting have been developed widely in recent years. In order to get the best efficiency of forecasting models, the authors gave four kinds of aggregative methods of group forecasting [2,3]. They are generalized weighted arithmetic mean combining forecasting model, generalized weighted logarithmic mean combining forecasting model, generalized weighted arithmetic proportional mean combining forecasting model and generalized weighted logarithmic proportional mean combining forecasting model [4,5].

2 Time Series Forecasting

Time series is always used in the forecasting of the life of system, the demand of spares, the cost required, and so on.

A time series maybe the combination of the follow situations, such as long time trend, season transformation, circularly change and irregularity change.

In general, T_t denotes the long time trend, S_t denotes the season change. C_t denotes the circularly change. , R_t denotes randomly turbulences.

For addition model,

$$y_t = T_t + S_t + C_t + R_t \tag{1}$$

For multiplication model,

$$y_t = T_t \cdot S_t \cdot C_t \cdot R_t \tag{2}$$

D. Jin and S. Lin (Eds.): Advances in MSEC Vol. 1, AISC 128, pp. 653–658.
springerlink.com

For combined model,

$$y_t = T_t \cdot S_t + R_t \tag{3}$$

$$y_t = S_t + T_t \cdot C_t \cdot R_t \tag{4}$$

Here, y_t is the records of the observed objective. $E(R_t) = 0$, $E(R_t^2) = \sigma^2$.

If there is no suddenly fluctuate, and the variance of the randomly turbulences is small during the forecasting time, some Empirical methods can be used.

3 Moving Average Method

3.1 Simple Moving Average Method

Suppose a series, $y_1, \cdots, y_T$. The average number of moving $N < T$. The one step simple moving average method is denoted as follows.

$$M_t^{(1)} = \frac{1}{N}(y_t + y_{t-1} + \cdots + y_{t-N+1}) = M_{t-1}^{(1)} + \frac{1}{N}(y_t - y_{t-N}) \tag{5}$$

When there is a Undulate along with the average, the one step simple moving average methods can be used.

$$\hat{y}_{t+1} = M_t^{(1)} = \frac{1}{N}(\hat{y}_t + \cdots + \hat{y}_{t-N+1}), \quad t = N, N+1, \cdots \tag{6}$$

The forecasting standard error is

$$S = \sqrt{\frac{\sum_{t=N+1}^{T} (\hat{y}_t - y_t)^2}{T - N}} \tag{7}$$

In generally, N is chosen during the range of: $5 \le N \le 200$. When the trend is not obviously and there are more random influence factors, N should be lager. Otherwise, N should be smaller.

3.2 Weighed Moving Average Method

Based on the simple moving average method, the recent data carries more information. And the importance is various. Then the weight is given to the different data according to the importance of the data.

Suppose the series. $y_1, y_2, \cdots, y_t, \cdots$

$$M_{tw} = \frac{w_1 y_t + w_2 y_2 + \cdots + w_N y_{t-N+1}}{w_1 + w_2 + \cdots + w_N} \quad t \ge N \tag{8}$$

Here M_{tw} is the weighed moving average for t periods. w_i is the weight of y_{t-i+1}, which denotes the importance of the information.

Then, the forecasting equation is as follows.

$$\hat{y}_{t+1} = M_{tw} \tag{9}$$

Here, the chosen of w_t is empirically. In generally, the recent data has lager weigh and the remote has smaller weight. The degree is determined by the knowledge of the data information.

3.3 Trend Moving Average Method

In the method the trend is used for forecasting. For one step moving average

$$M_t^{(1)} = \frac{1}{N}(y_t + y_{t-1} + \cdots + y_{t-N+1}) \tag{10}$$

Based on one step moving , two step moving can alsobe carried out. Thus

$$M_t^{(2)} = \frac{1}{N}(M_t^{(1)} + \cdots + M_{t-N+1}^{(1)}) = M_{t-1}^{(2)} + \frac{1}{N}(M_t^{(1)} - M_{t-N}^{(1)}) \tag{11}$$

When time series $\{y_t\}$ has a linear trend from some time, it is supposed the trend is linear from this time.

$$\hat{y}_{t+T} = a_t + b_t T \ , \quad T = 1, 2, \cdots \tag{12}$$

Here t is the current time. T is the forecasting time. a_t is intercept. b_t is gradient. Both are also called smooth indexes.

Obviously, we can get

$$\begin{aligned} a_t &= y_t \\ y_{t-1} &= y_t - b_t \\ y_{t-2} &= y_t - 2b_t \\ y_{t-N+1} &= y_t - (N-1)b_t \end{aligned} \tag{13}$$

As the result the smooth indexes are

$$\begin{cases} a_t = 2M_t^{(1)} - M_t^{(2)} \\ b_t = \dfrac{2}{N-1}(M_t^{(1)} - M_t^{(2)}) \end{cases} \tag{14}$$

It should be noted that the chosen of w_t is upon the analysis of the time series.

4 Exponential Smooth Method

4.1 One Step Exponential Smooth Method

For time series $y_1, y_2, \cdots, y_t, \cdots$, α is the weight. And $0 < \alpha < 1$.

Then one step exponential smooth equation is

$$S_t^{(1)} = \alpha y_t + (1-\alpha) S_{t-1}^{(1)} = S_{t-1}^{(1)} + \alpha (y_t - S_{t-1}^{(1)}) \tag{15}$$

And the deduce equation is

$$M_t^{(1)} = M_{t-1}^{(1)} + \frac{y_t - y_{t-N}}{N} \tag{16}$$

Develop the equation, we can get

$$S_t^{(1)} = \alpha y_t + (1-\alpha)[\alpha y_{t-1} + (1-\alpha) S_{t-2}^{(1)}] = \alpha \sum_{j=0}^{\infty} (1-\alpha)^j y_{t-j} \tag{17}$$

4.2 Two Step Exponential Smooth Method

$$S_t^{(1)} = \alpha y_t + (1-\alpha) S_{t-1}^{(1)}, \quad S_t^{(2)} = \alpha S_t^{(1)} + (1-\alpha) S_{t-1}^{(2)} \tag{18}$$

Here $S_t^{(1)}$ is one step smooth value, $S_t^{(2)}$ is two step exponential smooth value. For the current time series $\{y_t\}$

$$\hat{y}_{t+T} = a_t + b_t T, \quad T = 1, 2, \cdots \tag{19}$$

$$\begin{cases} a_t = 2S_t^{(1)} - S_t^{(2)} \\ b_t = \dfrac{\alpha}{1-\alpha}(S_t^{(1)} - S_t^{(2)}) \end{cases} \tag{20}$$

4.3 Three Step Exponential Smooth Method

$$\begin{cases} S_t^{(1)} = \alpha y_t + (1-\alpha) S_{t-1}^{(1)} \\ S_t^{(2)} = \alpha S_t^{(1)} + (1-\alpha) S_{t-1}^{(2)} \\ S_t^{(3)} = \alpha S_t^{(2)} + (1-\alpha) S_{t-1}^{(3)} \end{cases} \tag{21}$$

$$\hat{y}_{t+T} = a_t + b_t T + C_t T^2, \quad T = 1, 2, \cdots \tag{22}$$

It should be denoted that the weight is very important. The appropriate chosen of weight α is the key step for the accurate forecasting. Empirically, α can be chosen in the range of 0.1～0.3. if the time series is smoothly and the weight should be smaller, such as from 0.1 to 0.5. and if the turbulence is obviously , α should be larger , such as from 0.6 to 0.8. in practice, the trials are always carried out for the appropriate α .

5 Numerical Example

In a material warehouse, the consume quantity of the material is show in Table 1.

Table 1. Consume quantity of the material

Year	t	Initial value	One step smooth value	Two step Smooth value	Model value	Relative error%
1996	1	35	34.6500	34.5525	34.3929	1.73
1997	2	34	34.4550	34.5818	34.7893	-2.32
1998	3	33	34.0185	34.5437	34.2739	-3.86
1999	4	38	35.2129	34.3862	33.2682	12.45
2000	5	36	35.4491	34.6342	36.3939	-1.09
2001	6	35	35.3143	34.8787	36.6132	-4.61
2002	7	40	36.7200	35.0093	35.9366	10.16
2003	8	37	36.8040	35.5225	39.1639	-5.85
2004	9	39	37.4628	35.9070	38.6347	0.94
2005	10	42	38.8240	36.3737	39.6854	5.51
2006	11	43	40.0768	37.1088	42.3244	1.57
2007	12	40	40.0537	37.9992	44.3168	-10.79
2008	13	—	—	—	42.9887	—
2009	14	—	—	—	43.8692	—

From the data, we can fine the consume quantity is smoothly, thus the weight is chosen as $\alpha = 0.3$. Then the time series is established as follows.

$$y_t = (y_1, y_2, y_3, \cdots, y_{11}, y_{12}) = (35, 34, 33, \cdots, 43, 40) \tag{23}$$

The initial value $S_0^{(1)}$ is the average of the first two value. $S_0^{(2)}$ is the average of the first two value of one step smooth value. That is

$$S_0^{(1)} = \frac{y_1 + y_2}{2}, \quad S_0^{(2)} = \frac{S_0^{(1)} + S_1^{(1)}}{2} \tag{24}$$

From the calculation principle

$$S_t^{(1)} = \alpha y_t + (1-\alpha) S_{t-1}^{(1)}, \quad S_t^{(2)} = \alpha S_t^{(1)} + (1-\alpha) S_{t-1}^{(2)} \tag{25}$$

$S_t^{(1)}, S_t^{(2)}$ is calculated and shown in Table 1.

Thus

$$S_{12}^{(1)} = 40.0537, \quad S_{12}^{(2)} = 37.9992 \tag{26}$$

When $t = 12$

$$a_{12} = 2S_{12}^{(1)} - S_{12}^{(2)} = 42.1082, \quad b_{12} = \frac{\alpha}{1-\alpha}(S_{12}^{(1)} - S_{12}^{(2)}) = 0.8805 \tag{27}$$

So, we can get the trend equation with time $t = 12$.

$$\hat{y}_{12+T} = 42.1082 + 0.8805T \tag{28}$$

The forecasting values of year 2008 and 2009 are

$$\hat{y}_{2008} = \hat{y}_{13} = \hat{y}_{12+1} = 42.9887\,,\ \hat{y}_{2009} = \hat{y}_{14} = \hat{y}_{12+2} = 43.8692 \tag{29}$$

Let $T = 1$, then

$$\hat{y}_{t+1} = \left(1 + \frac{1}{1-\alpha}\right) S_t^{(1)} - \frac{1}{1-\alpha} S_t^{(2)} \tag{30}$$

Let $t = 1, 2, \cdots, 12$, the match values of different year can also be calculated. The result is shown in Figure 1.

From the result we can see that the max error appeared in year of 1999, and it is 12.45%. the minimum is 0.94%, in year 2004. the trend is smoothly as a whole.

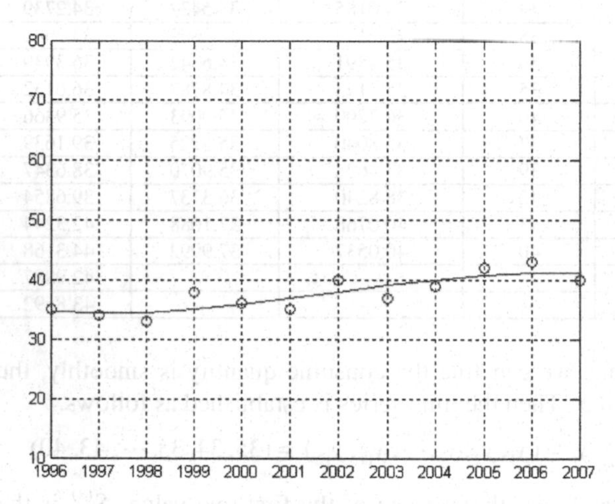

Fig. 1. Match value

6 Conclusion

In practice to forecast the accurate assumes quantity of the material is very important for equipment management. Exponential smooth method is appropriate for the forecasting of the material consume quantity. The numerical example result shows that the exponential smooth method is suitable and applicable.

References

1. Bates, J.M., Granger, C.W.J.: Combination of Forecasts. Operations Research Quarterly 20(4), 451–468 (1969)
2. Meeker, W.Q., Escobar, L.A.: Statistical Methods for Reliability Data. John Wiley & Sons, Inc., New York (1998)
3. Dickinson, J.P.: Some Comments on the Combination of Forecasts. Operational Research Quarterly 26(1), 205–210 (1975)
4. Clemen, R.T.: Combining Forecasts: A Review and Annotated Bibliography. International Journal of Forecasting 5(4), 559–563 (1989)
5. Coulson, N.E., Robins, R.P.: Forecast Combination in a Dynamic Setting. Journal of Forecasting 12(1), 63–67 (1993)

Study on Optimal Model and Algorithm of Equipment Spares

Qingtian Han, Wenjing Cao, and Wenqiang Li

Naval Aeronautical and Astronautical University
Yantai, China
hbluesky@163.com

Abstract. The goal of maintenance program is to optimize system performance through cost reduction and increased availability. A optimal model was established based on the reliability of the system and satisfactory rate of component spares. The genetic algorithm was designed and used to solve the model. Analysis was carried out for different parameters of the component spare, and the system reliability was different. The result shows that the model is suitable and applicable.

Keywords: optimal model, genetic algorithm, spares management.

1 Introduction

As systems become more complex and automation increases, there is a growing interest in the study of equipment maintenance and reliability. Increased dependence on industrial systems, military systems, and computer networks has emphasized the importance of maintaining systems to reduce unplanned downtime and maintenance costs. Especially for the military equipments, the army must pass an evaluation of operational suitability, which is measured in terms of reliability and maintainability performance. The goal of any maintenance program is to optimize system performance through cost reduction and increased availability. This is achieved by reducing the frequency of failures and the amount of downtime. In many situations the cost of a failure includes costs associated with bringing the system back to an operational state and the costs associated with the downtime incurred.[1,2] Downtime consists of the time it takes to discover that a failure exists, identify the problem, acquire the appropriate tools and parts, and perform the necessary maintenance actions.[3]

2 Calculation of the Spare Number

These are imperfect repair models, or non-homogeneous processes (NHP), and are more realistic for situations encountered in practice. The broader class of imperfect repair models consists of all models that allow repair actions to return the system to an intermediate condition between minimal repair and renewal. The homogeneous process is often a special case of an imperfect repair model. However, constructing models and obtaining analytical results under the assumption of an NHP is often

D. Jin and S. Lin (Eds.): Advances in MSEC Vol. 1, AISC 128, pp. 659–664.
springerlink.com

relatively difficult. This is particularly true when developing an expression for the point availability of an NHP.

For the same equipment spares with life time of Exponential distribution, and the reliability of detection and switch is 1, the reliability of system follows Poisson process, that is showed in Fig, 1..

$$R_s(t) = e^{-\lambda t}\left[1+\lambda t+\frac{(\lambda t)^2}{2!}+\frac{(\lambda t)^3}{3!}+\frac{(\lambda t)^4}{4!}+\ldots+\frac{(\lambda t)^n}{n!}\right] \quad (1)$$

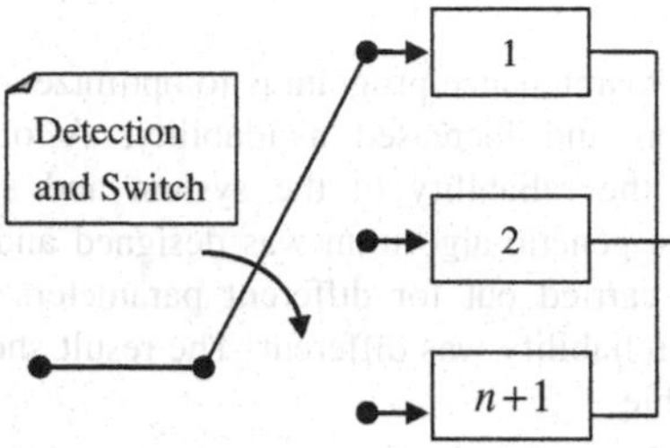

Fig. 1. Reliability of the system

If the probability of the spare is P , and the number of the spare is K, then

$$P = \frac{(k\lambda T)^n e^{-k\lambda t}}{n!} = \sum_{n=0}^{n=s}\left[\frac{R(-\ln R)^n}{n!}\right], n = 0,1,2,3,\cdots s. \quad (2)$$

2.1 Poisson Empirical Model

From (1), we can get the empirical model

$$S = \left(\lambda kt + \sqrt{\lambda kvt}\right) \quad (3)$$

Here, S is the desire spare number. K is the number of the basic spare number in an equipment. λ is the failure rate of the component. V is the logistic level. $k = U_p, U_p$ is the upper percent of the distribution. t is the use time of the equipment.

2.2 Forecasting Model Based on Reliability

The number of spares can also be forecasted based on the MTBF or MTBM.

$$Q_i = \frac{N \square K_i \square T}{T_{BF}} \quad (4)$$

Here, Q_i is the number of all the components installed in the equipments during a period. N is the number of the total equipments. k_i is the number of the components installed in a single equipment. T is the use time. T_{BF} is the MTBF.

2.3 Control Model of the Stores

For majority of general spare parts, the method of Economic Order Quantity (EOQ) is always used. The total cost of the stores C_B is determined by order cost C_A, storehouse management cost C_B and the purchase cost C_K. In generally, the cost are mostly closed by, and the average is used c_1, then the C_A is related to n with ratio. Storehouse management cost C_B includes sending, inspecting, custody and upkeep cost. For simply, the value of the components multiply the index c_2, then the management cost is gotten. The purchase cost C_K is related only to the price per unit c_k and the demanded quantity in a period R .the relation of the type of different cost is showed in Fig.2.

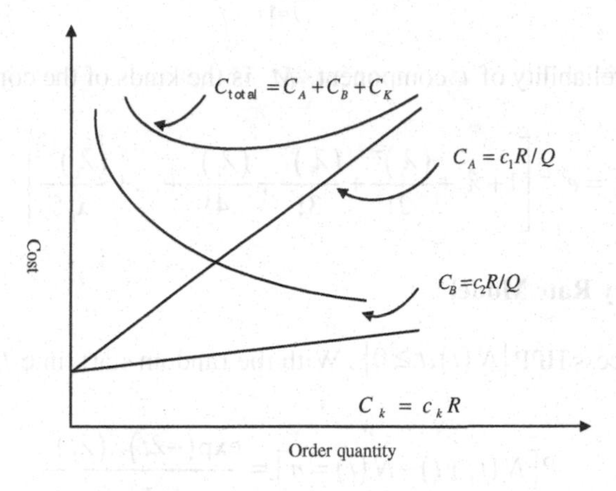

Fig. 2. One Cost of the spares instore

The total cost is as follows.

$$C_{total} = C_A + C_B + C_k = \frac{c_1 R}{Q} + \frac{c_2 R}{Q} + c_k R \tag{5}$$

$$n = \frac{R}{Q_0} = \sqrt{\frac{c_2 R}{2c_1}} \tag{6}$$

3 Study on Optimal Model

With the goal of the minimum of cost without lack of spares. The optimal model is as follows.

$$Max(z) = P\{N_1(T) \le m_1, \cdots, N_n(T) \le m_n\}$$

$$\begin{cases} P\{N_i(T) \le m_i\} \ge a_i \\ F \le w \\ m_i \ge 1 \end{cases} \tag{7}$$

Based on the system reliability the model is established as follows.

$$\begin{cases} \max R \\ s.t. \quad \sum_{i=1}^{M} x_i \cdot c_i \le C \quad x_i \ge 0 \end{cases} \tag{8}$$

3.1 System Reliability Model

System reliability can be denoted as follows.

$$R = \prod_{i=1}^{m} R_i \tag{9}$$

Here, R_i is the reliability of i component. M is the kinds of the component.

$$R_i = e^{-\lambda_i t}\left[1+\lambda_i+\frac{(\lambda_i)^2}{2!}+\frac{(\lambda_i)^3}{3!}+\frac{(\lambda_i)^4}{4!}+\ldots+\frac{(\lambda_i)^{x_i}}{x_i!}\right] \tag{10}$$

3.2 Satisfactory Rate Model

For Poisson process HPP$\{N(t), t \ge 0\}$. With the random start time t_0, during time t.

$$P\left[N(t_0+t)-N(t)=n\right]=\frac{\exp(-\lambda t)\times(\lambda t)^n}{n!} \tag{11}$$

During time t the expected number of spares is

$$E\left[N(t)\right]=\lambda t \tag{12}$$

If the equipment system is composed by L same components, the expectation of the failure number during time t is $L\lambda t$. Then we can get

$$P_i(k)=\sum_{k=0}^{n}\frac{\exp(-L\lambda t)\times(L\lambda t)^k}{k!} \tag{13}$$

3.3 System Spare Satisfactory Rate Model

The system spare satisfactory rate is an important beacon parameter that effects available degree of system and system task to run, the requirement overall considers each key to constitute life characteristic, task execution time and homologous spare parts of parts satisfies factors like rate, etc. for the components with exponential life time, the satisfactory rate is

$$P = \frac{\sum_{k=0}^{n} (\lambda_i t) P_i(k)}{\sum_{k=0}^{n} \lambda_i t} \tag{14}$$

Thus we can get the whole cost of n types of components.

$$F = \sum_{i=0}^{n} m_i (c_i + d_i T) - 0.5 \lambda_i d_i T^2 \tag{15}$$

And

$$F < w \tag{16}$$

Based on the analysis of the cost, allowed of the cost, the more spending, the more reliability of equipments. And the Non - linear programming model is as follows.

$$Max\, R$$
$$\begin{cases} F < w \\ FR \ge c \\ m_i \ge 1 \end{cases} \tag{17}$$

4 Genetic Algorithms Used for Calculation

Genetic Algorithms is a class of learning algorithms based on parallel search for an optimal solution. The algorithms as well as most of the terminology are inspired from evolutionary processes in nature. A Fitness Function defines what is to be optimized and the choice of fitness function is what defines the task to be solved.[4,5]

The parallel search is normally done synchronously in time steps that are called generations. In each generation, a number of search paths are maintained, and in analogy with evolution, these are called individuals. The whole set of individuals in a generation is referred to as the population. The central idea in genetic algorithms is to preserve and create variations of the individuals that seem most promising and remove the others. Both the variation and the selection can be done in a number of ways and what is best often turn out to be task specific.

The basic model of genetic is as follows.

$$\text{SGA} = \{ C, E, P_0, M, \Phi, \Gamma, \Psi, T \} \tag{18}$$

There are four parameters which should be predetermined, M, T, P_c, and P_m. M is the size of the worm. T is the generation number. P_c is the probability of cross, and is generally choused in the range of 0.4~0.99. P_m is the probability of variation, and is generally choused in the range of 0.0001~0.1.

The spare part number and the fitness are show in Fig. 3. for different number of the component type, reliabilities of the system are different.

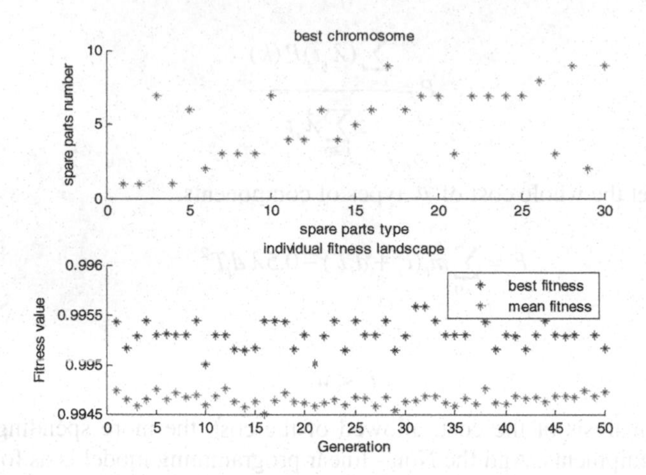

Fig. 3. Calculation result

5 Conclusion

The model was established based on the reliability of the system and satisfactory rate of component spares. For the calculation of the problem, genetic algorithm was used. For different parameters of the component spare, the system reliability is different. The result shows that the model is suitable and applicable.

References

1. Meeker, W.Q., Escobar, L.A.: Statistical Methods for Reliability Data. John Wiley & Sons, Inc., New York (1998)
2. Dhillon, B.S.: Systems Reliability, Maintainability and Management. Petrocelli Books, New York (1983)
3. Tohma, Y., Matsunaga, Y.: Application of hypergeometric distribution model to the hardware debugging process. Computer Systems Science Fd Engineering 9(1), 25–30 (1994)
4. Yah, Z.-Y., Kang, L.-S., Chen, Y.-P.: A New Multi-Objective Evolutionary Algorithm: Steady Elimination Evolutionary Algorithm. Journal ofWuhan University(Natural Science Edition), 33–38 (November 2003)
5. Kang, L.S., Ding, L.X.: An Orthogonal Multi-objective Evolutionary Algorithm for Multi-objective Optimization Problems with Constraints. Evolutionary Computation, 33–36 (December 2004)

Research on Role and Context-Based Usage Control Model

HaiYing Wu

Dept. of Communication Engineering, Engineering University of Armed Police Force, Xi'an ShaanXi 710086, China

Abstract. In order to meet the requirements of access control in pervasive environment, this paper adopts the UCON model as foundation and adds the concept of role and context as the decision factors to construct the Role and Context-based Usage Control ($RC\_UCON_{ABCC'}$) model. The context includes the three factors such as time, location and environment. The model family adds $RC\_UCON_{preC'}$ model and $RC\_UCON_{onC'}$ model. These two models are defined by Temporal Logic of Action (TLA). In RC_UCON model, roles are assigned to the Subject which simplifies the authorization process.

Keywords: Pervasive Computing, Access Control, Usage Control, Context, Temporal Logic of Action.

1 Introduction

Pervasive Computing is a new computing model which makes the information space and physical space closely inosculate, and provides transparent digital information services at anytime and anywhere[1]. Pervasive Computing makes the computer disappears from sight of people [2] which is called Openness. Openness of Pervasive Computing determines that the demand of security is outstanding in Pervasive Computing environment. The Access Control is an effective method which can ensure safety of equipments and services in Pervasive Computing environment.

The literature [3-4] have advanced that the most important feature is Context in Pervasive Computing, so that the Access Control of Pervasive Computing must consider the Context Information sufficiently; pervasive environment is dynamic and open system, thus authorization must be dynamic in Access Control, in addition must consider the continuity; because of openness of pervasive environment, the attributes need to be changed by the subject behavior and this kind of change must affect authorization at this time or next time. Thus UCON model is more adequate pervasive environment than Traditional Access Control model and RBAC model, but UCON model doesn't fully consider the context information. The system is huge because of the openness, and authorization management is great burden, and authorization management of the right become complex and difficult. This paper advances RC_UCON model to solve the access control in Pervasive Computing. In the paper, the context is introduced into the RC_UCON model to solve that the UCON model didn't consider the context. The concept of role can distribute the all rights which the subject need at a time.

D. Jin and S. Lin (Eds.): Advances in MSEC Vol. 1, AISC 128, pp. 665–670.
springerlink.com

2 RC_UCON Model

2.1 Specification of the Model

RC_UCON model adds the concept of role and judgment of context information and processes the Mutability and Continuity. The Continuity can satisfy the dynamic authorization. The Mutability can satisfy the adaptability of Pervasive Computing.

RC_UCON model is distributed into two parts. The first part, the assignments are shown as Fig. 1 which is subject-user assignment and role-subject assignment. The core is that the user is assigned corresponding subject according to the attributes of user (ATT (U)), then the subject is assigned corresponding role according to the attributes of subject (ATT(S)). These assignments make all rights to be assigned to the subject, and these rights belong to the role. The second part, the structures is shown as Fig. 2, the judgment of context is added into the decision factors. The structure denotes that the right can be executed on what condition. When the subject executes every right, the State Transition is shown as Fig. 3.

2.2 Introduction of the Model

Role is a set of rights which will be executed, and can be found by administrator. $Role$ is a set of role, $Role = \{role_1, role_2, \cdots, role_i\}$, $\forall role_i \in Role$, $role_i = \{r_1, r_2, \cdots, r_i\}$. In a role, the rights must abide by the principle of least privilege and separation of duties. There is Hierarchical Relationship (RH) between roles $\mathrm{RH} \subseteq Role \times Role$.

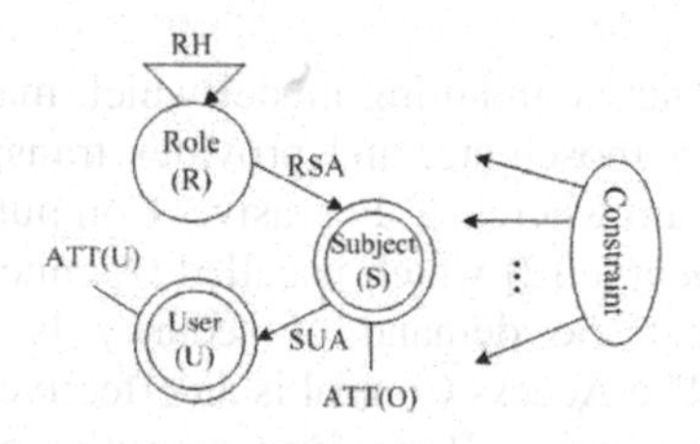

Fig. 1. SUA and RSA Assignment

User is sponsor of the operation. Attribute of the user structure the ATT (U). The set of user is denoted as $U = \{u_1, u_2, \cdots, u_i\}$.

Subject is entity which possesses some attributes ATT(S) and the rights to operate the objects. $S = \{s_1, s_2, \cdots, s_i\}$ denotes as a set of subject.

Subject-User Assignment (SUA) The subject is assigned to the user according to the ATT (*u*). The relationship of the subject and the user is many to many $\mathrm{SUA} \subseteq S \times U$.

Role-Subject Assignment (RSA) The role is assigned to the subject according to the ATT(*s*). The relationship of the role and the subject is many to many $\mathrm{RSA} \subseteq R \times S$.

Definition 1. Function $Assign(u, s)$ is denoted that the subject s is assigned to the user u; $Assign(s, role)$ is denoted that the $role$ is assigned to s.

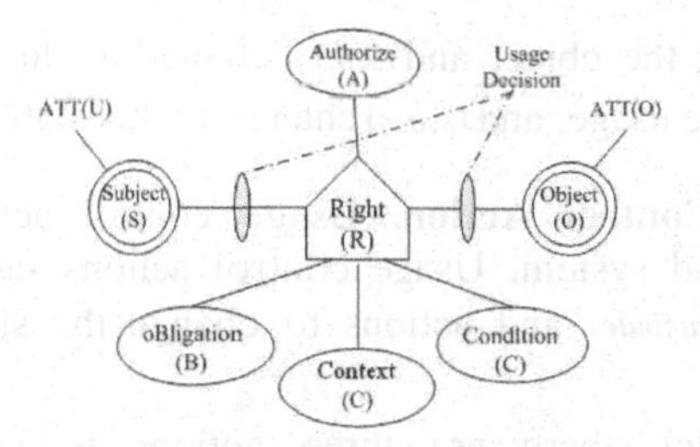

Fig. 2. RC_UCON Model Structure

In Fig. 2, the definitions of all the elements like literature [5]. The newly increased context information (C') is used to denote the state of entity, which is the important decision factor in access control of pervasive environment. In this paper, C' is divided into time, location, environment and identity [2] which is denoted as $(Time, Location, Environment, Identity)$. But the identity of subject is $s.identity \subseteq$ ATT(S) in UCON model, so that C' is denoted as $(Time, Location, Environment)$ in RC_UCON model.

Time Context Information is divided into point of time $curT$ and period of time $perT$. $curT$ is the time at which the subject request access. $perT$ is a time interval between t_i and t_j , $perT=\{[t_i, t_j] \mid i, j \in N, i<j\}$, $\forall t_s \in perT$, $\forall t_k \in perT$, $s<k$, $t_s < t_k$.

Location Context Information is divided into the location of subject l_s and the location of object l_o ; $l_s \cap l_o \neq \phi$ denote that the subject is in the location of the object.

Environment Context Information is divided into two states, $'free'$ and $'busy'$, which denote as environment in the busy state or free state.

Usage Decision (UD) has four decision factors such as A, B, C and C' . Definition of A, B, C like literature [5]. C' is context information. Judge to A, B, C and C' , if result is satisfied, the right *r* can be authorized to subject *s*. If result isn't satisfy, the right *r* can't be authorized to subject *s*.

2.3 Action and State of the Model

Definition 2. State A triple (s,o,r) denotes as an access request. $state(s,o,r)$ is a state which denotes that subject *s* access object *o* with right *r*.

In this model, a whole period of access is divided into seven different states which structure a set of state $\{initial, requesting, denied, accessing, waiting, revoked, end\}$. $state(s,o,r)$ is mapping from $\{(s,o,r)\}$ to $\{initial, requesting, denied, accessing, waiting, revoked, end\}$ .Semantics of each state are described below. Semantics of state $initial$ is that (s,o,r) has not been produce. Semantics of state $requesting$ is that (s,o,r) has been produce, but waiting for Usage Decision of system. Semantics of state $denied$ is that (s,o,r) is refused according to usage strategy before usage. Semantics of state $accessing$ is that the subject accesses the object immediately when the subject is permitted to access the object. The model adds state $waiting$ which denotes that (s,o,r) change to this state when the context isn't satisfied during access. State $revoked$ denotes that the system revoke this access when the subject is accessing the object and (s,o,r) change to this state. State *end* denotes that the subject finish the usage, and (s,o,r) change to this state.

Definition 3 Usage Control Action. Usage control actions are structured by operation of subject and system. Usage control actions include actions to update attribute values *Action(attribute)* and actions to change the status of a single access process $Action(s,o,r)$.

In RC_UCON model, there are three actions to update attribute values. *Preupdate(attribute)* denotes that system update the attribute values of subject or object before access or after access denied. *Onupdate(attribute)* denotes that system update the attribute values of subject or object during access. *Postupdate(attribute)* denotes that system update the attribute values of subject or object after access.

There are seven actions to change the states. $TryAccess(s,o,r)$ denotes that subject produces initial access request, and the state changes to *requesting* . $PermitAccess(s,o,r)$ denotes that system executes access request, and the states change to *accessing* . $WaitAccess(s,o,r)$ denotes that system executes waiting request and the state changes to *waiting* . $DenyAccess(s,o,r)$ denotes that system executes access denied and state changes to *denied* . $RevokeAccess(s,o,r)$ denotes that system executes access revoked and state changes to *revoked* . $EndAccess(s,o,r)$ denotes that subject executes ended access and state changes to *end* .

In $RC\_UCON_{ABCC'}$ model, actions to update attribute values and actions to change the status are shown as Figure 3.

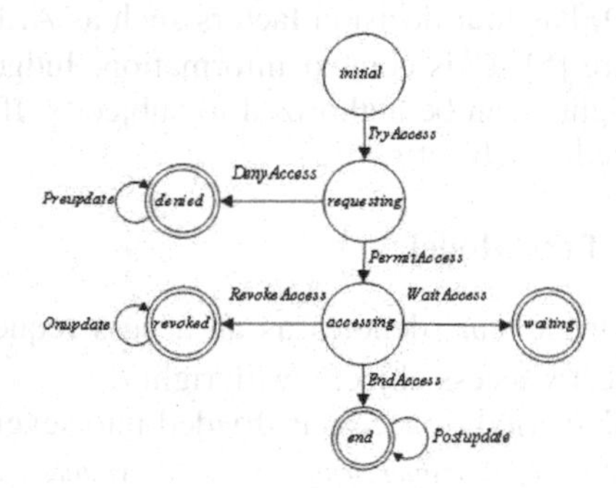

Fig. 3. $RC\_UCON_{ABCC'}$ Model State Transition

2.4 Temporal Logic of Actions of Model

Definition 3. Temporal Logic of Action (TLA) Extending temporal logic by introducing Boolean valued actions can be used to specify systems and their properties, especially for interactive and concurrent systems.

Temporal operator: $\square$denotes as *Always*, $\blacksquare$denotes as *Has – always – been*, $\diamond$denotes as *Eventually*, $\blacklozenge$denotes as *Once*, $\bigcirc$ denotes as *Next*, $\mathcal{U}$ denotes as *Until*, $\mathcal{S}$ denotes as *Since*, $\ominus$ denotes as *Previous*;

Logical connectors: →inclusion, ¬ false, ∧ conjunction, ∨ disjunction.
A logical formula in RC_UCON model is defined by the following grammar in BNF:

$$\beta ::= a \mid P(t_1,\cdots,t_n) \mid (\neg\beta) \mid (\beta\wedge\beta) \mid (\beta\rightarrow\beta) \mid \square\beta \mid \diamond\beta \mid \bigcirc\beta \mid \beta\ \mathcal{U}\ \beta \mid \blacksquare\beta \mid \blacklozenge\beta \mid \ominus\beta \mid \beta\ \mathcal{S}\ \beta \mid$$

"|"denotes as "or"," a "denotes as action. $P(t_1,\cdots,t_n)$ denotes as n dimension predicate, $t_1,\cdots,t_n$ is predicate terms.

The core models of RC_UCON model, such as $\mathrm{RC\_UCON_A}$ $\mathrm{RC\_UCON_B}$ $\mathrm{RC\_UCON_C}$, are similar to $\mathrm{UCON_A}$ $\mathrm{UCON_B}$ $\mathrm{UCON_C}$, therefore the definition likes the literature [7]. The corresponding core models of RC_UCON add the $\mathrm{RC\_UCON}_{C'}$ model, mainly include $\mathrm{RC\_UCON}_{preC'_0}$ and $\mathrm{RC\_UCON}_{onC'_0}$.

Rule of $preC'_0$ Model

In predicate $C'_1(t_s,t_o)$, t_s is divided into the current time $curT_s$ at which the subject advances the access request and the period of time $perT_s$ in which the subject access object. t_o is divided into the current time $curT_o$ at which the object is accessed and the period of time $perT_o$ in which the object can be accessed. Predicate $C'_1(t_s,t_o)$ denotes as $(curT_s \in perT_s)\wedge(curT_s \in perT_o)$ which illuminates the time condition which must be satisfy when the subject advances the access request. Predicate $C'_2(l_s,l_o)$ denotes as $l_s \cap l_o \neq \phi$ which illuminates that the subject must be in the location of the object. $C'_3(e)$ denotes as $e = 'free'$ which illuminates that the objects are requested in the free state.

$$PermitAccess(s,o,r) \rightarrow \blacklozenge\ (TryAccess(s,o,r)\wedge(C'_1(t_s,t_o)\wedge C'_2(l_s,l_o)\wedge C'_3(e)))$$

In this rule, $C'_1(t_s,t_s)\wedge C'_2(l_o,l_o)\wedge C'_3(e) = true$ denotes that the subject will use right to access the object and must be satisfy time context, space context at same time. The object must be in the free state.

Rule of onC'_0 Model

When (s,o,r) is in state *accessing*, as long as one predicate is not satisfy in these three predicates $C'_1(t_s,t_o)$, $C'_2(l_s,l_o)$, $C'_3(e)$ then $WaitAccess(s,o,r)$ is executed. Predicate $C'_4(t_s,t_o)$ denotes as $curT_o \geq \mathrm{MAX}(perT_s)$ which illuminates that $EndAccess(s,o,r)$ is executed after the subject has already accessed the object. Predicate $C'_5(t_s,t_o)$ denotes as $curT_o \geq \mathrm{MAX}(perT_s)$ which illuminates that $RevokeAccess(s,o,r)$ is executed if it's time to put on the access which the object can provide to the subject.

$$\Box(\neg(C_1'(t_s,t_o)\wedge C_2'(l_s,l_o)\wedge C_3'(e))\wedge(state(s,o,r)=accessing)\ \rightarrow WaitAccess(s,o,r))$$

$$\Box(C_4'(t_s,t_o)\wedge(state(s,o,r)=accessing)\ \rightarrow EndAccess(s,o,r))$$

$$\Box(\neg C_4'(t_s,t_o)\wedge C_5'(t_s,t_o)\wedge(state(s,o,r)=accessing)\ \rightarrow RevokeAccess(s,o,r))$$

3 Summarize

There are many shortages when $UCON_{ABC}$ model is directly used in Pervasive Computing Environment. In particular, the context information is not fully considered and one right is authorized to the subject at a time. Therefore the paper advances $RC\_UCON_{ABCC'}$ model which is aimed at access control of Pervasive Computing Environment. Based on $UCON_{ABC}$ model, RC_UCON model adds the concept of role and new decision factor of context. The context mainly considers the factors of time, location and environment. RC_UCON_{ABC} model adds $RC\_UCON_{preC_0'}$ model and $RC\_UCON_{onC_0'}$ model. At last, the paper gives the temporal logic description of the additional model.

Acknowledgments. This paper is supported by the Basic Research Fund Project of Engineering University of Armed Police Force (No.WJY201021).

References

1. Dou, W., Wang, X., Zhang, L.: New Fuzzy Role-based Access Control Model for Ubiquitous Computing. Computer Science 37(9), 63–67 (2010)
2. Xu, G., Shi, Y.: Pervasive/Ubiquitous Computing. Chinese Journal of Computers 26(9), 1042–1050 (2003)
3. Xin, Y., Luo, C.: Context-based RBAC model in pervasive computing. Computer Engineering and Design 31(8), 1693–1697 (2010)
4. Guo, Y., Li, R.: Research on access control for pervasive computing. Journal of Central China Normal University (Nat. Sci.) 40(4), 504–506 (2006)
5. Park, J., Sandhu, R.: The UCONABC Usage Control Model. ACM Transactions on Information and System Security 7(1), 128–174 (2004)
6. Xie, H., Zhang, B.: A Role-Based Usage Control Authorization Model. Microel Ectronics & Computer 27(6), 137–141 (2010)
7. Zhang, X., Sandhu, R.: Formal Model and Policy Specification of Usage Control. ACM Transactions on Information and Security 8(4), 351–387 (2005)

A No Interference Method for Image Encryption and Decryption by an Optical System of a Fractional Fourier Transformation and a Fourier Transformation

Huaisheng Wang

College of mathematics and physics, Shanghai University of electric power, 200090, Shanghai, China
wanghs11111@126.com

Abstract. A no interference encryption and decryption for a digital image is presented in this paper. The encrypting process consists of a fractional Fourier transformation and a Fourier transformation. A digital image is first coded with a random phase plate, then it takes a fractional Fourier transformation. The transformed field function is expanded 4 times and coded with another random phase mask. Meanwhile the system takes a Fourier transformation. The real part of the transformed function is taken as an encrypted image. For the decryption of the digital image, first the encrypted image undergoes an inverse Fourier transformation. Then the upper left corner of the transformed function is cut off. At last the cut-off function takes an inverse fractional Fourier transformation. The original digital image can be extracted from the final transformed function. Compared with optical holography, an interference process is not needed in our program. This will reduce the complexity of optical system.

Keywords: optical image security, Fourier transformation, fractional Fourier transformation.

1 Introduction

Optical information treatment has many characteristics such as high speed and high parallel treating. The information safety is an important research field. Optical image encryption and decryption [1-6] have important value in research and application.

In order to easily treat images in computer, we need a real function to express a encrypting image. In [7] a real-valued encryption of a digital image is presented. The optical system in this real-valued encryption is composed of a phase image coded by a random phase mask and the system undergoes a Fourier transformation. While in the decrypting process an interference process and an inverse Fourier transformation are adopted. Fractional Fourier transformation is an important tool both in mathematics and optics [8]. Fractional Fourier transformation combined with phase mask is widely used in optical image encryption [9-11].

We put forwards a no interference encryption of an image by an optical system which composed of a fractional Fourier transformation and a Fourier transform in this paper. The main characteristic of the system is that we do not need interference

D. Jin and S. Lin (Eds.): Advances in MSEC Vol. 1, AISC 128, pp. 671–676.
springerlink.com

operation in the decrypting process as compared with that in [7]. Because there are more keys in the encrypting system, the safety of the optical encryption is reinforced.

The rest of this paper is arranged as follows: section 2 put forwards a no interference encrypting optical system; section 3 gives the corresponding no interference decrypting system; section 4 makes a computer simulation; finally section 5 draws a conclusion.

2 A No Interference Encryption

The optical system of the no interference encrypting process for a digital image is shown in Fig.1.

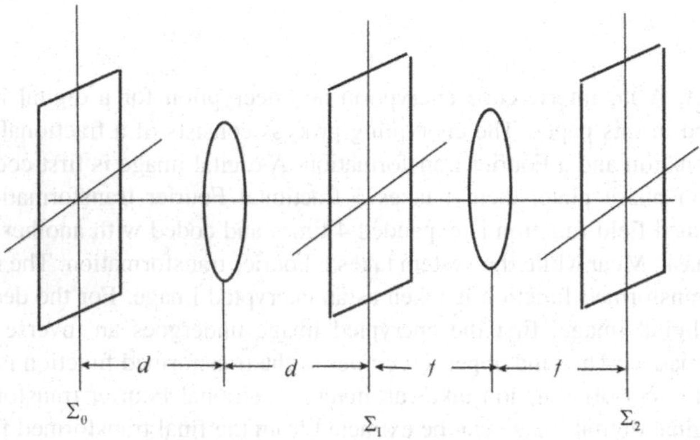

Fig. 1. The real-valued encrypting

The original digital image is $f(x, y)$ which coded with a random phase mask and is put in the plane Σ_0 . The function in plane Σ_0 is

$$u_0(x_0, y_0) = f(x_0, y_0)\exp(i\pi r_1(x_0, y_0)) , \tag{1}$$

where $r_1(x_0, y_0)$ is a random matrix. We assume that a monochromatic plane light incites the plan Σ_0 . The system first takes a p-order fractional Fourier transformation. The optical function in plane Σ_1 can be calculated

$$u(x_1, y_1) = c_1 \iint \exp(i\frac{\pi(x_1^2 + y_1^2 + x_0^2 + y_0^2)}{\lambda f_1 tg\alpha}) \times u_0(x_0, y_0)\exp(-i\frac{2\pi(x_0x_1 + y_0y_1)}{\lambda f_1 \sin\alpha})dx_0dy_0 \quad , \tag{2}$$

where c_1 is a constant, f_1 is the standard focal length. Based on the property of fractional Fourier, the following equations are effective: $\alpha = p\pi/2$, $f_1 = f'\sin\alpha$, $d = 2f'\sin^2(\alpha/2)$, where f' is the focal length of the lens of fractional Fourier transformation.

We enlarge the function $u(x_1, y_1)$ to $O(x_1, y_1)$

$$O(x_1, y_1) = \begin{bmatrix} u(x_1, y_1)\exp(i\pi r_2(x_1, y_1)) & 0 \\ 0 & 0 \end{bmatrix}, \quad (3)$$

where $r_2(x_1, y_1)$ is a random matrix. Through a Fourier transformation, the light field at plane Σ_2 is

$$v(x_2, y_2) = c_2 \iint \exp(-i2\pi[x_1x_2 + y_1y_2]/\lambda f) \times O(x_1, y_1)dx_1dy_1, \quad (4)$$

where c_2 is a constant. The real part of the $v(x_2, y_2)$ is known as the encrypting image.

3 A No Interference Decryption

The decrypting system is shown in Fig. 2.

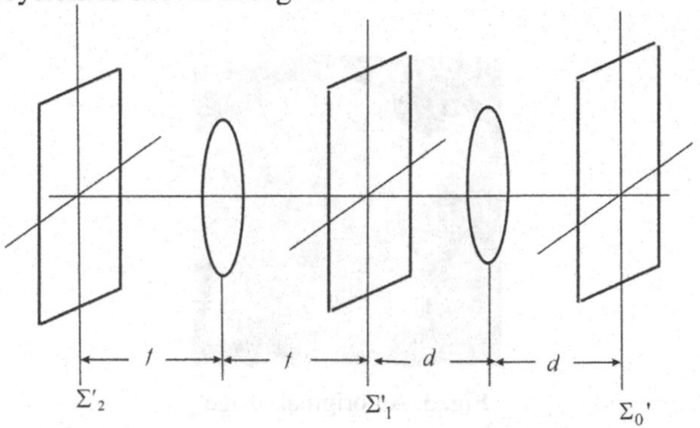

Fig. 2. The optical system for decryption

We just take the encrypted function as the initial field function at plane Σ_2'

$$v'(x_2', y_2') = (v(x_2', y_2') + v^*(x_2', y_2'))/2. \quad (5)$$

First the initial field function undergoes an inverse Fourier transformation. The optical light field at plane Σ_1' can be expressed as

$$u'(x_1', y_1') = c_3 \iint \exp(i2\pi[x_1'x_2' + y_1'y_2']/\lambda f) \times v'(x_2', y_2')dx_2'dy_2', \quad (6)$$

where c_3 is a constant.

We cut off the top left corner of the 1/4 part of the matrix $u'(x_1', y_1')$ as a function $u''(x_1', y_1')$. We make a function $U'(x_1', y_1')$ as follows

$$U'(x_1', y_1') = 2\exp(-i\pi r_2(x_1', y_1'))u''(x_1', y_1'). \quad (7)$$

At last the function $U'(x_1', y_1')$ goes through an inverse fractional Fourier transformation. The field function in plane Σ_0' can be calculated as

$$f'(x_0', y_0') = c_4 \iint \exp(-i\frac{\pi(x_1'^2 + y_1'^2 + x_0'^2 + y_0'^2)}{\lambda f_1 tg\alpha}) \times U'(x_1', y_1') \exp(i\frac{2\pi(x_0' x_1' + y_0' y_1')}{\lambda f_1 \sin\alpha}) dx_1' dy_1' \quad , \quad (8)$$

where c_4 is a constant. We can restore the original image from the function $f'(x_0', y_0')$.

4 Computer Simulation

In order to check the effect of the on interference encrypting and decrypting method, we take a 360×360 pixels original image as an example. Fig. 3 shows the original image.

Fig. 3. An original image

In our optical system the parameters are : $\lambda = 6.328\times 10^{-5}$ cm, the focal length of the fractional Fourier transformation $f' = 150$ cm, $P = 1/2$, $\alpha = P\pi/2$, $f_1 = f'\sin\alpha$, $d = f_1 tg(\alpha/2)$, $f = 106$ cm. The coded images at plane Σ_0 and Σ_2 are shown in Fig.4. and Fig.5 respectively.

Fig. 4. The coded image at plane Σ_0

Fig. 5. The coded image at plane Σ_2

The decrypting image is shown in Fig. 6. The effect of decrypting image is very good because there is no approximation in the decrypting process.

Fig. 6. The decoded images at plane Σ_0'

5 Conclusion

In this paper we put forward a no interference method for encryption and decryption of a digital image. Two phase plates are used in encrypting process. The encrypting process involves a fractional Fourier transformation and a Fourier transformation. The decrypting system contains a corresponding inverse Fourier transformation and a corresponding inverse fractional Fourier transformation. Because there is no interference operation in our system, the optical implement for the system is relatively easy. The real function of encrypting image is very convenient for printing and storage. There are more parameters in our encrypting system. This will increase the safety of encryption.

References

1. Refregier, P., Javidi, B.: Optical Image Encryption Based on Input Plane and Fourier Plane Random Encoding. Opt. Lett. 20(7), 767–769 (1995)
2. Barrera, J., Henao, R., Tebaldi, M., Torroba, R., Bolognina, N.: Digital Encryption with Undercover Multiplexing by Scaling the Encoding Mask. Optik. 120, 342–346 (2009)
3. Kumar, P., Joseph, J., Singh, K.: Optical Image Encryption Using a Jigsaw Transform for Silhouette Removal in Interference-based Methods and Decryption with a Single Spatial Light Modulator. Applied Optics 50(13), 1805–1811 (2011)
4. Chen, L., Zhao, D.: Optical Color Image Encryption by Wavelength Multiplexing and Lensless Fresnel Transform Holograms. Opt. Express 14, 8552–8560 (2006)
5. Tao, R., Xin, Y., Wang, Y.: Double Image Encryption Based on Random Phase Encoding in the Fractional Fourier Domain. Opt. Express 15, 16067–16079 (2007)
6. Chen, W., Chen, X., Sheppard, C.: Optical Image Encryption Based on Diffractive Imaging. Opt. Lett. 35, 3817–3819 (2010)
7. Rong, L., Ping, L.: Optical Image Security Based on Random Phase Real-value Encryption. Acta Photonica Sinica (in chinese) 33(5), 605–608 (2004)
8. Lohmann, A.: Image Rotation, Wigner Rotation, and the Fractional Fourier Transform. J. Opt. Soc. Am. A 10(10), 2181–2186 (1993)
9. Tao, R., Xin, Y., Wang, Y.: Double Image Encryption Based on Random Phase Encoding in the Fractional Fourier Domain. Opt. Express 15, 16067–16079 (2007)
10. Chen, L., Zhao, D.: Color Information Processing (Coding and Synthesis) with Fractional Fourier Transforms and Digital Holography. Opt. Express 15, 16080–16089 (2007)
11. Unnikrishnan, G., Joseph, J., Singh, H.: Optical Eecyption by Double Random Phase Encoding in the Fractional Fourier Domain. Opt. Lett. 25(12), 887–889 (2000)

A Minimum Spanning Tree Problem in Uncertain Networks

FangGuo He and GuiMing Shao

College of Mathematics and Computer Science, Huanggang Normal University, Huanggang, China
hfg0118@126.com

Abstract. This paper deals with a minimum spanning tree problem where each edge weight is a random variable. In order to solve the uncertain network optimization, the concept of the α-optimistic cost spanning tree is proposed and a stochastic optimization model is constructed according to the theory of stochastic programming. We adopted an efficient method to convert the stochastic optimization problem into the deterministic equivalent, and use Kruskal's algorithm to solve the problem. Finally, a numerical experiment is given to show the effectiveness of the proposed method.

Keywords: Minimum spanning tree, network optimization, uncertain programming.

1 Introduction

The minimum spanning tree (MST) problem is a classical combinatorial optimization problem in network. The problem is to find a least cost spanning tree in an edge weighted graph. This problem has been well studied and many efficient polynomial-time algorithms have been developed by Kruskal , Prim and Sollin [1]. In the real world, MST problem is applied in many fields. For instance, when designing a layout for communication system, if a decision maker wishes to minimize the cost for connection between cities, the problem is formulated as an MST problem. As other examples, the objective is to minimize the time for construction or to maximize the reliability. Most research papers with respect to MST problems dealt with the case where each weight is constant. However, real world problems not always have precise data. Problems parameters have uncertain nature. For example, links in a communication network can malfunction or degrade as a result of congestion, accidents, weather, etc. In order to investigate more realistic cases, it is necessary to consider the situation that the each edge weight is not a constant but an uncertain value such as a fuzzy or random variable.

Several authors have examined the MST problems where the edge weights are uncertainty. Katagiri [2] examined the case where the edge weights are fuzzy random variables, he introduced a fuzzy goal for the objective function by fuzzy theorem. Tiago and Akebo studied the MST problem with fuzzy parameters and proposed an exact algorithm to solve it. Kevin and Douglas [3] considered the problem in networks with varying edge weights and use the algebraic structure to describe the relationship

D. Jin and S. Lin (Eds.): Advances in MSEC Vol. 1, AISC 128, pp. 677–683.
springerlink.com

between different edge-weight realizations of the network. Furthermore, some researchers [4] proposed fuzzy random programming models.

In this research we consider networks in which the edge weights are normally distributed random variables. We propose the concept of stochastic minimum spanning tree and formulate the model for our problem according to the theory of stochastic programming. In order to solve the model, we propose a method to convert the stochastic optimization into the deterministic equivalent, and make the uncertain optimization problem become a classical minimum spanning tree. Finally, a numerical experiment is given to show the effectiveness of the proposed method.

2 Problem Description

Given a connected, undirected graph, a spanning tree of that graph is a subgraph which is a tree and connects all the vertices together. A single graph can have many different spanning trees. We can also assign a weight to each edge, which is a number representing how unfavorable it is, and use this to assign a weight to a spanning tree by computing the sum of the weights of the edges in that spanning tree. A minimum spanning tree or minimum weight spanning tree is then a spanning tree with weight less than or equal to the weight of every other spanning tree. Considering a graph $G = (V, E)$, we let $V = \{1, 2, ..., n\}$be a finite set of vertices representing terminals or telecommunication satiations etc., and $E = \{ e_1, e_2, \ldots, e_m \}$ be a finite set of edges representing connections between these terminals or stations, in which the lengths of the edges are assumed to be stochastic. Each edge is denoted by an ordered pair (i, j), where $(i, j) \in E$. For a subset of nodes S, we define $E(S) = \{(i, j) | i, j \in S\}$ to be the edges whose endpoints are both in S. We define the following binary decision variables for all edges $(i, j) \in E$:

$$x_{ij} = \begin{cases} 1 & \text{if edge } (i, j) \text{ is selected in the spaning tree} \\ 0 & \text{otherwise} \end{cases}$$

It has been proved that $x = \{x_{ij} | (i, j) \in E\}$ is a spanning tree if and only if

$$\begin{aligned} & \sum_{i=1}^{n} \sum_{j=1}^{n} x_{ij} = n - 1 \\ & \sum_{i,j \in S} x_{ij} \le |S| - 1, \quad \forall S \subseteq V, S \ne \varnothing \\ & x_{ij} \in \{0,1\}, i, j \in V \end{aligned} \tag{1}$$

Let ξ_{ij} be the weights of edges $(i, j) \in E$, where the ξ_{ij} are normally distributed random variables. Then the sum of the weights of a spanning tree of G is

$$T(x, \xi) = \sum_{i=1}^{n} \sum_{j=1}^{n} \xi_{ij} x_{ij} \tag{2}$$

In the following sections, stochastic programming models will be provided for the minimum spanning tree problem with stochastic edge weights.

3 The Models of Stochastic Optimization

The first type of stochastic programming is the so-called expected value model [5]. The essential idea of expected value model is to optimize the expected values of objective functions subject to some expected constraint.

When the weights of edges are stochastic, the spanning tree $T(x, \xi)$ becomes stochastic, too. It is unmeaning to find a least cost spanning tree when the weights are random variables. Then we consider minimizing the expected value of $T(x, \xi)$. We provide a idea to formulate the MST problem by expected value model as follows,

$$\min E(T(x,\xi)) = E(\sum_{i=1}^{n}\sum_{j=1}^{n}\xi_{ij}x_{ij}) \tag{3}$$

subject to Eqs. (1)

where $T(x, \xi)$ is defined by Eqs. (2), and $E(T(x, \xi))$ is the expected value of $T(x, \xi)$. A solution x is feasible if and only if x satisfy the formula (1). A feasible solution x^* is an optimal solution to (3) if $E(T(x^*, \xi)) \le E(T(x, \xi))$ for any feasible solution x.

As the second type of stochastic programming developed by Charnes and Cooper, chance-constrained programming offers a powerful means of modeling stochastic decision systems with assumption that the stochastic constraints will hold at least α of time, where α is referred to as the confidence level provided as an appropriate safety margin by the decision-maker. After that, Liu [6] generalized chance-constrained programming to the case with not only stochastic constraints but also stochastic objectives. Now we construct a chance-constrained programming model of minimum spanning tree by the uncertainty theory of Liu.

Since $minT(x, \xi)$ is meaningless if $T(x, \xi)$ is stochastic, a natural idea is to provide a confidence level α at which it is desired the $T(x,\xi) \le \overline{T}$, where the confidence level α is provided by the decision-maker. So the objective is to minimize $\overline{T}$ instead of $T(x, \xi)$ with a chance constraint as follows,

$$P\{T(x,\xi) \le \overline{T}\} \ge \alpha \tag{4}$$

where $\overline{T}$ is referred to as the α-optimistic value to $T(x, \xi)$. A solution x is called feasible if and only if the probability measure of the event $T(x,\xi) \le \overline{T}$ is at least α. In other words, the constraint will be violated at most (1-α) of time. Hence we have the following chance-constrained programming model on minimum spanning tree,

$$\min \overline{T} \tag{5}$$

subject to $P\{T(x,\xi) \le \overline{T}\} \ge \alpha$ and Eqs. (1)

where $\min \overline{T}$ is called the α-optimistic cost spanning tree and $T(x, \xi)$ is defined by Eqs. (2).

4 Equivalent Formulation and Algorithm

The models formulated in section 3 are stochastic programming problems. The traditional solution methods require conversion of the uncertain models to their respective deterministic equivalents. As we know, this process is usually hard to perform and only successful for some special cases. In this paper we propose an approach to convert the chance constraint (4) into the deterministic equivalent.

Theorem 1: Assume that the stochastic vector $\xi = (\xi_1, \xi_2, \cdots, \xi_n)$ and the function $g(x,\xi) = \xi_1 x_1 + \xi_2 x_2 + \cdots + \xi_n x_n$. If ξ_j ($j = 1, 2,\ldots n$) are assumed to be independently normally distributed random variables, i.e $\xi_j \sim (\mu_j, \sigma_j^2)$, then $P(g(x,\xi) \le b) \ge \alpha$ if and only if

$$\sum_{j=1}^{n} \mu_j x_j + \Phi^{-1}(\alpha)\sqrt{\sum_{j=1}^{n} \sigma_j^2 x_j^2} \le b$$

where Φ is the standardized normal distribution.

Proof: Since $\xi_j \sim (\mu_j, \sigma_j^2)$($j$=1,...$n$) are assumed to be independently normally distributed random variables, the quantity

$$\tau = \sum_j \xi_j x_j - b$$

is also normally distributed with the following expected value and variance,

$$E(\tau) = \sum_j \mu_j x_j - b \quad V(\tau) = \sum_j \sigma_j^2 x_j^2$$

We note that

$$\eta = \frac{\tau - E(\tau)}{\sqrt{V(\tau)}} = \frac{\sum \xi_j x_j - b - (\sum \mu_j x_j - b)}{\sqrt{\sum \sigma_j^2 x_j^2}}$$

must be standardized normally distributed. Since the inequality $\sum_{j=1}^{n} \xi_j x_j \le b$ is equivalent to

$$\frac{\sum \xi_j x_j - b - (\sum \mu_j x_j - b)}{\sqrt{\sum \sigma_j^2 x_j^2}} \le -\frac{\sum \mu_j x_j - b}{\sqrt{\sum \sigma_j^2 x_j^2}}$$

Therefore, we have

$$\mathrm{P}(\sum \xi_j x_j \le b) \ge \alpha \Leftrightarrow \mathrm{P}(\eta \le -\frac{\sum \mu_j x_j - b}{\sqrt{\sum \sigma_j^2 x_j^2}}) \ge \alpha \tag{6}$$

where η is the standardized normally distributed variable, and let Φ be its distributed function. Then the formula (6) holds if and only if

$$\Phi^{-1}(\alpha) \le -\frac{\sum \mu_j x_j - b}{\sqrt{\sum \sigma_j^2 x_j^2}}$$

Thus, the theorem is proved.

Theorem 2: If a_i (i =1, 2, …, l) are positive number, then $\sqrt{\sum_{i=1}^{l} a_i^2} \geq \frac{1}{\sqrt{l}} \sum_{i=1}^{l} a_i$, and the equality holds when a_i (i =1, 2, …, l) are the same.

This theorem can be proved by mathematical induction. After the chance constraint (4) is converted to its deterministic equivalent by theorem 1, it becomes a nonlinear constraint. We expected to turn it into linear constraint. In fact, the purpose can be achieved according to the theorem 2, and the nonlinear problem is relaxed to a linear problem.

Let m be the number of edge in a graph. By the theorem 1 and 2, the optimization problem (5) discussed in section 3 can be given in the form:

$$\min \sum_i \sum_j (E(\xi_{ij}) + \Phi^{-1}(\alpha) \frac{V(\xi_{ij})}{\sqrt{m}}) x_{ij} \tag{7}$$

subject to Eqs. (1)

Where $E(\xi_{ij})$ and $V(\xi_{ij})$ denote expected value and variance of random variable ξ_{ij}, respectively.

For the optimization problem (3), we can conclude the following formula in terms of the properties of expected value.

$$\min E(\sum_i \sum_j \xi_{ij} x_{ij}) = \sum_i \sum_j E(\xi_{ij}) x_{ij} \tag{8}$$

subject to Eqs. (1)

It is very clear that the problems (7) and (8) are classical minimum spanning tree problems. The problems can be solved by many different algorithms. It is the topic of some very recent research [7-9]. There are several "best" algorithms, depending on the assumptions you make. In this paper we use Kruskal's algorithm to solve our problem, and the step of the algorithm is as follows:

Step 1: Find the cheapest edge in the graph (if there is more than one, pick one at random). Mark it with any given colour, say red.

Step 2: Find the cheapest unmarked (uncoloured) edge in the graph that doesn't close a coloured or red circuit. Mark this edge red.

Step 3: Repeat Step 2 until you reach out to every vertex of the graph (or you have N−1 coloured edges, where N is the number of vertices.) The red edges form the desired minimum spanning tree.

5 An Illustrative Example

In order to illustrate the method developed here, we consider an 8-node complete graph. The edge weights of the problem are represented as normally distributed random variables, and their distribution functions are shown in the following in the edge-weight matrix W , where $N(\mu, \sigma^2)$ means normally distribution with mean μ and standard deviation σ. For simplicity, we let σ =1.

$$W = \begin{bmatrix} 0 & N(10,1) & N(3,1) & N(4,1) & N(17,1) & N(14,1) & N(8,1) & N(10,1) \\ & 0 & N(2,1) & N(16,1) & N(18,1) & N(14,1) & N(5,1) & N(19,1) \\ & & 0 & N(4,1) & N(9,1) & N(6,1) & N(13,1) & N(8,1) \\ & & & 0 & N(11,1) & N(14,1) & N(15,1) & N(9,1) \\ & & & & 0 & N(12,1) & N(15,1) & N(7,1) \\ & & & & & 0 & N(14,1) & N(5,1) \\ & & & & & & 0 & N(9,1) \\ & & & & & & & 0 \end{bmatrix}$$

By the theorem 1 and theorem 2, the minimum spanning tree with stochastic edge weights can be converted to the deterministic form. To solve the model (3), we use Kruskal's algorithm to calculate the expected value of minimum spanning tree, and gain the solution $x^* = (e_{12}, e_{14}, e_{25}, e_{36}, e_{37}, e_{56}, e_{78})$, and the expected value is 32. Similarly, we use Kruskal's algorithm to solve the model (5) with different confidence levels α.

The solution of 0.95-optimistic cost spanning tree is $x = (e_{13}, e_{14}, e_{23}, e_{27}, e_{36}, e_{56}, e_{68})$, and 0.95-optimistic cost is 34.17; the solution of 0.7-optimistic cost spanning tree is $x = (e_{12}, e_{14}, e_{25}, e_{36}, e_{37}, e_{56}, e_{78})$, and 0.7-optimistic cost is 32.7.

6 Conclusion

In this paper, we considered the problem of minimum spanning trees in uncertain networks in which the edge weights are random variables. The concept of the α-optimistic cost spanning tree is proposed and the models of expected value and chance-constrained programming are formulated. Based on uncertain theory, we propose a method to convert the stochastic optimization into the deterministic equivalent, and make the uncertain optimization problem become a classical minimum spanning tree. Finally, we use Kruskal's algorithm to solve the problem, and give a numerical experiment to show the effectiveness of the proposed method.

Acknowledgments. This work is supported by the fund of department of education of Hubei Province (No. D20102904), the research fund for doctor in Huanggang Normal University (09cd158).

References

1. Ahuja, R.K., Magnati, T.L.: Network flows: theory, algorithms and applications. Prentice-Hall, Englewood Cliffs (1993)
2. Katagiri, H., Mermri, E.B., Sakawa, M., Kato, K.: A Study on Fuzzy Random Minimum Spanning Tree Problems through Possibilistic Programming and the Expectation Optimization Model. In: The 47th IEEE International Midwest Symposium on Circuits and Systems (2004)
3. Hutson, K.R., Shier, D.R.: Minimum spanning trees in networks with varying edge weights. Springer Science + Business Media, LLC (2006) (published online)
4. Katagiri, H., Sakawa, M., Ishii, H.: Fuzzy random bottleneck spanning tree problems using possibility and necessity measures. European Journal of Operational Research 152, 88–95 (2004)
5. Liu, B.: Uncertainty Theory, 2nd edn. Springer, Berlin (2008)

6. Liu, B.: Theory and Practice of Uncertain Programming, 1st edn. Physica-Verlag, Heidelberg (2002)
7. Petrica, C.: POP: New Models of the Generalized Minimum Spanning Tree Problem. Journal of Mathematical Modelling and Algorithms 3, 153–166 (2004)
8. Dooms, G., Katriel, I.: The *Minimum Spanning Tree* Constraint. In: Benhamou, F. (ed.) CP 2006. LNCS, vol. 4204, pp. 152–166. Springer, Heidelberg (2006)
9. Feremans, C., Labbe, M., Laporte, G.: The generalized minimum spanning tree problem: polyhedral analysis and branch-and-cut algorithm. Networks 43(2), 71–86 (2004)

Author Index

An, Fengjie 569
Ao, Qinyun 471

Bai, QiuJu 429
Bo, Zhao 97, 105

Cai, Guoliang 447
Cai, Li 459
Cao, CaiFeng 497
Cao, Jian 251
Cao, Lei 401, 409
Cao, Wenjing 653, 659
Chen, Cheng 337
Chen, Gang 323
Chen, Haiyan 587
Chen, Jian 471
Chen, Jianbao 13
Chen, Qisong 615
Chen, Riyuan 227
Chen, Shangping 291
Chen, Xiaowei 615
Chen, Yan 375
Chen, YanQiu 177
Cheng, Tingting 13
Cheng, Yu 271
Cheng, Yun 191

Deng, Wanjun 219
Diao, Lixin 453
Ding, Juan 447
Ding, Muhua 277
Ding, Shoucheng 213
Do, Hyun-Lark 51, 61, 65, 75, 93, 129
Dong, Wanxin 305
Du, Huiying 515

Fan, Xinghua 447
Feng, Xian 355

Gao, Yu 383
Gu, Guoai 219
Guo, Jun 343, 349
Guo, Lei 7
Guo, Yecai 343, 349

Han, Kezhong 561
Han, Qingtian 653, 659
Han, Wei 159, 165
He, FangGuo 677
He, Hong 389, 395
He, Lin 389, 395
He, Qian 415
He, Yihui 409
Hu, WenLong 607
Hu, Xuezhi 139
Hua, Delin 545
Hua, Li 587
Huang, Haiquan 55
Huang, Jing 553
Huang, Jun Steed 361
Huang, Lei 7

Jan, Yih-Haw 623
Jiang, Chunlan 233
Jiang, DeDong 497
Jin, Quan 199

Li, De-dong 251
Li, Ming 233
Li, Tao 389

Li, Wei 607
Li, Wenqiang 653, 659
Li, Xin 305
Li, YaoLong 311
Li, YiMing 285
Li, Yiqun 199
Liang, Jie 251
Liang, Shuang 145
Lin, Nan 375
Lin, Qin-ying 7
Lin, Yaning 569
Liu, Gang 491
Liu, Guangfu 509
Liu, GuiFei 429
Liu, Houcheng 227
Liu, Jinhua 139
Liu, Ni 375
Liu, Qiang 453, 485
Liu, Xiangnan 601
Liu, Xian-mei 85
Liu, Xiaoming 337
Liu, Xin 79
Liu, Yuanchao 133
Liu, Zhe 575
Liu, Zhenhua 245
Lu, Donghua 587
Lu, Linlin 453, 485
Lu, ManSha 69
Lu, YangHua 581
Luo, Qiang 79
Lv, Bingqing 553
Lv, Jie 601
Lv, XiaoQi 79

Mi, YongHong 355

Na, Wang 423
Nan, Jian-guo 7

Pan, Hailan 471
Pan, Lian 337
Pan, Wei 441
Pang, Yanxia 423
Peng, GuoXing 111

Qin, Xinghong 465
Qin, Yanhong 465
Qu, ZhengHui 435

Ru, Yi 153

Shang, FuHua 1
Shao, GuiMing 677
Shen, He 355
Shen, Xu-kun 85
Shen, Yan 123
Song, Jianshe 19, 25, 33, 39
Song, Li-xia 525
Song, Shiwei 227
Song, Zhumei 629, 635
Su, DongHai 435
Su, JianYing 459
Sun, Chunling 641
Sun, Guangwen 227
Sun, Haitao 515
Sun, PeiLli 177
Sun, Shuangshuang 123

Tan, Huashan 477
Tao, Fengyang 509
Tian, Feng 85
Tian, Xianzhi 171
Tong, Qi 111
Tong, SanHong 191
Tu, Guangcan 453, 485

Wang, Chao 233
Wang, Chongwen 367
Wang, Chundong 133
Wang, Feng 191
Wang, Huadong 317
Wang, Huaibin 45, 133
Wang, Huaisheng 671
Wang, Jianda 183
Wang, Jin 259
Wang, Jun 401, 409
Wang, Min 311
Wang, Ruihua 25, 33, 39
Wang, Wen-Xiang 525
Wang, Xiuchen 575
Wang, Yanchun 183
Wang, Yanli 191
Wang, Zaicheng 233
Wen, Xinling 153
Wu, HaiYing 665
Wu, Hao 521
Wu, Jiande 375
Wu, Maonian 615
Wu, ShuQin 329
Wu, Yibing 33, 39

Xi, Gaowen 539
Xi, Wenjing 509
Xie, HongTao 1
Xie, Xiao-fang 251
Xing, TingYan 355
Xiong, Ying 271
Xu, Fang 343, 349
Xu, Hui 647
Xu, Jiangchun 375
Xu, JunHui 19
Xu, Ke 429
Xu, Wencai 349
Xue, Xiaojun 375

Yan, Qisheng 277
Yang, Chao 491
Yang, Fei 401
Yang, Tong 389, 395
Yang, Xiang 441
Yang, Yanlong 615
Yang, Yong-hua 383
Yang, You 477
Yang, Zheng 521
Yao, Linlin 415
Yao, Qin 447
Yao, Wenjuan 291
Yi, Lingyan 227
Yi, ZhaoXiang 19, 25
Yin, Xin 395
You, Houxing 265
You, Xine 593
Yu, Airong 401, 409
Yu, Ping 477
Yu, Yan-Hua 525
Yuan, Jixuan 629, 635
Yue, Xian-Fang 533

Zhai, Wenting 647
Zhang, Bo 123
Zhang, Hongbo 117
Zhang, Jianhua 245
Zhang, Jing 299
Zhang, Lina 647
Zhang, Lingfeng 593
Zhang, XiaoLin 581
Zhang, Xiongmei 19, 25, 33, 39
Zhang, Xuezheng 199
Zhang, YiWen 607
Zhang, Yuanyi 441
Zhang, Zhijie 515
Zhao, Xiaoyan 205
Zheng, ZhongMei 191
Zhou, Haiyun 45
Zhou, Jun 415
Zhou, Kai 85
Zhou, Ming 503
Zhou, ShengXuan 355
Zhou, Shiqiong 629, 635
Zhou, YaDong 1
Zhu, BoTao 361
Zhu, Huanjun 13
Zhu, JingWen 361
Zhu, Shengqing 291
Zhu, XiJun 429
Zou, Xing 239